Introductory Plant Biology

~~Ch 19~~

Ex I lect 1-8
pgs 1) 3-12
      86-87 98-100

2) 23-36
3) 36-40
4) 191-207
5) 205-216
6) 87-91 96-98
              64-69
7) 76-77 109-129
8) 50-60

Third Edition
Introductory Plant Biology
Kingsley R. Stern
California State University—Chico

Wm. C. Brown Publishers
Dubuque, Iowa

Book Team

John D. Stout
Senior Editor
Mary J. Porter
Assistant Editor
Natalie Gould
Production Editor

Barbara J. Grantham
Designer
Faye M. Schilling
Photo Research Editor
Mavis M. Oeth
Permissions Editor

Wm. C. Brown
Chairman of the Board
Mark C. Falb
President and Chief Executive Officer

Wm. C. Brown Publishers, College Division

Lawrence E. Cremer
President
James L. Romig
Vice-President, Product Development
David A. Corona
Vice-President, Production and Design
E. F. Jogerst
Vice-President, Cost Analyst
Bob McLaughlin
National Sales Manager

Marcia H. Stout
Marketing Manager
Craig S. Marty
Director of Marketing Research
Eugenia M. Collins
Production Editorial Manager
Marilyn A. Phelps
Manager of Design
Mary M. Heller
Photo Research Manager

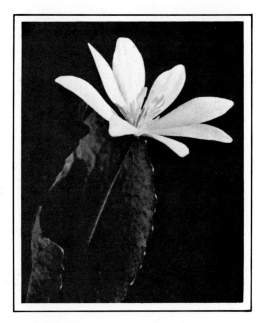

The cover photograph is of bloodroot (*Sanguinaria canadensis*), an early spring-flowering herb that occurs in rich woods throughout eastern and midwestern North America. The name stems from the presence of a bright red juice that is produced throughout the plant but particularly in the rhizome. The red juice contains the alkaloid *sanguinarine*, which was used in the past in the treatment of skin and other cancers, and is reportedly again being used for that purpose. Native Americans used bloodroot juice as a dye, in the treatment of rheumatism, ringworm, and sore throat, and as an agent to induce vomiting.

Contents

Preface   xiii

1

The Development of Plant Study   3
Introduction   4
Early History and Development of
    Plant Study   5
Summary   11
Review Questions   12
Discussion Questions   12
Additional Reading   12

2

Nature of Life   13
Introduction   14
Attributes of Living
    Organisms   14
Chemical and Physical Bases of
    Life   16
Energy   18
Chemical Components of
    Protoplasm   18
Summary   20
Review Questions   22
Discussion Questions   22
Additional Reading   22

3

Cells   23
Introduction   24
Cells   24
Cell Wall   29
Protoplasm   29
Cellular Reproduction   36
Communication in Cells   40
DNA   41
RNA and Synthesis of
    Proteins   44
Summary   47
Review Questions   49
Discussion Questions   49
Additional Reading   49

4

Tissues   50
Introduction   51
Meristematic Tissues   51
Permanent Tissues   52
Summary   60
Review Questions   61
Discussion Questions   61
Additional Reading   61

5

Roots and Soils   62
Introduction   63
Root Structure   64
Specialized Roots   70
Human Relevance of Roots   76
Soils   77
Summary   82
Review Questions   83
Discussion Questions   83
Additional Reading   84

6

Stems   85
Introduction   86
External Form of a Woody
   Twig   86
Origin and Development of
   Stems   87
Tissue Patterns in Stems   90
Specialized Stems   98
Applied Uses of Wood   101
Summary   106
Review Questions   107
Discussion Questions   108
Additional Reading   108

7

Leaves   109
Introduction   110
Leaf Arrangements and
   Types   110
Internal Structure of Leaves   113
Specialized Leaves   116
Why Leaves Change Color and
   Fall in Autumn   125
Human and Ecological Relevance
   of Leaves   126
Summary   128
Review Questions   129
Discussion Questions   129
Additional Reading   129

8

Water in Plants   130
Introduction   131
Molecular Movement   131
Water and Its Movement Through
   the Plant   135
Transport of Food Substances in
   Solution   137
Regulation of Transpiration   138
Summary   140
Review Questions   141
Discussion Questions   141
Additional Reading   141

9

Metabolism in Plants   142
Introduction   143
Photosynthesis   143
Respiration   153
Digestion and Assimilation   159
Summary   160
Review Questions   162
Discussion Questions   162
Additional Reading   163

10

Growth   164
Introduction   165
Seed Germination   165
Plant Hormones   167
Mineral Requirements for
   Growth   173
Plant Movements   174
Phytochrome   183
Photoperiodism   184
Temperature and Growth   186
Dormancy   187
Summary   188
Review Questions   190
Discussion Questions   191
Additional Reading   191

11

Meiosis and Alternation of Generations 192
Introduction 193
Phases of Meiosis 196
Alternation of Generations 201
Summary 203
Review Questions 203
Discussion Questions 204
Additional Reading 204

12

Genetics and Evolution 205
Introduction 206
Genetics 206
Evolution 216
Mechanisms of Evolution 219
Summary 223
Review Questions 225
Discussion Questions 226
Additional Reading 227

13

Plant Names and Classification 228
Introduction 229
Development of the Binomial
  System of Nomenclature 230
Development of the Kingdom
  Concept 232
Classification of Major
  Groups 234
Synoptic Key to Major Groups of
  Organisms 236
Summary 238
Review Questions 239
Discussion Questions 240
Additional Reading 240

14

Kingdom Monera and Viruses 241
Features of the Kingdom 242
Division Schizophyta 242
  Class Schizomycetes—
  Bacteria 242
Compost and Composting 251
Bacteria and Disease 253
Koch's Postulates 256
Bacteria Useful to Humans 257
Division Schizophyta 259
  Class Cyanophyceae—
  Blue-Green Algae 259
Human and Ecological Relevance
  of Blue-Green Algae 263
Division Prochlorophyta—
  Prochlorophytes 265
Viruses 265
Human Relevance of
  Viruses 269
Summary 269
Review Questions 272
Discussion Questions 272
Additional Reading 273

15

Kingdom Protista 274
Features of the Kingdom 275
Introduction to the Algae 275
Division Chlorophyta—Green
  Algae 275
Division Euglenophyta—
  Euglenoids 285
Division Chrysophyta—Golden-
  Brown Algae 286
Division Pyrrophyta—
  Dinoflagellates 288
Division Phaeophyta—Brown
  Algae 288
Division Rhodophyta—Red
  Algae 292
Human and Ecological Relevance
  of the Algae 296
Division Myxomycota—Slime
  Molds 299
Human and Ecological Relevance
  of the Slime Molds 303
Other Members of the Kingdom
  Protista 303
Summary 303
Review Questions 305
Discussion Questions 305
Additional Reading 305

16

Kingdom Fungi and Lichens 307

Features of the Kingdom 308
Introduction to the Fungi 308
Division Mycota—True Fungi 309
  Class Chytridiomycetes—Chytrids 309
  Class Oomycetes—Water Molds 310
  Class Zygomycetes—Bread Molds 312
Human and Ecological Relevance of the Bread Molds 314
  Class Ascomycetes—Cup or Sac Fungi 315
Human and Ecological Relevance of the Cup or Sac Fungi 317
  Class Basidiomycetes—Club Fungi 320
Human and Ecological Relevance of the Club Fungi 328
Class Deuteromycetes—Imperfect Fungi 332
Lichens 334
Human and Ecological Relevance of Lichens 338
Summary 338
Review Questions 340
Discussion Questions 341
Additional Reading 341

17

Introduction to the Plant Kingdom: Bryophytes 343

Features of the Kingdom 344
Introduction to the Bryophytes 344
Division Bryophyta 346
  Class Hepaticae—Liverworts 346
  Class Anthocerotae—Hornworts 351
  Class Musci—Mosses 352
Human and Ecological Relevance of Bryophytes 356
Summary 357
Review Questions 358
Discussion Questions 359
Additional Reading 359

18

Introduction to Vascular Plants: Ferns and Their Relatives 360

Introduction 361
Division Psilophyta—Whisk Ferns 361
Division Lycophyta—Club Mosses and Quillworts 363
Human and Ecological Relevance of Club Mosses and Quillworts 369
Division Sphenophyta—Horsetails and Scouring Rushes 370
Human and Ecological Relevance of Horsetails and Scouring Rushes 374
Division Pterophyta—Ferns 376
Human and Ecological Relevance of Ferns 381
Fossils 384
Summary 386
Review Questions 388
Discussion Questions 388
Additional Reading 388

19

Introduction to Seed Plants: Gymnosperms 389

Introduction 390
Division Coniferophyta—Conifers 391
Other Gymnosperms 397
Division Cycadophyta—Cycads 397
Division Ginkgophyta—Ginkgoes 398
Division Gnetophyta—Gnetophytes 399
Human and Ecological Relevance of Gymnosperms 401
Summary 408
Review Questions 409
Discussion Questions 409
Additional Reading 410

20

Flowering Plants   411
Introduction   412
Division Anthophyta—Flowering
   Plants   413
Structure and Reproduction of
   Flowering Plants   415
Trends of Specialization and
   Classification in Flowering
   Plants   422
Herbaria and Plant
   Preservation   425
Summary   429
Review Questions   430
Discussion Questions   430
Additional Reading   431

21

Fruits and Seeds   432
Introduction   433
Fruits   433
Kinds of Fruits   434
Fruit and Seed Dispersal   443
Seed Structure   447
Germination   448
Longevity of Seeds   450
Summary   451
Review Questions   453
Discussion Questions   453
Additional Reading   453

22

**Flowering Plants and
Civilization   454**
Introduction   455
Origin of Cultivated Plants   455
Selected Families of Flowering
   Plants   457
Buttercup Family
   (Ranunculaceae)   458
Laurel Family (Lauraceae)   460
Poppy Family
   (Papaveraceae)   461
Mustard Family
   (Brassicaceae)   463
Rose Family (Rosaceae)   464
Legume Family (Fabaceae)   466
Spurge Family
   (Euphorbiaceae)   467

Cactus Family (Cactaceae)   469
Mint Family (Lamiaceae)   471
Nightshade Family
   (Solanaceae)   474
Carrot Family (Apiaceae)   475
Pumpkin Family
   (Cucurbitaceae)   476
Sunflower Family
   (Asteraceae)   478
Grass Family (Poaceae)   479
Lily Family (Liliaceae)   480
Orchid Family
   (Orchidaceae)   482
Summary   483
Review Questions   484
Discussion Questions   484
Additional Reading   484

23

Ecology   486
Introduction   487
Populations   488
Communities   488
Ecosystems   489
Humans in the Ecosystem   492
The Nitrogen Cycle   493
The Carbon Cycle   495
Succession   496
Fire Ecology   499
Major Biomes of North
   America   501
Special Topic: Pollination
   Ecology   509
Summary   514
Review Questions   516
Discussion Questions   516
Additional Reading   517

Appendixes

1
Scientific Names of Organisms Mentioned in the Text   A–1

2
Biological Controls   A–27
General Controls   A–27
Specific Controls   A–31
Companion Planting   A–33
Some Sources of Herb Plants and
   Seeds   A–35
Additional Reading   A–35

3
Useful Plants and Poisonous Plants   A–37
Wild Edible Plants   A–37
Poisonous Plants   A–41
Medicinal Plants   A–44
Hallucinogenic Plants   A–51
Spice Plants   A–52
Dye Plants   A–55
Additional Reading   A–57

4
Home Gardening and Houseplants   A–59
Vegetative Propagation   A–59
Major Types of Grafting   A–66
Pruning   A–73
Growing Vegetables   A–76
Common Vegetables and Their
   Nutritional Values   A–77
Growing Houseplants   A–83
Common Houseplants   A–85
Additional Reading   A–95

5
Metric Equivalents   A–97

Glossary   G–1
Index   I–1

Preface

This book is designed as an introductory text in botany. It assumes little knowledge of the sciences on the part of the student. It includes sufficient information for some shorter introductory botany courses open to both majors and nonmajors, but it is arranged so that certain sections—for example, "Early History and Development of Plant Study," "Soils," "Division Psilophyta"—can be omitted without disrupting the overall continuity of the course.

Botany instructors vary greatly in their opinions as to the depth of coverage needed for the topics of photosynthesis and respiration in a text of this type. Some feel that nonmajors, in particular, should have a brief introduction only; whereas others consider a more detailed discussion essential. In this text photosynthesis and respiration are discussed at three levels. Some may find one or two levels sufficient, and others may wish their students to become familiar with the processes at all three levels.

Despite eye-catching chapter titles and headings, many texts for majors and nonmajors relegate the current interests of a significant number of students to comparative obscurity. This text emphasizes current interests without giving short shrift to botanical principles. Present interests of students include subjects such as organic gardening, Native American and pioneer uses of plants, pollution and recycling, houseplants, backyard vegetable gardens, natural dye plants, poisonous and hallucinogenic plants, and the nutritional values of edible plants. The rather perfunctory coverage or absence of such topics in many botany texts has occurred partly because botanists have tended to believe that the topics are more appropriately covered in anthropology and horticulture courses. I have found, however, that both majors and nonmajors in botany who may be initially disinterested in the subject matter of a required course frequently become engrossed if the material is repeatedly related to such topics. Accordingly, a considerable amount of ecological and ethnobotanical material has been included with traditional botany throughout the book—without, however, resorting to excessive use of technical terms.

Organization of the Text

A relatively conventional sequence of botanical subjects is included. Chapters 1 and 2 cover introductory and background information; chapters 3 through 10 deal with structure and function; chapters 11 and 12 introduce meiosis, genetics, and evolution. Chapter 13 presents a five-kingdom system of classification; chapters 14 through 21 stress, in phylogenetic sequence, the diversity of organisms traditionally regarded as plants; and chapter 22 deals with ethnobotanical aspects and information of general interest pertaining to sixteen major families of flowering plants. Chapter 23 is an overview of the vast topic of ecology, although ecological topics and applied botany are included in most of the preceding chapters as well. Some of these subjects are broached in anecdotes that introduce the chapters, whereas others are mentioned in the human and ecological relevance sections (with which most of the chapters in the latter half of the book conclude).

Aids to the Reader

Review questions, discussion questions, and additional reading lists are provided for each chapter. New terms are defined as they are introduced, and those used more than once are boldfaced and included in a pronouncing glossary. The use of scientific names throughout the body of the text has been held to a minimum, but a complete list of the scientific names of all organisms mentioned is given in appendix 1. Appendix 2 deals with biological controls and companion planting; appendix 3 lists wild edible plants, poisonous plants, hallucinogenic plants, spices, and natural dye plants. Appendix 4 discusses vegetative propagation and pruning, gives horticultural information on houseplants and on the cultivation and nutritional value of vegetables. Appendix 5 gives some metric equivalents.

New to This Edition

The topics of cells, tissues, water in plants, and metabolism in plants are now treated in four chapters. Other revisions include an expansion of the discussion of the pressure-flow hypothesis (chapter 8), a brief discussion of the cell cycle added to the section on cellular reproduction (chapter 3), an augmented section on evolution (chapter 12), and a revised synoptic key to major groups of organisms (chapter 13), refinements in illustrations of life cycles in chapter 16 through 20, discussion of the nitrogen and carbon cycles in ecology (chapter 23), an updated appendix on biological controls, and a new appendix on metric equivalents.

Additional Aids

Instructor's Manual

The Instructor's Manual available with *Introductory Plant Biology* offers a variety of course schedules while providing overviews, goals, suggested answers, film sources, and examination questions for each text chapter.

Laboratory Manual

The Laboratory Manual to accompany *Introductory Plant Biology* is written for the student entering the study of Botany for the first time. The exercises utilize plants to introduce biological principles and the scientific method. The exercises are written to allow for maximum flexibility in sequencing.

Acknowledgments

The contributions of many individuals to the development of this book are gratefully acknowledged. Critical reviewers, who provided many valuable suggestions for improving and updating the text, include:

Lydia C. Arciszewski, Bergen Community College
Linda R. Berg, University of Maryland
Maynard C. Bowers, Northern Michigan University
Hayle Buchanan, Weber State College
Laura Chunosoff, Pace University
Michael Gardiner, University of Puget Sound
J. M. Herr, Jr., University of South Carolina
Robert C. Lommasson, University of Nebraska
James E. Marler, Louisiana State University
Robert S. Mellor, University of Arizona
P. C. Pendse, California Polytechnic State University, San Luis Obispo
Robert R. Robbins, University of Wisconsin
James L. Seago, State University of New York, Oswego
H. B. Ward, University of Mississippi
Henry Webert, Nicholls State University
Terry F. Werner, Harris-Stowe State College

Additional persons who read parts of the manuscript and made many helpful criticisms and suggestions include Richard S. Demaree, Jr., Eugene D. Thomas, Robert B. McNairn, George E. Corson, Jr., William L. Stephens, Donald T. Kowalski, and Robert A. Schlising. Others whose encouragement and contributions are deeply appreciated include W. T. Stearn, E. Gibbes Patton, Lorraine Wiley, Donald B. Joley, Isabella A. Abbott, Paul C. Silva, Donald E. Brink, Jr., Suzanne Costanza, the faculty and staff of the Department of Biological Sciences, California State University, Chico (especially Robert I. Ediger and Timothy B. Devine); my many inspiring students; the Lyon Arboretum, University of Hawaii; the editorial, production, and design staffs of the Wm. C. Brown Company Publishers; and most of all my wife Janet, and my children, Kevin and Sharon, who sacrificed a great deal to bring about the book's publication. Special thanks are due the artists, Denise Robertson Devine and Elly Simmons.

Kingsley R. Stern
Chico, California

Introductory Plant Biology

The Development of Plant Study

Overview

This chapter introduces you to botany: what it is, how it developed, how it relates to our everyday lives, and what its potential is for the future. The discussion includes a brief introduction to the beginning of plant study in ancient civilizations, an examination of the scientific method, and a look at the development of botany after the invention of the microscope. It concludes with a brief survey of the major disciplines within the field of botany.

Some Suggested Student Learning Goals

1. Explain briefly what the scientific method is and what hypotheses are.

2. Name or identify a contribution to the development of botany as a science made by each of the following: Theophrastus, Leeuwenhoek, Malpighi, Grew, van Helmont, Linnaeus.

3. Define the Doctrine of Signatures, and (as associated with the Doctrine of Signatures) a herbal.

4. List the major botanical disciplines and indicate briefly the particular aspect of botany with which each is concerned.

Introduction

Plants may not often make the front page of the newspaper, but hardly a day goes by without their at least being mentioned by the time one reaches the sports section. Some recent headlines I have noted, for example, include, "Herbs Blamed in Accidental Poisonings," "Ugly Desert Jojoba May Keep Bay Area Transit Buses Rolling," "Hot Future Seen for Algae," "Hydroseeding Changes the Face of Arizona," "State Owes Bushels to 'U' Plant Breeders," and "Plants and Flowers Make You Sexy."

Actually, all of us have been totally dependent on green plants since before we were born. An extravagant statement? Consider that green plants are the sole organisms capable of transforming the sun's energy into forms useable by animal life and that such energy is vital to the very existence of animal life. Consider further that virtually all the oxygen in the air we breathe is produced by green plants and that they are the only organisms that can remove the carbon dioxide waste we give off into the atmosphere. If some universal plant disease were to kill off all or a large portion of the vegetation on earth (including that found in the oceans), all the animals on land, in the sea, or in the air could conceivably suffocate, although they undoubtedly would starve first.

Apart from these aspects, plants and plant products are so interwoven with the fabric of human society that we largely take them for granted. We are aware, of course, that corn, potatoes, and other vegetables are plants; but all foods, including meat, fish, poultry, eggs, cheese, and milk, to mention but a few, owe their existence to plants. Condiments such as spices, mustard, and pepper and luxuries such as perfumes are produced by plants, as are dyes, adhesives, digestible surgical "gut," food stabilizers, beverages, and emulsifiers. Our houses are constructed with lumber from trees, which also furnish the cellulose for paper, cardboard, and synthetic fibers. Our clothing, camping equipment, bedding, draperies, and other textile goods are made from fibers of many different plant families. Coal is fossilized plant material, and oil comes from animals that were plant eaters either directly or indirectly. All medicines and drugs at one time came from plants, fungi, or bacteria, and many important ones, including some of the antibiotics, still do. Microscopic organisms are responsible for recycling both plant and animal wastes and for aiding in the building of healthy soils. Others are responsible for human diseases and allergies.

As in the past, primitive peoples use plants not only for food, shelter, clothing, and medicine but also in hunting and fishing. Today small teams of botanists, anthropologists, and medical doctors are interviewing witch doctors and healers in remote tropical regions and identifying plants with healing properties before the primitive lore is lost to modern civilization and the plants become extinct.

What of plants and the future? It is estimated that by the year 2000, the population of the earth will double, approaching 7 billion persons, every one of whom will need food, clothing, and shelter in order to survive. To ensure survival, it is conceivable that every man, woman, and child may eventually need to learn not only how to cultivate food plants but also how to use plants in rendering polluted water and land productive again and how to use them in renewing urban areas; in addition, everyone may need to learn how to reverse the trend of plants shrinking in number, a trend caused by the staggering increase in the number of earth inhabitants.

Some have suggested that the odds are against humanity saving itself from itself and have even indicated that it might become necessary to emigrate to another planet. If so, plants will undoubtedly play a vital role in space exploration. For a number of years, experiments with portable oxygen "generators" have been in progress. Tanks of water teeming with tiny green algae are taken aboard a spacecraft and installed so that they will be exposed to light for at least part of the time. The algae not only produce oxygen for the spacecraft inhabitants to breathe, but they also utilize the waste carbon dioxide end product of respiration. In addition, as they multiply, excess algae can be fed to a special kind of shrimp, which in turn multiply and become food for the space travellers. Other wastes are recycled by different microscopic plants. When this self-supporting arrangement, called a *closed system,* is perfected, the range of spacecraft should greatly increase because heavy oxygen tanks will not be necessary, and the amount of food reserves needed will be reduced.

As supplies of fossil fuels (especially oil) dwindle, increased attention is being paid to alternative energy sources. Methane gas, which can be used as a substitute for natural gas, has been produced from animal manures and decomposed plants in numerous villages in India and elsewhere for many years, and after several years of trial on a small scale in the United States, plans are now under way for production of methane on a larger scale from human sewage. Potatoes, grains, and other sources of carbohydrates are currently used in the manufacture of alcohols, some of which are being blended with gasoline ("gasohol"), and such uses are destined to increase in the future.

Genetic engineering, which involves the introduction of desirable genes from one organism into another, holds enormous potential for improvements in the quality and yield of crop plants and in the suppression or elimination of human defects and diseases. Plant breeders—who have already given us a wide variety of hybrid plants with greater vigor, disease resistance, and yield than the parent plants—are busily engaged in trying to produce plants that will inhibit weeds, grow in areas that are inhospitable to crop plants presently available, and repel insects. Plant breeders at Cornell University, for example, are working on developing a potato plant with sticky hairs that will trap insects, while other scientists are seeking to produce plants with proportionately higher protein and vitamin content. The next decade or two will undoubtedly see plants of various properties that our grandparents never dreamed were possible.

Early History and Development of Plant Study

On a number of occasions I have received visits from anxious mothers or from pediatricians wanting to know if parts of backyard plants that young children have consumed are poisonous. I have also been consulted by various law enforcement officials, landscape architects, pollen collectors, students interested in wild flowers, vegetarians, organic and traditional gardeners, and a variety of other professional and amateur persons with a wide range of interest in plant life. Some have wanted plants identified; others have wanted suggestions for treating diseased plants. Still others have wanted to know about grafting techniques, the effect of "gray water" on plants, the suitability of plants for specific locations, the preservation of plants, the edibility of wild plants, and a host of other plant-related subjects.

Knowledge about plant life throughout the world has now become so vast that it is impossible for any person to be an authority on more than a tiny fraction of it. But our libraries contain thousands of books dealing with virtually every facet of botanical investigation, and research journals publish the latest discoveries from around the world on an almost daily basis. Why and how has all this knowledge accumulated? To answer this question we need to take a brief look at the early history and development of plant study.

The Chinese have been cultivating medicinal and other useful plants for at least forty-five hundred years. Records from that far back in history are fragmentary, and it is often impossible to separate fact from legend; some authorities agree, however, that the founder of Chinese agriculture was an emperor by the name of Shen Nung, who was born in 2737 B.C. According to tradition, he is supposed to have invented the plow and to have established an annual seed-sowing ceremony, during which seeds of soybeans, wheat, rice, millet, and sorghum were planted by royalty. He was believed to have been an authority on poisons and antidotes, and he wrote a book on drugs and medicines that was incorporated into the *Pun-tsao,* a Chinese *pharmacopoeia* (an officially recognized book describing drugs and medicines) of forty volumes, published during the seventeenth century. During the Han dynasties, which lasted from about 200 B.C. until the birth of Christ, gardens became very extensive in China and many ornamental plants were cultivated. Plants such as primroses, poppies, and chrysanthemums were brought to the Western world from China over two thousand years ago.

The Chinese, however, were probably neither the first people nor even the first civilized people to practice forms of agriculture. Information gleaned from the walls of tombs, mummy wrappings, hieroglyphics, cave paintings, and carvings indicates that the Assyrians and Egyptians were cultivating fruits such as pomegranates, grapes, quinces, olives, and figs, as well as cereals, by 4000 B.C. One of their most important tree crops was the date, which was eaten fresh or dried, and the sap of date palms was fermented for wine. Although they knew nothing of the details, the Assyrians were aware of sexuality in the date palm and pollinated the female trees by hand. By the seventh century B.C., they had produced a systematically arranged list of medicinal plants, which suggests that the physicians and pharmacists of the day had a noteworthy knowledge of plants and their uses.

The Egyptians cultivated primitive forms of wheat and barley from about 5000 to 3400 B.C., although some authorities place the cultivation of these two cereals as far back as 15,000 to 10,000 B.C. More modern forms of cereals, such as six-rowed barley, may have been under cultivation by 2000 B.C. By the fifth century B.C., the Egyptians apparently were brewing *booza,* a beer, from barley.

The study of plants, called **botany** from the French word *botanique* (botanical) and three Greek words *botanikos* (botanical), *botanē* (plant or herb), and *boskein* (to feed), appears to have had its origins with Stone Age peoples who sought to modify their surroundings and feed themselves. Initially, the primary interest in plants was practical, centering around how plants might provide food, fibers, fuel, and medicine. Eventually, however, an intellectual interest arose; individuals became curious about how plants reproduced and how they were put together. This inquisitiveness led to plant study becoming a **science,** which broadly defined is simply "a search for knowledge of the natural world."

A science may be distinguished from other fields of intellectual endeavor by several features. It involves the observation, recording, organization, and classification of facts, and more importantly, it involves what is done with the facts. Scientific procedure involves experimentation, observation, and the verifying or discarding of information, chiefly through inductive reasoning from known samples. There is no universal agreement on the precise details of the process. A few decades ago, scientific procedure was considered to involve a routine series of steps. This series of steps came to be known as the *scientific method,* and there are still instances where such a structured approach works well. In general, however, the scientific method now describes the procedures of assuming and testing hypotheses. A **hypothesis** is simply a tentative explanation for something that has been observed. It may not be the correct explanation—testing will determine whether it is correct or incorrect. The nature of the testing will vary according to the circumstances and materials. For example, we may *observe* that apples are red fruits that taste sweet. We may then make the *hypothesis* that all red fruits taste sweet. We may *test* the hypothesis by tasting red fruits, and as a result of our testing (since red crab apples are bitter), we may *modify the hypothesis* to state that only some red fruits are sweet.

However one defines *science,* it is clear that plant science existed in ancient Greece. As in even older cultures, the study of plants started when the early Greeks developed a practical interest in food and drug plants, but they slowly became curious as well about the structure and function of plants. As the physicians and the pharmacists of the era gathered and used medicinal plants, they studied the variations and forms and came to recognize apparent relationships among them.

One of these Greek herbal physicians had a son in 384 B.C., who became one of the most renowned philosophers of all time—Aristotle the Stagirite. Although Aristotle is perhaps better known for his philosophical works, he was an accomplished mathematician and also acquired an extensive knowledge of nearly all aspects of natural history. In fact, he combined philosophical and scientific interests as few other philosophers have done. At the age of seventeen, Aristotle went to Athens, where he met and became a pupil of Plato. He left Athens after Plato's death in 347 B.C. and studied marine animals at a coastal area for several years, eventually returning to Athens to found the first botanical garden of which there is any record.

Figure 1.1 Theophrastus. (Courtesy National Library of Medicine)

When Aristotle died, he willed the botanical garden and its associated library to his pupil and assistant Theophrastus of Eresus (figure 1.1). Theophrastus was an extraordinary man, who soon acquired all the knowledge Aristotle had accumulated on plants and added prodigiously his own observations. It is said that he had two thousand disciples and wrote two hundred treatises. The most important of the latter to have survived are two books entitled *History of Plants* and *Causes of Plants.* So great were his contributions to botany as a science that the famous eighteenth-century Swedish botanist Linnaeus gave him the title "Father of Botany"; few, if any, dispute his right to the honor.

With the advent of the printing press in the middle of the fifteenth century, numerous printed botanical works called **herbals** (figure 1.2) appeared. They were primarily the products of German botanists, although some Italian and English botanists made their own contributions between 1470 and 1670. These *herbalists,* as they were called, were mostly concerned with medicinal plants, which they studied in the botanical gardens that had become numerous and extensive in Europe by this time. They produced elaborate and intriguing illustrations for the herbals, accompanied by sometimes outlandish stories and descriptions. Some of the stories became legends and developed into the *Doctrine of Signatures.* According to this doctrine, if a part of a plant, such as the root, had the shape of a part of the human body, it would be useful in treating a disease of the human part it most closely resembled. Thus, the meat of a walnut, which somewhat resembles a miniature brain, was used in treating brain diseases and hepatica leaves, which have lobes reminiscent of those of the liver, were used to treat ailments of that organ. One of the more famous herbalists was Otto Brunfels, who published a three-volume herbal in 1530. His work had excellent illustrations and is considered to be a link between ancient and modern botany.

The microscope had and continues to have a profound effect not only on plant studies but on the biological sciences and related fields. In 1590, two Dutch spectacle makers, Zacharias Janssen and his father, Hans, produced the first instrument for magnifying minute objects. The device contained two simple glass lenses. Zacharias is often referred to as the inventor of the compound microscope. A Dutch draper by the name of Anton van Leeuwenhoek (1632–1723), who ground lenses and made microscopes in his spare time, is best known for his development of primitive microscopes. Leeuwenhoek was the first to describe bacteria, sperms, and other tiny cells— some of his microscopes could magnify as much as two hundred times. In his will, he left 26 of his 400 handmade microscopes to the Royal Society of London.

Before the invention of the microscope, plant study had been dominated by investigations based primarily on the external features of plants. The magnification of the early microscopes was not very great by present standards, but these instruments nevertheless led to the discovery of cells (discussed in chapter 3) and opened up whole new areas of study. **Plant anatomy,** concerned chiefly with the internal structure of plants, was established through the efforts of several scientific pioneers. Early plant anatomists of note included Marcello Malpighi (1628–1694), an Italian who discovered various tissues in stems and roots, and Nehemiah Grew (1628–1711), an Englishman who described the structure of wood more precisely than any of his predecessors.

Plant physiology, which is concerned with plant function, was established by J. B. van Helmont (1577–1644), a Flemish physician and chemist, who was the first to demonstrate that plants do not have the same nutritional needs as animals. In a classic experiment, van Helmont planted a willow branch weighing 5 pounds in a tub filled with 200 pounds of dry soil. He covered the soil to prevent dust settling on it from the air, and after five years he reweighed the willow and the soil. He found that the soil weighed only 2 ounces less than it had at the beginning of the experiment, but that the willow had gained 164 pounds. He concluded that the tree had added to its bulk and size from the water it had absorbed. We know now that most of the weight came as a result of photosynthetic activity, but van Helmont deserves credit for landmark experimentation.

The seventeenth century saw a marked increase in botanical explorations to various parts of the globe. In the fifteenth century when Columbus visited Cuba, he found local Indian tribes cultivating corn (maize). This important food plant had apparently been in use by the pre-Incas of Peru some five thousand years earlier. By the time explorers ventured into the Americas in the 1600s, they found that the maize culture had spread from Argentina in the south to the St.

Figure 1.3 Linnaeus. (Courtesy National Library of Medicine)

Lawrence River area in Canada; American Indians had also domesticated the white potato; and in Mexico Indians were cultivating flowers and medicinal plants.

The explorers took large numbers of plants back to Europe with them, and it soon became clear to those working with the plants that some sort of formalized system was necessary just to keep the collections straight. Several *plant taxonomists* (botanists who specialize in the identifying, naming, and classifying of plants) proposed ways of accomplishing this, but we owe our present system of naming and classifying plants to the Swedish botanist Carolus Linnaeus (1707–1778) (figure 1.3). **Plant taxonomy** (also called *plant systematics*), which is the oldest branch of plant study, began in antiquity, but Linnaeus did more for the field than any other person in history. Thousands of plant names in use today are those originally recorded by this remarkable man in his book *Species Plantarum,* published in 1753. An expanded account of Linnaeus and his system of classification is given in chapter 13.

The nineteenth century saw the development of **plant ecology,** the study of the interaction of plants with one another and with their environment, and of the allied field of **plant geography,** which is the study of how and why plants are distributed where they are. **Plant morphology,** the study of the form and the structure of plants, also developed during this period.

Among the most noteworthy of the early plant geographers were two natives of Berlin, Germany, Carl Willdenow (1765–1812) and Alexander von Humboldt (1769–1859), who published books on the relationship of seed dispersal to plant distribution and on the associations of various plants with one another in tropical and temperate climates. These studies were brought to a climax by Sir Joseph D. Hooker (1817–1911), who eventually became director of the Royal Botanic Gardens in Kew, England (figure 1.4). Hooker traveled widely in the northern and southern hemispheres studying plant life, and he published *floras* (accounts of the plants of a specific region) of India and Antarctica. Charles Darwin, whose books revolutionized some basic biological concepts of the adaptation of organisms to their environment, said of Hooker's *Flora Antarctica,* "It is by far the grandest and most interesting essay on subjects of nature I have ever read."

By the beginning of the twentieth century much of the basic information incorporated in the plant sciences today had been discovered and elucidated, and the number of scientists engaged in investigating plants had increased conspicuously. **Genetics,** the science of heredity, had been

Figure 1.4 Sir Joseph D. Hooker. (Courtesy National Library of Medicine)

founded by the Austrian monk Gregor Mendel (1822–1884) through his classic experiments with peas; and **cytology,** the science of cell structure and function (now called **cell biology**), had received great impetus from the discovery of how cells multiply and function in sexual reproduction. Many more discoveries have been made in the twentieth century; these are discussed at appropriate junctures throughout the following chapters.

There is still a vast amount of information to be discovered, and new discoveries continue to be made daily. As far back as 1938, for example, an investigator by the name of Wellensiek noted that eleven thousand papers on botanical subjects had been published in that year alone; the number in recent years is undoubtedly many times greater. Further, it appears probable that at least one-third of all the organisms traditionally regarded as plants (particularly algae and fungi) have yet to be named, let alone thoroughly investigated.

Chapter 1

Efforts must be upgraded to save wild plants from extinction as civilization relentlessly invades their habitats. Also, both basic and applied research in botany must be given additional support if the earth's burgeoning human population is to continue to be fed, clothed, and housed.

Summary

We are totally dependent on green plants because they alone can convert the sun's energy into forms that are useable by, and vital to the very existence of, animal life. Plants and plant products are so interwoven with the fabric of human society that we largely take them for granted. We know that vegetables are plants, but animals, animal products, many luxuries and condiments, and other useful substances such as fibers, lumber, coal, medicines, and drugs either depend on plants or are produced by them. Today, teams of scientists are interviewing witch doctors in the tropics to try to discover hitherto little-known plants used by primitive peoples before the plants become extinct.

In a future of burgeoning world populations, all persons, to ensure human survival, may need to acquire some knowledge of plants and how to use them. If it should become necessary to emigrate to another planet, plants will undoubtedly play a vital role in space exploration as portable oxygen generators.

Our libraries contain thousands of books dealing with botanical subjects. All this knowledge has accumulated over the span of human existence. The Chinese practiced primitive agriculture at least forty-five hundred years ago, and the Assyrians and Egyptians were cultivating fruits and cereals by 4000 B.C. Botany, the study of plants, apparently began with Stone Age peoples' practical uses of plants. Eventually botany became a science, as intellectual curiosity about plants arose.

A science involves observation, recording, organization, and classification of facts. The verifying or discarding of facts is done chiefly from known samples through inductive reasoning. The scientific method involves specifically following a routine series of steps and generally assuming and testing hypotheses (tentative explanations for something that has been observed). Botanists believe that plant science existed in ancient Greece. A third-century Greek, Theophrastus, especially contributed much to the field of botany.

Numerous printed botanical works called herbals appeared after the invention of the printing press in the fifteenth century. The herbals were illustrated and sometimes contained bizarre stories about plants; these stories resulted in the development of the Doctrine of Signatures. The doctrine held that plant parts resembling human body parts were useful in treating ailments of those particular human organs.

The compound microscope, which was first produced by Zacharias and Hans Janssen in 1590 and developed by Leeuwenhoek in the 1600s, had a profound effect on studies in the biological sciences and led to the discovery of cells. Plant anatomy and plant physiology developed during the seventeenth century. Van Helmont was the first to demonstrate that plants have nutritional needs different from those of animals. He did so by careful experimentation and observation of a willow grown in a tub. Also during the seventeenth century, Europeans engaged in botanical exploration on other continents and took plants back to Europe. During the eighteenth century the Swedish botanist Linnaeus produced the elements of our present system of naming and classifying plants. During the nineteenth century plant geography and plant morphology developed, and by the beginning of the twentieth century genetics and cell biology had received their great impetus from the discovery of how cells multiply and function in sexual reproduction. Much remains yet to be discovered and investigated.

Review Questions

1. Briefly indicate contributions to plant science made by the following: Shen Nung, the ancient Egyptians, Theophrastus, Grew, Linnaeus.

2. What is meant by the *scientific method?*

3. Distinguish among herb, herbal, and herbalist.

4. What was the significance of van Helmont's experiment with the willow tree?

5. What is the oldest branch of botany, and why did it precede other branches?

6. What is the thrust of each of the other branches of botany?

7. What is the Doctrine of Signatures?

Discussion Questions

1. Since humans survived on wild plants for thousands of years, might it be desirable to return to that practice?

2. On the basis of what you have read, would you single out any one individual as having contributed the most to the development of plant study? Why?

3. How would you guess that Stone Age peoples discovered medicinal uses for plants?

4. Many of the early botanists were also doctors. Why do you suppose this is no longer so?

5. Consider the following hypothesis: "The majority of mushrooms that grow in grassy areas are poisonous." How could you go about testing this hypothesis scientifically?

Additional Reading

Anderson, E. *Plants, man and life.* Rev. ed. Berkeley, CA: University of California Press; 1967.

Anderson, F. J. *An illustrated history of the herbals.* New York: Cambridge University Press; 1977.

Arber, A. *Herbals, their origin and evolution.* 2d ed. Cambridge, England: Cambridge University Press; 1938. 1970 reprint of 1953 ed. Riverside, NJ: Hafner Publishing Company.

Baker, H. G. *Plants and civilization.* 3d ed. Belmont, CA: Wadsworth Publishing Co.; 1978.

Baumel, H. B. *Biology: its historical development.* New York: Philosophical Library; 1978.

Coleman, W. *Biology in the 19th century.* New York: Cambridge University Press; 1978.

Ewan, J., ed. *A short history of botany in the United States.* New York: Hafner Publishing Co.; 1969.

Gardner, E. J. *History of Biology.* 3d ed. Minneapolis: Burgess Publishing Co.; 1972.

Greene, E. L. *Landmarks of botanical history.* Washington, D.C.: Smithsonian Miscellaneous Collections. Vol. 54; 1909.

Journal of the History of Biology. Vol. 1, *et seq.* Cambridge, MA: Belknap Press, Harvard University; 1968.

Locy, W. A. *Growth of biology.* New York: Henry Holt and Co.; 1925.

McKenzie, A. E. *The major achievements of science.* New York: Simon and Schuster; 1973.

Nordenskiold, E. *The history of biology.* New York: Tudor Publishing Co.; 1935.

Peattie, D. C. *Cargoes and harvests.* Englewood Cliffs, NJ: Prentice-Hall; 1926.

———. *Green laurels, the lives and achievements of the great naturalists.* New York: Simon and Schuster; 1936.

Reed, H. S. *A short history of the plant sciences.* Waltham, MA: Chronica Botanica Co.; 1942.

Sirks, M. J.; Zirkle, C. *The evolution of biology.* New York: Ronald Press Co.; 1964.

Swift, L. H. *Botanical bibliographies: a guide to bibliographic materials applicable to botany.* Minneapolis: Burgess Publishing Co.; 1970.

Nature of Life 2

Overview

This chapter begins with a discussion of the attributes of living organisms (e.g., growth, reproduction, response to stimuli, metabolism, movement, complexity of organization, and adaptation to the environment). Then it examines the chemical and physical bases of life. A brief look at the elements and their atoms and molecules is followed by a discussion of bonds, compounds, mixtures, acids, bases, and salts. Forms of energy and the chemical components of protoplasm are examined next. The chapter concludes with an introduction to carbohydrates, lipids, and proteins.

Some Suggested Student Learning Goals

1. List the attributes of living organisms.

2. Define matter; describe its basic state.

3. Distinguish compounds from mixtures and describe acids, bases, and salts.

4. List the various forms of energy.

5. Name the elements found in protoplasm.

6. Describe carbohydrates, lipids, and proteins.

Introduction

Have you ever dropped a pellet of dry ice (frozen carbon dioxide) into a pan of water and watched what happens? As the solid pellet is rapidly converted to a gas, due to its contact with the warmer water, it darts randomly about the surface looking like a highly energetic bug waterskiing. Does all that motion make the dry ice alive? Hardly; yet one of the attributes of living things is movement. But if living things move, what about plants? Is a tree not alive because it does not crawl down the sidewalk? Again the answer is no, but these questions do serve to point out some of the difficulties encountered in defining *life*. In fact, some contend that there is no such thing as life—only living organisms—and that life is a concept based on the collective attributes of living organisms.

The activities of living organisms emanate from *protoplasm,* the physical basis, or the "stuff of life," contained in tiny structural units called *cells,* which are unique to living things. Both protoplasm and cells are discussed in chapter 3. Other attributes, or features, of living things are discussed in the following sections.

Attributes of Living Organisms

Growth

The complex phenomenon of growth has been described simply as an increase in volume. Growth, which results primarily from the production of new protoplasm, includes variations in *form*—some the result of inheritance, some the result of environmental response. As an example of environmental response, consider what might happen if you were to plant two apple seeds of the same variety in poor soil and subsequently give them unequal treatment. If you were to give one just barely enough water to allow it to germinate and grow, while you not only gave the other an ample water supply but also worked fertilizers and conditioners into the soil around it, you might expect the second one to grow much larger and be much more productive than the first. In other words, although your two apple trees grew from the same variety of apple seed, they would differ in form, following patterns of growth dictated by the protoplasm and the environment.

Reproduction

Dinosaurs were abundant 160 million years ago, but none exists today. Numerous mammals, birds, reptiles, plants, and other organisms are now on lists of endangered or threatened species, and many species are doomed to extinction within the next decade or two. All these once-living or living things have one feature in common: it became impossible or it has become difficult for them to reproduce. **Reproduction** is such an obvious feature of living organisms that we take it for granted—until it is lost.

When reproduction occurs, the offspring are always similar to the parents: guppies never have puppies, just more guppies; and a petunia seed, when planted, will not develop into a polyanthus. In addition, offspring of one kind tend to resemble their parents more than they do other individuals of the same kind. The laws governing these aspects of inheritance are discussed in chapter 12.

Response to Stimuli

If you stick a pin into a pillow, you certainly do not expect any reaction from the pillow, but if you were to stick the same pin into a friend, you know the reaction would be instantaneous (assuming the friend was conscious) because response to stimuli is a characteristic of all living things. You might argue, however, that when you stuck a pin into your geranium nothing happened, even though you were fairly certain the plant was alive. What you might not have been aware of is that the geranium did indeed respond, but in a manner very different from that of a human. Plant responses to stimuli are generally much slower than those of animals and usually are of a different nature. If the geranium's food-conducting tissue was pierced, it probably responded by producing a plugging substance called **callose** in the affected cells. Some studies have shown that callose may form within as little as five seconds after wounding. In addition, an unorganized tissue called **callus,** which forms much more slowly, may be produced at the site of the wound. Responses of plants to injury and to other stimuli such as light, temperature, and gravity are discussed in chapters 8 through 10.

Metabolism

Metabolism has been defined as the "sum total of all the biochemical reactions taking place within an organism." All living organisms undergo various metabolic activities that include the production of new protoplasm, the repair of damage, and normal maintenance. The most important activities include **respiration,** an energy-releasing process that is the most universal and definitive feature of all living cells; **photosynthesis,** an energy-harnessing process in green cells that is, in turn, associated with energy storage; **digestion,** the conversion of large or insoluble food molecules to smaller soluble ones; and **assimilation,** the conversion of raw materials into protoplasm and other cell substances. These topics are discussed in chapters 8 through 10.

Movement

As observed at the beginning of this chapter, plants generally do not move from one place to another (although their reproductive cells may do so). This does not mean, however, that plants do not exhibit movement, a universal characteristic of living things. Sensitive plants fold their leaves within a few seconds after being disturbed or subjected to sudden environmental changes, and bladderworts snap their tiny underwater traps shut in less than one-hundredth of a second. But most plant movements are slow and imperceptible, when compared with those of animals, and are primarily related to growth phenomena. They become obvious only when demonstrated experimentally or when shown by time-lapse photography. The latter often reveals many types and directions of motion, particularly in young organs. Movement is not confined to the organism

as a whole but occurs down to the cellular level. **Cyclosis,** a streaming motion of protoplasm, occurs constantly in living cells. The streaming tends to resemble a river flowing clockwise or counterclockwise within the boundaries of each cell, but movement may actually be in various directions.

Complexity of Organization

Living organisms are composed of large numbers of **molecules** (the smallest unit of an element or compound retaining its own identity). The molecules are not simply mixed, like the ingredients of a cake or the concrete in a sidewalk, but are organized into cells and tissues. Even the most complex nonliving object has only a tiny fraction of the types of molecules of the simplest living organism, and in the living organism the arrangements of these molecules are highly structured and complex. Bacteria and blue-green algae, for example, are considered to have the simplest cells known, yet each cell contains a minimum of six hundred different kinds of protein in addition to hundreds of other substances, and each component has a specific place or structure within the cell. When larger living objects such as flowering plants are examined, the complexity of organization is overwhelming, and the number of molecule types can run into the millions.

Adaptation to the Environment

Assume that you skip a flat stone across a body of water and it lands on the opposite shore. The stone is not at all affected by the change from air to water to land; it is totally incapable of responding to its environment. Living organisms, however, do respond to the air, light, water, and soil of their environment. As will be explained in later chapters, they are genetically adapted to their environment in many subtle ways. Some weeds (e.g., dandelions) can thrive in a wide variety of soils and climates, whereas many species now threatened with extinction have adaptations to their environment that are so specific they cannot tolerate even relatively minor changes.

Chemical and Physical Bases of Life

The Elements: Units of Matter

The basic "stuff of the universe" is called *matter*. On earth, matter occurs in three states; namely, *solid, liquid,* and *gas.* In simple terms, matter's characteristics are as follows:

1. It occupies space.

2. It has mass (weight).

3. It is composed of **elements,** of which there are at present 107 known (103 officially recognized)—17 have been produced artificially and 90 occur naturally on our planet. Only a few of the latter (e.g., nitrogen, oxygen, gold, silver, copper) are found in pure form in nature; the others are found combined together chemically in various ways.

The smallest subdivision of an element that can exist is called an **atom.** An atom can be described in terms of a **nucleus,** consisting of particles with positive electric charges called **protons** and uncharged particles called **neutrons.** Each kind of atom has a specific number of protons in its nucleus, ranging from one in the lightest element (hydrogen) to ninety-two in the heaviest element (uranium). The *atomic number* of an element reflects the number of protons in the nucleus of one atom. The *atomic mass* of an element is determined by the number of protons and neutrons present in the nucleus of one atom (see table 2.1).

Negative electric charges called **electrons** whirl around the atom's nucleus. The mass of an electron is so infinitesimal that it is generally disregarded. Electrons equal the protons in number, so that the positive electric charges of the protons and the negative electric charges of the electrons balance each other. The number of neutrons in the atoms of an element can vary slightly so that

Table 2.1 Atomic numbers and masses of some elements found in plants

Element	Atomic Number	Usual Atomic Mass
Hydrogen (H)	1	1
Helium (He)	2	4
Boron (B)	5	11
Carbon (C)	6	12
Nitrogen (N)	7	14
Oxygen (O)	8	16
Sodium (Na)	11	23
Magnesium (Mg)	12	24
Phosphorus (P)	15	31
Sulphur (S)	16	32
Chlorine (Cl)	17	35
Potassium (K)	19	39
Iron (Fe)	26	56

the element can occur in forms having different weights but with all the forms behaving alike chemically. Such variations of an element are called **isotopes.** The element oxygen, for example, has seven known isotopes. The nucleus of one of these isotopes contains eight protons and eight neutrons; another isotope holds eight protons and ten neutrons in the nucleus. An atom can be converted to a charged particle called an **ion** by the removal or addition of one or more electrons.

The atoms of most elements have the property of binding to other atoms to form new combinations; indeed, most elements do not exist independently as single atoms. The smallest independently existing particle of a compound (compound is defined in the next section), consisting of two or more atoms bound together, is called a *molecule.* One of the simplest molecules is that of the element hydrogen, which has two atoms bound together. (Figure 2.1 shows a model of a hydrogen molecule, as well as models of oxygen and water molecules.)

Molecules: Combinations of Elements

The **bonds** that hold the atoms together do so in various ways. Sodium chloride (table salt), for example, is formed by *ionic bonding* between an atom of sodium (Na) and an atom of chlorine (Cl). The sodium becomes a positively charged

Figure 2.1 Models of oxygen, water, and hydrogen molecules

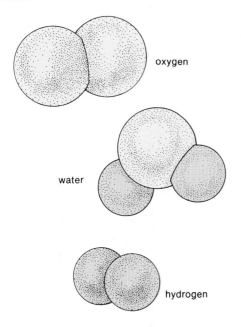

oxygen

water

hydrogen

ion when it loses one of its electrons, which is gained by an atom of chlorine. This extra electron makes the chlorine atom negatively charged, and the sodium ion and chlorine ion become bonded together by the force of opposite charge.

Covalent bonds, which are somewhat weaker than ionic bonds, are formed when two or more atoms share one, two, three, or more pairs of electrons. In some instances negatively charged oxygen and nitrogen molecules may form an attracting force between slightly positively charged hydrogen atoms of other molecules, thus forming a weak bond. Since such bonds are always associated with hydrogen, they are called *hydrogen bonds.* Hydrogen bonds have only about 5% of the strength of covalent bonds involving other elements.

The combining capacities of the atoms or ions are called *valence.* Atoms of the element calcium, for example, have a valence of two, while those of the element chlorine have a valence of one. In order for the atoms of these two elements to combine to form a molecule, there must be a balance between electrons lost or gained (i.e., the valences must balance); thus it takes *two* chlorine atoms to combine with *one* calcium atom.

When two or more elements are united in a definite ratio by chemical bonds, the substance is called a **compound.** The compound formed by the union of calcium and chlorine is called *calcium chloride.* It is customary to use standard abbreviations taken from the Latin names of the elements when giving chemical formulas or equations. Thus, calcium chloride is $CaCl_2$, the Ca indicating one atom of calcium and the Cl_2 indicating two atoms of chlorine. A **mixture** differs from a compound in not having all of its atoms or molecules chemically united in definite ratios—the proportions may vary. For example, granite is composed of several different materials that vary in proportion to one another throughout the rock; thus, granite constitutes a mixture.

Acids, bases, and salts are three of the most common kinds of compounds. An *acid* is defined as a compound that releases positively charged hydrogen ions (shown as H^+) when dissolved in water. An acid has a sour taste, like the citric acid in lemon juice. A *base* is defined as a compound that releases hydroxyl ions (negatively charged combinations of an oxygen and a hydrogen atom, shown as OH^-) when dissolved in water. Bases are also referred to as alkaline compounds and may have a soapy feel to them, although they can be quite corrosive. When an acid and a base are mixed, the H^+ ion of the acid unites with the OH^- ion of the base to form water (H_2O). The remainders unite to form a *salt.* For example, if hydrochloric acid (HCl) is mixed with sodium hydroxide (NaOH), which is a base, then water (H_2O) and sodium chloride (NaCl), which is a salt, are formed. The reaction is represented by symbols in an equation that shows what occurs:

$$HCl + NaOH \longrightarrow H_2O + NaCl$$

Nature of Life

17

Energy

The activities of all living things, from molecules to cells to whole organisms, require **energy** to take place. Energy can be defined as the "ability to produce a change in motion or matter," or simply as the "ability to do work." It takes a number of forms and may be converted from one form to another. Among its forms are chemical, electrical, heat, and light energy. Other forms include kinetic (motion) and potential energy. Potential energy is defined as the "capacity to do work owing to the position or state of a particle." For example, a ball resting at the top of a hill possesses potential energy that is converted to kinetic energy if the ball rolls down the hill. Some of the numerous energy exchanges occurring in living cells are discussed in later chapters.

Chemical Components of Protoplasm

The living substance of all cells is called **protoplasm.** Protoplasm is organized into numerous bodies of various sizes, most of which are discussed in chapter 3. About 96% of protoplasm is composed of the elements carbon, hydrogen, oxygen, and nitrogen; 3% consists of phosphorus and potassium. The remaining 1% includes calcium, iron, magnesium, sodium, chlorine, copper, manganese, cobalt, zinc, and minute quantities of other elements. When a plant first absorbs these elements from the soil or atmosphere, or when it utilizes breakdown products within the cell, the elements are in the form of simple molecules or ions. These simple forms may be converted to very large, complex molecules through the metabolism of the cells. The large molecules invariably have chains of carbon atoms within them and are said to be **organic,** in contrast to **inorganic** molecules, which usually contain no carbon atoms. The name organic was given to most of the chemicals of living things when it was believed that only living organisms could produce molecules containing carbon. Today, many organic compounds can be produced artificially in the labo-

Figure 2.2 Structural formula and model of a glucose molecule. The molecule can exist either as a straight chain (as shown) or in the form of a ring (not shown).

ratory, and scientists sometimes hesitate to classify as either organic or inorganic some of the 4 million carbon-containing compounds thus far identified. Most scientists, nevertheless, agree that inorganic compounds usually do not contain carbon. Four of the most important classes of organic compounds found in protoplasm are carbohydrates, lipids, proteins, and nucleic acids.

Carbohydrates

Carbohydrates have a number of roles in living organisms, but the two most important are as energy sources and structural components. Carbohydrates contain carbon, hydrogen, and oxygen in, or close to, a ratio of 1 C:2 H:1 O. Examples of carbohydrates include monosaccharides (simple sugars), disaccharides, and polysaccharides. A monosaccharide such as glucose has a skeleton chain of six carbon atoms, with hydrogen and oxygen atoms attached, giving it the formula $C_6H_{12}O_6$ (figure 2.2). Disaccharides consist of two simple sugars joined together. Sucrose, or common table sugar, is a disaccharide made up of the two simple sugars glucose and fructose; sucrose is the form in which sugar is usually transported by plants. Polysaccharides sometimes consist of hundreds of simple sugars attached to one another. Examples of polysaccharides are starches, which consist of 300 to 1,000 glucose units attached to one another, and the structural substance cellulose, which consists of about 2,000 slightly different glucose units attached to one

Figure 2.3 Partial structural formula and partial model of a fat molecule. The two fatty acids not depicted are similar to fatty acid A.

another. When glucose molecules are attached to one another to form a starch molecule, each of the glucose molecules gives up one molecule of water. The n in the formula for starch—$(C_6H_{10}O_5)n$—represents many units. In order for a starch molecule to become available as an energy source in cells, it has to be broken up into its individual simple sugar units through the restoration of the water molecule for each unit. This addition of water molecules is called **hydrolysis;** it takes place during the process of **digestion.**

Lipids

Lipids are fatty or oily substances that are insoluble in water. Because they can store more energy than carbohydrates can, they play an important role in the energy reserves and structural components of cells. Like carbohydrates, their molecules contain carbon, hydrogen, and oxygen, but proportionately there is usually much less oxygen present. Examples of lipids include **fats** (figure 2.3) and **oils,** whose molecules are manufactured from sugars and are composed of a unit of glycerol (an alcohol) with three fatty acids attached. The fatty acids have carbon atoms to which hydrogen atoms can become attached. If hydrogen atoms are attached to every available attachment point of these fatty acid carbon atoms, the fat is said to be *saturated;* if, however, there are very few places for the hydrogen atoms to attach themselves, as is the case with some highly advertised vegetable oils, the fat is said to be *polyunsaturated.* Like polysaccharides and proteins (discussed in the next section), lipids are broken down by hydrolysis. Other examples of lipids are *waxes,* which have the fatty acids attached to an alcohol molecule other than glycerol, and *phospholipids,* which are similar in structure to fats but have a phosphorus atom replacing one of the three fatty acids. Phospholipids are important components of cell membranes.

Proteins

A single cell may contain several thousand different kinds of **proteins.** These are important molecules that are usually very large and are composed of subunits called **amino acids** (figure 2.4). Proteins consist of carbon, hydrogen, oxygen, and nitrogen atoms, and sometimes also sulphur atoms. Many structural components of protoplasm are proteins.

There are about twenty different kinds of amino acids, and from fifty to fifty thousand of them are present in various combinations in each protein molecule. Each amino acid has two special groups of atoms: an amino group, NH_2, and a carboxyl group (the acid portion) COOH, both of which are attached to the same carbon atom. The amino (NH_2) group of an amino acid has the capacity to combine with other acids, while the carboxyl (COOH) group of the same amino acid has the capacity to combine with bases.

The NH_2 and the COOH groups of a protein's amino acid building blocks bond together to form chains. The bonds are called **peptide bonds,** and the chains they produce are called *polypeptide chains.*

The polypeptide chains of a protein frequently coil in a *helix;* the helix itself, which resembles a spiral staircase, may fold into a globular form, as illustrated by the protein cytochrome *c* shown in figure 2.5.

Since proteins vary in the total number of amino acids they contain, in the relative number of different kinds of amino acids present, and in the sequence of the amino acids in their polypeptide chains, the number of different kinds of protein possible staggers the imagination. Among living things it is no wonder that every single species of the millions in existence possesses at least one unique protein.

Enzymes, which are produced in all living cells, are complex proteins that function as organic catalysts. Even at very low concentrations they speed up specific chemical reactions in the cells where they are produced, without being used up in the reactions. They are essential to the normal metabolic activities of cells.

Figure 2.4 Structural formula and model of the amino acid glycine

Nucleic Acids

Nucleic acids are complex substances associated primarily with the nuclei of cells. Their structure and function are discussed in chapter 3.

Summary

Activities of living organisms stem from protoplasm, the physical basis, or the "stuff of life." Growth is among the attributes of living organisms. Growth has been described as an increase in volume; it results primarily from the production of new protoplasm. Variations in form may be inherited or result from response to the environment. Reproduction involves offspring that are always similar in form to their parents; if reproduction ceases, the organisms become extinct. Plants generally respond to stimuli more slowly and in a different fashion from animals. All living organisms exhibit metabolic activities, including production of new protoplasm, respiration, photosynthesis, digestion, and assimilation; they also all exhibit movement to varying degrees. Cyclosis is the streaming motion of protoplasm within living cells. Living organisms have a much more complex structure than nonliving objects and are adapted to their individual environments.

The basic "stuff of the universe" is called matter. It is composed of elements, the smallest subdivision of which is an atom. Atoms contain protons, neutrons, and electrons. Isotopes are forms of elements that have slight variations in the number of neutrons in their atoms. The earth's matter occurs in three states: solid, liquid, and gas. Atoms of most elements can bond to other

Chapter 2

Figure 2.5 Model of cytochrome *c*, a relatively small protein. (From Richard E. Dickerson and Irving Geis. 1982. *Proteins: Structure, Function and Evolution.* 2d ed. Benjamin Cummings Publishing Co., Menlo Park, Calif. Illustration copyright Irving Geis)

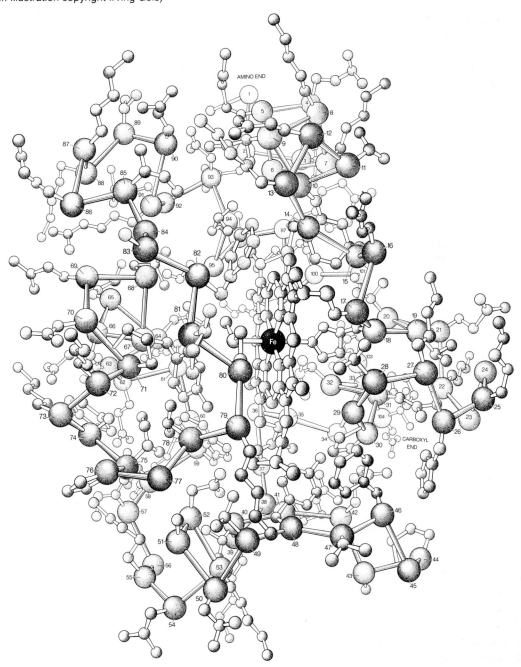

atoms, and those of many elements do not exist independently; molecules are the smallest independently existing particles. The combining capacities of atoms or ions are called valence. Compounds are substances composed of two or more elements combined in a definite ratio by chemical bonds; the atoms of mixtures are not all chemically united in definite ratios. Acids release positively charged hydrogen ions when dissolved in water. Bases release negatively charged hydroxyl ions when dissolved in water. Salts and water are formed when acids and bases are mixed.

Energy can be defined as "ability to produce a change in motion or matter" or simply as "ability to do work." Its forms include chemical, electrical, heat, and light, and kinetic and potential.

Protoplasm is composed primarily of carbon, hydrogen, oxygen, and nitrogen, with a little phosphorus and potassium, plus trace amounts of other elements. A plant may convert the simple molecules or ions it recycles or absorbs from the soil to very large, complex molecules. Organic molecules are usually large and have chains of carbon atoms within them. Carbohydrates contain carbon, hydrogen, and oxygen in a ratio of 1 C:2 H:1 O. Carbohydrates occur as monosaccharides (simple sugars, e.g., glucose); disaccharides, consisting of two simple sugars joined together (e.g., sucrose); and polysaccharides, consisting of many simple sugars joined together (e.g., starch, cellulose). Simple sugars each give up a molecule of water when they are attached to one another to form starch. Hydrolysis is the process of restoring a water molecule to each simple sugar when starch is broken down during digestion. Lipids (e.g., fats and oils) consist of a unit of glycerol with three fatty acids attached. They are insoluble in water and contain carbon, hydrogen, and oxygen, with proportionately much less oxygen than carbohydrates. Saturated fats are those that have hydrogen atoms attached to every available attachment point of their carbon atoms; if there are very few places for hydrogen atoms to attach themselves, the fat is said to be polyunsaturated. Proteins are usually large molecules that are composed of subunits called *amino acids*. Each amino acid has two special groups of atoms: an amino group (NH_2) and a carboxyl group (COOH); these groups bond amino acids together to form polypeptide chains; the bonds are called *peptide bonds*.

Review Questions

1. What is meant by the term *organic?*

2. How are acids, bases, and salts distinguished from one another?

3. Distinguish among carbohydrates, lipids, and proteins.

4. What distinguishes a living organism from a nonliving object such as a rock or a tin can?

5. What is energy, and what forms does it take?

Discussion Questions

1. Can part of an organism be alive while another part is dead? Explain.

2. What is the difference between inherited form and form resulting from response to the environment?

Additional Reading

Alder, I. *How life began.* New York: New American Library (Signet Books); 1957.

Baker, J. J. W.; Allen, G. E. *Matter, energy and life.* Reading, MA: Addison-Wesley Publishing Co.; 1965.

Raven, P. H.; Evert, R. F.; Curtis, H. *Biology of plants.* 3d ed. New York: Worth Publishers; 1981.

Cells 3

Overview

This chapter gives a brief review of the history of the discovery of cells and the development of the cell theory. After observations on cell size and structure, it discusses each of a cell's particulates, beginning with the cell wall. Included are the plasma membrane, mitochondria, endoplasmic reticulum, ribosomes, Golgi bodies, plastids, microtubules, lysosomes, microbodies, vacuoles, vacuolar membranes, and the nucleus. Distinctions between plant and animal cells, and those between prokaryotic and eukaryotic cells are then given. The chapter next discusses mitosis and cytokinesis; this is followed by a simple review of intracellular and intercellular communication, DNA structure and replication, and types of RNA and their functions.

Some Suggested Student Learning Goals

1. List five historical figures associated with the development of the cell theory and name their contributions.

2. Identify the following cell structures and organelles, and indicate the function of each: plasma membrane, mitochondria, endoplasmic reticulum, ribosomes, Golgi bodies, plastids, vacuoles.

3. Describe the components of a nucleus and the function of each component.

4. Contrast plant cells with animal cells.

5. Describe mitosis and the events that take place in each phase.

Introduction

All living organisms—aardvarks or apple trees, zebras or zamias, or even microscopic microbes—are composed of **cells.** Each of us began life as a single cell; this cell divided repeatedly until it developed into an organism consisting of billions of cells. During the first few hours of this organism's development the cells all resembled each other, but changes soon occurred, not only in their appearance but also in their function. The modifications of some, for example, permitted them to serve as conducting cells, while others came to function in secretion or support. Some cells live and function for many years; others mature and degenerate in just a few days. Even as you read this, millions of new cells are being produced in your body. Some cells add to your total body mass (if you have not yet stopped growing), but most replace the millions of older cells that are destroyed every second you remain alive. The variety and form of cells seems almost infinite, but certain features are shared by most of them. A discussion of these features forms the body of this chapter.

Cells

History

The discovery of cells is associated with the development of the microscope in the seventeenth century. Despite the fact that some of the optical properties of curved surfaces were known as early as the third century B.C., it was not until the end of the sixteenth century that the first microscope was produced by Zacharias Janssen and then developed by Leeuwenhoek into an instrument that could magnify up to two hundred times (see chapter 1). In 1665, the English physicist Robert Hooke, using such a microscope (figure 3.1), examined thin slices of cork that he had cut with a sharp penknife. Hooke described the boxlike

Figure 3.1 Robert Hooke's microscope as illustrated in one of his works. (Courtesy National Library of Medicine)

compartments he saw as resembling the surface of a honeycomb; he is credited with coining the term *cell.* What Hooke saw in the cork were really only the walls of dead cells, but he also observed "juices" in living cells of elderberry plants and thought he had found something similar to the veins and arteries of animals.

During the next fifty years Leeuwenhoek, along with Hooke's compatriot Nehemiah Grew, reported at frequent intervals on the existence of cells in a variety of plant tissues and on the form and structure of single-celled organisms that were referred to as "animalcules."

After this period, little more was reported on cells until the early 1800s. This lack of progress was due in large part to the imperfections of the primitive microscopes and also to the crude methods of tissue preparation used. Both microscopes and tissue preparations slowly improved, however. In 1809, the famous French biologist Lamarck concluded that "no body can have life if its constituent parts are not cellular tissue or are not formed by cellular tissue." In 1824, another Frenchman, R. J. H. Dutrochet, reinforced

Chapter 3

Figure 3.2 Pasteur's experiment. Nutrient broths were sterilized by boiling after they had been added to both straight-necked and swan-necked flasks. The broths were then left in their openmouthed containers for several weeks. Microorganisms such as bacteria entered the flasks from the atmosphere, but moisture in the bends of those flasks with curved necks trapped the organisms, causing the broth to remain sterile, while the broth in those flasks with straight necks became contaminated.

organisms trapped here

Lamarck's conclusions that all animal and plant tissues are composed of cells of various kinds, but neither of them realized that each cell could, in most cases, reproduce itself and exist independently. In 1833, the Englishman Robert Brown discovered that all cells contain a relatively large body that he called the *nucleus;* shortly thereafter, the German botanist Matthias Schleiden observed a smaller body within the nucleus that he called the *nucleolus.* Schleiden and a German zoologist, Theodor Schwann, were not the first to grasp the significance of cells, but they explained them with greater clarity and perception than others before them had done. They are generally credited with developing the cell theory, beginning with their publications of 1838–39. In essence, this theory holds that all living organisms are composed of cells and that cells form a unifying structural basis of organization.

In 1858 an important augmentation of the cell theory appeared in a classic textbook by another German scientist, Rudolf Virchow. He argued cogently that every cell comes from a preexisting cell ("*omnis cellula e cellula*") and that there is no spontaneous generation of cells. Virchow's publication stirred up a great controversy, because prior to this time there was a widespread belief among scientists and nonscientists alike that animals could originate spontaneously from dust. Many who had microscopes were thoroughly convinced they could see "animalcules" appearing in decomposing substances. The controversy became so heated that in 1860, the Paris Academy of Sciences offered a prize to anyone who could, through experiments, shed light on the matter. Just two years later the brilliant French scientist Louis Pasteur was awarded the prize. Pasteur, using swan-necked flasks, demonstrated conclusively that boiled media remained sterile indefinitely if microorganisms from the air were excluded from the media (figure 3.2).

In 1871 Pasteur proved that natural alcoholic fermentation always involves the activity of yeast cells. In 1897 the German scientist Eduard Buchner accidentally discovered that the yeast cells did not need to be alive for fermentation to occur. Much to everyone's surprise, Buchner found that extracts from the yeast cells would convert sugar to alcohol. This discovery quickly led to the identification and description of enzymes, the organic catalysts (substances that aid chemical reactions without themselves being changed) found in all living cells; it also led to the belief that cells were little more than miniature packets of enzymes. During the first half of the twentieth century, however, great advances were made in the refinement of microscopes and in tissue preparation techniques. Numerous structures and bodies, in addition to the nucleus, were observed in cells, and the relationship between structure and function came to be realized and understood.

The best light microscopes in use today can produce useful magnifications of up to fifteen hundred times under ideal conditions. Such microscopes will continue to be useful, particularly for observing living cells, into the foreseeable future. The advent of modern electron microscopes (figure 3.3), however, has permitted magnifications of two hundred thousand or more times, and scientists can now see a myriad of details that were unknown before the 1950s. Material to be viewed with a high-powered electron microscope is placed in a vacuum and must be sliced so thinly that living material cannot be observed. The techniques of preservation of materials for such observation have become so refined that even the preserved material can appear exceptionally lifelike. Cell biologists all over the world, using both electron and light microscopes, are presently engaged in research that will undoubtedly lead to further significant discoveries in the next few years.

Cell Size and Structure

Most plant cells, and the vast majority of animal cells, are so tiny they are invisible to the unaided eye. Cells of higher plants generally vary in length between 10 and 100 micrometers.[1] Since there are roughly twenty-five thousand micrometers to the inch, it would take about five hundred average-sized cells to extend across 1 inch of space; thirty of them could easily be placed across the head of a pin! Some bacterial cells are less than 0.5 micrometer in diameter, while cells of the green alga mermaid's wineglass are between 6 and 9 centimeters in length, and fiber cells of some nettles are about 20 centimeters long.

Because cells are so minute, the numbers occurring in full-grown organisms are astronomical. For example, it has been calculated that a

---

1. See appendix 5 for metric conversion tables.

Figure 3.3 Electron microscope

single mature leaf of a pear tree contains 50 million cells and that the total number of cells in the roots, stem, branches, leaves, and fruit of a full-grown pear tree exceeds 15 trillion. Can you imagine how many cells there are in one of the three-thousand-year-old redwood trees of California, which reach heights of 90 meters (300 feet) and measure up to 4.5 meters (15 feet) in diameter near the base?

Some cells are boxlike with six walls, but others assume a wide variety of shapes depending on their location and function. The most abundant cells in the younger parts of plants and fruits are packed together in such a way that they commonly have fourteen sides. These are discussed in the next chapter.

Chapter 3

Figure 3.4 Cross section through leaf mesophyll cells of a garden bean seedling seen with the aid of an electron microscope. (Courtesy Jean Whatley)

nuclear envelope

nucleolus

cell wall

nucleus

plasmodesma

mitochondria

vacuole

chloroplasts

ribosomes

intercellular space

endoplasmic reticulum

older chloroplast containing starch

The living part of the cell within the wall is called *protoplasm.* Two main components of protoplasm are recognized: the *nucleus,* which controls the cell's activities, and the **cytoplasm,** which is a souplike fluid containing water, dissolved substances, and many, small **organelles** (membrane-bounded bodies of various sizes and shapes). The organelles are the sites of many different activities that take place within the cell. A brief examination of each of the various cell components follows. (Figures 3.4 and 3.5 show these various components.)

Figure 3.5 *A.* Leaf cell diagrammed with the aid of a light microscope. *B.* The same cell greatly enlarged to show submicroscopic features.

nucleus
nucleolus
vacuole
cytoplasm
chloroplast
cell wall

A.

plasma membrane
ribosome cluster
cytoplasm

nucleus
nucleolus
primary cell wall
vacuole
crystal
vacuolar membrane (tonoplast)
mitochondrion
Golgi body
chloroplast
rough endoplasmic reticulum
smooth endoplasmic reticulum
plasmodesma
middle lamella

B.

Ribosomes

Cell Wall

A popular song of two or three decades ago had several verses listing food items the author purportedly disliked, with each verse ending, "But I like bananas, because they have no bones!" Indeed, bananas and all parts of plants differ from animals in that they have no bones or similar internal skeletal structures. Yet large trees support branches and leaves weighing many tons. They are able to do this because most plant cells have semirigid or rigid walls that perform the functions of bones; that is, they provide strength and support for the plants. When millions of these cells work together as a **tissue,** their collective strength is enormous. The largest trees alive today, the redwoods, exceed the mass or volume of the largest land animals, the elephants, by more than a hundred times. The wood of one tree could support the combined weight of a thousand elephants.

When new **cell walls** are first formed, a **middle lamella,** consisting primarily of **pectin** (the complex organic material that gives stiffness to fruit jellies), appears. This middle lamella is normally shared by two adjacent cells and is so thin that it may not be visible with an ordinary light microscope unless it is specially stained. A fine network of **cellulose** (a substance related to starch and sugar and used in the production of paper and other products) is laid down on either side of the middle lamella. Sometimes the cellulose is deposited in two stages, forming a *primary cell wall* and then a *secondary cell wall* inside the primary wall. When this happens, the secondary cell wall is usually the more extensive of the two structures. Most of the cell wall consists of cellulose and, depending on the type of cell involved, other substances such as sugars and **lignin** may be impregnated into the wall. Lignin is the substance from which the controversial pain reliever dimethyl sulfoxide (DMSO) is manufactured. The thickness of the wall can also vary, occupying more than 95% of the volume of the cell in some instances and less than 5% in others. Cells that function in food storage or manufacture usually have thin walls, while those primarily involved in support usually have walls of moderate to extensive thickness.

Protoplasm

Plasma Membrane

The outer boundary of the living part of the cell, the **plasma membrane** is roughly eight-millionths of a millimeter thick. To get an idea of how incredibly thin that is, consider that it would take 12,500 such membranes neatly stacked in a pile to achieve the thickness of an ordinary piece of writing paper. Yet this delicate structure is of vital importance in regulating what substances enter and leave the cell. Evidence obtained since the early 1970s indicates that this and other cell membranes are mosaics composed of lipids, with proteins interspersed throughout (figure 3.6). When these membranes are stained and magnified by an electron microscope as much as a million times, they appear as two dark parallel lines. Some proteins extend across the entire width of the membrane, whereas others apparently are loosely bound to the outer surface. The remainder of the cell contents usually push the plasma membrane up against the cell wall because of pressures developed by **osmosis** (see chapter 8), but the membrane is quite flexible and often forms folds, which may in turn become little hollow balls or vesicles that float off into the cell. The membrane may even shrink away from the wall temporarily. Thus, it behaves more like a fluid than a solid, but if it ever ruptures, the cell soon dies.

Mitochondria

Mitochondria are often referred to as the "powerhouses" of the cell, for it is within them that energy is released from sugar by the process of respiration (the role of mitochondria in respiration is further discussed in chapter 9). This energy is needed to keep the individual cells and the

Figure 3.6 Model of a small portion of a plasma membrane, showing its fluid mosaic nature. The proteins shown here as blocks (they are actually coiled chains of polypeptides), occur either on the surfaces or are embedded. Some of the embedded proteins extend all the way through the continuous part of the membrane, which is a double layer of lipids. The pairs of vertical lines extending inward from each lipid represent long chain fatty acids.

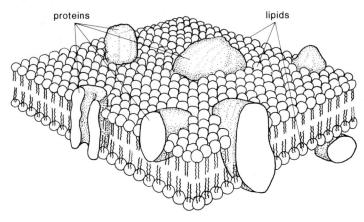

proteins   lipids

Figure 3.7 Mitochondrion greatly enlarged and cut away to show the cristae (folds of the interior membrane).

outer membrane

inner membrane

cristae

plant functioning as a whole. Mitochondria are numerous and tiny, typically measuring from 1 to 3 or more micrometers in length and having a width of roughly one-half micrometer. They assume various shapes, but are most frequently seen shaped as a gherkin or a paddle. Each is bounded by a double membrane, with the inner membrane forming numerous platelike folds called *cristae*. A sectioned mitochondrion resembles a scooped-out watermelon with inward extensions of the rind forming mostly incomplete partitions at right angles to the surface (figure 3.7). These partitions (cristae) greatly increase the amount of surface

Chapter 3

Figure 3.8 Endoplasmic reticulum and ribosomes in a young leaf cell of corn (*Zea mays*). (Courtesy Jean Whatley)

endoplasmic reticulum

ribosomes

area in contact with the enzymes involved in releasing the energy. Mitochondria seldom remain in one location for very long; they tend to congregate in those areas of the cell needing energy to perform a particular function.

Endoplasmic Reticulum

The **endoplasmic reticulum** (figure 3.8) is a complex system of membrane channels that occurs throughout the cytoplasm, with the amount and form varying considerably from cell to cell. It appears, in section, as a series of parallel membranes that resemble long, narrow bags, sacs, or tubes. The nucleus, which directs the various activities of the cell, is connected to the endoplasmic reticulum, and many of the most important activities occur either on the surface of the endoplasmic reticulum or between its compartments. One of the endoplasmic reticulum's chief functions appears to be the storing and transporting of protein and other cell products. Many enzymes involved in the process of respiration are synthesized on the endoplasmic reticulum and transported through it to the

Cells

mitochondria. Small granular particles called *ribosomes* (figure 3.8) may line the outer surfaces of the endoplasmic reticulum. Such endoplasmic reticulum is said to be "rough"; this is in contrast to "smooth" endoplasmic reticulum, which has no ribosomes lining the surface. Both types of endoplasmic reticulum can occur in the same cell.

Ribosomes

As indicated, ribosomes may line the endoplasmic reticulum, but they may also occur free in the cytoplasm and in the nucleus or chloroplasts. They are tiny, averaging about 20 nanometers in diameter in most plant cells. The unattached ribosomes often occur in clusters of 5 to 100, particularly when they are involved in performing their function of linking amino acids together to form the large, complex protein molecules that are a basic part of all living organisms. Ribosomes are roughly elliptical in shape and are composed of two subunits, which in turn are made up of RNA and anywhere from 50 to 150 different kinds of protein. Because they are not bounded by membranes, they technically are not true organelles.

Golgi Bodies

Golgi bodies (figure 3.9), often referred to as **dictyosomes** in plant cells, appear as groups of flat, roundish sacs, frequently bounded by branching tubules. Numbers of them are usually scattered throughout the cytoplasm of a living cell. The sacs are often organized into stacks of five to eight, but up to thirty or more are not uncommon in simpler organisms. Carbohydrates such as cellulose apparently collect in small *vesicles* (blisterlike bodies) that are pinched off from the margins of the Golgi body, or dictyosome, sacs. These vesicles migrate to the plasma membrane and fuse with it. There they discharge their contents to the outside, where they become a part of the cell wall. The enzymes needed for the "packaging" process are produced or contained within the Golgi bodies, or dictyosomes, themselves. Thus, one might describe them as collecting and packaging centers.

Figure 3.9 Golgi body (dictyosome) in a cell of an adder's tongue fern. (Courtesy B. E. Juniper)

Plastids

Several kinds of **plastids** are commonly found in living cells. In green plant cells, **chloroplasts** (figure 3.10), which are one kind of plastid, are usually the most conspicuous organelles seen. They occur in a variety of shapes and sizes, from the beautiful corkscrewlike ribbons found in cells of the green alga *Spirogyra* (see figure 15.5), to the star-shaped chloroplasts of other green algae such as *Zygnema* (see figure 15.1D). The chloroplasts of higher plants, however, tend to be shaped somewhat like two Frisbees glued together along their edges, and when they are sliced in median section they resemble the outline of a football.

Chapter 3

Figure 3.10 *A.* Chloroplast of a leaf cell of a garden bean. (Courtesy Jean Whatley) *B.* Interlocking thylakoids. (After Weier, T. E., T. Bisalputra, and A. Harrison. 1966. "Subunits in chloroplast membranes of *Scenedesmus quadricola." Journal of Ultrastructure Research* 15: 38–56. Copyright 1966 by Academic Press, Inc.)

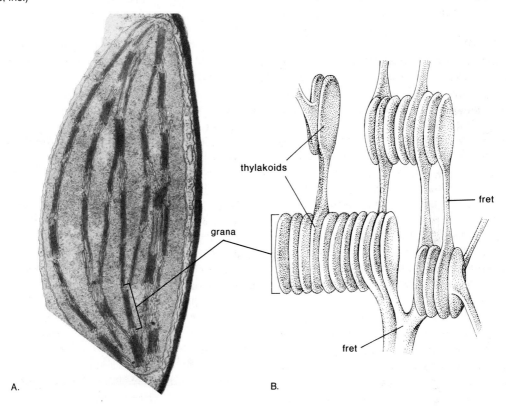

A.

B.

Although a number of the algae and a few other plants have only 1 or 2 chloroplasts per cell, in a green cell of higher plants the number of chloroplasts is usually much greater—100 to 150 is quite common. These may be from 2 to 10 micrometers in diameter, and each is bounded by two delicate membranes that enclose a colorless, enzyme-containing matrix, the **stroma.**

Suspended in the stroma are **grana** (singular: **granum**), which are stacks of coin-shaped membranes called **thylakoids;** these contain green **chlorophyll** pigments. As we shall see later, these "coin stacks," or grana, are vital to life as we know

it on our planet today, for it is within the thylakoids that the first steps of the all-important process of *photosynthesis* (see chapter 9) occurs. In photosynthesis, green plants convert carbon dioxide (from the air) and water to simple food substances, harnessing energy from the sun in the process. Thus, the existence of humans and all other animal life depends on the activities of the chloroplasts.

The grana within a chloroplast are linked together by arms called *frets*. There are usually about forty to sixty grana per chloroplast, and each granum may contain anywhere from 2 or 3 to more than 100 stacked thylakoids (see figure 3.10). There are usually four or five *starch grains* in the stroma, in addition to the enzymes mentioned.

Another type of plastid found in cells of more complex plants is the **chromoplast.** Similar to chloroplasts in size, chromoplasts vary considerably in shape, often being somewhat angular. They can develop from chloroplasts through internal changes, including the disappearance of chlorophyll. Chromoplasts are yellow, orange, or red in color due to the presence of carotenoid pigments, which they manufacture and accumulate. They are most abundant in the orange or red parts of plants, such as in ripe tomatoes, red peppers, or flower petals.

Leucoplasts are a third type of plastid common to cells of higher plants. They are essentially colorless and include *amyloplasts,* which are known to synthesize starches, and *elaioplasts,* which synthesize oils. If exposed to light, some leucoplasts will develop into chloroplasts.

Plastids of all types develop from **proplastids,** which are organelles having roughly the size and form of mitochondria. They are simpler in internal structure than plastids and have fewer thylakoids, which are not arranged in grana stacks. Proplastids frequently divide and are distributed throughout the cell so that when a cell itself divides, each daughter cell has its share. Plastids may also arise through the division of existing mature plastids.

Microtubules

Microtubules are unbranched, thin, hollow, tubelike structures of varying lengths that tend to be between 15 and 25 nanometers in diameter. They are most commonly found just inside the plasma membrane, where they apparently control the addition of cellulose to the cell wall. Other functions of microtubules include steering the vesicles from the Golgi bodies to the cell wall and aiding movement within the cell. Microtubules are also found in the special fibers that form the *spindle* of dividing cells discussed later in this chapter.

Nucleus

The nucleus is frequently the most conspicuous organelle in a living cell, although it may be obscured by chloroplasts in green cells when they are observed with the aid of a light microscope. In cells without chloroplasts, the nucleus may appear as a grayish lump, somewhat spherical or elliptical in shape and often lying against the plasma membrane to one side of the cell or in a corner. Some nuclei are irregular in form and, like most other organelles, they can vary greatly in size; however, they are generally from 2 to 15 micrometers or larger in diameter. Certain fungi and algae have numerous nuclei within a single extensively branched cell, but more complex plants usually have a single nucleus located within the cytoplasm of each living cell.

The nucleus is the control center that not only directs the myriad activities of the complex cell "factory" but also stores hereditary information, which is passed from cell to cell as new cells are formed. Each nucleus is bounded by two membranes, which are called the **nuclear envelope.** Up to one-third of the total surface area of the nuclear envelope is taken up with structurally complex pores (figure 3.11). These pores apparently are selective as to the kinds of molecules they permit to pass between the nucleus and the cytoplasm. The outer membrane of the nuclear envelope is connected to parts of the endoplasmic reticulum.

Within the nuclear envelope is a granular-appearing fluid called the *nucleoplasm.* The nucleoplasm is packed with short fibers that are about 10 nanometers in diameter, and several different larger bodies are suspended within it. The most noticeable of these larger bodies are **nucleoli** (singular: **nucleolus**), which superficially resemble miniature nuclei; there may be from one

Chapter 3

Figure 3.11 Glancing section through the nuclear envelope of a cell nucleus showing the nuclear pores in face view. (Courtesy Jean Whatley)

to several nucleoli in each nucleus. Nucleoli are composed primarily of protein and are not bounded by membranes. They are involved with the synthesis and exportation of a form of ribonucleic acid (**RNA**) to other parts of the cell. RNA is a complex substance vital to communication between the nucleus and the ribosomes. This communication is discussed in a later section of this chapter.

Other important nuclear structures, which are not apparent unless the cell is stained or is in the process of dividing, are the thin strands of **chromatin.** When a nucleus divides, the chromatin strands become shorter and thicker, and in their condensed condition, they are called **chromosomes.** Chromatin is composed of protein and deoxyribonucleic acid (**DNA**), another vital cell substance. DNA consists of units of heredity called **genes;** it is related to RNA. Each species of organism, plant and animal, has its own fixed number of chromosomes in each of its cells; the cells involved in sexual reproduction have half the number found in other cells of the same organism. The number of chromosomes present in a nucleus normally bears no relation to the size and complexity of the organism. Each body cell of a radish, for example, has eighteen chromosomes in its nucleus, while a cell of one species of goldenweed has four, and a cell of a tropical adder's tongue fern has over one thousand.

Other Organelles

Various small bodies distributed throughout the cytoplasm tend to give it a granular appearance. Included among such bodies are **lysosomes,** which are tiny spherical organelles containing enzymes; the enzymes play a role in regulating digestion in the cell. Previously it was believed that lysosomes were confined to the cells of animals and of fungi, but recent evidence suggests that small spherical bodies found in higher plant cells are actually lysosomes. Two other types of small spherical organelles found in the cytoplasm (one containing enzymes involved in one phase of photosynthesis and the other containing enzymes that aid in the conversion of fats to carbohydrates) are called *microbodies.*

Vacuoles

In a mature living plant cell, as much as 90% or more of the volume may be taken up by one or two large central **vacuoles** that are bounded by **vacuolar membranes** (tonoplasts). The latter, which constitute the inner boundaries of the living part of the cell, are similar in structure and function to plasma membranes.

The vacuole was evidently so called because of a belief that it was just an empty space; hence, its name has the same root as the word *vacuum.* Vacuoles, however, are filled with a watery fluid called **cell sap** and play a role in maintaining pressures within the cell (see the discussion in chapter 8). Cell sap contains dissolved substances such as salts, sugars, organic acids, and small quantities of soluble proteins. It also frequently contains water-soluble pigments called **anthocyanins.** Anthocyanins are responsible for many of the red, blue, or purple colors of flowers and some leaves. In some instances, anthocyanins accumulate to a greater extent in response to cold temperatures in the fall. They should not be confused, however, with the red and orange carotenoid pigments confined to the chromoplasts. These latter pigments are not soluble in water at all, but they also play a role in fall leaf coloration (see chapter 7).

Sometimes large crystals of waste products form within the cell sap after certain ions have become concentrated there. Vacuoles in newly formed cells are usually tiny and numerous. They increase in size and unite as the cell matures. In addition to accumulating the various substances and ions mentioned above, vacuoles are apparently also involved in the recycling of certain materials within the cell.

Differences between Plant Cells and Animal Cells

All animals have either internal or external support for their tissues from a skeleton of some kind. The type of support afforded by rigid cellulose structures is not found in animal cells; in fact, animal cells do not have cell walls. Instead, the plasma membrane, called the *cell membrane* by most *zoologists* (animal scientists), forms the outer boundary of animal cells. Plant cells have minute strands of cytoplasm called **plasmodesmata** (see the discussion on communication between cells on page 47) connecting the protoplasts of cells with each other through microscopic holes in the walls, whereas animal cells do not.

Other differences, which are minor, pertain to the presence or absence of certain organelles. *Centrioles,* for example, which are tiny paired organelles found just outside the nucleus, occur in all animal cells but are generally absent from higher plant cells. Centrioles are discussed in the section dealing with mitosis, which begins on page 38. *Plastids,* which are common in plant cells, are not found in animal cells. *Vacuoles,* which are often large in plant cells, are either small or absent in animal cells.

Eukaryotic Versus Prokaryotic Cells

With the few exceptions just mentioned, nearly all plant and animal cells exhibit the various features discussed so far. There are some very primitive organisms, however, whose cells lack a number of these features (e.g., membrane-bound nuclei and organelles). Such cells, called **prokaryotic,** to distinguish them from the typical **eukaryotic** cells just discussed, are covered in some detail in chapter 14.

Cyclosis

Cyclosis, or **cytoplasmic streaming,** occurs in living cells. When a living cell is examined with the light microscope, a current can be seen flowing within the protoplasm, carrying the organelles around the cell. This streaming probably facilitates exchanges of materials within the cell and plays a role in the movement of substances from cell to cell. The precise nature and origin of cyclosis is still not known, but there is evidence that bundles of tiny cytoplasmic fibers called *microfilaments* may be responsible for it; other evidence suggests that it may be related to the transport of cellular substances by microtubules.

Cellular Reproduction

The Cell Cycle

In order for growth, repair, reproduction, and other related activities to occur in living organisms, cells divide, producing new cells. The dividing cells go through an orderly series of events known as the **cell cycle.** This cycle is usually divided into *interphase* and *mitosis,* mitosis itself being subdivided into four phases.

Figure 3.12 The phases of mitosis as seen in onion root-tip cells. (Photomicrographs by Carolina Biological Supply Company)

A. Cells in interphase

B. Cell (center) in prophase

C. Cell (center) in metaphase

D. Cell in anaphase

E. Cell (center) in late anaphase

F. Cell in telophase

G. Cell in telophase. The nuclear envelopes are reforming.

H. Cell in telophase. The nucleoli have reappeared.

I. Recently divided cells. Cytokinesis is complete and the nuclei have reentered interphase.

Interphase (figure 3.12A)

Cells that are not dividing are said to be in *interphase* (figure 3.12A), and it is such cells that have been discussed up to this point.

For many years cells were considered to be "resting" when they were not actually dividing, but we know now that three consecutive periods of intense activity take place during interphase.

Immediately after a nucleus has divided, a period designated as G_1 begins. During this G_1 period, substances that either stimulate or inhibit the next period, designated S, are produced, and various organelles divide or grow. During the S period the unique process of DNA replication (duplication) takes place. Details of this process and of DNA

Cells

structure are discussed on pages 41–43. The final period of interphase, designated as G_2, sees the formation of substances and structures involved directly in mitosis.

The length of the cell cycle varies with the kind of organism involved, the type of cell within an organism, and with temperature and other environmental factors. In most instances, however, interphase may occupy up to 90% or more of the time it takes to complete the cycle.

Mitosis (figure 3.12B–H)

All organisms begin life as a single cell. Almost immediately, however, this initial cell may begin to divide, producing new cells; these, in turn, divide and produce more cells. The process of dividing, called **mitosis** (see figure 3.12), goes on in at least parts of the organism until death. Strictly speaking, mitosis refers to the division of the nucleus alone, but the division of the remainder of the cell, called **cytokinesis,** usually accompanies or follows mitosis; both processes will be considered together here.

In higher plants such as ferns, conifers, and flowering plants, mitosis occurs in specific regions, or tissues, called **meristems** (figure 3.13). Meristems are found in the root and stem tips and also in a thin, perforated and branching cylinder of tissue called the **vascular cambium** (often referred to simply as the **cambium**), located in the interior of stems a short distance from the surface. In some herbaceous and most woody plants, a second meristem similar in form to the cambium lies between the cambium and the outer bark. This second meristem is called the **cork cambium.** These specific tissues are discussed in the next chapter and also in chapters 5 and 6.

When mitosis takes place, it makes no difference how many chromosomes are contained in the nucleus. The daughter cells that result from the process have exactly the same number as the parent cell. It is a continuous process, which may take as little as five minutes or as long as several hours from start to finish. Typically, however, it

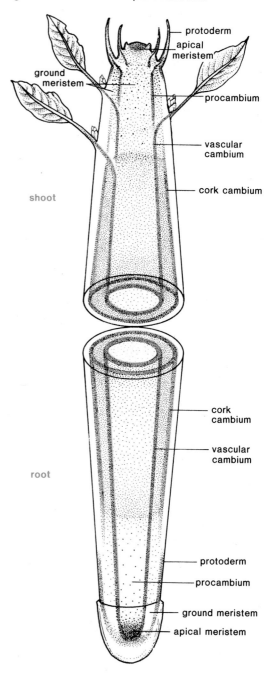

Figure 3.13 Locations of plant meristems

protoderm
apical meristem
ground meristem
procambium
vascular cambium
cork cambium
shoot

cork cambium
vascular cambium
root
protoderm
procambium
ground meristem
apical meristem

Chapter 3

Figure 3.14 Parts of a chromosome

centromere

chromatids

takes from thirty minutes to two or three hours. The process is usually divided into four arbitrary phases, primarily for convenience. Descriptions of the phases follow.

Prophase (figure 3.12B)

The main features of *prophase* (figure 3.12B) are (1) the chromosomes become shorter and thicker, and their double nature becomes apparent; and (2) the nucleolus and the nuclear envelope disappear.

Prophase takes up about as much time as the remaining three phases combined. The beginning of this phase is marked by the appearance of the chromosomes as faint threads in the nucleus. These chromosomes gradually coil or fold into thicker and shorter structures, and soon two threads, or **chromatids,** can be distinguished for each chromosome. The chromatids are themselves independently coiled; these coils appear to tighten and condense until the chromosomes have become relatively short, thick, and rodlike, with constrictions called **centromeres** (some call the constrictions **kinetochores**) holding each pair of chromatids together (see figure 3.14). When examined with the aid of a light microscope, the centromeres appear to be single structures, but they actually have become double during interphase and simply function as a single unit at this stage. Sometimes other constrictions may appear on individual chromosomes, giving them the appearance of having extra knobs. These knobs are

referred to as *satellites;* the constrictions at the base of the satellites have no known function, but the satellites themselves are useful in helping to distinguish certain chromosomes from others in a nucleus.

As prophase progresses, the nucleolus becomes smaller and eventually breaks down and disappears. The end of prophase is marked by the disappearance of the nuclear envelope.

In certain simpler organisms such as fungi and algae, and in virtually all animal cells, the cytoplasm just outside the nucleolus contains dual pairs of tiny keg-shaped organelles called *centrioles.* The centrioles are surrounded by microtubules, which radiate out from them and arrange cytoplasmic particles in the vicinity into starlike rays, collectively called an *aster.* At the beginning of prophase, the aster divides into two parts; one part remains at its original location, while the other part migrates around the nuclear envelope to the opposite side. Centrioles and asters have not been detected in the cells of most of the more complex members of the Plant Kingdom.

Metaphase (figure 3.12C)

The main features of *metaphase* (figure 3.12C) are (1) a **spindle,** made up of *spindle fibers,* forms in the area previously occupied by the nucleus; and (2) the chromosomes become aligned in the center of the cell.

After the nuclear envelope breaks down, spindle fibers can be seen in the area previously occupied by the nucleus. These are oriented so that they extend in arcs between the *poles* of the cell. The spindle fibers consist of groups of microtubules and are referred to collectively as the spindle. The chromosomes migrate to the center of the spindle and become aligned so that their centromeres are in a plane roughly in the center of the cell. This invisible line is called the *equator.* During their movement to the equator each chromosome becomes attached at its centromere to an individual spindle fiber. Most of the spindle fibers remain independently extended in arcs between the poles.

Anaphase (figure 3.12D and E)

The main feature of *anaphase* (figure 3.12D and E) is that the two chromatids of each chromosome separate and move to opposite poles. At the end of metaphase the centromeres of the chromatids of each chromosome are still joined together. Anaphase begins with these centromeres separating lengthwise and continues with the chromatids of each chromosome separating in unison and moving toward the poles. The chromatids give the appearance of being pulled toward the poles by the contraction of the spindle fibers attached to their centromeres. If the centromere is in the center of the chromatid, it leads the way, with the rest of the chromatid assuming a V shape as it appears to drag in the cytoplasm. All of the chromatids separate and move at the same time. Although experiments have shown that a chromatid will not migrate to a pole if the fiber attached to its centromere is severed, other experiments have shown that the chromatids will separate from one another but not move to the poles, even if no spindle is present. The forces behind both the separation of the chromatids and their movement to the poles have been the subject of much speculation, but these forces have not yet been positively identified.

Telophase (figure 3.12F–H)

The main features of *telophase* (figure 3.12 F–H) are (1) each group of chromatids becomes surrounded by a new nuclear envelope; (2) the chromatids become longer and thinner and are finally indistinguishable; (3) new nucleoli appear; (4) many of the spindle fibers disappear; and (5) a cell plate forms.

The transition from anaphase to telophase is not distinct, but telophase is definitely in progress when elements of the nuclear envelope appear around each group of chromatids at the poles. These elements gradually form an intact envelope as the chromatids return to the diffuse, indistinct threads seen at the onset of prophase. The new nucleoli appear on specific regions of certain chromosomes.

During telophase the spindle gradually disappears and a set of shorter fibers, or fibrils, composed of microtubules develops in the region of the equator. This set of fibrils is called a **phragmoplast.** Small vesicles, which have the appearance of droplets when viewed with a light microscope, appear along the plane of the equator on the phragmoplast fibrils. These vesicles, which are derived from Golgi bodies, evidently contain pectin. The vesicles increase in number from the center outward and merge, forming a disc-shaped **cell plate,** which continues to grow outward until it reaches the walls of the cell. Plasmodesmata (see figures 3.4 and 3.5), which are minute strands of cytoplasm that extend between cells through the walls, are apparently formed as portions of the endoplasmic reticulum are "trapped" between fusing vesicles of the cell plate.

The mature cell plate becomes the middle lamella between the daughter cells. New plasma membranes develop on either side of the cell plate as it forms, and the protoplasm of each daughter cell deposits new cell wall materials between the middle lamella and the plasma membranes. These new walls are relatively flexible and remain so until the cells increase to their mature size. At that time additional cellulose and other substances may be added, forming a secondary cell wall interior to the primary wall.

Communication in Cells

Amoebae and other microscopic creatures sometimes are referred to as "simple one-celled organisms." After our brief introduction to cells, it is apparent that microorganisms are simple only with regard to their overall structure and that cells are anything but simple within themselves. Indeed, as suggested in our discussion of the nucleus, cells are very complex "factories," directed by a control center. How does this control center, the nucleus, direct all the varied and intricate activities of its "factory?" The answer to this question is still being investigated by cell biologists in various parts of the world, but since the 1950s a great deal has been learned about cell communications.

In some ways, the nucleus functions like a combination of DNA-programmed computer and dispatcher. The rest of the cell receives coded messages or "blueprints" from the nucleus and assembles the items called for from the raw materials available to it. These raw materials are either absorbed by the plant from the soil or recycled from other areas.

In the cell, all these activities take place on a chemical level, and to follow such activities we now need to take a closer look at DNA and RNA.

DNA

Deoxyribonucleic acid (DNA) molecules of chromosomes consist of double helical (spiral) coils of repeating subunits called *nucleotides,* each composed of a base, a sugar, and a phosphate. The molecules are very large and contain, in heredity units known as genes, the coded information that precisely determines the nature and proportions of the myriad substances found in cells and also the ultimate form and structure of the organism itself. If this coded information were written out, it would fill over one thousand books of three hundred pages each—at least for the more complex organisms. The amount of DNA in the cells of a given species is constant, although certain cells produced during a life cycle will have half as much DNA as other cells of the same life cycle (see chapter 11). DNA molecules can replicate (duplicate themselves) in precise fashion. When a cell divides, the hereditary information contained in the DNA of the new cells is an exact copy of the original and can be passed on from generation to generation without change, except in the event of a **mutation** (discussed in chapter 12).

Structure of DNA

The thousands of atoms in a DNA molecule are found in chains of building blocks called **nucleotides.** Each nucleotide consists of three parts: (1) a nitrogenous base; (2) a 5-carbon sugar, deoxyribose; and (3) a phosphate group. Both the nitrogenous base and the phosphate group are bonded to the sugar (see chapter 2 and figure

Figure 3.15 Structure of a DNA molecule. In this enlargement of a small portion of a DNA molecule, the rungs of the twisted ladder formed by the two entwined spiraling strands consist of alternating units of sugar (S) and phosphate (P) molecules. The purines adenine or guanine (A or G) occur opposite the pyrimidines thymine or cytosine (T or C). The purines and the pyrimidines opposite each other are held together by hydrogen bonds linking the nitrogenous bases of the paired molecules. (After drawing from Volpe, E. Peter. 1981. *Understanding Evolution,* 4th ed. Wm. C. Brown Company Publishers, Dubuque, Iowa)

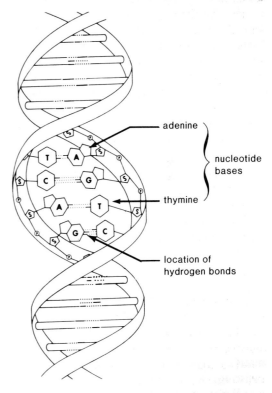

3.15). Four kinds of nucleotides occur in DNA. Each kind has a unique nitrogenous base, but all have the same types of phosphate groups and sugar. Two of the nucleotide bases—*adenine* and *guanine*—are called *purines;* they have a molecular structure that resembles two linked rings. The other two nucleotide bases—*cytosine* and *thymine*—are called *pyrimidines;* they have a molecular structure consisting of a single ring.

Each nucleotide in a DNA molecule is bonded to the next one in such a way that the nucleotides form a chain, with the sugar of one attached to the phosphate group of the next; the nitrogenous bases are oriented as side groups on the chain. Each species has its own unique DNA, with the nucleotides occurring in a different sequence for each kind of DNA molecule. The possibilities for variety in the sequences seem virtually unlimited (figure 3.16).

In 1951 two cell biologists, J. D. Watson and Francis Crick, initiated a series of brilliant investigations that eventually led to the construction of a model of a DNA molecule. When they began their work they had certain information on which to start building their conclusions. They knew, for example, that Linus Pauling had postulated that the structure of DNA might be similar to the structure of protein, and that he had shown part of the structure of some proteins to be helical and maintained by hydrogen bonds between the amino acids. They also knew, from studies by E. Chargraff and others at Columbia University the previous year, that there was a ratio of 1:1 between nucleotides with adenine and nucleotides with thymine in DNA molecules, and that there was also a ratio of 1:1 between nucleotides with guanine and those with cytosine. Furthermore, they knew, from studies the same year by Rosalind Franklin and M. H. F. Williams in which specialized X-ray equipment was utilized, that a DNA molecule was composed of regularly repeating units that appeared to be helically arranged.

When they had pieced together all the facts, Watson and Crick concluded that a DNA molecule consisted of a huge, entwined, double helix exactly 20 angstrom units wide, the strands of which appeared to be wrapped around an invisible pole—one strand spiraling in one direction and the other strand spiraling at the same angle in the opposite direction. They also concluded that

Figure 3.16 How nucleotides are paired in a tiny portion of a strand of DNA. The variations in sequences of pairs are virtually unlimited.

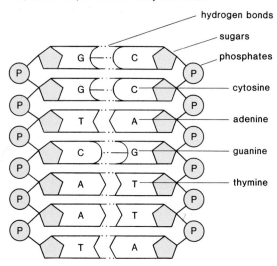

the nucleotides were linked in ladderlike fashion between the two strands. They knew, however, that if purines were linked across the strands of the helix, the size of the combined molecules would exceed the 20 angstrom width of the DNA molecule, and that two linked pyrimidines would not be wide enough to reach all the way across. If a purine and a pyrimidine were linked to each other, however, the width would fit the model perfectly. Accordingly, they concluded that the ladder rungs had to consist exclusively of purine–pyrimidine pairs. This meant that adenine–cytosine and guanine–thymine pairings could not occur, but adenine–thymine (or thymine–adenine) and guanine–cytosine (or cytosine–guanine) could occur. This nicely explained the 1:1 ratios shown by Chargraff and his colleagues, and such a model was proposed in 1953. Watson and Crick received a Nobel prize for their work, and subsequent research has continued to support their conclusions to the extent that their model of a DNA molecule is now generally accepted as an authentic representation.

Figure 3.17 Replication (duplication) of DNA. A molecule replicates by "unzipping" along the hydrogen bonds, which link the pairs of bases in the middle, and then each half serves as a template for bases available in the cell. As the new bases fit into the appropriate places and are linked together, a double helix is re-formed for each original half.

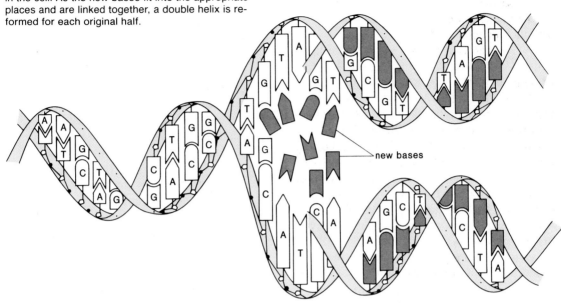

new bases

Replication (Duplication) of DNA

If DNA is indeed the determiner of inheritable characteristics, it must be able to duplicate itself precisely in order to pass along information from generation to generation. Since all DNA molecules have only four different nucleotides, the differences between the DNA of one organism and that of another must lie in the sequence of the four possible types of ladder rungs and their total number, as well as in the ratios of adenine and thymine to guanine and cytosine. If these four nucleotides are synthesized in living cells, what tells the cell exactly how to put them together to form the appropriate DNA molecules? Watson and Crick observed that if the two strands of DNA are "unzipped" by an enzyme breaking the bonds between the nucleotides down the middle of the ladder, each separated chain provides all the information needed for putting together a new ladder (figure 3.17). In other words, since guanine can pair only with cytosine and vice versa, a guanine nucleotide synthesized by the cell can lock on, or bond, only to a cytosine, and an adenine can bond only to a thymine. Thus, if a cell "unzips" the two strands in its DNA molecules, separate nucleotides can line up next to each of the single chains in precise sequence and be bonded together to form a new chain, with the separate chains functioning as molds, or templates, for self-duplication. This theory of Watson and Crick, also proposed in 1953, has since been supported by evidence obtained by the use of nucleotides containing isotopes of carbon and nitrogen atoms.

RNA and Synthesis of Proteins

Once the chemical nature of DNA and the way in which it duplicates itself became understood, attention was focused on the manner in which DNA determines a cell's or an organism's function and structure.

It was known that proteins are large complex molecules composed of twenty different kinds of amino acids linked together. If each of the four nucleotides in DNA coded only one amino acid, there could be not more than four amino acids. If all possible combinations of nucleotide pairs (sixteen) were used to code amino acids, the number would still not be enough to code the twenty known amino acids. If a minimum of three nucleotides were involved for each amino acid, however, up to sixty-four combinations would be possible—more than enough to code twenty different amino acids. Such an arrangement was postulated and later confirmed; each trio of nucleotides along a DNA strand does indeed code a specific amino acid. It was also known that most proteins are synthesized in the cytoplasm, whereas DNA is largely confined to the nucleus. Obviously, then, the DNA could not function as a direct template for protein synthesis. Some intermediary agent or agents had to be involved.

In the early 1940s, it was suspected that ribonucleic acid (RNA) was involved in protein synthesis because it had been found that cells producing large amounts of protein contained large amounts of RNA and vice versa. RNA is similar to DNA, but it differs in that its sugar component has an additional oxygen atom and it has a different pyrimidine, *uracil,* instead of thymine. RNA usually occurs as a single strand and only very rarely as a completely double strand.

Specific Types of RNA and Their Functions

Messenger RNA (mRNA)

Messenger RNA is a large molecule composed of several hundred to several thousand nucleotides. As with a new strand of DNA, a strand of mRNA forms along a portion of a DNA molecule like a row of jigsaw pieces fitting together with another row, with uracil taking the place of thymine. A sequence of three nucleotide bases, known as a *codon,* codes one amino acid. Once the mRNA molecule has been formed within the nucleus, enzymes split off and modify portions of the mRNA molecule. The enzymes are then carried out into the cytoplasm (probably through a nuclear pore). In the cytoplasm each mRNA unit becomes attached by one end to a ribosome, which is itself composed of another form of RNA and protein. Usually several ribosomes become associated with the mRNA molecule and move along in sequence, "reading" the coded information and participating in the bonding together of amino acids available in the cytoplasm. One mRNA molecule apparently is "read" by two or three such groups of ribosomes and then destroyed. It has been estimated that the average life of an mRNA molecule in the much-studied, digestive-tract bacterium *Escherichia coli* is about two minutes.

Transfer RNA (tRNA)

A molecule of tRNA consists of seventy-seven nucleotides bonded together in a single strand that doubles back on itself. There is at least one form of tRNA for each of the twenty amino acids. Each specific form of tRNA has an *anticodon loop,* where the strand doubles back on itself. This anticodon is a sequence of three nucleotides that will "recognize" a trio of specific amino acid nucleotides on a strand of mRNA. The amino acid is bound on the open, or *acceptor end,* of the tRNA strand with the aid of an enzyme that evidently identifies with both a specific tRNA strand and a specific amino acid (figure 3.18).

Figure 3.18 How a protein is synthesized. *1.* A strand of DNA in the nucleus serves as a template for the synthesis of a strand of mRNA. *2.* The mRNA moves out into the cytoplasm where it becomes associated with ribosomes. *3.* The tRNA strands each pick up a specific amino acid from the supply available in the cytoplasm. *4.* The tRNA briefly couples with the mRNA at a precise spot where its anticodon complements a codon on the mRNA. *5.* A ribosome moves along the mRNA, adding amino acids to the polypeptide chain of the protein being synthesized. *6.* The tRNA can now repeat the process of picking up an amino acid from the supply available. (Note: The molecules and the structures are not drawn to scale with respect to one another or the cell.)

Once the tRNA molecule has "captured" an amino acid, the tRNA becomes bound to a ribosome at the same point to which an mRNA molecule has become attached. Then, as the ribosome moves along the mRNA strand, the tRNA releases its amino acid and becomes detached, with another tRNA taking its place. In this way, the amino acids are lined up in sequence according to the mRNA code, and each is bonded to the amino acid of the next tRNA molecule until a complete protein molecule has been assembled. The tRNA thus relays amino acids for a specific "blueprint" coded by the DNA and carried by the mRNA. The amino acids of the various proteins of the cell are attached to each other precisely according to the information delivered from the nucleus, with specific enzymes ensuring that the right amino acids become designated proteins.

After a tRNA strand becomes detached and releases its amino acid, it is then available for reuse in "capturing" another amino acid molecule. Apparently tRNA molecules can repeat this cycle indefinitely.

Ribosomal RNA (rRNA)

Ribosomal RNA is formed along one strand of the DNA helix in the same manner as mRNA and tRNA molecules. Its role in protein synthesis is not clear, but it is known that ribosomes consist of two more-or-less spherical particles, one larger than the other, and that the ribosome contains a compact mass of RNA and protein. The larger of the two particles contains a "slot" into which a tRNA molecule can fit. How one tRNA molecule replaces another in this slot has not yet been determined. Ribosomal RNA, like transfer RNA, has a long life.

Figure 3.19 Pits are depressions or cavities in cell walls where the secondary wall does not form. There may be from one or two to several thousand in a cell. They always occur in pairs, with one on each side of the middle lamella. Some, called *bordered pits* (left), bulge out from the wall and resemble doughnuts in surface view, while others, called *simple pits* (right), do not bulge.

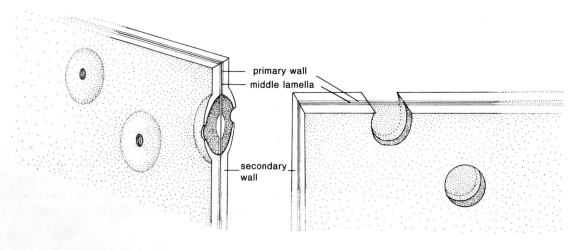

primary wall
middle lamella
secondary wall

Chapter 3

Communication Between Cells

Although cells are independent units capable of carrying on complex activities, it is essential that all living cells be coordinated in the activities of the organism as a whole through some system of communication. In plants, living cells are in contact with one another via the fine strands of cytoplasm called *plasmodesmata* (see figures 3.4 and 3.5), which were mentioned on page 36. These extend between adjacent cells through the walls. The minute hole through which the plasmodesma extends is often located within a pair of wall depressions called **pits** (figure 3.19). The translocation of sugars, amino acids, and other substances between cells apparently takes place by means of the plasmodesmata.

Summary

All living organisms are composed of cells. Cells are modified according to the functions they perform; some cells live for a few days whereas others live for many years.

The discovery of cells is associated with the development of the microscope. Robert Hooke was the first to use the word *cells* for boxlike compartments he saw in cork with the aid of a simple microscope in 1665. Leeuwenhoek and Grew reported frequently during the next fifty years on the existence of cells in a variety of tissues. In 1809 Lamarck concluded that all living tissue is composed of cells, and in 1824 Dutrochet reinforced Lamarck's conclusions. In 1833 Brown discovered that all cells contain a nucleus, and shortly thereafter Schleiden observed a nucleolus within a nucleus. Schleiden and Schwann are generally credited with developing the cell theory in 1838–39. The theory holds not only that all living organisms are composed of cells but that cells also form a unifying structural basis of organization. In 1858 Virchow argued cogently that every cell comes from a preexisting cell and that there is no spontaneous generation of cells from dust, as was the belief of the time. Pasteur confirmed Virchow's contentions in 1862, with experiments involving microorganisms in swan-necked flasks. Pasteur later proved that fermentation involves activity of yeast cells, and Buchner, in 1897, discovered that yeast cells do not need to be alive for fermentation to occur. This led to the discovery of organic catalysts called enzymes.

Cells are minute; most vary in diameter between 10 and 100 micrometers. They number into the billions in larger organisms such as trees. Plant cells are bounded by walls. The living part of the cell within the walls, protoplasm, has two main components: the nucleus, which functions as a control center, and cytoplasm, a fluid in which various membrane-bounded organelles are suspended.

A pectic middle lamella is sandwiched between the primary cell walls of adjacent cells. The primary wall and also the secondary cell wall, often added inside the primary wall, are composed mostly of cellulose; but secondary cell walls may also contain lignin and other substances. A flexible plasma membrane, which often forms folds, constitutes the outer boundary of the protoplasm; it regulates the substances that enter and leave the cell. Mitochondria are tiny, numerous organelles that are bounded by a double membrane with inner platelike folds called cristae; they are associated with respiration. The endoplasmic reticulum is a system of membrane channels associated with the storing and transporting of protein and other cell products. Granular particles called ribosomes may line the outer surfaces of the endoplasmic reticulum; they function in protein synthesis. Golgi bodies, or dictyosomes, are organelles that appear as groups of sacs and that function as collecting and packaging centers for the cell. Plastids are larger organelles that are green (chloroplasts), orange or red (chromoplasts), or colorless (leucoplasts). Chloroplasts contain enzymes, in a matrix called the stroma, and grana, which are stacks of coin-shaped membranes (thylakoids) containing green chlorophyll pigments; photosynthesis occurs in the thylakoids. Plastids develop from proplastids, which divide frequently, and also arise from the division of mature plastids. Microtubules are tiny tube-like structures that are commonly found just inside the plasma membrane. They apparently control the addition of cellulose to the cell wall. The nucleus is bounded by a nuclear envelope, consisting of two membranes that are perforated by numerous pores. Within the nucleus is a fluid called nucleoplasm, one or more spherical nucleoli, and thin strands of chromatin, which condense to become chromosomes when nuclei divide. Each species of organism has a specific number of chromosomes in each cell.

As much as 90% of the volume of a mature cell may be taken up with a vacuole or vacuoles that are bounded by a vacuolar membrane and contain a watery fluid called cell sap. Cell sap contains dissolved substances and frequently also water-soluble pigments called anthocyanins, which are red or blue.

Animal cells differ from most plant cells in not having a wall, plastids, or large vacuoles. Also, they have keg-shaped centrioles in pairs just outside the nucleus. Prokaryotic cells, which lack some of the features of the eukaryotic cells (the subject of this chapter), are discussed in chapter 14. Movement of cytoplasm within a cell in a streaming motion is called cyclosis.

In the cell cycle, cells that are not dividing are said to be in interphase, which is subdivided into three periods of intense activity that precede mitosis or division of the nucleus. Mitosis is usually accompanied by division of the rest of the cell and takes place in plant regions known as meristems. Mitosis is arbitrarily divided into four phases: (1) prophase, in which the chromosomes become apparent and the nuclear envelope breaks down; (2) metaphase, in which the chromosomes become aligned at the equator of the cell, and a system of spindle fibers, some of which are attached to the chromosomes at their centromeres, is fully developed; (3) anaphase, in which the chromosomes each separate lengthwise (the half chromosomes are called chromatids), with each group of chromatids migrating to opposite poles of the cell; and (4) telophase, in which each group of chromatids becomes a new nucleus, and a wall dividing the daughter nuclei forms, creating two daughter cells. The development of the dividing wall is initiated by the appearance of a set of short fibrils constituting the phragmoplast. Droplets, or vesicles, of pectin merge, forming a cell plate that grows to become the middle lamella of the new cell wall.

Chromosomes contain helical coils of DNA, which contain coded information determining the nature and proportions of substances in cells and the ultimate form and structure of the organism. DNA is composed of building blocks called nucleotides, which occur in four forms. Watson and Crick were awarded a Nobel prize in 1953 for developing a model of DNA that has since come to be generally accepted as authentic. DNA replicates (duplicates itself) by "unzipping" down its nucleotide "ladder rungs" and having new nucleotides replace the old ones until two new chains of DNA have been formed. New proteins are formed precisely according to coded information contained in the DNA of the nucleus by means of RNA, which occurs in three forms. Messenger RNA forms along part of a DNA molecule and is carried out of the nucleus into the cytoplasm, where it becomes attached to a ribosome. Ribosomes "read" the coded information and participate in assembling amino acids together to form a protein, with the aid of transfer RNA molecules. Ribosomes themselves are largely composed of a third form of RNA called ribosomal RNA. The synthesis of proteins is closely regulated by various mechanisms. Living cells are in contact with one another via fine strands of cytoplasm called plasmodesmata, which often extend through minute holes in wall depressions called pits.

Review Questions

1. How is cellular structure beneficial to plants and animals?

2. Of what importance to the cell is the wall?

3. How can you distinguish between cytoplasm and vacuoles?

4. What are plasmodesmata, and what is their importance to living plants?

5. What are pits? Where are they located?

6. What is the difference between protoplasm and cytoplasm?

7. What is the function of a cell nucleus? How does it perform its function?

8. What are the differences and similarities between plant and animal cells?

9. Are prophase and telophase of mitosis exactly the reverse of one another?

10. How would you describe the basic structure of DNA?

11. Why is the nucleus referred to as the control center of the cell?

12. What are the differences between DNA and the three forms of RNA?

Discussion Questions

1. Would you consider any of more useful than another? W

2. After you have completed your introductory plant science course, do you believe you would be able to determine the function of each of a cell's organelles in a laboratory? Explain.

Additional Reading

Avers, C. J. *Basic cell biology.* New York: D. Van Nostrand Co.; 1978.

DeWitt, W. *Biology of the cell: an evolutionary approach.* Philadelphia: W. B. Saunders Co.; 1977.

Dyson, R. D. *Cell biology: a molecular approach.* 2d ed. Boston: Allyn and Bacon; 1978.

Jensen, W. A. *The plant cell.* 2d ed. Belmont, CA: Wadsworth Publishing Co.; 1970.

Karp, G. *Cell biology.* Hightstown, NJ: McGraw-Hill Book Co.; 1979.

Kimball, J. W. *Cell biology.* 2d ed. Reading, MA: Addison-Wesley Publishing Co.; 1978.

Ledbetter, M. C.; Porter, K. R. *Introduction to the fine structure of plant cells.* New York: Springer-Verlag; 1970.

Loewenstein, W. R. "Intercellular communication." *Scientific American,* May 1970 (Offprint 1178).

Loewy, A. G.; Siekevitz, P. *Cell structure and function.* 2d ed. New York: Holt, Rinehart and Winston; 1970.

Swanson, C. P.; Webster, P. *The cell.* 4th ed. Englewood Cliffs, NJ: Prentice-Hall; 1977.

Wolfe, S. L. *Biology of the cell.* Belmont, CA: Wadsworth Publishing Co.; 1972.

Tissues

Overview

A discussion of meristems, permanent tissues (epidermis, periderm, parenchyma, collenchyma, sclerenchyma, secretory tissues), and conducting tissues (xylem and phloem) forms the body of this chapter.

Some Suggested Student Learning Goals

1. Identify the meristems present in plants; indicate where they are found.

2. Describe the conducting tissues of plants and the function of each cell component.

3. Identify tissues of plants that are neither meristematic nor function in conduction at maturity.

Introduction

Bricks cemented together to form a wall have considerably more strength and utility than an unorganized pile of them, and sections of pipe joined together can conduct fluids over far greater distances than isolated units. Likewise, cells that are united or organized can perform a common function and, as such, are much more useful than individual, detached cells. Indeed in plants, virtually all cells are produced in such a way that at maturity large numbers of them perform together in integrated fashion, providing strength, support, storage areas, and supply lines, and functioning in various other ways in plants as a whole.

A group of cells performing a common function is referred to as a *tissue*. A plant organ such as a leaf or stem is composed of a number of different tissues. Tissues are classified according to their structure, origin, or function. The following are major kinds of tissues found in higher plants. The specific kinds of cells associated with each tissue, as well as illustrations of them, are included in the discussions that follow the classification.

I. Meristematic Tissues (see figure 3.13)
   A. *Apical meristems*
   B. *Vascular cambium*
   C. *Cork cambium*
   D. *Intercalary meristems*
II. Permanent Tissues
   A. *Surface tissues*
      1. Epidermis
      2. Periderm
   B. *Storage, support, and secretory tissues*
      1. Parenchyma
      2. Collenchyma
      3. Sclerenchyma
      4. Secretory tissues
   C. *Conducting tissues*
      1. Xylem
      2. Phloem

Meristematic Tissues

Meristems or *meristematic tissues* are tissues in which cells actively divide. As new cells are produced, many mature from typically small, six-sided boxlike structures, each with a proportionately large nucleus in the center and tiny vacuoles, or no vacuoles at all (figure 4.1), into any one of numerous shapes and sizes related to the cell's ultimate function.

Apical Meristems

Apical meristems are found at, or near, the tips of roots and shoots, which are increased in length by the cells that the apical meristems produce. Three primary meristems, the **protoderm,** the **ground meristem,** and the **procambium** develop from each apical meristem, as do embryo leaves and buds. The tissues they produce are called *primary tissues*. Note their locations in figure 3.13; they are discussed in chapters 5 and 6.

Vascular Cambium

The meristem called the *vascular cambium* forms a thin, often branching, cylinder that runs the lengths of the stems and roots (except for the tips) of most perennial plants and many herbaceous annuals. It is primarily responsible for the production of tissues that increase the girth of a plant. The individual self-perpetuating cells of the vascular cambium are referred to as *initials;* both the cambium and its initials are discussed in chapters 5 and 6.

Cork Cambium

The cork cambium, like the vascular cambium, forms a thin cylinder that runs the lengths of stems and roots of woody plants; it is located to the exterior of the vascular cambium, however, so that in cross section the vascular cambium appears as a tube of smaller diameter lying within the tube of the cork cambium. The cork cambium is primarily responsible for the outer bark of woody plants; it is discussed in chapters 5 and 6. Tissues produced by the vascular cambium and the cork cambium are called *secondary tissues.*

Figure 4.1 Typical meristematic cell

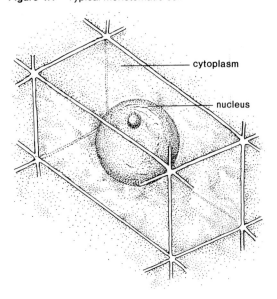

cytoplasm

nucleus

Intercalary Meristems

Grasses and related plants have neither a vascular cambium nor a cork cambium, but they do have apical meristems and other meristematic tissues, called *intercalary meristems,* in the vicinity of **nodes** (leaf attachment areas; see chapter 7), which occur at intervals along stems. Intercalary meristems, like apical meristems, produce increases in the lengths of stems.

Permanent Tissues

Permanent tissues are composed of cells that have been produced by meristems and have developed and matured into various shapes and sizes related to their functions. Permanent tissues that consist of only one kind of cell are generally referred to as *simple* tissues, while those consisting of more than one kind of cell are called *complex*. A discussion of the major kinds of permanent tissues, both simple and complex, follows.

Surface Tissues

Epidermis

The outermost layer of cells of all young plant organs is called the **epidermis**. Since it is in direct contact with the environment it is subject to modification by environmental factors and often includes several different kinds of cells. The epidermis is usually one cell thick, but in a few plants that produce aerial roots (e.g., orchids), the epidermis of the roots may be several cells thick. Such a multiple-layered epidermis, called *velamen,* also occurs in the leaves of some tropical figs and members of the pepper family. Most epidermal cells secrete a fatty substance called **cutin** within and on the surface of the outer walls. Cutin forms a protective layer called the **cuticle** (figure 4.2). The thickness of the cuticle to a large extent determines how much water is lost through the cell walls by evaporation. The cuticle is also exceptionally resistant to bacteria and other disease organisms and has been recovered from fossil plants millions of years old. Waxes are deposited on the cuticle in a number of plants (see figure 7.5), apparently reaching the surface through microscopic channels in the cell walls. The susceptibility of a plant to herbicides may depend on the thickness of these wax layers. Some wax deposits are extensive enough to have commercial value. Carnauba wax, for example, is deposited on the leaves of the wax palm. It and other waxes are harvested for use in polishes and phonograph records.

In leaves, the epidermal cell walls at right angles to the surface often assume bizarre shapes, with convolutions that, under the microscope, give them the appearance of a jigsaw puzzle. Epidermal cells of roots produce tubular extensions called *root hairs* (see figure 5.4) a short distance behind the growing tips; these greatly increase the absorptive area of the surface. Hairs of a different nature occur on the epidermis of above-ground parts of plants. These form outgrowths consisting of one to several cells. Leaves also have numerous small pores, the **stomata,** formed by pairs of specialized epidermal cells called **guard cells** (see

Figure 4.2 Small portion of a kaffir lily leaf, showing the thick cuticle secreted by the upper epidermis

cuticle

upper epidermis

parenchyma cells

figures 7.6 and 8.7). Guard cells differ in shape from other epidermal cells; they also differ in that chloroplasts are present within them. Stomata and guard cells are discussed in chapters 7 and 8. Some epidermal cells may be modified as **glands** that secrete protective or other substances (figure 4.3).

Periderm

In woody plants, the epidermis is sloughed off and replaced by a *periderm* after the cork cambium begins producing new tissues that increase the girth of the stem or root. The periderm constitutes the outer bark and is primarily composed of **cork** cells, which are somewhat rectangular and boxlike and are dead at maturity (figure 4.4). While the protoplasm of cork cells is still functioning, it secretes a fatty substance, **suberin,** into the walls. This makes cork cells waterproof and helps them protect the tissues beneath the bark. Some cork tissues, such as those produced by the cork oak, are harvested commercially and are used for bottle stoppers and in the manufacture of linoleum.

Some parts of a cork cambium form pockets of loosely arranged cells that are not impregnated with suberin. These pockets of tissue protrude through the surface of the periderm; they are called **lenticels** (figure 4.4) and function in gas exchange between the air and the interior of the stem. The fissures in the bark of trees have lenticels at their bases.

Storage, Support, and Secretory Tissues

Parenchyma

Parenchyma cells (figure 4.5) are the most abundant of the cell types, being found in almost all major parts of higher plants. They are roughly spherical in shape to begin with, but when all the spherical cells of parenchyma tissue push up against one another, their thin, pliable walls are forced into various shapes; the majority of the resulting cells have fourteen sides. They tend to have

Figure 4.4 Periderm and lenticel. Cross section through a small portion of elderberry periderm, showing a large lenticel. (Photomicrograph by Carolina Biological Supply Company)

lenticel

cork cells

Figure 4.5 Parenchyma cells in cross section

large vacuoles and may contain starch grains, oils, tannins (tanning or dyeing substances), crystals, and various other secretions. Spaces commonly occur between parenchyma cells; in fact, in water lilies and other aquatic plants, the intercellular spaces are quite extensive and form a network throughout the entire plant. This type of parenchyma tissue—with extensive connected air spaces—is referred to as *aerenchyma*, while parenchyma cells containing numerous chloroplasts (as found in leaves) form **chlorenchyma** tissue. The chief function of chlorenchyma tissue is photosynthesis, while parenchyma tissues without chloroplasts function primarily in food or water storage. Some parenchyma cells develop irregular extensions of the inner wall that greatly increase the surface area. Such cells are called *transfer cells;* they apparently play a role in transferring dissolved substances between adjacent cells. Many parenchyma cells live a long time; in some cacti, for example, they may live to be over one hundred years old.

Collenchyma

Collenchyma cells (figure 4.6), like parenchyma cells, have living protoplasm and may remain alive a long time. They are distinguished from parenchyma cells primarily by the thicker walls, which are also usually uneven in thickness. Collenchyma cells often occur just beneath the epidermis; typically they are longer than they are wide and their walls are pliable as well as strong. They provide flexible support for both growing organs and mature organs such as leaves and floral parts.

Sclerenchyma

Sclerenchyma tissue consists of cells that have thick, tough walls, normally impregnated with **lignin.** Most sclerenchyma cells are dead at maturity and function in support. Two types of sclerenchyma occur: **sclereids** and **fibers.** Sclereids (see figure 4.7 and plate 2A for illustrations

Figure 4.6 Collenchyma cells in cross section. Note the unevenly thickened walls.

Figure 4.7 Sclereids (stone cells) in cross section

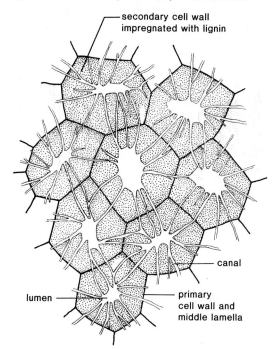

secondary cell wall impregnated with lignin

canal

lumen

primary cell wall and middle lamella

Chapter 4

Figure 4.8 Fibers. *Left:* Cross section through a portion of stem tissue from a linden tree. Note the thickness of the walls of the darker fibers. *Right:* Single fiber in longitudinal section

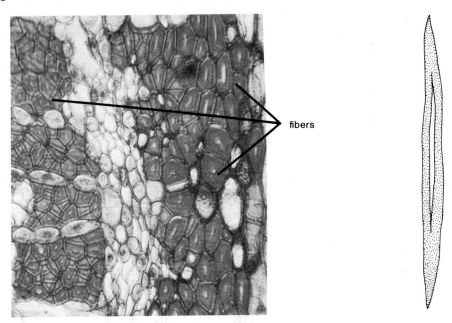

fibers

of two types of sclereids) may be randomly distributed in other tissues. For example, the slightly gritty texture of pears is due to the presence of groups of sclereids, or *stone cells,* as they are sometimes called. They tend to be about as long as they are wide, and sometimes occur in specific zones (e.g., the margins of camellia leaves) rather than being scattered within other tissues. Fibers (see figure 4.8 and plate 2C) may be found in association with a number of different tissues in roots, stems, leaves, and fruits. They are usually much longer than they are broad and have a proportionately tiny cavity, or *lumen,* in the center of the cell. At the present time, fibers from more than forty different families of plants are in commercial use in the manufacture of textile goods, ropes, string, canvas, and similar products. Archaeological evidence indicates that humans have been using fibers for at least ten thousand years.

Secretory Tissues

Secretory tissues or individual **secretory cells** release substances that have been produced within the protoplasm and moved to the outside of the cells producing them. Often the substances consist of waste products that are of no further use to the plant, but some substances such as **hormones** (see chapter 10) are vital to normal plant functions. Secretory tissues or cells can occur in a wide variety of places in a plant. Among the most common secretory tissues are those that secrete nectar in flowers; oils in citrus, mint, and many other leaves; mucilage in the glandular hairs of sundews and other insect-trapping plants; latex in members of several plant families such as the spurge family; and resins in coniferous plants such as pine trees. Latex and resins are usually secreted by cells lining tubelike ducts that form networks throughout certain plant species (see figure 6.10). Some plant secretions such as pine resin, rubber, mint oil, and opium have considerable commercial value.

Conducting Tissues

Conducting tissues are complex tissues that are primarily involved in moving water and soluble food substances throughout the plant. Some of these tissues are produced by apical meristems, but the bulk of such tissues in woody plants are produced by the vascular cambium and are often referred to as *vascular tissues.*

Xylem

An important part of the "plumbing" system of the plant, **xylem** conducts water and dissolved substances throughout all organs; it consists of a combination of parenchyma cells, fibers, *ray* cells, *vessels,* and *tracheids* (figure 4.9). **Vessels** are long tubes made up of individual cells called **vessel elements** that are open at each end, with barlike strips of wall material extending across the open areas in some instances. The cells are joined end to end, forming the tubes. **Tracheids,** which, like vessel elements, are dead at maturity, are tapered at each end and have no openings similar to those of vessels. The ends overlap with those of other tracheids, and wherever two tracheids are in contact with one another, pits (see figure 3.19)

Figure 4.9 Lengthwise views of water-conducting cells of xylem.
A. Tracheids. *B.* Vessel elements.

pit
A.

B.

are usually present. As indicated in chapter 3, pit pairs permit water to pass from cell to cell. In certain kinds of plants, such as cone-bearing trees, the xylem is composed almost entirely of tracheids. The walls of many tracheids have spiral thickenings on them that are easily seen with the light microscope. Most conduction is up and down, but some lateral (sideways) conduction also occurs. The lateral conduction takes place in the **rays;** ray cells are actually long-lived parenchyma cells.

Phloem

The components of **phloem** tissue (figure 4.10) of flowering plants include **sieve-tube elements** and **companion cells,** which are derived from common parent cells of the cambium and also usually ray cells, fibers, and parenchyma. Sieve-tube elements, like vessel elements, are laid end to end forming **sieve tubes,** but their end walls are not open. Instead, the walls are full of pores through which the cytoplasm extends from cell to cell. The

Figure 4.10 Phloem. A longitudinal section through a portion of stem phloem of tobacco. (After a drawing from Cronshaw, J. and Esau, K. *Journal of Cell Biology,* 38:298. July–September, 1968)

sieve-tube elements

phloem parenchyma cells

companion cell

sieve plate

porous regions of sieve elements are called **sieve plates.** Sieve-tube elements have no nuclei at maturity, despite the fact that their cytoplasm is very active in the phloem's major function of conducting food materials in solution throughout the plant. Apparently the adjacent companion cells form a very close relationship with the sieve tubes next to them, and function in some manner that brings about the conduction of the food.

Sieve cells, which are found in ferns and cone-bearing trees, are similar to sieve-tube elements but tend to overlap at their ends rather than form continuous tubes. Like sieve-tube elements, they have no nuclei at maturity, and they also have no adjacent companion cells. They do have equivalent adjacent parenchyma cells, called **albuminous cells,** which apparently function in the same manner as companion cells.

The various tissues discussed are shown as they occur in a woody stem in Plate 1.

Summary

A group of cells performing a common function is called a tissue. Apical meristems are found in the vicinity of the tips of stems and roots; the vascular cambium and the cork cambium occur as lengthwise cylinders within stems and roots; intercalary meristems occur in the vicinity of nodes of grasses and related plants.

Permanent tissues are produced by meristems. They include epidermis and periderm (surface tissues); parenchyma, collenchyma, sclerenchyma, and secretory tissues (storage, support, and secretory tissues); and xylem and phloem (conducting tissues).

Epidermis is usually one cell thick, with fatty cutin (forming the cuticle) within and on the surface of the outer walls; epidermis may include guard cells that form pores called stomata; root hairs, which are tubular extensions of single cells; other hairs that consist of one to several cells; and glands that secrete protective substances.

Periderm, which consists of cork cells and loosely arranged groups of cells comprising lenticels involved in gas exchange, constitutes the outer bark of woody plants.

Parenchyma cells are thin-walled, while collenchyma cells have unevenly thickened walls that provide flexible support for various plant organs. Two types of sclerenchyma—fibers (which are longer and tapering) and sclereids (which are short in length)—occur; both types have thick walls and are usually dead at maturity. Secretory tissues occur in various places in plants; they secrete substances such as nectar, oils, mucilage, latex, and resins.

Xylem conducts water and dissolved substances throughout the plant. It consists of a combination of parenchyma, fibers, ray cells (involved in lateral conduction), vessels (tubular channels composed of cells called vessel elements, which are attached end to end), and tracheids (cells with tapering end walls that overlap). Phloem conducts dissolved food materials throughout the plant. It is composed of sieve tubes (made up of cells called sieve-tube elements), companion cells (which apparently regulate adjacent sieve-tube elements), parenchyma, ray cells, and fibers. Sieve cells, which have overlapping end walls, and adjacent albuminous cells take the place of sieve-tube elements and companion cells in ferns and cone-bearing trees.

Review Questions

1. What is the function of meristems? Where are they located?

2. How are parenchyma, collenchyma and sclerenchyma distinguished from one another?

3. Distinguish between epidermis and periderm.

4. What types of substances do secretory cells secrete?

5. What are the functions of xylem and phloem? What cells are involved in their normal activities?

Discussion Questions

1. Most plant meristems are located at the tips of shoots and roots and in cylindrical layers within stems and roots. What could happen if they were present in leaves?

2. The cambium produces xylem toward the center of a tree and phloem toward the outside. Do you think it would make any difference if the positions of the xylem and phloem were reversed? Why?

Additional Reading

Cutler, D. F. *Applied plant anato[...]* Longman, Inc.; 1978.

Cutter, E. G. *Plant anatomy,* Part I: C[...] *tissues.* 2d ed. Reading, MA: Addiso[...] Publishing Co., Inc.; 1978.

Esau, K. *Anatomy of seed plants.* 2d ed. New [...] John Wiley & Sons, Inc.; 1977.

y: New York:

lls and

-Wesley

York:

Roots and Soils

Overview

This chapter discusses roots, beginning with the functions and development of roots from a seed. It covers the function and structure of the root cap, region of cell division, region of elongation, and region of maturation (with its tissues). The endodermis and pericycle are also discussed.

Specialized roots (food storage roots, water storage roots, propagative roots, pneumatophores, aerial roots, contractile roots, buttress roots, parasitic roots, mycorrhizae) are given brief treatment. This is followed by some observations on the economic importance of roots.

After a brief examination of soil horizons, the chapter concludes with a discussion of the development of soil, its texture, composition, structure, and its water.

Some Suggested Student Learning Goals

1. List the functions and forms of roots in general.

2. Describe the root cap, region of cell division, region of cell elongation, and region of maturation (including root hairs and all tissues), and indicate the function of each.

3. Identify the differences among the various types of specialized roots.

4. List at least ten practical uses of roots.

5. Describe how a good agricultural soil is developed from raw materials.

6. Contrast the various forms of soil water with regard to their specific location and availability to plants.

7. Compare sand, silt, and clay particles.

Introduction

You have probably seen pictures, at least, of the destruction caused by a tornado as it cut a swath through a village or a city, but have you ever seen what a twister can do to a forest? Large trees may be snapped off above the ground or knocked down, and branches may be stripped bare of leaves. Depending on the type of forest and the composition of the soil, however, you will probably see relatively few trees completely torn up by the roots and blown elsewhere. And in the tropics it is indeed rare to find healthy palm trees uprooted after a hurricane. Roots anchor trees firmly in the soil, usually by forming an extensive branching network that constitutes about one-third of the total dry weight of the plant. The roots of most plants do not usually extend down into the earth beyond a depth of 3 to 5 meters (10 to 16 feet); those of many herbaceous species are confined to the upper 6 to 9 decimeters (2 to 3 feet). A few plants such as alfalfa, however, frequently send their roots more than 6 meters (20 feet) into the earth. When the Suez Canal was being built, workers encountered roots of tamarisk at depths of nearly 30 meters (100 feet), and mesquite roots have been seen 53.4 meters (175 feet) deep in a pit mine in the southwestern United States. Some plants, such as cacti, form very shallow root systems but still effectively achieve *anchorage* by sending a densely branching mass of roots radiating out in all directions as far as 15 meters (50 feet) from the stem. In addition to anchorage, roots function extensively in the *absorption of water and minerals in solution.* The roots of some plants have specialized functions such as food or water storage. These functions are discussed in the section on specialized roots (page 70).

Although some aquatic plants (e.g., duckweeds) normally produce roots in water, and others (e.g., many orchids) produce aerial roots, the great majority of vascular plants develop their root systems in soils. Soils vary considerably in composition, texture, and other characteristics. These features of soils are discussed in the second half of this chapter.

When a seed germinates, a part of the **embryo** within it, the **radicle,** grows out and forms the first root. This may either develop into a *taproot,* from which branch roots develop, or it may give way to the development of numerous *adventitious* roots, forming a *fibrous* root system with roots of similar diameter (figure 5.1). **Adventitious** refers to structures arising at unusual places, such as roots growing from stems, or leaves or buds appearing at places other than leaf axils and tips of stems. Many mature plants exhibit a combination of taproot and fibrous root systems. The number of roots produced by a single plant may be prodigious. For example, a single mature ryegrass plant may have as many as 15 million individual roots and branchroots, with a combined length of 644 kilometers (400 miles) and a total surface area larger than a volleyball court, all contained within 57 cubic decimeters (2 cubic feet) of soil. This surface area is more than doubled by the presence of tiny *root hairs,* discussed in the next section.

Most *dicot* plants (plants with two "seed leaves") have taproot systems with one to several predominant roots from which secondary roots develop, while *monocots* (plants with one "seed leaf") have fibrous roots. Other types of roots, such as adventitious roots (see chapter 6), may develop in both dicots and monocots. In English and other ivies, lateral adventitious roots that aid in climbing appear along the aerial stems, and in certain plants with modified stems (e.g., rhizomes, corms, and bulbs; see figure 6.12), adventitious roots are the only kind produced.

Figure 5.1 Root systems. *A.* Fibrous root system of barley. *B.* Taproot system of a dandelion.

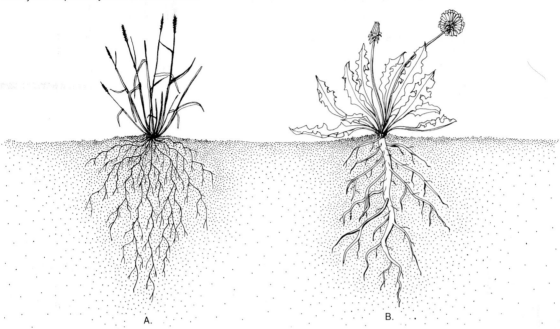

A.   B.

Root Structure

Botanists have traditionally recognized four regions or zones in developing young roots. Three of the regions are not sharply defined at their boundaries. The cells of each region gradually develop the form of those of the next region; the extent of each region varies considerably, depending on the species involved. These regions are called (1) the *root cap,* (2) the *region of cell division,* (3) the *region of elongation,* and (4) the *region of maturation* (figure 5.2).

Root Cap

The **root cap** is composed of a thimble-shaped mass of parenchyma cells covering the tip of each root. It is quite large and obvious in some plants, while in others it is nearly invisible. One of its functions is to *protect* from damage the delicate tissues behind it as the young root tip pushes through often angular and abrasive soil particles. The root cap has no equivalent in stems. Its outer cells secrete mucilage in the walls; these cells break down and are constantly being sloughed off (and replaced from the inside) forming a slimy lubricant that facilitates movement through the soil. The root cap can be slipped off or cut off, and when this is done, a new one is produced. Until it has been renewed, however, the root seems to grow randomly instead of downward, thus suggesting that the root cap has another function, namely, the *perception of gravity.* It is known that starch grains collect on the sides of the root cap cells facing the direction of gravitational force, and there is some evidence they may help influence the direction of root growth, although the actual manner in which they could accomplish this is not known.

Chapter 5

Figure 5.2 Longitudinal section through a dicot root tip. *Left:* Regions of the root. *Right:* Locations of the primary meristems of the root.

region of maturation

region of elongation

apical meristem (region of cell division)

root cap

protoderm

ground meristem

procambium

apical meristem (region of cell division)

root cap

Region of Cell Division

The root cap is produced by cells in the region of cell division, which is in the center of the root tip and surrounded by the root cap. It is composed of an **apical meristem** similar to that of stem tips. Most of the cell divisions take place at the edges of an inverted cup-shaped zone a short distance back of the actual base of the meristem just adjacent to the root cap. Here, the cells divide every 12 to 36 hours, while at the base of the meristem, they may divide only once in every 200 to 500 hours. Also, the divisions are often rhythmic, reaching a peak once or twice each day, usually toward noon and midnight, with relatively quiescent intermediate periods. Cells in this region are

mostly cuboidal in shape, with relatively large centrally located nuclei and few, if any, small vacuoles. As in stems, the apical meristem soon subdivides into three meristematic areas: (1) the **protoderm** gives rise to an outer layer of cells, the epidermis; (2) the **ground meristem,** to the inside of the protoderm, produces parenchyma cells of the cortex; (3) the **procambium,** which appears as a solid cylinder in the center of the root, produces primary xylem and phloem (figure 5.3). **Pith** tissue, which is seen in most stems, is absent in most dicot roots but is found in those of many monocots such as grasses.

Figure 5.3 Cross section through a portion of a root of corn (a monocot). (Photomicrograph by Carolina Biological Supply Company)

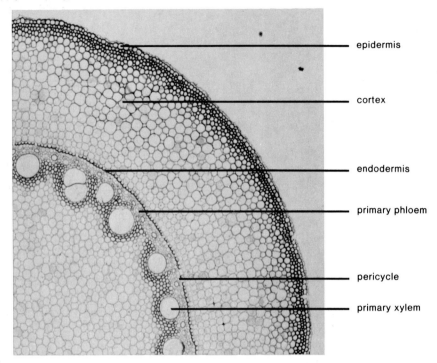

epidermis

cortex

endodermis

primary phloem

pericycle

primary xylem

Region of Elongation

The region of elongation, which merges with the apical meristem, usually extends about 1 centimeter (0.4 inch) or less from the tip of the root. Here the cells become several times their original length and also somewhat wider, while the tiny vacuoles merge and grow until one or two large vacuoles, occupying up to 90% or more of the volume of each cell, have been formed. No further increase in cell size takes place above this region, so only the root cap and apical meristem are actually pushing through the soil. Except for the tips and for those roots where there is gradual increase in girth through the addition of secondary tissues, the often very extensive remainder of each root remains stationary for the life of the plant.

Region of Maturation

Most of the cells mature into the various distinctive cell types of the primary tissues in this region, which is sometimes called the region of differentiation, or *root hair zone*. It is given this last name because of the numerous, hairlike, delicate protuberances that develop from many of the epidermal cells. These **root hairs,** as they are called, are not separate cells; in fact, the nucleus of the epidermal cell to which each is attached often moves out into the protuberance. They are so numerous that they appear as fine down to the naked eye, typically numbering more than 38 thousand per square centimeter (250 thousand per square inch) of surface area in roots of plants such

Chapter 5

Figure 5.4 Root hairs in contact with soil particles

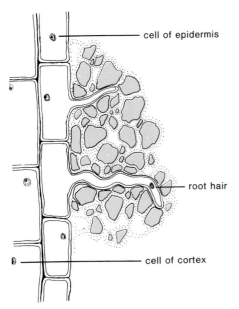

cell of epidermis

root hair

cell of cortex

as corn, and they seldom exceed 1 centimeter (0.4 inch) in length. A single ryegrass plant has been found to have more than 14 billion root hairs, with a total surface area almost the size of a football field. This obviously greatly increases the absorbing surface of the root, as the root hairs adhere tightly to soil particles with the aid of microscopic fibers that they produce (figure 5.4). When a seedling or plant is moved, many of the delicate root hairs are torn off or die within seconds if exposed to the sun, thus greatly reducing the plant's capacity to absorb water and minerals in solution. This is why plants should be watered and shaded after transplanting until new root hairs have formed. In any growing root, the extent of the root hair zone remains fairly constant, with new root hairs being formed toward the root cap and older root hairs dying back in the more mature regions. The life of the average root hair is usually not more than a few days, although a few live for a maximum of perhaps three weeks.

If the region of maturation is examined in cross section (figure 5.5), it will be seen that, with the exception of the pith in dicots, most of the tissues of young stems are also present in roots. This is hardly surprising, since roots and stems are connected to each other. However, there are some differences. The cuticle, which may be relatively thick on the epidermal cells of stems, is very thin on the root hairs and epidermal cells of roots. Presumably any significant amount of fatty substance present in or on the walls of water-absorbing cells would interfere with their function. The **cortex,** a tissue adjacent to the epidermis, functions primarily in food storage. It may be many cells thick and is similar to that of stems except for the presence of an **endodermis** at its inner boundary. The endodermis does not generally occur in stems, whereas it is nearly universally present in roots. It consists of a single layer of cells, most of which have thickened walls. Young endodermal cells also have a coating of fatty suberin, either covering all the inner wall faces or, more commonly, forming bands around the radial and transverse walls. These suberin bands are called **Casparian strips.** The protoplasts of the endodermal cells are fused to the Casparian strips, creating a barrier to the passage of water through the otherwise permeable cell walls. This barrier forces water to pass through the protoplasm of the endodermal cells and thus enables the root to regulate the movement of water and dissolved substances entering or leaving the central vascular cylinder. In some roots, the epidermis, the cortex, and the endodermis are sloughed off as girth increases, but in those roots where the endodermis is retained, all the inner walls of the cells eventually become coated with suberin. Thin-walled **passage cells** are scattered singly or in small groups throughout the thicker-walled endodermal cells found in the younger portions of the root.

Roots and Soils

Figure 5.5 Cross section through the region of maturation of a root of a buttercup (a dicot)

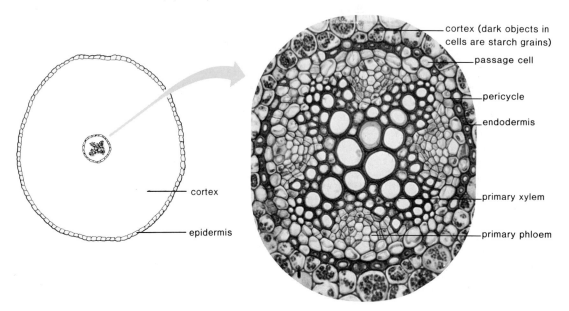

cortex (dark objects in cells are starch grains)

passage cell

pericycle

endodermis

primary xylem

primary phloem

cortex

epidermis

On the inside of the endodermis and immediately adjacent to it is a cylinder of parenchyma cells called the **pericycle.** This is usually one cell wide, although in some plants it may extend for several cells. It is a very important tissue, for the cells retain their capacity to divide even after they have matured, and it is within the pericycle that the *branch roots* and the *vascular cambium* of dicots arise (figure 5.6).

The *primary xylem,* with its water-conducting cells such as tracheids, forms a solid core in the center of most dicot and conifer roots, while in many monocot and some dicot roots it surrounds pith parenchyma cells. Root xylem, however, unlike stem xylem, first forms with short spirelike ridges, or arms, pointing toward the pericycle, at least in dicot roots. There are often

four of these arms, but there can be from two to several. *Primary phloem* forms in discrete patches between the xylem arms, and branch roots usually arise in the pericycle opposite the xylem arms. The primary phloem, containing food-conducting cells, also forms in patches in monocot roots, but in these roots the xylem arms may not be as strongly pronounced.

In woody dicots, some herbaceous dicots, and conifers, parts of the pericycle and parenchyma cells between the xylem arms and the phloem patches become a vascular cambium that starts producing secondary phloem to the outside and secondary xylem to the inside. Soon the cambium follows the outline of the primary xylem, and eventually, instead of appearing as patches and arms, the secondary conducting tissues appear as concentric cylinders; as is the case in stems, the primary tissues are crushed out of existence or sloughed off and lost. In woody plants, a cork

Chapter 5

Figure 5.6 Cross section through a water hyacinth (a dicot) root showing the origin of a branch root. (Photomicrograph by Carolina Biological Supply Company)

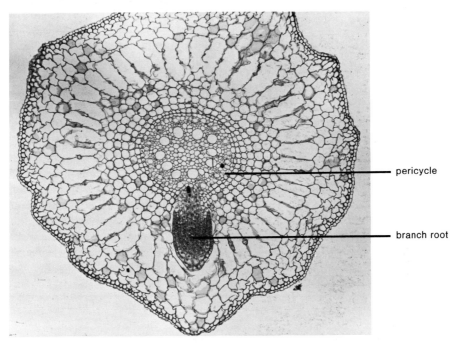

— pericycle

— branch root

cambium normally arises in the pericycle and gives rise to cork tissue (*periderm*) similar to that of stems. In fact, the cross section of a woody root several years old, with its annual rings of xylem along with its sapwood and heartwood, may be distinguished from that of a stem of the same species only by the relatively larger proportion of parenchyma cells present in the root. Although there are exceptions, monocot roots, like monocot stems, generally have no secondary meristems and therefore no secondary growth.

Natural grafting may occur between roots of different trees of the same species, particularly in the tropics. When the roots come in contact with one another, they unite through secondary growth, but the details of the uniting process are not yet known. One unfortunate aspect of this grafting is

that if one tree becomes diseased, the disease can be transmitted through these grafts to all the other trees connected to it.

In roots, there are no nodes or internodes, and branch or lateral roots arise internally in the pericycle, pushing their way out to the surface through the endodermis, cortex, and epidermis (see figure 5.6). This process is in contrast with that in stems, where branches develop from axillary buds. Some specialized roots, which at first glance may appear similar to specialized stems (see chapter 6) such as rhizomes, bulbs, and corms, can be distinguished by the absence of nodes and internodes.

Specialized Roots

As indicated earlier, most plants produce either a fibrous root system, a taproot system, or, more commonly, combinations of the two types. Some plants, however, have roots with modifications that permit specific functions in addition to the absorption of water and minerals in solution.

Food Storage Roots

Most roots and stems store some food, but in certain plants the roots are enlarged and store large quantities of starch and other carbohydrates (figure 5.7). In sweet potatoes and yams, for example, extra cambial cells develop in parts of the xylem of branch roots and produce numerous parenchyma cells that cause the organs to swell and provide storage areas for considerable quantities of carbohydrates. Similar food storage roots

are found in the deadly poisonous water hemlocks, in dandelions, and in salsify. Some food storage organs, such as those of carrots, beets, turnips, and radishes, are actually a combination of root and stem. In carrots, for example, approximately 2 centimeters (0.8 inch) at the top is derived from stem tissue, yet a casual inspection will not reveal any differences of origin.

Water Storage Roots

Some members of the pumpkin family, particularly those that grow in arid regions or in those areas where there may be no precipitation for several months of the year, produce huge water storage roots. In certain manroots, for example, roots weighing 30 kilograms (66 pounds) or more are frequently produced, and a major root of one calabazilla plant was found to weigh 72.12 kilograms (159 pounds). The water in the roots is apparently used by the plants when none is available in the soil.

Figure 5.7 Yams (food storage roots). (USDA photo)

Chapter 5

Propagative Roots

Many plants produce **adventitious** buds along the roots that grow near the surface of the ground. The buds develop into aerial stems called *suckers,* which have additional rootlets at their bases; these rooted suckers can be separated from the original root and grown individually. A number of fruit trees, such as cherries and pears, frequently produce such roots. The adventitious roots of horseradish and rice-paper plants can become a nuisance in gardens, the latter often producing propagative roots within a radius of 10 meters (33 feet) or more from the parent plant. Canada thistles and some other weeds have a remarkable facility to reproduce in this fashion as well as by means of seeds. This ability made it difficult to control them in the past, but some **biological controls** now being investigated (see appendix 2) may be an answer to the problem in the future.

Pneumatophores

Water, even after air has been bubbled through it, contains less than one-thirtieth of the amount of free oxygen found in the air. Thus, plants growing with their roots in water may have a problem in obtaining the oxygen necessary for respiration. Some swamp plants, such as the black mangrove and the yellow water weed, develop special "spongy" roots, called *pneumatophores,* which rise above the surface and facilitate gas exchange between the atmosphere and the subsurface roots to which they are connected (figure 5.8). Some older texts attributed a similar function to the woody "knees" of the bald cypress that occurs in southern swamps (see figure 19.18), but there is no conclusive evidence for this theory.

Figure 5.8 Pneumatophores of a tropical mangrove, rising above the mud of a coastal swamp. (Courtesy Lani Stemmerman)

Figure 5.9 Banyan tree, with numerous large prop roots that have developed from the branches

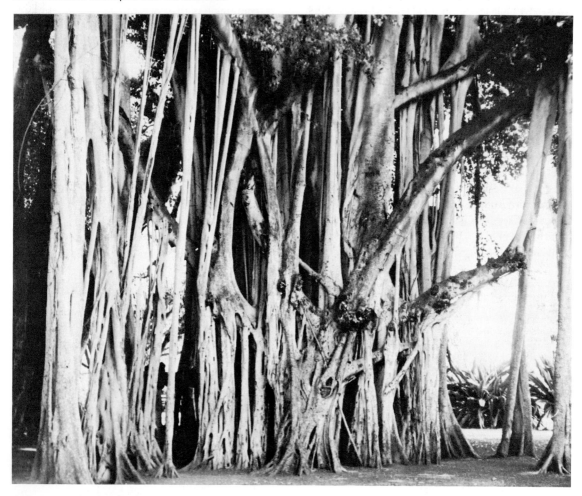

Aerial Roots

The various kinds of aerial roots produced by plants include the *velamen* roots of orchids, *prop* roots of corn and banyan trees (figure 5.9), *adventitious* roots of ivies, and *photosynthetic* roots of certain orchids. Velamen roots have an epidermis several cells thick. It has long been assumed that this thick epidermis aids in the absorption of rainwater, but some now suspect it may function more in preventing loss of moisture from the root. Corn prop roots, produced toward the base of the stems, support the plants in a high wind. A number of tropical plants, including the screw pines and various mangroves, produce sizeable prop roots extending for several feet above the surface of the ground or water. Debris collects between them and helps to create additional soil. Many of the tropical figs or banyan trees

Chapter 5

Figure 5.10 How a lily bulb, over three seasons, is withdrawn deeper into the soil through the action of contractile roots. *A.* A seed germinates. *B.* Contractile roots pull the newly formed bulb down several millimeters during the first season. *C.* The bulb is pulled down further the second season. *D.* The bulb is pulled down even further the third season. The bulb will continue to be pulled down in succeeding seasons until it reaches an area of relatively stable soil temperatures.

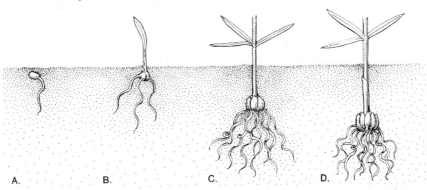

produce roots that grow down from the branches until they contact the soil. Once they are established, they continue secondary growth and look just like additional trunks. Some of these banyan trees attain great age and dimension. In India and southeast Asia, several such trees display almost one thousand root-trunks and have circumferences approaching 450 meters (1,476 feet). The oldest is estimated to be about two thousand years of age. The vanilla orchid, from which we obtain vanilla flavoring, produces chlorophyll in its aerial roots and thus can manufacture food with them through the process of photosynthesis. The adventitious roots of English ivy, Boston ivy, and Virginia creeper appear along the stem and aid the plants in climbing.

Contractile Roots

A number of herbaceous dicots and monocots have *contractile* roots that pull the plant deeper into the soil. Many lily bulbs are pulled a little deeper into the soil each year as additional sets of contractile roots are developed (figure 5.10).

The bulbs continue to be pulled down until an area of relatively stable temperatures is reached. Further, plants like dandelions always seem to have the leaves coming out of the ground as the top of the stem is pulled down a small amount each year when the root contracts. The contractile part of the root may lose as much as two-thirds of its length within a few weeks. Mechanisms of contraction include the thickening and constriction of parenchyma cells, causing the xylem elements to spiral somewhat like a corkscrew.

Buttress Roots

Some tropical trees produce huge buttresslike roots toward the base of the trunk, giving them great stability (figure 5.11). Except for their angular appearance, these roots look like a part of the trunk.

Figure 5.11 Buttress roots of a tropical fig tree

Parasitic Roots

A number of plants, such as the dodders, broom-rapes, and pinedrops, have no chlorophyll (necessary for photosynthesis) and have become dependent on plants with chlorophyll for their nutrition. They parasitize their host plants via somewhat rootlike projections called **haustoria** (singular: **haustorium**) (figure 5.12), which develop along the stem in contact with the host. The haustoria penetrate the outer tissues and establish connections with the water-conducting and food-conducting tissues. Some green plants, such as Indian warrior and the mistletoes, also form haustoria, but these apparently function primarily in obtaining water and dissolved minerals from the host plants, since the partially parasitic plants are capable of manufacturing at least some of their own food through photosynthesis.

Mycorrhizae

Members of a sizeable majority of flowering plant families have various fungi associated with their roots. The association is *mutualistic* (mutualism is a form of *symbiosis;* see page 242); that is, both the fungus and the root benefit from it but are also dependent upon the association for normal development. The fungus often forms a mantle of

Chapter 5

Figure 5.12 Twining, colorless stems of a dodder plant parasitizing a weed. Note the small, peglike haustoria that have formed but not yet penetrated the host plant's tissues.

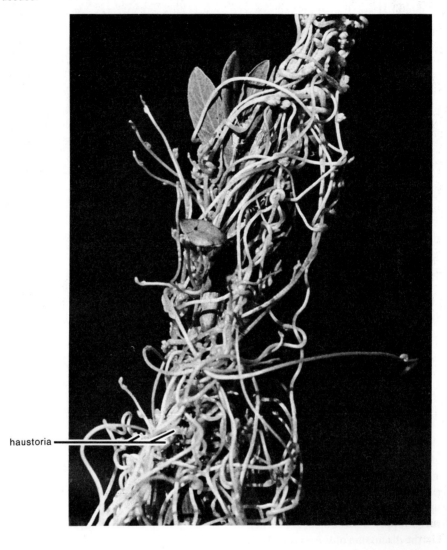

haustoria

millions of threadlike strands around the root and seldom penetrates deeper than the cortex. If fungal threads should happen to penetrate any deeper, they are apparently digested and broken down by the host plant. The fungus evidently breaks down complex molecules in the humus of the soil; the smaller molecules are absorbed by the root, while the root furnishes the fungus with food and amino acids. These "fungus-roots," or **mycorrhizae** (figure 5.13), are essential to the normal growth and development of some plants such as forest trees and orchids. Orchid seeds

Figure 5.13 Mycorrhizae *A.* Mycorrhizal root in cross section. *B.* Comparison, in longitudinal section, between a mycorrhizal root and one not associated with fungi.

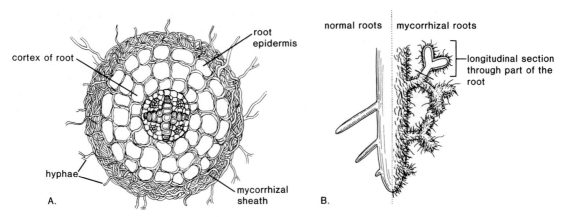

simply will not germinate until a mycorrhizal fungus invades their cells, and many plants seem to have considerable difficulty in absorbing essential elements such as phosphorus, even when the elements are abundant in the soil, if mycorrhizal fungi have been killed by fumigation or are otherwise absent. Other plants do not seem to need the mycorrhizae unless essential elements in the soil are present in amounts barely sufficient for healthy growth. Plants with mycorrhizae develop few root hairs as compared with those growing without an associated fungus.

Root Nodules

When legume plants (e.g., peas and beans), alders, and a few other plants are uprooted, numerous small swellings can be seen along the roots. These contain bacteria that supplement the plant's nitrogen supply. This subject is detailed in chapter 23 in the discussion of the nitrogen cycle. Root nodules should not be confused with root knots, which are also swellings and are seen in the roots of tomatoes and many other plants. Root knots develop in response to the invasion of tissue by small, parasitic roundworms (nematodes). Unlike bacterial nodules, root knots are not beneficial to the plant and the activities of the parasites within them can eventually lead to the premature death of the plant.

Human Relevance of Roots

Roots have been important sources of food for humans since prehistoric times, and some, such as the carrot, have been in cultivation in Europe for at least two thousand years. A number of cultivated root crops involve **biennials** (i.e., plants that complete their life cycles from seed to flowering and back to seed in two seasons). Such plants store food in a swollen taproot during the first year of growth and then utilize the stored food to produce flowers in the second season. Among the best-known biennial root crops are sugar beets, beets, turnips, rutabagas, parsnips, radishes, horseradishes, and carrots. Other important root crops include sweet potatoes, yams, and cassava. Cassava (figure 5.14), from which tapioca is made, forms a major part of the basic diet for millions of inhabitants of the tropics. It

Chapter 5

Figure 5.14 Cassava plant. (Courtesy Field Museum of Natural History, Chicago)

yields more starch per hectare (about 45 metric tons, the equivalent of 20 tons per acre) than any other cultivated crop, with a minimum of human labor. A number of minor root crops are cultivated in South America and other parts of the world. Among such plants are relatives of wild mustards, nasturtiums, and sorrel.

A number of spices, such as sassafras, sarsaparilla, licorice, and angelica, are obtained from roots. Sweet potatoes are used in the production of alcohol in Japan. Several important red to brownish dyes are obtained from roots of members of the Madder family, to which coffee plants belong. Drugs obtained from roots include aconite, ipecac, ginseng, gentian, and reserpine, a tranquilizer. A valuable insecticide, rotenone, is obtained from the barbasco plant, which has been cultivated for centuries as a "fish poison" by primitive South American tribes. When thrown into a dammed stream, the roots containing rotenone cause the fish to float but in no way poison them for human consumption. Other uses of roots are discussed in chapter 22.

Soils

The soil is a dynamic, complex, constantly changing part of the earth's crust that extends from a few inches deep in some places to hundreds of feet deep in others. It is essential not only to our existence but, at least indirectly, to the existence of most living organisms. It has a pronounced effect on the plants that grow in it, and they, in turn, have an effect on it.

If you were to dig up a shovelful of soil from your backyard and examine it, you would probably find a mixture of ingredients, including several grades of sand, rocks and pebbles, powdery silt, humus (a dark substance made up of incompletely decomposed organic materials), dead leaves and twigs, clods consisting of soil particles held together by clay and organic matter, plant roots, and small animals such as ants, pillbugs, millipedes, and earthworms. Also present, but not visible, would be hundreds of thousands of microorganisms, particularly bacteria, and, of course, air and water.

The soil became what it is today through the interaction of a number of factors: climate, parent material, topography of the area, living organisms, and time. Because there are thousands of ways in which these factors may interact, there are many thousands of different soils. The solid portion of a soil consists of mineral matter and organic matter. Pore spaces, shared by variable amounts of water and air, occur between the solid particles. The smaller pores often contain water, and the larger ones usually contain air. The sizes of the pores and the connections between them largely determine how well the soil is aerated.

If you were to dig down 1 or 2 meters (3 to 6 feet) in an undisturbed area, you would expose a soil profile of three intergrading regions that soil scientists refer to as *horizons* (figure 5.15). The horizons show the soil in different stages of development, and the composition varies accordingly. The upper layer, usually extending down 1 to 2 decimeters (4 to 8 inches), is called the A horizon, or *topsoil*. It is usually subdivided into a darker upper portion called the A_1 horizon and a lighter lower portion called the A_2 horizon. The A_1 portion in particular contains more organic matter than the layers below. The next 0.3 to 0.6 meter (1 to 2 feet) is called the B horizon, or *subsoil*. It usually contains more clay and is lighter in color than the topsoil. The C horizon at the bottom may vary from about a decimeter (4 inches) to several meters (6 to 10 feet or more) in depth; it may even be absent. It is commonly referred to as the *soil parent material* and extends down to *bedrock*.

Figure 5.15 Miami loam soil profile. (After Foth, H. D., L. V. Withee, H. S. Jacobs, and S. J. Thien. 1982. *Laboratory Manual for Introductory Soil Science.* 6th ed. Wm. C. Brown Company Publishers, Dubuque, Iowa)

A_1 horizon (dark loam)

A_2 horizon (light-colored loam)

B horizon (clay loam)

C horizon (parent material)

Chapter 5

Parent Material

The first step in the development of soil is the formation of parent material. This material accumulates through the weathering of rocks, which originate from volcanic activity (*igneous* rocks), from fragments of rocks deposited by glaciers, water, or wind (*sedimentary* rocks), or from changes in either igneous or sedimentary rocks brought about by great pressures, or heat, or both (*metamorphic* rocks).

Climate

Climate varies greatly throughout the globe, and its role in the weathering of rocks varies correspondingly. In desert areas, for example, there is little weathering by rain, and soils are poorly developed; whereas in areas of high rainfall, well-developed soils are common. In areas where there are great temperature ranges, rocks may split or crack as their outer surfaces expand or contract at different rates from the material beneath the surface. Low temperatures cause water in rock crevices to freeze, which, in turn, also causes cracks and splits.

Living Organisms and Organic Composition

There are numerous organisms of many kinds in the soil, in addition to plant roots. In the upper 3 decimeters (1 foot) of a good agricultural soil, living organisms constitute about one-thousandth of the total weight of the soil. This may not sound significant, but it amounts to approximately 6.73 metric tons per hectare (3 tons per acre). Bacteria and fungi present in the soil decompose all types of organic matter that accumulates when leaves fall and roots and animals die (this process is further discussed under the section on composting in chapter 14). Roots and all other living organisms present produce carbon dioxide, which combines with water to form an acid, thereby increasing the rate at which minerals dissolve. Ants and other insects, earthworms, burrowing animals, and birds all cultivate the soil through their activities and add to its organic increment either through wastes that they deposit or through the decomposition of their bodies when they die.

The total organic composition of a soil varies greatly. An average topsoil might consist of about 25% air, 25% water, 48% minerals, and 2% organic matter; however, soils in low wet areas, where a lack of oxygen prevents microorganisms from their normal activities, may contain as much as 90% organic matter. Except in legumes and a few other plants, almost all of the nitrogen utilized by growing plants, as well as much of the phosphorus and sulphur, comes from decomposing organic matter. In addition, as organic matter breaks down, it produces acids that, in turn, decompose minerals. Other roles of organic matter in the soil are discussed in the section on soil structure (page 80).

Topography of the Area

If the topography of an area is steep, soil may wash away or erode through the action of wind, water, and ice as soon as it is weathered from the parent material. It has been estimated that more than 20 metric tons of topsoil per hectare (8.2 tons per acre) is washed away annually from some prime croplands in the central United States. If an area is flat and poorly drained, however, pools and ponds may appear in slight depressions when it rains. If these bodies of water cannot drain relatively quickly, the activities of organisms in the soil are interrupted and the development of the soil is arrested. The ideal topography for the development of soil is one that permits drainage without erosion.

Table 5.1 Soil mineral components as classified by the United States Department of Agriculture

Mineral	Diameter (range in mm)	Comments
Stones	> 76 mm	do not support plant growth but affect permeability and erosion of the soil
Gravel	76 mm–2.0 mm	
Very coarse sand	2.0 mm–1.0 mm	
Coarse sand	1.0 mm–0.5 mm	
Medium sand	0.5 mm–0.25 mm	make soil feel gritty
Fine sand	0.25 mm–0.10 mm	
Very fine sand	0.10 mm–0.05 mm	
Silt	0.05 mm–0.002 mm	a lens or microscope needed to see any but the coarsest silt particles
Clay	< 0.002 mm	may absorb water, swell, and later shrink, causing the soil to crack as it dries

Soil Texture and Mineral Composition

Soil texture refers to the sizes of the individual soil particles (see table 5.1). The principal texture designations are sand, silt, and clay. Sands and gravels are usually composed of many small particles bound together chemically or by a cementing matrix material. Silt consists of particles that are mostly too small to be seen without a lens or a microscope. Clay particles are so tiny that even a powerful light microscope would not render them visible, although they can be seen with an electron microscope. Individual clay particles are called *micelles*; they are somewhat sheetlike, negatively charged, and held together chemically. They attract, exchange, or retain positively charged ions. Clay is plastic in nature because the water that tightly adheres to the surface of the particles acts both as a binding agent and a lubricant. Physically, clay is a **colloid** (that is, a suspension of particles that are larger than molecules but that do not settle out of a fluid medium).

"Light" soils have high sand and low clay content. "Heavy" soils have high clay content. Coarse-textured soils do not retain large amounts of water, whereas clay soils permit very little water to pass through.

Over half of the composition by weight of mineral matter is oxygen. Other common elements present include hydrogen, silicon, aluminum, iron, potassium, calcium, magnesium, and sodium, but soil inherits hundreds of different mineral combinations from parent material.

More plant nutrients are stored in the form of ions by clay and organic matter than by sand and silt. This is because the clay and organic particles, being smaller than those of sand and silt, have a greater total surface area per unit volume of material to which ions can become attached.

Soil Structure

Soil structure refers to the arrangement of the soil particles into groups of particles called *aggregates*. Aggregates in sands and gravels show little cohesion, but most agricultural soils have aggregates that stick together. Structure develops when colloidal particles clump together as a result of the activities of soil organisms, freezing and thawing, and so on. If the individual granules do not become coated with organic matter, they may continue to clump until they become clods.

Good agricultural soils are granular soils with pore spaces that occupy between 40% and 60% of the total volume of the soil. The pores contain air and water, and their sizes are more important than their total volume. Clay soils, for example, have more actual pore space than sandy soils, but the pores are so small that water and air are restricted in their movement through the soil. When the pores are full of water, air is kept out, which results in insufficient oxygen for root growth. Sandy soils have large pores that drain by gravity soon after they are filled. The water is replaced by air, but too much air speeds up nitrogen release by microorganisms. The plants cannot use nitrogen this quickly, and much of it is lost as a result.

Water itself can be harmful, because too much of it depletes mineral nutrients, slows the release of nitrogen, and otherwise interferes with plant growth. Countless numbers of houseplant enthusiasts and outdoor gardeners have stunted or killed the very plants they were trying to foster by "drowning" the objects of their affection with too much water or too frequent waterings. A general rule of thumb is to refrain from watering until the surface of the soil appears dry.

Water in the soil occurs in three forms. **Hygroscopic water** is chemically bound to the soil particles and is unavailable to plants. **Gravitational water** drains out of the pore spaces after a rain. If drainage is poor, it is this water that interferes with normal plant growth. Plants depend mainly on the third type, **capillary water,** for their needs. The structure and organic matter of the soil—which enable the soil to hold water against the force of gravity—the density and type of vegetational cover, and the location of underground water tables largely determine the amount of capillary water available to the plant. The ancient Incas of Peru knew that water would rise just so far in some areas. Where the water table was close to the surface, they removed the upper 0.6 meter (2 feet) of soil and planted their crops down in the hollowed-out areas so that the roots would be able to reach the capillary water. In some areas having sandy soils and low annual precipitation, soils can be compacted by a heavy roller to create finer capillaries to raise water from below. This technique is effective only if the available water is within 1.5 to 3.0 meters (5 to 10 feet) of the surface and if the soils do not contain much silt or clay.

After rain or irrigation, water in the soil drains away by gravity. The water remaining after such draining is referred to as the *field capacity* of the soil. Field capacity is mainly governed by the texture of the soil, but the structure and organic content also influence it to a certain extent. Plants readily absorb water from the soil when it is at, or near, field capacity; but as the soil dries, the film of water around each soil particle becomes thinner, and the force exerted by the plants to remove the water from the soil increases. If water is not added to the soil, a point is eventually reached at which the rate of absorption of water by the plant is insufficient for its needs, and the plant wilts permanently. The soil is then said to be at the *permanent wilting point*. In clay soils the permanent wilting point is reached when water content drops below 15%, while in sandy soils the permanent wilting point may be as low as 4%. *Available water* is soil water between field capacity and the permanent wilting point.

The pH (acidity or alkalinity) of a soil affects both the soil and the plants growing in it in various ways. A soil that is unusually acid or alkaline may be toxic to the roots, but these conditions do not normally affect plants directly nearly as much as they affect nutrient availability. For example, minerals such as copper, iron, and manganese become less available to plants as alkalinity increases, whereas highly acid soils inhibit the growth of nitrogen-fixing bacteria. Acid soils tend to be common in areas of high precipitation where significant amounts of bases are leached from the topsoil. It is a common agricultural practice to counteract soil acidity by adding compounds of calcium or magnesium in a process known as *liming*. Alkaline soils can be made more acid by the addition of sulphur, which is converted by bacteria to sulphuric acid; the addition of some nitrogenous fertilizers may have the same effect.

Summary

Roots function primarily in anchorage and in the absorption of water and minerals in solution. The radicle of a germinating seed develops into the first root, from which develops either a taproot with branch roots, or a fibrous root system in which the roots are of similar diameter; many plants have combinations of both systems, with some also developing adventitious roots as they mature. In developing young roots, four zones or regions are traditionally recognized: (1) A thimble-shaped root cap at the tip, which protects the root as it pushes through the soil and also aids in the perception of gravity. (2) A region of cell division that is surrounded by the root cap. It is composed of an apical meristem that subdivides into a protoderm, which produces the epidermis; a ground meristem that produces the cortex; and a procambium in the center that produces primary xylem and primary phloem. (3) A region of elongation in which the cells produced by the apical meristem become considerably longer and slightly wider. (4) A region of maturation in which the cells mature into the distinctive cell types of primary tissues.

Some of the epidermal cells in the region of maturation develop protuberances called root hairs; the root hairs greatly increase the absorbing surface of the root. The tissues that mature in the region of maturation are similar to those of stem tips, except that pith is absent in most dicot roots, and of a different origin (the procambium) in monocot roots. The cortex has an endodermis at its inner boundary. This consists of a cylinder, one cell wide, of cells that have suberin bands called Casparian strips around the radial and tranverse walls, and an occasional thin-walled passage cell interspersed throughout the cells with suberin. Immediately adjacent to the endodermis toward the center of the root are parenchyma cells constituting the pericycle. Branch roots and the vascular cambium arise in the pericycle. The primary xylem initially frequently forms a solid core in the center of the root; this core usually develops with two to several (usually four) arms, at least in dicot roots. At first, primary phloem is produced in discrete patches between the arms of the primary xylem, but as secondary tissues are added, the tissues eventually appear as concentric cylinders. In woody plants, a cork cambium usually arises in the pericycle and produces cork tissues similar to those of stems.

Roots may graft together naturally. There are no nodes or internodes in roots. Roots may be specialized as food storage or water storage roots; as pneumatophores, which facilitate respiration in swamps; as aerial roots, which function in an aerial habitat (velamen roots, prop roots, photosynthetic roots, adventitious roots that aid in climbing); as contractile roots, which pull a plant deeper into the soil; as buttress roots, which add stability; as haustoria, which have a parasitic function; and as mycorrhizal roots, which involve fungal associations. Some plants (e.g., legumes, alders) have nitrogen-fixing bacteria in small swellings (nodules) on their roots.

Root crops include sugar beets, beets, turnips, rutabagas, parsnips, radishes, carrots, sweet potatoes, yams, and cassava. Spices such as sassafras, sarsaparilla, licorice, and angelica are obtained from roots. Other uses of roots include the production of alcohol from sweet potatoes, and the extraction of dyes, drugs, insecticides, and poisons.

Soils contain a mixture of ingredients, including sands, rocks, silt, humus, dead organic matter, plant roots, small animals and microorganisms, plus air and water within pore spaces of various sizes.

A vertical column of soil exhibits intergrading regions called horizons; these show the soil in various stages of development. The A horizon (topsoil) is divided into two layers; the upper A_1 horizon is darker and contains more organic matter than the lower A_2 horizon. The B horizon (subsoil) usually contains more clay and is lighter in color than the A horizon. The C horizon (bottom portion) is referred to as the soil parent material. Parent material is developed through the weathering of rocks. The extent of weathering varies with the climate.

Living organisms in the soil decompose organic matter, producing carbon dioxide that combines with water to form an acid that increases the rate at which minerals dissolve and are decomposed themselves. Animals also cultivate the soil. The total amount of organic matter in a soil can vary greatly; most important plant nutrients come from decomposing organic matter. The topography of an area may cause the soil to erode as soon as it is produced or it may arrest its development.

Soil texture pertains to sizes of soil particles. Sands and gravels usually consist of many small particles bound together. Silt particles are too small to be seen without magnification, and clay particles (micelles) are visible only with the aid of an electron microscope. More than half of the composition by weight of mineral matter is oxygen. Soil structure refers to the arrangement of the soil particles into aggregates. Good soils are highly granular and have pore spaces that constitute about half the total volume, with the sizes of the pores being more important to aeration and water retention than the numbers present. Water in the soil occurs as hygroscopic water, which is chemically bound to the soil particles and unavailable to plants; gravitational water, which drains out of pores; and capillary water, which is water held against the force of gravity. The field capacity of the soil is the amount of water that remains after the rest of the water has drained away by gravity. Soil reaches the permanent wilting point when plants wilt permanently because they can no longer extract enough water from the soil for their needs. Available water is soil water between field capacity and the permanent wilting point.

Review Questions

1. Distinguish between a tiny root and a root hair. What is the function of a root hair?

2. What is the difference between parasitic roots and mycorrhizae?

3. If you were shown cross sections of a young root and a young stem from the same dicot plant, how could you tell them apart?

4. What is the function of the root cap and from which meristem does it originate?

5. How do endodermal cells differ from other types of cells?

6. Where do branch roots originate?

7. Indicate some spices and drugs obtained from roots.

8. What is soil parent material? Where does it come from, and how does it become soil?

9. What is the difference between soil texture and soil structure?

10. What types of soil water are recognized? What is *available* soil water?

Discussion Questions

1. Japanese gardeners regularly trim away parts of the root system to assist in dwarfing a plant. A plant's food is obtained by photosynthesis in the leaves; can you suggest why trimming the roots can cause dwarfing?

2. It was suggested that roots perceive gravity through the root cap. Would it really matter if roots grew randomly in the soil instead of responding to gravity?

3. From the viewpoint of the plant, can you suggest a practical reason for branch roots originating internally instead of at the surface?

4. When you eat a yam or a sweet potato, what kinds of compounds and cells are you consuming?

5. Persons associated with commercial nurseries and greenhouses often sterilize their soil by heating it to get rid of pests, but then they have to wait for a short time after it has cooled to use it. Why?

Additional Reading

Cronquist, A. *Introductory botany.* 2d ed. New York: Harper and Row, Publishers; 1971.

Cutter, E. G. *Plant anatomy: experiment and interpretation.* Part 2: *Organs.* Reading, MA: Addison-Wesley Publishing Co.; 1971.

Donahue, R. L.; Schickluna, J. C.; Robertson, L. S. *Soils: An introduction to soils and plant growth.* 4th ed. Englewood Cliffs, NJ: Prentice-Hall; 1977.

Epstein, E. *Mineral nutrition of plants: principles and perspectives.* New York: John Wiley and Sons; 1972.

———. Roots. *Scientific American* 228: 48–58; 1973.

Esau, K. *Anatomy of seed plants.* 2d ed. New York: John Wiley and Sons; 1976.

Fahn, A. *Plant anatomy.* 2d ed. Elmsford, NY: Pergamon Press; 1974.

Russell, R. S. *Plant root systems: their function and interaction with the soil.* New York: McGraw-Hill; 1977.

Chapter 5

Overview

After a brief introduction, this chapter discusses in general the origin and development of stems. Items such as the apical meristem and the tissues derived from it, leaf gaps, cambia, secondary tissues, and lenticels are included. This general discussion is followed by distinctions between herbaceous and woody dicot stems and monocot stems. This section covers annual rings, rays, heartwood and sapwood, resin canals, bark, laticifers, and vascular bundles.

Next comes a survey of specialized stems (rhizomes, stolons, tubers, bulbs, corms, cladophylls, and others). The chapter concludes with a discussion of the economic importance of wood and stems.

Some Suggested Student Learning Goals

1. Identify the tissues that develop from shoot apices and the meristems from which each tissue is derived. Distinguish between primary tissues and secondary tissues.

2. Describe the following and the function of each: vascular cambium, cork cambium, stomata, lenticels.

3. Contrast the stems of herbaceous and woody dicots with the stems of monocots.

4. Identify the cells that compose wood and describe annual rings, sapwood, heartwood, and bark. Explain how a log is sawed for commercial use.

5. Distinguish among rhizomes, stolons, tubers, bulbs, corms, cladophylls, and tendrils.

6. List at least ten human uses of wood and stems in general.

Introduction

Whether you order a dozen long-stemmed roses for someone special, use a toothpick or chopsticks, build a wood-framed house, sit in a wooden chair and read a newspaper, or graft one variety of fruit tree on to another, you may or may not be conscious of the fact that these and literally hundreds of other activities either directly or indirectly involve plant stems, and that these particular plant organs have been an integral part of human life since cave men first used clubs to kill for food.

One of the activities mentioned—*grafting*—usually involves artificially uniting stems, or parts of stems, of different but related varieties of plants. The careful matching of certain tissues is critical to its success, as is seen in the discussion of grafting in appendix 4. To understand how and why grafts may or may not be successful, however, and to identify which parts of stems are useful for food, clothing, construction lumber, furniture, paper, etc., we first need to examine the structure of stems and learn the basic functions of the various tissues.

Unlike animals, some plants can grow indefinitely, with the meristems at their tips increasing their length and other meristems increasing their girth for hundreds or even thousands of years. In stems, the cells produced by the meristems usually become the familiar, erect, aerial *shoot system* with branches and leaves. In certain plants, such as ferns or perennial grasses, however, this shoot system may develop horizontally beneath or at the surface of the ground; in other plants the stem may be so short and inconspicuous as to appear nonexistent. In a number of plants, modifications of the stem permit specialized functions such as climbing or the storage of food or water.

External Form of a Woody Twig

A woody twig consists basically of an axis with leaves attached (figure 6.1). The leaves may be attached to the twig in a spiral around the stem (they are then said to be *alternate,* or *alternately arranged*), or they may occur in twos (*opposite,* or *oppositely arranged*), or sometimes they occur in groups of three or more called *whorls* (**whorled arrangement**). The area, or region (not structure), of a stem where a leaf or leaves are attached is called a **node,** and a stem region between nodes is called an **internode.** A leaf usually has a flattened **blade** and a stalk, called the **petiole,** by which it is attached to the twig.

A bud occurs in the angle between a petiole and the stem. This angle is called an **axil,** and the bud that sits in the axil is an *axillary bud.* Axillary buds may become branches or they may contain tissues that will develop into the next season's flowers. Most, but not all, buds are protected by one or several *bud scales,* which fall off when bud tissue growth begins. In addition to axillary buds, a *terminal bud* is often present at the tip of each twig. The terminal bud usually resembles an axillary bud, but it is frequently a little larger. It usually produces tissues that extend the length of the twig during the growing season rather than forming a separate branch. The bud scales of a terminal bud leave tiny scars around the twig when they fall off in the spring. A twig's age can be determined by counting the number of groups of bud scale scars on it.

Sometimes scars with a different origin also occur on a twig. These scars originate from a leaf that has paired appendages called **stipules** at the base of the petiole. The stipules may remain throughout the life of the leaf, but in some plants they fall off as the buds expand in the spring, leaving tiny *stipule scars,* which may resemble a fine line encircling the twig or may be very inconspicuous small scars on either side of the petiole base.

Deciduous trees and shrubs (those that lose their leaves annually in the fall) have characteristic **leaf scars,** with dormant axillary buds above them after the leaves fall. Tiny **bundle scars,** which mark the location of the water-conducting and food-conducting tissues, can be seen within the leaf scars. There are frequently three bundle scars present, but the number can vary from one

Figure 6.1 Woody twig. *Left:* In its winter condition. *Right:* The same twig as it appeared the summer before.

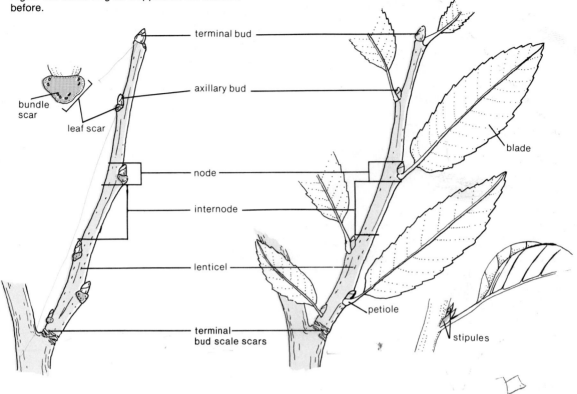

Figure 6.1 Woody twig. *Left:* In its winter condition. *Right:* The same twig as it appeared the summer before.

to many. The shape and size of the leaf scars, and the arrangement and numbers of the bundle scars, are characteristic of each species; it is often possible to determine the identity of a woody plant in its winter condition by means of these structures.

Origin and Development of Stems

As indicated in chapter 4, there is an apical meristem at the tip of each stem, and it is this meristem that produces the tissues resulting in the stem's increase in length. Before the onset of the growing season, the apical meristem is dormant; it is protected by bud scales of the bud in which it is located and also to a certain extent by leaf **primordia** (singular: **primordium**), the tiny embryo leaves that will develop into mature leaves after the bud scales drop off and growth begins.

The apical meristem in the embryo stem of a seed is also dormant until the seed begins to germinate.

When a bud begins to expand, or a seed germinates, the cells of the apical meristem undergo mitosis, and soon three primary meristems develop from it (see figure 3.13). The outermost of these primary meristems, the *protoderm*, gives rise to the epidermis. (As noted in chapter 4, the epidermis is usually one cell thick and usually becomes coated with a thin, fatty protective layer, the cuticle.) A cylinder of strands constituting the procambium appears to the interior of the protoderm. (The procambium produces water-conducting primary xylem cells and food-conducting primary phloem cells.) The remainder of the meristematic tissue, called ground meristem,

Figure 6.2 Longitudinal section through the tip of a *Coleus* stem. (Photomicrograph by Carolina Biological Supply Company)

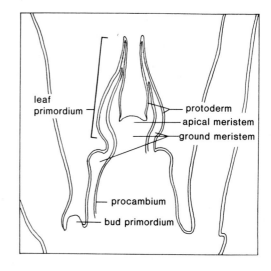

produces two tissues mostly composed of parenchyma cells. The parenchyma tissue in the center of the stem is the pith. (The cells of this area are very large and may break down shortly after they are formed, leaving a cylindrical hollow area; even if they do not break down early, they may eventually be crushed as new tissues produced by other meristems add to the girth of the stem, particularly in woody plants.) The other tissue produced by the ground meristem is the cortex. (The cortex may become more extensive than the pith, but in woody plants it, too, will eventually be crushed and replaced by new tissues produced from within.) The parenchyma of both the pith and the cortex function in storing food or sometimes, if chloroplasts are present, in manufacturing it.

All five of the tissues produced by this apical meristem complex (epidermis, primary xylem, primary phloem, pith, and cortex) arise while the stem is increasing in length and are called *primary tissues*. As these primary tissues are produced, the leaf primordia and the bud primordia

Figure 6.3 Portion of a young stem showing leaf gaps and bud gaps in the cylinder of vascular tissue. (After Esau, K. 1965. *Plant Anatomy.* 2d ed. John Wiley & Sons, Inc., New York. Copyright © 1965 John Wiley & Sons, Inc. Redrawn by permission of John Wiley & Sons, Inc.)

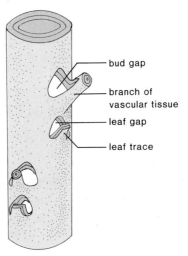

Chapter 6

Figure 6.4 Illustration of how a cell of the vascular cambium produces new secondary xylem cells to the inside and new secondary phloem cells to the outside. Note, in cross section, that the cambium gradually becomes shifted away from the center as new cells are produced.

surface of stem or root

center of stem or root

cell of cambium   immature xylem cell   immature phloem cell

(embryo buds in the axils of the leaf primordia) develop into mature leaves and buds (figure 6.2). As each leaf and each bud develops, a strand of xylem and phloem, called a *trace,* branches off from the cylinder of xylem and phloem extending up and down the stem and enters the leaf or the bud. As the traces branch from the main cylinder of xylem and phloem, each trace leaves a little thumbnail-shaped gap in the cylinder of tissue. These gaps are called **leaf gaps** and *bud gaps* (figure 6.3).

A narrow band of cells between the primary xylem and the primary phloem may retain its meristematic nature and become the vascular cambium, often referred to simply as the *cambium.* These cambial cells continue to divide indefinitely, with the divisions taking place mostly in a plane parallel to the surface of the plant; the secondary tissues produced thus add to the girth of the stem instead of to its length (figure 6.4).

If a nail is driven into the side of a tree and observed periodically over a number of years, differences between the activities of the apical meristem complex and those of the vascular cambium

will readily become apparent. The nail may eventually become embedded as the stem increases in girth, but it will always remain at the same height above the ground, as the cells that increase the length of a stem are produced only at the tips.

Cells produced by the vascular cambium become tracheids, vessel elements, fibers, or other components of *secondary xylem* (to the inside of the meristem), or become sieve-tube elements, companion cells, or other components of *secondary phloem* (to the outside of the meristem). The functions of these secondary tissues are the same as those of their primary counterparts—secondary xylem conducts water (and soluble nutrients), while secondary phloem conducts food (in soluble form) manufactured by the plant through photosynthesis.

In many plants, especially woody species, a second cambium arises within the cortex or, in some instances, develops from the epidermis or phloem. This is called the *cork cambium,* or **phellogen.** The cork cambium produces boxlike cork cells, which become impregnated with **suberin,** a fatty substance that makes the cells impervious to moisture. The cork cells, which are produced annually in cylindrical layers, die shortly after they are formed. The cork cambium may also produce parenchymalike *phelloderm* cells to the inside. Cork tissue makes up the outer bark of woody plants and functions in reducing water loss and in protecting the stem against mechanical injury. The tissue cuts off water and food supplies to the epidermis, which soon dies and is sloughed off. In fact, if the cork were to be formed as a solid cylinder covering the entire stem, vital gas exchange with the interior of the stem would not be possible. In young stems, such gas exchange takes place through the stomata located in the epidermis (see figures 7.6 and 8.7). As woody stems age, lenticels (see figure 4.4) develop beneath the stomata. As cork is produced, the parenchyma cells of the lenticels remain, so that exchange of gases (e.g., oxygen, carbon dioxide) can continue through spaces between the cells.

As indicated in chapter 4, lenticels occur in the fissures of the bark of older trees and often appear as small bumps on younger bark. In birch and cherry trees, the lenticels form conspicuous horizontal lines.

Tissue Patterns in Stems

Flowering plants develop from seeds that have either one or two "seed leaves" called **cotyledons** attached to the embryo stem (see chapters 10, 20, and 21), whereas the seeds of cone-bearing trees such as pines have several (usually eight) cotyledons. The cotyledons may function in storing food needed by the young seedling until its first true leaves can produce food themselves. Flowering plants that develop from seeds having two cotyledons are called **dicotyledons** (often abbreviated to *dicot*), while those developing from seeds with a single cotyledon are called **monocotyledons** (abbreviated to *monocot*). Dicots and

Figure 6.5 Cross section through a portion of a young sunflower stem, showing the arrangement of tissues typical of dicot herbaceous stems. (Courtesy Triarch, Inc.)

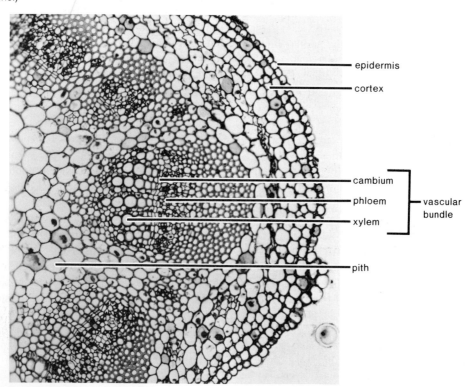

epidermis
cortex

cambium
phloem — vascular bundle
xylem

pith

monocots differ from one another in several other respects; differences in stem structure are noted in the following sections, and a summary of these and other differences is given in table 20.1.

Dicotyledonous Herbaceous Stems

In general, plants that complete their life cycles within one year (**annuals**) have green **herbaceous** (nonwoody) stems. Their tissues are largely primary, although cambia (plural of cambium) may develop some secondary tissues. Dicot herbaceous stems (figure 6.5) have discrete patches of xylem and phloem called **vascular bundles** which occur in a ring that separates the cortex from the pith, although in a few plants (e.g., foxgloves), the xylem and the phloem are produced as continuous rings. As previously noted, the procambium produces only primary xylem and phloem, but later a vascular cambium arises between these two primary tissues and adds secondary xylem and phloem to the vascular bundles. In some

plants the cambium extends between the vascular bundles, appearing as a narrow ring producing the conducting tissues within the bundles and the parenchyma cells between them. In other plants the cambium is confined to the bundles, each of which has its own small band of this meristematic tissue between the xylem and phloem.

Dicotyledonous Woody Stems

The arrangement of primary tissues in dicot woody stems (and also in those of cone-bearing trees) is very similar to that found in dicot herbaceous stems during the early stages of growth. As soon as the vascular cambium and the cork cambium start functioning, however, obvious differences begin to appear, the most conspicuous of which involve the secondary xylem, or *wood,* as it is best known (figure 6.6). Some tropical trees (e.g., ebony), in which both the vascular cambium and the cork cambium are active all year, produce an ungrained, uniform wood. The wood

Figure 6.6   Cross section through a portion of a young bittersweet stem

pith

xylem

phloem

cortex

phelloderm

cork cambium

cork tissue

vascular cambium

of most trees, however, exhibits seasonal growth. In trees of temperate climates, virtually all growth takes place during the spring and summer and then ceases for the year. When the vascular cambium of a typical broadleaf tree first becomes active in the spring, it usually produces relatively large vessel elements of secondary xylem; such xylem is referred to as *spring wood*. As the season progresses, the vascular cambium may produce vessel elements whose diameters become progressively smaller in each succeeding series of cells produced, or there may be fewer vessel elements in proportion to tracheids produced until tracheids (and sometimes fibers) predominate. This xylem, which is produced after the spring wood, and which has smaller or fewer vessel elements and larger numbers of tracheids, is referred to as *summer wood*. Over a period of years, the result of this type of switch between the early spring and the summer growth is a series of alternating concentric rings of light and dark cells. One year's growth of xylem is called an **annual ring.**

The vascular cambium produces more secondary xylem than it does phloem. In addition, xylem cells have stronger, more rigid walls than those of phloem cells and thus are less subject to collapse under pressure. As a result, the bulk of a tree trunk consists of annual rings of wood. The annual rings not only indicate the age of the tree (since normally only one is produced each year), but they can also tell something of the climate and other conditions occurring during the tree's lifetime (figure 6.7). For example, if the rainfall during a particular year is higher than normal, the annual ring for that year will be wider than usual. Sometimes caterpillars or locusts will strip the leaves of a tree shortly after they have appeared. This usually results in two annual rings being very close together, since very little growth can occur under such conditions. If a fire not resulting in the death of the tree occurs, it may be possible to determine the year of its occurrence,

Figure 6.7 Climatic history illustrated by a cross section through a 62-year-old tree. (Courtesy St. Regis Paper Company)

This tree is 62 years old. It's been through fire and drought, plague and plenty. And all of this is recorded in its rings.

Each spring and summer a tree adds new layers of wood to its trunk. The wood formed in spring grows fast, and is lighter because it consists of large cells. In summer, growth is slower; the wood has smaller cells and is darker. So when the tree is cut, the layers appear as alternating rings of light and dark wood.

Count the dark rings, and you know the tree's age. Study the rings, and you can learn much more. Many things affect the way the tree grows, and thus alter the shape, thickness, color and evenness of the rings.

For St. Regis these rings have a special significance. They record the steady accumulation of those fibers we use to create noteworthy printing papers, kraft paper and boards, fine papers, packaging products, building materials, and products for consumers.

Essentially, then, the life of the forest is St. Regis' life. That is why we—together with the other members of the forest products industry—are vitally concerned with maintaining the beauty and utility of America's forests for the generations to come.

ST REGIS

1904
The tree—a loblolly pine—is born.

1909
The tree grows rapidly, with no disturbance. There is abundant rainfall and sunshine in spring and summer. The rings are relatively broad, and are evenly spaced.

Chapter 6

1914
When the tree was 6 years old, something pushed against it, making it lean. The rings are now wider on the lower side, as the tree builds "reaction wood" to help support it.

1924
The tree is growing straight again. But its neighbors are growing too, and their crowns and root systems take much of the water and sunshine the tree needs.

1927
The surrounding trees are harvested. The larger trees are removed and there is once again ample nourishment and sunlight. The tree can now grow rapidly again.

1930
A fire sweeps through the forest. Fortunately, the tree is only scarred, and year by year more and more of the scar is covered over by newly formed wood.

1942
These narrow rings may have been caused by a prolonged dry spell. One or two dry summers would not have dried the ground enough to slow the tree's growth this much.

1957
Another series of narrow rings may have been caused by an insect like the larva of the sawfly. It eats the leaves and leafbuds of many kinds of coniferous trees.

since the burn scar may appear next to a given ring. The most recent season's growth is directly adjacent to the vascular cambium, and one need only count the rings back from the cambium to determine the actual year of the fire. It is not necessary to cut down a tree to determine its age. Botanists and foresters employ an *increment borer* for this purpose. This device removes a plug of wood from the tree at right angles to the trunk. The annual rings can then be counted in the plug; the small hole left in the tree can be treated with a disinfectant to prevent disease and covered up without harm to the tree.

A count of annual rings has produced some red faces on at least one occasion. Located in the community of Chico, California, the Hooker Oak,

named in honor of a famous British botanist who once examined it, was a huge, much-visited tree until its demise in 1977. Beneath the tree was a plaque indicating the tree to be over one thousand years old. A count of rings after its death, however, revealed it to be less than three hundred years old!

When a tree trunk is examined in transverse, or cross, section, lighter streaks or lines called *rays* can be seen radiating out from the center across the annual rings (see figure 6.7). That part of a ray within the xylem is called a *xylem ray;* its extension through the phloem is called a *phloem ray.* In basswood trees some of the phloem rays, when observed in cross section, flare out from a width of two or three cells near the cambium to many cells wide in the part next to the cortex (see

Figure 6.8 Three-dimensional, magnified view of a block of hardwood

transverse surface

vessel

tangential surface

ray

tracheid

vessel

fiber   cambium

sieve tube elements

ray

radial surface

plate 1). In longitudinal section, rays may be from two or three cells to fifty or more cells deep, but the majority of rays in both xylem and phloem are one or two cells wide. Rays consist of parenchyma cells that may remain alive for ten years or more. Their function is the lateral conduction of nutrients and water, with some cells also functioning in food storage. Ray cells can be observed in cross section if a woody stem is cut or split lengthwise along a ray (figure 6.8). Another view of rays (in tangential section) is obtained when the stem is cut at a tangent (i.e., cut lengthwise and off center).

As a tree ages, the protoplasts of some of the parenchyma cells that surround the vessels and tracheids grow through the pits in the walls of these conducting cells and balloon out into the cavities. The growth continues until much of the cavity of the vessel or tracheid has been filled. Such protrusions, called **tyloses** (singular: **tylosis**), prevent further conduction of water and dissolved substances. When this occurs, resins, gums, and tannins begin to accumulate, along with pigments that darken the color of the wood. This older, darker wood at the center is called **heartwood,** while the lighter, still functioning xylem closest to the cambium is called **sapwood.** Except for giving strength and support, the heartwood is not of much use to the tree since it can no longer conduct materials. Thus, a tree may live and function perfectly well after the heartwood has rotted away and left the interior hollow (see figure 6.9). It is even possible to remove part of the sapwood and other tissues and apparently not affect the tree very much, as has been done with giant trees such as the coastal redwoods of California, where holes big enough to drive a car through have been cut out without killing the trees.

Sapwood forms at approximately the same rate as heartwood develops, so there is always sufficient "plumbing" for the vital conducting functions. The relative widths of the two types of wood, however, vary considerably from species to species. For example, in the golden chain tree (a native of Europe and a member of the legume family), the sapwood is usually only one or two rings wide, while in several North American trees (e.g., maple, ash, and beech), the sapwood may be many rings wide.

Cone-bearing trees (see chapter 19) such as pine trees have xylem that consists primarily of tracheids; no fibers or vessel elements are produced. Since it has no fibers, the wood tends to be softer than that of trees with fibers. In many cone-bearing trees, **resin canals** are scattered not only through the xylem but throughout other tissues as well. These canals are tubelike and may

Figure 6.9 Australian baobab tree estimated to be over 2,000 years old. Its hollow trunk was once used for a town jail. (Courtesy *The Age,* Melbourne)

or may not be branched; they are lined with specialized cells that secrete resin into their cavities (figure 6.10). Although resin canals are commonly associated with cone-bearing trees, they are not confined to them. Tropical flowering plants such as olibanum and myrrh trees, for example, have resin ducts in the bark that produce the soft resins frankincense and myrrh of biblical note.

While the vascular cambium is producing secondary xylem to the inside, it is also producing secondary phloem to the outside. The term **bark** is usually applied to all the tissues outside the cambium, including the phloem. Some scientists distinguish between the *inner bark,* consisting of primary and secondary phloem, and the *outer bark (periderm),* consisting of cork tissue and cork cambium. Despite the presence of fibers, the thin-walled conducting cells of the phloem are not usually able to withstand for many seasons the pressure of thousands of new cells added to their interior, and the older layers become crushed and functionless. The parenchyma cells of the cortex to the outside of the phloem also function only briefly because they too become crushed or sloughed off. Before they disappear, however, the cork cambium begins its production of cork, and since new cambia arise inwardly in the phloem, the older bark may consist of alternating layers of crushed phloem and cork. The younger layers of phloem, nearest to the cambium, transport sugars and other substances in solution from their points of origin in the leaves to various parts of the plant, where they are either stored or used in the process of respiration. This sugar content of the phloem was recognized by North American Indian tribes. Some stripped the young phloem and cambium from Douglas fir trees and used the dried strips as food for winter and in emergencies.

Specialized cells or ducts called **laticifers** are found in about twenty families of herbaceous and woody flowering plants. These cells are most common in the phloem but occur throughout all parts of the plants. The laticifers, which resemble

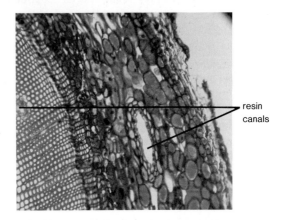

Figure 6.10 Resin canals in a portion of a pine stem

resin canals

vessels, form extensive branched networks of latex-secreting cells originating from rows of meristematic cells. Unlike vessels, however, the cells remain living and may have many nuclei. *Latex* is a thick fluid that is white, yellow, orange, or red in color and consists of gums, proteins, sugars, oils, salts, alkaloidal drugs, enzymes, and other substances. Its function in the plant is not clear, although some believe it aids in closing wounds. Some forms of latex have considerable commercial value (see the discussion under "Spurge Family" in chapter 22). Of these, rubber is the most important. Amazon Indians utilized rubber for making balls and containers hundreds of years before Pará rubber trees were cultivated for their latex. The chicle tree produces a latex used in the making of chewing gum. Several poppies, notably the opium poppy, produce a latex containing important medicinal drugs such as morphine and other drug complexes such as heroin. Other well-known latex producers include milkweeds, dogbanes, and dandelions.

Monocotyledonous Stems

Most monocots (e.g., grasses, lilies) are herbaceous plants that do not attain great size. The stems have neither a vascular cambium nor a cork cambium, and thus produce no secondary vascular tissues or cork. As in herbaceous dicots, the

Figure 6.11 *A.* Portion of a cross section through a monocot stem. *B.* Single vascular bundle (enlarged). (*B,* Photomicrograph by Carolina Biological Supply Company)

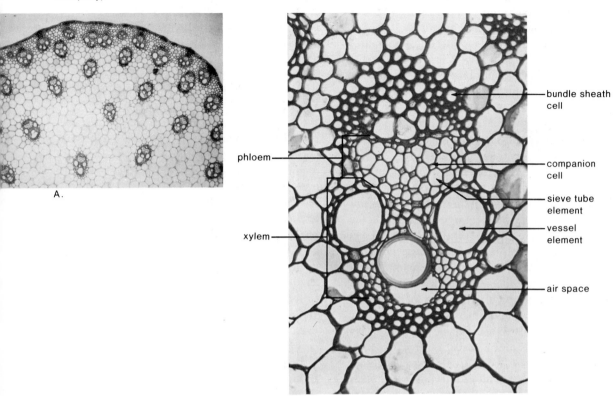

A.

phloem

xylem

bundle sheath cell

companion cell

sieve tube element

vessel element

air space

B.

surfaces of the stems are covered by an epidermis, but the xylem and phloem tissues produced by the procambium appear in cross section as discrete vascular bundles, scattered throughout the stem instead of being arranged in a ring (figure 6.11). Each bundle, regardless of its specific location, is oriented so that its xylem is closest to the center of the stem and its phloem is closest to the surface. In a typical monocot such as corn, a bundle's xylem usually contains two large vessels with several small vessels between them (figure 6.11). The first-formed xylem cells usually stretch and collapse under the stresses of early growth and leave an irregularly shaped air space toward the base of the bundle; the remnants of a vessel are often present in this air space. The phloem consists entirely of sieve tubes and companion cells, and the entire bundle is surrounded by a sheath of sclerenchyma cells. The parenchyma tissue between the vascular bundles is not separated into cortex and pith in monocots; it is usually referred to as *ground tissue,* or **fundamental tissue,** although its function and appearance are the same as those of the parenchyma cells in cortex and pith.

The bundles of a corn stem are more numerous just beneath the surface than they are toward the center. Also, a band of sclerenchyma cells, usually two or three cells thick, occurs immediately beneath the epidermis, and parenchyma cells in the area develop thicker walls as the stem matures. The concentration of bundles, each with its own sheath of sclerenchyma cells, the additional band of sclerenchyma cells beneath the epidermis, and the thicker-walled parenchyma cells all combine to give the stem the capacity to withstand stresses resulting from summer storms and the weight of the leaves and the ears of corn as they mature.

Palm trees differ from most monocots in that they attain considerable size, but they do so primarily as a result of their parenchyma cells continuing to divide and enlarge without a true cambium developing. Several popular houseplants (e.g., ti plants, *Dracaena, Sansevieria*) are monocots in which a secondary meristem develops as a cylinder that extends the length of the stem. Unlike the vascular cambium of dicots and conifers, this secondary meristem produces only parenchyma cells to the outside and secondary vascular bundles to the inside.

In wheat, rice, barley, oats, rye, and other grasses, there is an intercalary meristem (see chapter 4) at the base of each internode, in addition to the apical meristem. The stems of such plants elongate rapidly during the growing season, but because there is no vascular cambium producing tissues that would add to the girth of the stems, growth is columnar, with little difference in diameter between the top and the bottom.

Several commercially important cordage fibers (e.g., flax, hemp, jute, sisal) come from the stems and leaves of monocots, but the individual cells are not separated from one another by *retting* (a process that utilizes the rotting power of microorganisms thriving under moist conditions to break down thin-walled cells) as they are when fibers from dicots are obtained. Instead, during commercial preparation, entire vascular bundles are scraped free of the surrounding parenchyma cells by hand; the individual bundles then serve as unit "fibers." If such fibers are treated with chemicals or bleached, the cementing middle lamella between the cells breaks down; monocot fibers are not as strong or as durable as most dicot fibers.

Specialized Stems

While an erect shoot system is characteristic of most higher plants, many species have stems that perform specialized functions; these stems are modified accordingly (figure 6.12). Although the overall appearance of specialized stems may differ markedly from that of the stems discussed so far, they all have nodes, internodes, and axillary buds; these features distinguish them from roots and leaves, which do not have them. The leaves at the nodes of these specialized stems are often small and scalelike. They are seldom green, but full-size functioning leaves are also produced. Descriptions of some of the specialized stems follow.

Rhizomes

Rhizomes (figure 6.12) are horizontal stems that grow below ground, often near the surface of the soil. Superficially they resemble roots, but a close examination will reveal scalelike leaves and axillary buds at each node, at least during some stage of development, with short to long internodes in between. Adventitious roots are produced all along the rhizome, mainly on the lower surface. A rhizome may be a relatively thick, fleshy food-storage organ, as in irises, or it may be quite slender, as in many perennial grasses or some ferns.

Stolons

Stolons, more or less horizontal stems that are also called **runners,** differ from rhizomes in growing above ground, generally along the surface; they

Chapter 6

Figure 6.12 Types of specialized stems

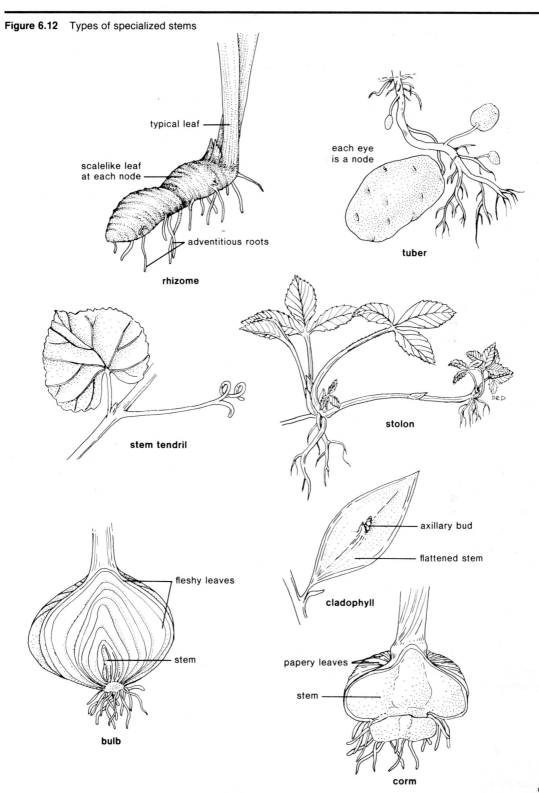

typical leaf

scalelike leaf
at each node

adventitious roots

rhizome

each eye
is a node

tuber

stem tendril

stolon

axillary bud

flattened stem

cladophyll

fleshy leaves

stem

bulb

papery leaves

stem

corm

also have long internodes (figure 6.12). In strawberries, stolons are usually produced after the first flowering of the season has occurred. Several may radiate out from the parent plant and grow along the surface of the ground, attaining lengths of up to 1 meter (3 feet) or more within a few weeks. Adventitious buds appear at alternate nodes along the stolons and develop into new strawberry plants, which can be separated and grown independently. In some houseplants, such as the saxifrages, stolons may grow out and hang over the edge of the pot, producing new plants at intervals. In Irish potatoes, stolons or rhizomes grow underground and produce potatoes at their tips.

Tubers

As food accumulates at the tips of underground stolons, such as those produced by Irish or white potato plants, several internodes swell, becoming **tubers** (figure 6.12). After the tuber is mature, the stolon dies, isolating it. The "eyes" of the potato are actually nodes formed in a spiral around the modified stem. Each eye consists of an axillary bud in the axil of a scalelike leaf, although the latter is visible only in very young tubers, the small ridge being a leaf scar.

Bulbs

Bulbs (figure 6.12) are actually large buds with a small stem at the lower end surrounded by numerous fleshy leaves. Adventitious roots grow from the bottom of the stem, but the fleshy leaves comprise the bulk of the bulb tissue, which functions in food storage. In onions, the fleshy leaves usually are surrounded by the scalelike leaf bases of long, green, aboveground leaves. Other plants producing bulbs include lilies, hyacinths, and tulips.

Corms

Corms, which superficially resemble bulbs, differ from them in being composed almost entirely of stem tissue except for the few papery scalelike

leaves sparsely covering the outside (figure 6.12). Adventitious roots are produced at the base, and corms, like bulbs, function in food storage. Well-known plants producing corms include crocuses and gladioli.

Cladophylls

In the butcher's-broom plant, stems are flattened and very leaflike in appearance. Such stems are called **cladophylls** (figure 6.12). In the center of each butcher's-broom cladophyll is a node bearing very small scalelike leaves with axillary buds. The feathery appearance of asparagus is due to numerous small cladophylls, which also occur in greenbriers and certain orchids.

Other Specialized Stems

In many cacti and in some of the spurges, the stems are stout and fleshy. Such stems are modified for water and food storage. Other stems may be modified in the form of thorns, as in the crab apple or honey locust, but not all thorns are modified stems. For example, the black locust has a pair of thorns at the base of each petiole of most leaves (these are parts of the leaf called *stipules,* mentioned in the discussion of twigs and discussed further in chapter 7). The prickles of raspberries, which originate from the epidermis, are technically not true thorns. Tiger lilies produce small aerial bulblets in the axils of their leaves.

Climbing plants have stems modified in various ways that adapt them for their growth habit. Some stems, called *ramblers,* simply rest on the tops of other plants, but many produce **tendrils.** These are specialized stems in the grape and Boston ivy but are modified leaves or leaf parts in plants like peas and cucumbers. In Boston ivy, the tendrils have adhesive disks. In English ivy, the stems climb with the aid of adventitious roots that arise along the sides of the stem and become embedded in the bark or other support material over which the plant is growing.

Applied Uses of Wood

The use of wood by humans dates back into antiquity, and present uses are so numerous that it would be impossible to list in a work of this type more than the most important ones. Before looking at the economic importance of wood, a brief discussion of its properties is in order.

In a living tree, up to 50% of the weight of the wood comes from the water content. Before the wood can be used, the moisture content is reduced to 10% or less through *seasoning,* either by air-drying the wood in ventilated piles or stacks or by drying it in special ovens known as *kilns.* The seasoning has to be done gradually and under carefully controlled conditions or the timber may warp and split along the rays, making it unfit for most uses. The dry part of wood is composed of 60 to 75% **cellulose** and about 15 to 25% **lignin,** an organic substance that is deposited in the walls of xylem cells and makes the walls tough and hard. Other substances present in smaller amounts include *resins, gums, oils, dyes, tannins,* and *starch.* The proportions and amounts of these and other substances determine how various woods will be used (figure 6.13).

Among the most important physical properties of wood is its *density.* Technically, the density is the weight per unit volume. The weight is compared with that of an equal volume of water and is stated as a fraction of 1.0. Because of the considerable air space within the cells, most woods have a *specific gravity,* as the comparative density is called, of less than 1.0. The range of specific gravities of known woods varies from 0.04 to 1.40, the lightest commercially used wood being balsa with a specific gravity of about 0.12. Woods with specific gravities of less than 0.50 are considered light; those with specific gravities of above 0.70 are considered heavy. Among the heaviest woods are the South American ironwood and lignum vitae, with specific gravities of over 1.25. Lignum vitae, obtained from West Indian trees, is extremely hard wood and is used instead of metal in the manufacture of main bearings for drive shafts of submarines because it is self-lubricating and less noisy.

A wood's ability to withstand decay organisms and insects is referred to as its *durability.* Most decay organisms need moisture to aid them in their enzymatic breakdown of cellulose and other wood substances, but the seasoning process usually reduces the moisture to a level below that necessary for the fungi and other decay organisms to survive. Other natural constituents of wood that repel decay organisms include tannins and oils. Wood with a tannin content of 15% or more may survive on a forest floor for many years after the tree has been felled by diseases of the phloem or other causes. Among the most durable of American woods are cedar, catalpa, black locust, red mulberry, and Osage orange. The least durable woods include cottonwood, willow, fir, and basswood.

Logs are usually cut longitudinally in one of two ways: along the radius or at right angles to the rays (figure 6.14). Radially cut or **quartersawed** boards show the annual rings in side view; they appear as longitudinal streaks and are the most conspicuous feature of the wood. Only a few perfect quartersawed boards can be obtained from a log, making them quite expensive. Boards cut at right angles to the rays (tangentially cut boards) are more common. In these, the annual rings appear as irregular bands of light and dark alternating streaks or patches, with the ends of the rays visible as narrower and less conspicuous vertical streaks. Lumber cut tangentially is referred to as being *plain-sawed,* or *slab cut* (figure 6.14). *Slabs* are the boards with rounded sides at the outside of the log; they are usually made into chips for pulping.

Knots are the bases of lost branches that have become covered, over a period of time, by new annual rings of wood produced by the cambium of the trunk. They are found in greater concentration in the older parts of the log toward the

Figure 6.13 Uses of some common North American woods. (Courtesy St. Regis Paper Company)

White Pine. A softwood widely used in home construction and for virtually everything from masts and matches to boxes and crates. Its soft, uniform texture and straight grain cuts easily in every direction, polishes well, and warps or swells little.

Hard Maple. In bowling pins and flooring for bowling alleys its uniform texture and hardness result in resistance to abrasion. The Romans used it for spears and lances. We turn it on lathes to make spools, bobbins, cue sticks, and croquet balls.

Black Walnut. A choice hardwood for fine furniture and interior paneling, because of the beauty of the heartwood grain, its ability to stay in place after seasoning, and its good machining properties. It is harder than oak, and shock-resistant.

White Oak. Makes good barrels because the wood is resilient, durable, and impermeable to liquids. This hardwood, which is about twice as dense as white pine, has many other uses ranging from flooring to fine cabinet work.

Baldcypress. Because it is weather-resistant without treatment, this wood was widely used for cross ties in the early days of railroading. Today it is used for water tanks and other applications requiring prolonged contact with water.

White Ash. Perfect for baseball bats, tennis racquets, oars and long tool handles. This hardwood's major virtues are straight grain, stiffness, strength, moderate weight, good bending qualities, and capacity for wearing smooth.

Red Spruce. A favorite for violin sounding boards because of its high resonant qualities. A softwood, it is easy to work, and is light in relation to its strength and stiffness. These qualities also make it eminently suitable for ladder rails, canoe paddles, and oars.

Hemlock. This relatively soft, light, straight-grained, resin-free wood, with its uniformly long fibers, is becoming one of the most important species for paper pulp. It is also used for structural lumber and plywood, and for boxes, barrels, and concrete forms.

Hickory. A hardwood unsurpassed for the handles of impact tools like axes and hammers, and for skis, because of its hardness, strength, toughness, and resiliency. In horse-and-buggy days it was widely used for wheel spokes and rims, singletrees, and buggy shafts.

Figure 6.14 How the surfaces of plain-sawed, quartersawed, and transversely sawed wood appear. (See figure 6.8)

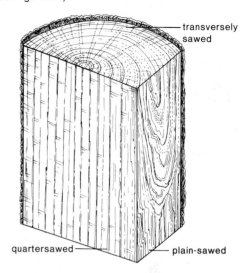

center, because in the forest the lowermost branches of a tree (produced while the trunk was small in girth) often die from lack of sufficient light. When a branch dies and falls off, the cambium at its base also dies, but the cambium of the trunk remains alive and increases the girth of the tree, slowly enveloping the dead tissue of the branch base until it may be completely buried and not visible from the surface. Knots usually weaken the boards in which they occur.

In the United States, about half of the wood produced is used as lumber, primarily for construction; the sawdust and other waste formed in processing the boards is converted to particle board and pulp. A considerable amount of lumber goes into the making of furniture, which may be constructed of solid wood or covered with a veneer. A *veneer* is a very thin sheet of desirable wood, carefully cut so as to produce the best possible view of the grain, that is glued to cheaper lumber (figure 6.15).

Figure 6.15 How a log is used. (Courtesy St. Regis Paper Company)

Will America run out of forests?
No! One reason is that we are making
every log work harder.

America's demand for lumber and paper grows by about 4 per cent each year. Can the forests meet this challenge? Yes. Good forest management helps us grow a continuous supply of the forest crop. And we get more from every log we use.

Sawmills have long known how to derive the maximum amount of lumber from a log. Nevertheless, at one time up to 50 per cent of some logs went unused. Today these "waste" pieces are routinely converted into chips for pulp and paper mills. Many pulp mills depend upon these chips for a good part of their supply. With millions of tons of chips so used each year, hundreds of thousands of acres of forest land need not be touched.

To St. Regis, conservation of the forest is a matter of prime concern. Its trees give us the wood for our products. We make printing papers, kraft papers and boards, fine papers, packaging products, building materials, and products for consumers.

Essentially, then, the life of the forest is St. Regis' life. That is why we—together with the other members of the forest products industry—are vitally concerned with maintaining the beauty and usefulness of America's forests for the generations to come.

ST REGIS

How the log is used

Debarking the log is essential to its full utilization because bark cannot be used for paper-making, and therefore any piece dropped in the chipper has to be free of bark. But the bark can be used for fuel and soil mulch.

The rounded sides of the log, called "slabs," are the first pieces sent to the chipper as the log goes through the saw-mill. This idealized picture shows the entire log being used for lumber, except for the slabs. Actually, as cutting continues, other pieces go to the chipper, including edgings, trim ends, and other parts of the log not usable as lumber. Each log presents different problems and can be handled differently.

The outer portions of the log have the fewest knots. This "clear" lumber is usually made into boards or planks varying in thickness from one to three inches.

Toward the center of the log, knots increase and the wood is less suitable for boards. Heavier planks, and square or rectangular beams are normally sawed from this section. The center of the log is used primarily for structural beams strong enough so that they are not weakened by knots. Knots are most frequent here because this is the oldest section of the tree. Branches that were removed during the early years of the tree's life left knots that were covered over as the tree grew outward.

Plywood is, in effect, a sandwich of thin wooden veneers. Veneer is made by "peeling," that is, holding a long blade against a rotating log. The wood is continuously peeled off, down to an eight-inch core. The core is then treated as though it were a small log. It can be made into lumber and, of course, the rounded portions go to the chipper.

Stems

The next most extensive use of wood is for pulp, which among other things is converted by various processes to paper, synthetic fibers, plastics, and linoleum. In recent years, it has been added as a filler to commercial ice cream and bread. Some hardwoods are treated chemically or heated under controlled conditions to yield a number of chemicals such as wood alcohol and acetic acid, but other sources of these products are now usually considered more economical than they were in the past. Charcoal, excelsior, cooperage (kegs, casks, and barrels), railroad ties, boxes and crates, musical instruments, bowling pins, tool handles, pilings, cellophane, photographic film, and Christmas trees are but a few of the additional wood products worth billions of dollars annually on the world market (see figure 6.13).

In developing countries, approximately half of the timber cut is used for fuel; in the United States, a little less than 10% is currently used for that purpose. In colonial times, wood was the almost exclusive source of heating energy. In the 1970s, Brazil's major cities were still using scrub timber from the surrounding forests to energize their utilities, but rapidly depleting supplies have pointed to the need for alternate sources of energy. Many types of coal are wood that has been compressed for millions of years until nearly pure carbon remains. The formation of coal and other fossils is discussed in chapter 18. Although the world's supply of coal is still plentiful, the rate at which this fossil fuel is being consumed makes it obvious that resources will eventually be exhausted unless our energy demands find renewable or less destructive alternatives.

Some of the vast array of secondary products from stems, including dyes, medicines, spices, and foods are discussed in later chapters and in the appendixes.

Summary

To understand the how and why of grafting, the propagation of plants from stems, and which stem parts are useful, an examination of stem structure and function is needed.

The shoot system of plants, with branches and leaves, is usually erect, but some stems may be horizontal or have modifications that permit climbing or storage of food or water. Woody twigs have leaves arranged alternately, oppositely, or in a whorl. Nodes are stem regions where leaves are attached; internodes occur between nodes. Most leaves have petioles and blades. Axillary buds occur in leaf axils. Most buds are protected by bud scales. Terminal buds occur at twig tips. When terminal bud scales fall off they leave groups of bud scale scars that can be used in determining the age of the twig. Stipules are paired appendages present at the base of some leaves; when they fall off, they leave small scars on the twig. When whole leaves fall they cause leaf scars on the twig, with tiny bundle scars within the leaf-scar surfaces.

Each stem has an apical meristem at its tip that produces tissues resulting in increase in length. Leaf primordia develop into mature leaves when growth begins. Three primary meristems develop from an apical meristem: the protoderm gives rise to the epidermis, which becomes coated with a cuticle; the procambium produces primary xylem and primary phloem; and the ground meristem produces pith and cortex. As each leaf and each bud develops from a primordium, a trace of xylem and phloem branches off from the main cylinder, leaving a leaf gap or a bud gap. A vascular cambium, producing secondary tissues, may arise between primary xylem and phloem. Secondary xylem cells include tracheids, vessel elements, and fibers. Secondary phloem cells include sieve-tube elements and companion cells.

In many plants, a cork cambium producing cork and phelloderm cells develops near the surface of the stem. Cork cells, which constitute the

Chapter 6

outer bark, have suberin in their walls; suberin is impervious to moisture and, therefore, the outer bark aids in protection. Lenticels in the bark permit gas exchange.

Dicotyledons (dicots) are plants whose seeds have two seed leaves (cotyledons), whereas monocotyledons (monocots) have seeds with one seed leaf. Herbaceous dicots have vascular bundles arranged in a ring in the stem. Woody dicots have most of their tissues arranged in concentric layers. The most conspicuous tissue is wood (secondary xylem). Spring wood usually has relatively large vessel elements, while summer wood has smaller vessels and/or a predominance of tracheids. An annual ring is one year's growth of xylem. A tree's age and other aspects of its history can be determined from annual rings. Rays, which function in lateral conduction, radiate out from the center of the trunk. Older wood toward the center (heartwood) ceases to function when its cells become plugged with tyloses. Younger, functioning wood (sapwood) is closer to the surface. A tree's functions are not particularly affected by the rotting of its heartwood. The wood of cone-bearing trees consists primarily of tracheids, and resin canals are often present.

In woody plants older tissues composed of thin-walled cells become crushed and functionless, and some are sloughed off. Laticifers are latex-secreting cells or ducts found in various flowering plants. The latex of some plants has considerable commercial value.

Monocot stems have neither a vascular cambium nor a cork cambium; their vascular bundles are scattered. The parenchyma tissue is not divided into pith and cortex; it is called ground, or fundamental tissue, instead. Each vascular bundle is surrounded by a sheath of sclerenchyma cells. The stems of monocots such as corn can withstand stresses because of the presence of numerous bundles and a band of sclerenchyma cells and thicker-walled parenchyma cells just beneath the surface.

Palm trees are monocots that become large because their parenchyma cells continue to divide. Other monocots develop a secondary meristem that produces parenchyma cells and secondary vascular bundles. Grasses have intercalary meristems at the base of each internode that contribute to rapid increases in length. Several commercially important cordage fibers are obtained from monocots.

Specialized stems include rhizomes, stolons, tubers, bulbs, corms, cladophylls, and tendrils. Such stems may have adventitious roots.

The dry part of wood consists primarily of cellulose and lignin. Resins, gums, oils, dyes, tannins, and starch are also present. Properties of wood that play a role in its use include density, specific gravity, and durability. Logs are usually cut longitudinally along the radius (quartersawed) or at right angles to the rays (tangentially, plain-sawed, or slab cut). Knots are bases of branches that have become covered over by new wood; they usually weaken the boards in which they occur.

About half the timber produced in the United States is used as lumber. Sawdust and waste is converted to particle board and pulp for paper, synthetics, and linoleum. Other timber is used for cooperage, charcoal, railroad ties, boxes, tool handles, etc. Developing countries use a greater proportion of their timber for fuel.

Review Questions

1. What is the function of bud scales?

2. How can you tell the age of a twig?

3. Distinguish among procambium, vascular cambium, and cork cambium.

4. How can you tell, when you look at a cross section of a young stem, whether it is a dicot or a monocot?

5. An Irish or white potato is a stem, but a sweet potato is a root. How can you tell?

6. If you were examining the top of a wooden desk, how could you tell if the wood had been radially or tangentially cut (quartersawed or plain-sawed)?

7. What differences are there between heartwood and sapwood?

8. Distinguish among corms, bulbs, and tubers.

9. What are laticifers?

10. What is meant by the specific gravity of wood?

Discussion Questions

1. If the cambium of a tropical tree were active all year long, how would its wood differ from that of a typical temperate climate tree?

2. It was mentioned that a nail driven into the side of a tree will remain at exactly the same distance from the ground for the life of the tree. Why?

3. Do climbing plants have any advantages over erect plants? Any disadvantages?

4. If two leaves are removed from a plant and one is coated with petroleum jelly while the other is not, the uncoated leaf will shrivel considerably sooner than the coated one. Would it be helpful to coat the stems of young trees with petroleum jelly? Explain.

5. Suggest some reasons for heartwood being preferred to sapwood for making furniture.

Additional Reading

Black, M.; Edelman, J. *Plant growth*. Cambridge, MA: Harvard University Press; 1970.

Core, H. A.; Côté, W. A.; Day, A. C. *Wood structure and identification*. 2d ed. Syracuse, NY: Syracuse University Press; 1979.

Cutter, E. G. *Plant anatomy: experiment and interpretation*. Part 1: *Cells and tissues*. 2d ed. Reading, MA: Addison-Wesley Publishing Co.; 1978.

————. *Plant anatomy: experiment and interpretation*. Part II: *Organs*. Reading, MA: Addison-Wesley Publishing Co.; 1971.

Esau, K. *Plant anatomy*. 2d ed. New York: John Wiley and Sons; 1965.

————. *Anatomy of seed plants*. 2d ed. New York: John Wiley and Sons; 1977.

Fahn, A. *Plant anatomy*. 2d ed. Elmsford, NY: Pergamon Press; 1974.

Fritts, H. C. *Tree rings and climate*. New York: Academic Press; 1976.

Metcalfe, C. R.; Chalk, L. *Anatomy of the dicotyledons*. Vols. 1 and 2. Oxford: Clarendon Press; 1950.

Meylan, B. A.; Butterfield, B. G. *Three-dimensional structure of wood*. London: Chapman and Hall; 1972.

O'Brien, T. P.; McCully, M. E. *Plant structure and development*. New York: Macmillan Co.; 1969.

Ray, P. M. *The living plant*. 2d ed. New York: Holt, Rinehart and Winston; 1972.

Schery, R. W. *Plants for man*. 2d ed. Englewood Cliffs, NJ: Prentice-Hall; 1972.

Steeves, T. A.; Sussex, I. M. *Patterns in plant development*. Englewood Cliffs, NJ: Prentice-Hall; 1972.

Zimmerman, M. H.; Brown, C. L. *Trees: structure and function*. New York: Springer-Verlag; 1971.

Leaves 7

Overview

This chapter introduces leaves by comparing them with solar panels and by discussing their general functions, morphology, and dimensions. This is followed by descriptive information on basic leaf types and specific forms and arrangements. The chapter next discusses the internal structure of leaves, including epidermis and cuticle, stomata, glands, mesophyll, and veins.

Specialized leaves, including tendrils, spines, window leaves, flower pot leaves, reproductive leaves, floral leaves, and insectivorous leaves, are then examined. The chapter concludes with an explanation of autumnal color changes and abscission and some observations on the human and ecological relevance of leaves.

Some Suggested Student Learning Goals

1. Describe the external forms and parts of leaves. Indicate the functions of a typical leaf, and name the specific tissues and cells that contribute to a leaf's normal functions.

2. Differentiate among pinnate, palmate, and dichotomous venation, and indicate differences between simple and compound leaves.

3. Contrast tendrils, spines, storage leaves, flower pot leaves, window leaves, reproductive leaves, floral leaves, and different types of insect-trapping leaves.

4. Explain why deciduous leaves turn various colors in the fall and how such leaves are shed.

5. Identify at least fifteen uses of leaves by humans.

Introduction

Most persons in industrialized countries are aware that fossil fuels, our largest source of energy, are limited and costly. One means of counteracting this problem, the use of solar energy, has been increasingly employed by the construction industry, particularly in the southwestern United States. A number of new houses are being built with flat panels and windows inclined at angles to maximize the amount of energy captured from the sun's rays. Some buildings have mechanical devices that slowly move the solar panels so that they will follow the sun in its daily course across the sky. Variations of such devices are undoubtedly destined to become commonplace in the future.

For aeons, before there was an "energy crisis," plants have had their own solar panels, which capture the sun's energy. The remarkably constructed structures called *leaves* are far more efficient than their comparatively clumsy counterparts on buildings. Their flattened surfaces, which are completely covered with a transparent protective layer of cells, the epidermis, admit light to all parts of the interior. Many leaves twist on their stalks, or petioles, without the aid of mechanical devices, so that their upper surfaces are inclined at right angles to the sun's rays throughout the day. Their lower surfaces (and sometimes their upper surfaces as well) are dotted with tiny openings called *stomata,* which not only permit air circulation but are also involved with cooling the interior of the leaf as water evaporates from the moist cell surfaces and passes out of them into the air in vapor form. Green leaves capture the light energy available to them by means of the most important process for life on earth, at least life as we know it today. This process, called *photosynthesis* (discussed in chapter 9), involves the trapping of energy in sugar molecules that are constructed from ordinary water and from carbon dioxide present in the atmosphere. All the energy needs of living organisms ultimately depend on photosynthesis, from the first day of their existence to the last.

Leaves also perform other functions. For example, all living cells *respire* (respiration is also discussed in chapter 9), and in the process of this and other metabolic activities, waste products are produced. These wastes are deposited outside the plant when the leaves are sealed off at the bases of the petioles and shed. The discarded leaves are then replaced with new ones the following season. Leaves are also involved in the movement of the water absorbed by the roots and transported throughout the plant. Most of the water reaching the leaves evaporates into the atmosphere by a process known as **transpiration** (discussed in chapter 8). In some plants, root pressure (see page 136) forces water out of *hydathodes,* which are special openings at the tips of leaf veins, usually at night when transpiration is not occurring (see page 139). The water thus expelled may contain ions secreted by root cells. Additional functions of leaves are discussed throughout this chapter.

Leaf Arrangements and Types

Most of the roughly 275,000 different kinds of plants that produce leaves can be distinguished from one another by their leaves alone. In addition, many plants such as lilies, ferns, and pine trees produce different forms of leaves on the same plant; some of these are reduced to tiny papery scales. The variety of shapes, sizes, and textures of leaves seems to be almost infinite. Some of the smaller duckweeds have leaves less than 1 millimeter (0.04 inch) wide. The leaves of the Seychelles Island palm attain spans of 6 meters (20 feet), and the floating leaves of a giant water lily, which reach 2 meters (6.5 feet) in diameter (figure 7.1), can support without sinking weights of more than 45 kilograms (100 pounds) distributed over their surface. Leaves have almost any form, including tubular, feathery, cup-shaped, spinelike, or needlelike (figure 7.2). They may be smooth or hairy, slippery or sticky, waxy or glossy, pleasantly fragrant or foul-smelling, edible or poisonous. Leaves may be of almost every color of the rainbow and, when viewed with a microscope, of exquisite beauty.

As indicated in chapter 6, all leaves originate as *primordia* in the buds, regardless of their ultimate size or form. At maturity most leaves consist of a stalk, called the **petiole,** and a flattened **blade,** or **lamina.** The blade has a network of veins

(vascular bundles), with the larger veins being parallel to one another in monocots (figure 7.3) and diverging from one another in various ways in dicots (see figure 7.7). A pair of appendages, called **stipules** (see chapter 6), are sometimes present at the base of the petiole. Leaves of flowering plants are associated with leaf gaps and may be simple or compound.

Figure 7.2 Types of leaves and leaf arrangements. *A.* Palmately compound buckeye leaf. *B.* Pinnately compound black walnut leaf. *C.* Alternate, simple but lobed leaves of tulip tree. *D.* Opposite, simple leaves of dogwood. *E.* Palmately veined maple leaf. *F.* Globe-shaped succulent leaves of string-of-pearls. *G.* Pinnately veined, lobed oak leaf. *H.* Grass leaf. *I.* Whorled leaves of bedstraw. *J.* Linear leaves of yew. *K.* Fan-shaped leaf of *Ginkgo,* showing dichotomous venation.

Figure 7.3 Portion of a monocot leaf showing the parallel veins

The arrangement of leaves may be in a *spiral,* or *alternate,* pattern on the stem, or two may be attached at the same node, providing an *opposite* arrangement. When three or more leaves occur at a node, they are said to be *whorled* (see chapter 6). The leaf itself may be a **simple leaf,** with an undivided blade, or it may be a **compound leaf,** with the blade divided into **leaflets** in various ways (see figure 7.2). **Pinnately compound** leaves have the leaflets in pairs along a central stalklike **rachis,** whereas **palmately compound** leaves have all the leaflets attached at the same point at the top of the petiole. Sometimes the leaflets of a pinnately compound leaf may be subdivided into still smaller leaflets, forming a *bipinnately compound* leaf. The arrangement of veins in a leaf or leaflet blade (*venation*) may also be either pinnate or palmate. In **pinnately veined** leaves there is a main vein, called the **midrib,** with secondary veins branching from it, but in **palmately veined** leaves several veins fan out from the base of the blade.

In a few leaves (e.g., *Ginkgo*) no midrib or other large veins are present; instead the veins fork evenly and progressively from the base of the blade to the opposite margin. This is called *dichotomous venation* (figure 7.2K).

Internal Structure of Leaves

If a typical leaf is cut transversely and examined with the aid of a microscope, three regions stand out: *epidermis, mesophyll,* and *veins* (referred to as *vascular bundles* in our discussion of roots and stems) (figure 7.4). The **epidermis** is a single layer of cells covering the entire surface of the leaf. The epidermis on the lower surface of the blade can sometimes be distinguished from the upper epidermis by the presence of stomata, which are discussed in the following paragraph. When seen from the top, the wavy, undulating epidermal cells frequently resemble pieces of a jigsaw puzzle fitted together. Except for guard cells, the upper epidermal cells contain no chloroplasts, their function being primarily protection of the delicate tissues to the interior. A thin coating of fatty **cutin** (the **cuticle**) is normally present, although it may not be visible with ordinary light microscopes without being specially stained. In addition to the cuticle, many plants produce other waxy substances on their surfaces (figure 7.5). In studies of the effects on plants of smog and auto exhaust fumes, it was found that these waxes may be produced in abnormal fashion on beet leaves within as little as twenty-four hours after exposure to the pollutants. Beet leaves also respond to aphid damage by producing wax around each tiny puncture. Occasionally the upper epidermal cells may contain crystals of waste materials. Different types of glands may also be present in the epidermis. **Glands** occur in the form of depressions, protuberances, or appendages either directly on the leaf surface or on the ends of hairs (see figure 4.3). Glands often secrete sticky substances.

Figure 7.4 Portion of a typical leaf

cuticle

upper epidermal cell

petiole

palisade
mesophyll cells

blade

spongy mesophyll cells

vascular
bundle
(vein)

stoma

guard cell

lower epidermis

air space

Figure 7.5 Small portion of the upper surface of a
Eucalyptus leaf. The light-colored starlike material is a
wax produced by the leaf. Magnification is
approximately 7,500×. (Courtesy B. E. Juniper)

Figure 7.6 Typical dicot stoma. *A.* Surface view. *B.* View in transverse section.

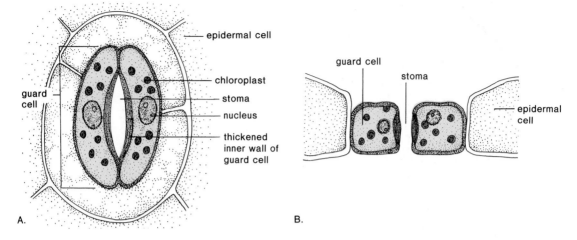

A.

B.

The lower epidermis usually resembles the upper epidermis in most respects, but it is perforated by numerous tiny pores, the **stomata** (figure 7.6), which also occur in the upper epidermis of some plants. Stomata are very numerous, ranging from about 1,000 to more than 1.2 million per square centimeter (6,300 to 8 million per square inch) of surface. An average-sized sunflower leaf has about 2 million of these pores throughout its lower epidermis. Each stoma is formed by two sausage- or dumbbell-shaped cells that usually are smaller than most of the neighboring epidermal cells. These are called **guard cells,** and unlike other cells of either epidermis, they contain chloroplasts. The amount of photosynthesis taking place in the guard cells aids in the functioning of the cells. The guard cell walls are flexible and thickened on the side adjacent to the pore. As the guard cells expand or contract with changes in the amount of water within the cells, their unique construction causes the stomata to open or close. When the guard cells are expanded, the stomata are open; when the water content of the guard cells decreases, the cells contract and the stomata close. Stomata function in permitting gas exchange between the interior of the leaf and the atmosphere and in regulating

the evaporation of most of the water entering the plant at the roots (see chapter 8). The stomata of water lilies and other plants with floating leaves are confined to the upper epidermis.

Most photosynthesis takes place in the **mesophyll** between the two epidermal layers. When two regions of the mesophyll are distinguishable, as is often the case, the uppermost consists of tightly packed, somewhat barrel-shaped parenchyma cells, commonly in two rows. This region is called the **palisade mesophyll** and may contain more than 80% of the leaf's chloroplasts. The lower region, consisting of loosely arranged parenchyma cells with abundant air spaces between them, is called the **spongy mesophyll.** Its cells are also rich in chloroplasts. Parenchyma tissue, because the cells contain numerous chloroplasts, is also referred to as *chlorenchyma* tissue. Besides being present in leaves, chlorenchyma tissue is found in the outer parts of the cortex in the stems of herbaceous plants. Inside the leaf, the surfaces of mesophyll cells in contact with the air are moist. If moisture decreases below a certain level, the stomata close, thus preventing further drying.

Figure 7.7   Net venation of a portion of a frangipanni leaf

Veins (vascular bundles) of various sizes are scattered throughout the mesophyll (figure 7.7). They consist of xylem and phloem tissues surrounded by a jacket of fibers called the **bundle sheath.** The veins give the leaf its "skeleton." The carbohydrates produced in the mesophyll cells are transported in solution throughout the plant by the phloem. Water, sometimes located more than 100 meters (330 feet) away in the ground below, is brought up to the leaf by the xylem, which, like the phloem, is part of a vast network of "plumbing" throughout the plant. Since veins run in all directions in the network, particularly in dicots, it is common, when examining a cross section of a leaf under the microscope, to see veins cut transversely, lengthwise, and at a tangent, all in the same section.

Monocot leaves, in addition to having parallel veins, usually do not have the mesophyll differentiated into palisade and spongy layers. Some monocot leaves (e.g., those of grasses) have large, thin-walled *bulliform cells* on either side of the main central vein toward the upper surface. The bulliform cells partly collapse under dry conditions, causing the leaf blade to fold or roll, thus reducing transpiration (see figure 10.16).

Specialized Leaves

If all leaves could function normally under any environmental conditions, special leaf modifications would provide no advantages to a plant. But the form and structure of plants of tropical rain forests do not permit them to thrive in a desert, and cacti soon die if planted in a creek because

Chapter 7

Figure 7.8 Pine needle in cross section

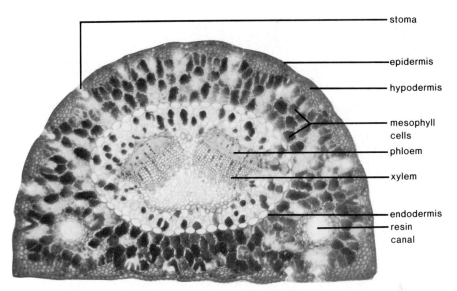

stoma

epidermis

hypodermis

mesophyll cells

phloem

xylem

endodermis
resin canal

their structure, form, and life cycles are attuned to specific combinations of environmental factors such as temperature, humidity, light, water, and soil conditions. In addition, the modifications of leaves occupying any single ecological niche may be very diverse, resulting in such a rich variety of leaf forms and specializations throughout the Plant Kingdom that only a few may be mentioned here.

Some modifications may be relatively minor and may occur along with unmodified leaves on the same plant. For example, leaves in the shade tend to be thinner and have fewer hairs than leaves on the same tree that are exposed to direct light. They also tend to be larger and to have fewer well-defined mesophyll layers and fewer chloroplasts than their counterparts in the sun.

In different climatic zones or habitats, however, the modifications are generally more pronounced. Thus, plants growing in arid regions may have thick and leathery leaves, fewer stomata or stomata sunken below the surface in special depressions, succulent water-retaining leaves

or no leaves at all (with the stems taking over the function of photosynthesis), or dense, hairy coverings. Pine trees, whose water supply may be severely restricted in the winter when the soil is frozen, have some leaf modifications similar to those of desert plants. The modifications include sunken stomata, a thick cuticle, and a layer of thick-walled cells (the **hypodermis**) beneath the epidermis (figure 7.8). The leaves of the compass plant face east and west, the blades at right angles to the ground, so that when the sun is overhead it strikes only the thin edge of the leaf, minimizing moisture loss. In plants that grow in water, the submerged leaves usually have considerably less xylem than phloem. Large spaces are found in the mesophyll, which is not differentiated into palisade and spongy layers. Other modifications are described in the following sections.

Leaves

Figure 7.9 Tendrils at the tip of a garden pea plant leaf

Tendrils

There are many plants whose leaves are partly or completely modified as **tendrils.** These leaves, when curled tightly around more rigid objects, help the plant in climbing or in supporting weak stems. In garden peas, the leaves are compound, and the upper leaflets are reduced to whiplike strands that, like all tendrils, are very sensitive to contact (figure 7.9). In fact, if you lightly stroke the lower half of a tendril, it will start curling in that direction within a minute or two.

In the yellow vetchling, the whole leaf is modified as a tendril, and photosynthesis is carried on by the leaflike stipules at the base. In the potato vine and the garden nasturtium, the petioles serve as tendrils, while in some greenbriers, stipules are modified as tendrils. In clematis, the rachises of some of the compound leaves serve very effectively as tendrils. Members of the pumpkin family, which includes squashes, melons, and cucumbers, produce tendrils that may be up to 3 decimeters (1 foot) in length.

As the tendrils develop, they become coiled like a spring. When contact with a support is made, the tip not only curls around it, but the direction of the coil reverses, and sclerenchyma cells develop in the vicinity of contact (see figure 10.11). This makes a very strong but flexible attachment that protects the plant from damage during high winds. It should be noted that the tendrils of many other plants, such as grapes, are not modified leaves but have developed from stems.

Spines

Many desert plants have their leaves modified as **spines.** This reduction in leaf surface correspondingly reduces water loss from the plants, and the spines also apparently protect the plants from browsing animals. Photosynthesis in such desert plants, which would otherwise take place in leaves, takes place in the green stems instead. These spines, along with those of barberries (figure 7.10B), are modifications of the whole leaf. In a number of woody plants, the stipules at the bases of the leaves are modified as short, paired spines

Figure 7.10 Spines. *A.* Stipules of a mesquite leaf modified as spines. *B.* Leaves of a common barberry modified as spines.

A.

B.

(figure 7.10A). As is the case with tendrils, not all spines are modified leaves; many are modified stems. The thorns of roses and raspberries originate at the surface of the stems.

Storage Leaves

As previously mentioned, desert plants may have succulent leaves with modifications that permit water storage. Just beneath the epidermis in such leaves are photosynthetic cells with chloroplasts, but in the interior there are large, thin-walled parenchyma cells without chloroplasts. These cells contain large vacuoles that can store proportionately large quantities of water. If removed from the plant and set aside, the leaves will often retain much of the water for several weeks. Other fleshy leaves, such as those that comprise the bulk of onion and lily bulbs, store large amounts of carbohydrates, which are utilized by the plant in the subsequent growing season.

Figure 7.11 Flower pot leaves of *Dischidia*. One leaf
is sliced lengthwise to reveal the root growing inside it.

Figure 7.12 Window plant. Note the transparent tips
of the window leaves.

Chapter 7

Figure 7.13 Leaves of an air plant, showing plantlets being produced along the margins

"Flower Pot" Leaves

Some of the leaves of *Dischidia* (figure 7.11), a greenhouse curiosity from tropical Australasia, develop into urnlike pouches that become the home of ant colonies. The ants carry in soil and add nitrogenous wastes, while moisture collects in the leaves through condensation of the water vapor coming through stomata from the mesophyll. This situation creates a good growing medium for adventitious roots, which develop at the same node as the leaf and grow down into the soil contained in the urnlike leaf. In other words, this extraordinary plant not only reproduces itself by conventional means but also, with the aid of ants, provides its own fertilized growing medium and flower pots and then produces special roots that "take advantage" of the situation.

Window Leaves

In the Kalahari desert of South Africa, there are at least three plants belonging to the carpetweed family that have unique adaptations to living in dry, sandy areas (figure 7.12). Their leaves, which are shaped like ice-cream cones, are about 3.75 centimeters (1.5 inches) long and are buried in the sand; only the dime-sized wide end of a leaf is exposed at the surface. This exposed end is covered with a thick epidermis and cuticle. It has a few stomata and is relatively transparent. Below the exposed end is a mass of tightly packed water-storage cells; these allow light coming through the "windows" to penetrate to the chloroplasts in the mesophyll, located all around the inside of the shell of the leaf. This arrangement, which keeps most of the plant buried and away from drying winds, allows the plant to thrive under circumstances that most other plants could not tolerate.

Reproductive Leaves

Some of the leaves of the walking fern are most unusual in that they produce new plants at their tips. Occasionally, three generations of plants may be found linked together. The *succulent* (designed to preserve moisture) leaves of air plants (figure 7.13) have little notches along the leaf margins in which tiny plantlets are produced, complete with roots and leaves, even after a leaf has been removed from the parent plant and pinned to a curtain or left on a counter. Each of the plantlets can develop into a mature plant if given the opportunity to do so.

Floral Leaves (Bracts)

Modified leaves known as **bracts** are found at the bases of flowers or flower stalks. In some plants such as the Christmas flower (poinsettia), the flowers themselves have no petals, but the brightly colored floral bracts that surround the small flowers make up for the absence of petals (figure 7.14). In other plants such as the dogwood, the tiny flowers in their buttonlike clusters do have inconspicuous petals, and the large white-to-pink bracts that surround the flower clusters, although they appear to the casual observer to be petals, are actually modified leaves.

Figure 7.14 Poinsettia "flower." There are several flowers without petals in the center; the most conspicuous parts of the "flower" are the colored bracts (modified leaves) surrounding the true flowers. (Courtesy Field Museum of Natural History, Chicago)

Insect-trapping Leaves

Highly specialized insect-trapping leaves have intrigued humans for hundreds of years. Almost two hundred species of flowering plants are known to have these leaves. Such plants occur mainly in swampy areas and bogs of tropical and temperate regions. In such an environment, certain elements may be deficient in the soil, or they may be in a form not readily available to the plants. Some of the needed elements are furnished when the soft parts of insects and other small organisms trapped by the specialized leaves are broken down and digested. All the plants contain chlorophyll and thus are able to make their own food. It has been demonstrated that they can develop in apparently normal fashion without insects if they are given the nutrients they need. The mechanisms of trapping the insects fall into four types, represented by the following plants (figure 7.15).

Pitcher Plants

Some of the leaves of pitcher plants have flattened blades and function like any other leaves, but others are formed like vases of various sizes and shapes, with or without umbrellalike flaps over the open ends. Some Asian pitcher plants are vines whose long petioles are twisted around branches for support and whose pitchers are formed at the tips of leaves. The plants give off a distinctive odor that attracts insects; these, upon arrival, seek out nectar-secreting glands located at the rim of the pitcher and usually fall into a pool of fluid at the bottom. If they try to climb

Figure 7.15 Insect-trapping leaves. *A.* Pitcher plant. *B.* Sundew. *C.* Venus flytrap. *D.* Bladderwort. (Photos by Carolina Biological Supply Company)

A.

B.

C.

D.

out, they find the walls highly polished and slippery. In fact, in some species of pitcher plants the walls are coated with wax, and as the insects struggle up the surface, their feet become coated with the wax, which builds up until the victims seem to have acquired heavy clodlike boots. Most insects never make it up the walls, but even if they do, they still face a formidable barricade of stiff downward-pointing hairs near the rim. Eventually they drown, and their soft parts are digested by bacteria and by enzymes secreted by digestive glands near the bottom of the plant. In North America, the pitcher plants produce their pitchers in erect clusters on the ground, but the mechanisms for trapping insects are similar.

Merchandisers have been shameless in their wholesale collection of these plants for sale, and pitcher plants may become extinct in the wild. The cobra plant, a pitcher plant restricted to a few swampy areas in California and Oregon, has been placed on official rare species lists.

Sundews

The tiny plants called *sundews* often do not measure more than 2.5 to 5.0 centimeters (1 to 2 inches) in diameter. The roundish to oval leaves are covered with about one hundred fifty to two hundred upright glandular hairs that look like miniature clubs. At the tip of each hair is a glistening clear drop of sticky fluid containing digestive enzymes. As the droplets sparkle in the sun, they may attract insects, which find themselves stuck if they alight. The hairs are exceptionally sensitive to contact, responding to weights of less than one-thousandth of a milligram, and bend inward, surrounding any trapped insect within the space of a few minutes. The digestive enzymes break down the soft parts of the insects, and after digestion has been completed (within a few days), the glandular hairs return to their original positions. If bits of nonliving debris happen to catch in the sticky fluid, the hairs barely respond, showing they can distinguish between protein and something inedible. Some sundew owners regularly feed their plants tiny bits of hamburger and

boiled egg white. Portuguese peasants use relatives of sundews with less specialized leaves in their homes as an effective substitute for flypaper. Another similar plant, the butterwort, rolls an edge of its specialized leaf over its victims.

Venus Flytraps

Found in nature only in wet areas of North Carolina and South Carolina, the Venus flytrap has leaves constructed along the lines of an old-fashioned steel trap. The two halves of the blade have the appearance of being hinged along the midrib, with stiff hairlike projections located along their margins. There are three tiny trigger hairs on the inner surface of each half. If two trigger hairs are touched simultaneously or if any one of them is touched twice within a few seconds, the blade halves suddenly snap together, trapping the insect or other small animal. As the organism struggles, the trap closes even more tightly. Digestive enzymes secreted by the leaf break down the soft parts of the insect, which are then absorbed. After digestion has been completed, the trap reopens, ready to repeat the process. As is the case with sundews, the traps do not normally close for bits of debris that might accidentally fall on the leaf, presumably because nonliving material does not move about and stimulate the trigger hairs.

Bladderworts

Bladderworts are plants with finely dissected leaves that are found submerged and floating in the shallow water along the margins of lakes and streams. Toward the bases of many of the leaves are tiny stomach-shaped bladders, each with a trapdoor over the opening at one end. The bladders, which are between 0.3 and 0.6 centimeter (0.125 to 0.25 inch) in diameter, trap aquatic insects and other small animals through a complex mechanism. Four stiff hairs located toward one end of the trapdoor act as triggers when an insect touches one of them. The trapdoor springs open, and water rushes into the bladder. The stream of water propels the victim into the trap, and the door snaps shut behind it. The action takes place in less than one-hundredth of a second and makes a distinct popping sound, which can be heard with the aid of a sensitive underwater microphone. The trapped insect eventually dies, is broken down by bacteria, and the breakdown products are absorbed by cells in the walls of the bladder.

Science fiction writers have contributed to superstitions and beliefs that deep in the tropical jungles there are plants capable of trapping humans and other large animals. No such plants have been proved to exist, however. The largest pitcher plants known hold possibly 1 liter (roughly 1 quart) of fluid in their pitchers, and small frogs have been known to decompose in them, but the trapping of anything larger than a mouse seems very unlikely.

Why Leaves Change Color and Fall in Autumn

Plants that seasonally drop their leaves are said to be **deciduous.** In temperate climates, new leaves are produced in the spring and are shed in the fall, but in the tropics the cycles coincide with wet and dry seasons rather than with temperature changes. Even so-called evergreen trees shed their leaves; they do so a few at a time, however, so that they never have the bare look of deciduous trees in their winter condition. The process by which the leaves are shed is called **abscission.**

Abscission occurs as a result of changes that take place in an *abscission zone* near the base of the petiole of each leaf (figure 7.16). Sometimes the abscission zone can be seen externally as a line of slightly different color on the petiole. In young leaves, hormones are produced that apparently inhibit the formation of the specialized layers of cells that permit abscission to occur. As the leaf ages, hormonal changes occur, and at least two layers of cells become differentiated. Closest to the stem, the cells of the *protective layer,* which may be several cells deep, become coated and impregnated with fatty **suberin.** On the leaf side, a *separation layer* develops in which the cells swell, sometimes divide, and also become gelatinous. In response to any of several environmental changes

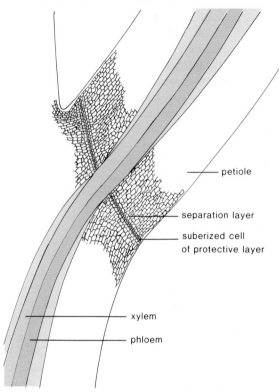

Figure 7.16   Abscission zone of a leaf

petiole

separation layer

suberized cell of protective layer

xylem

phloem

(such as lowering temperatures, decreasing day-lengths or light intensities, lack of adequate water, or damage to the leaf), the pectins in the middle lamella of the cells of the separation layer are broken down by enzymes. All that holds the leaf onto the stem at this point are some strands of xylem. Wind and rain then easily break the connecting strands, leaving tiny bundle scars within a leaf scar (see figure 6.1), and the leaf falls to the ground.

The chloroplasts of mature leaves contain several groups of pigments, including green *chlorophylls;* red, orange, or yellow *carotenes;* and pale yellow to brown *xanthophylls.* The first two of these groups play a role in photosynthesis. The chlorophylls are usually present in much larger quantities than the other pigments, and their intense green color masks or hides the presence of

Leaves

the carotenes and xanthophylls. In the fall, however, sugars accumulate in the leaves, causing the chlorophylls to break down, and other colors are revealed. In addition, the accumulation of sugars favors the formation of water-soluble anthocyanins and betacyanins in the vacuoles of the leaf cells. *Anthocyanins,* the more common of the two groups, are red if the cell sap is slightly acid and blue if it is slightly alkaline, and of intermediate shades if it is neutral. *Betacyanins* are usually red; they apparently are restricted to several plant families such as the cacti and the nightshades (the family to which potatoes and tomatoes belong). Some plants (e.g., birch trees) consistently exhibit a single shade of color in their fall leaves, but many (e.g., maple, ash, sumac) vary considerably from one locality to another or even from one leaf to another on the same tree, depending on the combinations of carotenes, xanthophylls, and other pigments present. Some of the most spectacular fall colors in North America occur in the Eastern Deciduous Forest, particularly in New England and the upper reaches of the Mississippi Valley. In parts of Wisconsin and Minnesota one can observe the brilliant reds, oranges, and golds of maples; the deep maroons (and also yellows) of ashes; the bright yellows of aspen, and the seemingly glowing reds of sumacs and wahoos, all in a single locality. Some fall coloration is found almost anywhere in temperate zones where deciduous trees and shrubs exist.

Human and Ecological Relevance of Leaves

We have already noted the supreme importance of photosynthesis to life as we know it, and the dependence of all living things on the two great products of the process, namely carbohydrates and oxygen. Leaves not only furnish the universally needed carbohydrate energy source but also provide vitamins and minerals, and along with their counterparts in the waters of the earth, they are at the beginning of every food chain (see chapter 23). Whether or not an animal eats leaves directly or eats animals that have grazed on vegetation, its existence depends on the activities of leaves.

Humans use shade trees and shrubs in landscaping for cooling as well as for aesthetic effects. The leaves of shade plants planted next to a dwelling can make a very significant difference in energy costs to the homeowner. Humans also use the leaves of cabbage, parsley, celery, lettuce, spinach, chard, and the petioles of rhubarb, to mention a few, as food. A large number of spices and flavorings are derived from leaves—thyme, marjoram, oregano, tarragon, peppermint, spearmint, wintergreen, basil, dill, sage, and savory. Many dyes (e.g., a yellow dye from bearberry, a reddish dye from henna, and a pale blue dye from blue ash) are extracted from leaves (see appendix 3), although nearly all commercial dyes are now derived from coal tar. Many cordage fibers for ropes and twines come from leaves, with various species of *Agave* (century plants) accounting for about 80% of the world's production. Bowstring fibers are obtained from a relative of the common houseplant *Sansevieria,* and Manila hemp fibers, which are used both in fine-quality cordage and in textiles, are obtained from the leaves of a close relative of the banana. Panama hats are made from the leaves of the panama hat palm, and palms and grasses are used in the tropics as thatching material for huts and other buildings.

In the high mountains of Chile and Peru, the leaves of the yareta plant are used for fuel. They produce a resin that causes the leaves to burn with an unusually hot flame. Leaves of many plants produce oils. Petitgrain oil from a variety of orange tree leaves and lavender, for example, are used for scenting soaps and colognes. Patchouli and lemongrass oils are used in perfumes, as is citronella oil, which was once the leading mosquito repellent before synthetic repellents gained favor. Eucalyptus oil, camphor, cajeput, and pennyroyal (figure 7.17) are all used medicinally.

Figure 7.17 Pennyroyal plant

Leaves are an important source of drugs used in medicine and also of narcotics and poisons. Cocaine, obtained from plants native to South America, is used medicinally as a local anesthetic and as a narcotic. Andean natives chew coca leaves while working and are reported capable of performing exceptional feats of labor with little or no food while under the influence of cocaine. Apparently the drug anesthetizes the nerves that convey hunger pangs to the brain. Belladonna is a drug complex obtained from leaves of the deadly nightshade, a native of Europe. The plant has been used in medicine for centuries, and several drugs are now isolated from belladonna. Included among the isolates are atropine, which is used in shock treatments, to dilate eyes, to relieve

pain locally, and to check secretions; and scopolamine, which is used in tranquilizers and sleeping aids. Another European plant, the foxglove, is the source of digitalis. This drug has been used for centuries in regulating blood circulation and heartbeat.

At present, more than 940 million kilograms (2 billion pounds) of tobacco for smoking, chewing, and use as snuff are produced annually around the world. Also at present, about one hundred thousand Americans die annually from lung cancer, and almost all of them have a history of cigarette smoking. Cigarette smoking is also considered a principal contributor to cardiovascular diseases. Though humans have long used tobacco, it has only been in the last two decades that its health threat has become appreciated.

Similarly, marijuana, the controversial plant widely used as an intoxicant, has been utilized in various ways for thousands of years. The active principle, tetrahydrocannabinol (THC), although found in the leaves, is concentrated in hair secretions among the female flowers. In recent decades marijuana has found increasing acceptance in the western hemisphere; however, it may not be the harmless intoxicant many thought it to be. A decision to legalize its prescription in pill form for the alleviation of nausea caused by chemotherapy treatments has been criticized by some medical authorities.

The drug lobeline sulphate is obtained from the leaves of a close relative of garden lobelias. It is used in compounds taken by those who are trying to stop smoking. The leaves of several species of *Aloe* yield a juice that is used to treat skin burns, particularly those accidentally received from X-ray equipment.

Several beverages are extracted by the brewing of leaves. Numerous teas have been obtained from a wide variety of plants, but most now in use come from a close relative of the garden camellia. Maté, the popular South American tea, is brewed from the leaves and twigs of a relative of holly. The alcoholic beverages pulque and tequila find their origin in the mashed leaves of *Agave* plants, and absinthe liqueur receives its unique flavor from the leaves of wormwood, a relative of western sagebrush, and other flavorings such as anise.

Insecticides of various types are also derived from leaves. A type of rotenone and a substance related to nicotine are obtained from tropical plants; both are effective against a variety of insects. Mexico's cockroach plant has leaves that when dried are highly effective in killing cockroaches, flies, fleas, and lice.

Carnauba and caussu waxes are obtained from the leaves of tropical palms. Leaves are used extensively by florists in floral arrangements and bouquets, and their uses for other aesthetic purposes are legion. Leaves may find more extensive use in the future as a direct source of food. It has been shown that a curd obtained by coagulating juice squeezed from alfalfa leaves contains more than 40% protein, and juices from other leaves have yielded better than 50% protein. Experiments are under way to make the leaf curd palatable for human consumption. Additional information on past and present uses of leaves is given in chapter 22 and in appendices 2, 3, and 4. The scientific names of the plants that are the sources of the various substances discussed are given in appendix 1.

Summary

Leaves are similar to solar panels in that they are covered with a transparent epidermis that admits light to the interior, and many twist on their petioles so that their flat surfaces are inclined to the sun throughout the day. Their lower and often their upper surfaces are dotted with stomata that permit air circulation and that function in other ways to facilitate photosynthesis, the primary function of leaves. Leaves also respire, accumulate wastes, and eliminate excess moisture via transpiration. The shapes, sizes, and textures of leaves vary greatly.

All leaves originate as primordia. Most leaves consist of a petiole and a blade. Some have paired stipules at the base of the petiole. Leaves of flowering plants may be simple or compound, and all are associated with leaf gaps. Both the venation and the compounding of the leaves may be either pinnate or palmate. In pinnate leaves, the main vein is the midrib in simple leaves and the rachis in compound leaves. A few leaves with evenly forking veins are said to have dichotomous venation.

Three groups of tissues occur in leaves. The epidermis is a single layer of cells covering the entire surface; it is coated with a thin cuticle, and may also have waxes, glands, hairs, and an occasional crystal within the cells. The lower epidermis usually contains numerous pores (stomata), each formed by a pair of guard cells that regulate both evaporation of water vapor from the leaf and gas exchange between the interior and the atmosphere. The mesophyll between the upper epidermis and the lower epidermis is divided into an upper palisade layer, which consists of rows of tightly packed parenchyma cells containing numerous chloroplasts, and a lower spongy layer in which the parenchyma cells are loosely arranged. The spongy layer contains a considerable amount of air space. Veins traverse the mesophyll.

Leaves may be specialized in many ways, each modification usually being associated with specific combinations of environmental factors. Plants in arid regions may have thick leaves, a reduction in the number of stomata, and water-storage leaves, or no leaves at all. Other specialized leaves include tendrils, which aid in the support of the plant; spines, which reduce leaf surface and aid in protection; food storage leaves; flower pot leaves; window leaves; reproductive leaves; floral leaves (bracts); and several types of insect-trapping leaves.

Leaves change color and fall in the autumn because of changes that occur in the abscission zone at the base of the petiole of each deciduous leaf. A protective layer of suberized cells and a separation layer of cells, which swell and become gelatinous, form adjacent to one another when hormones that inhibit the formation of these layers fall below a certain level in response to environmental changes. The green chlorophyll pigments break down, revealing pigments of other colors. In addition, sugars that accumulate in the leaves favor the formation of water-soluble anthocyanin and betacyanin pigments that may be red or blue in color. The leaves of some plants consistently turn a certain color in the fall, whereas leaves of other plants vary in color, depending on their locality or even on their position on the tree.

Review Questions

1. Leaves have no secondary xylem and phloem. Why not?

2. What are bracts?

3. Identify or define hydathode, transpiration, guard cells, mesophyll, venation, glands, compound leaf.

4. How can one distinguish between the upper and lower epidermis in most leaves?

5. What is the function of bundle sheaths?

6. How do leaves in shaded areas differ from leaves in sunlit areas on the same plant?

7. What leaf modifications are associated with dry areas, wet areas (e.g., lakes), climbing, and reproduction?

8. Why do leaves turn different colors in the fall?

Discussion Questions

1. Can you think of any advantages or disadvantages to a plant in having very tiny or very large leaves?

2. In chapter 3, it was noted that living cells are connected to one another by plasmodesmata holes in the walls. really need veins for distribution of materials

3. Is there any advantage to a le palisade mesophyll on the upper spongy mesophyll on the lower half

4. Pines and other plants that grow in very cold climates have sunken stomata just like plants of the desert, yet the annual precipitation where they grow may be very abundant. Is there any advantage to them in having such a modification?

5. Since tendrils and spines can be either modified leaves or modified stems, how might you determine the origin of a given specimen?

Additional Reading

Addicott, F. T. Physiology of abscission. *Handbuch der Pflanzenphysiologie* 15: 1094–1126; 1965.

Cutter, E. G. *Plant anatomy: experiment and interpretation.* Part 2: *Organs.* Reading, MA: Addison-Wesley Publishing Co.; 1971.

Darwin, C. *Insectivorous plants.* New York: Murray; 1884.

Esau, K. *Anatomy of seed plants.* 2d ed. New York: John Wiley and Sons; 1977.

Foster, A. S.; Gifford, E. M., Jr. *Comparative morphology of vascular plants.* 2d ed. San Francisco: W. H. Freeman and Co.; 1974.

Lloyd, F. E. *Carnivorous plants.* New York: Dover Publications, Inc.; 1976.

Philipson, W. R.; Balfour, E. E. Vascular patterns in dicotyledons. *Botanical Review* 29: 382–404; 1965.

Steeves, T. A.; Sussex, I. M. *Patterns in plant development.* Englewood Cliffs, NJ: Prentice-Hall; 1972.

Wardlaw, C. W. *Organization and evolution in plants.* London: Longmans; 1965.

Webster, B. D. Abscission. *McGraw-Hill yearbook of science and technology.* New York: McGraw-Hill Book Co.; 1969.

Water in Plants

Overview

This chapter begins by introducing diffusion through a comparison between balls in motion in a room and molecular activity. This introduction is followed by a discussion of osmosis, turgor, plasmolysis, active transport, and imbibition.

The chapter then takes up the entry of water into the plant, the movement of water through the plant, the evaporation of water into leaf air spaces, and transpiration.

Some Suggested Student Learning Goals

1. In simple terms explain diffusion, osmosis, turgor, imbibition, and active transport.

2. Describe the pathway, movement, and utilization of water in plants.

Introduction

Nearly everyone has had the experience of driving along a highway or city street when someone in the car says, "What's that smell?" Soon a bakery, or a paper mill, or perhaps a dead skunk comes into view, and the smell gets stronger. Then it fades away as the source is left behind. We take for granted the fact that there is a correlation between our proximity to an odor source and the intensity of the odor, but how and why does the odor reach us?

By way of an answer, let us imagine two adjacent rooms identical in height, width, and length, with no windows, doors, fixtures, or furniture. Now suppose we lift a small flap in the ceiling of one of them and drop in 100 tennis balls. These are ordinary tennis balls except for one extraordinary feature: they have perpetual motion motors in them that cause them to travel in any direction at 30 MPH. The tennis balls quickly make the room seem like a battlefield as they whiz around, bounce off the walls, floor, and ceiling, and also collide with one another, each time being deflected at a different angle. Shortly after they are introduced, the tennis balls will probably distribute themselves more or less evenly throughout the room.

Now what would you expect to happen if we opened up a small hole in the wall between the two rooms? Eventually a tennis ball should bounce or travel at just the right angle and go through the hole into the other room. Theoretically, it could then bounce straight back into the first room, but it seems unlikely that it would. Reason tells us that long before it might happen to strike the exact angle it needed to return, several more balls will come in from the first room. Given enough time, some balls might indeed bounce back into the first room, but in the long run, each of the two rooms would end up with roughly 50 tennis balls, with an occasional ball going between the rooms in either direction.

The situation just described is analogous to what takes place in nature on a molecular level.

Molecular Movement

Molecules and ions are constantly in random motion. Visual evidence of this can be seen with an ordinary light microscope. If a drop of India ink is diluted with water and observed through a microscope under high power, the tiny carbon particles of the ink appear to be in constant motion. This motion, known as *Brownian movement,* is the result of the bombardment of the visible particles by invisible water molecules, which are in constant motion themselves.

The differing intensity of smells discussed earlier involves molecules behaving something like the tennis balls. Through their random motion, molecules tend to distribute themselves evenly in the space available to them. Thus, if perfume molecules are kept in a bottle, they will distribute themselves evenly within the bottle, but if the stopper is removed, they will eventually disperse themselves evenly throughout the room, even if there is no fan or other device to move the air. This movement of molecules or ions from a region of higher concentration to a region of lower concentration is called **diffusion** (figure 8.1). Molecules that are moving from a region of higher concentration to a region of lower concentration are said to be moving *along a diffusion gradient,* while molecules going in the opposite direction are said to be going *against a diffusion gradient.* When the molecules, through their random movement in all directions, have distributed themselves evenly within the space available, they are considered to be in a state of *equilibrium.*

Figure 8.1 Diffusion of molecules from a crystal of dye placed on an agar plate. If the plate is left undisturbed, it may take years before the diffusion is complete.

A.   B.   C.   D.

Except within the area immediately surrounding the source, unaided diffusion requires a great deal of time because molecules and ions are infinitesimal. Something that is less than a forty-billionth of a millimeter in diameter is going to take a long time to move just one millimeter, even though the amount of movement may be great in proportion to the size of the particle concerned.

Large molecules move much more slowly than small molecules. If you added a tiny drop of a dye (which has relatively large molecules) to one end of a bathtub of water without disturbing the water in any way, it would take years for the dye molecules to diffuse throughout the tub and reach a state of equilibrium. In nature, however, wind and water currents distribute molecules much faster than they ever could be distributed by diffusion alone. Diffusion rates are also affected by the density of the molecules concerned and by other factors such as temperature.

Osmosis

Despite the fact that the protoplasts of living cells are bounded by membranes, it has been known for some time that water moves freely from cell to cell. This has led scientists to believe that plasma, vacuolar, and other membranes must have tiny holes in them, even though to the instruments presently available the holes are invisible. It is also known that these same membranes will not allow larger molecules or ions to pass through except under special circumstances. Such membranes have been described as **differentially permeable** or **semipermeable.**

Osmosis is the diffusion of water (or any other solvent) through a semipermeable membrane. The diffusing activity of the water molecules results in their moving from regions where they are most concentrated to regions where they are less concentrated. Osmosis is the primary means by which plants obtain water from their environment. In land plants, most water enters the cell walls and intercellular spaces of the epidermis and

Figure 8.2 Simple osmometer made by tying a differentially permeable membrane over the mouth of a thistle tube

fluid rising in tube

10% solution of sugar in water

differentially permeable membrane

pure water

the root hairs and travels along the walls until it reaches the endodermis. Here it crosses the differentially permeable membranes and protoplasts of the endodermal cells on its way to the xylem, where it flows to the leaves, evaporates within the leaf air spaces, and diffuses out (transpires) through the stomata into the atmosphere. Other forces involved in the movement of water through the plant are discussed in the next section.

A demonstration of osmosis can be made by tying a differentially permeable membrane over the mouth of a thistle tube that has been filled with a solution of 10% sugar in water (i.e., the solution consists of 10% sugar and 90% water). Fluid rises in the narrow part of the tube as osmosis occurs when the thistle tube is immersed in water (figure 8.2). Pressure applied to the top of the tube *could* prevent the rise of the fluid. *Potential* pressure such as this is called **osmotic potential,** or **osmotic pressure.** In other words osmotic or potential pressure is the pressure required to prevent osmosis from taking place. A solution in a beaker has an osmotic pressure determined by the conditions brought into effect for

Chapter 8

Figure 8.3 Portion of a leaf of the water weed *Elodea*.
A. Normal cells. *B.* Plasmolyzed cells.

A.

B.

that solution. The situation is somewhat analogous to that of a parked, new, high-powered sports car. The car has a potential top speed of 200 MPH. That is, a *skilled* driver could advance the car to a speed of 200 MPH *if* the driver had a straight, level highway, *if* there was no traffic, and so forth. An *inexperienced* driver on a city street at the 5:00 P.M. rush hour, however, might not be able to move the 200 MPH car at 25 MPH. Actual and potential pressures in cells are developed in similar fashion. Water enters a cell by osmosis until the osmotic pressure is balanced by the resistance of the cell wall to expansion. Water gained by osmosis keeps a cell firm, or **turgid,** and the pressure that develops against the walls as a result of water entering the vacuole of the cell is called **turgor pressure** *(pressure potential).* The release of turgor pressure can be heard each time you bite into a crisp celery stick or the leaf of a young head of lettuce. It is common practice to soak carrot sticks, celery, and lettuce in pure water to make them crisp.

If you placed turgid carrot and celery sticks in a 10% solution of salt in water, they would soon lose their rigidity and become limp enough to curl around your finger. The concentration of water within the cells would be greater than the concentration of water outside, and so there would be diffusion of water out of the cells into the salt solution. If you were to examine such cells with the aid of a microscope, you would see that the vacuoles, which are largely water, had disappeared and that the protoplasm had shrunk away from the walls and was clumped in the middle of the cell. Such cells are said to be *plasmolyzed.* This loss of water through osmosis, which is accompanied by the shrinkage of protoplasm away from the cell wall, is called **plasmolysis** (figure 8.3). If plasmolyzed cells are placed in fresh water before permanent damage is done, water reenters the cell by osmosis, and the cells become turgid once more.

Active Transport

Let us return for a moment to our two rooms with the tennis balls. Suppose that, in addition to the 100 tennis balls, we drop in 50 basketballs, also extraordinary in having perpetual motion motors that propel them in any direction at 12 MPH. They should also distribute themselves evenly throughout the room shortly after they are introduced. Assume, however, that the hole in the wall (which is large enough for the passage of a tennis ball) is too small to allow a basketball to pass through. The basketballs will then remain in the first room. But if we install a mechanical arm next to the hole in the second room, and if this arm can grab basketballs that come near the hole and squeeze them through, basketballs will be transported into both rooms *through the expenditure of energy.*

Plants expend energy, too. Plant cells generally contain a larger number of mineral molecules and ions than exist in the soil immediately adjacent to the root hairs. If it were not for the barriers imposed by the differentially permeable membranes, therefore, these molecules and ions would move from a region of higher concentration in the cells to a region of lower concentration in the soil. The plants absorb and retain these substances against a diffusion gradient, however, through the expenditure of energy. This process is called **active transport.** The precise mechanism of active transport is not fully understood, but it is believed that some sort of physiological pump in the plasma membrane is involved.

Imbibition

Osmosis is not the only force involved in the absorption of water by plants. *Colloidal* materials (i.e., materials that contain a permanent suspension of fine particles) and large molecules such as cellulose and starch usually develop electrical charges when they are wet, and they attract water molecules that adhere to the internal surfaces of the materials. This process, known as **imbibition,** results in the swelling of tissues, whether they are alive or dead, often to several times their original volume. Imbibition is the initial step in the germination of seeds (figure 8.4).

Figure 8.4 Pinto bean seeds before and after imbibition of water

The physical forces developed during germination can be tremendous, even to the point of causing a seed to split a rock weighing several tons. It has been found, for example, that a pressure of 42.2 kilograms per square centimeter (600 pounds per square inch) is needed to break the seedcoat of a fresh walnut from within, and that water being imbibed by a cocklebur seed develops a force of up to one thousand times that of normal atmospheric pressure. Yet walnuts germinate readily, and seeds that fall into the crevices of rocks or have boulders roll over on them also germinate when water and oxygen reach them, sometimes eventually splitting the rock from top to bottom (see figure 23.7). The huge stone blocks used in the construction of the pyramids of Egypt are believed to have been quarried by hammering rounded wooden stakes into holes made in the face of the stone and then soaking the stakes with water. As the stakes swelled, the force created by imbibition was sufficient to split the rock.

Water and Its Movement Through the Plant

If you were to place foil over the soil at the base of a plant, place the pot where it receives light, and then cover the potted plant with a bell jar, you would notice moisture accumulating on the inside of the jar within an hour or two. Because of the foil barrier, the water could not have come directly from the soil; it had to have come through the plant. More than 90% of the water entering a plant passes through and evaporates—primarily into leaf air spaces and then through the stomata into the atmosphere (see figure 8.6). (About 5 to 10% of the water lost escapes through the cuticle.) This process of water vapor loss from the internal leaf atmosphere is called **transpiration,** and the amount of water transpired by plants is greater than one might suspect. For example, mature corn plants each transpire about 30 liters (8 gallons) of water per week, while four-tenths of a hectare (1 acre) of corn may transpire more than 1,325,000 liters (350,000 gallons) in a 100-day growing season. A hardwood tree utilizes about 450 liters (120 gallons) of water while producing 0.45 kilogram (1 pound) of wood (or 1,800 liters while producing 0.45 kilogram of dry weight substance), and the 200,000 leaves of an average-sized birch tree will transpire from 750 to more than 3,785 liters (20 to 100 gallons) per day during the growing season. Humans recycle much of their water via the circulatory system, but if they were to have requirements similar to those of plants, each adult would have to drink well over 38 liters (10 gallons) per day.

Why do plants use so much water? Water constitutes about 90% of the weight of young cells. The thousands of enzyme actions and other chemical activities of cells take place in water, and a small amount is used in the process of photosynthesis. The exposed surfaces of the chlorenchyma cells within the leaf have to be moist at all times, for it is through this film of water that the carbon dioxide molecules needed for the process of photosynthesis enter the cell from the air. Water is also needed for cell turgor, which gives rigidity to herbaceous plants. Consider also what it must be like in the mesophyll of a flattened leaf that is fully exposed to the midsummer sun in areas where the air temperature soars to well over 38°C (100°F) in the shade. If it were not for the evaporation of water molecules from the moist surfaces, which brings about some cooling, and reradiation of energy by the leaf, the intense heat could damage the plant (see also the role of abscisic acid in this regard, discussed in chapter 10). Sometimes the transpiration is so rapid that the loss of water begins to exceed the intake, and the stomata may close, thus preventing wilting. When this occurs, the carbon dioxide supply from the atmosphere is cut off, but it can be partially compensated for from within the plant by the process of respiration, which produces carbon dioxide as a by-product.

How does water travel through the roots from 3 to 6 meters (10 to 20 feet) or more beneath the surface and then up the trunk to the topmost leaves of a tree that is more than 90 meters (300 feet) tall? We know that continuous tubular pathways of xylem run throughout the plant, extending from the young roots up through the stem and branches to the tiny veinlets of the leaves; we also know that the water gets to the start of this "plumbing system" by osmosis. But water is raised through the columns apparently by a combination of factors, and the process has been the subject of much debate for the past two hundred years.

One of the earliest explanations for the rise of water in a living plant was given by the English scientist Nehemiah Grew in 1682. He suggested that cells surrounding the xylem vessels and tracheids performed a pumping action that propelled the water along. This was questioned, however, when it was found that water will rise in lengths of dead stems also. Then the belief that capillary action moved the water became popular (figure 8.5). It is well known that water will rise in a narrow tube and that the heights attained are inversely proportional to the diameter of the tube. It is also known that this rise occurs through the forces involved in the forming of a concave *meniscus* (curved surface) at the top of the water column. But even though water can, indeed, rise 1 meter (3 feet) or more in a very narrow tube, air must be present above the column for the forces to work, and such is not the case in a plant. In fact, any air introduced into a water column in xylem interferes with the rise of water. Also, while capillarity might produce sufficient force to raise water 1 meter or 2, the diameter of the tubes is not small enough to raise it more than that.

Others have pointed to *root pressure* as the means whereby water is moved through plants. When some plants are pruned after growth has begun in the spring, water will exude from the cut ends. This is the result of root pressure, which

Figure 8.5 Capillarity in narrow tubes; the smaller the diameter of the tube, the greater the rise of the fluid

involves osmosis and perhaps other forces. Some plants do not "bleed" when they are pruned, however, and the force exerted by root pressure has been shown to be less than 30 grams per square centimeter (a few pounds per square inch), as a rule. This is considerably less than that needed to raise water to the tops of tall trees. Furthermore, root pressure seems to drop to negligible amounts in the summer, when the greatest amounts of water are moving through the plant.

The most satisfactory explanation of the rise of water in plants thus far suggested involves the **cohesion–tension–transpiration** theory. It is known that water molecules cohere to each other and adhere to capillary walls (e.g., those of xylem tracheids and vessels), permitting a certain amount of tension. It is possible, for example, to fill a glass with water, place a sheet of cardboard over the mouth, and invert the glass without the water spilling, because the cohesiveness of the water molecules to one another and their adhesion to the cardboard holds the cardboard against the rim of the glass. When water evaporates from the mesophyll cells in a leaf, these cells become less turgid. They then have a lower **water potential** (potential of cells to take up water until they are turgid) and receive more water by osmosis; in doing so they tend to become as turgid as they were before. The difference between the water potential in two areas generates the force to

PLATE 1. A. Cross section through a portion of a *Tilia* (basswood) stem. (Courtesy Triarch, Inc.)

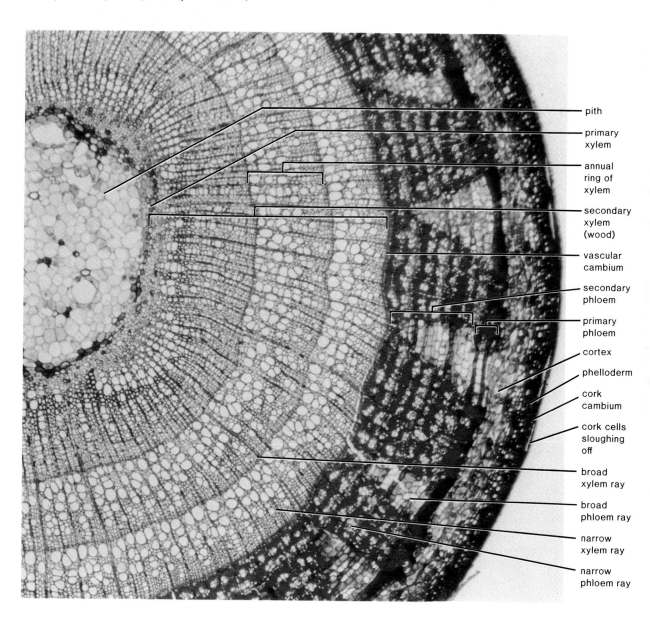

pith

primary xylem

annual ring of xylem

secondary xylem (wood)

vascular cambium

secondary phloem

primary phloem

cortex

phelloderm

cork cambium

cork cells sloughing off

broad xylem ray

broad phloem ray

narrow xylem ray

narrow phloem ray

PLATE 2.   A. Sclereids in the leaf of a wheel tree.
B. Hairs on the surface of a marijuana plant leaf.
C. Longitudinal section through fibers in a *Welwitschia*
leaf. D. Cross section through the apex of a stem of
coastal redwood. (Photomicrographs by G. S. Ellmore)

A

B.

C

D

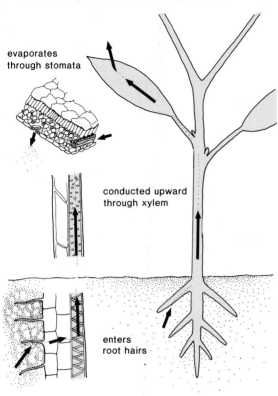

Figure 8.6 Pathway of water through a plant

evaporates
through stomata

conducted upward
through xylem

enters
root hairs

transport water in a plant. As transpiration occurs, a pull, or tension, results, drawing water from one molecule to another all the way through the xylem cells. Thus water "columns" move through the plant from roots to leaves, and the abundant water of a normally moist soil supplies these "columns" as the water continues to enter the root by osmosis (figure 8.6).

Transport of Food Substances in Solution

One of the most important functions of water in the plant involves the *translocation* (transportation) of food substances in solution by the phloem, a process that has only recently come to be better understood. Many of the studies that led to our present knowledge of the subject utilized aphids or radioactive tracers.

Most aphids feed on phloem by inserting their tiny tubelike mouthparts *(stylets)* through the leaf or stem tissues until a sieve tube is reached and punctured. The turgor pressure of the sieve tube then forces the fluid present in the tube through the aphid's digestive tract, and it emerges at the rear as a droplet of "honeydew." In the studies, research workers cut the stylets of feeding aphids so that much of the tiny tube remained where it had been inserted. The fluid exuding from the cut stylets was then collected (sometimes this continued for many hours afterwards) and analyzed. In other studies, by exposing a photosynthesizing leaf to carbon dioxide (a basic raw material of the process) that had been synthesized with radioactive carbon, the pathway of food substances manufactured by the leaf could be traced. Photographic film was used on which the radioactive substances produced an image corresponding to their location. Data obtained from such studies revealed that food substances in solution are confined entirely to the sieve tubes while they are being transported. At one time it was believed that ordinary diffusion and cyclosis (see chapter 3) were responsible for the movement of the substances from one sieve-tube element to the next, but it is now known that the substances move through the phloem at approximately 10 decimeters (almost 40 inches) per hour—far too rapid a movement to be accounted for by diffusion and cyclosis alone.

Pressure-flow Hypothesis

At present, the most widely accepted theory for movement of substances in the phloem is called the **pressure-flow hypothesis.** According to this theory, food substances in solution flow along concentration gradients between food-producing and storage regions (sources of the food) and *sinks*, which are places where the food is utilized (e.g., the growing tips of stems and roots). Turgor pressure, which develops as osmosis occurs, is responsible for driving the fluid through the sieve tubes toward the sinks. According to the theory, water is taken up by osmosis at the source (e.g., mesophyll tissue of a leaf) and is given up at the sink. As the food substances in solution are actively removed at the sink, the pressure in the adjacent sieve tubes is lowered, causing movement from the higher pressure at the source to the lower pressure at the sink. Most of the water diffuses back to the xylem, where it then returns to the source and is transpired or recirculated.

Regulation of Transpiration

As previously indicated, the stomata regulate transpiration and gas exchange. The guard cells forming each stoma have relatively elastic walls, and the part of the wall adjacent to the hole itself is considerably thicker than the remainder of the wall (figure 8.7). This thickness allows each stoma to be opened and closed by means of changes in the turgor of the guard cells. The stomatal pore is closed when turgor pressure is low, and open when turgor pressure is high. Changes in turgor pressures in the guard cells result from changes in solute (e.g., potassium ions, carbohydrates) concentrations brought about by osmosis and active transport between the epidermal cells and the guard cells. Guard cells expend energy to take up potassium ions from adjacent epidermal cells as the stomata open, and release potassium ions as the stomata close. With an increase in potassium ions, the water concentration in the guard cells is lowered, and the osmosis that takes place as a result brings in water that makes the cells turgid. Some believe that energy may also be expended to cause the potassium ions to leave, but whether the ions leave because of active transport or not, their departure also results in water leaving, making the cells less turgid, and closing the stomata (figure 8.8).

Figure 8.7 Portion of a strip of lily leaf epidermis showing stomata and surrounding epidermal cells. (Photomicrograph by Carolina Biological Supply Company)

Most plants have their stomata open during the day and closed at night, but water is conserved in a number of desert plants that have their stomata open only at night when there is less stress on the plants for water. Such plants have a specialized form of photosynthesis (see *CAM photosynthesis,* page 152) since the carbon dioxide they require for typical photosynthesis is shut off during daylight hours by the closing of the stomata. Some desert plants are able to store carbon dioxide taken in at night in the form of special substances that can be converted back to carbon dioxide during the daytime when photosynthesis occurs. Other desert plants have their stomata below the surface of the leaf or stem in little sunken chambers. These chambers often are overarched with hairs that further reduce water loss. Pine trees, which frequently occur in areas where the soil is frozen for part of the year and where limitations thus occur on available water,

also have sunken stomata (see figure 7.8). A few tropical plants that occur in damp, humid areas (e.g., ruellias) actually have stomata that are raised above the surface of the leaf, whereas plants that occur in water generally have no stomata on submerged surfaces.

Although the opening and the closing of the stomata regulate transpiration to a great extent, light, carbon dioxide concentration, and several other factors affect it as well. Air currents sweeping away the water molecules as they emerge from the stomata speed up the rates of evaporation within. Humidity plays an inverse role in transpiration rates: high humidity reduces transpiration, and low humidity accelerates it. A direct correlation also exists between temperature and the movement of the water molecules out of the leaf. The transpiration rate of a leaf at 30°C (86°F), for example, is about twice as great as it is for the same leaf at 20°C (68°F). The various adaptive modifications of leaves and their surfaces, which were mentioned in chapter 7, also may play important roles in influencing the amount of water transpired, as well as the availability of water to the roots.

If a cool night follows a warm humid day, some plants may produce water droplets at the tips of veins in their leaves through structures called *hydathodes.* This loss of water in liquid form is called **guttation** (figure 8.9). Although the droplets resemble dew, the two should not be confused. Dew is water that is condensed from the air, while guttation water is literally forced out of the plant by root pressures. As the sun strikes the droplets in the morning, they dry up, leaving a residue of salts and organic substances, one of which is used in the manufacture of commercial "flavor enhancers" (e.g., the monosodium glutamate in products such as Accent). In the tropics, the amount of water produced by guttation can be considerable. In taro plants, used by the Polynesians to make *poi,* a single leaf may produce as much as a cupful of water overnight through guttation.

Figure 8.8 How a stoma opens and closes. *A.* Closed stoma. *B.* As turgor pressure in the guard cells increases, the thinner outer walls stretch more than the thicker inner walls, causing the stoma to open.

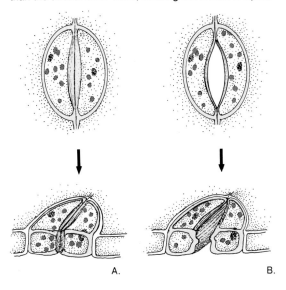

A.   B.

Figure 8.9 Droplets of guttation water on the margins of a tomato leaf. (Courtesy H. L. Dean)

Summary

Molecules and ions are in constant random motion and tend to distribute themselves evenly in the space available to them. In doing so they move from a region of higher concentration to a region of lower concentration by simple diffusion along a diffusion gradient, although they may also move against a diffusion gradient. Evenly distributed molecules are said to be in a state of equilibrium; diffusion rates are affected by temperature, the size of the molecules, the density of the molecules, and other factors such as wind and currents.

Osmosis is the diffusion of water through a differentially permeable membrane from a region of higher concentration to a region of lower concentration. It is the primary means by which plants obtain water from their environment. Osmotic pressure or potential is the pressure required to prevent osmosis from taking place. The pressure that develops in a cell as a result of water entering it is called turgor. Shrinkage of the protoplasm away from the cell wall as a result of osmosis taking place when the concentration of water inside the cell is greater than outside is called plasmolysis. Active transport is the expenditure of energy by a cell that results in molecules or ions entering the cell against a diffusion gradient. Imbibition is the attraction and adhesion of water molecules to the internal surfaces of materials; it results in swelling and is the initial step in the germination of seeds.

Most of the water that enters a plant passes through it and evaporates into the atmosphere through stomata by a process called transpiration. The water retained by the plant is used in photosynthesis and other metabolic activities of the cell; it is involved in the translocation of food substances in solution, and according to the pressure-flow hypothesis, such substances flow along concentration gradients between their sources and sinks (places where the food is utilized). According to the cohesion–tension–transpiration theory, water rises through plants because of a combination of factors including the adhesion of water molecules to the walls of the capillary conducting elements of the xylem, tension on the water columns created by the "pull" developed by transpiration, and the cohesion of water molecules.

Transpiration is regulated by the stomata, which open and close through changes in turgor pressure of the guard cells. These changes, which involve potassium ions, result from osmosis and active transport between the guard cells and the adjacent epidermal cells. Aquatic, desert, tropical, and some cold zone plants have modifications of stomata or specialized forms of photosynthesis that adapt them to their particular environments. Guttation is the loss of water in liquid form as it is forced out of the tips of leaf veins by root pressures through structures called hydathodes.

Review Questions

1. Distinguish among diffusion, osmosis, active transport, plasmolysis, and imbibition.

2. Why do living plants need a great deal of water for their activities?

3. Explain how a tall tree gets water to its tips without the aid of mechanical pumps.

4. What is the difference between transpiration and guttation?

5. Explain the pressure-flow hypothesis.

Discussion Questions

1. Why is salted meat less likely to spoil than unsalted meat?

2. Why would dye molecules in a bathtub of water take a long time to diffuse completely throughout the tub, but perfume molecules released in a closed room take considerably less time to do the same thing?

3. Why does osmosis not cause submerged water plants to swell up and burst?

4. Some bodies of water such as the Dead Sea have considerably higher salt concentrations than those of the human body. If you were swimming in such water, would you expect your cells to become plasmolyzed? Why?

Additional Rea

Crafts, A. S.; Crisp, C. E. _plants_. San Francisco: V 1971.

Epstein, E. _Mineral nutrition of p and perspectives_. New York: Joh Sons, Inc.; 1972.

Galston, A. W.; Davies, P. J.; Satter, R. L. _of the green plant_. 3d ed. Englewood Cli Prentice-Hall, Inc.; 1980.

Heath, O. V. S. _Stomata_. Oxford Biology Readers, No. 37. Burlington, NC: Carolina Biological Supply Co.; 1975.

Kluge, M.; Ting, I. P. _Crassulacean acid metabolism_. Ecological studies, Vol. 30. New York: Springer-Verlag; 1978.

Luettge, U.; Higinbotham, N. _Transport in plants_. New York: Springer-Verlag; 1979.

Meidner, H.; Sheriff, D. W. _Water and plants_. New York: John Wiley and Sons, Inc.; 1976.

Peel, A. J. _Transport of nutrients in plants_. New York: John Wiley and Sons, Inc.; 1974.

Rutter, A. J. _Transpiration_. Oxford Biology Readers, No. 24. Burlington, NC: Carolina Biological Supply Co.; 1972.

Salisbury, F. B.; Ross, C. W., eds. _Plant physiology_. 2d ed. Belmont, CA: Wadsworth Publishing Co., Inc.; 1978.

Sutcliffe, J. _Plants and water_. New York: St. Martin's Press, Inc.; 1968.

Phloem transport in
. H. Freeman and Co.;

nts: principles
n Wiley &

The life
ffs, NJ:

Metabolism in Plants

Overview

tosynthesis and respiration are
n presented at three different
ls: the essence of the process
is examined; the major steps are
introduced; the processes are
looked at in greater detail. One or
two levels may be sufficient for
some readers, while others will
want to explore all three. The
chapter discusses the importance
of the main features of each
process and summarizes the light
reactions, the dark reactions,
glycolysis, the Krebs cycle, and the
electron transport system. It
concludes with a tabular
comparison between
photosynthesis and respiration and
a few brief observations on
digestion and assimilation.

Some Suggested Student Learning Goals

1. Contrast the generalized
   equations of photosynthesis
   and respiration.

2. Describe what occurs in the
   light and dark reactions of
   photosynthesis, and list the
   principal products of the
   reactions.

3. Explain what occurs in
   glycolysis, the Krebs cycle, and
   the electron transport chain of
   respiration.

4. Distinguish between aerobic
   respiration and fermentation.

5. Compare digestion and
   assimilation.

Introduction

A number of years ago I had a student who claimed that her father was an alchemist, and that he was on the verge of discovering how to transform lead into gold. The belief that such a feat was possible was widespread in medieval times, but with the development of modern science and technology, such romantic yet totally unscientific ideas have been discarded by educated people in general and relegated to the history books. Yet some amazing chemical transformations of another kind take place all around us every day in green plants and have been taking place since long before humans appeared on this planet. By far the most important of these transformations, *photosynthesis,* involves little more than fresh air, water, a green pigment, and light, yet parts of the water and air are converted to a sugar without the aid of any cumbersome machinery. In addition, twenty-four hours a day, as long as any cell remains alive, stored energy is released from sugar by another process, *respiration.* These two processes take up the bulk of our discussion in this chapter.

Photosynthesis

Oil and coal today provide about 90% of the energy needed to power trains, trucks, ships, airplanes, factories, and a myriad of electrically energized appliances, computers, and communications systems. The energy within that oil and coal was originally "captured" from the sun by plants growing millions of years ago and transformed into fossil fuels by geological forces. The energy needs of transportation, industry, and the modern household seem insignificant, however, when compared with the combined energy requirements of all living organisms. Every living cell requires energy just to remain alive, and additional energy is needed for the cell to reproduce, grow, or do physical work as part of an organism. In addition, oxygen is vital to nearly all life in processes releasing that stored energy for use. Photosynthesis, at least indirectly, is not only the principal means of enabling a civilized society to function normally but also the sole means of sustaining life—except for a few bacteria that derive their energy from sulphur salts and other inorganic compounds. This unique manufacturing process of green plants furnishes raw material, energy, and oxygen. In photosynthesis, energy from the sun is harnessed and packed into sugar molecules made from water and carbon dioxide with the aid of chlorophyll; oxygen is given off as a by-product of the process.

The capacity of plants to satisfy the earth's energy needs probably will determine the ultimate size of human populations. Already in some quarters of the globe, starvation is commonplace where insufficient food is being produced to maintain the minimal energy requirements of overpopulated lands. While the Western world is still spending large sums of money on weight reduction as a result of the consumption of too much food, its populations are rapidly approaching the point at which they must either stabilize or they, too, will exceed the capacity of the plants to sustain them.

The food energy produced by plants is not necessarily the final limiting factor for human populations either, for pollution eventually could interfere with the principal by-product of photosynthesis to the extent of reducing the oxygen content of the atmosphere.

A great deal of photosynthesis occurs in organisms present in the oceans; consequently, it is estimated that more than 50% of the oxygen in the atmosphere originates in the seas. Laboratory tests have shown, however, that concentrations of DDT as low as 20 parts per billion are enough to stop many of the delicate marine algae from carrying on photosynthesis. Except for DDT concentrations of up to 7 or more parts per billion in estuaries, the concentration of such substances in ocean waters at present is thousands of times less than that. But, while the use of DDT and related chemicals has been curtailed in the United States in recent years, other countries are still using it, and residues are still washing off into rivers and on into the oceans. If the concentrations eventually were to build up to the point of stopping algal growth and reproduction, it is conceivable that the oxygen consumption of an overpopulated world could eventually exceed the capacity of the remaining plant life to replenish it, with suffocation of nearly all animal life the grisly result. In actuality, however, starvation would be far more likely to occur first.

It has been estimated that all of the world's green plants combined produce between 100 billion and 200 billion metric tons (110 billion and 220 billion tons) of sugar each year. However, a great deal of that total is produced in the oceans, where it is largely unavailable for human use. In addition, much of the sugar produced by land plants is used in the formation of wood, fibers, and other inedible plant materials. The immediate products of photosynthesis may be converted to storage forms of carbohydrates such as starch and other polysaccharides (chains of simple sugars) and disaccharides (e.g., sucrose) and others that the digestive activities of both plants and animals break down to smaller molecules. Photosynthetic sugars are also involved in the production of amino acids for proteins and a host of other cell constituents. In fact, photosynthesis produces more than 94% of the dry weight substance of plants, with the remainder coming from the soil.

Note to the Reader: Photosynthesis is undoubtedly the most important process on earth to life as we know it. It is also a complex process that can be summarized briefly or examined in detail. The following is a discussion of the subject at three different levels: (1) the essence of photosynthesis; (2) a brief introduction to the major steps of photosynthesis; and (3) a closer look at photosynthesis. The process of respiration, which is discussed after photosynthesis, is treated in similar fashion. The third level, in particular, contains detail that either may or may not be discussed in your course.

1. The Essence of Photosynthesis

Photosynthesis is an energy-storing process that takes place in leaves and other green parts of plants in the presence of light. The light energy is stored in a simple sugar molecule that is produced from carbon dioxide present in the air and water absorbed by the plant. When the carbon dioxide and the water are combined and form a sugar molecule in a chloroplast, oxygen gas is released as a by-product. The oxygen diffuses out into the atmosphere. The process is summed up in the following equation:

$$6CO_2 + 12H_2O + \text{light energy} \xrightarrow[\text{enzymes}]{\text{chlorophyll}} C_6H_{12}O_6 + 6O_2 + 6H_2O$$

carbon dioxide · water · · · glucose · oxygen · water

Photosynthesis takes place in chloroplasts (see figures 3.4, 3.5, and 3.10).

Let us briefly examine each of the principals in the process: *carbon dioxide, water, light,* and *chlorophyll.*

Carbon Dioxide

Our atmosphere consists of approximately 79% nitrogen and about 20% oxygen. The remaining 1% is made up of a mixture of less common gases, including 0.03% carbon dioxide and a little hydrogen, helium, argon, and neon. The 0.03% carbon dioxide in the air surrounding the leaves reaches the chloroplasts in the mesophyll cells by diffusing through the stomata, which are well adapted to the absorption of gases, and by going into solution in the thin film of water on the outside walls of the cells. By diffusion, it then passes through the walls into the cytoplasm and reaches the chloroplasts.

The amount of carbon dioxide constantly being taken from the atmosphere during daylight hours by all green plants is enormous. Just four-tenths of a hectare (1 acre) of corn (10,000 plants) accumulates more than 2,500 kilograms (5,512 pounds) of carbon from the atmosphere during a growing season. Over 10 metric tons (11 tons) of carbon dioxide are needed to furnish this much carbon. It has also been calculated that the total present atmospheric supply of carbon dioxide (more than 2.2 billion metric tons or about 50 metric tons over each hectare of the earth's surface) would be completely used up in about twenty-two years if it were not constantly being replenished. A large reservoir of carbon dioxide in the oceans maintains the atmosphere's percentage of the gas at an ambient 0.03, while plant and animal respiration, the burning of material containing carbon, volcanoes, and similar sources replace it at roughly the same rate at which it is removed during photosynthesis.

There is some evidence that atmospheric carbon dioxide concentration is increasing; in fact, a 1970 study by scientists at the Massachusetts Institute of Technology on critical environmental problems predicted a 20% increase by the year 2000. This could actually benefit plants, providing the level does not climb beyond 0.1 to 0.2% (which is extremely unlikely). Increases in yields of between 100% and 200% have been obtained by "fertilizing" plants with carbon dioxide, and some large commercial greenhouses have pipes over plant beds to supplement their natural supply. However, the changes in climate that would result from an increase in atmospheric carbon dioxide could well offset positive benefits.

Water

Less than 1% of all the water absorbed by plants is used in photosynthesis; most of the remainder is transpired and incorporated into protoplasm and other materials. The water used is the source of the oxygen given off as a by-product of photosynthesis, even though carbon dioxide also contains oxygen. This has been demonstrated by using oxygen isotopes to make both the carbon dioxide and the water used in photosynthesis. When the isotope is used only in the water, it appears in the oxygen gas given off. If, however, it is used only in the carbon dioxide, it does not appear in the oxygen gas but is confined to the sugar. Thus, it has been clearly shown that the water is the sole source of this important by-product. If it is in short supply, water may indirectly become a limiting factor in photosynthesis; in this case, the stomata would close and sharply reduce the carbon dioxide supply.

Light

Light exhibits properties of both waves and particles. Energy reaches the earth from the sun in waves of different lengths, the longest waves being radio waves, and the shortest being X rays. About 40% of the radiant energy we receive is in the form of visible light. If this visible light is passed through a glass prism, it splits up into its component colors. Reds are on the longer wave end and violets on the shorter wave end, with yellows, greens, and blues between (figure 9.1). Although nearly all of the visible light colors can be used in photosynthesis, those in the blue-violet and red-orange wavelengths are used most extensively; many in the green range are reflected. Leaves commonly absorb about 80% of the visible light that reaches them.

The intensity of light varies with the time of day, season of the year, altitude, latitude, and atmospheric conditions. On a clear summer day at noon in a temperate zone, sunlight attains an intensity in the vicinity of 10,000 footcandles. (A footcandle is the light cast by a standard candle at a distance of 1 foot.) In contrast, consider that a good reading lamp produces only about 25 footcandles. Plants vary considerably in the light intensities they need for photosynthesis to occur at optimal (ideal) rates, and other factors such as the amount of carbon dioxide available are related. For example, an increase in photosynthesis will not occur in some plants receiving more than 3,000 footcandles of light unless supplemental carbon dioxide is provided. With supplements, however, photosynthesis rates will continue to increase up to about 5,000 footcandles. Herbaceous plants on a forest floor can survive with less than 2% of full daylight, and some mosses are reported to thrive on intensities as low as 0.05 to 0.01%. Most land plants that naturally grow in the open need at least 30% of full daylight to thrive, and the optimal amount for some species of trees approaches full daylight. Shade plants often do well in 10% of full daylight, and too intense light may change the way in which some of

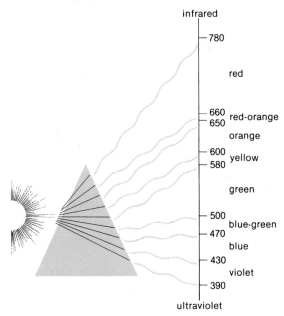

Figure 9.1 Visible light that is passed through a prism is broken up into individual colors with wavelengths ranging from 390 nanometers (violet) to 780 nanometers (red).

the metabolic activities of their cells takes place. For example, *photorespiration,* a type of respiration that depends on the rate of photosynthesis (see page 152), may occur in high light intensities. In this process, some of the chemical steps may differ from those that occur in the respiration of cells unaffected by the light. High light intensities may also cause an increase in transpiration, resulting in the closing of stomata. A sharp reduction in the available carbon dioxide supply inevitably follows.

Chlorophyll

There are several different types of chlorophyll molecules, all of which contain the element magnesium. They are otherwise similar in structure to hemoglobin, the red pigment in blood. Each molecule has a long lipid tail, which is believed to anchor the chlorophyll molecule in the lipid layers of the grana.

Chapter 9

Chloroplasts of most plants contain two kinds of chlorophyll associated with the thylakoid membranes. Chlorophyll *a* is bluish green in color and has the formula $C_{55}H_{72}O_5N_4Mg$. Chlorophyll *b* is yellowish green in color and has the formula $C_{55}H_{70}O_6N_4Mg$. Usually a chloroplast has about three times more chlorophyll *a* than *b*. When a molecule of chlorophyll *b* absorbs light, it transfers the energy to a molecule of chlorophyll *a*. Chlorophyll *b*, then, is an **accessory pigment;** that is, a pigment that makes it possible for photosynthesis to take place over a broader spectrum of light than would be possible with chlorophyll *a* alone. Other accessory pigments include carotenoids (yellowish to orange pigments found in all plants), phycobilins (blue or red pigments found in blue-green algae and red algae), and several other types of chlorophyll. Chlorophylls *c* and possibly *d* and *e* take the place of chlorophyll *b* in certain algae, and several other photosynthetic pigments are found in bacteria. The various chlorophylls are all closely related and differ from one another only slightly in the structure of their molecules.

In chloroplasts, about 250 to 400 pigment molecules are grouped as a **photosynthetic unit,** with countless numbers of these units in each granum. Two types of these photosynthetic units function together in the chloroplasts of green plants, bringing about the first phase of photosynthesis, the *light reactions* (discussed in the next section).

2. Major Steps of Photosynthesis

The process of photosynthesis takes place in two successive major steps called the **light reactions** and the **dark reactions.**

Light Reactions

In the light reactions, the first major phase of photosynthesis, light striking chlorophyll molecules that are embedded in the thylakoids of the grana in the chloroplasts (see page 33 and figure 3.10) initiates reactions that result in the conversion of some of the light energy to chemical energy. In the process, (1) water molecules are split apart, producing hydrogen ions and electrons, and oxygen gas is released; (2) energy-storing molecules commonly known as ATP are created;[1] and (3) the hydrogen from the split water molecules is involved in the creation of another substance known as $NADPH_2$ (nicotinamide adenine dinucleotide phosphate—more precisely shown as $NADPH + H^+$), which carries hydrogen and which is used in the second major phase of photosynthesis, the *dark reactions.*

Dark Reactions

The dark reactions are a series of reactions that take place outside of the grana in the stroma of the chloroplast (see page 33), regardless of whether or not light is present; the term *dark,* therefore, may be slightly misleading. The dark reactions may develop in different ways, depending on the particular kind of plant involved. In 1961, Dr. Melvin Calvin of the University of California received a Nobel prize for discovering how the most widespread type of dark reactions occur, and this process is now often referred to as the **Calvin cycle.**

In this cycle, carbon dioxide from the air is combined with a 5-carbon sugar, RuBP (ribulose bisphosphate), and then the combined molecules are converted, through several steps, to 6-carbon sugars. Energy and other items involved in these steps are furnished by the ATP molecules and $NADPH_2$ produced during the light reactions.

---

1. ATP stands for *adenosine triphosphate;* there are millions of ATP molecules in cells. When an ATP molecule releases its energy, it gives up one of three phosphorus "groups" and becomes ADP, or adenosine diphosphate. An ADP molecule becomes an ATP molecule again when it regains a third phosphorus "group" and stores energy.

Figure 9.2 Simple summary of photosynthetic reactions

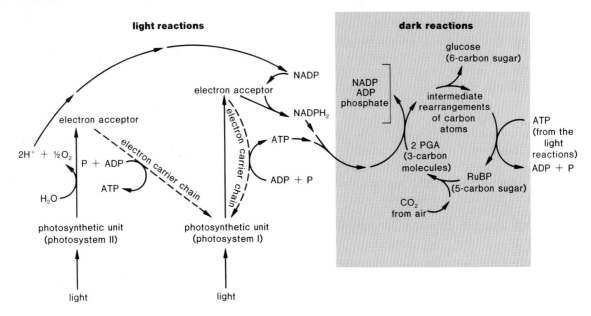

Some of the 6-carbon sugars that are produced during the dark reactions are recycled, while others are stored as starch or other polysaccharides (simple sugars strung together in chains). A simple summary of the photosynthetic reactions is portrayed in figure 9.2. More detailed diagrams of light and dark reactions are shown in figures 9.4 and 9.5.

Two molecules of a 3-carbon compound are shown as the first substance produced when carbon dioxide from the air and RuBP are combined and then converted during the dark reactions (figure 9.2). Many tropical plants produce a 4-carbon compound at this point instead. This 4-carbon pathway is discussed, along with a somewhat similar variation found in desert plants, in the next section.

3. A Closer Look at Photosynthesis

As indicated at the beginning of our discussion, photosynthesis is a very complex process. It involves several types of chemical reactions, including *oxidation-reduction reactions,* which are also vital to the process of respiration (discussed later in this chapter).[2]

A great deal has been learned about photosynthesis since Theodore de Saussure (1767–1845), a Swiss, first recognized and demonstrated that sunlit plants absorbed and utilized carbon from carbon dioxide and gave off oxygen, but much remains to be learned. The discussion that follows is a modest amplification of the details of the process. Those who wish more information are referred to the reading list at the end of the chapter.

---

2. Oxidation is the removal of electrons from a compound; *reduction* is the addition of electrons. When an electron is removed a proton may follow, with the result that hydrogen is often removed during oxidation and added during reduction. Oxygen usually is the oxidizing agent (i.e., the final acceptor of the electron), but oxidations can occur without oxygen being involved.

Chapter 9

Light Reactions Reexamined

We previously noted that light exhibits the dual characteristic of waves and particles. In the seventeenth century when Sir Isaac Newton first used a prism to produce a spectrum of colors from visible light, he postulated that light consisted of a series of discrete particles he called "corpuscles." His theory only partially explained light phenomena, however, and by the middle of the nineteenth century James Maxwell and others showed that light and all other parts of the vast electromagnetic spectrum travel in waves.

Before the beginning of the twentieth century, the wave theory of light also became inadequate to explain certain attributes of light, particularly in relation to the fact that a *photoelectric effect* can be produced in all metals. When a metal is exposed to radiation of a critical wavelength, either visible or invisible, it becomes positively charged because the radiation forces electrons out of the metal atoms. The ability of light to force electrons from metal atoms depends on its particular wavelength and not on its intensity or brightness. The shorter the wavelength, the greater the energy and vice versa.

When Einstein returned to the particulate nature of light in 1905 and proposed that light is composed of packets of energy he called **photons,** the wave theory of light was not discarded, however. Physicists gradually came to regard both theories of light as necessary for a complete description, and both waves and particles (photons) are now almost universally recognized as aspects of light. We also now recognize that the energy of a photon is not the same for all kinds of light; those of longer wavelengths have lower energy, and those of shorter wavelengths have proportionately higher energy.

As noted earlier, chlorophylls, the principal pigments of photosynthetic systems, absorb light primarily in the violet to blue and also in the red wavelengths; they reflect green wavelengths, which is why leaves appear green. Each pigment has its own distinctive pattern of light absorption;

Figure 9.3 Absorption spectrum of chlorophyll *a*

percentage of absorption

wavelength of visible radiation (in nanometers)

this is referred to as the *absorption spectrum* of the pigment (figure 9.3). When pigments absorb light, the energy levels of electrons are raised. When this occurs, the energy may be emitted immediately as light (a phenomenon called *fluorescence*). In chlorophyll, this is characteristically in the red part of the visible light spectrum, so that an extract of chlorophyll placed in light will appear red, or it may be emitted as light after a delay (a phenomenon called *phosphorescence*). The energy may otherwise be converted to heat, or it may be stored in a chemical bond as it is in photosynthesis.

Photosystems

The two types of photosynthetic units present in most chloroplasts collectively comprise **photosystems** designated as *photosystem I* and *photosystem II* (see figure 9.4). Each photosynthetic unit of photosystem I consists of approximately 200 molecules of chlorophyll *a;* about 50 molecules of

Figure 9.4 Summary of the light reactions of photosynthesis

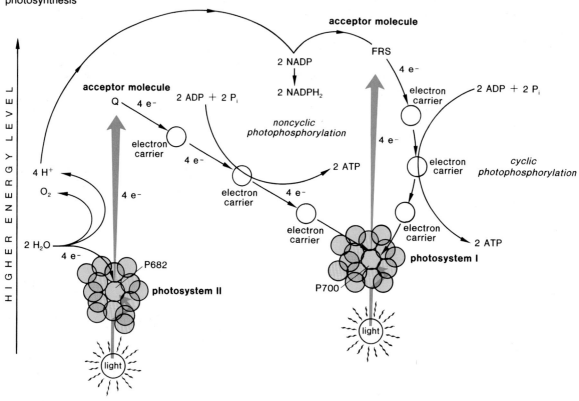

carotenoid pigment; and 1 special reactive, or "trap," molecule of chlorophyll *a* called P_{700}. A photosynthetic unit of photosystem II consists of approximately 200 molecules of chlorophyll *b* or other chlorophyll molecules; and 1 special reactive, or "trap," molecule of chlorophyll *a* called P_{682}. The numbers 700 and 682 of the "trap" molecules of chlorophyll *a* refer to peaks in the absorption spectra of light with wavelengths of 700 and 682 nanometers, respectively. These peaks differ slightly from those of the otherwise nearly identical chlorophyll *a* molecules of the photosynthetic units.

When a photon of light strikes a P_{682} molecule of a photosystem II photosynthetic unit, the energy boosts an electron out of place to a higher energy level. This is referred to as "exciting" an electron. The excited electron is picked up by the *acceptor molecule* Q. The departure of the electron leaves a "hole," which is filled by an electron from a special electron donor. After the special electron donor has donated four electrons, its own losses are replaced with electrons supplied by two molecules of water. As these water molecules are split, a molecule of oxygen and four protons are produced.

The acceptor molecule Q releases the excited electron to an electron transport system consisting of iron-containing pigments called *cytochromes* and other transfer molecules that pass the electron along to an electron donor for photosystem I. While the excited electron passes along the electron transport system, it loses some of its energy, a little at a time; this energy is stored in ATP molecules.

Chapter 9

A somewhat similar series of events occurs in photosystem I. When a photon of light strikes a P_{700} molecule in a photosynthetic unit, the energy excites an electron, which is transferred to an acceptor molecule of ferrodoxin-reducing substance (FRS). The excited electron may then follow one of two pathways: (1) it may be passed along an electron transport system in a cycle and returned to the P_{700} molecule, with ATP being generated in the process (this is called *cyclic photophosphorylation*), or (2) it may be passed via an intermediate compound to an important substance called *nicotinamide adenine dinucleotide phosphate* (NADP). Electrons from the initial splitting of water molecules in photosystem II are passed through the system, forming $NADPH_2$, and oxygen is released. $NADPH_2$ plays an important role in the second major stage of photosynthesis, the dark reactions. *Photophosphorylation* (making ATP using the energy of light) involving a flow of electrons in one direction from water to photosystem II to photosystem I to NADP is called *noncyclic*. A summary of the light reactions in the most widespread form of photosynthesis is shown in figure 9.4.

Dark Reactions Reexamined

We have seen how ATP and $NADPH_2$ are synthesized during the light reactions. Both of these substances play key roles in the synthesis of carbohydrate from carbon dioxide during the dark reactions. As indicated earlier, the dark reactions are really a whole series of reactions, each mediated by an enzyme in this major phase of photosynthesis. The reactions take place in the stroma of the chloroplasts and need no light to occur, although they normally take place during daylight hours.

There are three known mechanisms through which carbon dioxide is converted to carbohydrate during the dark reactions. The most widespread is the *Calvin cycle,* or *3-carbon pathway* (figure 9.5).

Figure 9.5 The Calvin cycle (dark reactions) of photosynthesis

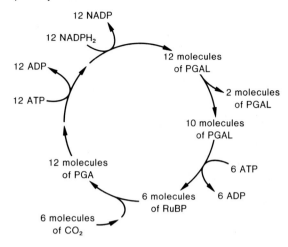

The main steps of the Calvin cycle (dark reactions) are as follows:

1. Six molecules of carbon dioxide from the air combine with six molecules of ribulose 1, 5-bisphosphate (RuBP, the 5-carbon sugar continually being formed while photosynthesis is occurring).
2. The resulting six 6-carbon molecules are immediately split into twelve 3-carbon molecules known as 3-phosphoglyceric acid (PGA).
3. The $NADPH_2$ (which has been temporarily holding the hydrogen ions and electrons released during the light reactions) and ATP (also from the light reactions) supply energy to convert the PGA to twelve molecules of phosphoglyceraldehyde (PGAL), a 3-carbon sugar-phosphate. These PGAL molecules are the stable end products of photosynthesis.
4. Ten of the twelve PGAL molecules are restructured and become six 5-carbon molecules of RuBP, the sugar with which the Calvin cycle was initiated.

The net gain in PGAL molecules can result either in an increase in the carbohydrate content of the plant (glucose, starch, cellulose, or related substances), or the PGAL can be used in pathways that lead to the net gain of lipids and amino acids.

4-Carbon Pathway

Plants of at least one hundred genera that occur primarily in the tropics produce a 4-carbon compound, oxaloacetic acid, instead of 3-carbon PGA during the initial steps of the dark reactions. Oxaloacetic acid is produced when a 3-carbon compound, phosphoenol pyruvate (PEP), and carbon dioxide combine with the aid of an enzyme, phosphoenol pyruvate carboxylase. Depending on the species, the PEP may then be converted to other 4-carbon acids, such as aspartate and malate. These organic acids are not, however, substitutes for PGA. They are converted to pyruvic acid and carbon dioxide. The pyruvic acid and ATP interact, forming additional PEP molecules, and the carbon dioxide combines with RuBP and is converted to PGAL in the Calvin cycle. Because the PEP system produces 4-carbon compounds, plants having this system are called 4-carbon, or C_4, plants, to distinguish them from plants that have only the 3-carbon, or C_3, system.

C_4 plants have several characteristic features. (1) They possess two forms of chloroplasts. There are large chloroplasts with few grana in the bundle sheath cells surrounding the veins. These large chloroplasts contain numerous starch grains and poorly developed grana. The much smaller, very numerous chloroplasts of the mesophyll cells usually lack starch grains and have well-developed grana. (2) High concentrations of PEP carboxylase and other enzymes are found in the bundle sheath cells surrounding the veins. This is significant because PEP carboxylase permits the conversion of carbon dioxide to carbohydrate at much lower concentrations than does RuBP carboxylase, the corresponding enzyme in the Calvin cycle. (3) The optimum temperatures for C_4 photosynthesis are much higher than those for C_3 photosynthesis, thus allowing C_4 plants to thrive under conditions that would adversely affect, or even kill, C_3 plants.

Obviously the C_4 pathway furnishes carbon dioxide to the Calvin cycle in a more roundabout way than does the C_3 pathway, but it should be noted that *photorespiration* (see page 146), which releases carbon dioxide and uses oxygen in the presence of light, occurs all the time the Calvin cycle is functioning. Both photorespiration (which reduces the efficiency of carbon fixation) and the Calvin cycle take place almost exclusively in the bundle sheath cells, where high concentrations of carbon dioxide and low concentrations of oxygen limit the amount of photorespiration that can occur. As a result, C_4 plants have photosynthetic rates that are two to three times higher than those of C_3 plants, not only because of the inhibition of photorespiration in the bundle sheath cells where RuBP carboxylase is functioning, but also because any carbon dioxide that escapes to the mesophyll cells can be fixed by the C_4 pathway there. In addition, PEP carboxylase is much more efficient in utilizing carbon dioxide at low concentrations than the RuBP carboxylase of the Calvin cycle.

CAM Photosynthesis

Crassulacean acid metabolism (CAM) photosynthesis is found in cacti, stonecrops, and many other succulent plants growing in regions of high light intensity. It is similar to C_4 photosynthesis in that 4-carbon compounds are produced during the dark reactions. In these plants, malic and isocitric acids accumulate in the chlorophyll-bearing tissues at night and are converted back to carbon dioxide during the day. A much larger amount of carbon dioxide can be converted to carbohydrate each day than would otherwise be possible, since the stomata of such plants generally admit very little carbon dioxide during the day because they are closed (closed stomata greatly reduce water loss). This arrangement permits the plants to function well under conditions of both limited water supply and high light intensity.

Respiration

The solar energy that is converted to chemical energy by the process of photosynthesis is stored in various organic compounds. If organic compounds such as wood or coal are burned, the energy is released very rapidly in the form of heat and light, and much of the useable energy is lost. Living organisms, however, "burn" their energy-containing compounds in numerous small steps that release tiny amounts of immediately useable energy, usually storing the released energy in ATP molecules and thus permitting available energy to be used more efficiently (table 9.1).

1. The Essence of Respiration

Respiration is an energy-releasing process that takes place in all living cells twenty-four hours a day, regardless of whether or not photosynthesis happens to be occurring at the same time in the cells. It is initiated in the cytoplasm and completed in the mitochondria. The energy is released from simple sugar molecules that are broken down during a series of steps controlled by enzymes. No oxygen is needed to initiate the process, but in **aerobic respiration** (the most widespread form of respiration) the process cannot be completed without oxygen gas; carbon dioxide and water are by-products. Aerobic respiration is summed up in the following equation:

Table 9.1 Summary comparison of photosynthesis and respiration

Photosynthesis	Respiration
1. Stores energy in sugar molecules	1. Releases energy from sugar molecules
2. Uses carbon dioxide and water	2. Releases carbon dioxide and water
3. Increases weight	3. Decreases weight
4. Occurs only in light	4. Occurs in either light or darkness
5. Occurs only in chlorophyll-containing cells	5. Occurs in all living cells
6. Occurs in chloroplasts (in eukaryotic cells)	6. Occurs in cytoplasm (glycolysis) and mitochondria (aerobic respiration)
7. Produces oxygen in green plants	7. Utilizes oxygen (aerobic respiration)
8. ATP produced through use of light energy	8. ATP produced with energy released from sugar

$$C_6H_{12}O_6 + 6O_2 + 6H_2O \xrightarrow{\text{enzymes}} 6CO_2 + 12H_2O + 36ATP$$

glucose oxygen water → carbon dioxide water

Anaerobic respiration and **fermentation** are two forms of respiration carried on by certain bacteria and other organisms in the absence of oxygen gas. These forms of respiration release much less energy than aerobic respiration; the two forms differ from one another in the manner in which hydrogen released from the glucose is combined with other substances (see discussion in a later section). Fermentation is very important industrially, particularly in the brewing industry. Two well-known forms of fermentation are illustrated by the following equations:

$$C_6H_{12}O_6 \xrightarrow{\text{enzymes}} 2C_2H_5OH + 2CO_2 + 2ATP$$

glucose   ethyl   carbon
               alcohol   dioxide

$$C_6H_{12}O_6 \xrightarrow{\text{enzymes}} 2CH_3CHOHCOOH + 2ATP$$

glucose   lactic acid

Energy released during respiration is stored in ATP molecules.

2. Major Steps of Respiration

Glycolysis

In all forms of respiration, the first major phase takes place in the cytoplasm and requires no oxygen gas (O_2). This phase, which is a series of small steps each controlled by an enzyme, is called **glycolysis**. During the process a small amount of energy is released, and some hydrogen atoms are removed from compounds derived from a glucose molecule. The essence of this complex series of steps is as follows:

1. In a series of reactions the glucose molecule becomes a fructose molecule carrying two phosphates.
2. This sugar (fructose) molecule is split into two 3-carbon fragments (PGAL).
3. Some hydrogen, energy, and water are removed from these 3-carbon fragments, leaving pyruvic acid (figure 9.6).

The energy needed to start the process of glycolysis is supplied by two ATP molecules. By the time pyruvic acid has been formed, however, four ATP molecules have been produced (for a net gain of two ATP molecules) from the energy released along the way; a great deal of the original energy in the glucose molecule remains in the **pyruvic acid.** The hydrogen ions and electrons released during the process are picked up and temporarily held by a hydrogen acceptor molecule called *nicotinamide adenine dinucleotide* (NAD). What happens to them next depends on the kind of respiration involved: *aerobic respiration,* true *anaerobic respiration,* or *fermentation.*

Aerobic Respiration

In aerobic respiration (the most common type of respiration), glycolysis is followed by two major stages: the *Krebs cycle* and the *electron transport chain.* Both stages occur in the mitochondria and involve many smaller steps, each controlled by enzymes (figure 9.6).

Krebs Cycle

The **Krebs cycle** was named after the British but German-born biochemist Hans Krebs, who received a Nobel prize in 1953 for his unraveling of many of the complex reactions that take place in respiration. This cycle is sometimes referred to as the *citric acid,* or *organic acid, cycle* because of the important role played by several organic acids during the process.

Before entering the Krebs cycle, the pyruvic acid produced by glycolysis loses carbon dioxide and is restructured to a 2-carbon acetyl group. This acetyl group combines with an acceptor called *coenzyme A* (CoA). This combination (called *acetyl CoA*) then enters the Krebs cycle, which is a series of organic acids that successively remove small amounts of energy and hydrogen until all the energy originally trapped in the glucose molecule has been transferred to energy storage compounds such as $NADH_2$, with most of the energy being transferred during the Krebs cycle. Carbon dioxide is produced as a by-product while the cycle is proceeding.

           Chapter 9

Figure 9.6   Summary of respiration

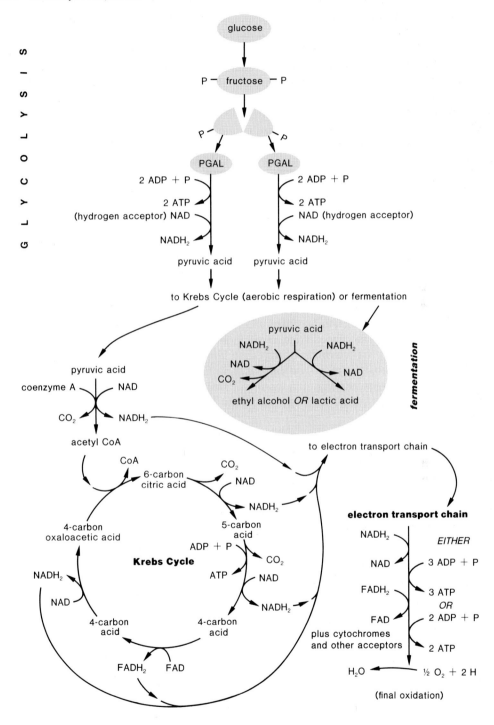

Metabolism in Plants

Electron Transport Chain

The hydrogen atoms removed during glycolysis and the Krebs cycle are picked up by the hydrogen acceptor molecule NAD (that becomes $NADH_2$), which passes the hydrogen atoms along the *electron transport chain*—a chain of special acceptor molecules (mostly iron-containing proteins called *cytochromes*)—and energy is released in small increments along the way. This energy is stored in ATP molecules. The hydrogen ions eventually are combined with oxygen gas, forming water. Thus, in the final step in aerobic respiration (see figure 9.6), oxygen acts as a "wastebasket" for hydrogen, and water is produced.

By the time the process is complete, the energy locked in a molecule of glucose has been released, with about 39% of it being stored in ATP molecules. This stored energy is then available for use in the synthesis of other molecules and for growth, active transport, and a host of other metabolic processes. Aerobic respiration produces a net gain of thirty-six ATP molecules from one glucose molecule, using up six molecules of oxygen and producing six molecules of carbon dioxide and twelve molecules of water. For each mole (180 grams) of glucose aerobically respired, 686 kcal of energy is released.

Anaerobic Respiration and Fermentation

Many bacteria carry on both fermentation and true anaerobic respiration simultaneously, making it difficult to distinguish between the two processes. Some texts use the terms *anaerobic respiration* and *fermentation* interchangeably to designate respiration occurring in the presence of little or no oxygen gas.

In many instances in the biological realm, glucose molecules may undergo glycolysis without sufficient oxygen being available to permit the completion of aerobic respiration. In such cases, the hydrogen released during glycolysis is simply transferred from the hydrogen acceptor molecules back to the pyruvic acid, after it has been formed, creating ethyl alcohol, lactic acid, or similar substances. A little energy is released during either fermentation or true anaerobic respiration, but most of it remains locked up in the alcohol, lactic acid, or other compounds produced.

In true anaerobic respiration, the hydrogen removed from the glucose molecule during glycolysis is combined with an inorganic ion, as is the case when sulphur bacteria (discussed in chapter 14) convert sulphate (SO_4) to sulphur or another sulphur compound or when certain cellulose bacteria produce methane gas (CH_4) by combining the hydrogen with carbon dioxide.

Oxygen gas is not required to make these compounds, but few organisms can live long without oxygen, and many that carry on fermentation can also respire aerobically. If oxygen becomes available, the remaining energy can be released by breaking these compounds down further. Only about 7% of the total energy in the glucose molecule is removed, and so much of that goes into the making of the alcohol or the lactic acid or is dissipated as heat that there is a net gain of only two ATP molecules (compared with thirty-six ATP molecules produced in aerobic respiration).

Living cells can tolerate only certain concentrations of alcohol. In media in which yeasts are fermenting sugars, for example, once the alcohol concentration builds up beyond 12%, the cells die and fermentation ceases. This is why most wines have an alcohol concentration of about 12%.

Factors Affecting the Rate of Respiration

Temperature

Temperature plays a major role in the rate at which the various respiratory reactions occur. For example, when air temperatures rise from 20°C (68°F) to 30°C (86°F), the respiration rates of plants double and sometimes even triple. The

Chapter 9

faster respiration occurs, the faster the energy is released from sugar molecules, with an accompanying decrease in weight. In growing plants, this weight loss is more than offset by the production of new sugar by photosynthesis. In harvested fruits, seeds, and vegetables, however, respiration continues without sugar replacement; some water loss also occurs. Because of this, most fresh foods are kept under refrigeration, not only to lower the respiration rate and water loss but also to dissipate the heat given off by respiring cells, which do not convert all the energy released into ATP but give off a certain amount in the form of heat as well. Keeping the temperatures down is also important to prevent the growth and reproduction of food-spoiling molds and bacteria, which may prefer warmer temperatures.

Heat inactivates most enzymes at temperatures above 35°C (95°F), but a few organisms such as various algae in the hot springs of Yellowstone National Park and similar places have adapted in such a way that they are able to thrive at temperatures of up to 60°C (140°F)—heat that will kill other organisms of comparable size almost instantly.

Water

Living cells often have a water content in excess of 90%, but mature seeds may have less than 10% water in their cells. When water content becomes this low, respiration does not cease, but it continues at a drastically reduced rate, resulting in only very tiny amounts of heat being released and of carbon dioxide being given off. Stored under dry conditions, seeds may remain viable (capable of germinating) for many years. If they come in contact with water, however, they swell by imbibition. Respiration rates then increase rapidly. If the wet seeds happen to be in an unrefrigerated storage bin, the temperature may increase to the point of killing the seeds. In fact, if fungi and bacteria begin to grow on the seeds, temperatures from the respiration can become so high that spontaneous combustion can sometimes occur.

Oxygen

If the roots of trees and houseplants have their oxygen supply sharply reduced due to flooding, their respiration rates may be decreased and their growth retarded; they may even die if the condition persists too long. In the case of stored foods, however, reducing the oxygen in the storage areas to bring about lower rates of respiration may be desirable, and consequently, it has become a common commercial practice to reduce the oxygen present in warehouses where crops are stored to as little as 1% to 3% by pumping in nitrogen gas, while maintaining low temperatures and humidity.

3. A Closer Look at Respiration

Respiration, like photosynthesis, is a very complex process, and, as with photosynthesis, it is beyond the scope of this book to explore the subject in great detail. The following amplification of information already discussed is modest, and those who wish further information are referred to the reading list at the end of the chapter.

Glycolysis Reexamined

As previously shown, this initial phase of all forms of respiration brings about the conversion of each 6-carbon glucose molecule to two 3-carbon pyruvic acid molecules via several small steps, each mediated by an enzyme. Energy needed to initiate the process is furnished by two ATP molecules, and as a result of the controlled energy release that takes place during the process, there is a net energy gain of two ATP molecules. Two $NADH_2$ (more precisely written $NADH + H^+$) molecules also are produced from each glucose molecule as it is converted to two pyruvic acid molecules, but much energy remains in the pyruvic acid.

The glucose molecule is prepared for the reaction by the addition of two phosphate groups furnished by ATP molecules. This restructured glucose molecule (now fructose 1,6-biphosphate) is split into two 3-carbon molecules, both of which eventually become phosphoglyceraldehyde (PGAL). Hydrogen protons and electrons are removed from the PGAL; these are picked up by a hydrogen acceptor molecule, nicotinamide adenine dinucleotide (NAD). After removal of the phosphate groups and water, **pyruvic acid** remains. Glycolysis, which requires no oxygen gas, is summarized in figure 9.6.

Krebs Cycle Reexamined

Before a pyruvic acid molecule enters the Krebs cycle, a molecule of carbon dioxide is removed and a molecule of $NADH_2$ is produced, leaving an acetyl fragment. The 2-carbon fragment is then bonded to a large molecule called *coenzyme A*. Coenzyme A consists of a combination of the B vitamin pantothenic acid and a nucleotide. The bonded acetyl fragment and coenzyme A molecule is referred to as *acetyl CoA*. The following equation summarizes the fate of the two pyruvic acid molecules produced during glycolysis:

2 pyruvic acid + 2 CoA + 2 NAD $\longrightarrow$

2 acetyl CoA + 2 $NADH_2$ + 2 CO_2

In addition to pyruvic acid, fats and amino acids can also be converted to acetyl CoA and enter the process at this point.

The $NADH_2$ molecules donate their hydrogen to an electron transport chain (discussed in the next section), six ATP molecules are formed, and the acetyl CoA enters the Krebs cycle (figure 9.6).

In the Krebs cycle, acetyl CoA is first combined with oxaloacetic acid, a 4-carbon compound, producing citric acid, a 6-carbon compound. The cycle then progresses in circular fashion. A carbon dioxide is soon removed, producing a 5-carbon compound. Then another carbon dioxide is removed, producing a 4-carbon compound. This 4-carbon compound, through additional steps, is converted back to oxaloacetic

acid, the substance with which the cycle began, and the cycle is repeated. Each full cycle uses up an acetyl group and regenerates an oxaloacetic acid molecule for the next turn of the cycle. Some hydrogen is removed during the process and is picked up by a flavoprotein acceptor molecule, flavin adenine dinucleotide (FAD). One molecule of ATP, three molecules of $NADH_2$, and one molecule of $FADH_2$ are produced for each turn of the cycle. The Krebs cycle may be summarized as follows:

oxaloacetic acid + acetyl CoA + ADP + P + 3 NAD + FAD $\longrightarrow$ oxaloacetic acid + CoA + ATP + 3 $NADH_2$ + $FADH_2$ + 2CO_2

Electron Transport Chain and Oxidative Phosphorylation

After completion of the Krebs cycle, the glucose molecule has been totally dismantled and some of its energy has been transferred to ATP molecules. A considerable portion of the energy remains, however, with the electrons that were picked up by the NAD and FAD as the glucose molecule was dismantled during glycolysis and the Krebs cycle. This energy is released as the electrons are passed along an *electron transport chain* in a process known as *oxidative phosphorylation.*

Most of the electron carriers in the system are cytochromes. They are very specific and, as electrons flow along the chain, they can transfer their electrons only to other specific acceptors. The cytochromes transfer the energy to ATP molecules. If hydrogen and electron transport begins with $NADH_2$, sufficient energy is produced to yield three ATP molecules from each $NADH_2$. Similarly, if hydrogen and electron transport begins with $FADH_2$, two ATP molecules are produced. When the electrons reach the end of the chain, they are picked up by oxygen, and combine with hydrogen ions, forming water.

If we retrace our steps through the entire process of aerobic respiration, we find that glycolysis yields two molecules of ATP and six molecules of $NADH_2$ (from which six more molecules of ATP are formed), for a total of eight ATP. Two of the ATP are "used" in the process, however, leaving a net gain of six ATP.

When pyruvic acid is converted to acetyl CoA, two more $NADH_2$ molecules (which are converted to six molecules of ATP) are produced, and the Krebs cycle yields two molecules of ATP, two molecules of $FADH_2$, and six molecules of $NADH_2$, making a total of twenty-four ATP and producing a grand total of thirty-six ATP for the aerobic respiration of one glucose molecule. The thirty-six ATP molecules represent about 39% of the energy originally present in the glucose molecule. The remaining energy is lost as heat or is unavailable to the process of aerobic respiration.

Digestion and Assimilation

After sugars are produced through photosynthesis they may undergo many transformations. Some sugars are used directly in respiration, but others not needed for that purpose may be transformed into lipids, proteins, or other carbohydrates. Among the most important carbohydrates produced from simple sugars are sucrose, starch, and cellulose. When photosynthesis is occurring in a leaf, sugar may be produced faster than it can be used or transported away to other parts of the plant. When this happens, the excess sugar may be converted to starch, temporarily stored in the chloroplasts, and then later changed back to a soluble form that is transported out of the leaf. The conversion of starch and other insoluble carbohydrates to soluble forms is called **digestion** (figure 9.7). The process is nearly always one of **hydrolysis,** in which water is taken up and, with the aid of enzymes, the links of the chains of simple sugars that comprise the molecules of starch and similar carbohydrates are broken. Ordinary starch, for example, is transformed to glucose with the aid of two enzymes and water as follows:

Figure 9.7 Starch grains. *A.* Before digestion. *B.* After partial digestion.

$$2(C_6H_{10}O_5)n + (H_2O)n-1 \xrightarrow[\text{(enzyme)}]{\text{amylase}} (C_{12}H_{22}O_{11})n \longrightarrow C_{12}H_{22}O_{11} + H_2O \xrightarrow[\text{(enzyme)}]{\text{maltase}} 2C_6H_{12}O_6$$

starch   water   maltose   maltose   water   glucose

Metabolism in Plants

In the preceding equation, *n* stands for a specific quantity, just as *x* stands for an unknown quantity in algebra. In the case of the starch, numerous simpler molecules are linked together. The *n* − 1 indicates that one less water molecule than the number of molecules in the starch chain is involved in the hydrolysis.

Fats are broken down to their component fatty acids and glycerol, and proteins to their amino acid building blocks in similar fashion. Digestion is carried on in any cell where there may be stored food, with very little energy being released in the process. In animals, special digestive organs also play a role in digestion, but plants have no such additional "help" in the process.

The process of digestion in plants and animals is similar except that in plants it takes place within the cells where the carbohydrate is stored, whereas in animals it usually occurs outside of the cells in the digestive tract.

Much of the sugar produced through photosynthesis is eventually used in the building of protoplasm and cell walls. This conversion process is called *assimilation.* Energy released through respiration is used in each building step, not only in the formation of fats, proteins, and complex carbohydrates but also in the manufacture of such products as resins, gums and oils, drugs, pigments, acids, and tannins. The capacity to assimilate is a fundamental attribute of living organisms, but much is still not known about the chemical changes involved in the process.

Summary

Photosynthesis is an energy-storing process that combines carbon dioxide and water in the presence of light with the aid of chlorophyll; oxygen is given off as a by-product. Virtually all life depends on photosynthesis, which takes place in the chloroplasts of both terrestrial and aquatic organisms. The carbon dioxide for the process occurs as a constant 0.03% of the atmosphere; less than 1% of water absorbed by a plant is used in photosynthesis, which utilizes light of primarily blue-violet and red-orange wavelengths. All types of chlorophyll contain a magnesium atom. Chlorophyll *a* is the principal chlorophyll pigment present in most leaves; chlorophyll *b* is an accessory pigment. An accessory pigment makes it possible for photosynthesis to occur over a broader spectrum of light than is possible with chlorophyll *a* alone. Other accessory pigments include chlorophylls *c, d,* and *e;* carotenoids; and phycobilins. In chloroplasts various pigment molecules are grouped into photosynthetic units that function in the light reactions, the first major phase of photosynthesis.

During the light reactions of photosynthesis, water molecules are split and oxygen gas is released, energy-storing ATP and $NADPH_2$ molecules are created, and hydrogen ions and electrons are released. In the dark reactions, carbon dioxide is combined with RuBP, and this combination is converted to 6-carbon sugars. The ATP and $NADPH_2$, produced during the light reactions, furnish the energy and materials needed for the dark reactions. The first substance produced in the dark reactions is usually a 3-carbon compound, but some tropical and desert plants produce a 4-carbon compound instead.

A closer look at photosynthesis reveals that there are two types of photosynthetic units, collectively comprising photosystems I and II, which are in most chloroplasts. Each photosystem contains a special reactive, or "trap," molecule of chlorophyll *a* that reacts to the energy of a photon of light by boosting an electron out of place to a higher energy level. In photosystem II, this "excited" electron is picked up by Q, an electron acceptor molecule, which releases it to an electron transport system consisting of cytochromes and other molecules; these pass it along to a special transfer molecule. The excited electron leaves a "hole" that is filled by an electron from an electron donor. The donor's losses are replaced with electrons from water molecules that are split, releasing oxygen. As an electron passes along the transport chain it loses small increments of energy, which are stored in ATP molecules. A similar series of events occurs in photosystem I, where the excited electron is either returned to the special reactive molecule (cyclic photophosphorylation) or passed on to NADP, which forms $NADPH_2$ with electrons from the original splitting of water molecules in photosystem II (noncyclic photophosphorylation).

In the dark reactions, which take place in the stroma of chloroplasts, the combination of RuBP and carbon dioxide is initially split into PGA molecules, then PGAL is formed with phosphorus from ATP. Ten of the twelve PGAL molecules are converted back to RuBP; the remaining two PGAL molecules are the basis for an increase in the carbohydrate content or net gain in lipids or amino acids of a plant. In plants that have a 4-carbon pathway, 4-carbon oxaloacetic acid is initially produced instead of 3-carbon PGA in the dark reactions. Four-carbon plants have large chloroplasts in the bundle sheaths and small chloroplasts in the mesophyll of their leaves, which contain higher concentrations of PEP carboxylase than those of C_3 plants. The high PEP carboxylase concentrations permit the conversion of carbon dioxide to carbohydrate at much lower concentrations than is possible in C_3 plants. In desert plants that have CAM photosynthesis, the 4-carbon compounds that accumulate in the chlorenchyma tissues at night are converted back to carbon dioxide during the day; this process is advantageous to plants whose stomata are closed during the day and thus admit little carbon dioxide.

Respiration is an energy-releasing process that takes place in the mitochondria and cytoplasm of all living cells. The energy is released from simple sugar molecules with the aid of enzymes. In aerobic respiration, oxygen is needed to release all the stored energy; carbon dioxide and water are by-products of the process. Anaerobic respiration and fermentation do not require oxygen gas, and much less energy is

released. The remaining energy is in the ethyl alcohol, lactic acid, or other such substances produced. Released energy is stored in ATP molecules. Temperatures, available water, and the oxygen content of the environment all affect respiration rates.

During the first step of respiration, glycolysis (which is common to all forms of respiration and requires no oxygen gas), two phosphates are added to a 6-carbon sugar molecule and the prepared molecule is split into two 3-carbon sugars (PGAL). Some hydrogen, energy, and water is removed from the PGAL, leaving pyruvic acid. There is a net gain of two ATP molecules. Hydrogen ions and electrons released during glycolysis are picked up by NAD, a hydrogen acceptor molecule, which becomes $NADH_2$.

In aerobic respiration, the pyruvic acid loses carbon dioxide and is restructured and combined with coenzyme A, becoming acetyl CoA. This enters the Krebs cycle, a cycle of organic acids in which energy, carbon dioxide, and hydrogen are removed. $NADH_2$ passes the hydrogen gained during glycolysis and the Krebs cycle along an electron transport chain; during the process small increments of energy are released and stored in ATP molecules; and the hydrogen is combined with oxygen gas, forming water in the final step of aerobic respiration (final oxidation).

Anaerobic respiration differs from fermentation in that hydrogen removed from glucose during glycolysis is combined with an inorganic ion in the anaerobic process, whereas in fermentation, the hydrogen is simply combined with the pyruvic acid or one of its derivatives; both processes occur in the absence of oxygen gas, with only about 7% of the total energy in the glucose molecule being removed, for a net gain of two ATP molecules.

A closer look at respiration reveals that two molecules of $NADH_2$ and two ATP molecules are gained during glycolysis when two 3-carbon pyruvic acid molecules are produced from a single glucose molecule. Another molecule of $NADH_2$ is produced when the pyruvic acid is restructured and combined with acetyl CoA prior to entry into the Krebs cycle. In the Krebs cycle, acetyl CoA combines with oxaloacetic acid, a 4-carbon compound, to produce first a 6-carbon compound, next a 5-carbon compound, and then several 4-carbon compounds. The last 4-carbon compound, after several steps, is oxaloacetic acid. Some hydrogen removed during the process is picked up by FAD, an acceptor molecule, and NAD; one molecule of ATP, three molecules of $NADH_2$, and one molecule of $FADH_2$ are produced during one complete cycle. Energy associated with electrons and/or with hydrogen picked up by NAD and FAD is gradually released as the electrons are passed along the electron transport chain during oxidative phosphorylation, and the energy is transferred to ATP molecules. Thirty-eight ATP molecules are produced altogether during the complete aerobic respiration of one glucose molecule, but two are used to "prime" the process so that there is a net gain of thirty-six ATP molecules.

Digestion takes place within plant cells with the aid of enzymes. During digestion large insoluble molecules are broken down by hydrolysis to smaller soluble forms that can be transported to other parts of the plant. The conversion of the sugar produced by photosynthesis to fats, proteins, complex carbohydrates, and other substances is termed assimilation.

Review Questions

1. What happens in the light reactions of photosynthesis?

2. What roles do water, light, carbon dioxide, and chlorophyll play in photosynthesis?

3. What is glycolysis?

4. What are the differences among aerobic respiration, anaerobic respiration, and fermentation?

5. How do temperature, water, and oxygen affect respiration?

6. Explain digestion and assimilation.

Discussion Questions

1. Since plants apparently benefit from small increases in the carbon dioxide available to them, would you suppose you would be doing your favorite houseplant a favor by breathing repeatedly on it?

2. If the dark reactions of photosynthesis take place in the light as well as in the dark, how did they come to be so named?

Additional Reading

Bjorkman, O.; Berry, J. High-efficiency photosynthesis. *Scientific American* 229:80–93; 1973.

Fogg, G. E. *Photosynthesis.* New York: American Elsevier Publishing Co.; 1972.

Gabriel, M. L.; Fogel, S. *Great experiments in biology.* Englewood Cliffs, NJ: Prentice-Hall; 1955.

Galston, A. W.; Davies, P. J.; Satter, R. L. *The life of the green plant.* 3d ed. Englewood Cliffs, NJ: Prentice-Hall; 1980.

Hall, D. O.; Rao, K. K. *Photosynthesis.* 3d ed. Baltimore: University Park Press; 1981.

Hatch, M. C.; Osmond, C. B.; Slatyer, R. O., eds. *Photosynthesis and photorespiration.* New York: Wiley-Interscience; 1971.

Lahetta, W. G. *Chemistry of respiration.* Lorain, OH: Dayton Labs; 1978.

Lehninger, A. L. *Biochemistry: the molecular basis of all structure and function.* 2d ed. New York: Worth Publishers, Inc.; 1975.

Levine, R. P. The mechanisms of photosynthesis. *Scientific American* 221:58–70; 1969.

Levitt, J. *Introduction to plant physiology.* 2d ed. St. Louis: C. V. Mosby Co.; 1974.

Meidner, H.; Mansfield, T. A. *Physiology of stomata.* New York: McGraw-Hill Book Co.; 1969.

Öpik, H. *The respiration of higher plants.* Baltimore: University Park Press; 1982.

Whittingham, C. *The mechanism of photosynthesis.* Baltimore: University Park Press; 1977.

Zelitch, I. *Photosynthesis, photorespiration and plant productivity.* New York: Academic Press; 1971.

10 Growth

Overview

This chapter introduces growth phenomena with a discussion of the distinctions among growth, differentiation, and development. This is followed by a brief introduction to seed germination and a discussion of plant hormones (auxins, gibberellins, cytokinins, abscisic acid, ethylene) and their roles in plant growth and development. A condensed view of mineral requirements for growth comes next.

The chapter explores plant movements, including spiraling, twining, contraction, nastic, tropisms, turgor, taxes, and miscellaneous other movements. The discovery and functions of phytochrome and photoperiodism are briefly surveyed. The chapter concludes with a discussion of the relationship of temperature to growth and dormancy.

Some Suggested Student Learning Goals

1. Define growth, differentiation, development, hormone, essential element. List the essential elements.

2. Identify the types of plant hormones and describe the major functions of each; list commercial applications for each.

3. Distinguish among the various types of plant movements and indicate the forces behind them.

4. Describe photoperiodism and make distinctions among short-day, long-day, intermediate-day, and day-neutral plants.

5. Explain what phytochrome is and how it functions.

6. Discuss the role of temperature in plant growth.

7. Define dormancy and stratification, and give examples.

Introduction

The word **growth** is used in a variety of contexts. If, for example, you observe a rubber balloon being inflated with gas, you may refer to its gradual increase in diameter as its growth in size; or if there is a leak in the roof you might remark that the puddle beneath is growing bigger. In the biological realm, growth is always associated with cells. It is defined simply as an "increase in volume due to the division and enlargement of cells," and may be applied to an organism as a whole or to any of its parts. As mentioned at the beginning of chapter 3, all living organisms begin as a single cell, which usually divides and keeps dividing until an entity consisting of possibly billions of cells is formed. As the various cells mature, they usually become larger and then *differentiate;* that is, they assume different forms adapted to specific functions such as conduction, support, or secretion of special substances. If **differentiation** did not occur, the result would be a shapeless blob with no distinct tissues and little, if any, coordination.

The term **development** is applied to change in form as a result of both growth and differentiation combined. What brings about growth and differentiation, and how do they occur? We already know they are controlled by the genes of the chromosomes, which not only dictate the kinds of cells that will develop but also their proportionate numbers and the ultimate size of the organism. The sizes of the cells themselves are limited by several factors. They must rely on molecular diffusion within for the transport of oxygen and food and for the removal of wastes, since they have no circulatory organs of their own. As a spherical cell enlarges, the total area of its surface, through which substances enter and leave, does not increase at the same rate as its volume. For example when a cell's surface area has increased fourfold, its volume has increased roughly six times, obviously putting a great strain on its capacity to obtain and exchange materials through its surface. Most plant cells have become adapted to the problem through various changes in shape (e.g., development of a convoluted outer surface; flattening of the cell) that increase the surface area and also through the development of large vacuoles, with the cytoplasm becoming a thin layer adjacent to the cell wall. These adaptations alone, however, do not increase the strength of the bounding membranes, and most cells divide before their enlarging becomes a significant problem.

Each metabolic step within cells is under the control of enzymes, whose development is likewise dictated by genes. In addition, many growth phenomena are influenced by **hormones.** These are chemical substances produced, usually in minute quantities, in one part of an organism and transported to another part where they have specific effects, such as causing stems to bend or initiating flowering. They should not be confused with enzymes, which are also produced in minute quantities. Enzymes are organic catalysts that speed up chemical reactions in the cells where they are produced; they are *not* transported from one part of an organism to another, and their chemical structure is different from that of hormones. **Vitamins** are organic molecules of varied structure that function in activating enzymes. They are essential in relatively small amounts for the normal growth and development of all organisms. Because some of their effects are similar to those of hormones, they are sometimes difficult to distinguish from hormones. Some scientists have suggested using the term *growth-regulating substances* for both hormones and vitamins to avoid having to distinguish between the two.

Seed Germination

In order for a seed to germinate, water must first be imbibed; changes in the embryo within the seed then begin to occur. First, a **radicle** (embryo root) emerges, followed by the **shoot.** How the shoot develops depends on the kind of seed concerned. In garden beans, for example, two **cotyledons** ("seed leaves") are attached to the embryo stem.

Figure 10.1 Germination of a bean seed

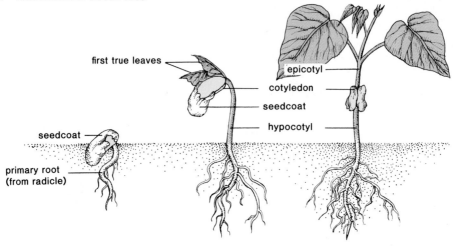

first true leaves
epicotyl
cotyledon
seedcoat
hypocotyl
seedcoat
primary root
(from radicle)

Figure 10.2 Germination of a corn grain

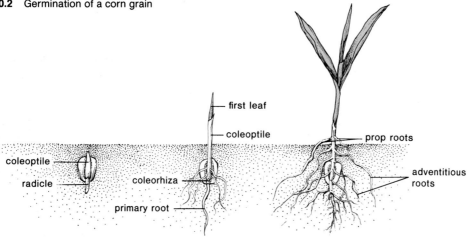

first leaf
coleoptile
prop roots
coleoptile
adventitious roots
radicle
coleorhiza
primary root

The part of the stem below the cotyledons, the **hypocotyl,** elongates and becomes arched. This arch pushes up above the ground and straightens out, carrying the cotyledons into the air. Then the **plumule,** which is an embryo bud consisting of the first true leaves, and the **epicotyl** (the part of the stem above the point of attachment of the cotyledons), begin to grow (figure 10.1; see also figure 21.21). In the scarlet runner bean, lima bean, and garden pea, the hypocotyl remains short so that the cotyledons stay beneath the surface while the epicotyl elongates.

Members of the grass family, of which corn is an example, have only one cotyledon, which also remains underground after germination (figure 10.2; see also figure 21.22). In addition, as these plants grow, the radicle pushes through a protective sheath, the **coleorhiza,** and the shoot emerges through a protective sheath, the **coleoptile.**

Chapter 10

Figure 10.3 Went's experiment with oat coleoptiles. *A.* Oat coleoptile. *B.* The tip of a coleoptile is cut off, placed on a small agar block, and left for an hour or two. *C.* If the agar block is then placed squarely on a decapitated coleoptile, growth is vertical. *D.* If, however, the block is placed off center, the tip bends away from the block, demonstrating that a substance affecting growth has diffused from the coleoptile tip into the agar and from the agar to the decapitated coleoptile.

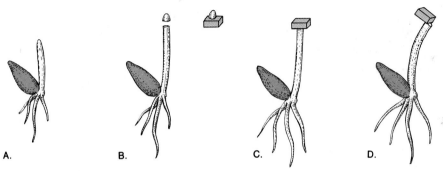

A. B. C. D.

Plant Hormones

In 1881 Charles Darwin and his son Francis were fascinated by the fact that coleoptiles bend toward a light source. They noticed that the bending did not occur if they placed little caps over the coleoptile tips and concluded that this part of the plant must be sensitive to light. In 1926 a young Dutch plant physiologist named Frits Went followed up the investigations of the Darwins and others by cutting off the tips of oat coleoptiles and placing the tips, cut surface down, on flat portions of **agar** (a gelatinlike substance obtained from marine algae). After a few hours, he removed the tips and cut the agar into little square blocks, which he then placed on decapitated coleoptiles. He found that if he placed a little block squarely over the cut top of a coleoptile, it grew straight up, but if he placed the block off center, the tip bent away from the side of the block (figure 10.3). He also found that the more coleoptile tips he placed on the agar to begin with, the more pronounced was the response. His experiments demonstrated conclusively that something produced in the coleoptile tips moved out into the agar and that this substance was responsible for the bending observed. Went named the substance **auxin** (from the Greek word *auxein,* meaning "to increase"), a term now in widespread use.

Auxins were the first plant hormones to be discovered; later several others were found. At least three major groups apparently promote the growth of plants. Some may also have an inhibitory effect, depending on the concentration. Other hormones and growth regulatory substances are known only for their growth-inhibiting features. All are produced in minute quantities and must be replenished, as they are constantly being used up during metabolic activities of cells. The following are the major known groups of hormones: *auxins, gibberellins, cytokinins, abscisic acid,* and *ethylene.*

Auxins

Auxins are produced by living protoplasm mainly in buds, young leaves, and other active young parts of plants. They stimulate the enlargement of cells, cause the Golgi bodies (dictyosomes) to increase the rates of secretion, play a role in controlling some phases of respiration, and are involved in numerous developmental aspects of growth, including flower initiation, sex determination, growth rate, fruit growth and ripening, abscission, rooting, aging, dormancy, and apical dominance (see the discussion of apical dominance on page 170). The effects of auxins combined with those of other hormones produce many of the growth phenomena.

Movement of auxins from the cells where they originate is *polar;* that is, they move away from their source, usually downward from a stem tip. They do not seem to be carried through the "plumbing" tubes but proceed from cell to cell, being carried faster by active transport than they would be by simple diffusion (see chapter 8). It was once believed that there were several naturally occurring auxins. It appears now, however, that there is only one active auxin, *indoleacetic acid,* usually referred to as IAA, and that other organic acids that act like auxins are actually converted to IAA in the plant where they perform their hormonal functions.

IAA and a number of organic acids that regulate growth have been produced synthetically. They are in widespread use in agriculture and horticulture. At the proper concentrations, IAA and other growth regulators stimulate the formation of roots on almost any plant organ. Accordingly, as noted in appendix 4, nurseries apply these substances in paste or powder form to the bases of cut pieces of stems to stimulate a much more vigorous production of roots than would occur naturally (figure 10.4).

Figure 10.4 Effect of auxin applied to the base of a *Gardenia* cutting four weeks after application. *Left:* Treated cutting. *Right:* Untreated cutting.

Chapter 10

Orchardists spray fruit trees with growth regulators to promote uniform flowering and fruit set, and they later spray the fruit to prevent the formation of abscission layers and subsequent premature fruit drop. A substantial saving in labor expenses results from being able to go through an orchard or a pineapple plantation and pick all of the fruit at one time. If growth regulators are applied to flowers before pollination occurs, seedless fruit can be formed and developed. Some orchardists use growth regulators for fruit thinning; that is, they control the number of fruits that will mature in order not to have too many small ones, and some even control the shapes of plants for sales appeal.

Other types of growth-regulating substances such as 2,4-D, 2,4,5-T, and 2,4,5-TP have been used as *herbicides* (vegetation killers). Prior to their discovery, weeds were controlled by hand labor and by the use of caustic or otherwise poisonous substances. Then it was found that these hormonelike compounds, when sprayed at low concentrations, kill some weeds. Broad-leaved plants like dandelions, plantains and others (for reasons that are not clearly understood) are more susceptible to low concentrations of 2,4-D and 2,4,5-T than narrow-leaved plants like grasses. Such growth regulators, therefore, have been used on lawns to kill weeds without noticeably affecting the grass.

Thus far there are no known directly adverse effects of 2,4-D on humans and other animal life, but such is not the case with 2,4,5-T, which was banned in 1979 by the Environmental Protection Agency for most uses in the United States. The controversy over the use of 2,4,5-T began in the 1960s during the Vietnam war when Agent Orange, a 1-to-1 mixture of 2,4-D and 2,4,5-T, was used to defoliate jungles. Subsequent tests and experiments with 2,4,5-T have produced leukemia, miscarriages, birth defects, and liver and lung diseases in laboratory animals. The diseases and defects are apparently caused by TCDD (2,3,7,8-tetrachlorodibenzoparadioxin), a dioxin contaminant that evidently is unavoidably produced in minute amounts during the manufacture of 2,4,5-T. TCDD kills laboratory animals in doses as low as a few parts per billion.

Gibberellins

In 1926 a Japanese scientist named E. Kurosawa reported the discovery of a new substance that was causing the "foolish seedling" effect on rice. Rice seedlings infected with a certain fungus had stems that grew twice as long as those of uninfected plants. Kurosawa found that extracts of the fungus applied to uninfected plants brought about the same growth stimulations as the fungal disease, but it was not until nine years later that another Japanese scientist was able to purify the substance, which he named **gibberellin** after the scientific name of the fungus that produced it. Today more than forty different gibberellins have been isolated from fungi of higher plants. Interest in their hormonal properties is high, for their ability to increase growth in plants is far more impressive than that of auxin alone, although traces of auxin apparently need to be present for gibberellins to produce their maximum effects. If a gibberellin, at the appropriate concentration, is applied to a cabbage, the plant may become 2 meters (6 feet) tall, and bush bean plants can become pole beans with a single application. There is not usually, however, a corresponding stimulation of root growth.

Gibberellins not only dramatically increase stem growth, but they are involved in nearly all of the same regulatory processes in plant development as auxins (figure 10.5). In some kinds of plants, flowering can be brought about by applications of gibberellins, and the dormancy of buds

Figure 10.5 Effect of gibberellin on growth. *Left:* An untreated chrysanthemum plant. *Right:* A similar chrysanthemum plant that has been treated with gibberellin. (USDA Photo)

and seeds can be broken. Some gibberellins appear to "lower the threshold of growth"; that is, plants may start growing at lower temperatures than usual after an application of gibberellin. This means an application could bring about a green lawn two or three weeks earlier in the spring. Gibberellin has been used to increase yields of sugar cane and hops and has revolutionized the production of seedless grapes through increasing the size of the fruit, eliminating hand-thinning, and helping to prevent fruit disease. It has also been used in navel orange orchards to delay the aging of the fruit's skin.

Field experiments have shown that if gibberellins are applied to plants at the same time herbicides are applied, the gibberellins may reverse the effects of the herbicides.

Cytokinins

By the close of the nineteenth century there was conjecture concerning the existence of a chemical that regulates cell division in plants, but it was not until 1964 that such a substance was identified in kernels of corn. Two decades before the corn substance was identified, something that stimulates embryo growth had been discovered in coconut milk. Such substances that promote cell division came to be known as **cytokinins.** Cytokinins also participate in the enlarging of cells, the differentiation of tissues, the development of chloroplasts, the stimulation of cotyledon growth, the delay of aging in leaves, and in many of the growth phenomena also brought about by auxins and gibberellins.

Experiments with living pith cells of tobacco plants have shown that such cells will enlarge when supplied with auxin and nutrients, but they will not divide unless small amounts of cytokinin are added. By varying the amounts of cytokinin it is possible to stimulate the pith cells to differentiate into roots or buds that will develop into stems. This last regulatory effect can be used to offset **apical dominance,** which is the suppression of the growth of the axillary or lateral buds brought about by auxin in a terminal bud. The extent of the suppression depends on how close the axillary buds are to the terminal bud and on the amount of inhibitor involved. Apical dominance is strong in trees with conical shapes and little branching toward the top (e.g., many pines, spruces, firs) and weak in trees with stronger branching above (e.g., elms, ashes, willows). When a terminal bud is deliberately or accidentally removed, the inhibiting auxin is also removed, and the nearest axillary buds begin to grow, developing inhibitors of their own (figure 10.6). Experiments with pea plants have shown that the axillary buds will begin to grow as little

Figure 10.6 The trunk of this bald cypress tree, which would normally be single, forked because of the earlier removal of the terminal bud.

Figure 10.6 The trunk of this bald cypress tree, which would normally be single, forked because of the earlier removal of the terminal bud.

as four hours after a terminal bud has been removed. If cytokinins are applied in appropriate concentrations to axillary buds, they will begin to grow even in the presence of a terminal bud.

Cytokinins are found most abundantly in meristems and other developing tissues, particularly those of young fruit. Most have not yet been approved for general agricultural use, but experiments have shown that they do prolong the life of vegetables in storage. Related synthetic compounds have been used extensively to regulate the height of ornamental shrubs, to keep harvested lettuce and mushrooms fresh, to shorten the straw length in wheat, so as to minimize the chances of the plants blowing over in a wind, and to lengthen the life of cut flowers.

Abscisic Acid

Another growth-inhibiting hormone, **abscisic acid** (ABA), was identified independently in the United States and England in 1964. It is found in many plant materials but is particularly common in fleshy fruits, where it apparently prevents seeds from germinating while they are still on the plant. ABA was once believed to promote the formation of abscission layers in leaves and fruits (see chapter 7), but despite its name it appears to have little influence on the process. When it is applied to active plant buds, the leaf primordia become bud scales and the bud goes into a dormant winter condition. Since nursery plants that are being shipped stand to suffer less damage in a dormant condition than in an active one, the application of ABA and similar growth inhibitors to such nursery stock has great practical value. The inhibiting effects of ABA can be reversed by the application of gibberellin.

ABA apparently helps leaves respond to excessive water loss. When the leaves wilt, ABA is produced in amounts several times greater than usual; this interferes with the transport or retention of potassium ions in the guard cells, causing the stomata to close. When the uptake of water again becomes sufficient for its needs, the ABA is broken down metabolically and the stomata reopen.

Ethylene

R. Gane discovered that **ethylene** was produced naturally by fruits in 1934, although it had been known for some time that artificially adding this simple gas to green fruits hastens their ripening. It is now known that ethylene is produced not only by fruits but also by flowers, seeds, leaves, and even roots, and that its regulatory effects make it a hormone in the broad sense of the word.

The production of ethylene by plant tissues varies considerably under different conditions. A surge of ethylene lasting for several hours becomes evident after various tissues, including

Figure 10.7 Effect of ethylene on holly twigs. Two similar twigs were placed under glass jars for a week; a ripe apple was at the same time placed under the jar on the right. Ethylene produced by the apple caused abscission of the leaves.

those of fruits, are bruised or cut, and applications of auxin can cause an increase in ethylene production of two to ten times. As pea seeds germinate, the seedlings produce a surge of ethylene when they meet interference with their growth through the soil. This apparently causes the stem tip to form a tighter crook, which may aid the seedling in pushing to the surface. Ethylene also apparently triggers its own production. If minute amounts are introduced to the tissues that produce the gas, a tremendous response by the tissues often results. These tissues may then produce so much ethylene that it can adversely affect the part concerned (e.g., flowers may fade in the presence of excessive amounts) (figure 10.7).

In ancient China, growers used to ripen fruits in rooms where incense was being burned, and citrus growers used to ripen their fruits in rooms equipped with a kerosene stove. In houses where artificial gas is used, occupants often experience great difficulty in growing houseplants, and greenhouse heaters using such fuel create a dilemma for owners attempting to promote plant growth. In all of these instances, minute amounts of ethylene gas resulting from the fuel combustion brought about the results, both good and bad. In fact, as little as one part of ethylene per ten million parts of air may be sufficient to trigger responses.

Commercially, ethylene is used extensively. It has been used for many years to ripen harvested green fruits such as bananas, mangoes, and

Chapter 10

Figure 10.8 Elements essential to growth

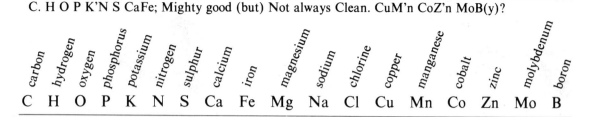

C. H O P K'N S CaFe; Mighty good (but) Not always Clean. CuM'n CoZ'n MoB(y)?

carbon	hydrogen	oxygen	phosphorus	potassium	nitrogen	sulphur	calcium	iron	magnesium	sodium	chlorine	copper	manganese	cobalt	zinc	molybdenum	boron
C	H	O	P	K	N	S	Ca	Fe	Mg	Na	Cl	Cu	Mn	Co	Zn	Mo	B

honeydew melons and to color up citrus fruits before marketing. (Some growers still use natural ethylene to ripen pears and peaches by wrapping each fruit individually in tissue paper. The paper retards the escape of ethylene from the fruit and hastens ripening.) "Resting" potato tubers will sprout following brief applications of ethylene, and seeds may be stimulated to germinate if given a short exposure to the gas just before sprouting, although treatment after sprouting inhibits growth. Ethylene is used in Hawaii to promote flowering in pineapples; it also causes members of the pumpkin family to produce more female flowers and thus more fruit. Because trees grown in containers in nurseries are usually crowded, they are often tall and spindly, but applications of small amounts of ethylene to the trunks while they are enclosed in plastic tubes causes a marked thickening, making the trees sturdier and less likely to break.

Mineral Requirements for Growth

Growth phenomena are not controlled by internal means alone. Light, temperature, soil structure and minerals, and other external factors all play a role. In fact, growth depends on a complex, interrelated combination of chemical and physical forces, both internal and external, all of which are in delicate balance with one another.

Plants may take up many elements from the soil, but besides the carbon, hydrogen, and oxygen obtained from carbon dioxide and water, only fifteen elements are essential to the growth of most plants. Sodium, a comparatively abundant element, is apparently required by some but not all plants. The **essential elements** can be remembered by a sentence (figure 10.8) that includes the symbols of the elements involved.

The mineral elements are usually put into two categories: the *macronutrients,* which are used by plants in greater amounts and constitute from about 0.5 to 3.0% of the dry weight of the plant; and the *micronutrients,* which are needed by the plant in very small amounts, often constituting only a few parts per million of the dry weight. The macronutrients (other than carbon, hydrogen and oxygen) are nitrogen, potassium, calcium, phosphorus, magnesium, and sulphur, with the first four usually making up about 99% of the nutrient total. The remaining elements, the micronutrients, are present in amounts ranging from bare traces—as in the case of sodium and cobalt, neither of which may actually be essential for some plants—to 1,500 parts per million of iron and manganese, and up to 10,000 parts per million of chlorine. In addition to these widely required elements, specific organisms may require others. For example, certain algae apparently also require the elements vanadium, silicon, or iodine, while some ferns utilize aluminum. One land plant is reported to require selenium, although many absorb this element without its being essential to their growth.

Figure 10.9 Tobacco plants grown in media deficient in various elements to show deficiency symptoms. *A.* Normal plant that has been supplied with all elements. The other plants have been grown in media deficient in specific nutrients, as follows: *B.* Nitrogen. *C.* Phosphorus. *D.* Potassium. *E.* Boron. *F.* Calcium. *G.* Magnesium. (USDA Photo)

When any of these elements is deficient in the soil, the plants exhibit characteristic symptoms of the deficiency, which disappear after the problem has been corrected (figure 10.9). Table 10.1 shows some of the uses of essential elements in plants and describes the symptoms of deficiency for each element.

Plant Movements

We noted in chapter 2 that all living organisms exhibit movement but that most plant movements are relatively slow and imperceptible unless seen in time-lapse photography or demonstrated experimentally. We now want to examine how and why plant movements occur.

First we need to distinguish among general types of movements. Some result from various growth phenomena, while others come about through changes in turgor pressures. Still others,

Chapter 10

Table 10.1 Uses of essential elements in plants

Element	Some Functions	Deficiency Symptoms
Nitrogen	part of proteins, nucleic acids, chlorophyll	relatively uniform loss of color in leaves, occurring first on the oldest ones
Potassium	activates enzymes; concentrates in meristems	yellowing of leaves, beginning at margins and continuing toward center; lower leaves mottled and often brown at tip
Calcium	essential part of middle lamella; involved in movement of substances through cell membranes	terminal bud often dead; young leaves often appearing hooked at tip; tips and margins of leaves withered; roots dead or dying
Phosphorus	necessary for respiration and cell division; high-energy cell compounds	plants stunted; leaves darker green than normal; lower leaves often purplish between veins
Magnesium	part of the chlorophyll molecule; activates enzymes	veins of leaves green but yellow between them with dead spots appearing suddenly; leaf margins curling
Sulphur	part of some amino acids	leaves pale green with dead spots; veins lighter in color than the rest of the leaf area
Iron	needed to make chlorophyll and in respiration	larger veins remaining green while rest of leaf yellows*
Manganese	activates some enzymes	dead spots scattered over leaf surface; all veins and veinlets remain green; effects confined to youngest leaves
Boron	influences utilization of calcium ions, but functions unknown	petioles and stems brittle; bases of young leaves break down

*NOTE: The symptoms of iron deficiency may be caused by several factors such as overwatering, cold temperatures, and nematodes (small roundworms) in the roots. The iron may be relatively abundant in the soil, but its uptake may be prevented or sharply reduced by these environmental conditions. Iron becomes more soluble under acid conditions—so much so that it can produce toxic conditions for most plants. Acid soil plants, on the other hand, have a much higher iron requirement than plants that require more alkaline environments. Iron *chelates,* when added to the soil or sprayed in solution on the foliage, furnish more soluble iron under nonacid conditions; thus, azaleas and other plants having high iron requirements can achieve normal growth in a nonacid soil.

It should be noted that all the micronutrients are harmful to plants when supplied in excessive quantities. Copper will kill algae in concentrations of one part per million, and boron has been used in weed killers. Even macronutrients are harmful if present in heavy amounts, although nonessential elements sometimes can counteract their toxicity.

usually involving reproductive cells or very primitive aquatic organisms (see chapters 14–20), occur when the whole cell or organism is propelled or pulled along by one or more little whiplike tails called **flagella** (singular: **flagellum**) or by rows of shorter beating hairs called **cilia** (singular: **cilium**). Finally, some (but not all) sudden movements, such as the splitting of a pod, occur because of structural response to changes in water content.

Growth Movements

Growth movements result from varying growth rates in different parts of an organ. They are principally related to young parts of a plant and are quite slow, as a rule taking at least two hours to become apparent although the plant may have begun microscopic changes within minutes of receiving a stimulus. The stimulus may be either internal or external.

Figure 10.10 Charles Darwin's demonstration of spiraling growth

paper blackened with carbon

sliver of glass glued to growing tip of plant

tip spirals as it grows, tracing its pattern on the paper

Figure 10.11 Typical twining of a tendril of a squash plant

Movements Resulting Primarily from Internal Stimuli

Spiraling Movements

Charles Darwin once attached a tiny sliver of glass to the tip of a plant growing in a pot. Then he suspended a piece of paper blackened with carbon over the tip and as the plant grew he raised the paper just enough to allow the tip to touch the paper without hurting the plant. He found that the growing point traced a spiral pattern in the blackened paper. We now know that such spiraling movements of growth are common to many plants (figure 10.10).

Nodding Movements

Members of the legume family such as garden beans, which form a crook in the hypocotyl upon germinating, exhibit a movement from side to side as the seedling pushes up through the soil. This nodding movement apparently facilitates the progress of the growing plant through the soil.

Twining Movements

Although twining movements are primarily internally stimulated, external forces such as gravity and contact may also play a role. These movements occur when cells in the stems of climbing plants such as morning glory elongate to differing extents, causing visible spiraling in growth (in contrast with the spiraling movements previously mentioned, which are not visible to the eye). Tendril twining, which is initiated by contact, results from an elongation of cells on one side of the stem and a shrinkage of cells on the opposite side, followed by differences in growth rates (figure 10.11). Some tendrils are stimulated to coil by auxin, while others are stimulated by ethylene. The process itself apparently involves osmosis and either light or ATP, with dissolved substances leaking out of the cells on the inside of the tendril.

Contraction Movements

In chapter 5 we noted that the bulbs of a number of dicots and monocots have contractile roots that pull them deeper into the ground. In lilies, for example, seeds germinating at the surface ultimately produce bulbs that end up 10 to 15 centimeters (4 to 6 inches) below ground level because of the activities of contractile roots. The aerial roots of some banyan trees straighten out by contraction after the roots have made contact with the ground. In contracting, cells of the cortex become compacted in parts of the root while cells in other parts swell, causing an accordion-like folding that "shrinks" the root. This "shrinking" has been shown to take place at the rate of 2.2 millimeters (0.1 inch) a day in sorrel. There is some evidence that temperature fluctuations at the surface determine how long the contracting will continue. When the bulb gets deep enough that the differences between daytime and nighttime temperatures are slight, the contractions cease.

Nastic Movements

When flattened plant organs such as leaves or flower petals first expand from buds, they characteristically alternate in bending down and then up as the cells in the lower part of the leaf enlarge faster than those in the upper part, followed by similar expansion in reverse. Such nondirectional movements (i.e., movements that do not result in an organ being oriented toward or away from the direction of a stimulus) are said to be *nastic*. In prayer plants (see figure 10.15) and others where the leaves or flowers fold up and reopen daily, turgor movements are also involved, as are the external stimuli of light and temperature; in a few instances nastic movements apparently are initiated by external stimuli alone.

Figure 10.12 This painted lady (stonecrop) plant evidences the bending of the stem that occurs when a box admitting light from only one side is kept over a plant for a month.

Movements Resulting from External Stimuli

Permanent, directed movements resulting from external stimuli are commonly referred to as **tropisms.**

Phototropism

The main shoots of most plants growing in the open tend to develop vertically, although the branches often grow horizontally. If a box is placed over a plant growing vertically and a hole is cut to admit light from one side, the tip of the plant will begin to bend toward the light within a few hours. If the box is later removed, a compensating bend develops, causing the tip to grow vertically again. Such a growth movement toward light is called a *positive phototropism* (figure 10.12), while a similar bending away from light is called a *negative phototropism*. The shoot

Figure 10.13 This *Coleus* plant was placed on its side the day before the photograph was taken. The stem tip bent upward within twenty-four hours of the pot being tipped over.

tips of most plants are positively phototropic, while roots are either insensitive to light or negatively phototropic. Leaves often twist on their petioles and become oriented at right angles to a light source in response to the illumination. Strictly speaking, this petiolar movement should be called *phototorsion* because no growth is involved as it is with phototropic responses of stems and roots. If one views a tree from directly overhead or observes a vine growing on a fence, it is perhaps surprising to note how little overlap of the leaves occurs; each leaf is placed so that it receives the maximum amount of light available.

We have already noted that if the tip of a coleoptile is covered or removed, the structure will not bend toward light, and that auxin is produced in the tip (figure 10.3). We have also noted that auxin promotes the elongation of cells, at least in certain concentrations. For some time it was believed that stem tips bent toward light because

auxin was destroyed or inactivated on the exposed side, leaving more growth-promoting hormone on the side away from the light and thus causing the cells there to elongate more and produce a bend. Careful experiments have shown, however, that stem tips growing in the open have the same total amount of auxin present as stem tips from the same species that have received light from one side. Other experiments indicate that the auxin migrates away from the light, accumulating in greater amounts on the opposite side, thus promoting greater elongation of cells on the "dark" side. Apparently an active transport system enables the auxin to migrate against a diffusion gradient.

Different intensities of light may bring about different phototropic responses. In Bermuda grass, for example, the stems tend to grow upright in the shade and parallel with the ground in the sun. In other plants, such as the European rock rose, which grows among rocks or on walls, the flowers are positively phototropic, but once

Figure 10.14 Clinostat

they are fertilized they become negatively phototropic. As the pedicels (stalks) elongate, the developing fruits are buried in cracks and crevices, where the seeds then germinate. Phototropic responses are not confined to flowering plants. A number of mushrooms, for example, show marked positive responses to light, and certain fungi that grow on horse dung evince striking phototropic movements, which are discussed in chapter 16.

Geotropism

Growth responses to the stimulus of gravity are called *geotropisms*. The primary roots of plants are positively geotropic, while shoots forming the main axis of plants are negatively geotropic (figure 10.13). As indicated in chapter 5, there is evidence that plant organs perceive gravity through the movement of starch grains in special cells of the root cap. These special cells, called *statocytes,* are also found in coleoptile tips and in the endodermis. When a potted plant is placed on its side, the starch grains in the statocytes will, within a few minutes, begin to float or tumble down until they come to rest on the side of the cells closest to the gravity stimulus. In roots, the cells on the side opposite the stimulus begin elongating within an hour or two, bringing about a downward bend, while the opposite occurs in stems. Some have suggested that mitochondria and Golgi bodies (dictyosomes) also respond to gravity, but precisely how either they or the starch grains bring about the response is the subject of much conjecture, although it is believed that auxin is the agent effecting the cell elongation while ABA evidently inhibits the elongation of cells on the side away from the stimulus. The stimulus of gravity can be cancelled by rotating a plant placed in a horizontal position. A simple device called a *clinostat* uses a motor to rotate a wheel to which a plant in a pot is attached (figure 10.14). As the plant rotates, both the stem and roots continue to grow horizontally instead of exhibiting characteristic geotropic responses. Obviously the starch grains in the statocytes do not settle while the plant is moving, and this apparently prevents auxin transport, which would bring about cell elongations that produce curvatures of root and stem.

Other Tropisms

It is well known that roots will enter cracked water pipes and sewers. In fact, roots have been known to grow upward for considerable distances in response to water leaks. Some have called such growth movements *hydrotropisms*. Other external stimuli that produce tropic responses include touch or contact *(thigmotropism)*, chemicals *(chemotropism)*, temperature *(thermotropism)*, wounding *(traumotropism)*, electricity *(electrotropism)*, and oxygen *(aerotropism)*. It has been noted that greater concentrations of roots occur on the north and south sides of wheat seedlings, and it has been suggested that magnetic forces may be involved; the term *geomagnetotropism* has been proposed for this phenomenon. Some of these tropisms have been induced primarily by artificial means, but others occur regularly in nature. For example, germinating pollen grains produce a long tube that follows a diffusion gradient of a chemical released within a flower; this is a chemotropism. Thermotropic responses to cold temperatures are seen in the shoots of many common weeds, which grow horizontally when cold temperatures prevail and return to erect growth when temperatures become warmer.

Figure 10.15 Prayer plant. *A.* The plant at noon. *B.* The same plant at 10 P.M., after "sleep" movements of its leaves had occurred.

A.

B.

Turgor Movements

Turgor movements result from changes in internal water pressures. The cells concerned may be in normal parenchyma tissue of the cortex, or they may be in special swellings called *pulvini* located at the bases of leaves or leaflets. Some turgor movements may be quite dramatic, taking place in a fraction of a second; others may require up to forty-five minutes to become visible.

Sleep Movements

About 90% of the members of the legume family, which includes peas and beans, and members of several other flowering plant families exhibit movements in which either leaves or petals fold as though going to sleep (figure 10.15). The folding and unfolding usually takes place in regular twenty-four-hour cycles, with closure most frequently taking place at dusk and opening occurring in the morning. Such cycles, which also occur in the Animal Kingdom, have come to be known as **circadian rhythms.** They appear to be controlled internally by the plants, although they are also geared to changing daylengths and sea-

sons; they do not, however, generally accelerate when temperatures increase. Although most plants exhibiting circadian rhythms manifest sleep movements at dusk and at dawn, the actual timing varies with the species; several open their flowers at different hours of the night.

About two hundred years ago, the famous Swedish botanist Linnaeus planted wedge-shaped segments of a circular garden with plants that exhibited sleep movements. The plants were arranged in successive order of their sleep movements throughout a twenty-four-hour day. One could tell the time simply by noting which part of the garden was "asleep" and which was "awake." A few others copied the garden clock idea, but the expense and difficulty of obtaining all the plants from different parts of the world and replanting them each year proved to be too great for the practice to be continued.

The movements are produced by the passage of water in and out of cells at the bases of the leaves or leaflets, causing turgor changes. The function of these movements is not clear, and the rhythms are not confined to sleep movements. Certain algae called *dinoflagellates* (see chapter 15) cause phosphorescence in warm ocean waters

Figure 10.16 Water conservation in a grass leaf brought about by the blade rolling up when sufficient water is not available to maintain normal turgor. *A.* Leaf when adequate water is available. *B.* Leaf that has rolled up. *C.* Cross section of a rolled leaf, enlarged to show the location of the large, thin-walled bulliform cells (arrows), which partly collapse under dry conditions and thus bring about the rolling of the leaf blade.

A.

B.

C.

by glowing in the dark. One species always glows brightly within two or three minutes of midnight, even if it is maintained in culture in continuous dim light.

Another type of rhythm is seen in the giant bamboos of Asia, which send up huge flowering stalks every thirty-three or sixty-six years, even if the plants have been transplanted to other continents and are growing under different conditions. These flowering stalks use up all the energy reserves of the bamboo, and they die shortly after appearing.

Water Conservation Movements

Leaves of many grasses have special thin-walled cells *(bulliform cells)* below parallel, lengthwise grooves in their surfaces. During periods when sufficient water is not available, these cells lose their turgor and the leaf rolls up or folds (figure 10.16). Experiments with certain prairie grasses have shown that the rolling effectively reduces transpiration to as low as between 5 and 10% of normal.

Contact Movements

The sudden movements of the Venus flytrap and bladderworts (discussed in chapter 7), involve turgor changes apparently triggered by electric charges released upon contact or as a result of variations in light or temperature. The sensitive plant has well-developed swellings *(pulvini)* at the bases of its many leaflets and a large pulvinus at the base of each leaf petiole (figure 10.17). When the leaf is stimulated, water rapidly leaves the cells in half of each pulvinus and goes into the intercellular spaces, causing sufficient loss of turgor to bring about a folding of both the leaflets and the leaf as a whole (figure 10.18). If a part of the stem of a sensitive plant is cut off and then immediately reattached with a water-filled piece of rubber tubing, it can be shown that something is transmitted from above the cut through the water. If a leaf above is stimulated to fold, after a few minutes a leaf or two below the cut will also fold. Potassium ions have been shown to leave

Figure 10.17 Longitudinal section through a pulvinus of a sensitive plant

Figure 10.18 Movement of a sensitive plant leaf. The leaf on the right has leaflets folded in response to being bumped; the remaining leaves are fully expanded.

cells in pulvini that are losing turgor, but whether the reaction is transmitted across water in the rubber tubing by these ions, by electrical charges, or by something else is not known. A normal response to a stimulus by a sensitive plant takes from a few seconds to less than two minutes.

Contact movements are not confined to leaves. Many flowers exhibit movements of stamens and other parts; such movements usually facilitate pollination. The pollen-receptive stigmas found in flowers of the African sausage tree have two lobes. These lobes fold inward to enclose pollen grains that have landed on them. The stigmas of monkey flowers and Asian cone flowers also exhibit turgor movements upon contact, while the pollen-bearing stamens of barberry and moss rose flowers jerk suddenly upon contact,

dusting visiting insects with pollen (figure 10.19). The most spectacular of all contact movements are seen in orchid plants (discussed in figure 23.23), in which little sacs of pollen become attached, sometimes forcibly, to the bodies of visiting insects, and in a few instances the benefactors receive dunkings in small buckets of fluid.

Taxes

The *taxis,* a type of movement that involves either the entire plant or its reproductive cells (see chapters 14–20), does not occur among flowering plants. In response to a stimulus, the cell or organism, either propelled or pushed by flagella or cilia, moves toward or away from the source of the stimulus. Stimuli include chemicals, light, and oxygen. In ferns, for example, the female reproductive structures manufacture a chemical that produces a *chemotaxy* in the male reproductive cells (sperms); that is, the sperms swim toward the source of the chemical. Certain one-celled algae are *phototactic;* they swim either toward or away from a light source. Other similar organisms exhibit *aerotactic* movements in response to changes in oxygen concentrations.

Figure 10.19 Action of the stamens of a barberry flower. *A.* Before contact. *B.* Upon contact.

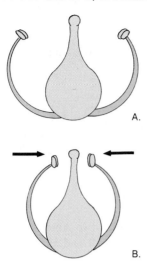

A.

B.

Miscellaneous Movements

Slime molds (discussed in chapter 15) inhabit damp logs and debris. During their active stages their protoplasm, which has no rigid cell walls, "flows" over dead leaves and other substrates like thin, slowly moving gelatin. Certain blue-green algae (e.g., *Oscillatoria*) wave slowly back and forth or slide up and down against each other in *gliding movements,* and other one-celled algae with thin glass shells (diatoms) give the appearance of being jet-propelled. It is not certain just how these movements occur, but there is evidence that submicroscopic fibrils produce rhythmic waves that bring about the motion. In the case of diatoms, some materials are apparently forced out of the cells, and the friction set up by the process may propel the organisms through the water in which they are found.

Dehydration movements do not involve living cells or hormones; the forces are purely physical, caused by imbibition or by the drying out of tissues or membranes. The individual fruitlets of filarees and other members of the geranium family have long, stiff, pointed extensions that are sensitive to changes in humidity. As humidity decreases during the day, they coil up, and then at night, when humidity increases, the coils relax. This alternating coiling and uncoiling results in the pointed fruitlets planting themselves into the ground in a corkscrewlike fashion (see figure 21.20).

A number of **fruits** that are podlike and dry at maturity split in various ways with explosive force, flinging seeds as far as 12 meters (39 feet) from the plant (see figure 21.9). Examples of such fruits include those of the garbanzo bean, witch hazel, vetches, and Mexican poppies. In ice plants and stonecrops, the converse is true; rain or dew causes the mature fruits to open due to the swelling of membranes. Pressures inside the squirting cucumber build up to the point where, upon abscission of the fruit, the seeds are expelled from the stalk in up to distances of more than 10 meters (33 feet). Dwarf mistletoes, which are parasitic on coniferous trees, produce tiny, sticky fruits that are explosively released and adhere to the trunks and branches of trees in the vicinity.

Phytochrome

Phytochrome is an extraordinary pale blue proteinaceous pigment that is associated with the absorption of light and that apparently occurs in all higher plants. Although its presence was suspected several years before 1959 when it was first isolated, it was not visible, and a special pigment-analysis instrument had to be constructed to detect it. Only minute amounts are produced, most found in meristematic tissues. It occurs in two stable forms, either of which can be converted to the other: P_{red} or P_r is a form that absorbs red light; $P_{far-red}$ or P_{fr} is a related form that absorbs the far-red light found at the edge of the visible light spectrum. When either form absorbs light, it is converted to the other form, so that P_r becomes P_{fr} when it absorbs red light, and P_{fr} becomes P_r when it absorbs far-red light. P_r is stable indefinitely in the dark. The normal effect of light in nature is to cause more P_r to become P_{fr} than vice versa. P_{fr} converts back to P_r in the dark over a period of several hours, or else becomes inactivated, but its conversion in the presence of appropriate light is instantaneous (figure 10.20).

Phytochrome pigments play a role in a great variety of plant responses: they are involved with several aspects of plant development, photoperiodism (discussed in the next section), changes in plastids, the production of anthocyanin pigments, and with the detection of shading by other plants. One of the most studied phenomena associated with phytochrome pigments involves the germination of seeds. Some seeds, for example, do not germinate in darkness, but the red part of the spectrum in sunlight converts P_r to P_{fr}, which in turn somehow unblocks the germinating mechanism. If such seeds (e.g., lettuce) are given suitable moisture and oxygen conditions and then exposed only to red light, they will germinate readily, but if they are given only far-red light they will not. Furthermore, if they are given alternating flashes of red and far-red light, the type of light in the final flash determines whether or not they will germinate; if the last flash, which may be of only a few seconds duration, is red light, the seeds will germinate, but if it is far-red light they will not.

When a seedling first emerges from the soil, light changes the P_r to P_{fr}, signaling the cells producing ethylene to reduce production. This in turn permits the crooks in the hypocotyls to relax, and the plant straightens up. Elongation of stems appears to be inhibited by P_{fr}. The mechanism is so sensitive that as little as three seconds of clear moonlight may produce a shortening of internodes of **etiolated** (spindly from having been grown in the dark) oat seedlings. When stems of a tree are shaded, however, more far-red light reaches them, causing the P_{fr} to be converted to P_r and thus lowering the inhibition so that the stems grow longer and out from under the shade.

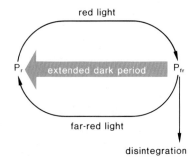

Figure 10.20 Phytochrome interconversions

Photoperiodism

In the early 1900s, two American plant physiologists, W. W. Garner and H. A. Allard, became curious about a variety of tobacco plant growing near Washington, D.C. The plants grew 3 to 4 meters (10 to 13 feet) tall during the summer, but they failed to flower unless they were brought into a greenhouse for the winter. If the plants were started in pots in the fall in a greenhouse, they grew only about a meter (3 feet) tall before flowering. Garner and Allard decided to see what would happen if they kept the tobacco plants, as well as some soybean seedlings that would not bloom before September, in Washington in complete darkness from 4 P.M. until 9 A.M., thus allowing them only seven hours of daylight. They found that all the plants flowered, and further investigations showed that the length of day (actually the length of the night) was directly related to the onset of flowering in many plants. They published the results of their investigations in 1920 and later called the phenomenon they had discovered **photoperiodism** (figure 10.21).

The critical length of day (i.e., the maximum or minimum length of day) for the initiation of flowering is frequently about twelve to fourteen hours, although it can vary considerably. Plants that will not flower unless the daylength is shorter than the critical length (e.g., in the fall or spring) are called **short-day plants.** They include asters,

Figure 10.21 Photoperiodism. Five cockscomb plants exposed to different daylengths. The plant on the left received 8 hours of light and 16 hours of dark; the plant on the right received 16 hours of light and 8 hours of dark; the plants between them (from left to right) received 10, 12, and 14 hours of light, respectively. The plants that received 14 or more hours of light did not flower. (USDA Photo)

chrysanthemums, dahlias, goldenrods, poinsettias, ragweeds, sorghums, salvias, strawberries, and violets. Plants that will not flower unless periods of light are longer than the critical length are called **long-day plants.** These include garden beets, larkspur, lettuce, potatoes, spinach, and wheat. Such plants usually flower in the summer but will also flower when left under continuous artificial illumination. Accordingly, leafy vegetables such as lettuce and spinach need to be harvested in the spring and grown again in the fall in temperate latitudes if "bolting" (producing flowering stalks) is to be avoided. Potato breeders in the United States grow their plants in northern states where the long summer days initiate flowering, but since the potatoes themselves are produced independently of flowering, the plants may be grown for crop purposes at any latitude where other conditions are favorable. Indian grass and several other grasses have two critical photoperiods; they will not flower if the days are too short and they also will not flower if the days are too long. Such species are referred to as **intermediate-day plants.** Other plants, particularly those of tropical origin, will flower under any length of day, providing, of course, they have received the minimum amount of light necessary for normal growth. Such plants are called **day-neutral plants** and include garden beans, calendulas, carnations, cyclamens, cotton, nasturtiums, roses, snapdragons, sunflowers, and tomatoes, as well as many common weeds such as dandelions.

With some plants, small differences in daylength may be critically important. Some varieties of soybeans, for example, will not flower when days are fourteen hours long but will flower if the daylength is increased to fourteen and a half hours. This difference could amount to less than 320 kilometers (200 miles) of latitude, so that certain varieties grown in the southern states might not produce fruit in the northern states and vice versa.

Commercial florists and nurserymen have made extensive use of photoperiods, manipulating with artificial light the flowering times of poinsettias, some lilies, and other plants so as to have them flower at times of the biggest demand, such as Christmas or Mother's Day. The light intensities used to lengthen the days artificially can be very low—often less than 10 footcandles (i.e., one-thousandth the intensity of full sunlight)—and can come from incandescent bulbs, which have more red wavelengths than fluorescent lamps and therefore have more effect on phytochromes, which are involved in photoperiodism.

A number of vegetative activities of plants are also affected by photoperiods. In the shortening days of the fall, for example, many woody plants will begin undergoing the changes that lead to the dormancy of buds, regardless of whether or not "Indian summer" temperatures may be prevailing. In the spring, certain seeds respond to photoperiods in their germination, both long-day and short-day species having been discovered. Plants that produce tubers (e.g., Jerusalem artichoke) may develop them only under short-day conditions, even though they are long-day plants with respect to flowering. Usually these photoperiodically controlled responses are valuable to the plants in that they prepare them for changes in the seasons and thus ensure their survival and perpetuation.

The precise mechanism of photoperiodism is not yet fully understood. After the existence of phytochrome had been demonstrated it was theorized that it might be involved in photoperiodism, with P_{fr} inhibiting flowering in short-day plants and promoting flowering in long-day plants. Presumably the P_{fr} would accumulate in the light in short-day plants and revert back to P_r or be broken down during dark periods. After the length of the night increased sufficiently, all or most of the P_r would disappear and flowering then would cease to be inhibited, with the converse taking place for long-day plants. P_{fr} has been demonstrated to disappear in many plants in as little as three or four hours of darkness, however, indicating that phytochrome interconversions alone cannot explain photoperiodism. Experiments have shown that a photoperiod is perceived by the leaves, and these are thought to produce a flowering hormone called *florigen,* which is then transported to the apical meristems where flower buds are initiated. Florigen, however, has not yet actually been isolated from a plant nor has it otherwise been proven to exist, leading to speculation that there may be no simple explanation for the phenomenon of photoperiodism.

Temperature and Growth

Each species of plant has an optimum temperature for growth, although the optimum may vary throughout the life of the plant, and a minimum temperature below which growth will not occur. Each species also has a maximum temperature above which injury may result (or, at least, growth will cease).

Dr. Frits Went, the discoverer of auxin, experimented with the growth of tomato plants under conditions in which night temperatures were lowered to 17°C (63°F) while day temperatures were maintained at 25°C (77°F). These plants were found to grow better than plants kept at a steady temperature around the clock. Went applied the term *thermoperiodism* to this phenomenon. It has since been shown that the optimal thermoperiod may change with the growth stage of the plant. Thus, young pepper plants have been shown to develop best when night temperatures are approximately 25°C (77°F), but as they grow older the optimum night temperature for their development drops more than 11°C (20°F). Lower night temperatures often result in a higher sugar content in plants and may also produce greater root growth, although some plants such as peas seem to be unaffected in this regard.

The growth of many field crops is roughly proportional to prevailing temperatures—making it possible to predict harvest times—although the number of days until maturity varies considerably with the locale. In 1855 the Swiss botanist A. P. de Candolle established a basis for the harvest dates of crops. His method, based on summing the temperature means for each day, makes it possible to follow a crop's progress with some precision. This method has been refined by multiplying the temperature means by the number of hours of daylight. In Hawaii a high degree of precision has been attained in predicting the date of the pineapple harvest by measuring the rate of growth within the temperature range at which the plants grow and multiplying the number of hours per day at each temperature. These methods, which work best for crops that are not closely regulated by photoperiods, do not, however, take into account the requirements of some plants for different optimum temperatures at different stages of their development.

Dormancy

Dormancy may be defined as a period of growth inactivity in seeds, buds, bulbs, and other plant organs. Growth is resumed only if environmental requirements of temperature, water, or daylength are met. We have already discussed the formation of abscission layers (see chapter 7) in the fall as plants change from an active condition to a *dormant* one. In temperate and cold climates, most plants and their seeds go through a dormant period lasting from a few days to several months, and in each case the dormancy nearly always has some survival value for the species. Many of the stone fruits (for example, cherries, peaches, and plums) need at least several weeks of rest at temperatures below 7°C (45°F) before the buds will develop into flowers or new growth,

and in some instances the buds will not develop before the tree has encountered a period of freezing temperatures. This dormancy adapts the plants to the conditions prevailing in the cooler temperate zones of the world. In response to prevailing conditions, seeds of desert plants germinate only after appreciable rain has fallen, and seeds of plants native to vernal pools and other wet areas may germinate only if under water.

The change from dormancy to a state in which germination will occur in seeds is controlled by a variety of factors, depending on the species concerned. Some plants become sensitive to photoperiods only after they have been subjected to a period of chilling, either while they are still seeds or shortly after germination. Growers often place moistened seeds of such plants in a refrigerator for several weeks. This process is known as *stratification* and is one of several practices employed to break dormancy. In some biennial plants, the cold treatment can be replaced by an application of gibberellin.

Some plants have their dormancies broken only after they have received a series of similar stimulations, so that a single stimulation such as an unseasonal rainstorm will not trigger growth. Other plants, such as poor man's pepper, require both certain temperatures and photoperiods for germination, and some weeds require specific thermoperiods plus some mechanical abrasion of the seedcoat to break dormancy. Still other seeds require specific enzyme action, acquired only by passing through the digestive tracts of birds, bats, or other animals, before they will germinate.

When a mechanical tomato picker was being developed in California during the early 1960s, tomato breeders wanted to have fruits that would be able to stand up to the bumping they would receive in the machinery, and so they set out to try to produce suitable varieties. Their search for new genes to use in their breeding programs led them to the wild tomatoes of the Galápagos Islands. When they brought the seeds back to California, however, they experienced difficulty in

Figure 10.22 The tree lupine has hard seeds, each with a moisture valve operated by specialized tissue ("counterpalisade" tissue) on either side of a pore through which water can enter the seed. When there is more moisture outside than inside, the specialized tissue swells and closes the pore. (After Hyde, E.O.C. 1954. "The function of the hilum in some Papilionaceae in relation to the ripening of the seed and the permeability of the testa." *Annals of Botany, N.S.* 18(70):244. © Annals of Botany Company; redrawn by permission.)

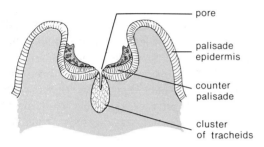

pore

palisade epidermis

counter palisade

cluster of tracheids

getting them to germinate, even after subjecting them to all known types of stratification treatments. Then they remembered that the giant tortoises of the Galápagos Islands ate these tomatoes, and they fed some to giant tortoises in a zoo. After passing through these animals, the seeds germinated promptly. Apparently something in the tortoises' digestive tracts broke the dormancy of the seeds.

Germination is prevented by a thick or restrictive seedcoat in the seeds of plants such as cocklebur, and the seeds of some legumes are almost stonelike because of the exceptional hardness of the outer "shell." Growers often nick or file seedcoats to allow water and gases to enter. In nature, various decay organisms break the seedcoats down, or the seeds themselves may secrete enzymes that aid in this process. The seeds of certain lupines have remarkable valves that restrict the uptake of water, preventing germination before the seedcoat is broken down by other means (figure 10.22).

Dormancy may be restricted to certain parts of the plant rather than occurring in the plant as a whole: in coffee, for example, only the flower buds may show dormancy. In other woody plants the onset of dormancy progresses from buds toward the base of the branch and on to the tip. Some seeds go through two cycles of dormancy before full growth occurs. In wake-robins, the radicle (embryo root) of the seed emerges and becomes established after being subjected to a period of cold, but the epicotyl does not emerge until a second season of cold has been encountered.

In many seeds, growth inhibitors such as ABA are present in the seedcoat, often in abundance. Before germination can occur, these inhibitors may need to be leached out by rain, broken down by enzymes, or have their effects overcome by other hormones. Studies have shown that gibberellin, which has the capacity to break dormancy when applied artificially, is produced by the germinating seeds of both cone-bearing and flowering plants. Ethylene, which is also effective in breaking dormancy, is produced by various germinating seeds, including those of oats, peanuts, and clover. Cytokinins do not generally appear to be involved in breaking dormancy, but a few instances are known of their doing so artificially. Although auxins are formed during the early stages of germination, they probably are not involved in the control of dormancy.

Many aspects of growth discussed in this chapter are applied to, or are involved in, vegetative propagation, which is surveyed in appendix 4.

Summary

Growth is defined as an "increase in volume due to the division and enlargement of cells." As cells mature they differentiate into forms adapted to specific functions. Development is a change in form as a result of growth and differentiation combined. Cells themselves assume different

Chapter 10

shapes and forms that adapt them to the problem involved in their total volume increasing at a greater rate than their total surface area.

Many growth phenomena are influenced by hormones, which are chemical substances produced in minute amounts in one part of an organism and transported to another part, where they have specific effects on growth, flowering, and other plant activities. Vitamins are organic molecules that function in activating enzymes; they are sometimes difficult to distinguish from hormones.

After water is imbibed by a seed, changes occur in its embryo. A radicle (which becomes the primary root) emerges first, followed by a shoot. In beans two cotyledons are attached to the embryo stem; the part of the stem below them is designated the hypocotyl and the part above them the epicotyl. A plumule is an embryo bud consisting of the first true leaves and the epicotyl. In some plants the cotyledons emerge above the ground; in others they remain beneath the surface. Members of the grass family have seeds with only one cotyledon. When such seeds germinate the radicle is at first protected by a sheathing coleorhiza and the stem by a sheathing coleoptile, through which each, respectively, emerges.

Darwin and his son noted in 1881 that coleoptiles bend toward a light source. Frits Went, in 1926, followed up on the Darwinian observations and demonstrated that something in coleoptile tips moved out into agar when decapitated tips were placed on it. He called the substance auxin and showed that it could cause coleoptiles to bend. Since Went's experiments, three groups of plant hormones have been found to promote the growth of plants; others that have been discovered are known only for their inhibitory effects. Auxins stimulate the enlargement of cells and are involved in many other growth phenomena. 2,4-D and 2,4,5-T are hormonelike compounds that have been used as weed killers and defoliants. 2,4,5-T has a highly toxic contaminant that has caused disease and defects in laboratory animals; it is now banned for most uses

in the United States. Gibberellins promote stem growth without corresponding root growth; cytokinins promote cell division, and can be used to offset apical dominance (the suppression of axillary buds brought about by auxin in a terminal bud); abscisic acid causes buds to become dormant and apparently helps leaves respond to excessive loss of water; ethylene gas hastens ripening of fruits and is used commercially to ripen green fruits.

Growth phenomena are controlled by both internal and external means and by chemical and physical forces in balance with one another. Besides carbon, hydrogen, and oxygen, fifteen of the elements taken up are essential to the growth of most plants. When any of the eighteen elements are deficient in the soil, characteristic deficiency symptoms appear in the plants.

Plant movements such as spiraling movements, nodding movements, twining movements, contraction movements, and some nastic movements are growth movements that result primarily from internal stimuli. Tropisms are permanent, directed growth movements that result from external stimuli such as light (phototropism), gravity (geotropism), water (hydrotropism), contact (thigmotropism), chemicals (chemotropism), etc. Turgor movements result from changes in internal water pressures; they may be very rapid or take up to forty-five minutes to beome visible. Turgor movements include sleep movements in which leaves or flowers fold daily in what is known as a circadian rhythm. Other types of plant movements include water conservation movements, contact movements, taxes, and dehydration movements.

Phytochrome is a pale blue pigment that occurs in all higher plants and plays a role in many different plant responses. It occurs in two forms, each of which can be converted to the other by the absorption of light. Daylight generally results in more P_r being converted to P_{fr} than vice versa. P_r absorbs red light and P_{fr} absorbs the far-red light found at the edge of the visible light spectrum. P_{fr} will convert back to P_r over a period of several hours in the dark; P_r is stable indefinitely in the dark. The P_r–P_{fr} mechanism is so sensitive that as little as three seconds of clear moonlight may produce enough P_{fr} to result in a shortening of the internodes of etiolated (spindly from having been grown in the dark) oat seedlings.

Photoperiodism is a response of plants to the duration of day or night. Short-day plants will not flower unless the daylength is shorter than a critical daylength. Long-day plants will not flower unless the daylength is longer than a critical daylength. Intermediate-day plants have two critical photoperiods and will not flower if the day is either too short or too long; day-neutral plants will flower under any daylength providing they have received a minimum amount of light.

The precise mechanism of photoperiodism is not yet fully understood. Phytochrome interconversions may play a role in photoperiodism but cannot be used alone to explain it. A flowering hormone called florigen is believed to be produced in leaves and to be involved in the initiation of flowering, but its actual existence has not yet been demonstrated.

Each species of plant has an optimum temperature for growth. Dormancy is a period of growth inactivity in seeds, buds, bulbs, and other plant organs. The change from dormancy to a state in which germination will occur in seeds is controlled by a variety of factors, including temperature, moisture, photoperiod, thickness of seedcoat, enzymes, and growth inhibitors. Stratification is the breaking down of dormancy in seeds by keeping them refrigerated and moist for several weeks.

Review Questions

1. What are the differences among hormones, enzymes, and vitamins?

2. Auxins, gibberellins, and cytokinins all promote growth. How does one distinguish among them?

3. How do hormonal herbicides function?

4. List several commercial applications of ethylene gas.

5. What are statocytes, and what role do they play in plant growth?

6. Distinguish among internally and externally induced growth movements, turgor movements, and dehydration movements.

7. How do day-neutral and intermediate-day photoperiod plants differ in their requirements?

8. What is dormancy, and how may it be artificially broken?

9. What are the requirements for a seed to germinate?

10. How does phytochrome differ from other plant pigments?

Discussion Questions

1. If green plants need light in order to manufacture food, why are seedlings that are germinated in the dark taller than those of the same age grown in the light?

2. If you remove a terminal bud from a stem, will growth in length stop altogether? Explain.

3. Would it be technically correct to say that some plants go to sleep at night? Explain.

4. Many plants produce seeds that require a period of dormancy before they will germinate. Of what value is the dormancy to the plant? Where might you expect to find plants with seeds that do not undergo dormancy?

5. The rapid turgor movements of the sensitive plant are rare in the Plant Kingdom. If all plants evolved such movements, would the phenomenon have possible survival value or other value to the plants? Explain.

Additional Reading

Briggs, W. R.; Green, P. B.; Jones, R. L., eds. *Annual review of plant physiology.* Vol. 27. Palo Alto, CA: Annual Reviews; 1976.

Hillman, W. S. *Photoperiodism in plants and animals.* Oxford Biology Readers, No. 107. Burlington, NC: Carolina Biological Supply Company; 1979.

Leopold, A. C.; Kriedemann, P. E. *Plant growth and development.* 2d ed. New York: McGraw-Hill Book Co.; 1975.

Luckwill, L. C. *Growth regulators and crop production.* Studies in Biology, No. 129. Baltimore: University Park Press; 1981.

Miller, G. T., Jr. *Living in the environment.* 2d ed. Belmont, CA: Wadsworth, Inc.; 1979.

Nickell, L. G. *Plant growth regulators—agricultural uses.* New York: Springer-Verlag; 1982.

Steeves, T. A.; Sussex, I. M. *Patterns in plant development.* Englewood Cliffs, NJ: Prentice-Hall; 1972.

Thimann, K. V. *Hormone action in the whole life of plants.* Amherst: University of Massachusetts Press; 1977.

Thomas, M.; Ranson, S. L.; Richardson, J. A. *Plant physiology.* 5th ed. London: Longman Group; 1973.

Torrey, J. G. *Development in flowering plants.* New York: Macmillan Co.; 1967.

Wareing, P. F.; Phillipps, I. D. *The control of growth and differentiation in plants.* 3d ed. Elmsford, NY: Pergamon Press, Inc.; 1981.

Weaver, R. J. *Plant growth substances in agriculture.* San Francisco: W. H. Freeman and Co.; 1972.

11 Meiosis and Alternation of Generations

Overview

After a comparison of asexual and sexual reproduction, this chapter explores the necessity for halving the chromosome number at some stage in the life cycle of sexually reproducing organisms. The phases of meiosis are systematically covered. The chapter also presents a discussion of alternation of generations and general rules for understanding the process.

Some Suggested Student Learning Goals

1. List all the phases of meiosis and briefly describe what occurs in each of them.

2. Indicate clearly what features meiosis and mitosis have in common and how they differ.

3. Explain the significance of crossing over to offspring.

4. In alternation of generations, indicate at what point each of the following occurs: change from N to $2N$; change from $2N$ to N; initiation of the gametophyte generation.

Figure 11.1 Contrast between asexual and sexual reproduction in a strawberry plant. A more detailed life cycle is shown in figure 11.7.

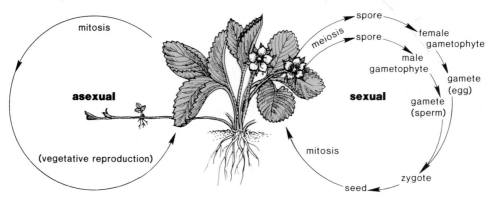

Introduction

As plants grow, the number of cells multiplies through the process of cell division discussed in chapter 3. Mitosis ensures, in very precise fashion, that the number of chromosomes and the nature of the DNA in the daughter cells will be identical with that in the parent cell. But living organisms do not grow indefinitely, although a few may remain alive for a long time. In order to perpetuate the species, they must reproduce or they will become extinct. This reproduction may take place through natural vegetative propagation (see appendix 4) or by means of special cells called *vegetative spores* (see chapters 16–20), which are produced by mitosis. Such processes, in which the cells involved are identical in their chromosomes to the cells from which they arise, are forms of **asexual reproduction** (the prefix *a* of *asexual* means "without," so asexual reproduction simply means reproduction without sex). Nearly all plants also undergo **sexual reproduction,** however, which in flowering and cone-bearing plants ultimately results in the formation of **seeds.** In sexual reproduction, special sex cells called **gametes** are produced. Two gametes, called **egg** and **sperm** in higher plants and animals, fuse together forming a single cell called a **zygote,** which is the first cell of a new individual (figure 11.1).

If gametes had the same number of chromosomes as all the other cells of the plant, then the zygote, since it results from the union of two gametes, would have double the number of its parents' chromosomes. If such zygotes were then to develop into mature plants, each of the new plants' cells would have double the original number of chromosomes, and when they produced gametes, the zygotes resulting from the union of this next generation of gametes would have four times the number of chromosomes of the grandparents. If that sort of thing were to continue for just twenty generations, a species having 16 chromosomes in each cell to begin with would end up with no fewer than 8,388,608 chromosomes per nucleus! The problem never actually arises, however, because at a certain stage in the life cycle of all organisms undergoing sexual reproduction, a unique process called **meiosis** occurs. This process brings about the development of gametes that have only half the number of chromosomes of any body developing from the zygote, and so when the gametes unite in pairs, forming zygotes, the original chromosome number is restored. Since some aspects of meiosis are similar to those of mitosis (see figure 11.2), it is strongly recommended that you review mitosis (chapter 3) before going through the phases of meiosis outlined in the following section.

Meiosis and Alternation of Generations

Figure 11.2 Contrast between mitosis and meiosis.

mitosis   prophase   metaphase   anaphase

interphase

meiosis   prophase I   metaphase I   anaphase I

telophase

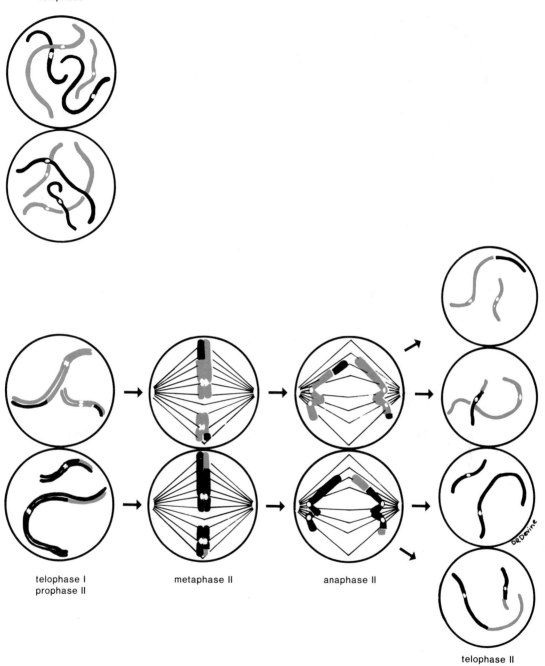

telophase I
prophase II

metaphase II

anaphase II

telophase II

When an interphase cell undergoes meiosis, four cells are produced as a result of two successive divisions, which take place without a "resting period" between them. Because of the remarkable events that occur during the process, the four cells, depending on the organism involved, are seldom to almost never identical with the original cell or with each other.

Phases of Meiosis

Like mitosis, meiosis is a continuous process that has been divided up into arbitrary phases for convenience. Bear in mind, however, that the lines between one phase and another are indistinct. Four main phases are recognized in each of the two divisions, and the initial phase is further subdivided. Since the main phases of the first division bear the same names as those of the second division, it has become customary to designate first division phases with a Roman numeral I and those of the second division with a Roman numeral II. The number of the chromosomes is reduced to half during the first division, but no further reduction in number takes place throughout the second division. Accordingly, some have referred to division I as a *reduction division* and to division II as an *equational division.*

Meiosis generally takes much longer than mitosis to be completed. In lilies, for example, mitosis runs its full course in about twenty-four hours, while meiosis takes about two weeks. In other organisms, meiosis may take months or even years to be completed. In human females the process sometimes takes as long as fifty years since it remains in a state of arrest for most of that time after it has begun.

Division I (Meiosis I)

Prophase I

The main features of prophase I are (1) the chromosomes pair up as they become shorter and thicker, and their double nature becomes apparent; (2) the nuclear envelope and the nucleolus disappear; and (3) the pairs of chromosomes often exchange parts with each other and then separate (figure 11.2; see figure 11.4 A, B).

As with prophase of mitosis, the beginning of this phase is marked by the appearance of chromosomes as faint threads in the nucleus. These threads gradually coil like a spring so that they become shorter in length, and the coil obviously becomes thicker than the thread alone. As the chromosomes become shorter and thicker they align themselves in pairs, and eventually two parallel threads, the **chromatids,** can be distinguished for each chromosome—four chromatids, therefore, in each pair of chromosomes. One of the chromosomes of each pair came from the male parent and the other from the female parent. They are identical with each other in length, in the amount of DNA present, and in having a centromere at the same precise location, but their genes may control contrasting characteristics (see chapter 12). Such chromosomes are called *homologues,* and associated pairs of **homologous chromosomes** are referred to as **bivalents.** The four chromatids of a bivalent each have their own centromere (figure 11.2; see figure 11.4 B).

As prophase I progresses, parts of the chromatids of the homologous chromosomes may break and be exchanged with each other. Depending on the length of the chromosomes, this exchange may occur at one to several points throughout the length of the paired homologous chromosomes, or if the chromosomes are short, it may not occur at all. The evidence for this exchange of parts appears a little later in prophase I, when the chromosomes of each bivalent appear to push each other apart. It can then be seen that

Figure 11.3 Chiasmata in chromosomes at prophase I of meiosis in crested wheat grass. (Photomicrograph by William Tai)

adjacent chromatids have crossed over each other at one to several points, and remain attached at these points. This phenomenon of **crossing over** brings about an exchange of some of the DNA contributed by the two parents and is the basis for some of the variability seen in the offspring. The crossovers are called **chiasmata,** and their relative positions provide a basis for the study of where various genes are located on the chromosomes (figure 11.3). After the chiasmata have appeared, the chromatids slowly tear apart, each remaining the same length as it was originally and with the same amount of DNA but now possessing "traded" material. The process is something like having two cars that are identical except for the color and the smell of the upholstery, and exchanging three of the fenders and the back seat between the two. When you were through with the exchange, you still would have structurally identical cars, but each would have a different combination of fenders and seats with respect to color and smell.

By the end of prophase I, the nuclear envelope has disintegrated, the nucleolus has disappeared, and spindle fibers are beginning to form. As in mitosis, some spindle fibers are attached to the centromeres of the chromosomes, whereas others extend from pole to pole.

Metaphase I

The main features of metaphase I are (1) the chromosomes line up at the equator of the cell in pairs and (2) the spindle becomes conspicuous and complete (figure 11.2; figure 11.4 C).

Meiosis and Alternation of Generations

Figure 11.4   Meiosis in crested wheat grass.
(Photomicrographs by William Tai)

A. **Early Prophase I.** Chromosomes begin to coil and contract and appear as threads.

B. **Prophase I.** Homologous chromosomes become aligned in pairs, and some crossing-over is visible.

F. **Prophase II.** The chromosomes coil and contract again; because of crossing over, the chromatids are no longer identical with each other.

G. **Metaphase II.** The chromosomes of each cell become aligned along their respective equators.

H. **Anaphase II.** The chromatids separate completely and migrate to the poles.

Chapter 11

C. **Metaphase I.** Bivalents become aligned along the equator.

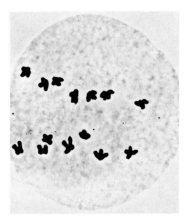

D. **Anaphase I.** Homologous chromosomes separate and migrate to opposite poles.

E. **Late Telophase I.** Chromosomes are at the poles, and the original cell becomes two cells.

I. **Telophase II.** The four groups of chromatids (now called chromosomes again) are at the poles; new cell walls begin to form.

J. **Formation of cell walls.** Cell walls form; the four cells will become pollen grains.

Meiosis and Alternation of Generations

Figure 11.5 Homologous chromosomes at metaphase I of meiosis.

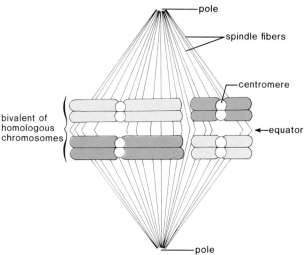

This phase resembles metaphase of mitosis except that when the chromosomes move to the equator, they line up along it in pairs, with homologous chromosomes on each side of the invisible "line" (figure 11.5). Also, as indicated in our discussion of mitosis in chapter 3, the two chromatids of a chromosome are joined together at their centromeres and function as a single unit at this stage.

Anaphase I

The main feature of anaphase I is that one whole chromosome (consisting of two chromatids) from each pair migrates to a pole (figure 11.2; figure 11.4 D).

In anaphase of mitosis, the two chromatids of each chromosome, which are being held together at their centromeres, separate as the centromeres become detached from each other, and a chromatid of each pair migrates to an opposite pole. Anaphase I of meiosis is fundamentally different in that the chromatids of each chromosome remain cohered at their centromeres and do

not separate from one another, a whole, two-stranded chromosome from each pair migrating to an opposite pole instead. Thus, when the chromosomes arrive at the poles, they still consist of two chromatids, but only half the total number of chromosomes is at each pole. If a particular chromosome was involved in crossing over in prophase I, one of the chromatids will consist of a mixture of chromatid material from a homologous chromosome and some of its original DNA.

Telophase I

Depending upon the species, the chromosomes may now either partially revert back to interphase by becoming longer and thinner, or they may proceed directly to division II (figure 11.2; figure 11.4 E). Normally no new nuclear envelopes are organized around the chromosomes, but the nucleoli usually do reappear.

Division II (Meiosis II)

In the second division of meiosis, the events of the various phases are similar to those of mitosis except that there is no duplication of DNA during any part of the interphase that may occur between the two divisions, and also the number of

Chapter 11

chromosomes is half the number that existed at the beginning of meiosis. In some organisms where there is no interphase, the process proceeds almost directly from telophase I to metaphase II.

Prophase II

The main feature of prophase II is the chromosomes become shorter and thicker and their double nature once more becomes apparent (figure 11.2; figure 11.4 F). Both groups of chromosomes undergo this stage simultaneously.

Metaphase II

The main features of metaphase II are (1) the chromosomes line up along the equator and (2) new spindles become conspicuous and complete.

The two spindles may be formed either at right angles to the ones that had formed in metaphase I or at the same angle. A spindle fiber is attached to each centromere, and other spindles extend from pole to pole (figure 11.2; figure 11.4 G).

Anaphase II

The main feature of anaphase II is the chromatids of each chromosome separate and migrate to opposite poles (figure 11.2; figure 11.4 H).

In mitosis, the centromeres separate during anaphase, but in meiosis the chromatids remain together during anaphase I and do not separate as their centromeres become detached until anaphase II.

Telophase II

The main features of telophase II are (1) the chromatids (now called chromosomes again) lengthen and become thinner and (2) new nuclear envelopes and nucleoli organize for each group of chromosomes (figure 11.2; figure 11.4 I).

This phase is accompanied by the formation of new cell walls between each of the four groups of chromosomes.

The set of chromosomes present in each of the four cells formed by the end of telophase II constitutes half the original number, and if crossing over and exchange of material between chromatids have occurred, none of the four cells will have exactly the same combination of DNA. In many plants, these four cells, which are called *sexual spores,*[1] develop into various structures or bodies, and it is within certain cells of these bodies that gametes are produced by mitosis.

Alternation of Generations

As we have noted, meiosis occurs at one stage of the life cycle of any organism undergoing sexual reproduction. The chromosomes of each cell that result from the process constitute a complete set, since one member of every original pair ends up in each cell. The original chromosomal complement, consisting of two complete sets of chromosomes, is restored when gametes unite, forming a zygote. Any cell having one set of chromosomes is said to be **haploid,** and any cell with two sets of chromosomes is said to be **diploid.** For example, the gametes of an organism are haploid, whereas a zygote of the same organism is diploid. This holds true regardless of the number of chromosomes peculiar to a given organism. Thus, we can state that an organism having N (a specified quantity) chromosomes in its haploid cells will have twice as many, or $2N$, chromosomes in its diploid cells.

In most animals the only haploid cells (i.e., cells with N or a single set of chromosomes) are the gametes (egg and sperm). In plants, however, this is generally not so. In a complete life cycle involving sexual reproduction, there is an alternation between a diploid ($2N$) **sporophyte** phase and a haploid (N) **gametophyte** phase. This is

---

1. Some use the term *meiospore* to distinguish a spore produced by meiosis from a vegetative spore produced by mitosis.

commonly referred to as **alternation of generations.** The diploid body itself is called a *sporophyte*. It develops from a zygote and eventually produces **spore mother cells,** each of which undergoes meiosis, producing four **spores.** The haploid bodies that develop from these spores are called **gametophytes.** These eventually form sex organs in which gametes are produced by mitosis. As we shall see in chapters to follow, the gametophytes of many primitive forms constitute a large part of the visible organism, but as we progress up through the Plant Kingdom to more complex plants, they become proportionately reduced in size until they may be only microscopic. The switch from one generation to the other takes place when spore mother cells undergo meiosis, producing spores, and again when **fertilization** (or **syngamy,** fusion of gametes, forming a zygote) occurs. Although the sporophyte generation of some primitive organisms may consist of a single cell (the zygote), the basic plan of alternation of generations can be seen in the Protistan, Fungal, and Plant Kingdoms. It becomes most conspicuous, however, in the Plant Kingdom, and it differs from one organism to the next in the forms of the various bodies and cells. If you will note the following six rules pertaining to alternation

of generations, along with the accompanying diagram (figure 11.6), you should have little trouble following the life cycle of any sexually reproducing organism discussed in this book.

1. The first cell of any *gametophyte generation* is normally a *spore (sexual spore* or *meiospore)*, and the last cell is normally a *gamete*.
2. Any cell of a gametophyte generation is *haploid (N)*.
3. The first cell of any *sporophyte generation* is normally a *zygote*, and the last cell is normally a *spore mother cell*.
4. Any cell of a sporophyte generation is *diploid (2N)*.
5. The change from a sporophyte to a gametophyte generation occurs as a result of *meiosis*.
6. The change from a gametophyte to a sporophyte generation occurs as a result of *fertilization* (fusion of gametes), which is also called *syngamy*.

The word *generation* as used in *alternation of generations* simply means *phase* of a life cycle and should not be confused with the more widespread use of the word pertaining to time or offspring.

Figure 11.6   Life cycle of a plant undergoing sexual reproduction

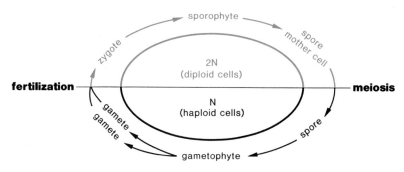

Summary

Reproduction may take place through natural vegetative propagation or by spores (asexual reproduction) or by sexual processes (sexual reproduction). In sexual reproduction two gametes unite, forming a zygote, which is the first cell of a new individual. The process of meiosis ensures that gametes will have half the chromosome number of the zygotes, and also ensures that offspring will not be identical with the parents in every respect.

Meiosis takes place by means of two successive divisions, each of which, like mitosis, is divided into arbitrary phases even though the process is continuous. In prophase I the chromosomes pair up, often exchange parts, and then separate. The similar chromosomes of each pair are referred to as being homologous, and associated pairs are referred to as bivalents. Exchange of parts of chromatids in bivalents may be effected by the parts initially crossing over one another and forming chiasmata, and then tearing apart again. In metaphase I the chromosomes become aligned at the equator in pairs and in anaphase I whole chromosomes from each pair migrate to opposite poles. In telophase I the chromosomes either partially revert to their interphase state or initiate the second division, which is essentially similar to mitosis. In prophase II the chromosomes of each of the two groups become shorter and thicker again, with both groups becoming aligned at their respective equators in metaphase II. In anaphase II the chromatids of each chromosome separate and migrate to opposite poles, and in telophase II the chromatids lengthen, and new nuclear envelopes and nucleoli appear for all four groups; new cell walls are produced between each of the four groups. If the chromatids have exchanged parts earlier, none of the four groups may have identical combinations of DNA, and each group has half the original number of chromosomes. Each of the four cells constitutes a spore (sexual spore or meiospore), which may develop into a body or structure within which gametes may be produced by mitosis.

Any cell having one set of chromosomes is said to be haploid or to have N chromosomes; any cell having two sets of chromosomes is said to be diploid or to have $2N$ chromosomes. In the life cycle of an organism that undergoes sexual reproduction, there is an alternation between a haploid phase and a diploid phase. The haploid body is called a gametophyte and the diploid body is called a sporophyte. The change from the haploid phase to the diploid phase occurs when two gametes (each N) unite, forming a zygote ($2N$) in the process of syngamy, or fertilization; the change from the diploid to the haploid phase occurs as a result of meiosis, when a $2N$ cell becomes four N cells. This switching of phases is referred to as alternation of generations.

Review Questions

1. Indicate when each of the following events occurs during meiosis: (a) crossing over; (b) chromatids separating at their centromeres and migrating to opposite poles; (c) chromosomes aligning themselves in pairs at the equator.

2. Is there any difference between mitosis and the second division phases of meiosis?

3. What is the significance of meiosis with respect to sexual reproduction?

4. What is meant by saying that the cells of a sporophyte phase are diploid or $2N$?

5. At what stage of a life cycle does the chromosome number of cells switch from N to $2N$?

Discussion Questions

1. If a plant reproduces very freely by asexual means, is there any advantage to its also reproducing sexually?

2. Is there any advantage to a plant in having an alternation of generations?

3. Would it make any difference if the events of the second division of meiosis occurred before the events of the first division?

4. Would it be correct to say that a bivalent has four times the amount of DNA present in the chromosomes of a cell in anaphase II?

5. Would you assume that the length of a chromosome might have anything to do with the number of crossovers it might form with its homologue?

6. Two mitotic divisions may take place in as little as a few hours, whereas meiosis may take much longer. Can you suggest any reasons for this?

Additional Reading

DeRobertis, E. D. P.; Saez, F. A.; DeRobertis, E. M. F., Jr. *Cell biology.* 6th ed. Philadelphia: W. B. Saunders Co.; 1975.

Dyson, R. D. *Cell biology: a molecular approach.* 2d ed. Boston: Allyn and Bacon, Inc.; 1978.

Raven, P. H.; Evert, R. F.; Curtis, H. *Biology of plants.* 3d ed. New York: Worth Publishers; 1981.

Wolfe, S. L. *Biology of the cell.* Belmont, CA: Wadsworth Publishing Co.; 1972.

Genetics and Evolution

Overview

In this chapter the story of Karpechenko's crossing of a radish and a cabbage introduces the topic of genetics. This is followed by a brief history of Mendel and his classic experiments with pea plants. Each of Mendel's principles or laws is discussed, and examples of monohybrid and dihybrid crosses are given. The chapter then introduces backcross, linkage, chromosomal mapping, and the Hardy–Weinberg law.

The discussion of evolution begins with an abbreviated history of Charles Darwin and the principles or tenets involved in the theory of evolution through natural selection. Mechanisms of evolution, including mutations, hybridization and introgression, polyploidy, apomixis, and reproductive isolation are discussed. The chapter concludes with an examination of the role of isolation and mutation in modifying Darwinism and with an examination of the evidence for evolution.

Some Suggested Student Learning Goals

1. Identify Mendel and explain the significance of his experiments with peas.

2. Describe or define dominance, phenotype, genotype, homozygous, heterozygous, monohybrid cross, dihybrid cross, backcross, linkage, chromosomal mapping, the Hardy–Weinberg law.

3. Give the ratios of the offspring in the first two generations from a monohybrid and a dihybrid cross. Describe the genotypes involved.

4. Describe how introgressive hybridization may lead to the development of new species.

5. List the contributions of Charles Darwin to theories of organic evolution and the tenets of natural selection as he understood them.

6. Explain the significance of mutation and reproductive isolation to evolution.

7. Give the reasons, past and present, for the controversy over evolutionary theory.

Introduction

When a plant is being grown as a crop, the more parts of the plant that are edible or otherwise useful, the more efficient the use of the land on which the crop is grown. Some parts of many crop plants are not used, however, except possibly as ingredients in compost. In lettuce, for example, the roots are not useful for either food or fuel; in tomatoes and grapes, only the fruits are edible; in potatoes, only the tubers are used; in cotton, the fibers and the seeds constitute the useful parts. In 1928 the Russian biologist G. D. Karpechenko reported on an experiment in which he tried to cross two food plants in the cabbage family—the radish, with an edible root, and the cabbage, with edible leaves. One might assume that such a cross, if successful, would produce an ideal crop plant. Karpechenko found that these two plants, although related, did not cross readily, but he did succeed in obtaining a few seeds from such crosses. When these seeds were planted, the results were dramatic. Huge plants grew and produced fertile seeds of their own. From a crop point of view, however, the cross was a dismal failure because these large plants had leaves mostly like those of a radish and roots like those of a cabbage—they were virtually useless!

Such crossing experiments are a part of **genetics,** the science that deals with heredity, or natural inheritance. Genetics is one of the youngest of the biological sciences, having developed after the details of meiosis and mitosis became understood at the turn of the century. It has since become a vast and very important field of study, having the potential for not only improving food production in a hungry world but also for solving various problems related to the inheritable human diseases, the effects of radiation, the control of population growth, and many other matters of direct human interest and concern.

Genetics

Genetics as a science originated about a generation before its significance became appreciated in the scientific community as a whole. An Austrian monk, Gregor Mendel (figure 12.1), taught between 1853 and 1868 in what is now the Czechoslovakian city of Brno. In the monastery there, he carried on a wide range of studies and experiments in physics, mathematics, and natural history. He also became an authority on bees and meteorology and kept notes on experiments involving two dozen different kinds of plants. Today he is best known for the studies he conducted with a number of varieties of pea plants.

Mendel published the results of his studies on pea plants in a biological journal in 1866, and he sent copies of his paper to leading European and American libraries. He also sent copies to at least two eminent botanists of the time. His work was, nevertheless, completely ignored or overlooked until 1900, when three botanists (Eric von Tschermak of Austria, Carl Correns of Germany, and Hugo de Vries of Holland), working independently in their own countries, reached the same conclusions as Mendel. Each, as a result of library research, also came across Mendel's original paper.

Peas, unlike most flowering plants, are self-pollinating and self-fertile. In other words, a pollen grain of a pea flower is capable of germinating on its own stigma, fertilizing one of its own ovules, and developing a fertile pea seed. Although Mendel did not know about mitosis and meiosis, he had observed that if a pea plant grew to a certain height, its offspring, if grown under the same conditions, would always reach approximately the same height, generation after generation. He wondered if he crossed a tall plant with a short one if the offspring would be intermediate in height. To make such a cross he would need to reach into a flower of one plant and remove the stamens before the pollen had matured, to prevent self-pollination. Then he could take

Chapter 12

Figure 12.1 Gregor Mendel. (Photo, ca. 1870s, Brünn, Moravia, courtesy of Brooklyn Botanic Gardens, New York; print at Hunt Institute, Pittsburgh, PA.)

pollen from another plant and apply it by hand to the stigma of the first flower. He also needed to do something to prevent insects from accidently bringing pollen from other flowers after the cross had been made (e.g., cover the experimental flowers with a cloth or paper bag).

When Mendel made such crosses, the results were astonishing. All of the offspring were tall.

There were no short or intermediate plants. He found, however, that if he allowed the offspring plants to pollinate themselves, their offspring, in turn, occurred in a ratio of approximately three tall plants to one short plant (figure 12.2).

Figure 12.2 Cross between a tall variety and a dwarf variety of peas

X

parents

F₁ generation (all tall)

F₂ generation (3 tall, 1 dwarf)

Mendel then tried crosses between peas with smooth seeds and those with wrinkled seeds. He also crossed green-seeded plants with yellow-seeded ones; he ultimately used seven different pairs of contrasting characteristics in all.

The original plants involved in making the crosses were referred to as the *parents,* or the *parental generation,* and their offspring were referred to as the *first filial generation* (*filial* is defined as "of or relating to a son or daughter"), usually abbreviated F_1. The offspring of F_1 plants were called the *second filial generation,* or F_2.

Chapter 12

Table 12.1 Summary of Mendel's data

Parents	F$_1$	F$_2$
Yellow × **green seeds**	all yellow	6022 yellow: 2001 green
Smooth × **wrinkled seeds**	all smooth	5474 smooth: 1850 wrinkled
Green pod × **yellow pod**	all green	428 green: 152 yellow
Long stem × **short stem**	all long	787 long: 277 short
Axial flowers × **terminal flowers**	all axial	651 axial: 207 terminal
Inflated pods × **constricted pods**	all inflated	882 inflated: 299 constricted
Red flowers × **white flowers**	all red	705 red: 224 white

Mendel counted the number of offspring in each cross, something none of his predecessors had done. In addition, he studied the inheritance of single pairs of contrasting characteristics and succeeded where others had failed because he did not try to interpret inheritance in terms of whole organisms. Besides being astute and meticulous in his work, it was also fortunate that he chose the plants and the characteristics he did, because we now know that many other characteristics are controlled by complex factors that do not give the simple, clear-cut results Mendel obtained. A summary of the data accumulated by Mendel in his work with peas is provided in table 12.1.

As Mendel began to collect data in his experiments, he realized that there must be something inside the pea plants that made them have yellow or green seeds and red or white flowers. He referred to this unknown agent as a *factor*. He also deduced that each plant must have two such factors for each characteristic, since even though all the offspring of an F$_1$ generation appeared more or less identical, the F$_2$ generation revealed some plants with one characteristic and others with the contrasting characteristic. These discoveries and deductions came to be a principle, or law, known as the *law of unit characters*. Stated simply, this law says that "factors, which always occur in pairs, control the inheritance of various characteristics." These factors later became known as **genes** (discussed in chapter 3); we know

now that there may be hundreds or even thousands of pairs of genes on each pair of homologous chromosomes (see chapter 11).

From his data and observations Mendel also deduced that one factor, or gene, of a pair may overcome or conceal the expression of the other. For example, in the F$_1$ generation of a cross involving yellow-seeded and green-seeded parents, all the plants had yellow seeds, yet both kinds of parents obviously had to contribute something to the cross since some of the F$_2$ plants had yellow seeds whereas others had green seeds. This deduction led to another principle known as the *law of dominance*. This principle says that "in any given pair of factors or genes, one may suppress or mask the expression of the other." The expressed factor is referred to as the **dominant** while the one not expressed is referred to as the **recessive.** In the yellow-seeded–green-seeded cross, the factor yellow is dominant and the factor green is recessive.

Mendel's crosses also made it clear that a plant's having yellow seeds did not indicate that both members of the pair of factors, or genes, controlling seed color were present, because the dominant could suppress or mask the presence of the recessive one. To distinguish between the appearance of an organism and its genetic components we use the terms *phenotype* and *genotype*.

Figure 12.3 Absence of dominance (incomplete dominance). A cross between a red-flowered and a white-flowered variety of snapdragons.

parents

red   X   white

F₁ (all pink)

pink

F₂

red   pink   pink   white

Phenotype refers to the physical appearance of the organism, and **genotype** refers to its genetic makeup. We customarily describe phenotypes with words and designate genotypes with letters. In Mendel's crosses, for example, seven pairs of phenotypes are shown as parents; these phenotypes include yellow seeds, green seeds, smooth seeds, wrinkled seeds, green pods, yellow pods, etc. In designating the genotypes of each of these plants, a capital letter is used to indicate the dominant member of a pair of genes; the *same letter* in lowercase is used to indicate the corresponding recessive. In the yellow-seeded–green-seeded pea cross, for example, yellow is dominant. If a plant has a yellow phenotype, its genotype could be either *YY* or *Yy*. The green genotype would be *yy*. The plant is said to be genotypically **homozygous** if both members of a pair of factors, or genes, are the same (e.g., *YY* or *yy*) and **heterozygous** if the pair includes both types (e.g., *Yy*).

Although the principle of **dominance** was amply demonstrated by Mendel's crosses, we now know that in many other instances neither member of a pair of factors, or genes, completely dominates the others; in other words, in some species there is an *absence of dominance*. In snapdragons, for example, the F₁ plants of a cross between a red-flowered parent and a white-flowered parent are all pink-flowered; the characteristics of both factors blend, yielding an intermediate condition. When an F₂ generation is produced from such a cross, the flowers of the offspring appear in a ratio of 1 red: 2 pink: 1 white. This phenomenon is also referred to as *incomplete dominance* (figure 12.3).

You will recall that the cells contributed by the parents in sexual reproduction (egg and sperm in higher organisms) are called *gametes* and that the offspring develop from *zygotes* formed when the gametes unite. You will also recall that gametes have only half the number of chromosomes that is present in the body cells of the organism.

Figure 12.4 Monohybrid cross between a yellow-seeded pea plant and a green-seeded pea plant. Yellow (**Y**) is dominant; green (**y**) is recessive.

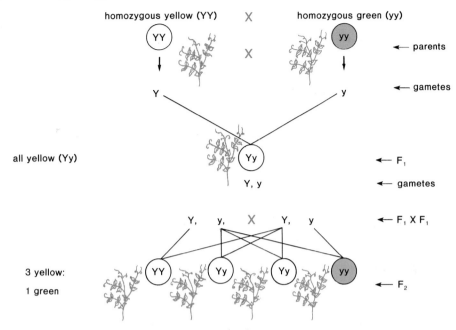

Each member of a pair of factors, or genes, controlling a given characteristic is located at a specific point on the chromosome, and the other member of the pair is found at the same precise point on a corresponding (homologous) chromosome in the cell. When meiosis occurs, the members of pairs of factors become separated and are not matched up again until a zygote is formed. Mendel called this separation of paired factors, or genes, into separate cells the *law of segregation*.

Monohybrid Cross

With these first three laws in mind, let us examine a cross between a homozygous yellow-seeded pea plant and a homozygous green-seeded pea plant (figure 12.4). The genotype of the homozygous dominant parent (yellow-seeded) is *YY;* the genotype of the homozygous recessive parent (green-seeded) is *yy.* You will recall from chapter 11 that gametes have only one member of a pair of homologous chromosomes. The gametes of the yellow-seeded parent (*YY*), therefore, will have a genotype of *Y;* the gametes of the green-seeded parent (*yy*) will have a genotype of *y.* No matter which egg unites with any sperm of the other parent, all the zygotes of this cross will be heterozygous, with the genotype *Yy.* This means that all the individuals of the F$_1$ generation will be yellow-seeded phenotypes, and all will be, like the zygotes, heterozygous. A cross of this type, involving two parents differing by only a single genetic trait, is called a **monohybrid** cross.

It is customary when making such crosses to try to produce an F$_2$ generation by crossing members of the F$_1$ generation with one another. Two kinds of gametes are produced in equal numbers by each F$_1$ parent. All of the F$_1$ plants involved are yellow-seeded phenotypes, and their genotypes are all *Yy.* When gametes are produced, half will be *Y* and half will be *y.* Thus, a *Y* egg may unite with either a *Y* sperm nucleus, producing a *YY* zygote, or, purely at random, the same egg may unite with a *y* sperm nucleus, producing a *Yy* zygote. The same type of random unions occur with *y* eggs so that either *Yy* or *yy* zygotes are produced. If all the offspring of a large number of such crosses are counted, it will be seen that they have been produced in a ratio of 1 *YY:* 2 *Yy:* 1 *yy* genotypes and the ratio will be approximately three yellow-seeded phenotypes to one green-seeded phenotype. These four equally possible F_2 offspring are shown at the bottom of figure 12.4.

Dihybrid Cross

Thus far, the crosses we have been discussing have involved single pairs of factors, or genes. But, as we have noted, homologous chromosomes may contain hundreds or even thousands of pairs of genes. If crosses are made involving parents that are homozygous for two or more pairs of genes, the nature of meiosis dictates that the genes will not necessarily be inherited together. In fact, the two or more sets of genes are separated when meiosis occurs and, if they are on separate chromosomes, they are combined again in the formation of zygotes. Another law of Mendel's, the *law of independent assortment,* recognizes this and states that the "factors (genes) controlling two or more contrasting pairs of characteristics segregate independently and the gametes combine at random." This law is illustrated in **dihybrid crosses,** which involve two pairs of genes located on separate pairs of chromosomes.

Let us assume that the factors for yellow seeds and green seeds are located on one pair of chromosomes while those for tall plants and dwarf plants are located on another pair. You will recall that tallness in peas is dominant and dwarfness is recessive. The phenotypes of the homozygous dominant parent will be yellow and tall while those of the recessive parent will be green and dwarf; their corresponding genotypes will be *YYTT* and *yytt.* All the gametes from the dominant parent will have a genotype of *YT*; all those

from the recessive parent will be *yt.* The dihybrid genotypes of the zygotes will all be *YyTt* and the phenotypes of the F_1 generation will all be yellow and tall. If these dihybrid members of the F_1 generation are then crossed with one another to produce an F_2 generation, four kinds of gamete genotypes will be produced in equal numbers by each parent because of the independent segregation that occurs during meiosis; these gamete genotypes are *YT, Yt, yT,* and *yt.* Since any one of the four kinds of gametes can unite randomly with any of the four kinds of gametes of the other parent, sixteen combinations are possible.

A *Punnett square* is used to determine the genotypes of the zygotes in order to avoid the confusion of having to remember all possible combinations of gametes. The diagram looks somewhat like a checkerboard, with one set of gametes across the top and the other set down one side (figure 12.5). If the genotypes of the zygotes are added up, it can be seen that there are nine different kinds in a ratio of 1 *YYTT:* 2 *YYTt:* 2 *YyTT:* 4 *YyTt:* 1 *YYtt:* 2 *Yytt:* 1 *yyTT:* 2 *yyTt:* 1 *yytt.* Four kinds of phenotypes are produced from these genotypes in a ratio of 9 yellow, tall: 3 yellow, dwarf: 3 green, tall: 1 green, dwarf. These ratios are expected when large numbers of individuals are involved. With small numbers, chance alone may not produce the expected ratios. Human populations, for example, are divided relatively equally into males and females, but in smaller groups such as families, the general population ratio of one male to one female may not be evident.

Backcross

If a scientific theory is valid, one should be able to test it experimentally. Mendel himself tested his predictions by means of **backcrosses.** He had found in pea flowers that red color is dominant and that white is recessive (as contrasted with snapdragons in which color dominance is lacking), so that all the F_1 offspring of a cross between a homozygous red-flowered parent and a homozygous white-flowered parent would be red. The genotypes of such a cross would be *RR* (red) $\times$ *rr* (white), and the F_1 generation would

Figure 12.5 Dihybrid cross between a yellow-seeded, tall pea plant and a green-seeded, dwarf pea plant. The F_2 offspring are shown in a Punnett square.

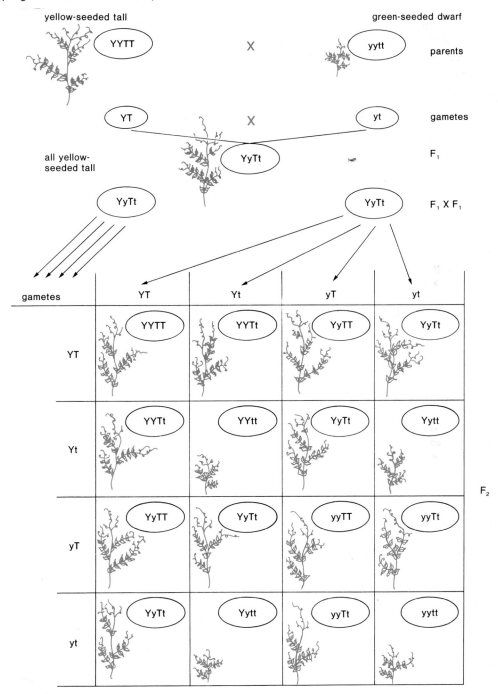

be all **Rr.** The gametes of such F_1 offspring would be either **R** or **r,** with each type being produced in equal numbers. Thus, if pollen from an F_1 hybrid is placed on the stigma of a white-flowered plant (which, since it shows the recessive trait, has to be homozygous), there is one chance in two that a dominant (**R**) gamete will form a zygote with a recessive (**r**) one and an equal chance that two recessive (**r**) gametes will form a zygote, with the offspring being produced in equal numbers of heterozygous red (**Rr**) and white (**rr**). Such backcrosses involving the homozygous recessive parent and the F_1 offspring are routinely made today, and invariably they yield the same results, namely, a 1:1 ratio of phenotypes. The same 1:1 ratios are produced even when there is no dominance. In snapdragons, for example, when the pink-flowered F_1 offspring are backcrossed with a red-flowered parent, a ratio of 1 red: 1 pink results. If the white parent is used, the backcross produces a ratio of 1 white: 1 pink.

Linkage

There may be thousands of genes on each chromosome, and the closer the genes are to one another on any given chromosome, the more likely they are to be inherited together. The tendency of genes on the same chromosome to be inherited together is called **linkage.**

In 1906, just a few years after the basic details of meiosis became known, W. Bateson and R. C. Punnett (after whom the Punnett square is named) of Cambridge University in England became the first to report on linkage in sweet peas. They knew from earlier work with sweet peas that purple flower color was dominant and red was recessive; they also knew that pollen grains with a shape that was oblong in outline ("long pollen") were dominant and that spherical pollen grains ("round pollen") were recessive. When they crossed a homozygous purple, long plant with a homozygous red, round plant, all the F_1 offspring were purple, long, as expected. When they crossed the F_1 plants with one another to obtain an F_2 generation, however, the F_2 offspring phenotypes

were produced in numbers that differed markedly from the expected 9:3:3:1 ratio. Puzzled by these results they tried a backcross, crossing F_1 plants (heterozygous purple, long) with the recessive parent (homozygous red, round). Again, the expected 1:1:1:1 ratio was not produced. Instead they obtained a ratio of 7 purple, long: 1 purple, round: 1 red, long: 7 red, round. Bateson and Punnett could not explain adequately what they had observed. In 1910, however, T. H. Morgan, who had observed similarly puzzling ratios in his experiments with fruit flies, theorized that linkage and crossing over were responsible, an explanation that is still considered valid today.

If the genes for purple and red, and long and round, were on separate pairs of chromosomes, the genotypes of the F_1 would all be **PpLl** and the homozygous recessive would be **ppll**. This should have produced 1 **PpLl:** 1 **Ppll:** 1 **ppLl:** 1 **ppll** instead of the 7:1:1:7 ratio actually obtained. Apparently the F_1 parent produced many more **PL** and **pl** gametes than **Pl** and **pL** gametes. If we designate the parents in the backcross $\frac{PL}{pl}$ and $\frac{pl}{pl}$ to indicate that **P** and **L** are on one chromosome and **p** and **l** are on another, we would expect the F_1 parent $\left(\frac{PL}{pl}\right)$ to have produced just two kinds of gametes, namely **PL** and **pl,** and the backcross to have produced only two kinds of phenotypes, namely, purple, long and red, round in equal numbers. But some purple, round and red, long phenotypes also were produced. The only plausible explanation for this seems to be that some **crossing over** occurs, with **P** and **L** genes occasionally being switched with **p** and **l** genes.

Chromosomal Mapping

As a result of thousands of experiments with many organisms, particularly those with fruit flies and with corn, it is now known that crossing over between particular pairs of linked genes occurs regularly. The frequency of crossing over varies widely, depending on the genes involved. We must conclude, therefore, that each gene has a specific location (*locus*) on a chromosome, and, indeed,

Figure 12.6 Salivary gland chromosomes of a common fruit fly. The staining bands seen on the chromosomes have, in some instances, been correlated with specific genes through chromosome mapping. (Photo © 1978 Charles P. Havel)

by calculating percentages of crossing over through counting large numbers of offspring phenotypes, we can effectively map the relative locations of genes on chromosomes. If, for example, we assume that the space within which 1% crossing over occurs is one unit of map distance between linked genes, then 7% crossing over between genes A and B would mean that they are 7 units of map distance apart on their chromosome.

A◄――― 7 units ―――►B

If we find there is 10% crossing over between A and a third linked gene C, and that B and C are shown by crossover percentages to be 17 units apart, we can deduce that A lies between B and C on its chromosome.

C◄―――10 units―――►A◄――7 units――►B
          17 units

However, if there is only 3% crossing over between B and C, we would have to conclude that B lies between A and C.

C◄――― 10 units ―――►A
◄3 units►B◄――― 7 units ――►

In Bateson and Punnett's backcross, two of every sixteen F$_2$ offspring were produced through recombinations due to crossing over; this two-sixteenths is 12.5%. We can deduce from this that the genes controlling flower color and pollen shape in sweet peas are 12.5 map units apart on the chromosomes. The locations of many hundreds of pairs of genes have been determined and mapped in corn and in fruit flies (figure 12.6), and the mapping of chromosomes in other organisms is currently being undertaken at institutions around the world.

The Hardy–Weinberg Law

Before Mendel's work became known, most biologists believed that inherited characteristics were a blending of those furnished by the parents, and it was difficult to understand why unusual characteristics did not eventually become so diluted that they essentially disappeared. After Mendel, biologists asked why dominant genes did not eventually completely eliminate recessive ones in breeding populations. G. H. Hardy, a British mathematician, and W. Weinberg, a German physician, pointed out the reason in 1908, and their observation became known as the *Hardy–Weinberg Law.*

The Hardy–Weinberg law states that "the proportions of dominant genes to recessive genes in a normally interbreeding population will be retained from generation to generation unless there are forces that change those proportions." The law operates, however, only when populations are large. In small populations, there can be a random loss of individual genotypes through, for example, a failure to breed. This in itself can lead to loss of a gene. The proportions can also be upset by a particular gene mutating to another form and then mutating back again to its original form at a slower rate. But *selection* is by far the most significant cause of exceptions to the Hardy–Weinberg law. The manner in which selection operates is discussed in the section on evolution, which follows.

Evolution

Few historical events within the past 126 years have had a greater impact on society in general and the biological sciences in particular than Charles Darwin's book, *On the Origin of Species.* Darwin's theory of evolution through "natural selection" has stimulated an enormous amount of thinking and research. It also created a great deal of controversy initially, and although the controversy has somewhat subsided, disagreement still

Figure 12.7 Charles Darwin. (Courtesy National Library of Medicine)

exists today. Part of the remaining controversy has stemmed from a failure of people in diverse fields to define terms and to distinguish between fact and theory in evolutionary matters. Evolution itself, for example, has been broadly defined by some as simply being synonymous with change. We are told that anything from cars to computers to cultures is evolving and that even our thought processes evolve. Thus, to distinguish between change in inanimate or intangible entities and change in living organisms, we need to refer to the latter as *organic evolution.* Even then we do not pin down the subject entirely, for there are varying theories of organic evolution. That which pervades and unifies most biological thought today, however, finds its origin in the observations of Charles Darwin (figure 12.7) and of a contemporary of his, Alfred Wallace, who independently arrived at the same conclusions as Darwin.

Charles Darwin was born in Shrewsbury, England, in 1809. His father, Dr. Robert Darwin, was a successful country physician and relatively wealthy. In 1825, at the age of sixteen, young Darwin was sent to the University of Edinburgh Medical School to follow in his father's footsteps. He did not do well in his studies, however, and dropped out after two years. The following year he went to Cambridge University to study for the ministry but did not excel academically in this field either. Part of the reason for his poor scholastic showing apparently was due to the fact that he spent much time in the countryside collecting beetles, rocks, and other specimens or talking at length with his Cambridge biology professors, who held him in high regard. He barely attained what today could be called a C average, but he managed to graduate with a degree in theology in 1831. He was then twenty-two years old but was still not really sure what he wanted to do with his life. While he was pondering such matters, the queen of England commissioned a sailing vessel, the H. M. S. *Beagle,* to undertake a voyage around the world for the purpose of charting coastlines, particularly those of South America. Young Darwin was recommended for the (unpaid) post of naturalist on the voyage by his Cambridge biology professors. His father was opposed to the idea, but his uncle sided with him and he accepted the position.

The voyage, which began December 27, 1831, and took five years, gave Darwin an opportunity to collect specimens on both sides of South America, as well as in the Galápagos Islands and along the coasts of Australia and New Zealand. He kept a daily journal and spent countless hours collecting alone on horseback. This gave him ample opportunity to think about the forms and distribution of the myriad new organisms he encountered. His thoughts slowly led to the development of ideas that later blossomed into his theory of evolution through natural selection.

Upon his return to England in 1836, Darwin obtained the financial support of his father and his cousin Emma, whom he married. Although still young, he retired to the country and began working on his collections and journal. He also carried on a voluminous correspondence with other biologists and made extensive investigations into pollinating mechanisms, earthworm ecology, geographical distributions of plants and animals, and several other areas of natural history. Throughout all of his activities he was guided by a concept that he had adapted from an essay written by Malthus in 1798 on human populations and food supplies. Darwin realized that although human beings might artificially improve or increase their food supply through selective breeding and cultivation, plants and animals could not do so and were therefore vulnerable to a process of selection in nature, which would explain changes in natural populations. He was reluctant to publish his ideas, however, and did not begin putting together his book on the origin of species until 1856.

Meanwhile, an English naturalist by the name of Alfred R. Wallace (1823–1913), who made major contributions to our knowledge of animal geography, independently concluded that natural selection contributed to the origin of new species and sent Darwin a brief essay on the topic in 1858. Urged by friends, Darwin published a statement of his views jointly with Wallace, and in 1859 Darwin's classic book *On the Origin of Species by Means of Natural Selection* was published.

In essence, Darwin's theory of evolution through natural selection is based on four principles, or tenets.

1. *Overproduction.* Many living organisms produce enormous quantities of reproductive cells or offspring. For example, a single maple tree produces thousands of seeds each year, each capable of becoming a new tree. Some fungi produce trillions of spores, each with the potential to become a new fungus.

2. *Struggle for Existence.* All the germinating seeds, spores, and other reproductive cells of living organisms "compete" for available moisture, light, nutrients, and space. The amounts of these elements available in nature are not sufficient to support all of these organisms, and many die as a result.

3. *Survival and Reproduction of the Fittest.* Those organisms best adapted to the environment have the best chance to survive and reproduce, while others less well adapted may succumb. A tree with thicker bark, for example, may survive cold temperatures; it hence may survive to reproductive age and therefore its offspring will bear the inherited features of the thicker-barked tree.

4. *Variation and Inheritance.* All living organisms vary. Those hereditary variations with survival value are inherited from generation to generation and accumulate in time, while other variations not important to the survival of the species are gradually eliminated.

One of the criticisms of Darwin's theory, published in 1859, was that it did not explain how hereditary variations originated and developed. It should be remembered, however, that Mendel's findings were not published until 1866, and the details of mitosis and meiosis did not become known until 1900–1906. Today, with our far greater knowledge of how variations occur and are inherited, we have come to understand much better the mechanisms of evolution in populations than was possible in Darwin's time.

Populations of organisms from aardvarks to zinnias are composed of a few to billions of individuals, yet even within very large populations it is virtually impossible to find two that are identical down to the last molecule. As humans, we are well aware of this within our own species, but variation exists in all living organisms, even though in simpler one-celled forms the differences may be much less obvious. Whether the differences are obvious or subtle, however, if the environment or other factors favor certain hereditary variations and bring about the elimination of others, the general characteristics of a population will eventually change.

In some instances, the environment may alter the phenotype slightly without affecting the genetic constitution of an organism. Plants that grow relatively tall at sea level may become dwarfed when they are transplanted to cooler areas in the mountains or drier areas near deserts, yet they are capable of breeding freely with plants at the original location if they are returned to that area. If one applies fertilizers to plants, stimulates their growth with hormones, or prunes them, the changes are not passed on to the offspring, for no permanent change occurs in a given population unless there is *inheritable variation*. The changes in transplanted, fertilized, or pruned plants are not transmitted to the offspring because the gametes of those plants will carry the same genetic information they would have carried if the transplanting, fertilizing, or pruning had not occurred.

Before the details of genetics became known, many prominent biologists believed that hereditary changes in populations over long periods of time (evolution) occurred as a result of the inheritance of acquired characteristics. One of the more prominent supporters of this widespread idea was Jean Baptiste Lamarck (1744–1829). He believed that giraffes attained their long necks over many generations as a result of the gradual increase in neck length as shorter-necked animals stretched to reach leaves on the branches of trees. As slight stretching of the neck occurred, this was supposed to have been passed on to the offspring, and eventually numerous tiny increases due to individual stretching added up to the present great neck-length of giraffes. If such were the case, one should be able to demonstrate it experimentally, and indeed, attempts were made, but all failed. For example, one such attempt involved the surgical removal of the tails of mice for many successive generations, but the mice of the last generation had tails of exactly the same average length as those of the first generation, because the repeated removal of the tails in no way affected the hereditary characteristics carried in the genes within the cells. Likewise, fruit trees that are pruned annually never produce even slightly dwarfed tree seeds as a result, even after many generations.

Mechanisms of Evolution

Despite the experiments just discussed, short-tailed mice and dwarf fruit trees do occasionally occur, but for reasons quite different from those proposed by Lamarck and his contemporaries. They come about as a result of a sudden change in a gene or chromosome. Such a change is called a **mutation,** a term introduced in 1901 by the Dutch botanist Hugo de Vries. Changes within chromosomes may occur in several ways. A part of a chromosome may break off and be lost (*deletion*), or a piece of a chromosome may become attached to another (*translocation*), or in some instances a part of a chromosome may break off and then become reattached after the piece has become inverted (*inversion*) (figure 12.8). A mutation of a gene itself may involve a change in one or more nucleotide pairs (see chapter 3). Mutation rates vary considerably from gene to gene, but mutations occur constantly in all living organisms at an average roughly estimated to be about 1 mutant gene for every 200,000 produced. Mutator genes that increase the rate of mutation in other genes have been discovered, but generally the mutation rate for a specific gene remains relatively constant unless changes in the environment (e.g., increase in cosmic radiation) occur. Most mutations are harmful, many times to the point of killing the cell. About 1% or fewer are either harmless or produce a characteristic that

Figure 12.8 Some types of chromosomal changes that can occur. *A.* **Deletion.** Part of a chromosome breaks off and is lost. *B.* **Translocation.** Part of one chromosome becomes attached to another. *C.* **Inversion.** Part of a chromosome breaks off, becomes inverted from its original position, and then is reattached.

A.

B.

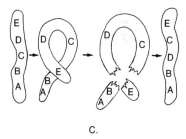

C.

may help the species survive changes in its environment. Modern theories of evolution extend Darwin's natural selection theory by recognizing that mutation and recombination of genes through hybridization are fundamental to bringing about the changes that result in new species. They also include in the process *reproductive isolation* (discussed in a later section).

Role of Hybridization in Evolution

Hybridization (the production of offspring by two parents that differ in one or more characteristics) seldom occurs naturally, but when it does occur the **hybrids** may be significant or important in evolutionary change, depending on how the characteristics of the parents were combined. If, for example, the environment changes (e.g., average temperatures drop or annual precipitation increases), hybrids may have gene combinations that are better suited to the new environment than those of either parent. If two species hybridize occasionally, *introgression* (backcrossing between the hybrids and the parents) may occur. If the backcrossing occurs repeatedly, some characteristics of the parents may disappear eventually from the population if the new combinations of genes in the offspring happen to be better suited to the environment than those of the parents, and as natural selection favors the offspring. Both parents and hybrids may, however, also evolve in other ways.

Occasionally—but rarely—during mitosis, a new cell wall may fail to develop between two daughter cells even though the chromosomes have divided. This situation can result in a cell with twice the original number of chromosomes. If mitosis were to occur normally after such a cell was formed and the cell divided repeatedly until a complete organism resulted, that organism would have double the original number of chromosomes in all its cells. This phenomenon, **polyploidy,** does occasionally occur in nature, and the individuals that result are known as *polyploids.*

In agriculture, polyploids are sometimes produced artificially by applying an alkaloidal drug called *colchicine* (derived from corms of the autumn crocus) to a field of plants. At the appropriate concentration many of the plants are killed, but in a few survivors the colchicine interferes with the separation of the chromatids during anaphase of mitosis and with the formation of the cell plate, doubling the chromosome number in the cells affected. The hybrids resulting from a cross between two species are often sterile because the chromosomes do not pair up properly in meiosis. If polyploidy does occur in such a hybrid, however, the extra set of chromosomes present from each parent presents an opportunity for any one chromosome to pair with its homologue in meiosis, and thus possibly the problem of sterility may be overcome. This type of polyploidy apparently occurred frequently in the past (in terms of geological time), and it is believed that more than one hundred thousand species of flowering plants that exist today originated in this fashion.

Sterile hybrids also may reproduce by asexual means. Asexual propagation, called *apomixis,* may include the development and production of seeds without fertilization, as well as other forms of vegetative reproduction discussed in chapters 4–7 and in appendix 4. When species reproduce principally by apomixis but sometimes also hybridize so that new combinations of genes are occasionally produced, they can be highly successful in nature. Dandelions and wild blackberries, for example, reproduce apomictically, as well as by other means, and are among the most successful plants known.

Reproductive Isolation

To visualize how mutation and isolation can bring about changes in populations, imagine that you are living in a broad valley bounded by ranges of high mountains. Imagine further that in the spring the valley is carpeted with millions of yellow daisies. In other words, a conspicuous element of the spring flora is a species of yellow daisy that is distributed throughout the valley floor.

Suppose a dominant mutation that results in red flowers occurs in one plant. The bees, when visiting this plant, become dusted with pollen, which they spread to other plants. The following season, when seeds from such plants germinate, there will be more red-flowered plants. Given enough time, red-flowered plants will appear throughout the entire valley.

Now suppose a major geological upheaval erects a new range of mountains through the valley, making two valleys. Then suppose a second mutation that occurs results in plants with long silky hairs. The bees spread the pollen carrying the mutation as before, but this time their activities are confined to one of the new valleys because the bees will not fly over the mountain range. Meanwhile, a third mutation that increases, for example, the length of the internodes occurs in the other valley, and then more mutations that affect leaf shape, physiology, or protein complements. In time, two different species, incapable of interbreeding with one another, could develop in the adjacent valleys as a result of the reproductive isolation brought about by the new mountain range and the mutations that occurred in the separated populations.

Isolation leading to the development of new species is not limited to physical barriers such as mountains or oceans. Climate and soils may play a role, or a mutant form within a population may flower at a different time, preventing exchange of genes between it and nonmutant forms. In the temperate deciduous woods of eastern North America, there are many populations of Dutchman's breeches, early spring-flowering herbs (discussed further in chapter 13). These have delicate, highly dissected leaves, which are nearly indistinguishable from those of related squirrel corn plants often found growing right among the Dutchman's breeches (see figure 13.1). So closely do the plants resemble each other vegetatively that early botanists and many lay persons assumed that they were the same species. It is believed, however, that at some point in geological history a mutation or mutations occurred that

caused some plants to begin flowering after other plants had already set seed. As a result, one group became reproductively isolated from the other group (figure 12.9). In due course, other mutations affecting the form and fragrance of the flowers and the shape and pigmentation of food storage bulblets beneath the surface also occurred, but the plants continued to occupy the same habitats. In short, we now have two closely related but distinct species growing together without interbreeding because it is no longer possible for them to do so.

Literally thousands of examples of such phenomena can be discerned or postulated in populations of living organisms in all parts of the world. Our knowledge of the mechanisms that bring about such changes is put to practical use in the development of new breeds of animals and a host of horticultural varieties of plants, many of which have proved invaluable in increasing food production through larger yields and disease resistance.

Discussion

If we now know and understand some of the mechanisms of evolution, why does there yet remain controversy over the subject? Obviously ignorance is a factor, but it is not the sole reason. Science deals with tangible facts and evidence that can be experimentally tested; beliefs stemming from metaphysics and religion are outside the realm of science to prove or disprove. Additional evidence in support of evolution is drawn from several areas, including the form and ecology of living organisms, and the way they are related to each other today; from the structure and relationships of molecules, and from fossils, which are remnants of previously living forms (figure 12.10). The simplest fossils are generally found in the oldest geological strata, while more complex forms tend to be found in younger strata. Still further evidence is drawn from the geographical distribution of organisms (see figure 12.9). Many

Figure 12.9 Geographical distribution of two species of *Dicentra*. *A.* Distribution of *D. canadensis* (squirrel corn). *B.* Distribution of *D. cucullaria* (Dutchman's breeches). (After Stern, K. R. 1961. "*Revision of Dicentra* (Fumariaceae)." Brittonia 13(1): 1–57. © The New York Botanical Garden. Reproduced by permission.)

A.

D. canadensis

B.

D. cucullaria

D. cucullaria
(western distribution)

Figure 12.10 Fossil fern

groups are confined to a single continent or island. In some instances where similar organisms occur on more than one land mass, there is evidence that the land masses concerned were once linked together, which would have permitted terrestrial migration. Still other conclusions are drawn from the physiology and chemistry of the organisms, and in animals, particularly, from the presence of vestigial organs (structures that are apparently now functionless but are presumed to represent previously functional organs in predecessor species). Among the best known such structures in humans are the appendix and wisdom teeth.

Most scientists who have studied the evidence feel that some form of evolutionary process is the only plausible explanation for the unity of life at the molecular and cellular level and the extraordinary diversity of the organisms now around us. There is little unanimity of thought, however, as to precisely how evolution proceeded in the past. One authority will be convinced that a certain group evolved from another, while other equally eminent authorities will maintain that the exact reverse occurred. Part of the reason for such paradoxes is that the historical record is quite incomplete. Although the fossil evidence for the evolution of a few organisms such as certain mollusks and the horse is fairly substantial, other fossil evidence is very fragmentary. Possibly fewer than one in each million organisms that once existed ever became a fossil, and there are probably no fossils at all of many herbaceous and soft-bodied organisms. "Missing link" fossils are frequently referred to, but often it is whole "chains" that are missing and science has recognized only a few isolated links. With such evidence, scientists can deal only in probabilities or possibilities, and it is inevitable that various, sometimes conflicting, interpretations result. Objective scientists freely acknowledge that numerous problems concerning the interpretation of the geological past and the pathways of organic evolution exist, but they ask their detractors to suggest more plausible alternatives. At this point, some apply the tenets of religious faith, which like history, is not subject to scientific experimentation. Some see no conflict between science and religion, and others are convinced that the two are mutually exclusive. Depending on individual points of view, an impasse may result, with persons of all persuasions—including, unfortunately, scientists—becoming dogmatic on the topic. Virtually no objective thinkers will deny, however, the extraordinary impact theories of evolution have had on modern peoples and on their concepts of the living world around them.

Summary

Karpechenko crossed a radish and a cabbage in an attempt to produce a better food crop, but the results were unsatisfactory; such crossing experiments are a part of genetics, the science that deals with heredity or natural inheritance. Genetics as a science originated with Gregor Mendel, an Austrian monk, who performed experiments with peas. Mendel made crosses with varieties of peas that had single pairs of contrasting characteristics, each pair being carried on a separate pair of homologous chromosomes. The original plants involved in making the crosses were referred to as the parents, or parental generation, while their offspring were designated as the F_1

generation and the offspring of crosses between members of the F_1 generation as the F_2 generation. F_1 plants were also called hybrids. Mendel differed from his predecessors in counting the number of all offspring and in being meticulous. He also was exceptionally lucky in the characteristics he chose to study. In the seven types of crosses he made, he found that the F_1 hybrids all resembled their parents in appearance, but the F_2 offspring were produced in a ratio of 3 plants resembling one parent to 1 plant resembling the other parent. He referred to agents for the characteristics inside the plants as "factors" and also deduced that each plant must have two factors for each characteristic. From this came the principle or law of unit characters, which states that "factors, which always occur in pairs, control the inheritance of various characteristics." The factors later became known as genes.

Mendel further deduced that one member of a pair of factors may suppress or mask the expression of the other member. The suppressing member was called a "dominant" and its counterpart a "recessive." To distinguish between the appearance of an organism and its genetic makeup, different terms are applied: "phenotype" for appearance and "genotype" for genetic makeup. Phenotypes are described with words while genotypes are designated with letters; capital letters are used for dominants and lowercase letters for recessives. If both members of a pair of factors, or genes, are identical, the plant is said to be homozygous; if the members of a pair of genes are contrasting, the plant is said to be heterozygous (for those factors). Sometimes the F_1 generation may phenotypically be more or less intermediate between the parents as a result of absence of dominance (incomplete dominance).

In a monohybrid cross the offspring are produced in a ratio of 3:1, when dominants and recessives are involved. In a dihybrid cross involving dominance and two pairs of genes carried on separate pairs of chromosomes, the offspring often are produced in a phenotypic ratio of 9:3:3:1.

Backcrosses are routinely made to check the results of genetic experiments. Such crosses involve $F_1 \times$ the recessive parent and normally produce a 1:1 (monohybrid) ratio or a 1:1:1:1 (dihybrid) ratio of phenotypes.

There are often thousands of genes on the same chromosome; genes that are inherited together are said to be linked. Bateson and Punnett discovered that flower color and pollen grain shape in sweet peas did not produce an expected 9:3:3:1 ratio of phenotypes in the F_2 generation. Morgan postulated that the 7:2:2:7 ratio they obtained instead was the result of linkage and occasional crossing over. Calculating crossover percentages has enabled workers to determine the relative position of genes on chromosomes, a process known as chromosomal mapping. The closer the genes are to each other on a chromosome, the less likely they are to be involved in a crossover, and vice versa.

Before Mendel's work became known, most biologists did not understand why recessive characteristics in a population did not eventually disappear. Hardy and Weinberg explained, in what is now known as the Hardy–Weinberg law, that the proportions of dominant genes to recessive genes in a normally interbreeding population will be retained from generation to generation unless there are forces that change those proportions. The most significant cause for exceptions to the Hardy–Weinberg law is selection.

The theory of organic evolution received its greatest impetus from Charles Darwin's *On the Origin of Species*. Darwin's theory is based on four principles, or tenets: (1) overproduction (many organisms produce enormous numbers of offspring); (2) "struggle for existence" (all living organisms "compete" for available nutrients, etc., but there is not enough for all and many die as a result); (3) "survival and reproduction of the fittest" (those organisms best adapted to the environment have the best chance to survive); and (4) variation and inheritance (those hereditary variations with survival value are inherited from generation to generation and accumulate in time, while other variations are eliminated).

Chapter 12

Before the details of genetics became known, many biologists believed that hereditary changes in populations over long periods of time (evolution) occurred as a result of the inheritance of acquired characteristics. One such biologist, Lamarck, believed giraffes had attained their long necks by constantly stretching for food and that the slight stretching in each generation had been passed on until in time the numerous small stretchings had added up to the present long necks. This theory was discredited experimentally by surgically shortening tails of mice for many generations and by pruning trees; none of the acquired characteristics were passed on.

Mechanisms of evolution include mutations in chromosomes or genes; hybridization (particularly introgression, the backcrossing of hybrids to parents in a natural population); polyploidy (having additional sets of chromosomes as a result of a cell's failure to produce a cell wall during mitosis and having the new cell develop into a new organism—those developing from sterile hybrids are especially significant); apomixis (asexual reproduction); and reproductive isolation.

If new dominant genes are produced in a freely interbreeding population, they will gradually be spread throughout the population, and the nature of the whole population will change in time. However, if a population is divided by a barrier such as a mountain range or an ocean, a dominant gene occurring in the one population will not spread throughout the isolated population as before. In time, because of the isolation, each new population may develop into separate species incapable of breeding with one another.

Additional evidence in support of evolution is drawn from the form, the ecology and relationships of living organisms, the molecular structure, the geographical distribution, fossils, and vestigial organs (structures that are apparently now functionless but are presumed to represent previously functional organs in predecessors). Most scientists feel some form of evolution is the only plausible explanation for the unity of life at the molecular and cellular level and the great diversity of life, but there is little agreement among them as to precisely how evolution proceeded in the past. This is partly due to the inadequacy of the fossil record, which is incomplete because proportionately so few organisms ever became fossils, and the fact that history is not subject to scientific experimentation. There are a variety of opinions and convictions on origins, but few advocates can deny the major impact that theories of evolution have had on modern peoples and on their concepts of life.

Review Questions

1. When did genetics originate as a science?

2. Why did Mendel succeed where others had failed?

3. Who were Tschermak, Correns, and de Vries?

4. Define hybrid, F_1 generation, phenotype, genotype, homozygous, heterozygous, dominant, recessive.

5. What are Mendel's laws?

6. In a variety of flowering plants, blue flowers and dwarf habit are controlled by dominant genes; white flowers and tall habit are the recessives. The gene pairs are carried on separate pairs of chromosomes. If a homozygous blue dwarf plant is crossed with a homozygous white tall plant,

   a. What are the phenotypes and genotypes of the F_1 generation?

   b. How many different *kinds* of genotypes could occur in the F_2 generation?

   c. What phenotypic ratio would you expect in the F_2 generation?

   d. What genotypes could result from the cross *BBDD* × *BbDd*?

   e. What phenotypes could result from the cross *bbDd* × *BBDd*?

7. In summer squashes, disk-shaped fruit is dominant over spherical fruit and white fruit color is dominant over yellow fruit color. If a homozygous white, disk-shaped variety is crossed with a homozygous yellow, spherical variety, how many different homozygous genotypes are possible in the F_2 generation?

8. A cross is made between a white-flowered plant and a red-flowered plant of the same variety. The F_2 generation consisted of 32 white-flowered plants, 64 pink-flowered plants, and 29 red-flowered plants. Explain.

9. In tomatoes, smooth skin is dominant and fuzzy skin is recessive; also, tallness is dominant and dwarfism is recessive. A homozygous smooth, tall variety is crossed with a homozygous fuzzy, dwarf variety, and the F_1 is backcrossed to a homozygous fuzzy, dwarf variety. The offspring of the backcross are 95 smooth, tall; 3 fuzzy, tall; 4 smooth, dwarf; and 96 fuzzy, dwarf. Is linkage involved? If so, what is the percentage of crossing over?

10. What is meant when genes are said to be linked?

11. What is the difference between organic evolution and other forms of evolution?

12. How did Darwin's theory of evolution differ from Lamarck's?

13. What basic modifications have been made in evolutionary theory since Darwin's time?

14. Why is a hybrid often sterile?

15. What is meant by a mechanism of evolution?

16. How are polyploids produced?

17. Why is reproductive isolation necessary for evolution to occur?

18. What is chromosomal mapping and how is it done?

19. What is the Hardy–Weinberg law?

20. What evidence is there to support modern concepts of evolution? Are there any problems with the evidence?

Discussion Questions

1. Mendel is said to have been exceptionally lucky in his discoveries. Do you think most scientists who make significant discoveries are lucky?

2. The peas Mendel worked with were largely self-fertile. Of what advantage or disadvantage to a species is such a phenomenon?

3. One of Darwin's tenets was that there is a struggle for existence among living organisms. Do plants struggle with one another to survive? If so, how do they do it?

4. Some populations change noticeably in form within a hundred years. If only 1 gene in every 200,000 per cell mutates, and if most mutations are harmful, how is such change possible?

5. Do you think there might be viable alternatives to organic evolutionary theory to explain the diversity of life around us?

Additional Reading

Anderson, E. *Plants, man and life.* Berkeley: University of California Press; 1968.

Ayala, F. J.; Valentine, J. W. *Evolving: the theory and processes of organic evolution.* Menlo Park, CA: Benjamin-Cummings Publishing Co.; 1979.

Darwin, C. *On the origin of species.* Facsimile 1st ed. of 1859. New York: Atheneum Publishers; 1967.

Ehrlich, P. R.; Holm, R. W.; Parnell, D. R. *The process of evolution.* 2d ed. New York: McGraw-Hill Book Co.; 1975.

Grant, V. *Plant speciation.* New York: Columbia University Press; 1971.

Kerkut, G. A. *Implications of evolution.* Oxford and New York: Pergamon Press; 1960.

Merrell, D. *Introduction to genetics.* New York: W. W. Norton and Co.; 1975.

Murray, J. *Genetic diversity and natural selection.* New York: Hafner Publishing Co.; 1972.

Srb, A.; Owen, K. D.; Edgar, R. S. *General genetics.* 2d ed. San Francisco: W. H. Freeman and Co.; 1965.

Stebbins, G. L. *Processes of organic evolution.* 3d ed. Englewood Cliffs, NJ: Prentice-Hall; 1977.

Wagner, R. P.; Judd, B. H.; Sanders, B. G.; Richardson, R. H. *Introduction to modern genetics.* New York: John Wiley & Sons, Inc.; 1980.

lant Names and
lassification

erview

chapter begins with a
ussion of the problems involved
... the use of common names for
plants, as illustrated by a survey of
such names for two related species
of American spring-flowering
perennials. It continues with a brief
historical account of the events that
led to the development and
acceptance of Linnaeus's Binomial
System of Nomenclature. The
history of the development of a
five-kingdom classification is
presented, along with a list of the
divisions and classes included in
the kingdoms covered in this text.
The chapter concludes with a
synoptic key to the kingdoms and
divisions of organisms.

Some Suggested
Student Learning Goals

1. List several problems
   associated with the application
   of common names to
   organisms.

2. Describe the Binomial System
   of Nomenclature, how it
   developed, and how it is
   currently used.

3. List several reasons for
   recognizing more than two
   kingdoms of living organisms.

4. Describe the bases for
   Whittaker's five-kingdom
   system.

5. Construct a dichotomous key
   to ten different objects (e.g.,
   pencils, golf balls).

Introduction

Biologists sometimes are thought by the general public to be slightly pompous when they refer to peanuts as *Arachis hypogaea* or an African hoopoe bird as *Upupa epops*. They are not, however, merely "putting on the dog" or trying to be difficult. Rather, they are identifying organisms by their *scientific names* through a system that has evolved over the last few centuries. Circumstances have dictated the necessity of distinguishing among the 10 million existing types of organisms, as well as those that have become extinct, in a manner that will identify them anywhere, regardless of the languages spoken locally.

At present, all living organisms are given a single two-part Latin scientific name, and most also have *common names*. Only one correct scientific name applies to all individuals of a specific kind of organism, no matter where they may be found, but many common names may be given to the same organism, and one common name may be applied to a number of different organisms. For example, the scientific name *Dicentra cucullaria* has been given to a pretty spring-flowering plant native to eastern North America and the Columbia River basin in the West (see distribution map in figure 12.9). Its unique flower shape, reminiscent of the baggy pants of a traditional Dutch costume, has resulted in its being given the common name Dutchman's breeches, but it also has the common names of little-boys' breeches, kitten's breeches, breeches-flower, boys-and-girls, Indian boys-and-girls, monkshood, white eardrops, soldier's cap, colicweed, little blue stagger, white hearts, butterfly banners, rice roots, and meadow bells. In addition to these English common names, the plant has Indian names, and in the Canadian province of Quebec it has French names. Often growing with it is a related plant, with similar leaves but flowers of a slightly different shape, having the scientific name *Dicentra canadensis*. Because of the similarities between the two plants and their close association in the woods where they occur, some people in the past assumed that they were merely two different forms of the same plant. This is reflected in the fact that some of the common names of *Dicentra cucullaria* (e.g., colicweed, Indian boys-and-girls, little blue stagger) have also been applied to *Dicentra canadensis* (figure 13.1), in addition to the names squirrel corn, turkey corn, turkey pea, wild hyacinth, fumitory, staggerweed, and trembling stagger. But the problem of the diversity and the overlapping of common names for these two plants does not stop there, for the names monkshood, soldier's cap, rice roots, meadow bells, and turkey corn have also been applied to completely unrelated plants.

In Europe, with its many languages, common names can become very numerous indeed. The widespread weed with the scientific name *Plantago major,* for example, is often called broad-leaved plantain in English, but it also has no fewer than 45 other English names, 11 French names, 75 Dutch names, 106 German names, and possibly as many as several hundred more names in other languages, with literally dozens of these common names also applying to quite different plants. If it were not for the early recognition by biologists and others of the urgent need for worldwide uniformity in naming and classifying all organisms, utter chaos eventually might have prevailed in communications concerning them.

Figure 13.1 A. *Dicentra cucullaria.* B. *Dicentra canadensis.*

A.

B.

Development of the Binomial System of Nomenclature

As was indicated in chapter 1, the first person of note to attempt to organize and classify plants was Theophrastus, the brilliant student of Aristotle and Plato. His third-century B.C. classification of nearly five hundred plants into trees, shrubs, and herbs, along with his distinctions between plants on the basis of leaf characteristics, was used for hundreds of years. Not until the thirteenth century A.D. was a distinction between monocots and dicots made on the basis of stem structure. After this, herbalists added considerably to Theophrastus's list, and by the beginning of the eighteenth century, details of fruit and flower structure were utilized in classification schemes in addition to form and habit. By this time European scholarly

institutions were bulging with thousands of plants that explorers were bringing back from around the world, and confusion over scientific and common names was multiplying. The use of Latin in schools and universities had become widespread, and it was then customary to use descriptive Latin phrase names for both plants and animals. All organisms were grouped into **genera** (singular: **genus**), with the first word of the Latin phrase indicating the particular genus to which the organisms belonged. For example, all known mints were given phrase names beginning with the word *Mentha,* the name of the genus. Likewise, the phrases for lupines began with *Lupinus,* and those for poplars began with *Populus.* A complete phrase name for spearmint read *Mentha floribus spicatis, foliis oblongis serratis.* Roughly translated, it means "Mentha with flowers in a spike (an elongated but compact flower cluster); leaves oblong, saw-toothed." A phrase name for the closely related peppermint read *Mentha floribus capitatus, foliis lanceolatis serratis subpetiolatis,* meaning "Mentha with flowers in a head; leaves lance-shaped, saw-toothed, with very short petioles."

It was at this point in history that the Swedish botanist, Carolus Linnaeus (1707–1778), entered the scene and established the basis for the present-day naming and classification of organisms. Linnaeus, who was nicknamed the "Little Botanist" at school, inherited his passion for plants from his father, who was a minister and an amateur gardener. He is said to have been much impressed at the age of four by his father's remarks about the uses of neighborhood plants in his home community of Råshult, located in southern Sweden. After a brief tenure as a student at the University of Lund, he spent most of his time making excursions to Lapland, Holland, France, and Germany. Eventually he became professor of botany and medicine at Uppsala, where he inspired large numbers of students, twenty-three of whom became professors themselves. It is said that he frequently led large field trips into the countryside, accompanied by a musical band.

When Linnaeus began his work, he set out to classify all known plants and animals according to their genera. In 1753, he published a two-volume work entitled *Species Plantarum,* which was later to become the most important of all works on plant names and classification. In this work, he not only included a referenced list of all the Latin phrase names previously given to the plants, but he also changed some of the phrases, when necessary, to reflect relationships, placing one to many specific kinds of organisms called **species**[1] in each genus. He limited each Latin phrase to a maximum of twelve words, and in the margin next to the phrase he listed a single word, which, when combined with the generic name, formed a convenient abbreviated designation for the species. The word in the margin for spearmint was *spicata,* and the word for peppermint was *piperita.* Thus, the abbreviated name for spearmint was *Mentha spicata* and for peppermint *Mentha piperita.* Although Linnaeus originally considered the phrase names the official names of the plants, he and those who followed him eventually replaced all the phrase names with abbreviated ones. Because of their two parts, these names became known as *binomials* and the method of naming became known as the Binomial System of Nomenclature. Today all organisms are named according to this system.

In addition to establishing the Binomial System of Nomenclature, Linnaeus sought to do more than merely publish a long list of plants. After all, such lists have limited usefulness if they cannot be used to identify the plants concerned. Linnaeus organized all known plants into twenty-four **classes,** which were based primarily on the number of stamens (pollen-bearing structures) in flowers. For example, all plants having five stamens per flower were placed in one class, while all having six stamens were placed in another; plants and other organisms that have no flowers at all (e.g., mosses, fungi) were assigned to a class of their own. His arrangement was for convenience and did not necessarily reflect natural relationships, but for the first time it became possible for a worker to identify an unknown plant. This arrangement was used in *Species Plantarum* and other works by Linnaeus (figure 13.2).

In 1867, more than one hundred years after *Species Plantarum* was published, about one hundred fifty European and American botanists

---

1. Note that *species* is like the word *sheep* in that it is spelled and pronounced the same way in either singular or plural usage. There is no such thing as a plant or animal "specie." Since Linnaeus's time, a species has been defined as "a population of individuals capable of freely interbreeding in nature but not generally interbreeding with members of another species."

```
124        TETRANDRIA DIGYNIA.
              HAMAMELIS.
virginiana. 1. HAMAMELIS, Gron. virg. 139. Cold. noveb. 18.
       Catesb. car. 3. p. 2. t. 2.
       Trilopus, Mitch. gen. 22.
       Piftacia virginiana nigra, coryli foliis. Pluk. alm. 298.
       Habitat in Virginia.

              CUSCUTA.
europæa.  1. CUSCUTA floribus feffilibus.
       Cufcuta nuda repens filiformis. Fl. fuec. 138. Mat med.
         55. Dalib. parif. 53.
       Cufcuta nuda repens fumiformis. Sauv. monfp. 11.
       Cufcuta major. Bauh. pin. 219.
       Cufcuta. Cam. epit. 984.
epithymum β. Cufcuta nuda repens filiformis. Sauv. monfp. 11.
       Epithymum f. Cufcuta minor. Bauh. pin. 219.
       Habitat in Plantis Europæ parafitica.
americana. 2. CUSCUTA floribus pedunculatis.
       Cufcuta caule aphyllo volubili repente. Gron. virg. 18.
       Cufcuta inter majorem & minorem media, filamentis
         longis & floribus late fuper arbores & campos fe ex-
         tendens, Sloan. jam. 85. hift. 1. p. 201. t. 128. f. 4.
       Habitat in Virginia.

              HYPECOUM.
procumbens. 1. HYPECOUM filiquis arcuatis compreffis articulatis.
       Hort. upf. 31.
       Hypecoum. Bauh. pin. 172. Dod. pempt. 449. Hort.
         cliff. 38. Roy. lugdb. 402.
       Habitat inter Archipelagi, Narbonæ & Salmanticenfes
         fegetes. ☉
pendulum. 2. HYPECOUM filiquis cernuis teretibus cylindricis.
       Hort. upf. 31. Sauv. monfp. 263.
       Hypecoum filiquis pendentibus non articulatis bivalvi-
         bus incurvis. Morif. hift. 2. p. 280.
       Cuminum fylveftre filiquatum ponæ. Dalech. hift. 698.
       Habitat in Galloprovincia ☉
erectum. 3. HYPECOUM filiquis erectis teretibus torulofis. Hort.
       upf. 32.
       Hypecoum tenuifolium, filiquis erectis teretibus. Amm.
         ruth. 58. t. 9.
       Habitat in Dauria.
                                        TRI-
```

met in Paris to try to standardize rules governing the naming and the classifying of plants. They agreed to use the works of Linnaeus as the starting point for all scientific names of plants and decided that his binomials, or the earliest ones published after him, would have priority over all others. International congresses of botanists have met at varying intervals since then and have revised and expanded these rules. Today the modified rules comprise what is known as the International Code of Botanical Nomenclature, which is a single book with a common index to its English, French, and German translations of the various rules and recommendations. It now

specifically recognizes Linnaeus's *Species Plantarum* as the starting point for scientific names of plants and spells out details of naming and classifying, which are followed by botanists of all nationalities. A similar code for animals has been developed and established by zoologists. It, too, uses a work of Linnaeus as the starting point for scientific names.

Development of the Kingdom Concept

If you were to ask the average person the differences between plants and animals, you would probably be told that animals move and plants do not, and that animals eat plants or other animals for food whereas plants manufacture their own food through photosynthesis. A distinction between plants and animals in general has probably always existed in the minds of intelligent beings, and it was natural when classification schemes were first developed that all living organisms would be placed, according to the highest category of **kingdom,** in either the *Plant Kingdom* or the *Animal Kingdom*. While this distinction still works well for the more complex plants and animals, it breaks down for some of the so-called simpler organisms. For example, there are more than three hundred species of single-celled organisms called *euglenoids* (see figure 15.11) inhabiting a variety of freshwater habitats. These little creatures have *flagella* that propel them through the water, and they also can ingest food particles through a groove called a *gullet*. Both of these features would be considered animallike, and so euglenoids are treated as animals in some textbooks. Many of these organisms also have chloroplasts, however, and if light is present, they can carry on photosynthesis efficiently enough not to need to ingest food. Accordingly, they have often been treated as plants in botany books. A similar problem has existed with slime molds, which resemble masses of protoplasm crawling across dead leaves and debris, feeding on bacteria and other substances as they go, and looking somewhat like amoebae, which botanists and zoologists alike have traditionally called animals.

Table **13.1** Four classifications of organisms into kingdoms

Two Kingdoms (Traditional)	Three Kingdoms (Hogg and Haeckel)	Four Kingdoms (Copeland)	Five Kingdoms (Whittaker)	
		Monera Bacteria Blue-green algae Prochlorophytes	**Monera** Bacteria Blue-green algae Prochlorophytes	cells prokaryotic
	Protista Bacteria Blue-green algae Prochlorophytes Algae Slime molds True fungi Protozoa Sponges	**Protista** Algae Slime molds True fungi Protozoa Sponges	**Protista** Algae Slime molds Protozoa Sponges	cells eukaryotic
			Fungi True fungi	absorb food in solution
Plantae Bacteria Blue-green algae Prochlorophytes Algae Slime molds True fungi Bryophytes Vascular plants	**Plantae** Bryophytes Vascular plants	**Plantae** Bryophytes Vascular plants	**Plantae** Bryophytes Vascular plants	obtain food via photosynthesis
Animalia Protozoa; sponges Multicellular animals	**Animalia** Multicellular animals	**Animalia** Multicellular animals	**Animalia** Multicellular animals	ingest solid food

When slime molds reproduce, however, dramatic changes take place. They become stationary and develop reproductive bodies that are distinctly plantlike, causing some to consider them plants.

In an attempt to overcome this problem, the biologists J. Hogg and E. Haeckel proposed a third kingdom in the 1860s. All organisms that did not develop complex tissues (e.g., algae, fungi, and sponges) were placed in a third kingdom called *Protoctista*, or *Protista*. This third kingdom included such a heterogeneous variety of organisms, however, that in 1938 another biologist by the name of H. F. Copeland proposed that it again be divided. He assigned the name *Monera* to all single-celled protists with prokaryotic cells (see chapters 3 and 14), leaving the algae, fungi,

and single-celled organisms with eukaryotic cells in the Kingdom Protoctista. Although many biologists considered Copeland's four-kingdom system of classification a definite improvement, it too had problems—particularly the fact that basic differences in the mode of nutrition existed among organisms within the Kingdom Protoctista. As a result, a sizable number of biologists today favor variations of a five-kingdom system proposed by R. H. Whittaker in 1969 (table 13.1). In Whittaker's system, three kingdoms of more complex organisms based on three basic forms of nutrition (photosynthesis, ingestion of solid food, and absorption of food in solution) are recognized, along with two kingdoms of protists, which

Table 13.2 Classification of organisms in five kingdoms

Kingdom Monera
Division Schizophyta
    Class Schizomycetes (bacteria)
    Class Cyanophyceae (blue-green algae)
Division Prochlorophyta (prochlorophytes)

Kingdom Protista
Division Chlorophyta (green algae)
Division Euglenophyta (euglenoids)
Division Chrysophyta
    Class Xanthophyceae (yellow-green algae)
    Class Chrysophyceae (golden-brown algae)
    Class Bacillariophyceae (diatoms)
Division Pyrrophyta (dinoflagellates)
Division Phaeophyta (brown algae)
Division Rhodophyta (red algae)
Division Charophyta (stoneworts)
Division Myxomycota
    Class Myxomycetes (true slime molds)
    Class Acrasiomycetes (cellular slime molds)
(Phylum Protozoa—protozoans)
(Phylum Porifera—sponges)

Kingdom Fungi
Division Mycota (true fungi)
    Class Chytridiomycetes (chytrids)
    Class Oomycetes (water molds)
    Class Zygomycetes (bread molds)
    Class Ascomycetes (cup fungi)
    Class Basidiomycetes (club fungi)
    Class Deuteromycetes (imperfect fungi)

Kingdom Plantae
Division Bryophyta
    Class Hepaticae (liverworts)
    Class Anthocerotae (hornworts)
    Class Musci (mosses)
Division Psilophyta (whisk ferns)
Division Lycophyta (club mosses)
Division Sphenophyta (horsetails)
Division Pterophyta (ferns)
Division Cycadophyta (cycads)
Division Ginkgophyta (ginkgoes)
Division Coniferophyta (conifers)
Division Gnetophyta (*Gnetum, Ephedra, Welwitschia*)
Division Anthophyta (flowering plants)
    Class Dicotyledonae (dicots)
    Class Monocotyledonae (monocots)

(Division Tracheophyta)

Kingdom Animalia (multicellular animals)

are distinguished on the basis of differences in cellular structure. The slime molds, which have no cell walls during their active state but do develop walls when they reproduce, still do not fit well into the five-kingdom system. Much information still needs to be accumulated to understand natural relationships, and until this occurs taxonomic categories will remain subject to different interpretations.

Classification of Major Groups

Since Linnaeus's time, a number of classification categories have been added between the levels of kingdom and genus. Genera are now grouped into **families,** families into **orders,** orders into **classes,** classes into **divisions,**[2] and divisions into kingdoms. Depending on which system of classification is used, there may be between twelve and nearly thirty divisions of "plants" recognized.

The following is a classification of major groups of living organisms utilizing a modification of Whittaker's five-kingdom system (table 13.2). It is the classification followed throughout this book. Hypothetical derivations and relationships are indicated in figure 13.3.

Viruses, which have neither cellular structure nor most of the other attributes of living organisms, do not fit into this classification but are discussed in chapter 14 after the bacteria. Lichens, each of which is a combination of an alga and a fungus in close association, are discussed in chapter 16.

---

2. *Division* is equivalent to the term *phylum,* which is used in animal classification. Although use of the word *phylum* for all living organisms has been proposed, the International Code of Botanical Nomenclature presently permits only the use of the word *division* for plants.

Figure 13.3 Hypothetical derivations and relationships among kingdoms and the major groups of organisms

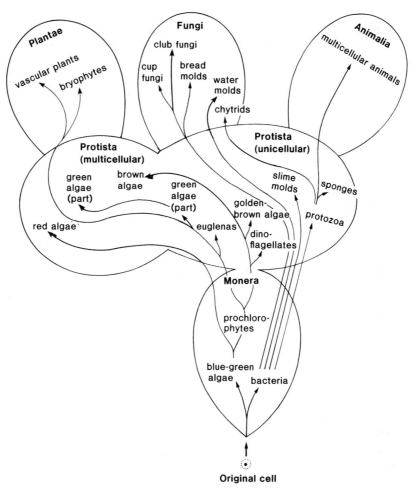

If we were to give a complete classification of an onion according to this particular arrangement, it would look like this:

Kingdom: Plantae
  Division: Anthophyta
    Class: Monocotyledonae
      Order: Liliales
        Family: Amaryllidaceae
          Genus: *Allium*
            Species: *Allium cepa* L.

The second part of the scientific name, the *specific epithet,* is followed by the name of the author, usually in abbreviated form. The author is the person (or persons) who originally named the plant or placed the species in a particular genus. It is customary to give binomials in italics (or to underline the words). In the case of *Allium cepa,* the L. following the name stands for Linnaeus, since this plant was one of the 7,300 species he included in his *Species Plantarum.*

In addition to these categories of classification, various "in-between" categories such as *subdivision, subclass,* and *suborder* have been used, and species themselves occasionally are broken down into *subspecies, varieties,* and *forms.* One popular classification arrangement recognizes only two divisions in Kingdom Plantae (Division Bryophyta for nonvascular plants and Division Tracheophyta for all vascular plants) and a number of subdivisions corresponding to those categories are given full division status.

Since human judgment has to enter into the compilation of classifications, there probably will never be complete agreement on them, but they are nevertheless useful, indeed necessary, for making some order of the diversity of plant life around us.

Those who specialize in identifying, naming, and classifying organisms and trying to sort out natural relationships are called *taxonomists* (see chapter 1). One of the activities of taxonomists is the construction of *keys* to help others identify organisms with which they may not be familiar. Most such keys are *dichotomous,* that is, they give the reader pairs of choices based on features of the organisms. If a key has been well constructed, we can arrive at an identification by careful observation of the organism, followed by elimination of the choices that do not apply. If, for example, we agree with the first of the two statements given for each number in the key, then we either use the identification at the end of the line or proceed to the *next indented statement beneath;* if we disagree with the first of the paired statements, then we look for the matching alternate statement and proceed from there. The following synoptic key, which deals with major groups of organisms, illustrates natural relationships. It covers only general features and does not indicate various details or occasional exceptions that would appear in keys for lesser groups. All organisms in any given group are presumed to be more closely related to each other than to organisms in another group.

Synoptic Key to Major Groups of Organisms (exclusive of Kingdom Animalia)

1. Unicellular, prokaryotic organisms with cell walls .... **Kingdom Monera**
    2. Pigments either absent or chlorophyll *a* and phycobilins present .... Division Schizophyta
    2. Chlorophylls *a* and *b* present; phycobilins absent .... Division Prochlorophyta
1. Unicellular, colonial, filamentous, or multicellular eukaryotic organisms, with or without cell walls.
    3. Organisms whose reproductive structures consist of a single cell or at least do not have a layer of sterile cells surrounding the reproductive structures; zygotes not developing into embryos.
        4. Cells with plastids, or if without plastids, having amoebalike naked protoplasts in the active state ....**Kingdom Protista**
      5. Cells with plastids, not amoebalike.
        6. Organisms essentially unicellular.
          7. Cells flexible; food reserves a carbohydrate called *paramylon* .... Division Euglenophyta[3]
          7. Cells not flexible; food reserves not paramylon.
            8. Cells with a flagellum at one end and another at right angles to it in a central groove .... Division Pyrrophyta

---

3. Only about a third of the euglenoids develop chloroplasts.

8. Cells either without flagella or having flagella not at right angles to each other.
    9. Plastids green; food reserves starch .... Division Chlorophyta (in part)
    9. Plastids having yellow, brown, or orange pigments prominent in relation to the chlorophylls; food reserves oils or carbohydrates other than starch .... Division Chrysophyta
6. Organisms multicellular.
    10. Organisms with accessory pigments essentially similar to those of higher plants; food reserves starch.
        11. Both male and female reproductive structures unicellular .... Division Chlorophyta (in part)
        11. Male reproductive structures multicellular ..... Division Charophyta
    10. Organisms with some accessory pigments differing from those of higher plants; food reserves carbohydrates other than ordinary starch.
        12. Organisms brownish in color due to presence of brown pigments; carbohydrate food reserve *laminarin* .... Division Phaeophyta
        12. Organisms reddish in color due to presence of red pigments; carbohydrate food reserve *floridean starch* .... Division Rhodophyta
5. Cells without plastids and without walls in the amoebalike active state .... Division Myxomycota
4. Cells without plastids but with definite cell walls .... **Kingdom Fungi**
3. Organisms with multicellular reproductive structures or the reproductive structures at least surrounded by a layer of sterile cells; zygotes developing into embryos .... **Kingdom Plantae**
13. Plants without xylem or phloem .... Division Bryophyta
13. Plants with xylem and phloem.
    14. Plants with dichotomously forking stems and *enations*[4] .... Division Psilophyta
    14. Plants with stems that branch in various ways or are unbranched; true leaves present.
        15. Plants with leaves having a single vein (microphylls).
            16. Stems not ribbed and not containing silica; leaves photosynthetic .... Division Lycophyta
            16. Stems ribbed and containing silica; leaves reduced to scales and non-photosynthetic .... Division Sphenophyta
        15. Plants with leaves usually having more than one vein (macrophylls).
            17. Plants reproducing by means of spores produced on the leaves .... Division Pterophyta
            17. Plants reproducing by means of seeds developed from ovules.
                18. Male reproductive cells (sperms) motile (i.e., capable of "swimming").

---

4. For a discussion of *enations*, see page 361.

19. Leaves pinnate and large, resembling those of palms ....Division Cycadophyta
19. Leaves smaller and fan-shaped, with numerous dichotomously forking veins ....Division Ginkgophyta
18. Male reproductive cells nonmotile.
20. Plants without flowers; seeds produced in cones or conelike structures.
21. Wood containing no vessels; resin canals present ....Division Coniferophyta
21. Wood containing vessels; resin canals absent ....Division Gnetophyta
20. Flowering plants ....Division Anthophyta

Once again, it should be emphasized that a key of this type, while calling attention to basic differences between major groups of organisms, may also oversimplify the differences, partly because it does not list all the characteristics of each member of a given group. Neither does it allow for the occasional exception to the rule or for some of the intermediate forms that frequently are transferred back and forth as research uncovers new evidence or as new lines of speculation develop. Keys at this level of classification also are not completely practical in that one sometimes needs to have specific stages in the life cycles of the plants available to be able to arrive at an identification. Nevertheless, such a synoptic key is generally preferable to a simple listing of groups and their characteristics for purposes of distinguishing among them.

Summary

All organisms are identified by a two-part Latin name (binomial), regardless of the language spoken in the area where the organisms occur; most also have one to many common names. The multiplicity of common names led to an early recognition by biologists of the need for worldwide uniformity in naming and classifying organisms.

Theophrastus (third century B.C.) was the first person of note to attempt to organize and classify plants. In the thirteenth century a distinction was made between monocots and dicots, and by the beginning of the eighteenth century herbalists had added many plants to Theophrastus's original list, with details of fruit and flower structure in addition to form and habit being utilized in classification schemes. By this time it had become customary in Europe to use descriptive Latin phrase names for all organisms, with the first word of the phrase indicating the genus (classification group) of each organism. Linnaeus compiled a comprehensive list of all known Latin phrase names for plants according to their genera in a work entitled *Species Plantarum,* published in 1753. In this work he placed one to many specific kinds of related organisms (species) in each genus; in the margin next to the Latin phrase he listed a single word, which, when combined with the name of the genus, formed an abbreviation for each species. This two-part Latin name (binomial) eventually replaced the phrase name, and the method of naming plants became known as the Binomial System of Nomenclature, a system still in use today. Linnaeus also organized all known plants into twenty-four classes, based primarily on the number of stamens in flowers.

In 1867, European and American botanists met in Paris and agreed to use Linnaeus's 1753 publication and binomials as the starting point for all scientific names of plants. The rules for naming and classifying plants drawn up at that meeting have periodically been revised and expanded, and today constitute the International Code of Botanical Nomenclature, a compendium of rules and recommendations now followed by botanists of all nationalities. A similar code governs the naming and classifying of animals.

When classification schemes were first developed, all living organisms were placed in either the Plant Kingdom or the Animal Kingdom. Many microscopic organisms have characteristics of both plants and animals, however, and in the 1860s Hogg and Haeckel proposed a third kingdom (Protista) for all organisms that did not develop complex tissues. In 1938 H. F. Copeland divided the Protista into two kingdoms, with organisms having prokaryotic cells being placed in the Kingdom Monera, and those with eukaryotic cells being left in the Kingdom Protoctista. Since there are differences in the mode of nutrition of organisms in Copeland's Kingdom Protoctista, many biologists now favor five kingdoms, as proposed by R. H. Whittaker in 1969. In Whittaker's system there are three kingdoms based on forms of nutrition (photosynthesis, ingestion of solid food, absorption of food in solution) and two kingdoms of protists based on differences in cellular structure. Slime molds, which have no cell walls at certain stages and do have walls at others, still do not fit well into the five-kingdom system, but this system appears to be the most satisfactory arrangement so far proposed.

Since Linnaeus's time several classifications have been added between the level of kingdom and genus. Genera are grouped into families, families into orders, orders into classes, classes into divisions, and divisions into kingdoms. Bacteria, blue-green algae, and prochlorophytes are placed in the Kingdom Monera; all algae, euglenoids, dinoflagellates, stoneworts, and slime molds are placed in the Kingdom Protista; all true fungi are placed in the Kingdom Fungi; liverworts, hornworts, mosses, and all vascular plants are placed in the Kingdom Plantae. Viruses, which lack most attributes of living organisms and are not classified within the five kingdoms, are discussed in chapter 14; lichens, which are a combination of an alga and a fungus, are discussed in chapter 16.

The second part of a binomial, the specific epithet, is followed by the name of the author (the one who named or classified the species), usually in abbreviated form. Subcategories (e.g., subspecies, suborder) are occasionally used in classification. Taxonomists may construct keys to aid in the identification of organisms. Most keys are dichotomous, that is, one arrives at an identification by making choices between pairs of statements given throughout the key. The key given to major groups of organisms deals only in generalities and does not indicate occasional exceptions or allow for intermediate forms that may be transferred back and forth between groups as new knowledge about them is gained.

Review Questions

1. What are the advantages and drawbacks of using scientific names as compared with common names?

2. What is the International Code of Botanical Nomenclature?

3. What is meant by the term *binomial?*

4. Which divisions include organisms found mainly in water?

5. List some characteristics by which the following may be distinguished from one another, and then construct a simple dichotomous key that could aid someone from another planet in identifying them: blondes, oranges, hammers, bicycles, rosebushes, automobiles, brunettes, cats, rats, raspberry bushes, dogs, wrenches, peaches.

Discussion Questions

1. Since phrase names are generally more descriptive of plants than binomials, do you think they should be revived?

2. After Linnaeus organized all the genera known to him into classes, a number of other categories of classification between the level of kingdom and genus were added. Are these other categories really useful?

3. Do you think biologists are "hairsplitting" when they divide the traditional Plant and Animal Kingdoms into five kingdoms? Explain.

Additional Reading

Bell, P. R.; Woodcock, C. L. F. *The diversity of green plants.* 2d ed. Reading, MA: Addison-Wesley Publishing Co.; 1972.

Bold, H. C.; Hundell, C. L. *The plant kingdom.* 4th ed. Englewood Cliffs, NJ: Prentice-Hall; 1977.

Cronquist, A. *The evolution and classification of flowering plants.* Boston: Houghton Mifflin Co.; 1968.

Ornduff, R., ed. *Papers on plant systematics.* Boston: Little, Brown and Co.; 1967.

Porter, C. L. *Taxonomy of flowering plants.* San Francisco: W. H. Freeman and Co.; 1967.

Scagel, R. F.; Bandoni, R. J.; Maze, J. R.; Rouse, G. E.; Schofield, W. B.; Stein, J. R. *Plants: an evolutionary survey.* Belmont, CA: Wadsworth Publishing Co.; 1984.

Stearn, W. T. *An introduction to the Species Plantarum and cognate botanical works of Carl Linnaeus.* Printed for the Ray Society and sold by Bernard Quaritch, London; 1957.

Kingdom Mone
and Viruse

The members of the Kingd
blue-green algae, and
prokaryotic cells. Su
velopes, but a lon
each cell. Oth
mitochondri
ticulum
membr
num

Overview

After a summary of the features of members of the Kingdom Monera, the chapter describes the luminescent bacteria associated with flashlight fish as an introduction to bacteria. A brief investigation of features of bacteria and their reproduction is followed by a discussion of genetic engineering and a look at representatives of heterotrophic and autotrophic bacteria. The chapter continues with a section on compost and composting, and then discusses various means of transmitting bacterial diseases and some of the diseases themselves. Next comes a listing of Koch's Postulates and a section on bacteria useful to humans.

The chapter discusses the similarities and differences between blue-green algae and bacteria and surveys the human relevance of blue-green algae. Prochlorophytes are introduced, and the chapter concludes with an overview of the nature, reproduction, and human relevance of viruses.

Some Suggested Student Learning Goals

1. Name the basic forms of bacteria. Explain how a prokaryotic cell differs from a eukaryotic cell and why prokaryotic organisms are difficult to classify.

2. Describe the forms of nutrition in bacteria.

3. Define genetic engineering and explain how it is accomplished.

4. List the various ways in which disease bacteria are transmitted; describe how each type of disease bacterium functions in causing disease.

5. List Koch's Postulates.

6. List at least ten bacteria useful to humans, and describe how they are useful.

7. Describe how bacteria, blue-green algae, prochlorophytes, and viruses differ from one another in form, pigmentation, and reproduction.

...om Monera (bacteria, ...prochlorophytes) all have ...ch cells have no nuclear en... ...g strand of DNA is present in ...er organelles such as plastids, ...a, Golgi bodies, and endoplasmic re... ...re also lacking, but ribosomes and ...anes are present (see figure 14.17). In a ...ber of members of the Kingdom Monera, ...lls may occur in a common matrix of gelatinous material as variously shaped colonies, or in the form of chains or **filaments** (threads), but each cell is completely independent—no protoplasmic connections exist between them. Some species are **motile** (capable of independent movement); simple **flagella** (whiplike tails that are often, but not always, in pairs) propel or pull a cell through the water. Several filamentous species exhibit a gliding motion in which the filaments glide back and forth independently or against each other, but most species are nonmotile. Nutrition is primarily by the absorption of food in solution through the cell wall, but some bacteria are *chemosynthetic* (capable of obtaining their energy through chemical reactions involving various compounds or elements) whereas a few others are photosynthetic; blue-green algae and prochlorophytes are generally photosynthetic. Reproduction is predominantly asexual, by means of **fission,** a form of cell division that does not involve mitosis since there are no nuclei or other organelles. The strand of DNA duplicates itself, however, and the DNA is distributed between the two new cells originating from the parent cell. Sexual reproduction is unknown, but genetic recombination occurs in several groups; it is facilitated by means of **pili** (minute tubes) that form connections between cells.

Division Schizophyta

Class Schizomycetes—Bacteria

The astonishing variety and beauty of undersea life, with representatives from all five kingdoms, has attracted the attention of snorkelers and scuba divers in ever-increasing numbers, particularly since Jacques Cousteau and other marine biologists have added to our knowledge of this vast domain with their sophisticated investigations and color cinematography. The shallow tropical waters of the Great Barrier Reef of Australia, the Virgin Islands National Park, Hanauma Bay of Hawaii, and countless other reefs teem with brilliantly colored marine life that may command the rapt attention of even the most blasé of observers for hours. Scuba divers who explore the oceans at night, however, in areas such as Indonesia's Banda Sea, the region just north of Darwin, Australia, or Israel's Gulf of Elat are frequently rewarded with a spectacular sight not seen by those who explore by day. They find themselves in the midst of a shimmering, constantly changing light display.

Countless numbers of lights, bright enough to read by, flicker, dart, glide slowly, or even congregate in glimmering groups. The lights come from large pouches beneath the eyes of small fish, which can turn them off at will by covering them with special folds of tissue. Unlike their deep-sea relatives, however, these flashlight fish, as they are called, do not produce the light within their own bodies. Instead, it comes from thousands of microscopic, luminescent *bacteria,* which enjoy a *symbiotic* relationship with the fish (figure 14.1). **Symbiosis** is simply an intimate association between two dissimilar organisms that benefits both of them. In this instance, the fish furnish some food and oxygen to the bacteria from numerous blood vessels in the pouches, while the bacteria produce the light used by the fish not only to search for food at night but also to attract live food and to confuse predators by flashing the lights on and off while swimming in erratic patterns.

Luminescent bacteria are but minor representatives of the most numerous of the earth's living organisms. The approximately seventeen hundred species of bacteria known today occur in astronomical numbers in almost every conceivable natural habitat. They are found on and in all plants and animals, in all soils, throughout both fresh and salt waters, in polar ice caps and bubbling hot springs, in coal and petroleum, throughout the atmosphere, in bottles of India ink, and almost anywhere else one may care to look for them. Just how many species there are is uncertain, as it is difficult to classify simple one-celled organisms on the basis of visible features alone. For example, one disease-causing bacterium with the scientific name of *Streptococcus pneumoniae* occurs in seventy-seven known "strains," all of which look alike but each of which causes a different form of pneumonia or related disease. Should each strain then be called a species? *Microbiologists,* who study microscopic organisms of all kinds, are not sure, but they tend to recognize clusters of strains based on what they do rather than how they look.

Although bacteria that cause disease and spoilage receive most of the publicity, more than 90% of bacterial species are either harmless or useful to humans. They are used in industry for a wide variety of purposes and are so important ecologically that life as we know it today would be vastly different without the activities of these tiny organisms. Both industrial uses and the role of bacteria in ecology are discussed in later sections.

Size, Form, and Classification

Although a few giant bacteria attain lengths of up to 40 micrometers, most do not exceed 2 or 3 micrometers in diameter, and a few of the smaller species approach 0.15 micrometer. The latter are so small that 6,500 of them arranged in a row would not quite extend across the head of a pin! Because of their small size, bacteria are not visible individually to the unaided eye and are studied in the laboratory with electron microscopes or with the highest magnifications of light microscopes. The thousands of different kinds occur

Figure 14.2 Three basic forms of bacteria. *A.* Cocci.
B. Bacilli. *C.* Spirilli. (Photomicrographs by Carolina
Biological Supply Company)

A.

B.

C.

primarily in three forms: *cocci* (singular: *coccus*), which are spherical or sometimes elliptical; *bacilli* (singular: *bacillus*), which are rod-shaped or cylindrical; and *spirilli* (singular: *spirillum*), which are in the form of a helix, or spiral (figure 14.2). Further classification is based on numerous visible features including, for example, presence of pigments, development of slimy or gummy capsulelike sheaths around cells, presence of hairlike or budlike appendages, development of internal thicker-walled *endospores,* and gliding movements by which threadlike groups of bacteria appear to slide back and forth. Some bacteria have slender flagella, usually about 5 to 10 micrometers in length, which propel them through fluid media. Others have somewhat shorter tubelike **pili** (singular: **pilus**), which resemble flagella but do not function in locomotion. These apparently enable bacteria to attach themselves to surfaces or to each other.

Bacteria are also grouped into two large categories based on their reaction to a dye. After a heat-fixed smear of cells has been stained blue-black with a violet dye, a dilute iodine solution, alcohol or acetone is added. When so treated, some species rapidly lose their color and are called *gram-negative;* others, called *gram-positive,* retain most of the color. Gram's stain, named after Christian Gram who discovered it in 1884, has many variations, but all the variations produce similar results.

Cellular Detail and Reproduction

Bacterial cells are *prokaryotic.* Although such cells have no nuclear envelopes or organelles, the complex, folded membranes apparently perform some of the functions of the organelles of eukaryotic cells. Bacterial cells have a single, long, very condensed DNA molecule in the form of a circle present in the cytoplasm. In addition, up to thirty or forty small, circular DNA molecules called *plasmids* may be present. The plasmids replicate independently of the large DNA molecule or chromosome. The chromosome and sometimes all or part of the plasmids replicate before the cell divides (figure 14.3). Mitosis as seen in eukaryotic cells does not occur. Instead, there is an internal reorganization of material during which the two DNA molecules migrate to opposite ends of the cell. Then, at approximately the middle of the cell, the plasma membrane and cell wall, as a result of forming a transverse wall, divide the cell in two. Finally, the two new cells separate and enlarge to their original size, or in some bacteria the cells remain attached to each other in chains. This simple form of asexual reproduction, **fission,** is found so universally in bacteria that they have been called *schizomycetes,* which means "fission fungi" (figure 14.4). Bacteria are not, however, currently believed to be closely related to the fungi, as was the case when the name was originally applied.

Under ideal conditions of moisture, food supply, and temperature, a bacterium may undergo fission every ten to twenty minutes. If a single bacterium of average size were to duplicate itself every twenty minutes for as little as thirty-six hours, a mass of bacteria numbering 322,981,536,679,200,000,000,000,000,000,000 and weighing 126,464,618,590 metric tons (137,438,953,472 tons) would be produced. If this mass were not compacted, its volume would exceed that of the earth. Of course, bacteria do not continue to reproduce at their maximum rate for very long because several factors prevent it. These factors include the exhaustion of food supplies and the accumulation of toxic wastes.

Bacteria do not produce gametes or zygotes, nor do they undergo meiosis; at least three forms of genetic recombination occur, however. In *conjugation,* part of the DNA strand or chromosome

Figure 14.3 Duplication of the DNA in a bacterial cell. The duplication begins (*A*) at a single point or bubble (upper right) that expands (*B*) until the two new rings of DNA separate (*C*). (Electron micrographs courtesy Tsuyoski Kakefuda)

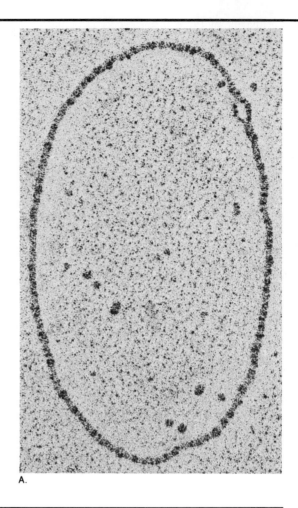

A.

Figure 14.4 Dividing bacterial cell. The new wall is growing inward. ($\times$ 48,750) (Courtesy R. G. E. Murray)

B.

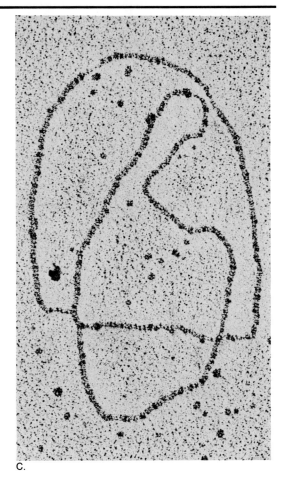

C.

Figure 14.5 Conjugation in bacteria. Part of the DNA strand of the donor cell migrates through the hollow, tubelike pilus to the recipient cell, where it is incorporated into the DNA of its new cell.

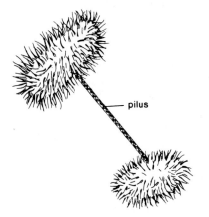

pilus

is transferred from a donor cell to a recipient cell through a tiny, hollow pilus while the two cells are in contact (figure 14.5). Once in the recipient cell the DNA fragment becomes part of the new cell. Bacterial cells that are capable of conjugating have different "sexual" forms of DNA, and only cells of different mating types may undergo conjugation. In *transformation,* a living cell picks up fragments of DNA released by dead cells into the medium in which it occurs and incorporates them into its own cell; in *transduction,* fragments of DNA are carried from one cell to another by viruses.

Figure 14.6 Splicing genes to bacterial plasmids

Genetic Engineering

Considerable controversy has arisen over the successful artificial addition of DNA to certain bacteria. Some fear that such tinkering by scientists could accidentally create uncontrollable monsters, while others hail such "genetic engineering" as the forerunner of eventual suppression or elimination of human defects and diseases.

The process of genetic engineering, or gene "splicing," begins with the isolation of pure bacterial DNA, which, as indicated in chapter 3, consists of two strands wound in a double helix. Next, the linkages between adjoining nucleotides are broken down by special bacterial enzymes known as restriction enzymes, so that the two strands are separated or "unzipped," leaving nucleotide "ends" that complement each other. Recombinant DNA is made by mixing large numbers of separated DNA strands from two different sources. When so mixed, a fragment from one source may associate with a fragment from the other source, and with the aid of a "repair" enzyme, the two fragments may link—in much the same fashion as recombination occurs during crossing over in meiosis—yielding a recombinant DNA molecule with parts of both original molecules. The recombinant plasmids are then inserted into bacterial cells where they replicate, thus introducing new traits into a particular strain of bacteria (figure 14.6).

Bacteria are easily propagated; thus, it is relatively simple to obtain large amounts of DNA for scientific studies with enormous potential. If, for example, genes for desirable traits in plants (e.g., superior nutritional quality of fruits or vegetables, superior metabolic efficiency, superior capacity for nitrogen fixation) could be isolated from selected plants and incorporated into bacterial plasmids, the bacteria could produce large amounts of recombinant DNA for introduction into plants that, in turn, could be propagated asexually if the desirable trait had been successfully incorporated. Viruses (see the final section of this chapter), some of which naturally infect plants, are also easily propagated and are subject to similar engineering techniques. It is possible that viruses will play a major role in the genetic manipulation of higher plants in the future.

Bacteria have been used for many years in the production of industrial chemicals, but genetic engineering may greatly increase such use in the future. More complex organisms have more complex genetic complements, however, and much additional research is needed before genetic engineering becomes commonplace in the envisioned elimination of diseases and defects in higher plants and animals. Pioneering experiments with higher animals and plants have already been performed. In 1980, for example, J. Gordon of Yale University squirted thousands of recombinant plasmids into newly fertilized mouse eggs that were in a dish, and then implanted the eggs in the oviducts of about ten female mice. The mice produced seventy-eight offspring, two of which proved to have traces of the new DNA in their cells. In 1981 scientists of the University of Wisconsin and the United States Department of Agriculture reported transferring a gene from a French bean seed into a cell of sunflower, and called the tissue that developed after the transfer "sunbean." The gene, which directs major protein production, is stable in the sunflower cell, and high levels of bean protein production are anticipated when technology is developed to regenerate a sunflower plant from the "sunbean" cells (see the section on mericloning and tissue culture in appendix 4).

Bacterial Nutrition and Metabolism

Heterotrophic Bacteria

Most bacteria are *heterotrophic* (organisms that cannot synthesize their own food and therefore depend on other organisms for it), and the majority of heterotrophic bacteria are *saprobes* (living organisms that obtain their food from dead organic matter). Saprobic bacteria, along with fungi, are primarily responsible for decay and for recycling all types of organic matter in the soil. Some of their recycling activities are discussed in the section on the nitrogen cycle in chapter 23. Other heterotrophic bacteria are *parasites* (living organisms that depend on other living organisms for their food); several parasites that cause important human diseases are discussed later in the chapter.

Autotrophic Bacteria

A few groups of bacteria are similar to green plants in being **autotrophic;** that is, they are capable of synthesizing organic compounds from simple inorganic substances. Some carry on photosynthesis without, however, producing oxygen as a by-product. Included in the photosynthetic bacteria are the *purple sulfur* bacteria, the *purple nonsulfur* bacteria, and the *green sulfur* bacteria. Cells of the first two groups appear purplish, or occasionally red to brown, because of the presence of a mixture of greenish, yellow, and red pigments. The greenish pigment *bacteriochlorophyll* is similar to the chlorophyll *a* of higher plants. The pigments are located in thylakoids or small spherical bodies, no plastids being present. In their photosynthesis, the purple bacteria substitute hydrogen sulfide, the bad-smelling gas given off by rotten eggs, for the water used by higher plants, so that the generalized equation for the process is as follows:

$$CO_2 + 2H_2S$$

carbon   hydrogen
dioxide   sulfide

$$\xrightarrow[\text{light}]{\text{bacteriochlorophyll}}$$

$$(CH_2O) + H_2O + 2S$$

carbo-   water   sulfur
hydrate

The equation for the purple nonsulfur bacteria is similar, but hydrogen from organic molecules is used instead of hydrogen sulfide. Green sulfur bacteria do use hydrogen sulfide, but their chlorophyll, called *chlorobium chlorophyll,* differs significantly in its chemistry from the chlorophylls of higher plants.

Other groups of bacteria are *chemoautotrophic;* that is, capable of obtaining the energy they require from various compounds or elements through chemical reactions involving the oxidation of inorganic molecules such as iron, sulfur, or nitrogen, or the oxidation of hydrogen gas. The following are among the better known groups of chemoautotrophic bacteria.

Figure 14.7 Scanning electron micrograph of nodules produced on the roots of a clover plant by nitrogen-fixing bacteria. Insert at upper right shows bacteria breaking out of the surface of a nodule. (Courtesy B. E. Juniper and D. Kerr)

Iron Bacteria

Iron bacteria transform soluble compounds of iron into insoluble substances that accumulate as deposits. Sometimes these bacteria become established in water pipes, and their deposits may eventually accumulate to such an extent that they reduce the rate of water flow.

Sulfur Bacteria

Unlike the purple bacteria, sulfur bacteria do not carry on photosynthesis. They can convert elemental sulfur to sulfuric acid, which may become concentrated enough to corrode metal pipes, and some are capable of producing and tolerating extreme acidity. They inhabit sulfur deposits, sulfur springs, oil wells, some coal mine waters, sewage, and soil.

Hydrogen Bacteria

Hydrogen bacteria flourish in the soil, where they combine hydrogen and oxygen molecules, forming water.

Nitrifying Bacteria

Nitrifying bacteria play a vital role in the earth's *nitrogen cycle* (discussed in chapter 23). One group converts ammonia (a compound composed of nitrogen and hydrogen) to nitrites, which are nitrogen-containing salts; another group converts nitrites to different salts called *nitrates*. The proteins of dead plant and animal tissues are the sources of the ammonia that is released by the activities of decay bacteria.

Nitrogen-fixing Bacteria

Most of the nitrogen supply of plants (and indirectly, therefore, of animals) is derived from the soil in the form of inorganic compounds taken in by the roots. These compounds include those that contain nitrogen chemically combined with oxygen or hydrogen. They are produced by bacteria and fungi as these break down the more complex molecules of dead plant and animal tissues to simpler ones. Some nitrogen from the air is also *fixed*, that is, converted to compounds such as ammonia, by various *nitrogen-fixing bacteria*.

These organisms gain access to various plants, particularly legumes (e.g., peas, beans, clover, alfalfa), through the root hairs. There is some evidence that the plants produce substances that attract the bacteria to them. Once in the cortex of the root they multiply, and as they do so, the root develops small nodules in which the bacteria thrive symbiotically with the plant (figure 14.7). There, with the aid of carbohydrates furnished by the host plant, they fix the nitrogen from the air, furnishing a significant part of the nitrogen requirements of the plant. Other free-living bacteria, including soil types, those that live on the surface of leaves, and blue-green algae, also are capable of fixing nitrogen. (See chapter 23 for further discussion of the role of nitrogen-fixing bacteria in the nitrogen cycle.)

Compost and Composting

Long before the existence of bacteria was known or even suspected, primitive agriculturists made rough piles of weeds, garbage, manure, and other wastes. They watched as the piles became reduced to a fraction of their original volume and changed into *compost,* a dark fluffy material that conditions and enriches soil when mixed with it. Today the compost pile is at the heart of the activities of organic gardeners and farmers, whose numbers increase daily. With many communities instituting ordinances forbidding the burning of leaves and refuse, and with space for waste disposal becoming scarce, a significant number of cities and towns have turned to the composting of street leaves and other materials on a large scale. They have found that they not only save space but are also producing a useful, ecologically compatible product (figure 14.8).

Because bacteria are ubiquitous, a single leaf or dead tissues in a pile of any size will eventually be decomposed. However, a compost pile is ideally 2 meters wide, 1.5 meters deep, and at a minimum, 2 meters long (6 feet wide, 3 feet deep, and at least 6 feet long). If the pile is of

Figure 14.8 Compost. *Left:* Garbage and leaves before composting. *Right:* After composting.

lesser length or breadth, or greater depth, the conditions that favor the breakdown of materials (created by the bacteria themselves) develop at a slower pace.

Any accumulation of household garbage, leaves, weeds, grass clippings, and/or manure may be heaped together, and it has been demonstrated that no so-called starter cultures of microorganisms are needed to initiate decomposition. Once the ever-present decay organisms have material to decompose, their numbers increase rapidly, and much heat is generated. In fact, the temperature in the center of the pile often rises to 70°C (158°F). As it rises, populations of organisms adapted to higher temperatures replace those not as well adapted. The

remains of the latter are then added to the compost. The high temperatures also kill many weed seeds and most disease organisms.

The proportion of carbon to nitrogen present in a compost pile largely determines the pace at which the microbial activities proceed, a ratio of 30 carbon to 1 nitrogen being optimal. Microbial growth stops if the proportion of carbon gets much greater, and nitrogen is lost in the form of ammonia if the ratio drops lower.

If the materials are kept moist and turned occasionally to aerate the pile, composting may be completed in as little as two weeks. Shredding the materials exposes much more surface area for the microorganisms to work on and speeds up the process. A few materials should be avoided when the compost is intended for use as a soil conditioner or crop fertilizer. Eucalyptus leaves and also walnut leaves and hulls contain substances that inhibit the growth of other plants. They should not be added to a compost pile unless they make up less than 20% of the total volume of material. Bamboo is particularly resistant to decay bacteria and will decompose much more slowly than other materials. Domestic cat manure may contain stages of a parasite that can infect humans, although the organism should be killed if proper composting techniques are employed. Generally, however, almost any organic materials are suitable.

While the value of compost to the soil is indisputable, its value as a fertilizer has certain limitations. The nitrogen content of good compost is about 2 to 3%, phosphorus 0.5 to 1%, and potash 1 to 2%, as compared with five to ten times those amounts in chemical fertilizers, which also require less labor to produce. Nevertheless, with the spiraling costs of chemical fertilizers, an increasing awareness of the problems accompanying their use, and greater public enlightenment concerning the accumulation and disposal of solid wastes, compost is undoubtedly destined to play a major role in pertinent agricultural practices of the future.

Bacteria and Disease

It has been calculated that plant diseases alone cause American farmers losses in excess of $4 billion per year. Although many plant diseases are caused by fungi, a number involve bacteria. The latter include diseases of pears, potatoes, tomatoes, squash, melons, carrots, citrus, cabbage, and cotton. Bacteria also cause huge losses of foodstuffs after they have been harvested or processed. They are even better known, however, as the culprits in many serious diseases of animals and humans.

Since they are so tiny and often incapable of independent movement, how are they transmitted, and what is it about their activities that enables them to fell creatures billions of times their size? We have learned much about their activities in the past century and know now that they can gain access to human tissues in a variety of ways.

Figure 14.9 Stop-action photograph of a sneeze. (Reprinted with the permission of Marshall W. Jennison, Syracuse University)

Access from the Air

Every time a person coughs, sneezes, or just speaks loudly, an invisible cloud of tiny saliva droplets is produced (figure 14.9). Each droplet contains bacteria or other microorganisms and a tiny amount of protein. The moisture usually evaporates almost immediately, leaving minute protein flakes to which live bacteria adhere. As normal breathing takes place, these soon may find their way into the respiratory tracts of other humans or animals, particularly if they have been confined within the air of a room. Fortunately, the natural resistance of most of those who acquire the bacteria in this way prevents the bacteria from multiplying to the point of causing a disease. When the resistance is not there, however, a number of diseases, including diphtheria, whooping cough, some forms of meningitis and pneumonia, and "strep throat" can develop. Psittacosis, a disease carried by birds, is caused by the inhalation of *chlamydias,* which are exceptionally minute organisms unable to manufacture their own ATP molecules. They are apparently "energy parasites," depending on their host cells for the energy needed to carry on their own functions.

Access Through Contamination of Food and Drink

When sewers were commonly open and food processing was done under less sanitary conditions than exist at present, a number of bacterial diseases, including cholera, dysentery, salmonella "food poisoning," and typhoid and paratyphoid fevers often reached epidemic proportions. Today, however, civilized countries rarely see such epidemics unless a natural disaster such as a flood or a typhoon disrupts normal sewage disposal, and the diseases are more often spread by "carriers" who handle food or by houseflies. These bacteria multiply in the intestinal tract or spread from there to other parts of the body.

Some bacteria (e.g., those responsible for Legionnaire's disease) are very common on algae in freshwater streams, lakes, and reservoirs. They usually pass through the human digestive tract harmlessly without multiplying, but on rare occasions, something unknown triggers their reproduction, often with fatal results.

Access Through Consumption of Bacterial Toxins

The most deadly of all known biological toxins is produced by a bacterium with the scientific name of *Clostridium botulinum* (figure 14.10). The name comes from *botulus,* the Latin word for sausage, since the bacterium was first discovered after people had died from eating some contaminated sausage at a picnic. Unlike salmonella "food poisoning," **botulism** is not an infection but is poisoning from a substance produced by bacteria that can grow and multiply anaerobically (in the absence of oxygen) in improperly processed or stored foods. Home-canned beans, beets, corn, and asparagus in particular have been known to permit the development of botulism bacteria. Just one gram (0.035 ounce) of the toxin is sufficient to kill 14 million adults, and a little more than half a kilogram (1.1 pound) could eliminate the human race!

The bacteria, which are present in most soils, produce unusually heat-resistant spores and are likely to be present on any soil-contaminated

Figure 14.10 Botulism bacteria. (Photo © Charles P. Havel)

foods. They may not be destroyed during canning unless the food is heated for thirty minutes at 80°C (176°F), or boiled for ten to fifteen minutes. They ordinarily will not grow in foods that have been preserved in brines containing at least 10% salt (sodium chloride) or in fairly acid media such as the juices produced by most stone fruits. The toxin is absorbed directly from the stomach and the intestines, affecting nerves and muscles and causing paralysis. While some antidotes are available, they are ineffective after symptoms have become advanced. Deaths reported from botulism in the United States reached a peak of about twenty-five per year in the 1930s but have declined to five or six per year since then, although there is recent evidence that deaths of infants due to hitherto unknown causes are actually due to botulism. For reasons that are not clear, botulism bacteria, which pass harmlessly through the digestive tracts of humans over the age of six months, may germinate in the intestines of infants between the ages of one to six months. One hundred and seventy cases of sudden infant death reported worldwide between 1976 and 1980 appear to have been due to botulism, and the evidence now suggests that botulism is responsible for about 5% of the eight thousand cases of sudden infant death that occur each year in the United States alone.

Chapter 14

Access Through Direct Contact

Bacteria responsible for diseases such as syphilis, gonorrhea, anthrax, and brucellosis enter the body through the skin or mucous membranes (i.e., those membranes lining tracts with openings to the exterior of the body). Both syphilis and gonorrhea are transmitted through sexual intercourse or other forms of direct contact, rarely through the use of public washroom towels or toilets. In 1976, gonorrhea accounted for the largest number of reported cases of any communicable disease in the United States, and currently more than 3 million Americans become infected every year. The symptoms of both syphilis and gonorrhea, which include persistent sores or discharges from the genitalia, sometimes disappear after a few weeks, leading victims to believe the body has healed itself. More often than not, however, symptoms reappear in different parts of the body at a later date. Both diseases are curable when treated promptly, but it is very important that such treatment be sought, since failure to do so can lead to sterility, blindness, and even death.

Anthrax, which is primarily a disease of cattle and other farm animals in addition to wild animals, is sometimes transmitted to humans, particularly workers in the wool and hide industries. Like syphilis and gonorrhea, it can be effectively eliminated if treated early enough but may be fatal if allowed to progress. Brucellosis, another disease of farm animals, is occasionally transmitted to humans through direct contact or through the consumption of contaminated milk. It is sometimes called "undulant fever" because of a daily rise and fall of temperature apparently associated with the release of toxins by the bacteria.

Access Through Wounds

When one steps on a dirty nail or is wounded in such a way that dirt is forced into body tissues, *tetanus* ("lockjaw") bacteria, which are common soil organisms, may gain access to dead or damaged cells. There they can multiply and produce a deadly toxin so powerful that 0.00025 gram (0.00000088175 ounce) is sufficient to kill an adult. In contrast, about 150 times that amount

Figure 14.11 Rat climbing over a barrier on a ship's mooring line. (Courtesy U.S. Public Health Service)

of strychnine is needed to achieve the same result. Control of tetanus through immunization is very effective and has become widespread. Several related bacteria that gain access to the body in the same way are responsible for potentially fatal *gas gangrene*, which used to be feared on the battlefields in the past but is now controlled through the use of **antibiotics** and aseptic techniques.

Access Through Bites of Insects and Other Organisms

Bubonic plague (the "Black Death") and tularemia are two bacterial diseases transmitted by fleas, deerflies, ticks, or lice that have been parasitizing infected animals, particularly rodents.

Rat fleas, found on infected rats that inhabit dumps, sewers, barnyards, and ships (figure 14.11), acquire the bacteria for plague and then pass them on to humans through their bites. The

Kingdom Monera and Viruses

disease has been found in ground squirrels and other rodents in the United States, particularly in the West, since 1900. In the past, bubonic plague has spread with great speed and reached devastating epidemic proportions. In 1665 in London, hundreds of thousands perished from the disease, and between 1347 and 1349 it is believed to have killed one-fourth of the entire population of Europe (about 25 million persons). Today plague is rare in North America, but it still occasionally manifests itself in port cities of Asia, Europe, and South America. Control depends on control of rats and fleas, which are virtually impossible to eradicate entirely, and the use of vaccines, which produce immunity for about six to twelve months.

Tularemia is primarily a disease of animals, but it may be transmitted to humans through the bites of infected ticks or deerflies. It is an occupational disease of meat handlers and is fatal in 5 to 8% of the cases. Ticks, lice, and fleas may also transmit *rickettsias,* which cause typhus and spotted fevers. Rickettsias are extremely tiny bacteria that live inside of eukaryotic cells. Pleuropneumonia-like organisms (PPLOs) are also minute bacteria that may be transmitted by various means. These have no cell walls and therefore are quite plastic. They are found in many plants, in hot springs, and in the moist surfaces of the respiratory and intestinal tracts of animals and humans. They are responsible for a form of human pneumonia and for numerous plant diseases. They are the only prokaryotes known that, as a group, are resistant to penicillin.

Koch's Postulates

Since there are so many bacteria present everywhere, how can one be certain that a given bacterium obtained from an infected person is actually the organism responsible for the observed disease? A German physician, Robert

Figure 14.12   Robert Koch. (Courtesy National Library of Medicine)

Koch (figure 14.12), became known during the latter half of the nineteenth century for his investigations of anthrax and tuberculosis. As a result of his work, Koch formulated rules for proving that a particular microorganism is the cause of a particular disease. His rules, with minor modifications, are still followed today. They have come to be known as *Koch's Postulates,* and their essence is as follows:

1. The microorganism must be present in all cases of the disease.
2. The microorganism must be isolated from the victim in pure culture (i.e., in a culture containing only that single kind of organism).
3. When the microorganism from the pure culture is injected into a susceptible host organism, it must produce the disease in the host.
4. The microorganism must be isolated from the experimentally infected host and grown in pure culture for comparison with that of the original culture.

Figure 14.13 Tomato hornworm before, and three days after, spraying with *Bacillus thuringiensis.*

Bacteria Useful to Humans

For many years, insect pests of food plants have been controlled primarily through the use of toxic sprays. Residues of the sprays remaining on the fruits and vegetables have accumulated in human tissues, however, often with adverse effects, while at the same time many organisms have become immune or resistant to the toxins. In addition, the sprays kill useful organisms, and precipitation runoff washes the toxins into streams, lakes, and oceans, harming or killing aquatic creatures. As an awareness of the undesirable effects of the use of toxic sprays has developed, alternative means of controlling crop predators have been sought. Today many harmful pests and even weeds can be limited to inconsequential numbers through the use of "biological controls" (see appendix 2).

Three such controls involve bacteria that have been registered for use by the United States Department of Agriculture. One, with the scientific name of *Bacillus thuringiensis* (often referred to as BT), has been remarkably effective against a wide range of caterpillars and worms, including peach tree borers, European corn borers, bollworms, cabbage worms and loopers, tomato and fruit hornworms (figure 14.13), tent caterpillars, fall webworms, leaf miners, alfalfa caterpillars, leaf rollers, gypsy moth larvae, and cankerworms. The bacteria, which are easily mass-produced by commercial companies, are sold in the form of a stable wettable dust containing millions of spores. When the spores are sprayed on food plants, they are absolutely harmless to humans, birds, animals, earthworms, or any living creatures other than moth or butterfly larvae. When a caterpillar ingests any tissue with BT spores on it, the bacteria quickly become active and multiply within the digestive tract, soon paralyzing the gut. The caterpillar stops feeding within two or three hours and slowly turns black, dropping off the plant in two to four days.

Kingdom Monera and Viruses

In 1983, a variety of *Bacillus thuringiensis* (var. *israelensis*) was introduced into the commercial market for the control of mosquitoes. Called BTI, the bacterium attacks only mosquito larvae ("wigglers") and one or two other pests. Six years of experiments performed on more than seventy species of fish, snails, shrimp, and insects demonstrated no adverse effects on either plants or animals other than mosquitoes (and a couple of lesser pests) even at dosages 100 times more powerful than those needed to kill mosquito larvae. Use of this bacterium to control mosquitoes in the future may constitute a significant step in lessening the ecological damage and disruption that so frequently accompanies the use of toxic chemicals for pest control.

Another bacterium, *Bacillus popilliae,* is also marketed in a powder form. When it is applied to soil where grubs of the highly destructive Japanese beetle are present, it causes what is known as "milky spore disease" in the grubs, which die in a few days. The spores are carried throughout the topsoil by rainwater, foraging grubs, and by organisms that feed on the grubs.

Bacteria play a major role in the dairy industry. Milk, which is composed of proteins, carbohydrates, fats, minerals, vitamins, and about 87% water, has exceptional nutritive value for animals and is also an excellent medium for the growth of many kinds of bacteria. Milk is sterile when secreted within the udder of the cow, but it picks up bacteria as it leaves the cow's body. It spoils rapidly if it is not obtained and stored under strictly sanitary conditions. Even after it has been pasteurized and refrigerated, the numbers of bacteria in it will increase the longer it stands. If milk is left in open containers in household refrigerators, for example, bacterial growth will not be kept in check for much longer than twenty-four hours.

Except for milk itself, either alone or in mixtures (e.g., ice cream), all dairy products are manufactured from raw or pasteurized milk by the controlled introduction of various bacteria. Such products include buttermilk, acidophilus milk, yogurt, sour cream, kefir, and cheese. Whey, the watery part of the milk separated from curd during cheesemaking, is utilized, along with starches and molasses, for the commercial production of lactic acid. Lactic acid is used extensively in the manufacture of textile and laundry products, in the leather tanning industry, as a solvent in lacquers, and in the treatment of calcium and iron deficiencies in humans.

In their metabolism of various sugars, proteins, and other organic substances, bacteria also produce waste products that have important industrial uses. Such products are often produced in large quantities by culturing bacteria in huge vats. These products include acetone, used in the manufacture of photographic film, explosives, and as a solvent (e.g., nail polish remover); butyl alcohol, another solvent used in the manufacture of synthetic lacquers; dextran, used as a food stabilizer and as a blood plasma substitute; sorbose, used in the manufacture of ascorbic acid (vitamin C); and citric acid, used as a lemon flavoring for foods. Some vitamin and medicinal preparations also involve bacterial synthesis.

Bacteria are used in the curing of vanilla pods, cocoa beans, coffee, and black tea and in the production of vinegar, sauerkraut, and dill pickles. Fibers for linen cloth are separated from flax stems by bacteria, and green plant materials are fermented in silos to produce ensilage for cattle feed. In recent years, the production of several important amino acids by bacteria has been exploited commercially. More than 6,800 metric tons (7,500 tons) of one amino acid, glutamic acid, are produced in the United States each year. This is in demand as a flavor-enhancing agent in the form of monosodium glutamate.

In the future, bacteria may be used on a large scale to furnish energy for engine fuels and for heating, cooking, and light. Certain anaerobic

Figure 14.14 Integrated organic digester operation for the production of methane gas. (Reprinted from *Producing Your Own Power* © 1974 by Rodale Press, Inc. Permission granted by Rodale Press, Inc., Emmaus, Pa. 18049)

bacteria produce *methane gas,* which can be substituted for natural gas. The methane is given off by the bacteria as they "digest" organic wastes in the absence of oxygen. Nine kilograms (20 pounds) of horse manure or 4.5 kilograms (10 pounds) of pig manure fed into a methane digester will produce all the gas needed per day for the average American adult's cooking needs. Considerably less green plant material is required to produce the same amount. The digester basically consists of an airtight drum connected by a pipe to a storage tank and a means of drawing off the sludge left after the gas has been produced (figure 14.14). The sludge itself makes an excellent fertilizer. In India, where cows produce over 812 million metric tons (800 million tons) of manure per year, numerous villages satisfy their fuel needs by means of manure-fed methane gas digesters, which have also been used in Europe since World War II and are now being employed in rural areas in the United States.

Division Schizophyta

Class Cyanophyceae— Blue-Green Algae

In the past, algae as a group have been distinguished from other organisms in being photosynthetic, in having relatively simple structures, and for those reproducing sexually, in having sex organs consisting of a single cell. However, as more has become known about cellular details, it appears that the differences between the organisms known as *blue-green algae* and other kinds of algae are quite basic. Blue-green algae, which along with bacteria and prochlorophytes comprise the Kingdom Monera, have prokaryotic cells, whereas all other algae, which are assigned to the Kingdom Protista, have eukaryotic cells.

Figure 14.15 Similarity of form between various genera of bacteria and blue-green algae. (After Pelczar, M. J., Jr., and R. D. Reid. 1972. *Microbiology,* 3d ed. Redrawn by permission of the McGraw-Hill Book Co.)

In fact, blue-green algae are so much like bacteria (figure 14.15) that there is a growing trend toward treating them as specialized bacteria. The only consistent distinctions between bacteria and blue-green algae appear to be that blue-green algae possess chlorophyll *a,* which is found in higher plants, and that oxygen is produced when they undergo photosynthesis; those bacteria capable of carrying on photosynthesis do not produce oxygen, and they do not have chlorophyll *a.*

Blue-green algae are found in almost as diverse a variety of habitats as bacteria. They are common in temporary pools or ditches, particularly if the water is polluted. They are not found in acidic waters, but they occur abundantly in other fresh and marine waters around the globe, from the frozen lakes of Antarctica to warm tropical seas, and in hot springs whose water temperatures approach 85°C (185°F). In Yellowstone National Park, a different species of blue-green algae is found in each temperature range of hot springs. There the algae precipitate chalky, insoluble carbonate deposits, which become a rocklike substance called *travertine.* The deposits accumulate at the rate of up to 2 to 4 millimeters per week, with other blue-green algae often forming brilliantly colored streaks in the travertine. Blue-green algae are frequently the first photosynthetic organisms to appear on bare lava after a volcanic eruption, and they occur in both desert and jungle soils. Some are found on the shells of turtles and snails, while others live symbiotically in various types of organisms, including amoebae, protozoans, diatoms, certain sea anemones and their relatives, some fungi, and in the roots of tropical palmlike plants called *cycads.* They also flourish in tiny pools of water formed at the bases of the leaves of tropical grasses and other plants, and they are well-known components of "compound" plants called *lichens* (see chapter 16).

When they occur symbiotically in other organisms, they commonly lack a cell wall and appear to function essentially as chloroplasts. When a eukaryotic cell containing chloroplasts divides, the chloroplasts divide at the same time. The cells of blue-green algae occurring within the cells of other organisms divide in similar fashion, leading to speculation that chloroplasts originated as blue-green algae living within other cells.

The cells of blue-green algae are minute and similar in size to those of bacteria. They do not have protoplasmic connections with each other, and each is capable of an independent existence. They often occur, however, in chains or hairlike filaments, which are sometimes branched. Several species occur in irregular, spherical, or plate-like colonies, the individual cells being held together by the gelatinous sheaths they secrete (figure 14.16). The sheaths may be colorless or pigmented with various shades of yellow, red, brown, green, blue, violet, or blue-black, which makes some colonies quite striking in appearance. The cells themselves appear blue-green in color in about half of the approximately fifteen hundred known species of blue-green algae. This color, which is due to the presence of chlorophyll *a* and a blue accessory pigment called *phycocyanin,* is often masked by the presence of other pigments. Several yellow or orange carotenoid pigments similar to those of higher plants are usually present, and varying amounts of a red pigment, *phycoerythrin* (in a form unique to the blue-green algae), may give the cells a distinct red color. The periodic appearances of large numbers of blue-green algae with considerable amounts of phycoerythrin are believed to have given the Red Sea its name.

The cell walls of blue-green algae do not contain cellulose, but are composed of polysaccharides similar to those found in the walls of bacterial cells. The organisms produce a carbohydrate called *cyanophycean starch.* This is probably identical with glycogen, the common storage form for sugar in higher animals and fungi. Blue-green algae also produce and store lipids and proteins.

Figure 14.16 Representative blue-green algae

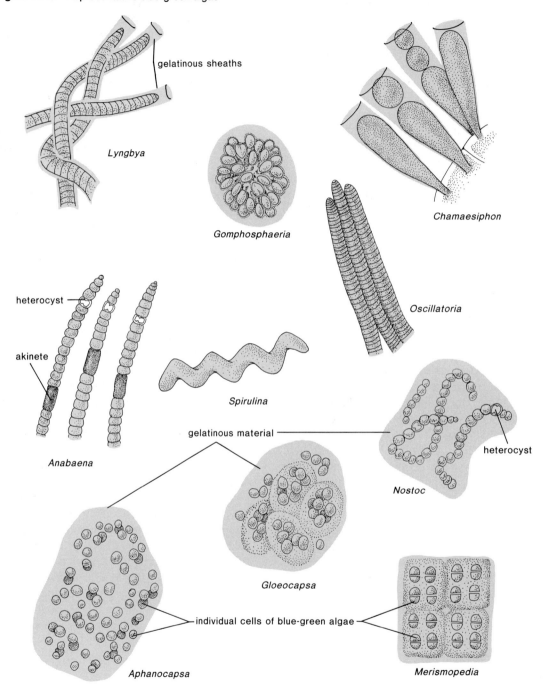

Chapter 14

Flagella are unknown in the blue-green algae, but some of these algae are nevertheless capable of movement. *Oscillatoria* (figure 14.16), for example, seem to rotate on axes and move in a gliding fashion, apparently by the twisting of minute fibrils inside the cell walls while secreting mucilage that creates friction. New cells are formed through fission, while new colonies or filaments may arise through fragmentation (breaking up) of older ones. In the common genera *Nostoc* and *Anabaena* (figure 14.16), which form chains of cells, fragmentation often occurs at special larger cells called **heterocysts** produced at intervals in the chains. It is believed that heterocysts may be important in the fixation of nitrogen from the air. Members of these two genera also may produce thick-walled cells called *akinetes*. These cells can resist freezing and other adverse conditions. This feature permits the species to survive cold or dry conditions, germinating, when favorable conditions return, and becoming new chains or filaments. Some akinetes have been known to germinate after lying dormant for more than eighty years. Blue-green algae, like bacteria, do not produce gametes or zygotes and do not undergo meiosis; however, genetic recombination has been reported—apparently taking place in similar fashion to that reported for bacteria—but its occurrence evidently is rare.

Human and Ecological Relevance of Blue-Green Algae

Blue-green algae are included among the many aquatic and photosynthetic organisms at the bottom of various food chains. They store energy in the sugar they produce through photosynthesis. Small fish and crustaceans feed on them, only to be eaten by larger fish, which then become food for other aquatic organisms or humans.

During warmer months various algae, and blue-green algae in particular, may become abundant in bodies of fresh water, especially if the water is polluted. A floating scum or mat called a *bloom* may cover extensive areas. In 1964, for example, one such bloom covered more than 2,000 square kilometers (800 square miles) of Lake Erie's surface. Such blooms often impart a fishy, or otherwise objectionable, odor or taste to the water, and toxic substances may be produced when the algae die and are decomposed by bacteria. There have been occasional reports of cattle, hogs, sheep, rabbits, dogs, and even poultry having been poisoned through drinking water that had an algal bloom. While the algae are alive, the oxygen content of the water increases temporarily, but later the decay bacteria decomposing their dead bodies may deplete the oxygen to the extent of killing fish and other organisms.

One species of blue-green algae, which occurs in abundance on the coral beaches of the Gulf of Mannar on the southeast coast of India, frequently causes the death of horses that feed on it. Most blue-green algae are not palatable to humans, but one species has been used for food by the natives of the Lake Chad region of central Africa for many years. Many poisonous fish become poisonous only after feeding on certain blue-green algae. A few colorless forms are mild parasites in humans and animals.

In human water supplies, blue-green algae frequently clog filters, corrode steel and concrete, cause natural softening of water, and produce undesirable odors or coloration in the water. Many communities control the blue-green algae in reservoirs through the addition of very dilute amounts of copper sulphate.

More than forty species of blue-green algae are known to fix nitrogen from the air at roughly the same rates as the nitrogen-fixing bacteria of leguminous plants. They may be more important than originally thought in this regard. In Southeast Asia, so much usable nitrogen is produced in the rice fields by naturally occurring blue-green algae that rice is often grown for many years on the same land without the addition of fertilizer.

Figure 14.17 Section through a cell of the prochlorophyte *Prochloron,* which lives on the surface of a marine animal called the sea squirt. Note the absence of a nucleus, or other organelles, and the concentric layers of membranes that perform some of the functions of organelles. (× 15,750) (Electron micrograph by Jean Whatley)

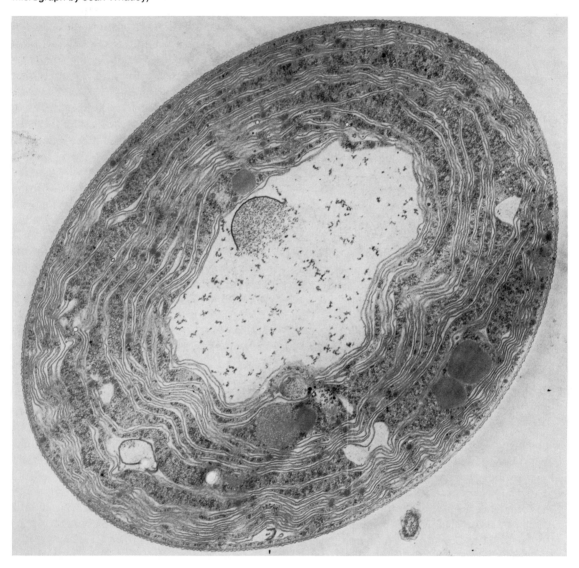

Division Prochlorophyta— Prochlorophytes

In 1976 R. A. Lewin of the Scripps Institute of Oceanography announced the discovery of unicellular, prokaryotic organisms with bright green cells that were living on marine animals called sea squirts. These organisms proved to have the chlorophylls *a* and *b* of higher plants, but no trace of the phycobilin accessory pigments associated with blue-green algae. Instead, their accessory pigments were confined to the carotenoid pigments found in higher plants. Lewin considered such a basic difference between blue-green algae and these bright green cells to be worthy of recognition at the division level, and he designated the new division as the Prochlorophyta. Some microbiologists who believe the blue-green algae are merely specialized bacteria are reluctant to recognize the prochlorophytes as belonging to a separate Moneran division because the basic prokaryotic cell structure is similar to that of blue-green algae (figure 14.17), but many botanists agree with Lewin's assessment of the significance of the pigment system.

The discovery of prochlorophytes adds weight to the theory that chloroplasts may have originated from such cells living within the cells of other organisms, especially since the pigments involved are identical with those of higher plants (see the discussion under blue-green algae, on page 261).

Viruses

Smallpox is an often fatal communicable disease that apparently has been widespread for thousands of years. Periodically throughout history it has killed countless numbers of individuals. As it develops in its victims, it manifests itself as blisterlike lesions on the skin, which later may become permanent pits or depressions. In 1901 an outbreak of smallpox in New York caused 720 deaths. Since that time, however, it has been eliminated in the United States through vaccination, and the World Health Organization now believes it has been eradicated from the remainder of the world.

Vaccination involves the introduction of a weakened form of disease agent to the body. The body's natural defenses, in fighting the agent, build up an immunity to the disease. The first known vaccinations were performed by two Englishmen, Benjamin Jesty, a farmer, and Edward Jenner, a country physician. They had both noticed that farmhands working with cows having cowpox, a comparatively mild disease related to smallpox, did not contract smallpox itself during the devastating epidemics of the disease that occurred in Europe from time to time. In 1796 Jenner scratched the skin of a boy with fluid he had obtained from a cowpox blister on the hand of a milkmaid. Six weeks later, he deliberately inoculated the boy with fluid from a blister of a smallpox victim, but the boy developed no symptoms of the disease. Jenner thus had performed a successful vaccination against a dread disease more than fifty years before Louis Pasteur developed the germ theory of disease.

We now know that smallpox is caused by agents considerably smaller in size than bacteria. During Pasteur's time virtually all infectious agents, including bacteria, protozoans, and yeast, were called *viruses*. One of Pasteur's associates, Charles Chamberland, discovered that porcelain filters would not allow bacteria to pass through but would permit the passage of an unseen agent that caused rabies, another serious disease of both animals and humans. Such agents of disease that could pass through filters became known as *filterable viruses*. Today we know that not only smallpox and rabies are caused by these filterable viruses but also measles, mumps, chicken pox, polio, yellow fever, influenza, fever blisters, warts, and the common cold. The word *filterable* is no longer used, and only those organisms that have certain unique features are now called viruses. These features, which include a complete lack of cellular structure, make viruses quite different

Figure 14.18 Papavoviruses in a human wart as seen with the aid of an electron microscope. (Courtesy Richard S. Demaree, Jr.)

from anything else in the five kingdoms of living organisms now recognized. In fact, there is some question as to whether or not viruses are even living. In 1946 Wendell Stanley, an American chemist, received a Nobel prize for demonstrating that a virus causing tobacco mosaic, a common plant disease, could be isolated, purified, and crystallized and that the crystals could be stored indefinitely but would always produce the disease in healthy plants at any time they were placed in contact with them. We also know that viruses do not grow by increasing in size, nor do they respond to external stimuli. They cannot move on their own, and they cannot carry on independent metabolism.

Chapter 14

Figure 14.19 Phage viruses as seen with the aid of an electron microscope. Insert drawing shows some of the detail. (Electron micrograph by Carolina Biological Supply Company)

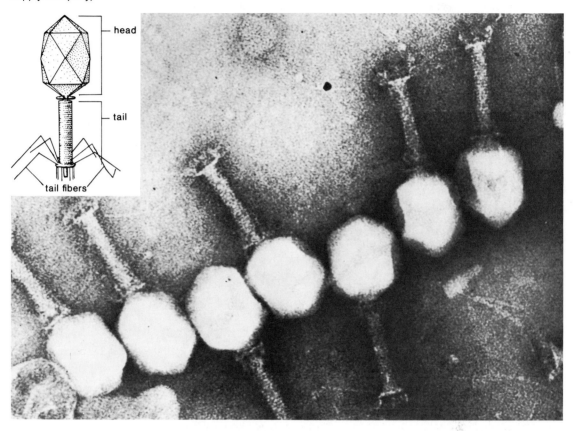

Viruses are about the size of large molecules, varying in diameter from about 15 to 300 nanometers (figure 14.18). Thousands of the smallest ones could fit inside a single bacterium of average size. Basically they consist of a nucleic acid core surrounded by a protein coat. The architecture of the protein coats varies considerably, but many have twenty sides and resemble tiny geodesic domes, while others have distinguishable head and tail regions. The nucleic acid core consists of either DNA or RNA—never both. Viruses have been classified in several ways. Originally they were grouped according to their hosts and the types of tissues or organs they affected. Now they are separated first according to the DNA or RNA in their cores; then, they are grouped according to size and shape, the nature of their protein coats, and the number of identical structural units in their cores.

Viruses that attack bacteria have been studied extensively. These are called **bacteriophages** or simply **phages** (figure 14.19). Some resemble the space exploration vehicles portrayed in the literature and films of science fiction. They consist of a head on top of a thin cylindrical core, which is surrounded by a sheathing coat. Toward the base of the core are six spiderlike fibril "legs," which anchor the virus in place.

Kingdom Monera and Viruses

Figure 14.20 Stages in the development of a phage virus within a bacillus bacterium. *A.* The virus becomes attached to the bacterium. *B.* The DNA of the virus enters the cell. *C.* Various components of the virus are synthesized from the DNA of the bacterium. *D.* The viral components are assembled into units. *E.* The assembled viruses are released as the bacterial wall breaks down. (After Pelczar, M. J., Jr., and R. D. Reid. 1972. *Microbiology,* 3d ed. Redrawn by permission of the McGraw-Hill Book Co.)

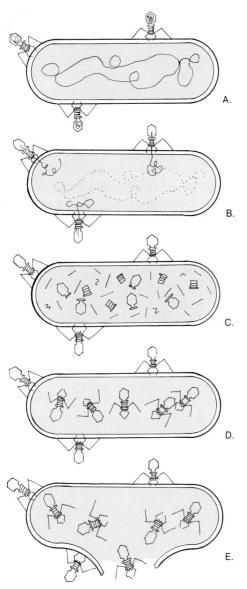

Viruses can replicate (reproduce themselves) only at the expense of their host cells. In doing so, they first become attached to a susceptible cell by their "legs." Then they penetrate to the interior, some types leaving their coats on the outside. Inside the cell, their DNA or RNA directs the synthesis of new virus molecules, which are then assembled into complete viruses. These are released from the host cell, usually as it dies (figure 14.20).

Some viruses (e.g., those causing influenza) can mutate (see chapter 12) very rapidly, allowing them to attack organisms that previously had been immune to them. As a result, new vaccines have to be developed constantly to combat new strains of viruses as they appear. Some viruses greatly affect the metabolism of their host cells. For example, in botulism bacteria, the botulism toxins are produced only if specific phages are present and active. Evidence is mounting that many forms of cancer, which usually involves abnormal cell growth, are caused by viruses. Scientists also suspect that all living organisms carry viruses in an inactive form in their cells, and they are trying to discover what causes the inactive viruses to become suddenly active.

Cells of higher animals that are invaded by viruses produce a protein called *interferon,* which is released into the fluid around the cells or into the bloodstream. Minute amounts of interferon in contact with cells cause the cells to produce a protective protein that prevents or inhibits the propagation of numerous types of viruses within the protected cells and also inhibits the capacity of viruses causing tumors to transform normal cells into tumor cells. Because of these properties of interferon, considerable research is currently being conducted on ways and means of producing it in much greater quantities than has hitherto been possible, for use in controlling cancers and many other viral diseases. Especially promising is the use of bacteria as hosts for donor DNA; this process, in effect, turns the bacteria into interferon synthesis centers (see figure 14.6), and

Chapter 14

in 1981, scientists at the University of Washington and the Genentech Corporation of San Francisco announced that they had succeeded in producing a form of interferon by splicing interferon genes into yeast cells. Because yeast cells are larger than bacteria, the process can potentially produce much larger quantities of interferon at considerably lower cost than is possible using bacteria.

Human Relevance of Viruses

The economic impact of viruses in both developing and industrialized countries is enormous. The annual loss in work time due to common cold and influenza viruses alone is virtually incalculable. While discomfort, adverse effects on employment, and even deaths due to viral diseases such as chicken pox, measles, German measles, mumps, and yellow fever have declined dramatically since immunizations against the diseases became widespread, they still take their toll; another viral disease, infectious hepatitis, periodically also still claims victims.

The production of vaccines for a number of diseases is undertaken by a flourishing worldwide multimillion dollar industry. Other mass-produced viruses are used to infect ticks, insects, and other disease organisms of both animals and plants. Some viruses cause great losses when they infest creamery vats during the manufacture of dairy products or culture vats during the production of antibiotics. One phage attacks nitrogen-fixing bacteria in the roots of leguminous plants, while other phages attack diphtheria and tuberculosis bacteria.

Recently it has been discovered that a number of insect pests of gardens can be controlled by making a dilute spray of the ground-up bodies of a quantity of the pests mixed with water in a blender. It has not been clearly demonstrated as yet, but it is suspected that some of the pests carry viruses that cause specific diseases in their own bodies. Such viruses would be well distributed by the blender and then effectively passed on to other insects in the garden when universally distributed through spraying.

Summary

The Kingdom Monera consists of prokaryotic organisms (bacteria, blue-green algae, and prochlorophytes) that occur as single cells, in colonies, or in the form of chains or filaments. Some cells may be motile (capable of movement) by means of whiplike flagella, or they may exhibit a gliding motion, but most are nonmotile. Nutrition is primarily by absorption of food in solution through the cell wall, but some members of the kingdom are photosynthetic or chemosynthetic (capable of obtaining their energy through chemical reactions with various compounds or elements). Reproduction is asexual by means of fission, a form of cell division that does not involve mitosis; some genetic recombination occurs in some groups by means of pili (minute tubes) that form connections between cells.

Luminescent bacteria are minor representatives of the most numerous of earth's living organisms. Bacteria are mostly 2 or 3 micrometers in diameter and are not visible individually except with the aid of a microscope. They occur primarily in the form of spheres (cocci), rods (bacilli), and spirals (spirilli) and are further classified on the basis of sheaths, appendages, and motion. Depending on their reaction to the blue-black Gram's stain, they are also classified as gram-positive or gram-negative.

Prokaryotic cells have no nuclear envelopes or organelles, but membranes present apparently perform some of the functions of organelles. Each cell has one long DNA molecule and sometimes up to thirty or forty small, circular DNA molecules called plasmids, which replicate independently of the large DNA molecule or chromosome. Mitosis does not occur, but a transverse wall forms near the middle of the cell and creates two cells after the DNA has replicated, by fission. Meiosis also does not take place, and gametes and zygotes are not produced. Genetic recombination occurs through conjugation (during which part of the DNA strand is transferred from a donor cell to a recipient cell through a hollow tubelike pilus); a living cell may also pick up and incorporate fragments of DNA released by dead cells in a process called transformation, or fragments of DNA may be carried from one cell to another by viruses in a process called transduction.

Genetic engineering involves the alteration of the DNA of a cell by artificially mixing large numbers of enzymatically "unzipped" DNA strands from two different sources; fragments from one source may associate with those from another source, creating recombinant DNA with parts of both original molecules. The recombinant plasmids are inserted into live bacterial cells where they replicate. Since bacteria are easy to propagate, large amounts of recombinant DNA may be obtained in this fashion. Such recombinant DNA may be used in the future for improving crops and possibly eliminating diseases of both plants and animals.

Heterotrophic bacteria (organisms that depend on other organisms for food) are mostly saprobes (organisms that obtain their food from dead organic matter), but some are parasites. Autotrophic bacteria are photosynthetic but do not produce oxygen; chemoautotrophic bacteria obtain their energy through oxidation of inorganic molecules such as iron, sulfur, nitrogen, and hydrogen.

Nitrifying bacteria play a role in the nitrogen cycle by converting ammonia to nitrites, and nitrites to nitrates; decay bacteria break down proteins with the release of ammonia. Nitrogen-fixing bacteria convert nitrogen from the air to forms that can be used by plants. Such bacteria inhabit the roots of legume plants, alders, and others.

Any dead tissues will eventually be decomposed to compost by bacteria and fungi, but piles of moderate size favor the sequence of bacteria usually involved in composting, which is increasingly being done more often by communities where ordinances against burning of leaves have been instituted. The proportion of carbon to nitrogen in a compost pile largely determines the pace at which decomposition proceeds, with a ratio of 30 carbon to 1 nitrogen being optimal. Compost is defininitely good for the soil but has limited value as a fertilizer.

Bacteria cause huge losses through plant diseases and food spoilage and many serious diseases in animals and humans. They gain access to their hosts from the air (e.g., diphtheria, whooping cough, some pneumonia forms), through contamination of food and drink (e.g., cholera, dysentery, salmonella food poisoning), through consumption of bacterial toxins (e.g., botulism), through direct contact (e.g., syphilis, gonorrhea, anthrax, brucellosis), through wounds (e.g., tetanus, gas gangrene), and through the bites of insects and other organisms (e.g., bubonic plague).

Koch formulated postulates (rules) for proving that a particular organism is the cause of a particular disease: The organism must be present in all cases of the disease; the organism must be isolated from the victim in pure culture; when the organism from the pure culture is injected into a susceptible host it must produce the disease in the host; the organism must be isolated from the experimentally infected host and grown in pure culture for comparison with that of the original culture.

Bacteria useful to humans include *Bacillus thuringiensis,* which attacks a wide range of caterpillar pests; *Bacillus thuringiensis israelensis,* which is highly destructive of mosquito larvae; and *Bacillus popilliae,* which attacks grubs of Japanese beetles. Bacteria also play a major role in the manufacture of dairy products such as yogurt, sour cream, kefir, and cheese; they are used in the manufacture of industrial chemicals, vitamins, flavorings, food stabilizers, and a blood plasma substitute; they play a role in the curing of vanilla, cocoa beans, coffee, and tea, and in the production of vinegar and sauerkraut; and they aid in the extraction of linen fibers from flax stems, in the production of ensilage for cattle feed, and in the production of several important amino acids. Bacteria from manures placed in digesters may, in the future, be used on a large scale for the production of methane gas for heating, cooking, and light.

Blue-green algae are similar in their cells to bacteria, but are distinguished from bacteria in having chlorophyll *a,* in producing oxygen when they undergo photosynthesis, and in having accessory pigments (blue and red) called phycobilins. They occur in fresh and marine water, in hot springs, on bare lava, on the shells of turtles and snails, and symbiotically within other organisms. Blue-green algae cells may have been the origin of chloroplasts, since they divide as chloroplasts do when their host cells divide. The cells may occur singly, in colonies within a gelatinous matrix, or as single or branched filaments. The blue-green color of the cells is due to the presence of a blue pigment, phycocyanin, but other pigments (e.g., red phycoerythrin) may mask the blue-green color. The carbohydrate cyanophycean starch is produced by blue-green algae. There are no flagella, but some species have gliding movements. Fragmentation may occur at heterocysts, which are special larger cells produced at intervals in the chains of some species. Thick-walled cells called akinetes may also be produced; akinetes can lie dormant for many years and are resistant to cold and drought.

Blue-green algae may become very abundant in bodies of polluted fresh water. Toxic substances are produced when the algae die and are decomposed by bacteria, and there are reports of various animals dying after drinking water in which blue-green algae have become abundant. At least forty species of blue-green algae are known to fix nitrogen.

Prochlorophytes are similar in form to blue-green algae but have pigmentation similar to that of higher plants and lack phycobilins.

Vaccination, which involves introducing to the body a weakened disease agent that in turn causes the body to build up an immunity to the disease while fighting the agent, was first performed by Jesty and Jenner in connection with smallpox. Smallpox, a disease caused by a virus, is now believed to have been eradicated.

Viruses, which have no cellular structure, are about the size of large molecules; some can be isolated, purified, and crystallized, yet remain virulent. They cannot grow or increase in size, and cannot be reproduced outside of a living cell, upon whose DNA they depend for duplication. Such attributes have led to speculation that they are not living organisms. Viruses consist of a core of nucleic acid surrounded by a protein coat; they are now classified on the basis of the DNA or RNA in their core, their size and shape, the number of identical structural units in their cores, and the nature of their protein coats.

Bacteriophages are viruses that attack bacteria. In replicating they become attached to a susceptible cell, which they penetrate, with their DNA or RNA directing the synthesis of new virus molecules from host material; the assembled new viruses are released when the host cell dies.

Cells of higher animals being invaded by viruses produce interferon, a protein that causes cells to produce a protective substance that inhibits duplication of viruses, and also inhibits the capacity of viruses to transform normal cells into tumor cells.

Viral diseases such as chicken pox, measles, mumps, and yellow fever have declined since immunizations against the diseases have become widespread. Mass-produced viruses are used to infect insects and other pests of both plants and animals. Some viruses cause losses in creamery vats. It is believed viruses may play a role in pest control when pests are liquefied in a blender and sprayed in gardens.

Review Questions

1. What is symbiosis? Give examples other than those mentioned in the text.

2. Why are bacteria not classified on the basis of visible features alone?

3. How do bacteria exchange DNA?

4. How does fission differ from mitosis?

5. Is photosynthesis the same in bacteria as it is in higher plants? Explain.

6. How do chemoautotrophic bacteria differ from photosynthetic bacteria?

7. What is the difference between nitrification and nitrogen fixation?

8. If decay bacteria use nitrogen, how does composting accumulate any nitrogen?

9. How are disease bacteria transmitted?

10. What are Koch's Postulates?

11. Why are many bacteria considered "useful"?

12. What do bacteria and blue-green algae have in common? How do they differ?

13. How do blue-green algae survive freezing and desiccation?

14. What is an algal bloom?

15. How do viruses differ from bacteria?

16. What is a vaccination?

17. What is a phage?

18. How do viruses multiply?

Discussion Questions

1. If a virulent phage were to eliminate all the bacteria in North America for one year, how would our lives be affected?

2. What would be the feasibility and the advantages or disadvantages of using only blue-green algae and nitrogen-fixing bacteria for our agricultural nitrogen needs?

3. Methane gas produced by bacteria is proving to be sufficient to meet all the fuel needs of villages in India. Do you think we could produce and use methane in a similar fashion in the United States?

4. If blue-green algae are capable only of asexual reproduction, does this mean that species of these organisms can never change in form?

5. As long as viruses can multiply, why should there be any question as to whether or not they are living?

Additional Reading

Brock, T. D. *Biology of microorganisms.* 3d ed. Englewood Cliffs, NJ: Prentice-Hall, Inc.; 1979.

Buchanan, R. E.; Gibbons, N. E., ed. *Bergey's manual of determinative microbiology.* 8th ed. Baltimore: Williams and Wilkins Co.; 1974.

Carr, N. G.; Whitton, B. A., ed. *The biology of the blue-green algae.* Oxford: Blackwell; 1973.

Chakrabarty, A. M. *Genetic engineering.* Boca Raton, FL: CRC Press; 1978.

Fenner, F.; McAuslan, B. R.; Mimms, C. A.; Sambrook, J.; White, D. O. *The biology of animal viruses.* 2d ed. New York: Academic Press; 1974.

Fogg, G. E.; Stewart, W. D. P.; Fay, P.; Walsby, A. E. *The blue-green algae.* New York: Academic Press; 1974.

Frobisher, M.; Hinsdill, R. D.; Crabtree, K. T.; Goodheart, C. R. *Fundamentals of microbiology.* 9th ed. Philadelphia: W. B. Saunders Co.; 1974.

Gibbs, A. J.; Harrison, B. D. *Plant virology: the principles.* New York: John Wiley and Sons, Inc.; 1979.

Luria, S. E., et al. *General Virology,* 3d ed. New York: John Wiley and Sons, Inc.; 1978.

Palmer, C. M. *Algae in water supplies.* Public Health Service Publication No. 657. Washington, DC: U.S. Government Printing Office; 1962.

Pelczar, M. J., Jr.; Reid, R. D.; Chan, E. C. S. *Microbiology.* 4th ed. New York: McGraw-Hill Book Co.; 1977.

Stanier, R. Y.; Adelberg, E. A.; Ingraham, J. *The microbial world.* 4th ed. Englewood Cliffs, NJ: Prentice-Hall; 1976.

Stoner, C. H., ed. *Producing your own power.* Emmaus, PA: Rodale Press; 1974.

ingdom Protista

erview

...ginning with a summary of the
...tures of members of the
...gdom Protista, this chapter
discusses each division of algae
and the slime molds. For the
Chlorophyta (green algae), asexual
and sexual reproduction in
*Chlamydomonas, Ulothrix,
Spirogyra,* and *Oedogonium* are
covered, and mention is made of
Chlorella, desmids, *Acetabularia,
Volvox,* and *Ulva.*

After a representative euglenoid
is described, the chapter discusses
the Division Chrysophyta (golden-
brown algae), with emphasis on
diatoms. It next mentions the
Division Pyrrophyta (dinoflagellates)
and the role of members of the
division in red tides and
phosphorescence. A brief treatment
of the Phaeophyta (brown algae)
and Rhodophyta (red algae)
includes a discussion of the life
cycle of a representative of each
division. This is followed by a table
summarizing divisional differences
with respect to food reserves,
special pigments, and flagella. The
chapter concludes with a digest of
the human and ecological relevance
of the algae and a discussion of the
Division Myxomycota (slime molds).

Some Suggested Student Learning Goals

1. Describe those features that
   the members of the Kingdom
   Protista share with one
   another, and note the basic
   ways in which they differ.

2. Diagram the life cycles of
   *Chlamydomonas, Ulothrix,
   Spirogyra,* and *Oedogonium;*
   indicate where meiosis and
   fertilization occur in each.

3. List at least two features that
   distinguish the Chlorophyta,
   Euglenophyta, Chrysophyta,
   Pyrrophyta, Phaeophyta, and
   Rhodophyta from one another.

4. Describe how a diatom differs
   in structure and form from
   other members of the Division
   Chrysophyta.

5. Name the function of each of
   the three thallus forms of a red
   alga such as *Polysiphonia.*

6. List at least twenty
   economically important uses of
   algae, in addition to the
   numerous uses of algin that
   were given.

7. Describe the life cycle of a
   slime mold, beginning and
   ending with a plasmomodium.

b

Features of the Kingdom

In contrast with the Kingdom Monera, whose members have prokaryotic cells, all members of the Kingdom Protista have eukaryotic cells. The organisms comprising this kingdom are very diverse and heterogenous, but none have the distinctive combinations of characteristics possessed by members of the Kingdoms Plantae, Fungi, or Animalia. Many, including the euglenoids, protozoans, and some algae, consist of a single cell, whereas other algae are multicellular or occur as colonies or filaments, and the slime molds consist of a mass of protoplasm usually containing numerous nuclei. Nutrition is equally varied, with the algae being photosynthetic, the protozoans and slime molds ingesting their food, and the euglenoids either carrying on photosynthesis or ingesting their food.

Individual life cycles vary considerably, but reproduction is generally by cell division and sexual processes. Many protistans are motile, usually by means of flagella, but some have *amoeboid movement* (a "crawling-flowing" of protoplasm that is not restricted by rigid cell walls) and others are nonmotile.

Introduction to the Algae

Children fortunate enough to have lived within access of ocean beaches where seaweeds are cast ashore by the surf, have often enjoyed stamping on the bladders ("floats") of kelps to make them "pop." Some have collected and pressed beautiful, feathery red seaweeds, and others who have waded around the shores of freshwater lakes or in slow-moving streams have encountered slimy-feeling pond scums. All who have kept tropical fish in glass tanks have sooner or later had to scrape a brownish or greenish film from the inner surfaces of the tank, and those who have lived in homes with their own swimming pools have learned that the white plaster of the pool soon acquires colored patches on its surface if chemicals are not regularly added to prevent them from appearing.

Algae, which have no flowers or leaves (although some of the seaweeds do have flattened, leaflike blades), are involved in our everyday lives in more ways than most people realize (see the discussion on the human and ecological relevance of the algae, which begins on page 296), and seaweeds, pond scums, fish tank films, and colored patches in swimming pools are but a few of the numerous kinds of algae all assigned to the Kingdom Protista. Based on the form of their reproductive cells and combinations of pigments and food reserves, the algae are grouped into several major divisions.

Division Chlorophyta—Green Algae

The approximately seventy-five hundred species of organisms comprising the Division Chlorophyta are commonly known as the *green algae*. They are a very diverse and cosmopolitan group. Some consist of microscopic unicellular forms or threadlike filaments, while others form platelike colonies, netlike tubes, or hollow balls. Some green algae are seaweeds, resembling lettuce leaves or green ropes up to several meters long. A number of unicellular forms grow in greenish patches or streaks on the bark of trees or help camouflage jungle animals, growing in large numbers while attached to their fur. Still others thrive in snowbanks, live in flatworms and sponges, or are found on the backs of turtles. They are the most common member in lichen "partnerships" (see chapter 16). The greatest variety, however, is found in freshwater ponds, lakes, and streams, with additional diversity evident in the oceans, where they are an important part of the *plankton* (free-floating, mostly microscopic organisms), and thus of food chains (figure 15.1).

The cells of green algae resemble those of higher plants in having the same kinds of chlorophylls and other pigments in their chloroplasts. They also store their food in the form of starch.

Figure 15.1 Representative green algae. A. *Volvox.* The cells form hollow, spherical colonies that spin on their axes as the flagella of each cell beat in such a way that the motion is coordinated. New colonies are formed within the old ones. B. *Scenedesmus.* The outer cells of each four-celled colony have spinelike extensions of the cell wall; the function of the spinelike extensions is unknown. C. *Pediastrum.* The cells form flattened, platelike, circular colonies. D. *Zygnema* (Portion of a filament). Each cell of this green alga contains two somewhat starlike chloroplasts. (Photomicrographs by Carolina Biological Supply Company)

A.

Most can undergo both asexual and sexual reproduction (see chapter 11). The manner in which they do so is illustrative of the forms of reproduction found in most of the organisms discussed in the chapters to follow, and so several different representative green algae will be examined in some detail here.

Chlamydomonas

A lively little alga, *Chlamydomonas* (figure 15.2), is a common inhabitant of quiet freshwater pools. It is considered to be one of the more primitive of eukaryotic organisms, with fossil relatives occurring in rock formations reported to be nearly 1 billion years old.

Chlamydomonas is unicellular, with a somewhat oval cell surrounded by a cellulose wall. A pair of whiplike flagella at one end enables the cell to move very rapidly. The flagella are, however, difficult to see with ordinary light microscopes, and the cell itself is usually less than 25 micrometers (one ten-thousandth of an inch) long. Near the base of the flagella are two or more vacuoles, which can contract and expand. They apparently regulate the water content of the cell. A dominant feature of each *Chlamydomonas* is a single, usually cup-shaped chloroplast, which at least partially hides the centrally located nucleus. Located in the chloroplast are one or two roundish glistening **pyrenoids,** which are structures of uncertain function but thought to be associated with the synthesis of starch. Most species also have a red **eyespot** on the chloroplast near the base of the flagella. Although it is merely part of an organelle within a single cell and nothing like an eye in structure, it has been demonstrated to be sensitive to light.

Chapter 15

B.

C.

D.

Figure 15.2 *Chlamydomonas*

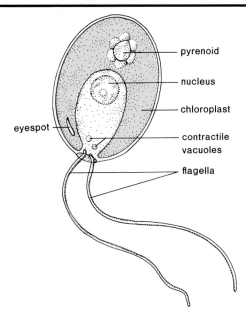

pyrenoid

nucleus

chloroplast

eyespot

contractile
vacuoles

flagella

Kingdom Protista

Asexual Reproduction

Before a *Chlamydomonas* reproduces asexually, the cell's flagella degenerate and drop off or are reabsorbed. Then the nucleus divides by mitosis, and the entire protoplasm becomes two cells within the cellulose wall. These two "daughter cells," which develop flagella, escape and swim away as the "parent" cell wall breaks down. Once they have grown to their full size, they may repeat the process. Sometimes mitosis occurs more than once, so that four, eight, or up to thirty-two little cells with flagella are produced inside the parent cell. Occasionally flagella do not develop, and the cells remain together in a colony. When growth conditions change, however, each cell of the colony may develop flagella and swim away. This type of reproduction brings about no changes in the number of chromosomes present in the nucleus, and all the cells remain *haploid* (see chapter 11).

Sexual Reproduction

Under certain environmental conditions, unknown forces within the organisms may bring about a congregating of many cells in populations of *Chlamydomonas*. Careful study of such events has revealed that pairs of cells appear to be attracted to each other by their flagella, and the protoplasts of these cells function as gametes. The cell walls break down as the protoplasts slowly emerge and fuse together, forming zygotes. A new wall, often relatively thick and ornamented with little bumps, forms around each zygote. This may remain dormant for several days, weeks, or even months, but under favorable conditions a dramatic change occurs. The protoplast, which is now *diploid,* undergoes meiosis, producing four haploid *Chlamydomonas* cells. These swim away and grow to their full size when the old zygote wall breaks down (figure 15.3).

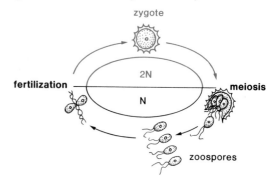

Figure 15.3 Sexual life cycle of *Chlamydomonas*

Ulothrix

Often an examination of dead twigs, rocks, and other debris in cold freshwater ponds, lakes, and streams reveals a threadlike alga called *Ulothrix* (figure 15.4), whose name is derived from the Greek words meaning "little hair." Each alga consists of a single row of cylindrical cells attached end to end and forming a thread or **filament**. The basal cell of each filament is slightly longer than the other cells and functions as an attachment cell, or **holdfast.** The nucleus of each cell is surrounded by a wide chloroplast that is shaped like a partial to nearly complete bracelet. Each chloroplast contains one to several pyrenoids. Filaments increase in length when individual cells divide, and any cell except the holdfast may do so.

Asexual Reproduction

The protoplast of any cell except the holdfast appears to clump and condense inside the rigid cell wall, divide by mitosis, and become *zoospores*. **Zoospores** are motile cells that do not fuse with other cells. Many different kinds of algae produce zoospores. The zoospores of *Ulothrix* are quite similar to *Chlamydomonas* cells in that they have contractile vacuoles and an eyespot, but they have four flagella instead of two.

Frequently a protoplast divides one to several times before becoming zoospores, but after zoospores are formed they usually escape from the parent cell through a pore in the wall. After

Chapter 15

Figure 15.4 Life cycle of *Ulothrix*

swimming about for a few hours to several days, they settle on submerged objects, shed their flagella, and divide. One of the first two daughter cells becomes a holdfast, while the other continues to divide and forms a new filament. In some instances, the protoplasts do not produce flagella after they have condensed and divided like developing zoospores, but they are otherwise capable of germinating and producing new filaments. Such cells, which are released when the parent cell wall breaks down, are called *aplanospores*.

Sexual Reproduction

Both asexual and sexual reproduction start out in the same way. The protoplast of any cell except the holdfast appears to condense and then divides. Up to sixty-four zoosporelike cells, each with two flagella, may be produced. When these cells escape from the parent cell walls, however, they function as gametes, uniting in pairs with gametes from other filaments and forming zygotes. The zygotes form thick walls and become dormant. Eventually their protoplasts undergo meiosis, giving rise to zoospores, which then can become new filaments.

Although the gametes of *Ulothrix* come from different filaments, they are identical in size and appearance. Sexual reproduction involving such gametes is called **isogamy**. It is found only in simpler organisms. As in *Chlamydomonas,* the zygotes of *Ulothrix* are the only diploid (2*N*) cells in the life cycle. All the other cells are haploid (*N*).

Spirogyra

One's initial reaction to a first encounter with "watersilk," as *Spirogyra* is called, may be less than ecstatic because of the slimy feel of the watery sheaths surrounding its filaments. The microscope, however, reveals a beautiful alga with some of the most striking chloroplasts known.

These common freshwater algae, which form unbranched filaments of cylindrical cells, are found most frequently floating in masses at the surface of quiet waters. The cells each contain a long, slightly frilly, ribbon-shaped chloroplast that looks as though it had been spirally wrapped around an invisible pole occupying most of the cell's interior. Some species of *Spirogyra* have as many as sixteen of these chloroplasts in each cell. Every one of these elegant green ribbons has pyrenoids at regular intervals along it.

Asexual Reproduction

Unlike the two algae just discussed, *Spirogyra* does not form any zoospores or other cells with flagella. Any cell is capable of dividing, but the only asexual reproduction resulting in new filaments is brought about through the breakup or fragmentation of existing filaments. This frequently occurs as a result of a storm or other disturbance.

Figure 15.5 *Spirogyra* (watersilk). *A.* Portion of a vegetative filament showing the ribbonlike chloroplasts spirally arranged in each cell. The centrally located darker object in each cell is a nucleus. *B.* From adjacent cells of two filaments close to each other, papillae have grown out opposite one another and formed conjugation tubes. *C.* The condensed protoplasts in the cells on the left are functioning as male gametes and migrating through the conjugation tubes to the stationary female gametes in the cells on the right. *D.* Zygotes have been produced in some of the cells on the right as a result of fusion of gametes. (Photomicrographs by Carolina Biological Supply Company)

A.   B.   C.   D.

Sexual Reproduction

In colonies of *Spirogyra,* the filaments usually are produced so close to each other that they may actually be touching. When sexual reproduction begins, the individual cells of adjacent filaments form little dome-shaped bumps, or **papillae** (singular: **papilla**), opposite each other. As these papillae grow, they force the filaments apart slightly, and then they fuse at their tips, forming small cylindrical **conjugation tubes** between each pair of cells. The condensed protoplasts then function as gametes. Usually those of one filament will seem to flow or crawl like amoebae through the conjugation tubes to the adjacent cells, where each fuses with the stationary gamete and forms a zygote. Each moving protoplast is considered a male gamete, while the stationary ones function as female gametes (figure 15.5).

The zygotes usually develop thick walls and remain dormant for some time, often over the winter. Thick-walled zygotes are characteristic of most freshwater green algae. Eventually their protoplasts undergo meiosis, producing four haploid cells. Three of these disintegrate, and a single new *Spirogyra* filament grows from the interior of the old zygote shell. The type of sexual reproduction shown by *Spirogyra* is called **conjugation.**

Oedogonium

Aquatic flowering plants and other algae often provide surfaces to which *Oedogonium* (pronounced ee-doh-goh'-nee-um), a filamentous green alga, may attach itself. It is, however, in no way parasitic. Algae and plants that attach themselves to other plants in such a manner are called **epiphytes.** The basal cells of the unbranched filaments form holdfasts, and the terminal cell of each filament is rounded. The remaining cells are cylindrical and attached end to end. The name *Oedogonium* comes from words meaning "swollen reproductive cell" and is quite apt, as the female reproductive cells do indeed bulge noticeably. Each cell contains a large netlike chloroplast, which rolls and forms a tube something like a loose wickerwork basket around and toward the periphery of each protoplast. There are pyrenoids at a number of the intersections of the net.

Asexual Reproduction

Akinetes, the thick-walled overwintering cells, may occasionally be formed, but more commonly zoospores are produced singly in cells at the tips of the filaments (figure 15.6). Unlike the zoospores of most other algae, those of *Oedogonium* have about 120 small flagella forming a fringe around the cell toward one end. The zoospores look like tiny balding faceless heads, and their rapid rotating movements can be quite entertaining when viewed through a microscope. After they escape from their filaments, they eventually settle and form new filaments in the same manner as *Ulothrix.*

Sexual Reproduction

Oedogonium shows more sexual specialization than any of the three algae previously discussed. Short boxlike cells called **antheridia** (singular: **antheridium**) are formed in the filaments alongside the ordinary vegetative cells. A pair of male gametes, or *sperms,* is produced in each antheridium. The sperms resemble the zoospores but are smaller. Certain cells become swollen and round to elliptical in outline. These cells, called **oogonia** (singular: **oogonium**), each contain a single female gamete or *egg,* which occupies nearly all of the cell. As the egg matures, a pore develops on the side of the oogonium. When sperms escape from the antheridia, they are attracted to the oogonia by a substance released by the eggs. One sperm eventually enters the oogonium through the pore and unites with the egg, forming a zygote. In some species of *Oedogonium,* oogonia are produced only on female filaments and antheridia only on male filaments. Sometimes the male filaments are dwarf. They attach themselves to the female filaments, and then both produce hormones that influence the other's development.

Figure 15.6 Life cycle of *Oedogonium*

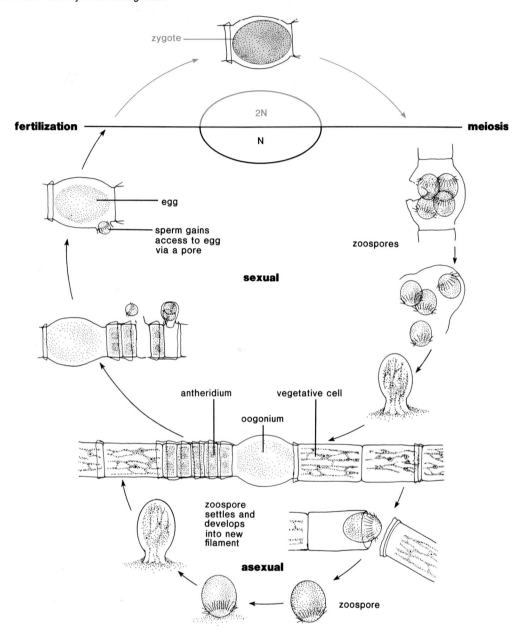

zygote

2N

N

fertilization

meiosis

egg

sperm gains
access to egg
via a pore

zoospores

sexual

antheridium

vegetative cell

oogonium

zoospore
settles and
develops
into new
filament

asexual

zoospore

Zygotes may remain dormant for a year or more, but they eventually undergo meiosis, producing four zoospores, each of which is capable of developing into a new filament. The sexual reproduction exhibited by *Oedogonium,* in which one gamete is motile (capable of spontaneous movement) while the other gamete is larger and stationary, is called **oogamy.**

Other Green Algae

As indicated earlier, the green algae constitute a very diverse group, with an extraordinary variety of forms and chloroplast shapes exhibited in their seventy-five hundred species. Each species obviously has to reproduce in order to perpetuate itself. Most undergo both sexual and asexual reproduction, but a few do not. For example, the worldwide algae that make parts of some tree trunks appear as though they had received a light brushing or spattering of green paint usually are unicellular or colonial forms that reproduce only asexually. These and *Chlorella,* another widespread green alga composed of tiny spherical cells, reproduce by forming either daughter cells or aplanospores through mitosis. The daughter cells often remain together in packets, while the aplanospores of *Chlorella,* which are formed inside the parent cells, grow to full size as the parent cell wall breaks down.

Chlorella is an alga very easy to culture and is a favorite organism of research scientists. Many major investigations in photosynthesis and respiration have utilized this alga, and in the future it may not only become important in human nutrition (see the section on human and ecological relevance of the algae, which begins on page 296), but it could also play a key role in long-range space exploration. Present exploration is severely limited by the weight of oxygen tanks and food supplies needed on a spacecraft, and so scientists have turned to *Chlorella* and similar algae as portable oxygen generators and food sources. Future spacecraft may be equipped with tanks of such algae. These would carry on photosynthesis, using available light and carbon dioxide given off

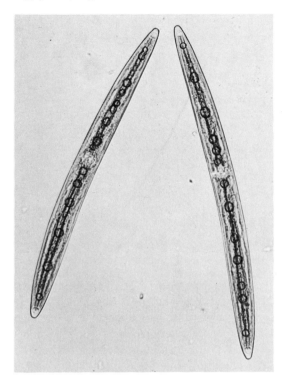

Figure 15.7 Two cells of the common desmid, *Closterium.* (Photomicrograph by Carolina Biological Supply Company)

by the astronauts, while furnishing them with oxygen. As the algae multiplied, the excess could either be eaten or fed to freshwater shrimp, which could, in turn, become food for the astronauts. Still other algae and bacteria could recycle other human wastes. Such a self-perpetuating "closed system," as it is called, has already been successfully tested with mice and other animals. Many algae, however, are known to produce traces of deadly carbon monoxide gas, and until this problem is resolved, humans will not be subjected to such research.

Desmids (figure 15.7), whose twenty-five hundred species of crescent-shaped, elliptical, and star-shaped cells are mostly free-floating and unicellular, reproduce by conjugation. In

Figure 15.8 Portion of a water net (*Hydrodictyon*), a green alga whose cells form a network in the shape of a tube

Figure 15.9 Mermaid's wineglass (*Acetabularia*)

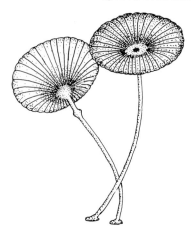

the beautiful *water nets* (*Hydrodictyon*) (figure 15.8), sexual reproduction is isogamous, with up to one hundred thousand flagellated gametes being produced in a single cell. Sexual reproduction is also isogamous in the mermaid's wineglass (*Acetabularia*), a marine alga consisting of a single huge cell shaped like a delicate mushroom (figure 15.9). Each cell is up to 5 centimeters (2 inches) long. This alga has been used in classic experiments demonstrating the influence of the nucleus on the form of the cell. If the "cap" of an alga is removed and the nucleus is replaced with a nucleus taken from another species, the base regenerates a cap identical with the previous one. If this new cap is also removed, however, the next cap that develops shows form characteristics of both species. If the intermediate cap is then removed, the next cap that develops is identical with that of the species from which the nucleus originally came. Clearly the original nucleus directed development of cytoplasmic substances regulating cap form, and when these are gone, the replacement nucleus exerts its own influence.

Volvox (see figure 15.1A) is representative of a line of green algae that forms colonies, apparently by means of single cells similar to those of *Chlamydomonas,* held together in a secretion of gelatinous material. In some colonies, the cells are actually connected to one another by cytoplasmic strands. The flagella of individual cells beat separately but pull the whole colony along. A *Volvox* colony may consist of several hundred to many thousands of cells arranged so that they resemble a hollow ball, which appears to spin on its axis as it moves. Reproduction may be either asexual or sexual, with smaller daughter colonies being formed inside the parent colony. The daughter colonies are released when the parent colony breaks apart.

Sea lettuce *(Ulva)* is a multicellular seaweed with flattened green blades that may attain lengths of 1 meter (3 feet) or more (figure 15.10). The blades are anchored to rocks by means of a holdfast at the base. Some blades are diploid; these produce spores that develop into haploid blades bearing gametangia. The gametes from these haploid blades fuse in pairs, forming zygotes, each of which can potentially grow into a new diploid blade. Except for the reproductive structures, the haploid and diploid blades of sea lettuce are indistinguishable from one another.

Chapter 15

Figure 15.10 Sea lettuce (*Ulva*)

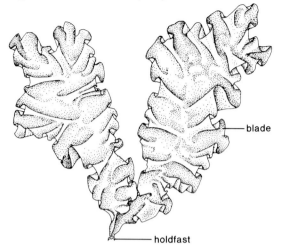

blade

holdfast

Figure 15.11 *Euglena*

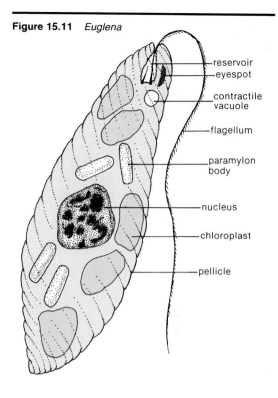

reservoir
eyespot
contractile vacuole
flagellum
paramylon body
nucleus
chloroplast
pellicle

Stoneworts, which loosely resemble horsetail plants (see chapter 18), are sometimes included with the green algae. Their sexual reproduction is oogamous, and they are placed in their own division (Charophyta) in this book because the antheridia in which the sperms are produced are multicellular and because other features of both their vegetative growth and reproduction are more complex than those of any of the green algae.

Division Euglenophyta—Euglenoids

Barnyard pools and sewage treatment ponds often develop a rich green bloom of algae. A superficial examination of water from such a pool with the aid of a microscope usually reveals what appear to be large numbers of active *Chlamydomonas* cells. A closer inspection, however, probably will reveal *euglenoids,* of which *Euglena* is a common example (figure 15.11).

Euglena differs from *Chlamydomonas* in a number of important respects. For one thing, the spindle-shaped cell has no rigid wall, and it can be seen to change shape even as the organism moves along. Just beneath the plasma membrane are fine strips that spiral around the cell parallel to one another. The strips and the plasma membrane together are called a *pellicle,* which is devoid of cellulose. A single flagellum, which has numerous tiny hairs along one side, pulls the cell through the water. A second very short flagellum is present within a *reservoir* at the base of the functional flagellum. Other features of *Euglena* include the presence of a *gullet,* or groove, through which food can be ingested, several to many mostly disc-shaped chloroplasts, and a red eyespot located in the cytoplasm near the base of the flagella. A unique carbohydrate food reserve called *paramylon* normally is present in the form of small whitish bodies of various shapes.

Reproduction is by cell division. The cell starts to divide at the flagellar end and eventually splits lengthwise, forming two complete cells. Sexual reproduction has been hinted at but has never been confirmed.

Some *Euglena* species can live in the dark if appropriate food and vitamins are present. Others are known to reproduce faster than their chloroplasts under certain circumstances, so that some chloroplast-free cells are formed. As long as a suitable environment is provided, these cells also can survive indefinitely. *Euglena's* clear manifestation of both photosynthetic capacity and animallike methods of obtaining nutriment caused it in the past, when only two kingdoms were recognized, to be treated as a plant in botany texts and as an animal in zoology texts.

Division Chrysophyta— Golden-Brown Algae

If the members of this division were not primarily microscopic, many undoubtedly would become collectors' items in the art and antique shops of the world because of their exquisite form and ornamentation (figures 15.12, 15.13 and 15.14). They are grouped into three classes: *Yellow-green algae* (Xanthophyceae), *golden-brown algae* (Chrysophyceae), and *diatoms* (Bacillariophyceae). Superficially the organisms of each class may appear unrelated to each other, but they do have several features in common, including food reserves, specialized pigments, and other cell characteristics. Some members of each class produce a unique "resting" cell called a *statospore* (figure 15.13). These cells resemble miniature glass apothecary bottles, complete with plugs that dissolve or "uncork," releasing the protoplast inside. Many statospores are striking in form, with finely sculptured ornamentations on the surface.

Diatoms (figure 15.14) are the best known and economically the most important members of the division. These mostly unicellular algae occur in astronomical numbers in both fresh and salt water but are particularly numerous in colder marine habitats. They also usually dominate the algal flora on damp cliffs, the bark of trees, bare soil, or the sides of buildings. Some of the more

Figure 15.12 Some representatives of the Division Chrysophyta. A. *Vaucheria.* B. *Stipitococcus.* C. *Dinobryon.*

than forty thousand known species can withstand extreme drought, and one species is known to have become active after lying dormant for forty-eight years in dry soil.

Diatoms look like little glass pillboxes, with half of the rigid crystal-clear wall fitting inside the other overlapping half. Most marine diatoms are circular in outline when viewed from the top, while freshwater species tend to resemble the outline of a boat when viewed from the top. The walls contain pectin and aluminum silicate, an ingredient of glass. They usually have exquisitely fine grooves and pores that are composed of exceptionally minute passageways that connect the protoplasm with the watery environment outside the shell. These fine grooves and pores are so uniformly spaced they have been used to test the resolution of microscope lenses.

Chapter 15

Figure 15.13 Statospores, which resemble apothecary bottles, are formed by many of the golden-brown algae. (After G. M. Smith. 1950. *The Freshwater Algae of the United States.* 2d ed. Redrawn by permission of the McGraw-Hill Book Co.)

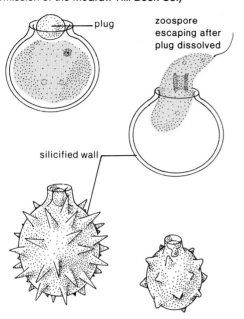

Figure 15.14 Diatoms. (Photomicrograph by Carolina Biological Supply Company)

Each diatom may have one, two, or many chloroplasts per cell. These frequently have a golden-brown color because of the dominance of **fucoxanthin,** a brownish pigment also found in the brown algae, and the accessory pigment chlorophyll *c* is typically present.

Many freshwater diatoms move backward and forward with somewhat jerky motions. It is believed that the movements occur in response to external stimuli. Extremely tiny fibrils are thought to take up water as they are discharged into a lengthwise groove or pores through which moving cytoplasm protrudes. When they come in contact with a surface, they stick and contract, moving the cell as the caterpillar treads on a tractor do and leaving a trail similar in some respects to that of a snail.

Reproduction in diatoms is unique. After a protoplast, which is diploid, has undergone mitosis and division, the two halves of the cell separate, with a daughter protoplast remaining in each half. Then a new half wall fitting *inside* the old half is formed. This occurs for a number of generations, with the result that some of the cells become progressively smaller. Eventually, however, a protoplast undergoes meiosis, producing four gametes, which then escape. These fuse with other gametes, becoming zygotes called *auxospores*. Auxospores are like any other zygotes except that they rapidly increase in size before their rigid "pillbox" walls are formed around them, thus restoring the original size of the diatom species.

Division Pyrrophyta— Dinoflagellates

If you are at an ocean beach in midsummer and ever notice a distinct reddish tint to the water, you are probably witnessing a phenomenon known as a **red tide.** If it is feasible for you to do so at the time, you should dip a cup of seawater and save it until you can examine it under a microscope. (If you add a few drops of formaldehyde or other preservative substance, the material should last indefinitely, although the organisms in the water will, of course, be dead.)

Red tides are caused by the sudden and not fully understood multiplication of unicellular organisms called *dinoflagellates* (figure 15.15). Your cup of seawater should contain a large number of these dinoflagellates, which are the best known representatives of the Division Pyrrophyta. Some resemble armor-plated spaceships, while others may be smooth or have fine lengthwise ribs. All have a distinctive arrangement of their two flagella, which are usually attached near each other in a groove. One, which acts as a rudder, trails behind the cell, while the other gives the cell a spinning motion as it undulates in another groove like a tiny snake. The latter groove encircles the cell at right angles to the other groove. Most species have two or more disc-shaped chloroplasts, which contain distinctive brown pigments in addition to various other pigments, including chlorophylls *a* and *c*. Some have an eyespot, and all have a unique nucleus in which the chromosomes remain condensed and clearly visible throughout the life of the cell.

Dinoflagellates occur in most types of fresh and salt water, but those that cause red tides have received the most publicity because they also produce powerful toxins that kill fish and shellfish. In fact, the toxins, which also paralyze the human nervous system, have been studied for possible use in chemical warfare. The havoc caused by major red tides to the fishing industry

Figure 15.15 Dinoflagellates. A. *Ceratium.* B. *Gonyaulax.*

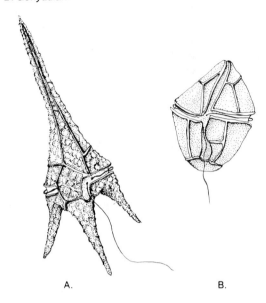

A.　　　　　　　　B.

is so great that several laboratories have been conducting research with dinoflagellates and their marine habitats to try to find a way of preventing such destruction in the future.

Some dinoflagellates are luminescent and play a major role in the phenomenon of *phosphorescence* often seen in tropical ocean waters. At night, if dinoflagellates are numerous, it sometimes looks as though every wave is glowing brightly as it breaks.

Reproduction is by cell division. Sexual reproduction appears to be rare.

Division Phaeophyta—Brown Algae

Most seaweeds that are brown to olive green in color are assigned to this division. Many are relatively large; none are unicellular or colonial. Very few occur in fresh water, the vast majority being found in colder ocean waters, usually in shallower areas, although the giant kelp (figure 15.16) may be found in water up to 30 meters (100 feet) or more deep.

Figure 15.16 Kelps. *A.* Giant kelp (*Macrocystis*). (Photo by Wm. E. Ferguson) *B.* Sea palms (*Postelsia*).

Figure 15.17 Parts of the brown alga *Nereocystis*, a kelp. The heaviest pounding surf seldom dislodges brown algae from rocks.

A.

B.

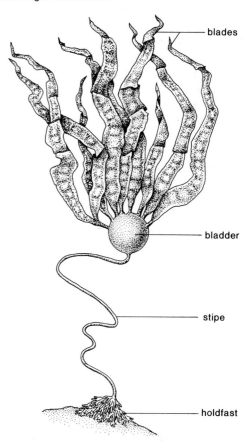

blades

bladder

stipe

holdfast

Many have a body that is differentiated into a *holdfast*, a *stipe*, and flattened leaflike *blades* (figure 15.17). The **holdfast** is a tough, sinewy structure resembling a mass of intertwined roots. It holds the seaweed to rocks so tenaciously that even the heaviest pounding of surf will not readily dislodge it. The **stipe** is a stalk, often hollow, which has a meristem either at its base or at the blade junctions. Since the meristem produces new tissue at the base, the oldest parts of the blades are at the tips. The **blades,** which like most of the rest of the body are photosynthetic, may have gas-filled floats called *bladders* toward their bases. Analysis of the bladder contents of some kelps has revealed the presence of more than 10% carbon monoxide, the deadly gas produced by internal combustion engines.

The color of the brown algae reflects the presence of varying amounts of the brown pigment *fucoxanthin,* in addition to chlorophylls *a* and *c* and several other pigments in the chloroplasts. The main food reserve is a carbohydrate called *laminarin.* **Algin** (see table 15.2 and the discussion under human and ecological relevance of the algae, page 296) occurs on or in the cell walls. Reproductive cells are unusual in that the two flagella are inserted laterally (i.e., on the side) instead of at the ends. The only motile cells in the brown algae are the reproductive cells.

In some localities off the coast of British Columbia and Washington, herring deposit spawn (eggs) in layers up to 2.5 centimeters (1 inch) thick on both surfaces of giant kelp blades in late spring. In the past, and to a limited extent at present, native North Americans have harvested these spawn-covered blades, sun-dried them, and used them for winter or feast food. Even today some school children are given small pieces of the dried or preserved material for lunchbox snacks.

The Sargasso Sea gets its name from a brown rockweed, *Sargassum,* which is washed up in large quantities on the shores along the Gulf of Mexico after tropical storms (figure 15.18). A species occurring in the Pacific Ocean has been used, in chopped form, as a poultice on cuts received from coral by native Hawaiians. This and several other brown algae reproduce asexually by fragmentation, while some produce aplanospores. Sexual reproduction takes several different forms, depending on the species, with the conspicuous phases of the life cycles usually being diploid.

Figure 15.18 *Sargassum,* a floating brown alga from which the Sargasso Sea got its name. It is also found in other marine waters. (Courtesy I. A. Abbott)

In the common rockweed *Fucus* (figure 15.19), separate male and female **thalli** (singular: **thallus,** the term for a body that is usually flattened, multicellular, and not organized into leaves, stems, and roots) are produced. Somewhat puffy fertile areas called **receptacles** develop at the tips of the branches of the thallus.

Chapter 15

Figure 15.19 Life cycle of the common rockweed *Fucus*

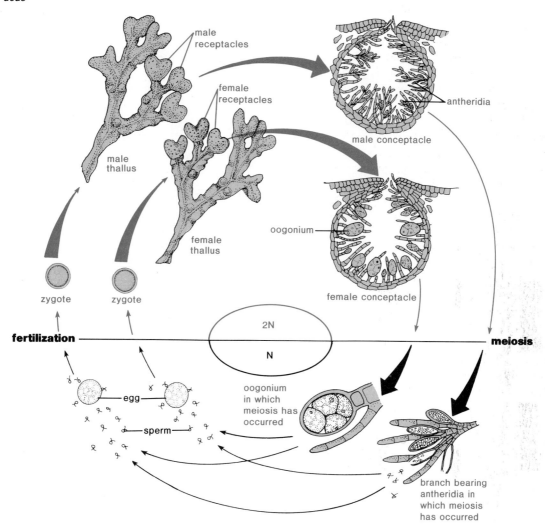

The surface of each receptacle is dotted with pores (visible to the naked eye) that open into special, spherical hollow chambers called *conceptacles*. Within the conceptacles **gametangia** (cells or structures in which gametes are produced) are formed. Eight eggs are produced in each *oogonium* (female gametangium) as a result of a single diploid nucleus undergoing meiosis followed by mitosis. Meiosis also occurs in each *antheridium* (male gametangium), but three mitotic divisions follow meiosis so that sixty-four sperms are produced. Eventually both eggs and sperms are released into the water, where fertilization takes place, and the zygotes develop into mature thalli, completing the life cycle.

Figure 15.20 Representative red algae.
A. *Botryocladia.* B. *Gigartina.* C. *Ahnfeltia.*
D. *Microcladia.*

A.

B.

Division Rhodophyta—Red Algae

Like the brown algae, most of the red algae are seaweeds (figure 15.20); however, they tend to occur in warmer and deeper waters than their brown counterparts. Some grow attached to rocks in intertidal zones, where they may be exposed at low tide. Others grow at depths of up to 175 meters (575 feet), where light barely reaches them. A few are unicellular, but most are filamentous. The filaments frequently are so tightly packed that the plants appear to have flattened blades or to form branching segments. Some develop as beautiful feathery structures that have the appearance of delicate works of art.[1] None match the large kelps in size, however; the largest species seldom exceed 9 or 10 decimeters (3 feet) in height.

1. Most seaweeds produce their own "glue" and are easy to mount on paper for display. Fresh specimens can be laid directly on clean, high-rag-content paper, covered with a layer or two of cheesecloth, and pressed between sheets of blotting paper for a day or two until they are dry. Feathery types will make better specimens if they are placed in a shallow pan of water so that their delicate structures float out, as they do in the ocean. Slide paper under them and carefully lift so that the seaweed spreads out naturally on the surface. Once the specimens are dry, remove the cheesecloth (which keeps them from sticking to blotting paper), and they remain glued to the paper. They can then be displayed or stored indefinitely. Green and brown seaweeds can be treated in similar fashion.

Certain marine algae, which form crusty growths or jointed-appearing upright structures on rocks, accumulate calcium salts as they grow and often contribute to the development of coral reefs. These *coralline* algae need no special treatment to be displayed, although some may lose their natural pinkish or purplish color when they die.

C.

D.

The red algae have relatively complex life cycles, often involving three different types of body structures. Meiosis usually occurs on a special body called a *tetrasporophyte,* while gametes are produced on separate male and female bodies. All of the reproductive cells are nonmotile and are carried passively by water currents. Zygotes may migrate from one cell to another through special tubes, which form bizarre loops in some species.

In *Polysiphonia* (figure 15.21), a feathery red alga that is widespread in marine waters, the three types of thalli (male gametophyte, female gametophyte, and tetrasporophyte) all outwardly resemble one another in being about 2 to 15 centimeters (1 to 6 inches) tall, and in being branched into many fine, threadlike segments. The male sex organs, called *spermatangia,* somewhat resemble dense clusters of tiny grapes on slender branches

of the male gametophyte thallus. Each spermatangium, as it is released from its branch, functions as a nonmotile male gamete or *spermatium.* The female sex organs, called *carpogonia,* are produced on the female gametophyte thallus. Each carpogonium looks something like a microscopic bottle with a very long neck called the *trichogyne.* A single nucleus at the base of the carpogonium functions as the female gamete, or egg. Since the spermatia have no flagella they cannot move of their own accord, but currents may carry them considerable distances. If a spermatium should brush against a trichogyne it may become attached, with the walls between the spermatium and the trichogyne breaking down and permitting the nucleus of the spermatium to migrate to the egg nucleus, with which it unites forming a zygote.

Figure 15.21   Life cycle of the red alga *Polysiphonia*

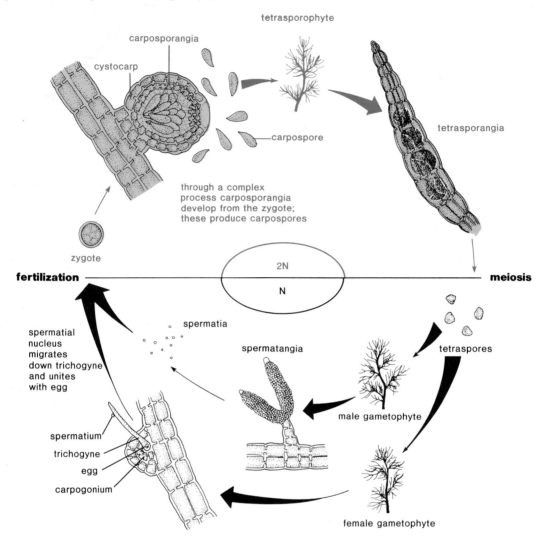

tetrasporophyte

carposporangia

cystocarp

tetrasporangia

carpospore

through a complex
process carposporangia
develop from the zygote;
these produce carpospores

zygote

2N

N

fertilization

meiosis

spermatia

spermatial
nucleus
migrates
down trichogyne
and unites
with egg

spermatangia

tetraspores

male gametophyte

spermatium

trichogyne

egg

carpogonium

female gametophyte

Table 15.1 Comparison of the divisions of eukaryotic algae

Division	Food Reserves	Special Pigments[1]	Flagella
Chlorophyta 7,500 species of green algae	starch	chlorophyll *b*	2 or more at the apex or none
Euglenophyta 700 species of euglenoids	paramylon and fats	chlorophyll *b*	1–3 near the apex
Chrysophyta 6,000 species of yellow-green algae, golden-brown algae, and diatoms	oils, chrysolaminarin	chlorophyll *c* a few with chlorophyll *e*	1–2, usually unequal, or none
Pyrrophyta 1,100 species; primarily dinoflagellates	starch, fats, oils	chlorophyll *a*	2: one trailing, one girdling
Phaeophyta 1,500 species of brown algae	laminarin, mannitol (an alcohol)	chlorophyll *c* fucoxanthin	2: unequal and lateral
Rhodophyta 3,900 species of red algae	floridean starch	chlorophyll *d* (in some species); phycobilins	none

[1]Note: All divisions of eukaryotic algae have chlorophyll *a* and yellow to orange pigments called *carotenes* and *xanthophylls,* although the combinations and specific types of the latter vary with the divisions. The fucoxanthin of the brown algae is a xanthophyll.

Next, the zygote develops into a cluster of clublike *carposporangia* toward the base of an urn-shaped body (the *cystocarp*) formed around the carposporangia by the female gametophyte thallus. Diploid asexual spores, called *carpospores,* are produced by the carposporangia; these, when released, are carried away by ocean currents. When a carpospore lodges in a suitable location (e.g., rock crevice or hull of a ship) it germinates and grows into a *tetrasporophyte,* which closely resembles the gametophyte thalli. *Tetrasporangia* are formed along the branches of the tetrasporophytes; each tetrasporangium undergoes meiosis, giving rise to four haploid *tetraspores.* When tetraspores germinate they develop into male or female gametophytes, and the life cycle is complete.

The red to purplish colors of most red algae are due to the presence of varying amounts of red and blue accessory pigments called *phycobilins,* which are closely related to those found in the blue-green algae and which have led to the belief that the red algae may have been derived from the blue-green algae. Several other pigments, including chlorophyll *a* and sometimes chlorophyll *d,* are also present in the chloroplasts. The reserve food is a carbohydrate called *floridean starch.* A number of red algae also produce **agar** and other important gelatinous substances discussed in the next section.

A comparative summary of the food reserves, special pigments, and flagella of the divisions of eukaryotic algae is given in table 15.1.

Table 15.2 Some Uses of Algin

Food
1. As a thickening agent in toppings, pastry fillings, meringues, potato salad, canned foods, gravies, dry mixes, bakery jellies, icings, dietetic foods, fountain syrups, candies, puddings.

2. As an emulsifier and suspension agent in soft drinks and concentrates, salad dressings, barbecue sauces, frozen food batters.

3. As a stabilizer in chocolate drinks, eggnog, ice cream, sherbets, sour cream, coffee creamers, party dips, buttermilk, dairy toppings, milkshakes, marshmallows.

Paper
1. Provides better ink and varnish holdout on paper surfaces; provides uniformity of ink acceptance, reduction in coating weight and improved holdout of oil, wax, and solvents in paperboard products. Makes improved coating for frozen food cartons.

2. Used for coating greaseproof papers.

Textiles
1. Thickens print paste and improves dye dispersal. Reduces weaving time and eliminates damage to printing rolls or screens.

Pharmaceuticals and Cosmetics
1. Thickening agent in weight control products, cough syrups, suppositories, ointments, toothpastes, shampoos, eye makeup.

2. Smoothing agent in lotions, creams, lubricating jellies. Binder in manufacture of pills. Blood anticoagulant.

3. Suspension agent for liquid vitamins, mineral oil emulsions, antibiotics. Gelling agent for facial beauty masks, dental impression compounds.

Industrial Uses
1. Used in manufacture of acidic cleaners, films, seed coverings, welding rod flux, ceramic glazes, boiler compounds (prevents minerals from precipitating on tubes), leather finishes, sizing, various rubber compounds (e.g., automobile tires, electric insulation, form cushions, baby pants), cleaners, polishes, latex paints, adhesives, tapes, patching plaster, crack fillers, wall joint cement, fiberglass battery plates, insecticides, resins, tungsten filaments for light bulbs, digestible surgical gut (disappears by time incision is healed), oil well-drilling mud. Used in clarification of beet sugar. Mixed with alfalfa and grain meals in dairy and poultry feeds.

Brewing
1. Helps create creamier beer foam with smaller, longer-lasting bubbles.

Human and Ecological Relevance of the Algae

We have already noted that numbers of blue-green algae are at the bottom of aquatic food chains, but members of all the protistan algal divisions play similar roles. Diatoms, for example, are consumed by fish that feed on plankton. The oils produced by diatoms are converted into cod-liver and other liver oils, which are rich sources of vitamins for humans. Diatoms also have other extensive and more direct industrial uses. In the past, they have apparently been at least as numerous as they are now. As billions upon billions of them have reproduced and died, their microscopic glassy shells have accumulated on the ocean floor, forming deposits of "diatomaceous earth." These deposits have accumulated to depths of hundreds of meters in some parts of the world and are quarried in several areas where past geological activity has raised them above sea level. At Lompoc, California, beds of diatomaceous earth are more than 200 meters (650 feet) deep, while in the Santa Maria oilfields of California, deposits reach a depth of 1,000 meters (3,280 feet).

Chapter 15

This light, porous, powdery-looking material contains about 6 billion diatom shells per liter (1.057 quarts), yet a liter weighs only eight-tenths of a kilogram (1.76 pounds). It also has an exceptionally high melting point of 1,750°C (3,182°F) and is insoluble in most acids and other liquids. These properties make it ideal for a variety of industrial and domestic uses, including many types of filtration. The sugar industry uses diatomaceous earth in sugar refining, and its use for swimming pool filters is widespread. It is also used in silver and other metal polishes, in toothpaste, and in the manufacture of light-reflecting paint used on highway markers and signs and on the automobile license plates of some states. It is packed as insulation around blast furnaces and boilers. It is also manufactured into the construction panels used in prefabricated housing.

A few green algae have occasionally been used for food, but members of this division have generally been used less by humans than those of other divisions. With dwindling world food supplies, this could soon change. Sea lettuce has been used for food on a limited scale in Asian countries for some time, and several countries are experimenting with the suitability of **plankton** for human consumption. Except for vitamin C, *Chlorella* contains most of the vitamins needed in human nutrition, and since it is so easy to culture, it may become an important food source in many parts of the globe. *Chlorella* has also been investigated as a potential oxygen source for atomic submarines, in addition to its possible use in space exploration.

Commercially produced ice cream, salad dressing, beer, jelly beans, latex paint, penicillin suspensions, paper, textiles, toothpaste, ceramics, and floor polish today all share a common ingredient, **algin,** produced by the giant kelps and other brown algae, and it is now used in so many products (table 15.2) that one might wonder how the world used to get along without it.

Algin has the unique ability to regulate water "behavior" in a wide variety of products. It can, for example, control the development of ice crystals in frozen foods, regulate the penetration of water into a porous surface, and generally stabilize any kind of suspension such as an ordinary milkshake or other thick fluid containing water. It is produced by several kinds of seaweeds, but a major source is the giant kelp found in the cooler ocean waters of the world, usually just offshore where there are strong currents (see figure 15.16A). This large seaweed, which has been known to attain a length of 92 meters (300 feet), sometimes grows at the rate of 3 to 6 decimeters (1 to 2 feet) per day. Specially equipped ocean-going vessels (figure 15.22) harvest the kelp by mowing off the top meter (3 feet) of growth, taking the chopped material aboard and then transferring it to processing centers onshore where it is extracted and refined.

Brown algae also produce a number of other useful substances. Many seaweeds, but particularly kelps, build up concentrations of iodine to as much as twenty thousand times that of the surrounding sea water. Although it is cheaper to obtain iodine from other sources in the United States, dried kelp has been used in the treatment of goiter, which results from iodine deficiency, in other parts of the world. Kelps are relatively high in nitrogen and potassium and have been used as fertilizer for many years. They have also been used as livestock feed in northern Europe and elsewhere. In the Orient, many marine algae are used for food in soups, confections, meat dishes, vegetable dishes, and beverages. In Japan, acetic acid is produced through fermentation of seaweeds.

During the Irish famine of 1845–1846, *dulse,* a red seaweed, became an important substitute for the potato crop that had been destroyed by blight. Dulse also occurs along both the Atlantic and Pacific coasts of North America, where some is still gathered for food. Another red seaweed, *purple laver* ("nori"), which occurs in both American and Asian waters, is used extensively for food, particularly in the Orient. In Japan, it is cultured

Figure 15.22 Specially equipped vessel harvesting kelp off the California coast. The ship moves backward through the kelp beds as the machinery at the stern mows the kelp and conveys it into the hold. (Courtesy Kelco Company)

on nets or bamboo stakes set out in shallow marine bays (figure 15.23). It is harvested when the thin, crinkly, gelatinous blades are several centimeters (2 to 3 inches) in diameter and is used in meat and macaroni dishes, soups, and dry spiced delicacies. *Irish moss* is another important edible red alga; it is also used in bulking laxatives, cosmetics, and pharmaceutical preparations. Blancmange is a dessert made from Irish moss and milk. Carrageenin is a mucilaginous substance extracted from Irish moss and used as a thickening agent in chocolate milk and other dairy products. Funori, obtained from yet another red alga, is used as a laundry starch, as an adhesive in hair dressings, and in some water-based paints.

One of the most important of all algal substances is **agar,** produced most abundantly by the red alga *Gelidium.* This gelatinous substance is used around the world in laboratories and medical institutions as a culture medium for the growth of bacteria. When various nutrients are added to it, it can also be used as a culture medium for the growth of both plant and animal cells. Recently, full-sized plants have been induced to develop from pollen grains sown on nutrient agar, orchid tissues are cultured commercially on it and induced to grow into full-sized plants (see the section on mericloning in appendix 4), and its use for the capsules containing drugs and vitamins is now worldwide. It is also used as an agent in bakery products to retain moistness, as a base for cosmetics, and as an agent in gelatin desserts to promote rapid setting.

Chapter 15

Current research involving red algae and other seaweeds indicates they contain a number of substances of potential medicinal value. More than twenty seaweeds have been used in preparations designed for the expulsion of digestive tract worms, control of diarrhea, and the treatment of cancer. Some have shown considerable potential as antibiotics and insecticides. Recently, chemical relatives of DDT have been found to be produced by certain red algae. The sea hare and other marine animals feed on such algae, and it is possible that such animals may degrade (break down) the DDT-like compounds to simpler substances—a feat unknown among land animals.

Division Myxomycota—Slime Molds

An item in an early edition of *Ripley's Believe It or Not* referred to "human hair growing on trees." The "hairs," while indeed resembling short human hair, were actually the reproductive structures of a species of *slime mold* (figure 15.24). These curious organisms are totally without chlorophyll and thus are incapable of producing their own food. They are a bit of a puzzle to biologists because they are distinctly animallike during much of their life cycle but just as distinctly plantlike or funguslike when they reproduce. At

Kingdom Protista

Figure 15.24 Reproductive bodies (sporangia) of the slime mold *Stemonitis*. (Courtesy Field Museum of Natural History, Chicago)

present, they are placed in the Kingdom Protista because they lack a cell wall for most of the life cycle and because they ingest their food like amoebae by engulfing it in food vacuoles, where it is digested. They also produce cells that resemble and move like amoebae at one stage of their life cycle, but few biologists feel completely comfortable in regarding them as protists, and, as our knowledge of the biological realm becomes refined, slime molds may well be placed in another kingdom.

The tiny roundish spores of the more than five hundred species of slime molds average only 10 to 12 micrometers in diameter and thus are individually invisible to the naked eye. Nevertheless, they are present nearly everywhere and are especially abundant in airborne dusts. If one places almost any dead leaf or piece of bark on a food source such as a few dry oatmeal flakes in a covered dish and adds a few drops of water, the chances are that any slime mold spores present will germinate. In some instances, they will germinate in as little as fifteen minutes, but germination usually takes several hours or longer.

Within a few days a curious glistening mass of active slime mold material somewhat resembling the netted venation of a leaf may appear. This material, whose "veins" tend to merge into the shape of a fan at its leading edges, is the slime mold *plasmodium* (figure 15.25). Inspection of a plasmodium with a hand lens or dissecting microscope reveals no cell walls. The protoplasm in the veins, particularly that located toward the center, flows very rapidly and rhythmically; brief pauses in the protoplasmic movement occur at regular intervals, and the flow reverses after each pause.

Plasmodia are often white, but they also may be brilliantly colored in shades of yellow, orange, blue, violet, or black. A few are colorless and transparent. They are found on damp forest debris, under logs, on old shelf or bracket fungi, sometimes on older mushrooms, and in other moist places where dead organic matter is present. They tend to creep forward at a rate of up to 2.5 centimeters (1 inch) or more per hour, often against slow moisture seepage, feeding on bacteria and other organic particles as they go. They contain many diploid nuclei, all of which divide

Chapter 15

Figure 15.25 Plasmodium of the slime mold *Physarum*. (Courtesy H. L. Dean)

Figure 15.26 Common slime mold sporangia. A. *Arcyria*. B. *Lycogala*. (B, Courtesy L. L. Steimley)

A.

B.

simultaneously and frequently as growth occurs. With an adequate food supply, a plasmodium may increase to twenty-five times its original size in just one week.

When the food supply dwindles or other features of the environment such as moisture or light change significantly, dramatic events occur. Usually the plasmodium heaps up into many separate small **sporangia** (singular: **sporangium**) (figure 15.26), each containing thousands of minute one-celled **spores.** The sporangia frequently are globe-shaped, but in some species they occur as long or wide stationary bodies of variable shape. Depending on the species, sporangia may be either

Kingdom Protista

Figure 15.27 Life cycle of a slime mold. (After Alexopoulos, C. J.; Mims, E. W. 1979. *Introductory mycology,* 3d ed. Copyright © 1979 John Wiley & Sons, Inc., New York. Redrawn by permission of John Wiley & Sons, Inc.)

sporangia begin to form

older plasmodium

young plasmodium

sporangia

zygote

fertilization

2N

N

spore

meiosis
(either in the sporangia or the spores)

union of myxamoebae

myxamoeba

germination

union of swarm cells

swarm cell

on slender stalks or without stalks. In other species, the whole plasmodium may form a single spore-bearing body that develops in a variety of shapes. In still others, combinations of body forms may occur. The spores are often dispersed throughout a jumbled mass of threads called a *capillitium.*

As a spore is formed, a single nucleus and a little cytoplasm become surrounded by a wall. Meiosis takes place in the spore, and three of the four resulting nuclei degenerate. When the spore germinates, one or more amoebalike cells called *myxamoebae* emerge. Sometimes these have flagella, in which case they are called *swarm cells.* Either form may become like the other through the development or loss of flagella. Myxamoebae, or swarm cells, at first feed on bacteria and other food particles. Sooner or later they function as gametes, fusing in pairs and forming zygotes. A new plasmodium usually develops from the zygote, although occasionally zygotes or small plasmodia may fuse and form larger plasmodia (figure 15.27).

The slime molds as a group are divided into two classes. The majority of species have typical plasmodia and follow the patterns just discussed. About two dozen species, however, are called *cellular slime molds* because they do not produce true plasmodia. Instead, individual amoebalike cells feed independently, dividing and producing new separate cells from time to time. When a population reaches a certain size, they stop feeding and clump together, forming a mass called a *pseudoplasmodium.* The pseudoplasmodium looks and crawls like a garden slug. It eventually becomes stationary and is transformed into a sporangiumlike mass of spores.

Human and Ecological Relevance of the Slime Molds

The slime molds perform an ecological role similar to that of bacteria in breaking down organic particles to simpler substances; they also reduce bacterial populations in their forest and woodland habitats. Except for occasionally causing lawns to become somewhat unsightly, they are of little economic significance.

Other Members of the Kingdom Protista

Protozoans (Phylum Protozoa) and sponges (Phylum Porifera) are included within the Kingdom Protista; they have, however, traditionally been regarded as animals and are not covered in this book.

Summary

The Kingdom Protista includes organisms that all have eukaryotic cells but are otherwise diverse. Members may be unicellular or multicellular, occur as colonies or filaments, and in the case of slime molds, consist in the vegetative state of a mass of protoplasm containing numerous nuclei. Modes of nutrition include photosynthesis, ingestion of food, or a combination of both. Many members are motile by means of flagella or by amoeboid movements, but some are nonmotile.

Green algae occur in a wide variety of aquatic habitats. Their cells have the same pigments and reserve food (starch) as those of higher plants. The unicellular green alga, *Chlamydomonas,* has a pair of flagella that enable the cell to move rapidly. Within the cell there is a chloroplast containing one or two pyrenoids, two or more vacuoles, and often a red eyespot. Asexual reproduction is by mitosis; sexual reproduction is by union of like gametes (i.e., it is isogamous).

Ulothrix is a filamentous green alga that attaches to various objects by means of a basal cell called a holdfast; each cell contains a braceletshaped chloroplast. Asexual reproduction is by zoospores, sexual reproduction is isogamous.

Spirogyra is a floating, filamentous green alga with spiral, ribbonlike chloroplasts. It does not produce flagellated cells of any kind. Asexual reproduction is by fragmentation (breaking of filaments with each fragment adding new cells by mitosis). Sexual reproduction is by conjugation.

Oedogonium is a filamentous green alga that is epiphytic (attached in a nonparasitic manner) on aquatic plants and other algae; it has cylindrical, netted chloroplasts. In asexual reproduction, zoospores are produced. In sexual reproduction short cells of a filament function as antheridia in which sperms are produced, and large swollen cells function as oogonia in each of which a single egg is produced. A sperm enters an oogonium through a pore and fertilizes the egg. This union of a small motile gamete and a large nonmotile gamete is called oogamy.

Other green algae include *Chlorella,* a tiny unicellular alga that is easy to culture and has been used in experiments to produce oxygen for space vehicles; desmids, which are free-floating unicellular algae that reproduce by conjugation; mermaid's wineglass, which produces huge mushroomlike cells; *Volvox,* a colonial alga that forms motile hollow balls of hundreds to thousands of cells; sea lettuce, whose blades are anchored to rocks by means of a holdfast; and many others. Stoneworts, which are sometimes included with the green algae are retained in a separate division (Charophyta) because they have multicellular antheridia and other features more complex than those of the green algae.

Euglenoids superficially resemble *Chlamydomonas* cells but differ in not having a rigid cell wall, and in having only one functional flagellum, a gullet, and a unique carbohydrate food reserve, paramylon. Reproduction is by cell division; sexual reproduction has not been confirmed.

The golden-brown algae are grouped into three classes that include the yellow-green algae, the true golden-brown algae, and the diatoms. Some members of each class produce statospores. Diatoms, which have a glassy shell that consists of two "halves" that fit together like a pillbox, are extremely abundant, particularly in colder marine waters. The shells are usually etched with fine grooves and pores through which the cytoplasm is in contact with the environment. Diatoms are often golden-brown in color due to the presence of the brownish pigment fucoxanthin in the one to many chloroplasts occurring in each cell. Diatoms move in caterpillar fashion by contact of the cytoplasm with a surface as it protrudes through the pores. In asexual reproduction the two "halves" of a cell separate after mitosis of the protoplast, and a new "half" forms within each original portion. An auxospore (zygote) is produced through a sexual process involving the fusion of gametes.

Dinoflagellates are unicellular organisms with two flagella. Some are known to cause red tides and when present in large numbers can kill fish and shellfish by releasing paralyzing toxins into the water. Dinoflagellates are also responsible for phosphorescence (a glowing of sea water when disturbed) in the tropics.

The brown algae include the largest of seaweeds. Many are differentiated into a stalk (stipe), flattened blades that may have bladders toward their bases, and a tough, sinewy holdfast that holds the seaweed to the rocks. Fucoxanthin is largely responsible for the color of brown algae; their main carbohydrate food reserve is laminarin; some produce algin, a useful gelatinous substance; reproductive cells have lateral flagella. In the common rockweed, *Fucus,* eggs (from oogonia) and sperms (from antheridia) are released into the water where fertilization occurs, and the zygotes develop into mature thalli.

The red algae, whose members tend to be smaller than those of the brown algae, have relatively complex life cycles involving three different types of thalli and nonmotile gametes. In *Polysiphonia* and other representatives, the three thalli are the tetrasporophyte, the male gametophyte, and the female gametophyte.

The colors of red algae are partially due to the presence of red and blue phycobilins, which are related to those of the blue-green algae; chlorophyll *d* may be present in the chloroplasts; the carbohydrate food reserve is floridean starch. Some red algae produce agar, an economically important gelatinous substance.

Algae are important in aquatic food chains and in numerous other ways. Diatom shells have accumulated for thousands of years on ocean floors and make up diatomaceous earth that is used for filtering, polishes, insulation, and reflectorized paint. *Chlorella* is a potentially important food and oxygen source. Algin is used as a stabilizer and thickening agent in hundreds of food products, paints, medicines, papers, ceramics, cosmetics, beer, etc. Brown algae are a source of fertilizer and iodine, and some serve as food for both livestock and humans. Red algae

Chapter 15

are a source of agar which is used as a culture medium for bacteria and other organisms or tissues; some are also used for human food; and substances produced by red algae (e.g., carrageenin, funori) are used as thickening agents, as laundry starch, and as adhesives. Some red algae have potential medicinal value.

Slime molds are animallike in their vegetative state; they consist of a multinucleate mass of protoplasm that flows over damp surfaces, ingesting food particles encountered. They are plantlike or funguslike when they reproduce, forming stationary sporangia containing spores, from which myxamoebae, or swarm spores, emerge upon germination. Myxamoebae, or swarm spores, function as gametes, with new plasmodia developing from the zygotes. Some members of this division are called cellular slime molds because they do not produce true plasmodia; instead, independent cells clump together, becoming a pseudoplasmodium that crawls like a slug and eventually is converted into a stationary, sporangiumlike mass of spores.

Review Questions

1. What forms of sexual and asexual reproduction occur in the green algae?

2. Which groups of algae produce the following important products: (1) agar, (2) algin, (3) nerve poisons, (4) abrasives for polishes?

3. Where and how is algin obtained?

4. How would you distinguish *Chlamydomonas* from *Euglena?*

5. *Spirogyra, Ulothrix,* and *Oedogonium* all form filaments. How can you tell them apart?

6. In the green algae studied, where in the life cycles does the chromosome number change from *N* to *2N* and vice versa?

7. How do cells of diatoms differ from those of other organisms?

8. Is there any difference in structure between the holdfasts of microscopic green algae and brown algae?

9. Why are some green algae red and some red algae green?

10. What is the difference between a plasmodium and a pseudoplasmodium?

Discussion Questions

1. Some algae are attached to solid objects or other organisms, while others are free-floating. What are the advantages and disadvantages of each type of growth?

2. The variety and sizes of algae found in the oceans are considerably greater than those of freshwater forms. Can you suggest reasons?

3. Seaweeds that grow in intertidal zones, where they may be exposed between tides, are often more gelatinous than their continually submerged counterparts. Explain.

4. Why do some algae grow on one side of a tree and not all around the trunk?

5. Should the bladders, or floats, of some of the kelps give these algae any advantage over those algae that do not possess such structures?

Additional Reading

Abbott, I. A.; Dawson, E. Y. *How to know the seaweeds.* 2d ed. Pictured Key Nature Series. Dubuque, IA: Wm. C. Brown Company Publishers; 1978.

Abbott, I. A.; Hollenberg, G. J. *Marine algae of California.* Stanford: Stanford University Press; 1976.

Abbott, I. A. *Limu: an ethnobotanical study of some edible Hawaiian seaweeds.* 3d ed. Lawai, HI: Pacific Tropical Botanical Garden; 1984.

Bold, H. C.; Wynne, M. J. *Introduction to the algae.* Englewood Cliffs, NJ: Prentice-Hall; 1978.

Bonner, J. T. *The cellular slime molds.* 2d ed. Princeton: Princeton University Press; 1968.

Buetow, D. E., ed. *The biology of Euglena. 1: General biology and ultrastructure.* New York: Academic Press; 1968.

Chapman, V. J.; Chapman, D. J. *Seaweeds and their uses.* 3d ed. London: Chapman & Hall; 1980.

Crowder, W. Marvels of the mycetozoa. *National Geographic Magazine* 49(4):421–44; 1926.

Dawes, C. J. *Marine botany.* New York: John Wiley & Sons; 1981.

Dixon, P. S. *Biology of the Rhodophyta.* Edinburgh: Oliver and Boyd; 1973.

Gibor, A. *Acetabularia:* a useful giant cell. *Scientific American* 215: 118–24; 1966.

Gray, W. D.; Alexopoulos, C. J. *Biology of the myxomycetes.* New York: Ronald Press Co.; 1968.

Jackson, D. F., ed. *Algae and man. Proceedings of the NATO Advanced Study Institute.* New York: Plenum Press; 1964.

Kelco Company. *Search in the aquasphere. The story of Kelco.* San Diego, CA: Kelco Co.; 1968.

Palmer, C. M. *Algae in water supplies.* U.S. Department of Health, Education and Welfare. Public Health Service. Washington, DC: Superintendent of Government Documents; 1962.

Picket-Heaps, J. D. *Green algae: structure, reproduction and evolution in selected genera.* Sunderland, MA: Sinauer Associates; 1975.

Prescott, G. W. *How to know the freshwater algae.* 3d ed. Pictured Key Nature Series. Dubuque, IA: Wm. C. Brown Company Publishers; 1979.

Round, F. E. *The biology of the algae.* 2d ed. New York: St. Martin's Press; 1973.

Scagel, R. F.; Bandoni, R. J.; Maze, J. R.; Rouse, G. E.; Schofield, W. B.; Stein, J. R. *Plants: an evolutionary survey.* Belmont, CA: Wadsworth Publishing Co.; 1984.

Smith, G. M. *Freshwater algae of the United States.* 2d ed. New York: McGraw-Hill Book Co.; 1950.

Turner, N. J. *Food plants of British Columbia Indians.* Part 1: *Coastal peoples.* Victoria: British Columbia Provincial Museum; 1974.

Kingdom Fun
and Lichen

Features of the Kingdom

In the past the true fungi, sl
teria were all placed in
Plant Kingdom. Onc
ences between pro
became appar
placed in th
it also b
othe

Overview

The chapter opens with a summary of the features of the Kingdom Fungi and a review of how the kingdom came to be recognized. The true fungi are treated as belonging to six classes: Chytridiomycetes, Oomycetes, Zygomycetes, Ascomycetes, Basidiomycetes, and Deuteromycetes. Selected members of each class and representative life cycles are presented, along with discussion of their economic importance. Among the topics or fungi discussed are nematode-trapping fungi, *Pilobolus,* truffles, morels, ergot, yeasts, stinkhorns, puffballs, bracket fungi, bird's-nest fungi, smuts, rusts, poisonous and hallucinogenic fungi, Black Forest mushrooms, mushroom culture, antibiotics, industrial products obtained from fungi, and fungi in nature. The chapter concludes with a description and discussion of various forms of lichens. Natural dyeing is discussed in a footnote, and the economic importance of lichens is reviewed.

Some Suggested Student Learning Goals

1. Describe the general features that distinguish the Kingdom Fungi from the other kingdoms.

2. Distinguish the classes of fungi from one another on the basis of their hyphae or cells, and their reproduction.

3. Define sporangium, conidium, coenocytic, mycelium, dikaryotic, zygospore, ascus, basidium.

4. Assign each of the following to its fungal class: athlete's foot, Dutch elm disease, *Pilobolus,* penicillin mold, stinkhorn, chytrid, yeast, ergot, chestnut blight fungus, puffball, smut.

5. List five economically important fungi in each of four different classes of fungi.

6. Describe how lichens are classified and identified.

7. Describe the basic structure of a lichen.

me molds, and bac-
single division of the
the fundamental differ-
karyotic and eukaryotic cells
nt, however, the bacteria were
prokaryotic Kingdom Monera. Then
came evident that the reproductive and
features of slime molds, such as the absence
cell walls in the vegetative state and the amoeboid movement of their plasmodia and myxamoebae, removed them far from the fungi, which
are all filamentous or unicellular saprobic or parasitic decomposers. Accordingly, the slime molds
were assigned to the heterogeneous Kingdom
Protista.

It has become increasingly apparent that the
metabolism, reproduction, and general lines of
diversity of fungi are quite different from those
of plants; fungi evidently were independently derived from ancestral unicellular eukaryotic organisms. Thus, they are now placed in their own
kingdom, which consists of a single division with
several classes.

The filamentous fungi all produce threadlike
hyphae (singular: **hypha**), which grow at their
tips, usually branching and often *anastomosing*
(forming a network) into a mass known as a **mycelium.** Structures such as mushrooms consist of
hyphae tightly packed together. All fungi except
the water molds (Class: Oomycetes) have cell
walls that consist primarily of *chitin,* a material
also found in the shells of arthropods (e.g., insects, crabs). Their primary food reserve is glycogen, which is also a food reserve of animals. The
cell walls of water molds contain a polysaccharide related to cellulose. These molds are probably not directly related to the other fungi—they
apparently have been derived independently from

protists. Most fungi are nonmotile, but some produce reproductive cells with flagella. Most fungi
produce spores and exhibit a variety of forms of
sexual reproduction. Fungi cannot engulf and ingest food; instead, they absorb substances through
the cell walls. The food materials are often broken down with the aid of enzymes secreted to the
outside by the cells. Because of the great variety
of form and reproduction throughout the Kingdom Fungi, a neat pigeonholing of all the members into distinct groups is difficult; broad classes
can, however, be recognized.

Introduction to the Fungi

To many, the word *fungus* evokes visions of
mushrooms or some sort of creeping infestation.
While mushrooms are indeed fungi, and while
many fungi do give the appearance of creeping
along the ground, the true fungi assume a great
variety of forms. There are about one hundred
thousand known species of mushrooms, rusts,
smuts, mildews, molds, stinkhorns, puffballs, truffles, and other organisms assigned to the Kingdom Fungi, and more than one thousand new
species are described each year. Possibly as many
as two hundred thousand species still await to be
described or discovered. The hyphae of fungi
(figure 16.1) grow so rapidly if appropriate food
is available that a single day's growth could extend more than a kilometer (0.6 mile) if all the
hyphae produced were laid end to end. Some fungi
thrive in freezers if the temperature is not lower
than $-5°C(23°F)$, while others freely reproduce under temperature conditions of $55°C$
$(131°F)$ or higher. Along with bacteria, fungi are
vital to the natural recycling of dead organic material, but they also cause huge economic losses
through food spoilage and disease (see the sections on the human and ecological relevance of
fungi later in the chapter). Scientists who study
fungi are known as *mycologists,* and consumers
of fungi are called *mycophagists* (from the Greek
word *myketos,* a fungus).

Chapter 16

Figure 16.1 Typical fungal hyphae, as seen with the aid of a dissecting microscope

Division Mycota—True Fungi

Class Chytridiomycetes—Chytrids

If you were to immerse dead leaves or flowers, old onion bulb scales, dead beetle wing covers, or other organic material in water that has been mixed with a little soil, it is very probable that within a day or two, thousands of microscopic *chytrids* would appear on the surfaces of the immersed objects (figure 16.2). These simplest of fungi include many parasites of aquatic flowering plants and algae and also numerous saprophytic species. A chytrid consists mostly of a spherical cell with colorless branching threads called **rhizoids** at one end. The rhizoids anchor the fungus

Figure 16.2 Chytrids on a dead leaf that is under water

to its food source. Many chytrid species reproduce only asexually, producing zoospores within a spherical cell. When the zoospores, each of which has a single flagellum, are released, they settle and grow into new chytrids. Some species undergo sexual reproduction by means of the fusion of two motile gametes or by the union of two nonmotile cells whose zygote nuclei undergo meiosis. The origin of chytrids is unknown, but the presence of flagella on the motile cells has led some authorities to suggest they may have originated from the protozoa; some species have cellulose present in their cell walls, and they do not appear to be directly related to the other fungi.

Class Oomycetes—Water Molds

Water mold fungi are familiar to those who have kept tropical fish aquaria or have seen salmon at the end of a spawning run. They appear as cottony growths wherever cuts and bruises have occurred on a fish's body, and they frequently attack the eyes of ailing fish. They also grow on dead insects and readily develop on marijuana seeds placed in water.

The hyphae of water molds may branch repeatedly and form large mycelia. No individual cells are formed during growth, however, and large numbers of nuclei are scattered throughout the mycelium. Such multinucleate mycelia without crosswalls are said to be **coenocytic** (pronounced see-no-sitt-ik) (figure 16.3). In asexual reproduction, tips of certain hyphae are separated from the remainder of the mycelium through the formation of crosswalls. Numerous biflagellated zoospores are produced in these special tips. The zoospores eventually give rise to new water mold mycelia after emerging through a terminal pore. Sexual reproduction involves oogonia and antheridia, which arise on side branches under the influence of hormones produced by the fungus. The mycelium is diploid, and meiosis takes place in the oogonia and antheridia. Zygotes formed in the oogonia eventually give rise to new mycelia (figure 16.4).

Figure 16.3 Coenocytic hyphae of a water mold

nuclei

Two very important members of this class of fungi are responsible for serious diseases of higher plants. Neither grows under water, although dew or rainwater is necessary for their reproduction. One, called *downy mildew of grapes,* develops and undergoes its life cycle on grape leaves, usually killing the leaves and consequently the vine if not controlled. This disease seriously threatened the French wine industry in the latter part of the nineteenth century after it had been introduced into the vineyards on imported American cuttings. It was controlled within a few years when it was discovered that Bordeaux mixture, a combination of copper sulphate and lime, inhibited the growth of downy mildew. This unappetizing-looking mixture, which is the first substance known to have been used as a fungicide, was originally sprayed on grape vines to discourage passersby from picking the grapes.

Chapter 16

Figure 16.4 Life cycle of the water mold *Saprolegnia*

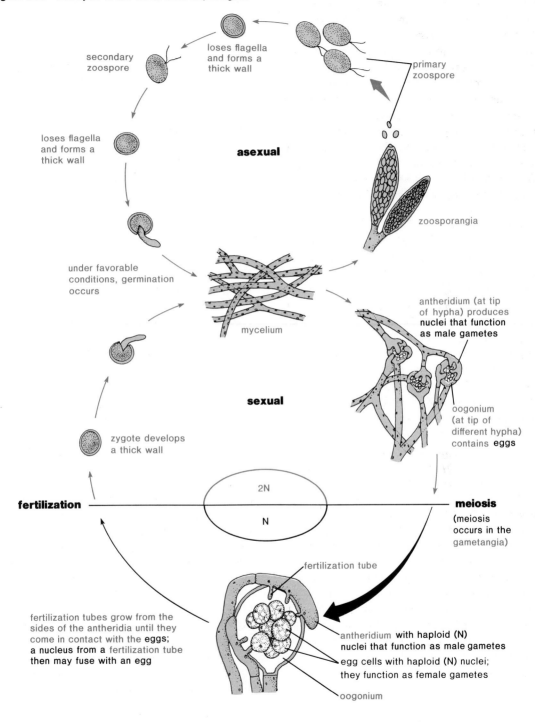

secondary
zoospore

loses flagella
and forms a
thick wall

primary
zoospore

loses flagella
and forms a
thick wall

asexual

zoosporangia

under favorable
conditions, germination
occurs

mycelium

antheridium (at tip
of hypha) produces
**nuclei that function
as male gametes**

sexual

oogonium
(at tip of
different hypha)
contains **eggs**

zygote develops
a thick wall

2N

N

fertilization

meiosis

(meiosis
occurs in the
gametangia)

fertilization tube

fertilization tubes grow from the
sides of the antheridia until they
come in contact with the **eggs**;
a nucleus from a fertilization tube
then may fuse with an egg

antheridium with haploid (N)
nuclei that function as male gametes

egg cells with haploid (N) nuclei;
they function as female gametes

oogonium

The other important disease-causing oomycete is called *late blight of potato*. This virulent organism plagued potato farmers to such an extent in the past that at times it altered the course of history. In the summer of 1846, the entire potato crop of Ireland was wiped out in one week, bringing about a famine during which a million persons starved to death and many more emigrated to the United States. The fungus produces its mycelium inside the leaves, and so it is relatively immune to spraying once its spores land on the surface and send hyphae into the tissues beneath. Development of the fungus is checked, however, if the leaves are sprayed with Bordeaux mixture or similar substances before the spores have germinated.

Class Zygomycetes—Bread Molds

Although black bread molds are the best known members of this class, they are not the only fungi that grow on bread; in fact, so many organisms can contribute to bread spoilage that nearly all commercially baked goods have, in the past, had chemicals such as calcium propionate added to the dough to retard the growth of such organisms. There is now a trend toward eliminating preservatives from bread, pies, and other bakery items since the chemicals that have made the goods a less suitable medium for the growth of fungi apparently are unhealthy for humans after prolonged use, and alternative ways of retarding fungal growth and spoilage are being sought.

Rhizopus (figure 16.5), a well-known representative black bread mold in this class, has spores that are exceedingly common everywhere. They have been found in the air above the North Pole, over jungles, on the inside and outside of buildings, in soils, clothing, automobiles, and hundreds of kilometers out to sea, easily carried there by prevailing winds and breezes. When a spore lands in a suitable growing area, it germinates and soon produces an extensive mycelium, which, like that of the water molds, is *coenocytic*

(without crosswalls). It contains numerous haploid nuclei. After the mycelium has developed, certain hyphae called **sporangiophores** grow upright and produce globe-shaped *sporangia* at their tips. Numerous black spores are formed within each sporangium. When these spores are released through the breakdown of the sporangium wall, they may blow away and repeat the cycle.

Such reproduction, involving no union of gametes, is asexual, but black bread molds also reproduce sexually in a unique fashion. Although there is no visible difference in form, black bread mold mycelia occur in two different mating strains. When a hypha of one strain encounters a hypha of the other, swellings called **progametangia** develop opposite each other on the hyphae. These protuberances grow toward each other until they touch. A crosswall is formed a short distance behind each tip, and the two **gametangia** merge, becoming a single large multinucleate cell in which the nuclei of the two strains fuse in pairs. A thick ornamented wall then develops around this cell with its numerous diploid nuclei. This structure, which may lie dormant for months, is called a **zygospore.** Eventually the zygospore cracks open, and one or more sporangiophores with sporangia at their tips emerge. Meiosis apparently takes place just before this occurs.

No members of this class produce zoospores, but some produce the spores externally on hyphae instead of forming them in a sporangium. One interesting group of such fungi in this class parasitizes protozoans and other small animals in various ways. Some develop an unbranched body inside their victim and slowly absorb nutrients until the host dies. They then produce a chain of spores that may stick to or be eaten by another victim. Others capture their prey on sticky hyphae to which passing amoebae adhere. One soil-dwelling group captures nematodes (eelworms) in hyphal rings or loops akin to the tripknots mischievous children sometimes tie in long grasses. In some, the rings are a little smaller than the circumference of a nematode, which tapers at both ends. When a nematode randomly happens

Figure 16.5 Life cycle of the black bread mold *Rhizopus*

zygote

thick-walled zygospore

2N

N

fertilization

meiosis

gametangia

germination

sporangium with spores

sexual

+progametangium

+mycelium

spores germinate forming mycelium

−progametangium

−mycelium

asexual

asexual

sporangium

spores

sporangiophore

spore germinates

spore on bread germinates

mycelium

+ strain

− strain

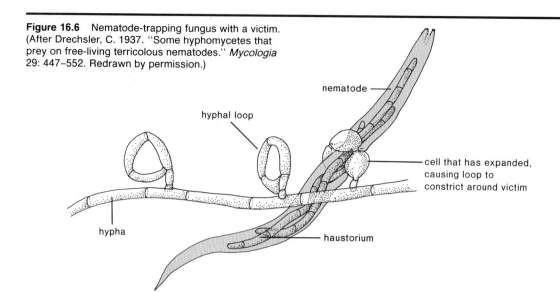

Figure 16.6 Nematode-trapping fungus with a victim. (After Drechsler, C. 1937. "Some hyphomycetes that prey on free-living terricolous nematodes." *Mycologia* 29: 447–552. Redrawn by permission.)

to stick its head through such a loop, it frequently tries to struggle forward instead of backing out and becomes trapped (figure 16.6). The fungus then produces small rhizoidlike outgrowths called **haustoria** which grow into the worm's body and digest it. More specialized species in this group have loops that constrict around the worm less than a tenth of a second after contact. The loops are spaced so that sometimes the tail of the worm gets caught in a second loop while the worm is thrashing around after being caught by the head. A number of these worm-trapping fungi can easily be grown on agar media, but they do not form loops unless an eelworm is placed in the dish. Evidently the eelworms produce a substance that promotes the development of the loops.

One dung-inhabiting genus of fungi in this class has the scientific name of *Pilobolus,* derived from two Greek words meaning "cap-thrower." The name is quite appropriate as the mature sporangia are forcibly "shot" a distance of up to 8 meters (26 feet), where they adhere to grass or other vegetation (figure 16.7). When the vegetation is ingested by animals, the spores pass unharmed through the digestive tract and germinate in the dung. These fungi are very sensitive to light and release their sporangia with force precisely

in its direction. This action can be demonstrated by placing some horse dung in a glass dish with a lid and then covering it with black paper. Cut an opening in the paper (the pattern of the opening can be elaborately frilly or angular if desired), and set the dish where it will receive sufficient light for a few days. If *Pilobolus* is present and sporangia have been produced and forcibly discharged, a black path corresponding closely to the cut-out area will appear on the glass.

Human and Ecological Relevance of the Bread Molds

One species of bread mold is used in Indonesia and adjacent areas to produce a food called *tempeh*. Basically, tempeh consists of boiled skinless soybeans that have been inoculated with a bread mold and allowed to sit for twenty-four hours. The mycelium that develops holds the soybeans together and produces enzymes that increase the content level of several of the B vitamins. The tempeh is fried, roasted, or diced for soup and is prepared fresh daily. Other bread mold species are used with soybeans to make a Chinese cheese called *sufu*.

At least two bread mold species are now used commercially to carry out important steps in the

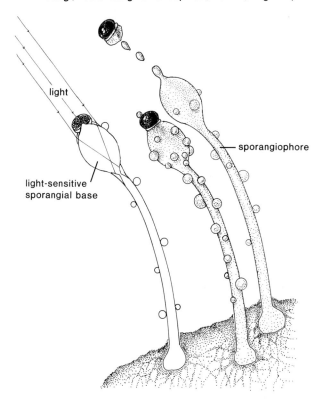

Figure 16.7 The dung fungus *Pilobolus,* which releases its spores with force toward a light source. (After A. H. Reginald Buller. 1934. *Researches on Fungi,* vol. 6. Longman Group Ltd., Harlow, England.)

light

sporangiophore

light-sensitive sporangial base

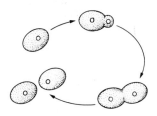

Figure 16.8 How a yeast cell reproduces asexually by budding

manufacture of birth control pills and anesthetics, while others are utilized in the production of industrial alcohols and as a meat tenderizer. Yet another species produces a yellow agent used for coloring margarine.

Class Ascomycetes— Cup or Sac Fungi

Should you happen one summer to find yourself in the south of France, you might be startled to see men and women with coils of rope slung around their shoulders pushing pigs in wheelbarrows. They are not on their way to a hog-hanging festival but are instead wheeling the pigs into the woods to help them find *truffles.* These are gourmet "mushrooms" that grow between 2.5 and 15 centimeters (1 and 6 inches) beneath the surface

of the ground, usually near oak trees. They are somewhat prunelike in appearance and may be more than 10 centimeters (4 inches) in diameter, although most are smaller. They give off a tantalizing aroma that pigs can detect from more than 15 meters (50 feet) away, and the animals, in an attempt to get to them, strain at the ropes that are tied around their necks. The owners dig up the truffles and reward the pigs with acorns or other less interesting food.

Truffles are the reproductive bodies of representatives of a large and varied class of fungi called **cup** or **sac fungi.** More than twenty thousand species, including yeasts, powdery mildews, brown fruit rots, ergot, morels, microscopic parasites of insects, canned fruit fungi, and many others, have thus far been described. Most produce mycelia superficially similar to those of the previous two classes, but the hyphae are divided up into individual cylindrical cells. Each crosswall has one or more pores through which the single to many tiny nuclei can pass. Little bodies that serve as plugs are located near the pores. They seal off individual cells if they become damaged.

Asexual reproduction is by means of **conidia** (singular: **conidium**). Conidia are spores that are produced externally—outside of a sporangium— either singly or in chains at the tips of hyphae called *conidiophores* (see figure 16.32). In the case of yeasts, asexual reproduction is by **budding** (figure 16.8). When a yeast cell buds, a small protuberance appears to balloon out slowly from the cell and become pinched off as it grows to full size.

Figure 16.9 Life cycle of a cup fungus. When hyphae of two different strains of a cup fungus become closely associated, male *antheridia* may be formed on one and female *ascogonia* on the other. Male nuclei migrate into an ascogonium and pair but do not fuse with the female nuclei present. Then new hyphae (ascogenous hyphae), whose cells each contain a pair of nuclei, grow from the ascogonium. In a process involving fusion of the pairs of nuclei (followed by meiosis), fingerlike *asci*, each containing four or eight haploid nuclei, are formed in a layer called a *hymenium* lining an ascocarp. The haploid nuclei become *ascospores*, which are discharged into the air. They are potentially capable of initiating new mycelia and repeating the process. (After Sharp, L. W. 1943. *Fundamentals of Cytology*. Redrawn by permission of the McGraw-Hill Book Co.)

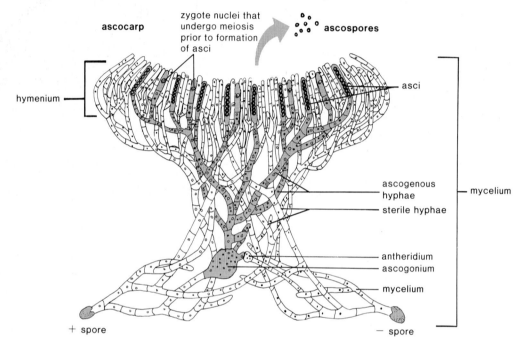

Sexual reproduction always involves the formation of sacs called **asci** (singular: **ascus**). In the more complex cup or sac fungi, ascus formation is preceded by the development of special hyphae (ascogenous hyphae) into which pairs of nuclei migrate. One nucleus of each pair originates from a "male" mycelium or cell that has fused with a "female" organ or cell. As the ascogenous hyphae grow, the paired nuclei divide in a unique way so that there is one pair of nuclei in each cell. These ascogenous hyphae become several cells in number at maturity, and long tubular asci develop following a process of nuclear fusion, which occurs at their tips. The diploid zygote nucleus undergoes meiosis, and the resulting four haploid nuclei usually divide once more by mitosis so that there is a row of eight nuclei in each ascus. These nuclei become *ascospores* as they are walled off from one another with a little cytoplasm (figure 16.9). Thousands of asci are often packed together into a structure called an *ascocarp*. This may be cup-shaped (figure 16.10), completely enclosed, or flask-shaped with a little opening at the top. Truffles are really enclosed ascocarps.

Chapter 16

Figure 16.10 Ascocarps of a cup fungus. (Courtesy Robert A. Schlising)

Figure 16.11 Morels. (Photograph by Carolina Biological Supply Company)

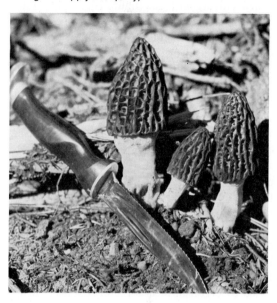

Cup-shaped ascocarps may be several centimeters (2 to 3 inches) in diameter and may be brilliantly colored on the inside. When ascospores are mature, they are often released with force from the asci. If an open ascocarp is jarred at the right time, it may appear to belch fine puffs of "smoke" consisting of thousands of ascospores.

When an ascospore lands in a suitable area, it may germinate, producing a new mycelium, and then repeat the process. In many instances, however, a number of asexual generations involving conidia are produced between the sexual cycles.

Human and Ecological Relevance of the Cup or Sac Fungi

Morels, which have been called by some "the world's most delicious mushrooms," and truffles have been prized as food for centuries. Neither is a true mushroom; mushrooms are included in the club fungi discussed in the next section. Wealthy Romans and Greeks used to insist on preparing them personally according to various recipes they had concocted, and they are still prized as a gourmet food today. Numerous unsuccessful attempts have been made to grow them commercially, but morel mycelium has been cultured and used for flavoring soups. Morels (figure 16.11) are the size of a medium to smallish mushroom, tan in color, and have a wrinkled, slightly conelike top on a stalk that resembles a miniature tree trunk. The numerous depressions between the wrinkles each contain thousands of asci. Although morels are perfectly edible by themselves, there are unconfirmed reports that some persons have become ill after consuming them with alcohol; caution in this regard is advised.

A related "mushroom" called a *false morel* or *beefsteak morel* is considered a delicacy by many but has caused death in others. This points up the peculiar unpredictability of the interaction between the human metabolism and certain fungi. For some, the fungi are quite safe to eat; for others they are not—a literal example of "one man's meat being another man's poison."

Kingdom Fungi and Lichens

317

When rye, and to a lesser extent, other grains come into flower in a field, they may become infected with ergot fungus (figure 16.12). This fungus seldom causes serious damage to the crop, but as it develops in the maturing grain, it produces several powerful drugs. If the infected grain is harvested and milled, a disease called **ergotism** may occur in those who eat the contaminated bread. The disease can affect the central nervous system, often causing hysteria, convulsions, and sometimes death. Another form of ergotism causes gangrene of the limbs; the affected limb may drop off as a result. It can also be a serious problem for cattle grazing in infected fields. They frequently abort their fetal calves and may succumb themselves.

Ergotism was common in Europe in the Middle Ages. Known then as "St. Anthony's Fire," it killed 40,000 people in an epidemic in A.D. 994. In 1722, the cavalry of Czar Peter the Great was felled by ergotism just as the czar was about to conquer Turkey, and the conquest never took place. As recently as 1951, an outbreak hospitalized 150 victims in a French village; 5 persons died and 30 became temporarily insane, imagining they were being chased by snakes and demons.

In small controlled doses, ergot drugs have proved medically useful. They stimulate contraction of the uterus to initiate childbirth and have been used in abortions and in the treatment of migraine headaches. Ergot is also an initial source for the manufacture of the hallucinogenic drug lysergic acid diethylamide, popularly known as LSD.

Fungi in this class play a basic role in the preparation of baked goods and alcohol. Enzymes produced by yeast aid in fermentation (see chapter 9), producing ethyl alcohol and carbon dioxide in the process. The carbon dioxide produced in bread dough forms bubbles of gas, which are trapped, causing the dough to rise and giving bread its porous texture. Part of the flavor of individual wines, beers, ciders, *sake,* and other alcoholic beverages is imparted by the species, or

Figure 16.12 Ergotized rye. The prominent darker objects scattered throughout the ears of rye are kernels infected with ergot fungus. (USDA Photo)

strain, of yeast used to ferment the fruits or grains.

Yeasts are at least indirectly involved in the manufacture of a number of other important products. Ephedrine, a drug also produced by leafless western desert shrubs, is obtained commercially from certain yeasts. The drug is used in nose drops and in the treatment of asthma. Yeasts are also a rich source of B vitamins and

Chapter 16

Figure 16.13 Peach leaves infected with peach leaf curl

are used in the production of glycerol for explosives. Ethyl alcohol is used in industry as a solvent and in the manufacture of synthetic rubber, acetic acid, and vitamin D. Yeast contains about 50% protein and makes a nourishing feed for livestock. Its enzyme, invertase, is used to soften the centers of chocolate candies after the chocolate coating has been applied.

Several very important plant diseases are found in this class of fungi. Dutch elm disease, a disease originally described in Holland, has devastated the once numerous and stately elm trees in many towns and cities in the midwestern and eastern United States and has now spread to the west and south. When Dutch elm disease was first discovered, limited control of the disease was achieved by spraying the trees with DDT to kill the elm bark beetles that spread the disease from tree to tree. DDT killed many useful organisms as well, however, and its use for such purposes was eventually banned. Other sprays have since been used, again with limited success, and biological controls are now being sought. One such control was reported in 1980 by G. Strobel of Montana State University. He injected pseudomonad bacteria into twenty diseased trees and saved seven of them when the bacteria multiplied and killed the Dutch elm disease fungus. Whether or not this control technique will meet with widespread success remains to be seen, but it is the type of control that is much to be preferred to poisonous sprays that upset delicate ecological balances (see appendix 2, which deals with biological controls).

Chestnut blight has virtually eliminated the once numerous American chestnut trees from the eastern deciduous forests. Attempts to control the disease have met with very limited success thus far. Much better success has been obtained in controlling peach leaf curl, a disease that attacks the leaves of some stone fruits, especially peach trees (figure 16.13). Sprays that contain copper or zinc salts apparently inhibit the germination of spores of peach leaf curl, and have been effective in preventing serious damage when applied to trees before the dormant buds swell and open in the spring.

Class Basidiomycetes—Club Fungi

A few years ago I received a telephone call at my office from a woman previously unknown to me. She said she had had an operation some months before that had left her with a peculiar halitosis. She also said she frequently needed to expectorate, and that she had noticed each time she did so out of her kitchen window, a "growth" would come up out of the ground, smelling like her halitosis! She wanted to know if I would be interested in seeing one of the "growths." I assured her that I was, not having any idea what to expect. In due course, the woman arrived with a large club fungus known as a *stinkhorn* (figure 16.14). It somewhat resembled a morel with a spongy stalk, but the top was partly covered with a slimy, evil-smelling mass of spores. Any connections, real or imagined, between the halitosis, the smell of the stinkhorn, and the location of the fungi were, of course, purely coincidental. Undoubtedly some time earlier a fly, attracted by the odor of a stinkhorn elsewhere, had carried spores adhering to it to the ground by the woman's kitchen window, and the fungus had gone through its life cycle there.

Stinkhorns are interesting but relatively unimportant representatives of another large class of fungi, the club fungi. Other members of this class include mushrooms (plate 3D) or toadstools (the only distinctions between mushrooms and toadstools are based on folklore or tradition—botanically there is no difference), puffballs, earth stars (plate 3B), shelf or bracket fungi (plate 3C), rusts, smuts, jelly fungi, and bird's-nest fungi. They are called club fungi because in sexual reproduction they produce their spores at the tips of swollen hyphae that often resemble baseball bats or clubs. These swollen hyphal tips are called **basidia** (singular: **basidium**). The hyphae, like those of the cup or sac fungi and blue-green molds, are divided up into individual cells; however, these cells have either a single nucleus or,

Figure 16.14 Common stinkhorn. Note the slimy mass of spores on the cap. (Courtesy Leland Shanor)

in some stages, two nuclei. The crosswalls have a central pore that is surrounded by a swelling, and both the pore and swelling are covered by a cap. This cap, with some exceptions, blocks passage of nuclei between cells but allows cytoplasm and small organelles to pass through.

Asexual reproduction is much less frequent in club fungi than in the other classes of fungi. When it does occur, it is mainly by means of conidia, although a few species produce buds similar to those of yeasts, and others have hyphae that fragment into individual cells, each functioning like a spore and forming a new mycelium after germination.

Figure 16.15 Development of a clamp connection

Sexual reproduction in many mushrooms begins in the same way as it does for members of the two classes previously discussed. When a spore lands in a suitable place—often an area with good organic material and humus in the soil—it germinates and produces a mycelium just beneath the surface. The mycelium soon becomes divided into cells, each containing a single haploid nucleus. Such a mycelium is said to be **monokaryotic.** Monokaryotic mycelia of club fungi often occur in four mating types, usually designated simply as types 1, 2, 3, and 4. Only types 1 and 3, or types 2 and 4, can mate with each other. If the growth of the hyphae of compatible mating types happens to bring them close together, cells of each mycelium may unite, initiating a new mycelium in which each cell has two nuclei. Such a mycelium is said to be **dikaryotic.** Dikaryotic mycelia usually have little walled-off bypass loops called **clamp connections** between cells on the surface of the hyphae (figure 16.15). The clamp connections develop as a result of a unique type of mitosis that ensures that each cell will have one nucleus of each original mating type within it. When a cell at the tip of a dikaryotic hypha is ready to divide, a small looping protuberance appears on the side of the cell and starts to grow backward. As it does so, one nucleus migrates into it, and both nuclei then undergo mitosis simultaneously. The protuberance fuses back to the cell, and a transverse wall forms across it between the daughter nuclei. Another crosswall forms in the original cell just above the point at which the loop or clamp connection has united with it. This latter crosswall forms between the two daughter nuclei of the second original nucleus. When the process has been completed, there are two cells, each with two nuclei, and a walled-off clamp connection to one side.

After developing for a while, the dikaryotic mycelium may become very dense and form a compact, solid-looking mass called a *button*. This pushes above the surface and expands into a *basidiocarp*, commonly called a mushroom (figure 16.16). Most mushrooms have an expanded umbrellalike *cap* and a *stalk*. Some have a ring called an **annulus** on the stalk. It is the remnant of a membrane that extended from the cap to the stalk and tore as the cap expanded. Some mushrooms, such as the notorious death angel mushroom, also have a cup called a **volva** at the base (figure 16.17). Thin, fleshy-looking plates called **gills** radiate out from the stalk on the underside of the cap. Microscopic examination of a gill reveals the compacted hyphae of which it is composed and innumerable **basidia** oriented at right angles to the flat surfaces of the gill.

As each basidium matures, the two nuclei unite and then undergo meiosis. The four nuclei that result from meiosis become *basidiospores* when they migrate through four (in a few species, two) tiny pegs at the tip of the basidium and form their own walls around them. The tiny pegs, called *sterigmata* (singular: *sterigma*), serve as stalks for the basidiospores. One large mushroom

Figure 16.16 Life cycle of a typical mushroom

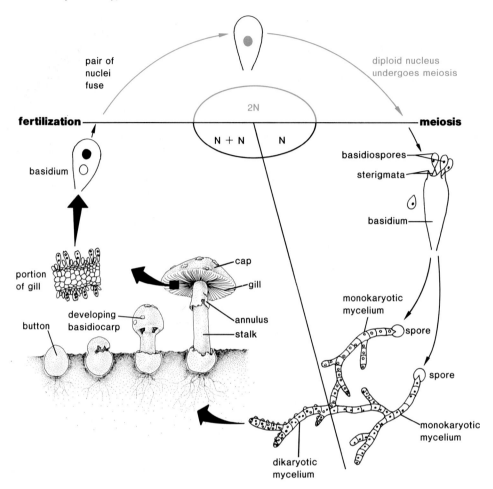

may produce several billion basidiospores within a few days. These are forcibly discharged into the air between the gills, and they blow away with the slightest draft. If you remove the stalk and place a mushroom cap gill-side down on a piece of paper, covering it with a dish to eliminate air currents, the spores will fall and adhere to the paper in a pattern perfectly reflecting the arrangement of the gills. The dish and cap can be removed in a day or two, and the spore print (figure 16.18) can be made more or less permanent with the application of a little clear varnish or shellac. Professional mycologists use such spore

Chapter 16

Figure 16.17 Death angel mushroom. Note the egglike volva at the base. (USDA Photo)

Figure 16.18 Spore print

Figure 16.19 *Boletus*, a mushroom that has pores instead of gills beneath its cap. (USDA Photo)

prints as an aid to identification, employing white paper for dark-colored spores and black paper for white or light-colored spores.

In nature, some of the basidiospores eventually repeat the reproductive cycle. Often a dikaryotic mycelium radiates out in an ever-expanding circle from its starting point, periodically producing basidiocarps in so-called fairy rings. If conditions are favorable, the mycelium continues to grow at the edges for many years while dying in the center as food resources are depleted. Some mycelia have been known to grow in this fashion for over five hundred years.

Some mushrooms produce their spores on the surfaces of thousands of tiny pores instead of on gills (figure 16.19).

Kingdom Fungi and Lichens

Shelf or *bracket fungi* (figure 16.20) grow out horizontally from the bark or dead wood to which they have become attached, some adding a new layer of growth each year. Perennial species can become large enough and so securely attached that they can support the weight of an adult.

Other members of this class produce spores within parchmentlike membranes, forming somewhat ball-like basidiocarps called *puffballs.* Puffballs, which are generally edible, range in diameter from a few millimeters to 1.2 meters (0.125 inch to 4 feet) (figure 16.21). They have no stalks and rest in contact with the ground. Literally trillions of spores may be produced by a large puffball. These are released through a pore at the top or from random locations when the outer membrane breaks down. *Earth stars,* which are similar to puffballs, differ from them in having a ring of appendages at the base that look like a set of woody petals around a flower.

Bird's-nest fungi (figure 16.22), which grow on wood or manure, form nestlike cavities in which "eggs" containing basidiospores are produced. In some species, each "egg" has a sticky thread attached to it. When raindrops fall in the nests, the eggs may be splashed out, and as they fly through the air the sticky threads catch on nearby vegetation, whipping the eggs around it (figure 16.23). When animals graze on the vegetation, the spores pass unharmed through the intestinal tract.

Figure 16.20 Oyster mushrooms. These are shelf, or bracket, fungi with gills; many bracket fungi have pores instead of gills. (Courtesy Perry J. Reynolds)

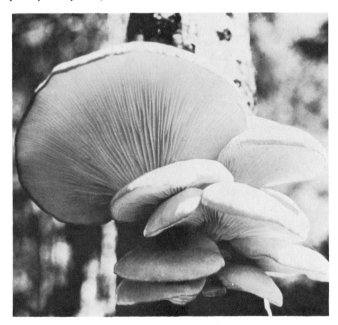

Chapter 16

Figure 16.21 Giant puffballs. The largest weighed 6.12 kilograms (13.5 pounds). (Courtesy Louise White, Redding, California, *Record-Searchlight*)

Figure 16.22 Bird's-nest fungus

Figure 16.23 How the "eggs" in a bird's-nest fungus are dispersed. (After Brodie, Harold J. 1951. "The splash-cup dispersal mechanism in plants." *Canadian Journal of Botany* 29: 224–234. Reproduced by permission of the National Research Council of Canada from the *Canadian Journal of Botany,* vol. 29, pp. 224–34, 1951.)

Raindrops splash the "eggs" out of the nest. The tail of an "egg" whips around a blade of grass. When an animal grazes, the "egg" passes unharmed through the digestive tract and germinates in the dung.

Kingdom Fungi and Lichens

Smuts are parasitic club fungi that do considerable damage to corn, wheat, and other grain crops. In corn smut (figure 16.24), the mycelium grows between the cells of the host. The hyphae absorb nourishment from these cells and also secrete substances that stimulate them to divide and enlarge, forming tumors on the surfaces of the corn kernels. These eventually break open, revealing millions of sooty black spores, which are blown away by the wind. Some smuts affect only the flowering heads or grains, while others infect the whole plant.

Rusts, which also are parasites, attack a wide variety of plants. Some rusts grow and reproduce on only one species of flowering or cone-bearing plant. Others, however, require two or more different hosts to complete their life cycles. *Black stem rust of wheat,* which has reduced wheat yields by millions of bushels in a single year in the United States alone, has plagued farmers ever since wheat was first cultivated thousands of years ago. More than three hundred races of black stem rust are now known. This rust requires both common barberry plants and wheat to complete its life cycle (figure 16.25).

Chapter 16

Figure 16.25 Life cycle of black stem rust of wheat

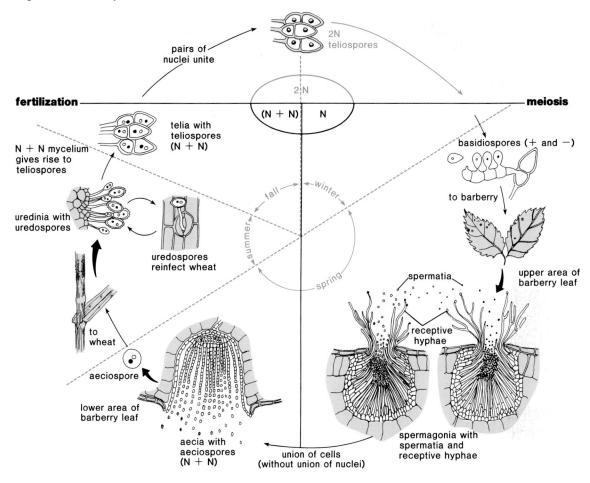

pairs of nuclei unite

2N teliospores

2N

(N + N)   N

fertilization

meiosis

telia with teliospores (N + N)

basidiospores (+ and −)

N + N mycelium gives rise to teliospores

to barberry

uredinia with uredospores

fall

winter

summer

spring

uredospores reinfect wheat

upper area of barberry leaf

spermatia

receptive hyphae

to wheat

aeciospore

lower area of barberry leaf

aecia with aeciospores (N + N)

union of cells (without union of nuclei)

spermagonia with spermatia and receptive hyphae

Since two hosts are necessary for black stem rust of wheat to complete its life cycle, partial control of the disease has been accomplished through an attempt to eradicate common barberry bushes. An estimated 600 million such plants have been destroyed in the United States since 1918. Producing rust-resistant strains of wheat has also helped, but even as new strains are developed, the rusts themselves hybridize or mutate (see chapter 12), producing new races capable of attacking previously immune varieties of cereals.

Another serious rust with two hosts is the *white pine blister rust,* which has caused huge losses of valuable timber trees in both the eastern and western United States. Basidiospores infect the pine trees; when the basidiospores germinate, other types of spores are produced. These different spores in turn infect currant and gooseberry bushes, and the spores formed on the currants and gooseberries eventually give rise to new basidiospores, completing the cycle. The U.S. Forest Service has had a program of gooseberry bush eradication in operation for many years in an attempt to alleviate the problem. Spraying programs have been implemented more recently as an alternative, with some success. Other rusts with two hosts include *apple rust* (alternate host: cedar trees), *poplar leaf spot* (alternate host: larch or tamarack trees), and *corn rust* (alternate host: sorrel).

Human and Ecological Relevance of the Club Fungi

Of the approximately twenty-five thousand described species of club fungi, fewer than seventy-five are known to be poisonous. Many of the latter are, however, not readily distinguishable by amateurs from edible species; and some edible forms such as the inky cap and shaggy mane mushrooms, which cause no problems by themselves, may make one very ill if consumed with alcohol. Few of the poisonous forms normally are fatal, but unfortunately some—such as the death angel, which causes 90% of the fatalities attributed to mushroom poisoning—are relatively common. Poisoning from death angels and similar species usually takes from six to twenty-four hours to manifest itself, and until recently, successful treatment was impossible by the time the intense stomachache, blurred vision, violent vomiting, and other symptoms occurred. Hope for survival of the victims now lies with the administration of a drug known as thioctic acid; however, either the drug or the mushroom poison leave the patient with hypoglycemia (a blood sugar deficiency). Some wild mushroom lovers have fed a part of their collection to a dog or cat, and when nothing happened to the animal after an hour or two, they have eaten the mushrooms themselves; both they and the animals later succumbed. Others have mistakenly believed that the toxic substances are destroyed by cooking. Records show, however, that before the discovery of thioctic acid as an antidote, death ensued in 50 to 90% of those

Figure 16.26 Teonanacatl mushrooms. (Courtesy Field Museum of Natural History, Chicago)

who had eaten just one or two of the deadly mushrooms, cooked or raw, and that even as little as 1 cubic centimeter (less than a 0.5 cubic inch) can be fatal.

A number of widespread but very unreliable beliefs exist concerning distinctions between edible and poisonous mushrooms. One holds that a silver coin placed in the cooking pan will turn black while the mushrooms are cooking if any poison is present. Some members of both edible and poisonous species will turn such a coin black, and some will not. Another superstition holds that edible species can be peeled while poisonous ones cannot. Again this is a fallacy, for death angels peel quite easily. Still other erroneous beliefs maintain that poisonous mushrooms appear only in the fall or the early spring, or that all mushrooms eaten by snails and beetles are edible, or that all purplish-colored mushrooms are poisonous, or that all mushrooms growing in grassy areas are edible. Again, these notions simply do not agree with the facts. Wild mushrooms can be eaten with confidence *only* if they have been identified by a knowledgeable authority. It is foolhardy for anyone to do otherwise.

Some poisonous mushrooms cause hallucinations in those that eat them. During the Mayan civilization in Central America a number of *teonanacatl* ("God's flesh") sacred mushrooms (figure 16.26) were used in religious ceremonies.

Figure 16.27 Fly agaric mushrooms. (USDA Photo)

The consumption of these mushrooms, which has continued among native groups in Mexico and Central America to the present, results in sharply focused, vividly colored visions. Similar use of the striking, fly agaric mushroom (figure 16.27) in Russia, and for a while in the Indus valley of India, dates back to many centuries B.C. Users appear to go into a state of intoxication. Related species that occur in the United States have not produced the same effects but have, instead, caused the user to become nauseated and to vomit. In Siberia, users have noted that the intoxicating principle is passed out in the urine; as a result,

some persons have adopted the practice of drinking the urine of fly agaric users. Reindeer, incidentally, are reported to be obsessed with both fly agaric and human urine.

More than 90% of a mushroom is water, and mushrooms generally contain smaller quantities of nutritionally valuable substances than most foods. An apparent exception is the legendary Black Forest *(Shiitake)* mushroom grown for centuries in China and now cultured in the United States. It has more than double the protein of ordinary, commercially grown mushrooms and is very rich in calcium, phosphorous, and iron. It also contains significant amounts of B vitamins, vitamin D_2 (ergosterol), and vitamin C and has excellent flavor. Ancient Chinese royalty believed that eating Black Forest mushrooms would promote healthful vigor and retard the aging process. It appears there could be some substance to this belief, as recent research suggests that mushrooms in general may be one of the richer sources of RNA, which has been shown to retard the aging process in cells through its conversion to ATP by the metabolism. Lentinacin, an agent capable of lowering human cholesterol levels, has been obtained from Black Forest mushrooms, and purified extracts from spores of the mushrooms have demonstrated antiviral activity against influenza and polio viruses in laboratory animals. The extracts evidently induce the formation of interferon, a virus-combating substance produced by animal cells.

Many types of mushrooms have been cultivated for food since ancient times. In the second century B.C., a Greek doctor by the name of Nicandros taught people how to grow mushrooms underneath fig trees on manure dug into the ground. Andrea Cesalpino, a noted Italian botanist and physician of the sixteenth century, cultivated mushrooms by scattering pulverized poplar bark on very rich soil near poplar trees with which the mushrooms were associated. In more recent times, Italians have cultivated mushrooms on waste material from olives, on coffee grounds, on remnants of oak leaves after tannins have been extracted for leather tanning, and on laurel berries. Today jelly fungi and various mushrooms are cultivated in the Orient on media composed of wood and manure.

Only one species of mushroom *(Agaricus campestris)* is cultivated to any extent in the United States. It has been grown in basements,

Figure 16.29 Fungi in a termite fungus garden.
A. The conspicuous objects are conidia. *B.* Close-up
of a conidium with numerous spores on its surface.
(Courtesy D. H. Kistner and P. Fleischman)

A.

B.

caves, and abandoned mines, but contrary to popular belief, light does not affect its growth at all.
Large-scale operations generally use windowless
warehouses with stacked rows of shallow planting beds because temperatures, humidity, and
other climatic factors are easier to control in such
buildings (figure 16.28). The mushrooms are
grown on compost made from straw and on horse
or chicken manure. Prior to use, the compost is
pasteurized for a week to destroy insects and their
eggs. Then it is inoculated with *spawn,* which is
compact mycelium grown from germinated basidiospores sown on bran of wheat or other grains.
After inoculation, the spawn is covered with a thin
layer of soil. The moisture content of the compost
is controlled with regular, light waterings, and a
humidity of about 75% is maintained. The mycelium grows in temperatures ranging from just
above freezing to 33°C (91.4°F), but commercial growers try to keep the temperatures between 9°C and 13°C (48°F to 55°F) because the
mushrooms are less subject to disease or insect
attacks and are also firmer when grown under cool
conditions. The mushroom buttons appear within
a week to ten days after spawn is planted and
continue appearing for six months or more. About

1 kilogram (2.2 pounds) of mushrooms is obtained from each 10 square decimeters (slightly
over 1 square foot) of growing area. An estimated 59,000 metric tons (65,000 tons) of mushrooms are produced annually in the United States.

As indicated in the introduction to fungi,
fungi of all classes and bacteria are the most important organisms involved in natural recycling
processes. Fungi are constantly breaking down
dead wood and debris and returning the components to the soil where they can be reused.
Sometimes, as we have seen, they can be very destructive from a human viewpoint, attacking
everything from living plants and harvested or
processed food to shoe leather, paper, cloth, construction timbers, paint, petroleum products, upholstery, and even glass, particularly in warmer
climates. Others play a nondestructive ecological
role. For example, several species of fungi are
found in ant and termite nests. The insects cultivate their fungus gardens by bringing in bits of
leaves, other plant debris, caterpillar droppings,
and their own feces to form a rich growing medium (figure 16.29). They also lick the hyphae

and constantly probe them, using them for food as they grow. Some wood-boring beetles have pouches in their bodies that function as fungus spore containers. The spores germinate in the wood tunnels, their mycelia forming a lining that produces yeastlike cells on which the beetles and their larvae feed. Many higher plants have **mycorrhizal fungi** (see chapter 5) associated with their roots. These fungi greatly increase the absorptive surface area around the roots, and may be far more important than root hairs in this regard, particularly in mature roots.

Class Deuteromycetes— Imperfect Fungi

Any fungus for which a sexual stage has not been observed is classified as an imperfect fungus. Many otherwise unrelated fungi are grouped together in this artificial class, which includes several well-known disease organisms as well as fungi important in disease control and food processing. If a member of this group is studied further and a sexual stage is discovered, the fungus is reassigned to its appropriate class.

All imperfect fungi reproduce by means of conidia. In one such group, which is found growing on dead leaves and debris at the bottom of streams with rapidly moving water, the spores often have four long extensions on them. These arms are arranged in such a way that three may catch on a flat surface like a tripod, keeping the fungus from being washed downstream.

Among the best known of the imperfect fungi are the *Penicillium* molds (figure 16.30), which produce penicillin, a well-known and widely used *antibiotic* (a substance produced by a living organism that interferes with the normal metabolism of another living organism). Sir Alexander Fleming, a British scientist, noticed in 1929 that certain bacteria would not grow in the vicinity of the mycelium of a *Penicillium* mold and gave the name penicillin to the element in the mold that

Figure 16.30 *Penicillium* colonies. The tiny droplets of fluid on the surface contain penicillin.

prevented the bacterial growth. His findings did not particularly excite the medical profession until the outbreak of World War II some ten years later. At that time, a team of British and American scientists at Northern Regional Laboratories in Peoria, Illinois, set out to see if they could coax *Penicillium* molds into producing more of this antibacterial substance, primarily because war casualties created a need for greater quantities of more effective medicines that could keep wounds from becoming infected. They began with cultures from the original mold observed by Fleming, but the amounts of penicillin it produced were so small that it was very expensive to obtain significant quantities for human use. They appealed to the general public, asking them to send in any material they found with a greenish or bluish mold on it. They received huge quantities of moldy trash from all over the United States; however, the breakthrough in the research came when a different species of *Penicillium* mold, one that yielded twenty-five times the penicillin produced by the original culture, was found on a moldy cantaloupe from a local market. The scientists set to work germinating individual spores

Figure 16.31 Roquefort cheese. (Photo by Bob Coyle)

of this new mold on culture media, and by careful selection they eventually were able to isolate a strain that produced more than eighty times the original quantity of penicillin. Later, when this strain was subjected to X-radiation, still other forms were produced that upped the penicillin output to 225 times that of Fleming's mold. Today most of the penicillin produced around the world comes from descendants of that cantaloupe mold. Literally hundreds of other antibiotics effective in combating human and animal diseases have been discovered since the close of World War II, and the production of these drugs is a vast worldwide industry.

Penicillium molds are also useful for purposes other than the manufacture of penicillin. Some are introduced into the milk of cows, sheep, and goats at stages in the production of "smelly"

cheeses such as Roquefort (figure 16.31), Camembert, blue, Gorgonzola, and Stilton. The molds produce enzymes that break down proteins and fats in the milk, giving the cheeses their characteristic flavors.

Aspergillus is a genus of imperfect fungi whose species produce dark brown to blackish or yellow spores. It is closely related to the *Penicillium* molds, and is extensively used in industry. One or more species are used commercially for the production from sugar of citric acid, a substance for flavoring foods and for the manufacture of effervescent salts that was originally obtained from oranges. Citric acid is also used in the manufacture of inks and in medicines, and it is even used as a chicle substitute in some chewing gums. These same fungi also produce gallic acid used in photographic developers, dyes, and indelible black ink. Other species are used in the production of artificial flavoring and perfume substances, chlorine, alcohols, and several acids. Further uses are in the manufacture of plastics, toothpaste, and soap, and in the silvering of mirrors.

One species of *Aspergillus* is used in the Orient and elsewhere to make soy sauce, or *shoyu,* by fermenting soybeans with the fungus. A Japanese food called *miso* is made by fermenting soybeans, salt, and rice with the same fungus. More than one-half million tons of miso are consumed annually.

A number of diseases of both man and animals are caused by *Aspergillus* species. Called *aspergilloses,* the diseases attack the respiratory tract after the spores have been inhaled (figure 16.32). One type thrives on and in human ears. Other imperfect fungi are responsible for the irritating and worldwide problem known as "athlete's foot," for white piedra (a mild disease of beards and mustaches), and for tropical diseases of the hands and feet that cause the limbs to swell in grotesque fashion. The highly contagious ringworm diseases are also caused by imperfect fungi.

Figure 16.32 Tip of a conidiophore of *Aspergillus* as seen with the aid of a microscope. Conidia are radiating out around the swollen tip. Several species of *Aspergillus* are responsible for human respiratory and skin diseases called *aspergilloses*. (Courtesy Turtox/Cambosco)

Lichens

A student who was interested in natural dyeing came to me a few years ago and asked if she could experiment with the dye potential of local plants as a special project. In the course of her experimentation, she obtained beautiful shades of yellow, brown, and green from two dozen common local plants, using simple "recipes."[1] During the

This usually takes from one to several hours. The solid waste is then strained out and discarded. A *mordant* (substance that helps fibers take up dye permanently) is then often added to the liquid in amounts varying from one teaspoon to one-half cup. Commonly used mordants include alum, detergent ammonia, copper sulphate, tin (stannous chloride), and white vinegar. White wool or other fibers are washed in warm water and detergent, rinsed in warm water, and then left to soak in hot water for up to an hour. The fibers are then quickly transferred to the hot dyeing liquid and simmered for an hour or left overnight. The dyed material is then rinsed in warm water until no more dye diffuses out, washed again with detergent, and dried. Reaction of the dyed material to light can be tested by placing it in direct sunlight for several days. (See appendix 3 for a number of sources of natural dyes.)

---

1. Most natural dye "recipes" call for simmering at least a liter (approximately 1 quart) or two of loosely packed fresh or dry material covered with water in a large enamel kettle until most of the coloring appears to be in the water.

following summer she extended the project to include lichens growing on the trees and rocks at a camp where she served as a counsellor. The rich colors she obtained from the lichens were even more spectacular, which is perhaps not surprising since these organisms are also used in commercial dyeing.

Lichens have traditionally been referred to as prime examples of symbiotic relationships. Each consists of a fungus and an alga intimately associated in a spongy body called a **thallus.** The thallus can range in diameter from less than 1 millimeter to more than 2 meters (0.04 inch to 6.5 feet). The alga supplies the food for both organisms, while the fungus protects the alga from harmful light intensities, produces a substance that accelerates photosynthesis in the alga, and absorbs and retains water and minerals for both organisms. The evidence suggests, however, that it would probably be more correct to say that the fungus parasitizes the alga in a controlled fashion, actually destroying algal cells in some instances.

There are about twenty-five thousand known species of lichens. The algal component is either a green or a blue-green alga, and a few lichens have two species of algae present. One species of alga may be found in many different lichens, but each lichen has its own unique species of fungus. With the exception of about twenty tropical species of lichens, which have a club fungus, all lichens have members of the cup or sac fungi and blue-green molds class for their fungal components. It is possible to isolate and culture the components separately. When this is done, however, the fungus takes on a very different, compact but indefinite shape, and the algae grow faster than they do when they are part of a lichen. The fungal component is very rarely found growing independently in nature, whereas the algal component may do so; lichen species, therefore, are identified according to the fungus present.

Lichens grow very slowly, at a maximum rate of 1 centimeter (0.4 inch) and a minimum of as little as 0.1 millimeter (.004 inch) per year. They are capable of living to an age of forty-five hundred years or more and are tolerant of environmental conditions that kill most other forms of life. They are found on bare rocks in the blazing sun or bitter cold in deserts, in both arctic and antarctic regions, on trees, and just below the permanent snow line of high mountains where nothing else will grow. They even attach themselves to such man-made substances as glass, concrete, and asbestos. Part of the reason for this wide range of tolerance is the presence in the lichens' thalli of gelatinous substances that enable them to withstand periods of rapid drying alternating with wet spells. While they are dry they do not carry on photosynthesis and temporarily become dormant. In this state, most environmental extremes do not affect them at all. One environmental factor from modern civilization does, however, have a marked effect on them. They are exceptionally sensitive to pollution, particularly that of sulphur dioxide. In fact, their sensitivity to this air pollutant is so great that it has been possible to calculate the amount of sulphur dioxide present in the air solely by mapping the occurrence or disappearance of certain lichens in a given area. Such studies have shown that some species of lichens have disappeared entirely from industrial areas and are now facing extinction due to air pollution.

The lichen thallus generally consists of three or four layers of cells or hyphae (figure 16.33). At the surface is a protective layer constituting the *upper cortex*. The hyphae are so compressed they resemble parenchyma cells, and it is here that gelatinous substances are present that aid in the layer's protective function. Below the upper cortex is an *algal layer* in which algal cells are

Figure 16.33   Section through a foliose lichen

scattered among strands of hyphae. Next is a *me-dulla* consisting of loosely packed hyphae, which occupies at least half the volume of the thallus. A number of substances produced by the lichen are stored here. A fourth layer, called the *lower cortex,* is frequently but not always present. It resembles the upper cortex but is usually thinner and is often covered with anchoring strands of hyphae called *rhizines.*

Lichens have been loosely grouped into three major growth forms, which have no basis in their natural relationships but are convenient as a first step in their identification (figure 16.34). *Crus-tose* lichens are attached to or embedded in their substrate over their entire lower surface. They often form brightly colored crusty patches on bare rocks and tree bark. Some that grow on sedi-mentary rocks send their hyphae as much as 1 centimeter (0.4 inch) into the rock. Others grow

just beneath the cuticle of the leaves of tropical hardwood trees, with no apparent harm to the leaves. *Foliose* lichens have somewhat leaflike thalli, which often overlap one another. They are weakly attached to the substrate. The edges are frequently crinkly or divided into lobes. *Fruti-cose* lichens may resemble miniature shrubs, or they may hang down in festoons from branches. Their thalli, which are usually branched, are ba-sically cylindrical in form and are attached at one point.

It should be stressed that while lichens may be attached to trees or other plants, they in no way parasitize them.

Although the fungal component of a lichen reproduces sexually, lichens are naturally dis-persed in nature primarily by asexual means. In about a third of the species, small powdery clus-ters of hyphae and algae called **soredia** (singular: **soredium**) are formed and cut off from the thallus

Figure 16.34 Three types of lichen thalli. *A.* Crustose lichens on the surface of a rock (see plate 3A). *B.* Foliose lichen. *C.* Fruticose lichen.

A.

C.

B.

in a set pattern as it grows. Rain, wind, running water, and animals act as agents of dispersal. In other lichens, specialized parts of the thallus may break off or be separated by decay. Sexual reproduction is similar to that of the cup or sac fungi and blue-green molds, except that the ascocarps produce spores continuously for many years. No one has ever observed the initiation of a new thallus in nature, but it is believed that they arise after ascospores carried by the wind come in contact with independently living algae, germinate, and parasitize them. Lichen algae reproduce by mitosis and simple cell division.

Human and Ecological Relevance of Lichens

Lichens provide food for many lower animals as well as large mammals. Reindeer and caribou can survive exclusively on fruticose lichens in Lapland when other food is unavailable, while North African sheep graze on a crustose lichen. By themselves lichens do not make good food for human consumption, but they have been used as a food supplement (e.g., in soups) in parts of Europe. Most have acids that make them unpalatable, and some (e.g., rock tripe) have had harsh laxative effects on those who have tried them.

More than half of the lichens investigated have antibiotic properties. One lichen substance has been used in Europe in combination with another antibiotic in the treatment of tuberculosis. Europeans have also used lichen antibiotics to produce salves that have been effective in treating cuts and skin diseases.

Lichens were used for dyes by the Greeks and Romans, and a lichen dye industry persisted for many centuries in Europe. Native Americans and others also used lichens for dyes. Coal tar dyes have now largely replaced those of lichens, but lichens are still used in the manufacture of Scottish tweeds and in the production of litmus paper. Litmus paper is used in elementary chemistry laboratories as an acid-alkaline indicator, turning red under acid conditions and blue under alkaline ones.

Soaps are scented with extracts from lichens, and such extracts are still used in the manufacture of some European perfumes. Because of their resemblance to miniature trees and shrubs, some fruticose lichens are used by toy makers for the scenery of model railroads and car tracks. The importance of lichens in initiating soil formation is discussed in chapter 23.

Summary

Members of the Kingdom Fungi were grouped in the past with the slime molds and bacteria in a single division of the Plant Kingdom, but bacteria later were placed in the Kingdom Monera because of their prokaryotic cells, and the slime molds were placed in the Kingdom Protista because of the nature of their plasmodia, their reproduction, and other considerations. The reproduction, metabolism, and lines of diversity of the fungi were sufficiently different from those of members of the Plant Kingdom to warrant their being placed in a kingdom of their own.

True fungi are all filamentous or unicellular with chitinous walls (except for the water molds, whose walls have a polysaccharide related to cellulose); their food reserve is glycogen. Most fungi are nonmotile but some produce cells with flagella; most produce spores. Sexual reproduction is varied. They absorb their food in solution through their walls. Several broad classes of fungi are recognized.

About two-thirds of the possibly three hundred thousand species of fungi still await description or discovery. Filamentous fungi produce threads called hyphae that collectively constitute a mycelium. Fungi may thrive at temperatures below freezing or well above human comfort levels. They are major natural decomposers and cause huge economic losses through food spoilage and diseases.

The Class Chytridiomycetes (chytrids) includes simple fungi that consist mostly of spherical cells with rhizoids at one end. Reproduction is primarily asexual, by means of zoospores.

Chapter 16

The Class Oomycetes (water molds) includes fungi that cause diseases of fish and other aquatic organisms, and grow on dead insects and submerged organic material. They also cause two major diseases of flowering plants (downy mildew of grapes, which seriously threatened the French wine industry until it was controlled with Bordeaux mixture; and late blight of potato, which ruined the Irish economy in the mid-nineteenth century and brought about starvation and death of large numbers of Irish peasants). Water molds have coenocytic mycelia (i.e., the hyphae have no crosswalls). They reproduce asexually by means of zoospores, and sexually by means of gametes produced in oogonia and antheridia produced in response to hormones secreted by the fungi.

The Class Zygomycetes (bread molds)—represented by *Rhizopus,* which has exceptionally widespread spores—is characterized by coenocytic mycelia that produce upright hyphae (sporangiophores) with sporangia containing numerous spores at their tips (in asexual reproduction), and sexually by gametangia produced on hyphae of opposite strains that merge, creating a single large cell in which nuclei fuse in pairs. This large cell becomes a zygospore in which meiosis occurs prior to its cracking open and sporangiophores and sporangia emerging. Some spores, called conidia, are produced externally on hyphae in this class, but conidia production is more common in other fungal classes. Some conidia-producing zygomycetes trap eelworms and digest them. Another, *Pilobolus,* releases its sporangia with force toward a light source. Bread molds are used to make tempeh and sufu, and are used for various industrial purposes.

Members of the Class Ascomycetes (cup or sac fungi) include truffles, yeasts, powdery mildews, brown fruit rots, ergot, morels, insect parasites, and others. Their hyphae have crosswalls with pores. Asexual reproduction is by conidia or budding. Sexual reproduction involves the formation of sacs called asci following union of male and female structures. In some groups ascogenous hyphae develop after the union, and these, in turn, develop into asci contained in cup-shaped or enclosed ascocarps. Ascospores are produced, following meiosis, in the asci. Upon release, each ascospore may germinate, forming a new mycelium.

Morels are prized edible members of this class. Ergotism is a disease produced when rye infected with ergot is ingested; the disease has killed thousands in the past. Yeasts produce carbon dioxide and alcohol used in baking and brewing; the drug ephedrine; and proteins and vitamins used for human and livestock consumption. Dutch elm disease, chestnut blight, and peach leaf curl are caused by fungi in this class.

Stinkhorns are relatively unimportant representatives of the Class Basidiomycetes (club fungi), which includes mushrooms, bracket fungi, rusts, smuts, jelly fungi, and bird's-nest fungi. The hyphae are divided into cells that have either one nucleus or two. Asexual reproduction, which is less common than in other classes, is usually by means of conidia. Sexual reproduction involves the union of cells of two mating types of monokaryotic hyphae (hyphae whose cells have a single nucleus); this initiates the development of a dikaryotic mycelium whose cells each have two nuclei, and usually have small external protuberances called clamp connections between each cell. A dikaryotic mycelium may become very dense and push above the surface, forming a button that expands into a basidiocarp or mushroom. The expanded cap of the mushroom is usually elevated on a stalk that may have a ring called an annulus around it and sometimes a cup, the volva, at the base. Gills radiate out from the stalk under the cap. The gills consist of hyphae that produce club-shaped basidia, in each of which fusion of two nuclei occurs, followed by meiosis; four basidiospores usually develop on the outside of each basidium. The basidiospores may germinate and repeat the cycle. Puffballs are stalkless club fungi that may each produce trillions of spores. Shelf fungi may add a new layer of growth each year. Bird's-nest fungi produce their basidiospores in "eggs" that have a sticky thread attached. When splashed by rain the eggs may be displaced and become attached to vegetation ingested by animals; the eggs pass through the intestinal tract of the animals unharmed. Smuts infest cereals; rusts, which may have more than one host, infest cereals, white pines, apple trees, and other important plants. Rusts may be partially controlled by eradicating alternate hosts.

Fewer than seventy-five species of the approximately twenty-five thousand described club fungi are known to be poisonous, with the death angel mushroom responsible for about 90% of the fatalities attributed to mushroom poisoning. Thioctic acid has been found to be an antidote to mushroom poisoning. Many unreliable beliefs are associated with attempts to distinguish between poisonous and edible mushrooms; the only safe way, however, is to have a mushroom identified by an authority. Some mushrooms have been used for intoxication and hallucinatory purposes. Mushrooms are not high in nutritional value, with the exception of the Black Forest mushroom, which also has produced substances capable of lowering cholesterol levels and inhibiting viruses. Many types of mushrooms have been cultivated for food but only one species is commercially cultivated in the United States.

Fungi and bacteria play a major role in natural recycling processes. They are also destructive of many economically important substances; they are cultivated by ants and termites in their nests, and form mycorrhizal associations with the roots of higher plants.

Members of the Class Deuteromycetes (imperfect fungi) all lack known sexual reproductive phases but otherwise resemble members of the other classes of fungi. Among the best known members of the class are the *Penicillium* molds, which produce penicillin, and *Aspergillus,* some species of which are used commercially in the production of a wide variety of products, including shoyu and miso. *Aspergillus* and other imperfect fungi also cause diseases in humans and animals, including athlete's foot and ringworm.

Lichens, whose bodies (thalli) consist of an alga and a fungus in intimate symbiotic relationship, are used commercially as a source of dyes. The algal component is usually a green alga or a blue-green alga, which may occur in many different species of lichens, but each lichen species has its own unique fungal component. Except for a few tropical species with club fungi, all lichens have a cup or sac fungus for this component, and it is by the fungus that lichens are identified.

Lichens grow very slowly; they can stand great temperature ranges and may live to be thousands of years old. They are, however, exceptionally sensitive to industrial pollution. The lichen thallus usually consists of several layers of hyphae, including a protective layer (the upper cortex), an algal layer containing algal cells, a medulla consisting of loosely packed hyphae, and a lower cortex, which is often covered with anchoring rhizines, but which may not always be present.

Lichens occur in three growth forms: crustose lichens, which are attached to their substrate over their whole lower surface; foliose lichens, which are somewhat leaflike and weakly attached; and fruticose lichens, which are erect or pendent. Lichens reproduce sexually but are naturally dispersed primarily by asexual means. Some produce small powdery clusters called soredia, which become detached and are dispersed by rain, wind, water, and animals.

Lichens provide food for animals, especially in cold areas; most are unpalatable to humans. Many lichens have antibiotic properties; some lichens are used for perfume, litmus paper, dyeing Scottish tweeds, and as scenery for model railroads.

Review Questions

1. Why are fungi placed in a kingdom separate from slime molds and bacteria?

2. How do the cells of fungi differ from those of other organisms?

3. What does coenocytic mean? To which classes of fungi does it apply?

4. What means of asexual reproduction do the fungi exhibit?

5. If you were looking at hyphae of a bread mold, a cup fungus, and a club fungus through a microscope, could you distinguish among the three? How?

6. Which fungi produce zoospores?

7. Define conidia, haustoria, zygospore, sporangiophore, ascus, mycelium, ascocarp, monokaryotic, dikaryotic, clamp connection.

8. Is there any one best way to tell a poisonous mushroom from an edible one?

9. What is ergotism?

10. Discuss the human relevance of the cup or sac fungi and blue-green molds.

11. What is a spore print?

12. To which class of fungi do each of the following belong: truffles, puffballs, LSD fungi, stinkhorns, peach leaf curl, late blight of potato, bird's-nest fungi, jelly fungi, downy mildew of grape, death angel, shelf or bracket fungi?

13. What is a fairy ring, and why does it form?

14. How do rusts and smuts attack their hosts?

15. Under what conditions do commercially produced mushrooms grow best?

16. Why are some fungi referred to as being "imperfect"?

17. Discuss the economic importance of imperfect fungi.

18. What is the relationship between the alga and the fungus in a lichen?

19. What makes up the basic structure of a typical lichen?

20. How is it possible for lichens to live to such great ages?

Discussion Questions

1. Since fungi produce so many trillions of spores, why is it we are not overrun by them?

2. Slime molds were once treated as fungi in elementary botany books. Do you think their being assigned to a different kingdom is justified?

3. Why are fungus diseases much more common now in the United States than they were at the time of the Declaration of Independence?

4. If all rusts had only one host, would they (theoretically, at least) be easier to control?

5. Since most lichens are not good for food and are not very important economically, should we be as concerned about some of them becoming extinct as we are about Africa's big game animals disappearing?

Additional Reading

Ahmadjian, V.; Hale, M. E. *The lichens.* New York: Academic Press; 1973.

Ainsworth, G. C.; James, P. W.; Hawksworth, D. L. *Ainsworth and Bisby's dictionary of the fungi.* Kew, Surrey, England: Commonwealth Mycological Institute; 1973.

Alexopoulos, C. J.; Mims, C. W. *Introductory mycology.* 3d ed. New York: John Wiley and Sons; 1979.

Batra, S. W. T.; Batra, L. R. The fungus gardens of insects. *Scientific American* 217(5): 112–20; 1967.

Bessey, E. A. *Morphology and taxonomy of the fungi.* Philadelphia: The Blakiston Co.; 1950.

Burnett, J. H. *Fundamentals of mycology.* 2d ed. New York: Crane, Russak and Co., Inc.; 1977.

Christensen, C. M. *The molds and man: an introduction to the fungi.* 3d ed. Minneapolis: University of Minnesota Press; 1965.

———. *Molds, mushrooms and mycotoxins.* Minneapolis: University of Minnesota Press; 1975.

Cooke, R. C. *Fungi, man and his environment.* London: Longman; 1978.

Duddington, C. L. *Beginner's guide to the fungi.* New York: Drake Publishers; 1972.

Ferry, B. W.; Baddeley, M. S.; Hawksworth, D. L., eds. *Air pollution and lichens.* Toronto: University of Toronto Press; 1973.

Hale, M. E., Jr. *The biology of lichens.* 2d ed. Baltimore: University Park Press; 1977.

————. *How to know the lichens.* 2d ed. Dubuque, IA: Wm. C. Brown Company Publishers; 1979.

Johnson, T. W., Jr.; Sparrow, F. K. *Fungi in oceans and estuaries.* Monticello, NY: Lubrecht and Cramer; 1970.

Kramer, J. *Natural dyes: plants and processes.* New York: Charles Scribner's Sons; 1972.

Lamb, I. M. Lichens. *Scientific American,* October 1959, pp. 2–9.

Large, E. C. *The advance of the fungi.* New York: Dover Publications; 1962.

Miller, O. K., Jr. *Mushrooms of North America.* New York: E. P. Dutton and Co.; 1972.

Moore-Landecker, E. *Fundamentals of the fungi.* Englewood Cliffs, NJ: Prentice-Hall; 1972.

Rinaldi, A.; Tyndalo, V. *The complete book of mushrooms.* New York: Crown Publishers; 1974.

Smith, A. H. *The mushroom field hunter's guide.* Ann Arbor: University of Michigan Press; 1963.

Sparrow, F. K., Jr. *Aquatic phycomycetes.* Ann Arbor: University of Michigan Press; 1960.

Webster, J. *Introduction to fungi.* New York: Cambridge University Press; 1977.

Introduction to th[e]
Plant Kingdo[m]
Bryophyte[s]

Features of the Kingd[om]

Plants have several ma[in]
[ch]lorophyll *a*, chlorophyll
those of green alg[ae]
have starch as [...]
[cel]lulose in th[...]
algae; an[...]
cell p[...]
Be[...]

Overview

This chapter opens with a discussion of the features of the Plant Kingdom and its members, and then introduces bryophytes by noting that birds use them in constructing their nests. Following a discussion of the habitats and general life history of bryophytes, liverworts are examined, with an emphasis on *Marchantia*. Next, the chapter examines hornworts and their life cycles, and then discusses mosses in greater detail. The chapter concludes with observations on the human and ecological relevance of bryophytes.

Some Suggested Student Learning Goals

1. List the features that distinguish the Plant Kingdom from other kingdoms.

2. Describe how bryophytes as a group differ from other plants.

3. Describe the basic differences between thalloid liverworts and "leafy" liverworts.

4. Explain how a liverwort thallus can be distinguished from that of a hornwort.

5. Describe the structures involved in the life cycle of a moss and where meiosis and fertilization occur.

6. List the features liverworts, hornworts, and mosses have in common, and describe how their sporophytes differ.

7. List five uses of bryophytes by humans.

...om

...or pigments (e.g., chlo-... *b*, carotenoids) similar to ...e (see chapter 15). Plants also ...heir primary food reserve and cel-...eir walls—again, features of green ...d all plants produce a phragmoplast and ...ate (see chapter 3) when their cells divide. ...sides one brown alga, the only other organisms known to produce such structures during cell division are a few green algae. These features shared by plants and green algae suggest they were derived from a common ancestor.

Fossil records indicate that plants first appeared about 400 million years ago, so that any hypothetical ancestor probably made the transition from an aquatic environment to the land even earlier. By the time plants became established on land, several features that prevented them from drying out had developed. Most plants have a fatty cuticle on their surfaces and also a sterile jacket layer of cells surrounding both the gametangia and the spore-producing cells (sporangia). In addition, the zygote develops into an *embryo* within tissues that originally surround the egg.

Unlike members of the Kingdoms Monera, Protista, and Fungi, which (with the exception of stoneworts) have unicellular gametangia, all plants produce their gametes in multicellular gametangia. Members of the Plant Kingdom exhibit greater structural organization and diversity than those of other kingdoms, with corresponding specialization in the tissues involved in photosynthesis, conduction, support, anchorage, and protection. Reproduction is primarily sexual, with diploid phases of the life cycles assuming the predominant roles in the more advanced members.

The multicellular embryos of plants, mentioned earlier, are unknown in the kingdoms studied in the preceding chapters. This was recognized in past classifications when all organisms were placed in either the Plant or Animal Kingdoms; the Plant Kingdom was then divided into two subkingdoms: the monerans, protists, and fungi were placed in the Subkingdom Thallophyta ("thallus plants") and all other organisms except animals were placed in the Subkingdom Embryophyta ("embryo plants").

One division of nonvascular plants, the *bryophytes,* and several divisions of vascular plants are included in the Kingdom Plantae. The bryophytes are discussed in this chapter and the vascular plants are discussed throughout the remaining chapters of the book.

Introduction to the Bryophytes

As a youth growing up in Africa, I spent many blissful hours watching birds and other wildlife in undisturbed areas during my school vacations. I vividly recall marveling at the way weaver birds constructed their nests from grasses and reeds, with long tubular entrances that dangled down from twigs over water, making it difficult for tree snakes and other predators to get to the eggs or the young. I was more than intrigued by the incredibly soft cobweb lining of the tiny nests of sunbirds, which are something like our hummingbirds; the sloppy stick and grass nests of the noisy Hadeda ibises; or the foul-smelling nests of hoopoes, often located in hollow logs or abandoned termite mounds.

Birds use a great variety of materials in constructing and lining their nests, much of them originating in local plant life. In the summer of 1975, two American biologists, David Breil and Susan Moyle, working at a field station in the Appalachian Mountains in Virginia, examined fifty-eight different nests of native birds. The nests had been constructed by a dozen species, including robins, phoebes, cedar waxwings, slate-colored juncos, and barn swallows. The scientists found no fewer than sixty-five different species of bryophytes distributed in varying degrees throughout all except one of the nests.

Figure 17.1 Mosses and ferns on a tree trunk

Figure 17.2 *Grimmia,* a rock moss that survives on bare rocks, often in scorching sun

Bryophytes, which include *mosses, liverworts,* and *hornworts,* frequently make a soft and cool-looking green covering on damp banks, trees, and logs that are shaded for at least a part of the day (figure 17.1). Some that can withstand long periods of desiccation are also found, however, on bare rocks in the scorching sun (figure 17.2), and others occur on frozen alpine slopes. The bryophytes also frequent habitats ranging from sea level near ocean beaches to altitudes up to 5,486 meters (18,000 feet) on high mountains. A number are very specific indeed as to their growth sites. For example, some species are found only on the antlers and bones of dead reindeer. Others are confined to the dung of herbivorous animals, while still others can grow only on the dung of carnivores. A few found in the tropics attach themselves to large insects. The pygmy mosses, which appear annually on bare soil after rains, are only 1 to 2 millimeters (0.04 to 0.08 inch) tall and can go through their whole life cycle in a few weeks. The widespread peat mosses, which are very important ecologically in bogs and in their transformation to dry land (see chapter 23), sometimes form floating mats over water and keep conditions so acid that the growth of bacteria and fungi is inhibited. Dead organisms deposited in such waters or bogs are often preserved for hundreds and even thousands of years. The luminous moss, which appears to glow in reflected light with an eerie golden green radiance, is found near the entrances in caves and in other dark, damp places. The upper surfaces of its cells are curved slightly, and each functions as a tiny magnifying glass, concentrating the dim light on the chloroplasts at the base and thus adapting the plant to photosynthesis under conditions that would not support the survival of other organisms (figure 17.3).

Water, at least in the form of dew or rain, is essential for the reproduction of all bryophytes. Many haircap mosses have lignin in the walls of special cells in the centers of their stems, but none of the bryophytes has xylem or phloem; hence most are soft and pliable. It is not surprising, therefore, that birds often use them to line their nests.

Bryophytes and ferns (discussed in chapter 18) exhibit an *alternation of generations* (see chapter 11) more conspicuously than do most organisms. In mosses, for example, the "leafy" plant is a major part of the *gametophyte* generation, which produces the gametes; periodically the *sporophyte* generation, which produces the *spores,* emerges from a "leafy" gametophyte tip, usually in the form of a tiny capped urn on the end of a slender stalk.

The bryophytes are divided into three classes on the basis of structure and reproductive features.

Figure 17.3 Lenslike cells of the protonema of the luminous moss. The cells concentrate light on the chloroplasts at their bases. (After von Denffer, D., Schumacher, W., Magdefrau, K. and Ehrendorfer, F. 1976. *Strasburger's Textbook of Botany,* 30th German ed., translated by Bell, P. and D. Coombe, 1976. Redrawn by permission of Gustav Fischer Verlag, Stuttgart, and Longman Group, London and New York.)

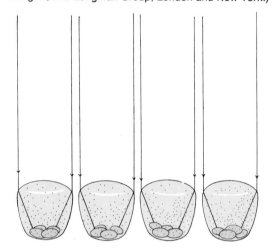

Division Bryophyta

Class Hepaticae—Liverworts

The word *wort* simply means plant or herb. In medieval times, when the Doctrine of Signatures (see chapter 1) was promulgated, the herbalists of the day thought that some of these bryophytes—those that have flattened bodies with lobes shaped somewhat like those of the human liver (see figure 17.4)—were useful for treating ailments of that organ. No such efficacy has been demonstrated, but the name liverwort is still universally used today.

Liverworts that have flattened lobed bodies *(thalli),* although widespread, actually constitute only about 20% of the approximately ninety-five hundred known species. The remainder are "leafy" and superficially look more like true mosses. Liverworts differ from mosses (discussed in a later section) in several details and are considered less complex than mosses. They develop

Chapter 17

Figure 17.4 *Marchantia* (a thalloid liverwort).
A. Thallus with male gametophores. *B.* Thallus with
female gametophores. (Courtesy Field Museum of
Natural History, Chicago)

A.

B.

their thalli or "leafy" stages (gametophytes) almost directly from spores, and they have distinct
top and bottom surfaces, with one-celled *rhizoids*
on the lower surface. The rhizoids resemble tiny
roots and function in anchorage. Many liverworts
have cells with special thickenings in the walls at
the corners, and all have prostrate as opposed to
an upright form of growth.

Probably the best known of the non-"leafy"
liverworts is one with the scientific name of *Marchantia* (named in honor of a French botanist, N.
Marchant). The flattened lobed but leafless gametophyte body is called a *thallus,* and all similar liverworts are said to be *thalloid* (figure 17.4).
The most widespread species of *Marchantia* is
often found on damp soil, particularly after a fire
has burned over the area.

The thallus forks dichotomously as it grows.
Each branch has a central lengthwise groove and
a notch at the apex. Meristematic cells at these
notches add new growth to the thalli. Older tissues at the rear decay as the new growth is added.
The upper surface of the thallus is divided into
diamond-shaped or polygonal segments, the lines
marking the limits of the chambers below. Each
segment has a small bordered pore opening into
the interior. Examining a section through a thallus with a microscope is something like looking
at a series of covered prickly pear cactus gardens
sitting on a decorative rock wall (figure 17.5). The
"rock wall," which may make up the bulk of the

Introduction to the Plant Kingdom: Bryophytes

Figure 17.5 Section through a portion of a
Marchantia thallus

thallus, consists of parenchyma cells that contain few, if any, chloroplasts. The tissue apparently functions in the storage of substances manufactured by other cells. The bottom layer of cells constitutes an epidermis from which rhizoids and scales arise. The "cactus gardens" consist of upright branching rows of chlorenchyma cells in an air space. The individual "gardens" are enclosed by vertical walls and covered by a slightly dome-shaped layer of epidermal cells. In the center of each "roof," the conspicuous pore looks something like a short, suspended, open-ended barrel. The pore remains open at all times.

Marchantia reproduces asexually by means of **gemmae** (singular: **gemma**). These are groups of cells that form tiny lens-shaped pieces of tissue that become detached from the thallus. They are produced in small cups formed at intervals on the upper surface of the liverwort gametophyte (see figure 17.6). Raindrops may splash the gemmae as much as 1 meter (3 feet) away. Each gemma is capable of growing into a new thallus. In addition, parts of an older thallus may die, isolating patches of active tissue that may then continue to grow independently.

In sexual reproduction, *Marchantia* develops more specialized structures than are found in other liverworts. Male and female organs are produced on separate gametophytes, with both male and female gametangia being formed on **gametophores** (figure 17.6). These are somewhat umbrellalike structures rising from the thallus on slender stalks above the central grooves of the thallus. The top of the male gametophore, or **antheridiophore**, is disclike with a scalloped margin, whereas that of the female gametophore, or **archegoniophore**, resembles the hub and spokes of a wagon wheel. **Antheridia** (club-shaped male sex organs each containing numerous sperms) are produced in rows just beneath the upper surface of the antheridiophore, whereas **archegonia** (flasklike female sex organs each containing a single egg), which are also produced in rows, hang neck downward beneath the spokes of the archegoniophore. Raindrops sometimes splash the released sperms more than one-half meter (1.5 feet) away. Fertilization may occur before the stalks of the archegoniophores have finished growing.

After fertilization, the zygote develops into a multicellular **embryo** (an immature *sporophyte*). The sporophyte (diploid spore-producing

Figure 17.6 Life cycle of the thalloid liverwort
Marchantia

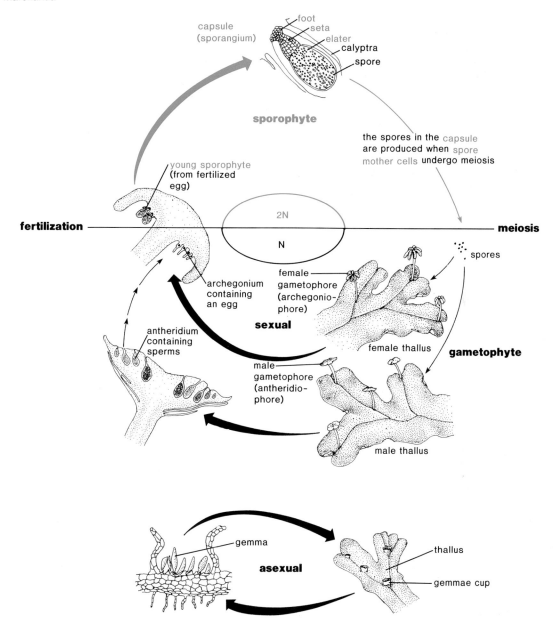

Figure 17.7 Longitudinal section through a sporophyte of *Marchantia*

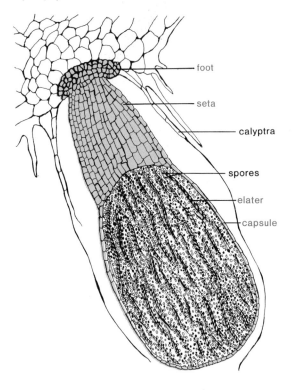

- foot
- seta
- calyptra
- spores
- elater
- capsule

Figure 17.8 "Leafy" liverwort

phase) is anchored in the tissues of the archegoniophore by means of a knoblike **foot.** The sporophyte resembles a tiny watermelon suspended from the embedded foot by a short thick stalk called the **seta.** The main part of the sporophyte, in which different types of tissues develop, is called a *capsule.* One type of tissue in the capsule consists of **spore mother cells** that undergo meiosis, producing spores, whereas the cells of another type of tissue do not undergo meiosis but develop instead into long, pointed **elaters** (figure 17.7). Elaters, which have spiral thickenings, are sensitive to changes in humidity, and they twist and untwist rapidly, thereby aiding in the dispersal of spores. Membranes covering the capsule, and a caplike tissue, the **calyptra,** which grows out from the gametophyte, protect the young sporophyte until it is mature. Then the

capsule splits, and the spores are carried away by air currents. Under favorable conditions the spores germinate, producing new gametophytes.

Other thalloid forms such as the floating or amphibious liverworts do not produce gametophores. Instead the archegonia and antheridia develop within the thallus beneath the central grooves, where the sporophytes also are formed. The spores are liberated from the submerged sporophytes as the thallus decays.

"Leafy" liverworts (figure 17.8), which are often found abundantly in tropical jungles and in fog belts, always have two rows of partially overlapping "leaves" whose cells contain distinctive oil bodies. The "leaves" have no midribs, and unlike the "leaves" of mosses, they often have folds and lobes. In the tropics, the lobes form little water pockets in which tiny animals are nearly always present. It has been suggested that these water pockets may function like the pitchers of pitcher plants (see chapter 7). On the underside of "leafy" liverworts is frequently a third row of "underleaves," which are smaller in size than the other "leaves" and not visible from the top. A few rhizoids, which anchor the plants, develop from the stemlike axis at the base of the "underleaves."

Members of all classes of bryophytes often have mycorrhizal fungi associated with their rhizoids. In some instances, the fungi are apparently at least partially parasitic. One species of liverwort, which lives underneath mosses, is completely colorless and depends on its fungal associate for its existence.

Figure 17.9 *A.* Hornworts. (Photograph by Carolina Biological Supply Company) *B.* Detail of a hornwort sporophyte.

A.

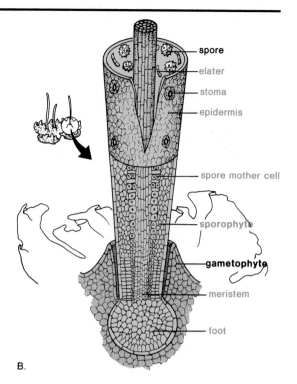

spore
elater
stoma
epidermis
spore mother cell
sporophyte
gametophyte
meristem
foot

B.

The archegonia and antheridia of the "leafy" liverworts are produced in cuplike structures composed of a few modified "leaves," either in the axils of "leaves" or on separate branches. At maturity, the sporophyte capsule may be pushed out from among the "leaves" by elongation of the seta. When a spore germinates, it produces a short filament consisting of a few photosynthetic cells (**protonema**) that soon develops into a new gametophyte plant.

Class Anthocerotae—Hornworts

Hornworts, whose rounded gametophytes resemble those of thalloid liverworts, received their common name from their sporophytes, which resemble miniature cattle horns when they are mature. They seldom exceed 2 centimeters (0.4 inch) in diameter and are found mostly on moist earth in shaded areas, although some occur on trees. There are about one hundred species distributed around the world, but they are uncommon in arctic regions. They differ from liverworts and mosses in that they have only one large chloroplast per

cell, and that each chloroplast has pyrenoids similar to those of green algae. The thalli have pores and cavities filled with mucilage, in contrast with the pores and cavities of thalloid liverworts, which are filled with air. Nitrogen-fixing blue-green algae often grow in the mucilage. Rhizoids anchor the plants to the surface.

Asexual reproduction in hornworts occurs through fragmentation or separation of lobes from the main part of the thallus. In a few instances, tiny tubers capable of becoming new gametophytes are formed.

In sexual reproduction, archegonia and antheridia are produced in rows just beneath the upper surfaces of the gametophytes. As is the case with both mosses and liverworts, some species of hornworts have **unisexual** plants whereas other species are bisexual. Sporophytes of hornworts are most distinctive. They have no *setae* (stalks) and look like tiny green broom handles or horns rising through a basal sheath from a foot beneath the surface of the thallus (figure 17.9). A meristem above the foot continually increases the length of the sporophyte from the base when conditions are favorable. As growth occurs, spore mother cells surrounding a central rodlike axis in the sporophyte undergo meiosis, producing spores. The tip of the sporophyte horn splits into two or three segments, releasing the spores, and the segments continue to peel back as long as the meristem is producing new tissue at the base.

Class Musci—Mosses

Many different organisms have been called mosses. In fact, almost any greenish covering or growth on tree trunks and forest floors has probably been called "moss" at one time or another. Examples include lichens (e.g., reindeer moss), red algae (e.g., Irish moss), flowering plants (e.g., Spanish moss), and club mosses. The latter superficially resemble large true mosses but have xylem and phloem and thus are vascular plants. About twenty thousand species of mosses are currently known. These are divided into three different groups, called *peat mosses, true mosses,*

and *rock mosses.* As observed in the following discussion, mosses are distinct, both in form and reproduction, from any other groups of organisms.

The "leaves" of moss gametophytes differ from those of more complex plants in that they have no mesophyll tissue, stomata, or veins; also their cells are all haploid. The blades are nearly always only one cell thick, and they are never lobed or divided, nor do they have a petiole. Some have a *costa,* a relatively thick rib that is darker than the surrounding cells, running lengthwise down the middle. The costa occasionally projects beyond the tip in the form of a hair or spine. The "leaf" cells usually contain numerous lens-shaped chloroplasts except at the midrib. The "leaves" of peat mosses, however, have large transparent water storage cells (without chloroplasts) that adapt them to water absorption and storage, with small green photosynthetic cells sandwiched between the large cells (figure 17.10). The "leaves" are usually arranged in a spiral or alternately on an axis. The axis is somewhat stemlike but also has no xylem or phloem, although cells at the

Figure 17.10 Enlargement of a portion of a peat moss *(Sphagnum)* leaf. *A.* Surface view. *B.* Further enlarged cross section of living and dead cells.

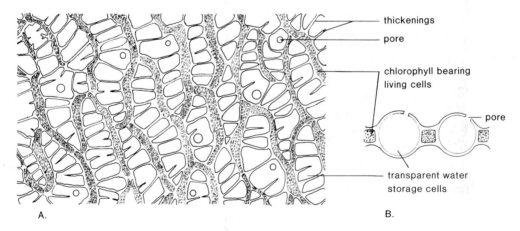

thickenings

pore

chlorophyll bearing living cells

pore

transparent water storage cells

A.

B.

Chapter 17

center may appear to form a strand that differs slightly in appearance from that of the surrounding tissue. At the base are rootlike *rhizoids,* consisting of several rows of colorless cells. Their primary function is to anchor the plant. Some water absorbed by rhizoids rises up the central strand, but most water used by the plant apparently travels up the outside of the plant by means of capillarity. The closely packed habit of many mosses and the fact that they rarely extend more than a few centimeters (2 to 3 inches) into the air favors such outside movements of water, which is absorbed directly through the plant surfaces.

Mosses initiate sexual reproduction with the formation of multicellular gametangia, usually at the apices of the "leafy" shoots (see figure 17.13), although many produce them on special separate branches. In many mosses, both male and female sex organs are produced on the same plant; in others, they are produced on separate plants. The female gametangium is a cylindrical projection, the *archegonium* (figure 17.11). The cylinder is slightly swollen near the middle, where certain cells break down, leaving a cavity called a **venter** in which a single egg cell is produced. The part of the archegonium above the venter is called the *neck.* It may taper toward the tip and contains a narrow canal. The canal is at first plugged with cells, but these break down as the archegonium matures, leaving an opening to the outside at the top. A number of archegonia usually are produced at the same time, with sterile hairlike rows of cells called *paraphyses* (singular: *paraphysis*) among them.

Male gametangia, which also have paraphyses among them, are sausage-shaped to roundish with walls that are one cell thick. These *antheridia* (figure 17.12) each are borne on a short stalk. Inside is a mass of tissue that develops into coiled or comma-shaped sperm cells, each with a pair of flagella. This mass of sperms is forced out of the top of the antheridium when it absorbs water and swells. After release, the sperm mass breaks up into individual cells, aided, it is believed, in some cases by fats produced by the moss, while in other instances rain splash is responsible.

Figure 17.11   Longitudinal section through the tip of a female gametophyte of the moss *Mnium*

arche-
gonium

neck canal

paraphysis

venter

egg

Figure 17.12 Longitudinal section through the tip of a male gametophyte of the moss *Mnium*

Archegonia release sugars, proteins, acids, or other substances that attract the sperm, and eventually a sperm, after swimming down the neck of an archegonium, unites with the egg, forming a *zygote* (figure 17.13). The zygote usually grows rapidly into a spindle-shaped *embryo,* which breaks down the cells at the base of the archegonium and firmly establishes itself in the tissues of the stem by means of a swollen knob called a *foot.* As the embryo grows, cells around the venter divide, thus accommodating its increasing size. The length of the embryo soon exceeds the length of the venter cavity. The top of the latter is split off and is left sitting like a pixie cap on top of the embryo. By this time, the embryo is a developing *sporophyte* whose cells are all diploid. The pixie cap, called a *calyptra,* remains until the sporophyte is mature. In one genus known as the "extinguisher" mosses, the calyptra looks just like a little candlesnuffer, and in the haircap mosses it resembles a miniature, pointed, goatskin cap such as might be worn by a Shakespearean actor.

The cells of the sporophyte become photosynthetic as it develops, remaining so until maturity. The sporophyte, however, depends on the gametophyte to varying degrees for some of its carbohydrate needs as well as for at least a part of its water and minerals, which are absorbed through the foot.

The mature sporophyte consists of a *sporangium* or *capsule* located at the tip of a slender stalk, the *seta.* Depending on the species, the seta may be less than 1 millimeter (0.04 inch) long, or it may be up to 15 centimeters (6 inches) long; most are less than 5 centimeters (2 inches) long. The capsule, which may resemble a tiny apple, a pear, an urn, a box, or a wingtip fuel tank of an airplane, usually has from 3 or 4 to over 200 stomata at or near its surface. Unless extremely dry conditions prevail, the stomata normally remain open until the capsule begins to age, and then they close permanently. The free end of the capsule is usually protected by a little rimmed lid, the **operculum,** which falls off at maturity.

As the capsule matures, *spore mother cells* inside undergo meiosis, producing *spores.* These spores are released from the free end of the capsule after the operculum has fallen off, usually

Figure 17.13   Life cycle of a moss

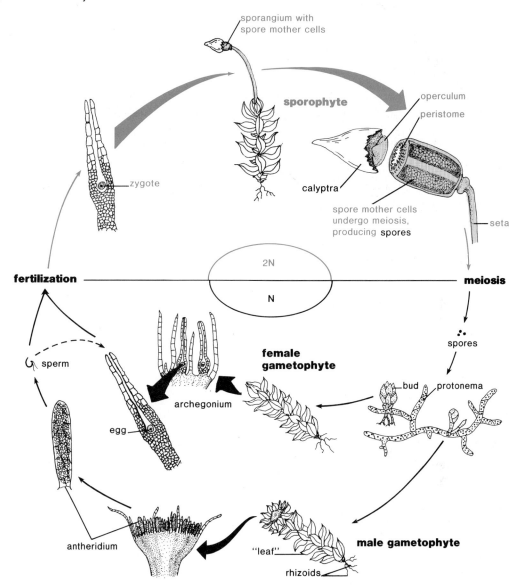

sporangium with spore mother cells

sporophyte

operculum

peristome

calyptra

spore mother cells undergo meiosis, producing **spores**

seta

zygote

fertilization

2N

N

meiosis

sperm

female gametophyte

spores

bud   protonema

archegonium

egg

antheridium

"leaf"

male gametophyte

rhizoids

through a structure called a **peristome.** Most peristomes consist of sixteen narrowly triangular and membranous teeth arranged in a circle around the rim of the capsule. They are frequently colored orange or red and are often beautifully sculptured with bars and fringes. They are sensitive to changes in humidity and open or close in response to such changes. In a few species of mosses, the peristome is a cone-shaped structure with pores through which the spores are released. Certain rock mosses have neither a peristome nor an operculum. The spores in these mosses are released when the capsule splits longitudinally along four lines.

In the dung mosses, a putrid odor is given off when the spores are ready for release. Some of the spores adhere to the legs and bodies of flies, which are attracted by the odor, and are disseminated as the insects clean themselves. Most moss spores are simply blown away by the wind, however; and if they fall in a suitable damp location, they usually germinate relatively quickly. In the majority of mosses, fine green tubular threads consisting of single rows of cells with chloroplasts first emerge from the spores. These soon branch and grow, forming an algalike *protonema*. The protonema can be distinguished from a filamentous green alga by the crosswalls of its cells, which are usually formed at oblique angles instead of right angles, and by the chloroplasts, which are all lens-shaped. If light and other conditions are favorable, tiny "leafy" buds appear at intervals along the protonemal filaments after about two to four weeks of growth. These "leafy" buds develop rhizoids at the base and grow into new "leafy" gametophytes, thus completing the cycle.

Some reproduction in mosses does not depend on such a sexual cycle and its alternation of generations. It has been demonstrated under laboratory conditions that cells of archegonia and antheridia, paraphyses, leaves, stems, and rhizoids all can develop protonemata. Two American biologists, Norton Miller and Howe Ambrose, collected bryophyte gametophyte fragments from a snowbed in the Canadian high arctic during a recent summer and found that 12% of their samples resumed growth in various ways when cultured in the laboratory. On this basis, they calculated that at their study site there were over four thousand bryophyte fragments per cubic meter (1.3 cubic yards) of snow capable of becoming new plants. They suggested that wind dispersal by fragments may be routine in arctic regions. Such dispersal and vegetative reproduction also occur widely in more temperate areas.

Human and Ecological Relevance of Bryophytes

Certain bryophytes and lichens are pioneers on bare rock after volcanic eruptions or other geological upheavals and slowly convert the rock to soil, which can then be inhabited or utilized by other organisms (see chapter 5). Mosses in particular retain moisture, slowly releasing it into the soil. They reduce flooding and erosion and contribute to humus formation. They are indicators of the presence or absence of calcium in the soil, of soil salinity, and of acid conditions. The presence of one genus in a dry area indicates that running water occurs there during a part of the year. A few mosses are occasionally a problem in water reservoirs, where they may plug entrances to pipes. A few bryophytes are reported to be grazed, along with lichens, by foraging mammals in arctic regions, but bryophytes are not generally edible.

Some mosses have been used for packing dishes and stuffing furniture, and Native Americans are reported to have used mosses under splints when setting broken limbs. By far the most important bryophytes to humans, however, are the peat mosses. When allowed to absorb water, 1 kilogram (2.2 pounds) of dry peat moss will take up 25 kilograms (55 pounds) of water. Its extraordinary absorptive capacity has made it very useful as a soil conditioner in nurseries and as a component of potting mixtures. Live shellfish and other organisms are shipped in it, and its natural acidity, which inhibits bacterial and fungal

Chapter 17

growth, gives it antiseptic properties. This latter feature combined with its absorbency, which is greater than that of cotton, has made it a useful poultice material for application to wounds. It was used for this purpose during the Crimean War and on an emergency basis during World War I. Extensive peat deposits have been formed from the remains of peat mosses that flourished in past eras. Peat, like the undecomposed peat mosses, is used around the world as a soil conditioner and as a fuel. It is also used in the curing of malt for Scotch whiskey (see appendix 1 for the scientific names of all the bryophytes discussed).

Summary

Members of the Plant Kingdom have a cuticle and produce their gametes in multicellular organs surrounded by a sterile jacket of cells; their zygotes develop into embryos, and tissues specialized for photosynthesis, conduction, support, anchorage, protection, and reproduction are produced. Diploid phases of the life cycles are predominant in more advanced members. Cell plates and phragmoplasts are formed when plant cells divide; cell plates occur only in certain green algae and one brown alga outside of the Plant Kingdom. The similarity in pigments, food reserve (starch), and occurrence of cell plates suggests a common ancestor for the green algae and plants, which, according to fossil records, may have originated more than 400 million years ago. The Plant Kingdom includes one division of bryophytes and several divisions of vascular plants.

Birds use many materials in constructing their nests. Breil and Moyle found sixty-five species of bryophytes used in fifty-eight nests of native birds in Virginia. Bryophytes (liverworts, hornworts, mosses) are found in habitats ranging from bare rocks in hot sun to frozen slopes, and from sea level to high altitudes; some, such as those found on bones or dung or insects are very specific in their habitats. The luminous moss is adapted to existing under conditions of very low light. Having external water present is essential to reproduction in bryophytes. Bryophytes exhibit a marked alternation of generations, with the gametophyte being more conspicuous than the sporophyte.

The best known liverworts have gametophytes with flattened, dichotomously forking thalli, but about 80% of the liverwort species are "leafy." Liverworts have distinct top and bottom surfaces, with one-celled rhizoids that function in anchorage on the lower surface.

The thallus of the well-known liverwort *Marchantia* has a central lengthwise groove along its upper surface, which is divided into polygonal segments that each contain a pore. Beneath each pore is a chamber containing upright branching rows of chlorenchyma cells arising from brick-shaped parenchyma cells in wall-like layers at their bases. Rhizoids and scales arise from the bottom row of parenchyma cells. *Marchantia* reproduces asexually by means of pieces of tissue called gemmae that are splashed by rain out of cups in which they are formed and also by fragmentation of the thallus. In sexual reproduction, *Marchantia* produces flasklike archegonia beneath the "spokes" of an umbrella-shaped archegoniophore that arises from the thallus, and club-shaped antheridia in disc-shaped tops of antheridiophores, also arising from the thallus. Each archegonium contains an egg and each antheridium contains many sperms. The zygote formed from the union of egg and sperm develops into a sporophyte that is anchored to the archegoniophore by a knoblike foot, from which is suspended a capsule connected to the foot by a stalk, the seta. The capsule contains spore mother cells, which undergo meiosis and produce spores, and pointed elaters, which do not undergo meiosis. The elaters, which twist with changes in humidity, aid in spore dispersal. The calyptra, which grows out from the gametophyte, and other membranes protect the spores until the capsule splits, releasing them; the spores may then develop into new gametophytes.

"Leafy" liverworts have two rows of overlapping "leaves" and frequently a third row of "underleaves" not visible from above. The "leaves" often have lobes that retain rain water.

Hornworts, which have one chloroplast with pyrenoids per cell, resemble liverworts in their gametophytes, but their sporophytes are hornlike and have a meristem above the foot. Hornwort thalli have pores and cavities filled with mucilage in which blue-green algae often grow. Asexual reproduction involves the separation of lobes from the main part of the thallus. Archegonia and antheridia are produced in rows beneath the upper surface of a thallus. The tip of the hornlike sporophyte splits vertically, releasing the spores.

The moss gametophyte consists of an axis to which "leaves" are attached, with rhizoids at the base. The "leaves" are haploid and have no mesophyll, stomata, or veins. Water is absorbed primarily directly through the plant surfaces.

Multicellular archegonia and antheridia are produced at the tips of "leafy" shoots. Each archegonium has a cavity, the venter, containing a single egg and a neck through which a sperm gains access to the egg. Sperms are produced in sausage-shaped antheridia. After fertilization the zygote grows into an embryo, which is attached to the gametophyte by means of a foot embedded in its tissues. The embryo develops into a sporophyte consisting of a capsule and a seta, with a caplike calyptra partially covering the capsule. Spore mother cells in the capsule undergo meiosis, producing spores that are released through the teeth of the peristome, a structure at the tip of the capsule. The peristome is initially protected by a lid, the operculum, which falls off when the spores are mature. When moss spores germinate, they develop into threadlike protonemata on which "leafy" buds appear. The buds grow into new gametophytes. Protonemata have been demonstrated to develop from cells of various other moss tissues besides spores.

Mosses may be pioneers, along with lichens, on bare rocks. They are indicators of soil calcium, salinity, and acidity. Mosses are not generally edible, although a few are grazed in arctic regions. Some mosses are used for packing material, but the most significant use is that of peat mosses for soil conditioners. Peat mosses can absorb and retain large amounts of water, and their natural acidity gives them antiseptic properties. Peat deposits, from peat mosses that flourished in past eras, are used for fuel and also as a soil conditioner.

Review Questions

1. What basic features distinguish members of the Plant Kingdom from those of other kingdoms?

2. What features distinguish the bryophytes discussed in this chapter from other plants?

3. How could you tell a hornwort thallus from that of a thalloid liverwort?

4. Contrast the sporophytes of mosses, liverworts, and hornworts.

5. What is a protonema? Do all bryophytes have them? How would you tell a protonema from a green alga?

6. What adaptations do bryophytes have for their particular habitats?

7. Which parts of the life cycles of bryophytes have haploid (N) cells? Which parts have diploid (2N) cells?

8. Why is a bryophyte "leaf" technically not the same as a flowering plant leaf?

9. Define calyptra, operculum, capsule, peristome, paraphysis, foot, seta, archegoniophore, thallus, underleaves, elaters.

10. If you were to single out one bryophyte as being the most important member of its division from a human viewpoint, which would you choose? Why?

Discussion Questions

1. Very few fossils of bryophytes have been found. Suggest reasons for this.

2. Do the multicellular sex organs of plants give them any advantages over other organisms?

3. After reading about the characteristics and uses of peat mosses, can you suggest some possible new uses of these plants?

4. Some bryophytes produce unisexual gametophytes, while others produce bisexual gametophytes. Should one type have any survival or adaptive advantage of the other? Explain.

Additional Reading

Doyle, W. T. *The biology of higher cryptogams.* New York: Macmillan Co.; 1970.

Grout, A. J. *Mosses with hand-lens and microscope.* Published in five parts by the author in Brooklyn, NY; 1903–1910.

Parihar, N. S. *An introduction to Embryophyta.* Vol. 1: *Bryophyta.* 5th ed. Allahabad, India: Central Book Depot; 1965.

Redfearn, P. L., Jr.; Conard, H. S. *How to know the mosses and liverworts.* 2d ed. Pictured Key Nature Series. Dubuque, IA: Wm. C. Brown Company Publishers; 1979.

Richardson, D. H. S. *The biology of mosses.* New York: John Wiley & Sons; 1981.

Schuster, R. M. *The Hepaticae and Anthocerotae of North America.* Vol. 1. New York: Columbia University Press; 1966.

Watson, E. V. *The structure and life of bryophytes.* 3d ed. London: Hutchinson and Co.; 1971.

troduction to
ascular Plants
rns and Their Relatives

rview

chapter opens with a brief
v of the features that
guish the major groups of
seedless vascular plants from one
another and from the bryophytes
and then discusses representatives
of each division. Included in the
discussion are whisk ferns
(Psilotum), club mosses
(Lycopodium, Selaginella),
quillworts *(Isoetes)*, horsetails
(Equisetum), and ferns. A digest of
the human and ecological relevance
of each group is given, and life
cycles of representatives are
illustrated. The chapter concludes
with an examination of fossils, and
each type is briefly described. A
table showing the geologic time
scale is provided.

Some Suggested
Student Learning Goals

1. List the basic structural
   differences between
   bryophytes and vascular
   plants.

2. Distinguish the four divisions of
   seedless vascular plants from
   one another.

3. Describe the differences in the
   life cycles of ground pines
   (Lycopodium) and spike
   mosses *(Selaginella)*.

4. Describe the structural features
   of horsetail *(Equisetum)*
   sporophytes.

5. Indicate all the structures
   involved in alternation of
   generations in a fern.

6. List ten important uses of
   seedless vascular plants.

7. Explain what a fossil is and
   distinguish among the various
   types of fossils.

Introduction

Our survey of all organisms traditionally regarded as plants thus far has taken us from simple one-celled prokaryotic organisms, through more specialized eukaryotic protists and fungi, and on to the bryophytes. We now take up several divisions of plants whose members have developed internal conducting tissues but have not produced seeds.

As indicated in the introduction to the Plant Kingdom in the preceding chapter, the hypothetical ancestor of the bryophytes and the vascular plants was probably a multicellular green alga that became established on the land over 400 million years ago. The shift to terrestrial habitats was accompanied by the appearance of features such as sterile jackets of cells around gametangia, development of embryos within protective tissues, and cutin—all significant in preventing the drying out of vital parts. The bryophytes, however, had little if any internal conducting tissue, and water was absorbed directly through the aboveground parts of the plants. In addition, external water was essential to the reproductive processes. Then internal conducting tissues (xylem and phloem) began to evolve, true leaves appeared, and roots that function in anchorage and absorption developed. At the same time gametophytes became progressively smaller and more dependent on and protected by proportionately larger sporophytes. Unlike conifers and flowering plants, the primitive vascular plants discussed in this chapter do not produce seeds. They include the ferns and a number of their relatives often referred to as "fern allies."

Four divisions of seedless vascular plants are recognized: Division Psilophyta (whisk ferns), whose members are unique among vascular plants in having sporophytes that have neither true leaves nor roots and consist of stems and rhizomes that fork evenly; Division Lycophyta (club mosses and quillworts), whose stems are covered with mostly photosynthetic **microphylls** (leaves with a single vein whose trace is not associated with a leaf gap—see chapter 7); Division Sphenophyta (horsetails and scouring rushes), whose sporophytes have ribbed stems containing silica deposits and whorled scalelike microphylls that lack chlorophyll; and Division Pterophyta (ferns), whose sporophytes have **megaphylls** (leaves with more than one vein and a leaf trace that is associated with a leaf gap—see chapter 7) that are often large and much divided. Fossils, discussed at the end of the chapter, give us clues to the ancestry of some of these plants and also to the derivations of seed plants discussed in the chapters to follow.

Division Psilophyta—Whisk Ferns

There is nothing very fernlike in the appearance of *whisk ferns,* more commonly referred to by their scientific name of *Psilotum,* but they do loosely resemble small green whisk brooms. Exactly where these plants fit in the Plant Kingdom is not clear. Traditionally they have been associated with a number of extinct plants called *psilophytes,* which flourished perhaps 400 million years ago, but there is a general lack of fossil evidence to substantiate the link. Some have classified them with the true ferns because their gametophytes share features with those of true ferns. By most criteria presently in use they are, however, among the simplest of all living vascular plants and thus will be the first discussed.

Psilotum sporophytes consist of evenly forking stems that commonly approach 30 centimeters (1 foot) in height, occasionally growing to a height of 1 meter (3 feet) or more (figure 18.1). The visible stems arise from short branching rhizomes just beneath the surface of the ground. Tiny green protuberances called **enations** are spirally arranged at intervals along the stems, but no leaves or roots are present. Photosynthesis takes place in the outer cells of the stem, which contains a central cylinder of xylem surrounded by phloem. The xylem is star-shaped in cross section. The rhizomes, which have short rhizoids over

Figure 18.1 *Psilotum* sporophyte

are terminated by small *sporangia* fused together in threes and resembling miniature yellow pumpkins. Spores released from the sporangia germinate slowly in the soil, on the bark of tree ferns, or in similar habitats. The gametophytes that arise from the spores are easily overlooked as they are colorless, seldom more than 2 millimeters (0.08 inch) long, and found beneath the surface of the soil. These gametophytes, which are cylindrical and may fork like the stems, sometimes resemble tiny dog bones. They are *saprophytic,* obtaining their nutrients through symbiotic association with fungi. Archegonia and antheridia are produced randomly over the surface of the same gametophyte. After a sperm unites with an egg in an archegonium, the zygote develops a foot and a rhizome. As soon as the rhizome establishes itself with the aid of mycorrhizal fungi, upright stems are produced and the rhizome separates from the foot (figure 18.2).

Whisk ferns grow naturally primarily in tropical and subtropical regions. In the United States, they are found in Florida, Louisiana, Texas, Arizona, and Hawaii. They are extensively cultivated in Japan and in greenhouses around the world. A close relative, *Tmesipteris,* with leaflike appendages, occurs in Australia and the South Pacific. Fossil plants resembling whisk ferns in many respects have been found in Silurian geological formations (see table 18.1). As intimated in the introduction to the whisk ferns, these geological formations are estimated to be as much as 400 million years old. One group of these fossil plants, exemplified by *Rhynia,* had naked stems and spindle-shaped terminal sporangia. A second group of fossils, represented by *Zosterophyllum,* had somewhat rounded sporangia produced along the upper parts of naked stems. *Zosterophyllum* and its relatives first appeared during the Devonian period; they are considered to be likely ancestors of the club mosses discussed in the next section.

Whisk ferns are of little economic importance. Their spores, which have a slightly oily feel, were once used by Hawaiian men to reduce loincloth irritations of the skin; Hawaiians also made a laxative liquid by boiling whisk ferns in water.

their surfaces, perform the functions of roots with the aid of mycorrhizal fungi, some of which establish themselves in the cells of the cortex just beneath the epidermis.

Very short stubby branches appear at times along the upper parts of the angular stems. These

Chapter 18

Figure 18.2 Life cycle of *Psilotum*

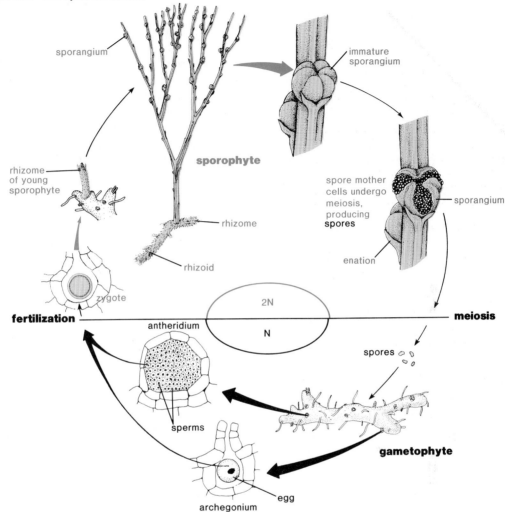

sporangium

immature sporangium

sporophyte

rhizome

spore mother cells undergo meiosis, producing **spores**

sporangium

rhizome of young sporophyte

rhizoid

enation

zygote

2N

N

fertilization

meiosis

antheridium

spores

sperms

gametophyte

archegonium

egg

Division Lycophyta—Club Mosses and Quillworts

On one occasion during my youth, my father was interviewed by a newspaper reporter in a large South American city. At the end of the interview, which took place in a hotel room with the rest of the family present, the reporter set up a press camera and said he wanted to take a picture of the whole family. After posing us, he took a vial from his pocket and poured a little powder along a metal bar attached to one end of a T-shaped device that otherwise looked like a flashlight. He then told us to smile, and a moment later there was a blinding flash accompanied by a large billow of smoke. I learned later that the flash powder (forerunner of flashbulbs and strobe lights) he had used contained millions of spores of primitive vascular plants called *club mosses.*

The approximately four hundred known species of club mosses superficially look enough like large true mosses to have caused the great Swedish botanist Linnaeus to lump both together in a single class. Once details of the structure and the form of club mosses were known, however, it became obvious that they are quite unrelated to true mosses.

Two major genera of club mosses exist today (several others became extinct about 270 million years ago). The sporophytes of all have small leaves called **microphylls** (see chapter 7), true stems, and true roots.

The two genera of club mosses with living members are *Lycopodium* and *Selaginella*. Species of *Lycopodium* are frequently called *ground pines,* partly because they often grow on forest floors and resemble little Christmas trees, complete with "cones" that are usually upright but that in some species hang down (figure 18.3).

They are distributed throughout the world, although they are more abundant in the tropics and wetter temperate areas. Ground pine sporophytes have either simple or branched stems, mostly less than 3 decimeters (1 foot) tall although tropical species may attain heights of 1.5 meters (5 feet) or more. The upright, or sometimes pendent, stems develop from rhizomes, which also branch. Their leaves may be whorled or in a tight spiral and rarely exceed 1 centimeter (0.4 inch) in length. Adventitious roots arise along the rhizomes. Many of the epidermal cells of the roots produce root hairs (see chapter 5).

At maturity, some species of ground pines develop sporangia shaped like kidney beans on short stalks in the axils of several of the leaves. Such sporangium-bearing leaves are called **sporophylls.** In other species, the sporophylls are reduced in size, lack chlorophyll, and are clustered in terminal conelike structures called **strobili** (singular: **strobilus**).

Spore mother cells in the sporangia undergo meiosis, producing spores that are carried away by air currents upon their release. Those of some species germinate in a few days if they land in a suitable location. Other species produce spores that may not germinate for up to several years. Upon germination, the spores produce independent gametophytes that occur in various shapes,

Figure 18.3 Species of *Lycopodium* native to northern and eastern North America

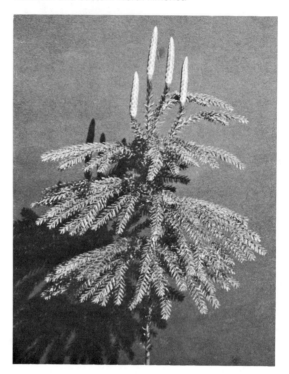

Figure 18.3 Species of *Lycopodium* native to northern and eastern North America

some resembling tiny carrots, with most or all of the body developing in the ground. Other gametophytes develop primarily on the surface, assisted by mycorrhizal fungi. Surface gametophytes, or the parts of them exposed to light, are green. All types produce both antheridia and archegonia on the same gametophyte. Zygotes develop embryos with a foot, stem, and leaves, and these develop into mature sporophytes (figure 18.4). If the gametophyte is underground, the young sporophyte does not develop chlorophyll until it emerges into the light. Several sporophytes may be produced from a single gametophyte. A number of ground pines also reproduce asexually by means of small *bulbils* (bulb produced in the axils of leaves), each of which is capable of developing into a new sporophyte.

Members of *Selaginella,* the second major genus of living club mosses, are sometimes called *spike mosses* (figure 18.5). The approximately seven hundred species are widely scattered around

Chapter 18

Figure 18.4 Life cycle of the ground pine *Lycopodium*

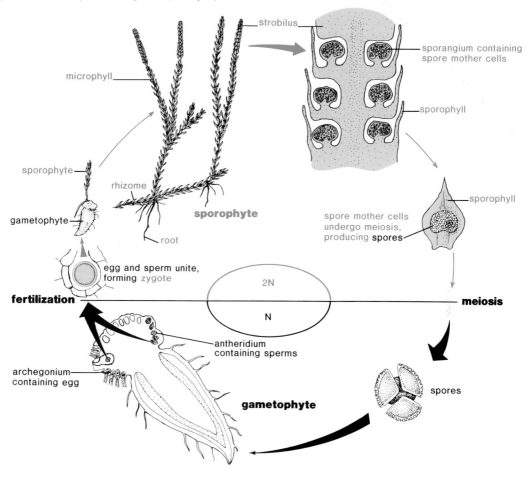

strobilus

sporangium containing spore mother cells

microphyll

sporophyll

sporophyll

sporophyte

rhizome

sporophyte

root

spore mother cells undergo meiosis, producing **spores**

gametophyte

egg and sperm unite, forming zygote

2N

fertilization

N

meiosis

antheridium containing sperms

archegonium containing egg

gametophyte

spores

Figure 18.5 *Selaginella*, a spike moss

Introduction to Vascular Plants

Figure 18.6 Life cycle of the spike moss *Selaginella*

the world in wetter areas, but they are especially abundant in the tropics. One or two are common "weeds" in greenhouses. They tend to branch more freely than ground pines, from which they differ in several respects. The two most obvious differences are (1) their leaves each have a tiny extra appendage, or tongue, called a **ligule,** on the

upper surface near the base, and (2) they produce two different kinds of spores and gametophytes.

Sexual reproduction of spike mosses begins in similar fashion to that of ground pines. Sporangia are produced on sporophylls called either *microsporophylls* or *megasporophylls* (figure 18.6). **Microsporophylls** bear sporangia containing numerous **microspore mother cells** that

Figure 18.7 Quillwort (*Isoetes*) sporophytes. (Courtesy Robert A. Schlising)

undergo meiosis, producing tiny **microspores,** whereas the sporangia of **megasporophylls** usually contain a **megaspore mother cell** that becomes four comparatively large **megaspores** upon undergoing meiosis. Each microspore has the potential to develop into a male gametophyte, which consists simply of a somewhat spherical antheridium surrounded by a sterile jacket of cells. Either 128 or 256 sperm cells with flagella are produced in each antheridium. A megaspore develops into a female gametophyte, which is also relatively simple in structure. By the time this gametophyte is mature, however, it consists of many cells, which have been produced inside the megaspore. As it increases in size, it eventually ruptures its thickened spore wall, and several archegonia are produced in the exposed seams. The development of both the male and the female gametophytes often occurs, or is at least initiated,

before the spores are released from their sporangia. Fertilization and development of new sporophytes are similar to those of ground pines.

There are about fifty species of *quillworts* (all in the genus *Isoetes*) (figure 18.7). Most are found in areas where they are at least partially submerged in water for part of the year. Their leaves (microphylls) are reminiscent of green porcupine quills, although they are not as stiff and rigid as their animal counterparts. The leaves, which are somewhat spoon-shaped at the base, are arranged in a tight spiral on a stubby stem that resembles the corm of a gladiola or a crocus. Ligules similar to those of spike mosses occur toward the leaf bases. The corms have a vascular cambium and may live for many years. Wading birds and muskrats often include the corms in their diet. The plants are generally less than 1 decimeter (4

Introduction to Vascular Plants

Figure 18.8 Life cycle of a quillwort (*Isoetes*)

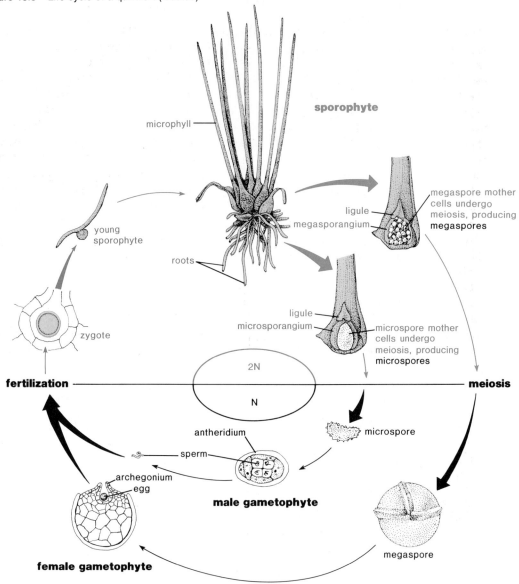

inches) tall, but one species has leaves that reach a length of 6 decimeters (2 feet). Reproduction is similar to that of spike mosses, with both types of sporangia being produced at the bases of the leaves (figure 18.8). Up to 1 million microspores may occur in a single microsporangium. Development of gametophytes does not begin until the spores have been released.

Some of the ancient and extinct relatives of the club mosses were large and treelike, reaching heights of 30 meters (100 feet) and having

Figure 18.9 Small portion of the surface of the fossil lycopod, *Lepidodendron*, showing leaf (microphyll) bases similar to those seen in modern lycopods. (Courtesy University of Illinois Paleobotany Laboratory)

Human and Ecological Relevance of Club Mosses and Quillworts

Like the whisk ferns, club mosses and quillworts are of little economic importance today. As was mentioned at the beginning of the discussion, large numbers of club moss spores produce a flash of light when ignited. This characteristic was exploited at one time in the manufacture of theatrical explosives and photographic flashlight powders. Druggists also used to mix spore powder with pills and tablets to prevent them from sticking to one another. The spore powder itself has been used for centuries in folk medicine, particularly for the treatment of urinary disorders and stomach upsets. Members of some Native American tribes used it as a talcum powder for babies, snuffed it to arrest nosebleeds, and applied it following childbirth to staunch hemorrhaging. Sometimes it was also used to stop bleeding from wounds. Club moss extracts have been used in several countries in the past to reduce fevers, but medicinal use has now largely been abandoned, partly because of undesirable side effects.

Native Americans of Washington, Oregon, and British Columbia used to experience a mild form of intoxication after chewing parts of one local species of club moss. It is reported that they became unconscious if they chewed too much.

A species of spike moss native to Mexico and to the southwestern United States is sometimes sold in stores as a novelty. Known as the resurrection plant, it shrivels and rolls up in a ball when dry, appearing to all intents and purposes completely dead, but quickly unfolds and turns green when sprinkled with water. Other spike mosses have been placed on shelves indoors without water for nearly three years and then have resumed growth when given water. Club mosses and spike mosses have been used ornamentally indoors and outdoors as ground covers. Some club mosses are

trunks up to 1 meter (3 feet 3 inches) in diameter. They were dominant members of the forests and swamps of the Carboniferous period, which reached its peak some 325 million years ago (see table 18.1). Quillworts first appeared in the Cretaceous period, about 130 million years ago. Their fossils reveal that even the oldest were remarkably similar to their present-day relatives (figure 18.9). A knowledge of the life histories and ancestry of various members of this division enhances our understanding of the development and diversity of the Plant Kingdom as a whole.

Figure 18.10 Horsetails (*Equisetum*). *A.* Unbranched species. *B.* Branched species. (*B,* Courtesy Field Museum of Natural History, Chicago)

A.

B.

spray painted and used as Christmas ornaments or in floral wreaths. Several species of *Lycopodium* have been exploited to the extent that they are now on rare and endangered species lists; they should no longer be collected.

Quillwort corms have been eaten by domestic and wild animals, waterfowl, and humans.

Division Sphenophyta—Horsetails and Scouring Rushes

Many backpackers and campers have become aware of a unique use of a relatively common and widespread genus of plants (*Equisetum*) known

as *horsetails or scouring rushes.* The plants accumulate significant deposits of silica on the inner walls of the epidermal cells of the stems and make excellent scouring material for dirty metal pots and pans. Native Americans were aware of this scouring property of the plants and used them extensively.

About twenty-five species of horsetails (the name usually applied to branching forms, which look a little like a horse's tail) and scouring rushes (unbranched forms) (figure 18.10) are distributed sporadically throughout all continents except Australia. They usually grow less than 12 decimeters (4 feet) tall, but some in the tropics and coastal redwood forests of California exceed 4.6 meters (15 feet) in height. Where branches

Chapter 18

Figure 18.11 Horsetail (*Equisetum*) stem in cross section

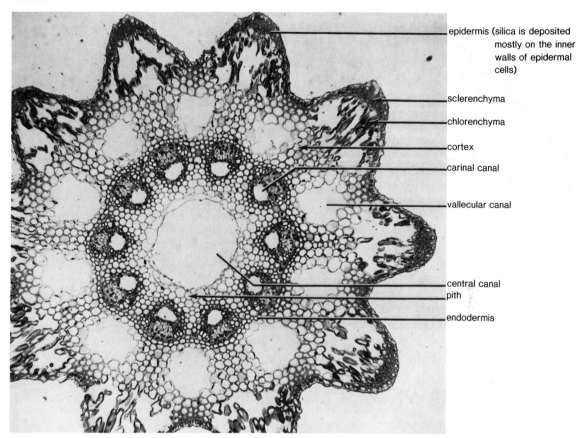

epidermis (silica is deposited mostly on the inner walls of epidermal cells)

sclerenchyma

chlorenchyma

cortex

carinal canal

vallecular canal

central canal
pith
endodermis

occur, they are normally in whorls at regular intervals along the jointed stems. Both branched and unbranched species have tiny scalelike leaves (microphylls) in whorls at the nodes. These leaves are fused together at their bases, forming a collar. They are green when they first appear, but they soon wither and bleach, and virtually all photosynthesis occurs in the stems. The stems are distinctly ribbed and exhibit well-defined nodes and internodes, with numerous stomata in the grooves between the ribs. When a stem is viewed in cross section (figure 18.11), it can be seen that the pith in the center breaks down at maturity, leaving a hollow central canal. Outside of the pith are two rings of smaller canals. The inner ring consists of canals that are aligned opposite the ribs of the stem. These are called *carinal canals* and function in water conduction. Each has a patch of xylem and phloem to the outside. The canals of the outer ring, called *vallecular canals,* contain air. They are larger than the carinal canals and are aligned opposite the "valleys" between the ribs. The stems arise from horizontal rhizomes, which, like the aerial stems, are ribbed and have regular nodes and internodes. In some species, the rhizomes may form extensive branching systems as much as 2 meters (6.5 feet) below the surface. They also have adventitious roots. Spe-

Figure 18.12 Life cycle of a horsetail (*Equisetum*) that
has separate vegetative and reproductive shoots

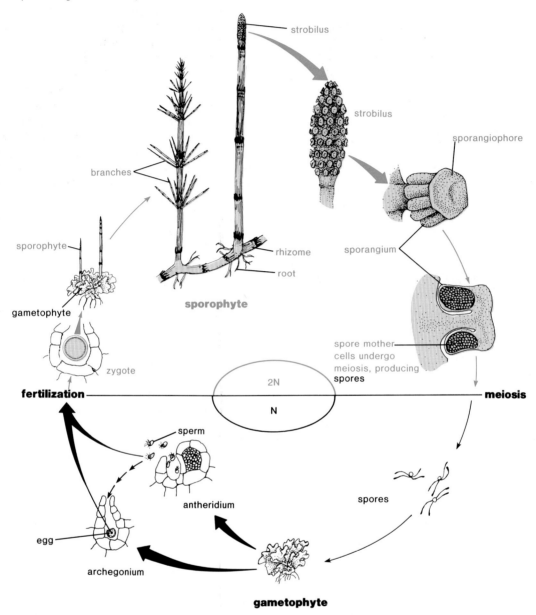

strobilus

strobilus

sporangiophore

branches

sporophyte

gametophyte

rhizome

root

sporophyte

sporangium

spore mother
cells undergo
meiosis, producing
spores

zygote

fertilization

2N

N

meiosis

sperm

spores

antheridium

egg

archegonium

gametophyte

Figure 18.13 Horsetail spores. *A.* With elaters coiled. *B.* With elaters spread.

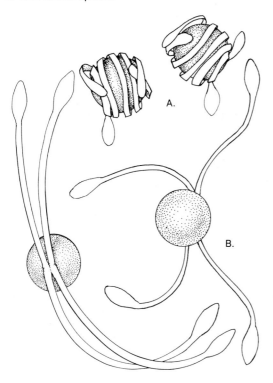

A.

B.

marks the top of a *sporangiophore,* which will have five to ten elongate *sporangia* attached at the rim. The sporangia, which surround the stalk of the sporangiophore, point inward toward the center of the strobilus to which the stalks of each sporangiophore are attached. The sporangia are not visible until maturity. At that time, the sporangiophores separate slightly, revealing the sporangia beneath, and the spores are released.

Spore mother cells in the sporangia undergo meiosis, producing the spores, which are distinctive in appearance. They are green and at one pole have four ribbonlike appendages that are slightly expanded at the tips (figure 18.13). The appendages are called **elaters** (not related to the elaters of liverworts). The elaters are very sensitive to changes in humidity, and they aid in spore dispersal. While spores are being carried aloft by a breeze, the elaters are more or less extended like wings. If they encounter a pocket of humidity, indicating a damp area below, the elaters coil around the spore, causing it to drop in an area where the probability of moisture being available for germination is increased.

Upon germination, which usually occurs within a week, spores produce pinhead-sized, lobed, cushionlike green gametophytes that seldom grow to more than 8 millimeters (0.36 inch) in diameter. Rhizoids anchor them to the surface. At first, about half of the gametophytes are male with antheridia, and the other half are female with archegonia. After a month or two, however, the female gametophytes become bisexual in most species, producing only antheridia from then on. When water contacts mature antheridia, the sperms produced within appear to be explosively ejected by sudden changes in water pressures. Several eggs on a female or bisexual gametophyte may be fertilized, and the development of more than one sporophyte is common.

Horsetails and scouring rushes belong to the single remaining genus *(Equisetum)* of a division of several different orders that flourished in the

cies of horsetails may be distinguished from one another on the basis of internal stem structure as well as by means of external features.

If stems or rhizomes are broken up by a disturbance such as a storm or foraging animals, the segments are potentially capable of growing into new sporophyte plants. Reproduction in most species usually involves a sexual process, however. In some species, special cream- to buff-colored nonphotosynthetic stems are produced from rhizomes in the spring (figure 18.12). **Strobili** (singular: **strobilus**) in the shape of small cones are produced at the tips of these special stems, or at the tips of regular photosynthetic stems in other species. These strobili are usually about 2 to 4 centimeters (0.75 to 1.5 inches) long, and their surfaces look like a honeycomb, with dovetailing hexagonal plates. Each hexagon

Introduction to Vascular Plants

373

Figure 18.14 Reconstruction of a coal age (Carboniferous) forest. (Courtesy Field Museum of Natural History, Chicago)

Carboniferous period some 300 million years ago. At that time, some members of this division, like the ancient club mosses, were treelike (figure 18.14) and reached heights of over 15 meters (50 feet), while others may have been vinelike in appearance. A number were similar to present-day horsetails in having leaves in whorls (figure 18.15), jointed stems with internal canals, and sporangiophores with sporangia hanging down from the rims.

Human and Ecological Relevance of Horsetails and Scouring Rushes

The seventh-century Romans are reported to have eaten the young strobili of field horsetails boiled like asparagus or fried after mixing with flour. Members of a number of Native American tribes peeled off the silica-impregnated outer parts of young stems and ate the inner portions either raw or boiled, while others cooked parts of the rhizomes of the giant horsetail for food. Cows, goats, muskrats, bears, and geese also have been known to eat horsetails. Hopi Indians ground dried stems of another species to flour, which they mixed with cornmeal to make bread or mush. Field horsetails have occasionally been reported to be poisonous to horses, however, and they are not recommended for human consumption unless they are cooked in three or four changes of water to eliminate or reduce their acid and drug content.

Native Americans and Asians had several other uses for these plants. One tribe drank the water from the carinal canals of the stem, and another thought the shoots were "good for the blood." It is not known if there was any basis for this latter idea, but there is an old unconfirmed report that field horsetail consumption "produces a decided increase in blood corpuscles." At least one or two species are known to have a mild diuretic effect (a **diuretic** is a substance that increases the flow of urine), and they have been used in the past in folk medicinal treatment of urinary and bladder disorders. Some have also been used as an antacid or an astringent (an astringent is a substance that arrests discharges, particularly of blood). One species was used in the treatment of gonorrhea, and others were used for tuberculosis.

At least two Native American tribes burned the stems and used the ashes to alleviate sore mouths or applied the ashes to severe burns. Members of another tribe ate the strobili of a widespread scouring rush to cure diarrhea, and still others boiled stems in water to make a hairwash for the control of lice, fleas, and mites.

At one time, the use of scouring rush stems for scouring and sharpening was widespread. They were used not only for cleaning pots and pans but also for polishing brass, hardwood furniture, and flooring and for honing mussel shells to a fine edge. Scouring rushes are still in limited use for these purposes today.

Some species of horsetails accumulate certain minerals in addition to silica. Veins of such minerals have been located beneath populations of horsetails by analyzing the plants' mineral contents. This process of analysis involves a chemical treatment of the tissues followed by the use of X-ray equipment (for the scientific names of species discussed, see appendix 1).

In the geological past, the giant horsetails and club mosses were a significant part of the vegetation growing in vast swampy areas. In some instances, the swamps were stagnant and slowly sinking, permitting the gradual accumulation of plant remains, which, because of the lack of oxygen in the water, were not readily attacked by decay bacteria. Such circumstances, over aeons of time, were ideal for the formation of coal. Today, if you section a lump of coal thinly enough and examine it under the microscope, you can still see bits of tissue and spores of plants that were living hundreds of millions of years ago. One soft coal known as *cannel* consists primarily of the spores of giant horsetails and club mosses that through the ages were reduced to carbon.

Figure 18.16 Sporophytes of a common fern with typically dissected fronds. (Courtesy Robert A. Schlising)

Division Pterophyta—Ferns

If one were to take a worldwide poll of opinion concerning ornamental plants, *ferns* undoubtedly would be at or near the top of the popularity list. Their leaves display such an infinite and aesthetically pleasing variety of form that Thoreau was once moved to state, "God made ferns to show what He could do with leaves."

The approximately eleven thousand known species of ferns vary in size from the tiny floating forms less than 1 centimeter (0.4 inch) in diameter to giant tropical tree ferns up to 25 meters (82 feet) tall. Although fern leaves, commonly referred to as **fronds,** are typically dissected and feathery in appearance (figure 18.16),

some are undivided, pleated, or tonguelike, and others resemble a four-leaf clover or grow in such a way as to form "nests." The latter ferns often accumulate enough humus in the tropics to provide food and shelter for huge earthworms that are up to 6 decimeters (2 feet) long. Since external water is essential to their sexual reproduction, ferns are most abundant in wetter tropical and temperate habitats, but a few are adapted to survival in drier areas.

As in all the plants discussed in this chapter, the sporophyte is the conspicuous phase of the life cycle (figure 18.17). It consists of the fronds, a stem in the form of a rhizome, and adventitious roots that arise along the rhizome. The fronds, regardless of their ultimate form, usually first appear tightly coiled at their tips. These **croziers,**

Figure 18.17 Life cycle of a fern

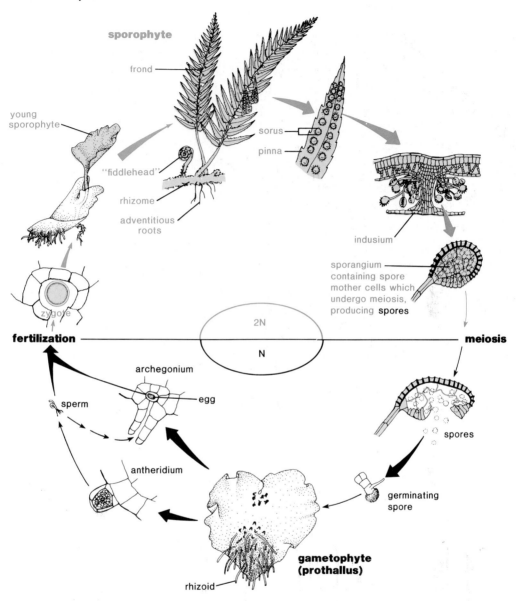

sporophyte

frond

young
sporophyte

sorus

pinna

"fiddlehead"

rhizome

adventitious
roots

indusium

sporangium
containing spore
mother cells which
undergo meiosis,
producing **spores**

zygote

fertilization

2N

N

meiosis

archegonium

egg

sperm

spores

antheridium

germinating
spore

rhizoid

**gametophyte
(prothallus)**

Figure 18.18 Croziers ("fiddleheads") of a tropical tree fern.

or "fiddleheads" (figure 18.18), then unroll and expand, revealing the blades. At maturity, the blades are frequently divided into segments called **pinnae** (singular: **pinna**), which are attached to a midrib, or *rachis*. A stalk, or *petiole,* is usually present at the base.

When the fronds have expanded, small, often circular, rust-colored patches of powdery looking material may appear on the lower surfaces of some or all of the blades. Because these patches appear similar to fungal rusts, some fern owners have thought that their plants were diseased and have turned to sprays to deal with the "problem," or have even carefully scraped the patches off.

Their appearance is normal and healthy, however, and examination with a hand lens or dissecting microscope will reveal that the patches are actually clusters of *sporangia* (figure 18.19). The sporangia may be scattered evenly over the lower surfaces of the fronds, but more often they are confined to the margins; very frequently they are found in numerous discrete clusters called **sori** (singular: **sorus**). In many species of ferns, the sori are protected by individual flaps of tissue called **indusia** (singular: **indusium**) while they are developing. As the sporangia mature, the indusium, which is often attached at its base in umbrella fashion to the frond surface, shrivels and exposes the sporangia beneath.

Chapter 18

Figure 18.19 Pinnae of fern fronds, showing some types of arrangements of sporangia. A. *Cheilanthes.* B. *Polypodium.* C. *Cystopteris.* D. *Cibotium.* E. *Davallia.* F. *Cyathea.* (*B,* Courtesy Robert A. Schlising)

margin forms a false indusium

hair

sporangium

A.

B.

midrib

indusium

C.

indusium

margin

vein

blade

sporangium

D.

F.

blade

indusium of sorus

opening to "pocket" formed by indusium over sorus

vein

E.

margin

vein

blade

young sorus

indusium splitting open revealing sporangia

Figure 18.20 Release of spores from a fern sporangium. *A.* Intact sporangium. *B.* Spores being ejected as the sporangium splits and the annulus snaps back.

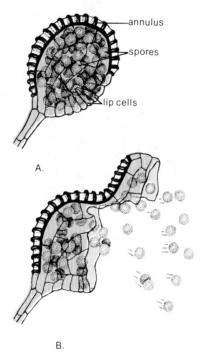

A.

B.

Most of these microscopic sporangia are on stalks and look something like tiny transparent baby rattles with a conspicuous row of heavy-walled brownish cells along the edge. This row of cells, somewhat resembling a millipede, is called an **annulus.** It functions in catapulting spores out of the sporangium with a distinct snapping action influenced by moisture changes in the cells (figure 18.20).

Spore mother cells undergo meiosis in the sporangia, producing spores usually totaling either 48 or 64 per sporangium. Sporangia of some of the primitive adder's tongue ferns may have up to 15,000 spores, however, and the number of sporangia is often so great that it has been estimated that a single beech fern plant will produce a total of 50 million spores. In certain aquatic or

Figure 18.21 Fern prothalli growing on the surface of a clay pot

amphibious ferns, two kinds of spores are produced; single large megaspores are produced in some sporangia, whereas numerous tiny microspores are produced in others. The vast majority of fern species, however, produce only one kind of spore.

After the spores have been dispersed, initially by being flung out of their sporangia and then carried by wind, relatively few end up in habitats suitable for their survival. Such habitats include shady wet ledges or moist soil (the spores can also easily be encouraged to germinate on damp clay flowerpots in the home or greenhouse). Those that do come to rest in favorable locations germinate, producing little "Irish valentines," or **prothalli** (singular: **prothallus**), as the green heart-shaped gametophytes of these and other seedless vascular plants are more properly called (figure 18.21). These structures, which often curl slightly at their edges, may be 5 to 6 millimeters (0.25 inch) in diameter and are readily seen without a microscope. They are only one cell thick, except toward the middle where they are slightly thicker. Rhizoids are produced in this central area on the lower surface, with antheridia interspersed among them. Archegonia are also produced, usually closer to the notch of the heart-shaped gametophyte. The archegonia are somewhat flask-shaped, with curving necks

Chapter 18

Figure 18.22 Artist's restoration of ancient preferns. A. *Rhynia.* B. *Psilophyton.*

A.   B.

Figure 18.23 Fossil ferns. (Courtesy Field Museum of Natural History, Chicago)

that protrude above the surface, while the antheridia are more spherical and often elevated above the surface on short stalks. From thirty-two up to several hundred sperms with several to numerous flagella may be produced in a single antheridium.

Only one zygote develops through an embryo stage into a young sporophyte on any prothallus, regardless of the number of eggs that may be fertilized. This sporophyte usually has smaller, simpler fronds during its first season of growth, but these give way to typical full-sized fronds as growth continues.

Fossil remains of ferns and plants thought to be ancestors of ferns abound in ancient deposits. The ancestors of ferns, sometimes referred to as *preferns* (figure 18.22) are found in Devonian formations (see table 18.1) estimated to be 375 million years old. Most resemble ferns in habit but otherwise are more like whisk ferns than true ferns in that they lack broad fronds. Well-preserved fossils of tree ferns related to large tropical ferns of the present day are found in geological strata dating back to between 250 and 320

million years ago (figure 18.23). These became so abundant during the latter part of the Carboniferous period some 300 million years ago that this stage in geological history used to be referred to as the "Age of Ferns." However, the twentieth-century discovery that a number of these fernlike plants produced seeds on their fronds raised questions as to the true nature of the fossils, and the name "Age of Ferns" was dropped.

Human and Ecological Relevance of Ferns

In some areas of the United States, it is unusual to find a house without at least a Boston fern growing within. Ferns make ideal houseplants in that many of them are adapted to growing in low light, and most are not as susceptible as other plants to aphids, mites, mealybugs, and similar

Figure 18.24 Orchid plant growing on a slab of tree fern bark

pests. Outdoors, ferns are equally popular as ornamentals. Invariably some eventually become subjects for the landscape artist's brush or the natural history photographer's camera.

A number of ferns also have practical value. Commercial growers of the brilliantly colored anthuriums in Hawaii and elsewhere have found that native tree ferns provide the ideal amount of shade and other environmental conditions needed for bringing their flowers to perfection. Tree fern "bark" and rhizome "bark" of certain species has long been a favorite medium of orchid growers (figure 18.24). Its texture is eminently suited to the growth of the orchids' aerial roots, and as the bark slowly decomposes, rainwater trickling over its surface picks up nutrients that are particularly appropriate for these plants. The demand for fern "bark" for orchids has exceeded the supply for a number of years, and it has become very expensive on the market.

As the young "fiddleheads" of many species of ferns unroll, a dense covering of hairs is visible on the petiole and rachis. In the past, the silky hairs of some of the larger tropical tree ferns (figure 18.25) were stripped and used for upholstery, pillow, and mattress stuffing. During the late 1800s over 1,900 metric tons (2,094 tons) of this material was shipped from Hawaii to the mainland, and if it were not for the eventual substitution of alternative materials, these magnificent plants might have been totally destroyed. Some tropical hummingbirds use these hairs along with scales of other ferns to line their nests, and at one time Polynesians used them for embalming their dead. The trunks of tree ferns have been used in the construction of small houses in the tropics. Parts of one Hawaiian species of small tree fern and the fronds of an Asian fern yield red pigments used for dyeing cloth.

The bracken fern, which is distributed worldwide, has been used and even cultivated for human food for many years, particularly in Japan and New Zealand, even though it has also long been known to be mildly poisonous to livestock. Recent research in both Europe and Japan has shown conclusively that bracken fronds fed to experimental animals produce intestinal tumors, and because of this the consumption of these fronds for food should be actively discouraged.

Chapter 18

Figure 18.25 Tropical tree ferns in Java. (Photo by W. H. Hodge)

Indigenous peoples of many areas where ferns occur used the rhizomes and young fronds of a number of species of ferns for food in the past. Native Americans often baked the rhizomes of sword ferns, lady ferns, and others in stone-lined pits, removed the outer layers, and ate the starchy inner material. Similarly, native Hawaiians ate the starchy core of their tree ferns as emergency food. In Asia, the oriental water fern is still sometimes grown for food in rice paddies and used as a raw or cooked vegetable. In Malaya, a relative of the lady fern is frequently used as a vegetable.

Uses of ferns in folk medicine abound. They have been used in the treatment of diarrhea, dysentery, rickets, diabetes, fevers, eye diseases, burns, wounds, eczema and other skin problems, leprosy, coughs, stings and insect bites, as a poison antidote, for labor pains, constipation, dandruff, and a host of other maladies. The male fern, which is more common in Europe than the United States, contains a drug that is effective in expelling intestinal worms (e.g., tapeworms). Its use for this purpose dates back to ancient times, and it is still occasionally used, although synthetic medicines have now largely replaced it. The licorice fern, which was used by Native Americans of the Pacific Northwest in the treatment of sore throats and coughs, was also used as a flavoring agent and a sugar substitute. Members of a tribe in California chewed stalks of goldback ferns to quell toothaches and snuffed a liquid made from the fronds of the bird's-foot fern to arrest nosebleeds.

The fronds of bracken and other ferns have been used in the past for thatching houses. Anyone who has placed such fronds in compost piles knows that they break down much more slowly than the leaves of other plants. Bracken fronds are still occasionally used as an overnight bedding base by fishermen and hunters. A substance extracted from these fronds has been used in the preparation of chamois leather, and the rhizome is used in northern Europe in the brewing of ale.

The chain fern, which has large fronds up to 2 meters (6.5 feet) long, has two flexible leathery strands in the petiole and rachis of each frond. Native Americans and others have gathered the fronds for many years to strip these strands for use in basketry and weaving. They do so by gently cracking the long axis with stones to expose the strands, which are then easily removed. The glossy black petioles of the five-finger fern have also been long used in intricate basketry patterns by Native Americans. In Southeast Asia, the climbing fern, whose rachis may attain lengths of up to 12 meters (40 feet), is still a favorite material (when available) for the weaving of baskets.

One of the floating water ferns, which forms tiny plants little bigger than duckweeds, is spread over wide areas where the climate is relatively mild. It sometimes forms such dense floating mats that it is believed to suffocate mosquito larvae, which periodically need to reach the surface for air; it is thus known as the mosquito fern in some

Introduction to Vascular Plants

Table 18.1 Geologic Time Scale

Era	Period	Age	Earliest Evidence of Plants	Duration (millions of years)	Million Years Before Present
Cenozoic	Quaternary	Age		2	
		of			2
	Tertiary	Modern		63	
		Seed			65
Mesozoic	Cretaceous	Plants		65	
			grasses		130
	Jurassic	Age	flowering plants	45	
		of			175
	Triassic			45	
		Ancient			220
Paleozoic	Permian	Seed	ginkgoes	50	
			conifers		270
	Carboniferous	Plants		80	
		Age of			
		Spore-bearing			350
	Devonian		seed ferns	50	
		Plants	lycopods		
			horsetails		400
	Silurian	Age		40	
					440
	Ordovician	of		60	
					500
	Cambrian	Bacteria	mosses	100	
		and	marine algae		600
	Precambrian	Algae		1700?	2300?

parts of the United States. This same fern frequently has blue-green algae living symbiotically in cavities between cells. The blue-green algae fix nitrogen, and it has been shown experimentally that plants without these organisms do not grow as well as those with them (scientific names of ferns discussed are given in appendix 1).

Fossils

Several references to **fossils** have been made in this and preceding chapters, and other references occur in the chapters that follow. We have become increasingly aware of worldwide limitations of energy sources, and we now frequently hear references to "fossil fuels." Exactly what is a fossil, and how did it come into existence?

Originally the word *fossil* was applied to anything unusual found in rocks, but its use is now more or less confined to any recognizable prehistoric object of an organic nature or its impression, which has been preserved from past geological ages in the earth's crust (table 18.1). Such a definition includes teeth marks, borings, impressions of footprints and tracks, dung deposits, and deposits of chemicals, which are evidence of the activities of algae and bacteria. Age itself does not make a fossil. Human remains thousands of years old that have been found in caves and tombs are not regarded as fossils, but it is conceivable that other remains of the same age could be preserved in rocks and regarded as fossils.

Fossils are formed in a number of different ways. The conditions for their formation almost

Chapter 18

PLATE 3. *A.* Crustose lichens. *B.* Earth stars. *C.* A bracket fungus. *D.* Common woods mushrooms. (*B*, Courtesy Perry J. Reynolds)

A

B

C

D

PLATE 4. *A.* Cinnamon fern. *B.* Maidenhair fern.
C. Holly fern. *D.* 'Ama'uma'u, a small Hawaiian tree
fern. (*C*, Courtesy Perry J. Reynolds)

A

B

C

D

invariably include the accumulation of sediments in an area where plants, animals, or other organisms are present or to which they have been transported. Such areas are found in swamps, oceans, lakes, or other bodies of water. Hard parts such as wood or bones are more likely to be preserved than soft parts. Other environmental factors such as the presence of salt or antiseptic chemicals or stagnation resulting in low oxygen content of the water, all favor fossil formation. Another factor enhancing the chances of organic material becoming fossilized is quick burial. Such burial might result if an organism was unable to avoid a rain of ash from a volcanic eruption or was trapped in a cave or sudden sandstorm. Other forms of quick burial occur when an organism falls into quicksand, a body of water, or asphalt.

Descriptions of several common types of fossils follow.

Molds, Casts, Compressions, and Imprints

After silt or other sediment has buried an object and hardened into rock, the organic material may be slowly washed away by water percolating through the pores of the rock. This leaves a space that may then be filled in with silica deposits. If only a space is left, it is called a *mold;* if it is filled in with silica, it is called a *cast.* Artificial casts of plaster or wax compounds may be made from molds of the original objects. When objects such as leaves are buried by layers of sediment, the

sheer weight of the overlying material may compress them to as little as 5% of their original thickness. When this occurs, all that is left is a thin film of organic material and an outline showing some surface details. Virtually no preservation of cells or other internal structure takes place. Such fossils, which are very common, are called *compressions* (figure 18.26). The image of a compression, like the details of a foot or a hand pressed into wet cement, is called an *imprint.* Coal is a special type of compression, often involving different plant parts that are thought to have been subjected to enormous pressures after the fallen plants slowly accumulated in a swamp.

Petrifactions

Petrifactions (figure 18.27) are uncompressed rocklike materials in which the original cell structure has been preserved. About twenty different mineral substances, including silica and the salts of several metals, are known to bring about petrifaction. At one time it was believed that the process occurred through the replacement of the plant parts by minerals in solution, one molecule at a time. It is now believed, however, that chemicals in solution infiltrate the cells and cell walls, where they crystallize and harden, preserving the original material permanently. Petrifactions can be studied by cutting thin sections with a diamond or carborundum saw and polishing the material with extremely fine grit powder until it is thin enough for light to pass through. Then it can be examined with a compound microscope. Another simpler method of studying petrifactions involves etching the cut surface with a dilute acid and then applying a plastic or similar film. As soon as the film hardens, it can be peeled off. Such *peels* display lifelike microscopic details of the surface with which they were in contact. They are commonly used by *paleobotanists* (botanists who study fossil plant materials) in their research.

Coprolites

Coprolites are the fossilized dungs of prehistoric animals and humans. They may contain pollen grains and other plant and animal parts that provide clues to the food and feeding habits of past organisms and cultures.

Unaltered Fossils

Some plants or animals fell into bodies of oil or water that, because of substances present or a nearly total lack of oxygen, did not permit decay to occur. Some animals died in snowfields, and their bodies where permanently frozen. In such instances, preservation in the unaltered state occurred. *Unaltered fossils,* particularly frozen ones, are, however, very rare.

Summary

Four divisions of seedless vascular plants are recognized: Division Psilophyta members (whisk ferns) have neither leaves nor roots, but have stems that fork evenly; Division Lycophyta members (club mosses and quillworts) have mostly photosynthetic microphylls (leaves with one vein whose traces have no leaf gaps); Division Sphenophyta members (horsetails and scouring rushes) have ribbed stems and scalelike nonphotosynthetic leaves; and Division Pterophyta members (ferns) have megaphylls (leaves with more than one vein and traces associated with leaf gaps).

Whisk ferns are the simplest of all living vascular plants, consisting of evenly forking green stems that have small protuberances called enations, but no leaves; roots are also lacking. The stem contains a central cylinder of xylem and phloem. Spores produced in sporangia formed along the stems germinate into tiny saprophytic gametophytes over the surface of which archegonia and antheridia are scattered. After fertilization the zygote develops a foot and a rhizome; upright stems are produced when the rhizome separates from the foot. An Australian relative of whisk ferns, *Tmesipteris,* has leaflike appendages. Fossil plants resembling whisk ferns have been found in Silurian geological formations.

Club mosses presently have living members of two genera. Ground pines *(Lycopodium)* develop sporangia in the axils of sporophylls (sporangium-bearing leaves); spore mother cells in the sporangia undergo meiosis, producing spores that germinate into gametophytes containing both archegonia and antheridia. Several sporophytes may be produced from one gametophyte. Spike mosses *(Selaginella)* have a ligule (tiny tonguelike appendage) on each microphyll; they also produce microsporophylls with microsporangia in which microspores are formed, and megasporophylls with megasporangia in which fewer, larger megaspores are formed. The microspores develop into male gametophytes with antheridia and the megaspores develop into female gametophytes with archegonia.

Quillworts have quill-like microphylls that arise from a cormlike base. The corms have a cambium that remains active for many years.

Some fossil relatives of club mosses were large dominant members of the forests and swamps of the Carboniferous period.

Club moss spores have been used for flash powder, medicinal purposes, as talcum powder,

Chapter 18

and to staunch bleeding. The plants themselves have been used as ornamentals, novelty items, Christmas ornaments, and for intoxicating purposes.

Horsetails and scouring rushes *(Equisetum)* accumulate deposits of silica in their epidermal cells and have made good scouring material for Native Americans and others. They occur in both unbranched and branched forms. The stems are jointed and ribbed and have tiny scalelike leaves in whorls at each joint; they are also hollow in the center and contain two rings of canals. Carinal canals (inner ring) conduct water; vallecular canals (outer ring) contain air. The stems arise from rhizomes that branch extensively below the surface of the ground. In some species, nonphotosynthetic stems with conelike strobili are produced in the spring; in other species strobili are produced on photosynthetic stems. The strobili are composed of sporangiophores that support sporangia in which spore mother cells undergo meiosis, producing spores with ribbonlike elaters that are sensitive to changes in humidity. Equal numbers of male and female gametophytes are produced, but female gametophytes may become bisexual after a month or two; the development of more than one sporophyte from a gametophyte is common.

Ancient relatives of horsetails were the size of trees when they flourished in the Carboniferous period of 300 million years ago.

Horsetails have been used for food after the parts containing silica were removed, but they are not recommended for human consumption. They have also been used medicinally as a diuretic, as an astringent, and in the treatment of venereal disease and tuberculosis; other uses include as a hairwash, a mineral indicator, and a metal polish. Cannel coal consists primarily of spores of giant horsetails that were reduced to carbon.

Fern leaves (fronds) are typically dissected and feathery in appearance but vary greatly in form. They usually first appear as croziers (tightly coiled immature fronds) that unroll and expand. Rust-colored patches of sporangia appear on the lower surfaces of fronds. The sporangia may be arranged in various ways but commonly occur in discrete clusters called sori, which may be protected by flaps of tissue called indusia. Each sporangium has a row of heavy-walled cells, the annulus, that functions in catapulting mature spores out of the sporangium after they have developed from spore mother cells following meiosis. Heart-shaped gametophytes called *prothalli* develop after spores germinate; the prothalli contain both archegonia and antheridia; only one zygote develops into a sporophyte. Preferns, the ancestors of ferns, are found in Devonian deposits estimated to be 375 million years old, and ferns (especially tree ferns) became so abundant during the latter part of the Carboniferous period that this era was once referred to as the "Age of Ferns," but the discovery of seeds on some of them raised questions about relationships and the term was dropped.

Ferns are used as ornamentals, a source of "bark" for growing orchids and other plants, a source of stuffing materials for bedding, in tropical construction, as food, and in numerous folk medicinal applications. Other uses include as a source of material for basketry and weaving, as an ingredient in brewing ale, and as an ingredient in the preparation of chamois leather. One floating fern, which forms dense mats, is believed to suffocate mosquito larvae.

Fossils, which are recognizable prehistoric objects of an organic nature, are formed in different ways. Molds, casts, compressions, and imprints are formed when material buried by silt or other sediment has hardened into rock and the organic material has slowly been washed away by water. Petrifactions are uncompressed rocklike materials in which the original cell structure has been preserved. Coprolites are fossilized dungs that may contain pollen grains and other plant and animal parts. Unaltered fossils are those of plants or animals that may have fallen into bodies of oil or water or snowfields and were not subjected to decay.

Review Questions

1. What basic features do the ferns and their relatives possess that distinguish them from any organisms thus far studied?

2. How does the gametophyte of a whisk fern differ from that of a true fern?

3. Which of the fern relatives have significantly functional leaves? In those without conspicuous leaves, how are the carbohydrate needs of the plants met?

4. How does one distinguish among ground pines, spike mosses, and quillworts?

5. How did the ancient ground pines differ from those of the present day?

6. How do the spores and the female gametophyes of horsetails differ from those of any other plants studied thus far?

7. What is the location and function of carinal canals?

8. In your opinion, which have the most human relevance today: club mosses and horsetails of the present or those of the geological past? Why?

9. Define crozier, rachis, pinna, indusium, prothallus, prefern.

10. Diagram the life cycle of a typical fern.

11. Summarize present and past human use of ferns.

12. How are fossils formed, and what different types are recognized?

Discussion Questions

1. Would you assume that there is any significance to the fact that both the sporophytes and the gametophytes of whisk ferns branch in the same manner?

2. Do spike mosses have any advantage over ground pines in producing two kinds of spores and gametophytes?

3. Some gametophytes of fern relatives develop underground, while others develop at the surface. If you were to be a gametophyte, which would you prefer? Why?

4. After looking at the internal structure of a horsetail stem, can you suggest a function for the silica in the ribs?

5. How would we be affected if all ferns were to become extinct in a few years?

Additional Reading

Bierhorst, D. W. *Morphology of vascular plants.* New York: Macmillan Co.; 1971.

Cobb, B. *A field guide to the ferns and their related families.* Boston: Houghton Mifflin Co.; 1977.

Grillos, S. J. *Ferns and fern allies of California.* Berkeley: University of California Press; 1966.

Hubbard, D. H. *Ferns of Hawaii National Park.* Volcano, HI: Hawaii National Park and Hawaii Natural History Association; 1952.

Jermy, A. C.; Crabbe, J. A.; Thomas, B. A., eds. *The phylogeny and classification of the ferns.* New York: Academic Press, for the Linnaean Society of London; 1973.

Lloyd, R. M. Ethnobotanical uses of California pteridophytes by western American Indians. *American Fern Journal* 54: 76–82; 1964.

Mickel, J. *How to know the ferns and fern allies.* Pictured Key Nature Series. Dubuque, IA: Wm. C. Brown Company Publishers; 1979.

Puri, H. S. Indian pteridophytes used in folk remedies. *American Fern Journal* 60: 137–43; 1970.

Sporne, K. R. *The morphology of pteridophytes. The structure of ferns and allied plants.* 4th ed. Atlantic Highlands, NJ: Humanities Press, Inc.; 1975.

Turner, N. J. *Food plants of British Columbia Indians.* Part 1: *Coastal peoples.* Victoria: British Columbia Provincial Museum; 1975.

Verdoorn, F., ed. *Manual of pteridology.* The Hague, Netherlands: Martinus Nijhoff; 1938.

Introduction
Seed Plar
Gymnosper

Introduction

Some of the giant tree fern
graceful leaves held abov
superficially resemble
who has sat at the
one of its massi
been abrupt
ence bet
palm
be

Overview

The chapter begins by exploring the differences between ferns and seed plants, and then briefly relates the geological history of gymnosperms. Since conifers are the most significant of the four divisions of gymnosperms having living representatives, pine trees are used to discuss the life cycle of a typical gymnosperm. Features of the leaves, roots, and stems of pines are discussed, and the life cycle is portrayed. Other conifers such as yews, podocarps, junipers, and redwoods are given brief mention, and a short discussion of other gymnosperms such as cycads, ginkgoes, *Ephedra, Gnetum,* and *Welwitschia* follows. The chapter concludes with a digest of the human and ecological relevance of gymnosperms, with particular emphasis on the conifers.

Some Suggested Student Learning Goals

1. Describe typical male and female conifer strobili, what features they have in commo and how they differ.

2. Describe what distinguishes the four divisions of living gymnosperms from one another.

3. List the modifications of pine leaves that adapt them to a harsh environment.

4. Indicate which of the following structures occur in the life cycle of a pine tree: archegonia, antheridia, eggs, sperms with flagella, male and female gametophytes, sporophytes that are dependent on gametophytes for nutrition, integuments, vessels, spore mother cells, embryo, pollen grains.

5. Explain the function of each of the following: resin canals, albuminous cells, mycorrhizal fungi, nucellus, generative cell, megaspore.

6. List a use for each of at least ten different conifers.

7. Identify three major representatives of the Division Gnetophyta.

s, which have large ... an unbranching trunk, ... coconut palms. Anyone ... base of a coconut palm when ... ve fruits has fallen, however, has ... y reminded of a fundamental differ- ... ween ferns and coconut palms. The ..., which are flowering plants, and cone- ... aring plants such as pine trees reproduce primarily by means of **seeds,** whereas the ferns do not.

The oldest known seeds were produced by plants that appeared late in the Devonian period, more than 350 million years ago. Seeds provided a significant adaptation for plants that had invaded the land. Unlike spores, seeds have a protective seedcoat and a supply of food for the embryo, which may be capable of lying dormant through lengthy periods of freezing weather, drought, and in some instances may even survive fire. This survival value of seeds undoubtedly played a major role in seed plants becoming the dominant vegetation on earth today.

The first seed plants were so fernlike in appearance that, as indicated in chapter 18, they were originally classified as ferns. When fossils with obvious seeds on the fronds were discovered, however, these *pteridosperms* ("seed ferns") were reclassified as *gymnosperms*.

The name **gymnosperm** is derived from two Greek words: *gymnos,* meaning naked, and *sperma,* a seed. The name refers to the exposed nature of the seeds, which are produced on the surface of sporophylls or similar structures instead of being enclosed within a fruit as they generally are in the flowering plants (figure 19.1). The seed-bearing sporophylls are often spirally arranged in female strobili (cones), which along with smaller male strobili that produce *pollen grains* develop on the sporophyte. The female gametophyte is produced within an **ovule** containing a fleshy, nutritive diploid **nucellus** that is itself

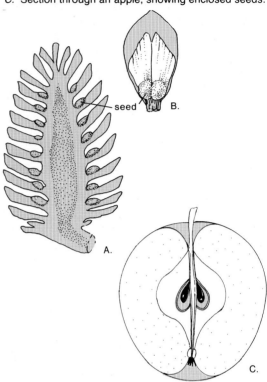

Figure 19.1 Comparison between exposed (gymnosperm) seeds and enclosed (angiosperm) seeds. *A.* Exposed seeds on a female (woody) cone. *B.* Single female cone scale with two seeds. *C.* Section through an apple, showing enclosed seeds.

enclosed within one or more layers of diploid tissue. These additional layers of tissue constitute an **integument,** which becomes a **seedcoat** after the fertilization and development of an embryo has occurred.

Gymnosperm sporophytes are primarily trees and shrubs, although a few are vinelike in habit. The gametophytes are even more reduced proportionally than they are in ferns and their relatives. Unlike the gametophytes of members of the plant kingdom discussed to this point, they are not free-living but develop within structures that, as indicated, are enclosed by parts of the sporophyte.

Four divisions of gymnosperms have living representatives at this time. The **conifers** (Division Coniferophyta) constitute the largest and most significant group by far, totaling some 575 species. Pine trees, firs, spruces, hemlocks, redwoods, cedars, and others belong to this division. Fossils of some conifers extend back 290 million years to the late Carboniferous era. Other gymnosperms include the cycads (Division Cycadophyta), which have palmlike leaves but produce their seeds in cones; ginkgoes (Division Ginkgophyta), which have smaller, fan-shaped leaves and seeds enclosed in a fleshy covering; and gnetophytes (Division Gnetophyta), whose wood contains vessels—a structural element unknown in other gymnosperms.

Division Coniferophyta—Conifers

The largest genus of conifers, *Pinus* (pines) has over 100 living species. They are the predominant trees in the vast coniferous forests of the northern hemisphere. They have also been planted extensively in the southern hemisphere; but only the Merkus pine occurs there naturally, and its distribution barely extends south of the equator. They include the world's oldest known living organisms, the bristlecone pines, which occur in the White Mountains of eastern central California and in the Snake range on the central Nevada-Utah border. Some trees still standing are about forty-six hundred years old, and one that was, unfortunately, cut down in 1964 was found to have been about forty-nine hundred years old.

Pine leaves, which are needlelike, are arranged in clusters or bundles of two to five leaves each (a handful of species have as many as eight or as few as one to a cluster). Regardless of the number of leaves, each cluster (*fascicle*) forms a cylindrical rod if the leaves are held together (figure 19.2). Pine trees often occur in areas where the topsoil is frozen for a part of the year, making it difficult for the roots to obtain water. In addition, the leaves may be subjected to high winds and bitterly cold temperatures. Accordingly, they have several modifications that enable them to withstand harsh environments. A **hypodermis,** consisting of a layer or two of thick-walled cells, occurs just below the epidermis. The epidermis itself is coated with a thick cuticle. The stomata,

Figure 19.2 Cluster (fascicle) of pine needles in cross section, showing how the needles form a cylindrical rod. (Photomicrograph by Carolina Biological Supply Company)

instead of being at the surface, are recessed or sunken in small cavities. The veins and their associated tissues are surrounded by an endodermis similar to that of roots (see chapter 5), and the mesophyll cells do not have the obvious air spaces typical of the spongy mesophyll of the leaves of flowering plants (see figures 7.4 and 7.8). Conspicuous **resin canals** occur in the mesophyll. These resin canals, which occur throughout other parts of the plant as well, consist of tubes lined with special cells that secrete resin. This substance, which is aromatic and antiseptic, prevents the development of fungi and deters certain insects from attacking the tree. In other conifers, resin canals are apparently produced primarily in response to injury.

Unlike the xylem of woody dicots (see chapter 6), that of pine trees and many other gymnosperms consists primarily of tracheids and contains no vessel elements or fibers. Because of the absence of fibers, pine wood is said to be *soft* while that of broadleaf trees is described as *hard.*

Figure 19.3 Portion of a cross section through the stem of a pine tree showing distinct annual rings

cork
resin canal
phloem
vascular cambium
xylem
pith

The degree of hardness varies considerably from species to species, however. In most conifers, the xylem is proportionately very wide, with distinct annual rings (figure 19.3) as a result of a relatively rapid rate of growth during the growing season. Resin canals are formed both vertically and horizontally throughout various tissues. The bark, which includes the secondary phloem, often becomes relatively thick. It frequently attains a width of 7.5 centimeters (3 inches) or more, and in extreme cases, such as the giant redwood (which is in a different family of conifers), it may even reach 6 decimeters (2 feet) in extent. The phloem does not have companion cells, but certain other cells called **albuminous cells** apparently perform the same function.

Roots of pine trees are invariably associated with mycorrhizal fungi. In fact, pine seedlings germinated in sterilized soil do not grow well at all until the fungi are introduced or permitted to develop. The roots of pine trees adjacent to one another often interweave. New England pioneers made use of this characteristic in eastern white pines when clearing land. They would tip over the stumps of felled trees with the roots still attached and use them for fences; the fences survived for many years.

Reproduction

As in the spike mosses, quillworts, and a few of the ferns, two kinds of spores are produced in pines. *Microspores* occur in microsporangia that develop in pairs toward the bases of papery or

Chapter 19

Figure 19.4 Cluster of male pine cones. (Photograph by Carolina Biological Supply Company)

Figure 19.5 Pollen grains of a pine tree, as seen with the aid of a microscope. The pairs of dark objects attached to each pollen grain are air sacs that provide added buoyancy.

Figure 19.6 Longitudinal section through a pine ovule

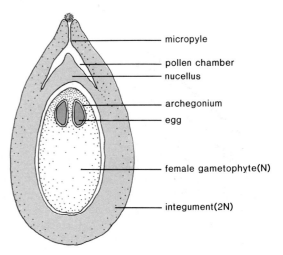

- micropyle
- pollen chamber
- nucellus
- archegonium
- egg
- female gametophyte(N)
- integument(2N)

inch) long. They become shriveled and spent within a few weeks and then fall from the trees. Microspore mother cells in the microsporangia each undergo meiosis, producing the four haploid *microspores*. These then develop into **pollen grains,** each consisting of four cells and a pair of external air sacs that look something like tiny water wings (figure 19.5). These sacs give the pollen grains added buoyancy and contribute to the fact that the grains frequently are carried great distances by the wind. Pine trees produce pollen grains in astronomical numbers. The grains accumulate as a fine yellow dust on cars, shrubbery, or anything else in the vicinity, and they often form an obvious scum on pools and puddles.

Megaspores are produced in *ovules* at the bases of the *female cone* scales. The female cones (also called "seed cones"), or strobili, are much larger than the male cones, attaining lengths of 6 decimeters (2 feet) in sugar pines, and weighing as much as 2.3 kilograms (5 pounds) in Coulter pines. They have woody scales, with inconspicuous bracts between them, arranged in a spiral around an axis. They are produced on the upper branches of the same tree on which the male cones appear. The ovules (figure 19.6),

membranous scales arranged in a spiral or in whorls around an axis, forming a strobilus or *male cone.* Male cones (also called "pollen cones"), which are usually produced in the spring in clusters of up to fifty or more toward the tips of the lower branches (figure 19.4), are commonly not more than 1 to 2 centimeters (about 0.5 to 0.75

Introduction to Seed Plants: Gymnosperms

which occur in pairs toward the base of each scale, are larger and more complex than their microsporangial counterparts in the male cones. Each contains a multicellular nutritive tissue called the **nucellus,** which is surrounded and enclosed by a thick, layered **integument.** The integument has a channel or pore called a **micropyle** that opens toward the center of the cone. One of the integument layers later becomes the **seedcoat** of the seed. A single spore mother cell within the nucellus of each ovule undergoes meiosis, producing a row of relatively large *megaspores.* All except one of the megaspores soon degenerate. The remaining one slowly develops over a period of months into a female gametophyte, which ultimately may consist of several thousand cells. Toward the end of the gametophyte development, two to six *archegonia* become differentiated at the end facing the micropyle. Each archegonium contains a single large *egg.* Often when one examines stained thin sections of a pine ovule under the microscope, only one archegonium or egg may be seen, or the micropyle may appear to be missing. This frequently occurs because the knife that cut the section may not have sliced precisely through the middle of all the structures.

Female cones, which are green at first, commonly take two seasons to mature into the woody structures that are familiar to nearly everyone. During the first spring the cone scales separate, and pollen grains carried by the wind sift down between the scales. There they catch in sticky drops of fluid (pollen drops) oozing out of the micropyles. As the fluid evaporates, the pollen is drawn down through the micropyle to the top of the nucellus. After pollination the scales grow together, protecting the developing ovule. It is not until about a month after pollination that meiosis occurs and the megaspores are formed. The female gametophyte is not mature with archegonia for more than a year after that. Meanwhile the pollen grain forms an outgrowth called a **pollen tube,** which slowly digests its way through the nucellus to the area where the archegonia develop. While the pollen tube is growing, two of the original four cells in the pollen grain enter it. One of these, called the **generative cell,** divides and forms two more cells, called the *sterile cell* and the *spermatogenous cell.* The latter divides again, producing two male gametes, or *sperms.* These sperms, unlike those encountered up to this point in the Plant Kingdom (and also in some other gymnosperms), have no flagella. This germinated pollen grain, with its pollen tube and two sperms, constitutes the *male gametophyte.* Notice that no antheridium has been formed (figure 19.7).

About fifteen months after pollination the pollen tube reaches the archegonium. The contents of the pollen tube are discharged, and one sperm unites with the egg, forming a zygote. The other sperm and remaining cells of the pollen grain degenerate. The sperms of other pollen grains present may unite with the eggs of other archegonia, and each zygote begins to develop into an embryo. This may be likened to the development of fraternal twins or triplets in animals. At a later stage, an embryo may divide in such a way as to produce the equivalent of identical twins or quadruplets in animals. Normally, however, only one embryo completes development. While this development is occurring, one of the layers of the integument hardens, becoming a seedcoat. A thin membranous layer of the cone scale becomes a "wing" on each seed. This wing aids in the seed's dispersal, which may also be helped along by squirrels and other animals breaking open the cones (see figure 19.7 for a diagram of the life cycle of a pine).

In other species, such as the knobcone pine and the jack pine, the cones remain on the tree with the scales closed until they are seared by fire. Sometimes these cones are slowly engulfed by tissue and buried as the cambium increases the girth of the branch. Seeds of such engulfed cones have been reported to germinate after they have been dug out of the stem.

Figure 19.7   Life cycle of a pine tree

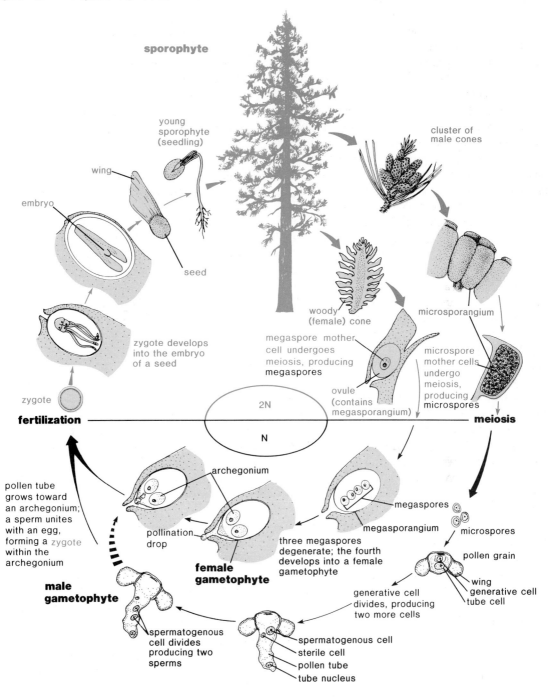

sporophyte

young
sporophyte
(seedling)

wing

embryo

seed

cluster of
male cones

woody
(female) cone

microsporangium

zygote develops
into the embryo
of a seed

megaspore mother
cell undergoes
meiosis, producing
megaspores

microspore
mother cells
undergo
meiosis,
producing
microspores

ovule
(contains
megasporangium)

2N

N

zygote

fertilization

meiosis

archegonium

megaspores

megasporangium

microspores

pollen tube
grows toward
an archegonium;
a sperm unites
with an egg,
forming a zygote
within the
archegonium

pollination
drop

three megaspores
degenerate; the fourth
develops into a female
gametophyte

pollen grain

wing
generative cell
tube cell

**female
gametophyte**

generative cell
divides, producing
two more cells

**male
gametophyte**

spermatogenous
cell divides
producing two
sperms

spermatogenous cell
sterile cell
pollen tube
tube nucleus

Introduction to Seed Plants: Gymnosperms

Figure 19.8 One-seeded cones of a podocarp (*Podocarpus*). Note the fleshy covering and large appendage at the base of each cone.

Other Conifers

Not all conifers produce woody female cones with conspicuous scales, nor do all produce both male and female cones on the same tree. In the yews and California nutmeg, for example, ovules are produced singly at the tips of short axillary shoots. Each ovule is at least partially surrounded by a fleshy, cuplike covering called an **aril.** In yews, this is bright red and open at one end, giving the "cone" the appearance of a small red hors d'oeuvre olive with its stuffing removed. The fleshy "cones" are produced only on female trees, while the pollen-bearing strobili, or male cones, are produced only on male trees. Podocarps, which are conifers of the southern hemisphere, are widely planted as ornamentals in regions with milder climates. Their one-seeded "cones" are similar to those of yews, but they are not open at one end and an additional larger structure is at the base of the otherwise yewlike, fleshy-coated seed (figure 19.8).

Chapter 19

The scales of the female cones of junipers tend to be fleshy at maturity, so that they look more like berries than cones. Juniper pollen, as well as that of a number of other conifers, does not have air sacs. Redwood female cone scales are flattened at the tips and narrow at the base, where they do not overlap one another as do pine cone scales.

The two species of living redwoods found in California are both renowned for their size, height, and longevity. It is not uncommon for coastal redwoods to attain a height of 90 meters (295 feet), and one tree in Humboldt County, California, with a height of 111.6 meters (366.2 feet) is believed to be the tallest tree in the world. The other species, usually referred to as "Big Trees" or giant redwoods and confined to the eastern slopes of California's Sierra Nevada range, does not become quite as tall as the coastal redwood but exceeds it in total size. The "General Sherman" tree in Sequoia National Park, for example, is 31 meters (101.5 feet) in circumference at the base and over 24 meters (79 feet) in circumference 1.5 meters (5 feet) above the ground; it weighs an estimated 5,594 metric tons (6,167 tons). The total timber in this single tree exceeds 600,000 board feet, sufficient to build more than 75 five-room houses or enough to make 20 billion toothpicks. It is over 3,500 years old.

Other Gymnosperms

Members of the other three divisions of gymnosperms with living representatives are not as numerous or as well known as the conifers and outwardly do not resemble them at all. In fact, some of them look more like leftover props from a science-fiction movie set. They include the *cycads,* the *ginkgoes,* and the *gnetophytes* (pronounced nee-toe-fytes).

Division Cycadophyta—Cycads

Cycads are slow-growing plants of the tropics and subtropics that look like a cross between a tree fern and a palm. They have unbranched trunks

Figure 19.9 Cycad that is producing male strobili

that grow to more than 15 meters (50 feet) tall in a few species, with a crown of large pinnately divided leaves. Extinct members of this division, known as cycadeoids, were abundant during the Mesozoic era. Several of the approximately one hundred known living species are presently facing extinction. Their life cycles are similar to those of pine trees; their sperms, however, have numerous flagella. The male and female strobili, which are sometimes massive, are produced on separate plants (figure 19.9). The female strobili of some species are covered with feltlike or woolly hairs and can become as much as 1 meter (3 feet, 3 inches) long.

Figure 19.10 Ginkgo. *A.* Mature tree. (Courtesy Robert A. Schlising) *B.* Seeds and leaves. (Courtesy Field Museum of Natural History, Chicago)

A.

B.

Division Ginkgophyta—Ginkgoes

There is only one living species of *Ginkgo* (figure 19.10), whose name is derived from Chinese words meaning "silver apricot." The fossil record indicates *Ginkgo* and other members of its family (Ginkgoaceae) were once widely distributed, especially in the northern hemisphere. Despite isolated reports to the contrary, there are doubts that ginkgoes now exist anywhere they have not been cultivated, and the plant has often been called a living fossil.

Ginkgoes are frequently referred to as maidenhair trees because of the similarity of the notched, broad, fan-shaped leaves to the individual pinnae of maidenhair ferns. They are widely cultivated in the United States and are popular street trees in some areas. The leaves, which are produced in a spiral on short, slow-growing spurs, have no midrib or prominent veins. Instead, hairlike veins branch dichotomously (fork evenly) and are relatively uniform in their width. They are deciduous and turn a bright golden yellow before abscission in the fall. The life cycle of ginkgoes is also similar to that of pines. The male and female reproductive structures are produced on separate trees, and the sperms have flagella. Mature seeds are enclosed in a fleshy covering and resemble a small plum. These seeds have a rank odor and are irritating to the skin of some individuals. Because of this, male trees, which are propagated from cuttings, are preferred for ornamental purposes.

Chapter 19

Figure 19.11 Joint fir (*Ephedra*). *A.* Single plant. *B.* Branch bearing male strobili. *C.* Branch bearing female strobili.

Figure 19.12 *Gnetum.* (Courtesy Field Museum of Natural History, Chicago)

Division Gnetophyta—Gnetophytes

The gnetophytes comprise a heterogeneous group of some seventy species of plants distributed among three genera. They are unique among the gymnosperms in having vessels in the xylem. The genus *Ephedra,* whose members are sometimes called *joint firs* (figure 19.11), accounts for more than half of the species. These plants are somewhat shrubby, inhabit drier regions, and superficially resemble horsetails. The leaves, which are produced in twos or threes at a node, are very tiny and turn brown soon after they appear. The slightly ribbed branches, which are green when they are young and carry on most of the photosynthesis, are often whorled. A tubular extension, like the neck of a tiny bottle, extends into the air from the integument of an ovule at the time of pollination. Sticky fluid oozes out of this

extension, which constitutes the micropyle, and airborne pollen catches in it. Male and female strobili may be produced on the same plant or on different ones, depending on the species.

Most of the remaining species in this division are in the genus *Gnetum* (figure 19.12), which has not been given an English common name. Its members, which have broad leaves similar to those of flowering plants, occur in the tropics of South America, Africa, and Southeast Asia. Most are vinelike, but the best known species is a tree that grows up to 10 meters (33 feet) tall.

The third genus, *Welwitschia* (figure 19.13), has only one species, which is confined to the temperate Namib and Mossamedes deserts of southwestern Africa. Here the average annual rainfall is only 2.5 centimeters (1 inch), and some years it does not rain at all. The plants apparently survive much of the time on fog and dew water absorbed through the surfaces of the leaves.

Welwitschias are also truly extraordinary plants
in appearance. The stem, which rises only a short
distance above the surface, is in the form of a large
shallow cup that tapers at the base into a long
taproot. At maturity the plants, which may live
to be 100 years old, have a crusty, barklike covering on the surface of the stem cup. The stems
may exceed 1 meter (3 feet, 3 inches) in diameter. Throughout the life of the plant only two
leaves are produced. These are wide and straplike, each with a meristem at the base. The meristems constantly add to the length of the leaves,
but as the leaves flap about in the wind they become tattered and split, and they wear off at the
tips so that they seldom exceed 2 meters (6.5 feet)
in actual length. Male and female strobili, which
are produced on separate plants, are on axes that
emerge from the axils of the leaves so that they
appear to be growing around the rim of the stem
cup.

Chapter 19

Human and Ecological Relevance of Gymnosperms

What do we plant when we plant the tree?
We plant the ship, which will cross the sea.
We plant the mast to carry the sails;
We plant the planks to withstand the gales—
The keel, the keelson, the beam, the knee;
We plant the ship when we plant the tree.

What do we plant when we plant the tree?
We plant the houses for you and me.
We plant the rafters, the shingles, the floors,
We plant the studding, the lath, the doors,
The beams, the siding, all parts that be;
We plant the house when we plant the tree.

What do we plant when we plant the tree?
A thousand things that we daily see;
We plant the spire that out-towers the crag,
We plant the staff for our country's flag,
We plant the shade, from the hot sun free;
We plant all these when we plant the tree.

Henry Abbey

As a group, the gymnosperms are second only to the flowering plants in their impact on our daily lives. Space does not permit a detailed account of all they contribute, but the following sections provide an overview of some of their uses, past and present.

Conifers

In the early 1970s the late author–naturalist Euell Gibbons filmed a series of television commercials in which he mentioned uses of several wild plants for food. In one of the commercials, he rhetorically inquired if his audience had ever eaten a pine tree and added, "many parts are edible." The edibility of parts of many conifers was known to Native Americans long before Europeans set foot on the North American continent. In fact, early explorers found large numbers of pine trees stripped of their bark; for centuries these inner parts (phloem, cambium) had been used for emergency food. The Adirondack Mountains of New

York are believed to have received their name from a Mohawk Indian word meaning "tree-eater," in reference to Native American use of the inner bark of eastern white pine trees. This material (specifically the phloem) contains sugars that make it sweet to the taste. Some tribes ate the material raw, some dried it and ground it to flour, and others boiled it or stored dried strips for winter food. Early settlers in New England candied strips of eastern white pine inner bark, and both they and local Native Americans drank a tea made of the needles, which are rich in vitamin C, to prevent scurvy.

The seeds of nearly all species of pines are edible, but those of western North America include the larger and better tasting varieties. The protein content of those analyzed generally ranges between 15 and 30%, with much of the remainder consisting of oils. California Indians relished digger pine seeds in particular, but even the small seeds of ponderosa pine were eaten raw or made into a meal for soups and bread. Cones of pinyons were collected by tribes of the Southwest and thrown on a fire to loosen the seeds. These were then pounded and made into cakes or soup. The latter was often fed to infants. In Siberia, peasants obtain a nutritious oil by crushing the seeds of the Siberian white pine, but its use has declined since corn and cottonseed oils became available. In Italy and other parts of Europe, "pignolias," the seeds of the stone pine, are cooked in stews and soups. They are also used in cakes and cookies, and some are exported to the United States for this purpose. Many of the so-called nuts used by commercial American bakers in cakes and confectionery, however, are really seeds from the east Himalayan chilghoza pine. Other sources include the Mexican stone pine and a few pinyons.

Eastern white pines were favored for use as masts in sailing vessels. In colonial days, the royal surveyors marked certain trees for the use of the

Figure 19.14 Southern longleaf pine from which resin is being extracted. Part of the bark was cut away and a receptacle to collect the resin for commercial use was placed at the bottom of the cut area. (Courtesy FRP Company)

Crown, and severe penalties were imposed on colonists who ignored the ban on the use of any white pine not growing on private land. Eastern white pine wood contains less resin than that of other species and was extensively used for crates, boxes, matchsticks, furniture, flooring, and paneling. By the end of the nineteenth century, eastern white pines, which originally occurred over vast tracts of the northeastern United States and Canada, had been decimated by wholesale logging done with no thought to conservation. White pine blister rust also took its toll. Although new growth is now being encouraged, most white pine lumber used today originates in large stands of western white pine in the Pacific Northwest. In California, smog has severely damaged ponderosa and other native pines. The U.S. Forest Service has been experimenting with Afghanistan pine, a smog- and drought-resistant pine from Russia and adjacent areas, as a replacement for native trees. In tests thus far growth rates have been very rapid. Rapid growth is a desirable commercial feature, since considerably more timber can be produced in the same time needed to obtain it from slower-growing species.

The resin produced in the resin canals of conifers is a combination of a liquid solvent called *turpentine* and a waxy substance called *rosin.* When a conifer tree is wounded or damaged by insects, resin usually covers the area, sometimes trapping the insects. Out in the air, the turpentine evaporates quickly, leaving a protective layer of rosin, which prevents water loss and fungal attacks. Both turpentine and rosin are very useful products, and a large industry centered in the southern United States (figure 19.14) and in the south of France is devoted to their extraction and refinement. They are often referred to as *naval stores,* a term that originated when the British Royal Navy used large quantities for caulking and sealing their sailing ships. Most naval stores (and a third or more of the lumber used in the United States) today come from a group of southern yellow pines, particularly slash pine. Pitch pine,

which was also a source of naval stores before slash and other yellow pines proved more profitable, was used in the past for the construction of water wheels for mills. The wood of this pine was also used as fuel for steam engines, as it produces considerable heat when it burns.

Turpentine is considered a premier paint and varnish solvent, whereas rosin is used by musicians on violin bows. Baseball pitchers use rosin to improve their grip on the ball, and batters apply pine tar to the handles of bats to minimize slippage. In the past, pine pitch was used by Native Americans for patching canoes, and it is believed that Noah's Ark was sealed with pitch from aleppo pine. Pine resin was used for purifying wine in the first century A.D., and Greeks still add it to certain wines today. *Colofonia* is the Spanish word for a type of resin that was produced in abundance by the numerous Monterey pines around the old Spanish capital of Monterey. The early California priest, Padre Arroyo, suggested during the first half of the nineteenth century that California received its name from this Spanish word.

The huge kauri pines of New Zealand, which are in a genus different from that of true pines, are the source of a mixture of resins called *dammar,* which are used in high-quality colorless varnishes. This was also the resin originally used in the manufacture of linoleum. The substance, also called *amber,* is obtained primarily in fossil form from former or present kauri pine forest areas. It occurs as lumps of translucent material with a deep orange-yellow tint. Some of the lumps weigh up to 45 kilograms (100 pounds). The supply, which was at its peak at the turn of the century, is now nearing exhaustion. Remarkably lifelike preservations of prehistoric insects millions of years old have occurred in amber (figure 19.15).

Other products refined from resin are used in the manufacture of menthol for cigarettes (menthol also occurs naturally in members of the mint family), floor waxes, printer's ink, paper coatings, varnishes, and perfumes.

Figure 19.15 Prehistoric insect preserved in Baltic amber. (Courtesy Field Museum of Natural History, Chicago)

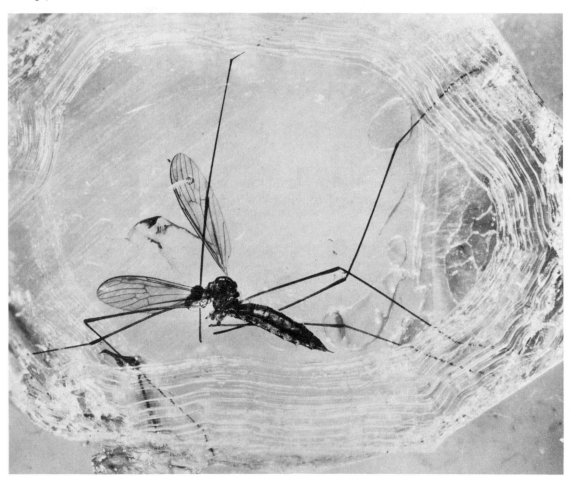

The chief source of pulpwood for newsprint (figure 19.16) and other paper in North America is the white spruce. Enormous quantities are used every day. A single midweek issue of a large metropolitan newspaper may use an entire year's growth of 50 hectares (123 acres) of these trees, and that amount may double for weekend editions. A large American publishing company, in an attempt to find ways of reducing paper consumption in the United States, tried trimming 2.5 centimeters (1 inch) from the width of all rolls of toilet paper in their building facilities. They found that the employees still used the same number of rolls per month as they had previously. From this it was calculated that if all rolls of toilet paper were similarly trimmed in width, 1 million trees would be saved each year in the United States alone.

The roots of the white spruce are quite pliable and were used in split form by Native Americans for lashing canoes and for basketry. Spruce

Chapter 19

beer, brewed from young twigs and leaves with a sugar source such as honey or molasses added, was once used as a remedy for scurvy. Resin of white, red, black, and sitka spruces was used as a type of chewing gum by Native Americans, who sometimes hardened it slightly in cold water. Europeans who have tried it report that it has to be of the right consistency to be enjoyable, since it behaves like unhardened taffy if it is too soft and is bitter if it is too hard.

The tracheids of spruces have spiral thickenings on the inner walls. These apparently are responsible for giving the wood a resonance that makes it ideal for use as soundboards for musical instruments (figure 19.17). Sitka spruce of the Pacific Northwest produces a strong resilient wood that is a favorite material of manufacturers of light airplanes.

Figure 19.17 Violin with soundboard made from red spruce. (Courtesy Scherl & Roth, Inc.)

Introduction to Seed Plants: Gymnosperms

Larches, which along with the dawn redwood and bald cypress are exceptions to the rule that conifers are evergreens, have some of the toughest of all conifer woods. Fence posts of larch are known to last twenty years; in the southern and southwestern United States, posts of juniper wood and bald cypress last even longer, some remaining useable for forty to fifty years or more. The resin of the western larch has been used in the manufacture of baking powder, and the European larch is the source of a special type of turpentine.

There are about forty species of true firs, which are widely used in the construction, plastic, and paper industries, as ornamentals, and as Christmas trees. The balsam fir produces on its bark blisters containing a clear resin. This resin, known as Canada balsam, was used in the past for cementing optical lenses and is still used for making permanent mounts on microscope slides. It has medicinal properties, too, and was used by New England colonists in sore throat medications. Douglas fir, found in the mountains of the West, is not a true fir. In the Pacific Northwest, it grows into giant trees that are second only to the redwoods in size. It is probably the most desired timber tree in the world today. The wood, which is strong and relatively free of knots as a result of rapid growth with less branching than most other conifers, is heavily used in plywoods and is a major source of large beams; a useful wax is extracted from the bark. Exploitation has nearly eliminated virgin stands, but large numbers of new trees are being grown in managed forests.

Coastal redwoods are also prized for their wood, which contains substances that inhibit the growth of fungi and bacteria. The wood is lightweight, strong, and soft, but it splits easily. It is used for construction, furniture, posts, greenhouse benches, and for many other purposes. California wineries use it extensively for wine barrels. The giant redwoods are so huge that a double bowling alley was built on the log of the

Figure 19.18 Bald cypress trees in a southern swamp. Note the "knees" protruding above the surface in the foreground. (Courtesy Field Museum of Natural History, Chicago)

first specimen cut down. They are no longer logged and serve now almost exclusively as tourist attractions.

Wood of the bald cypress, found in southern swamps, is like that of the redwood in being very resistant to decay. It is used for railroad ties, coffins, and shingles and was used in the past for guttering. The trees are well known for their "knees," which are tapering growths from the roots that rise above the level of the water (figure 19.18). At one time it was widely believed that these were a means of admitting oxygen to the roots, but this is now in doubt. They are favored for making knickknacks such as wall ornaments and lamp bases. The leaves of the bald cypress yield a red dye.

Chapter 19

A dull red dye can be obtained from the younger bark of the eastern hemlock, which is also a source of tannins for shoe leather. The wood of these small trees contains exceptionally hard knots, which can chip an axe blade. It sputters and throws out sparks freely when placed on a fire. Native Americans made a poultice for scrapes and cuts by pounding the inner bark. British Columbia Indians used to scrape out the cambium of both the western and mountain hemlocks for food.

Eastern white cedar, also known as arborvitae, is a favorite ornamental in temperate areas. The wood is pliable, and several Native American tribes used it for canoes. The Atlantic or southern white cedar was the first tree to be used for the construction of pipes for pipe organs in North America. During World War II old logs of this species found in a swamp in New Jersey were milled and used in the construction of patrol torpedo boats.

The fleshy red aril surrounding the seeds of yews is sweet and edible, but the seeds themselves and other parts of the plants are definitely poisonous. The wood, which is tough and resilient, is favored for making bows. A red or purple dye can be obtained from the bark and roots. Podocarps, two species of which are valuable timber trees in New Zealand, have edible seeds.

The wood of incense cedars is used in the manufacture of venetian blinds and pencils. Red cedar wood is also used for pencils, as well as for cedar chests, closet lining, fence posts, and cigar boxes. It was used at one time in Germany for smoking hams. An aromatic oil used in floor-sweeping compounds is extracted from red cedar wood, and the "berries" of this and related junipers are widely eaten by birds. Many Native American tribes used the berries and inner bark as survival food during bleak winters. Some roasted the dried berries and brewed a beverage from them. Berries of the dwarf juniper are used to flavor gin. Some authorities indicate that the word *gin* may have been derived from *genievre*, the French word for juniper berry.

Other Gymnosperms

Florida arrowroot starch was once obtained from the extensive cortex and pith of a species of cycad whose northernmost distribution occurs in Florida. Before it could be used for human consumption, however, a poisonous substance had to be leached out. Since cycads grow too slowly to make continued preparation of the arrowroot starch profitable, the practice has been abandoned. Today cycads that are seen outside of their natural habitats are being grown primarily for ornamental or educational purposes. In Louisiana, however, the large compound leaves are used in Palm Sunday religious services. Some species, which are nearing extinction in the wild, may soon be known only in botanical gardens and conservatories.

Despite the foul smell of the fleshy integument of seeds of *Ginkgo*, the starchy food reserves of the seeds themselves are edible. In the Orient, the substance is widely used for food, either boiled or roasted, and is to be found imported in canned form in Chinese food stores of large metropolitan areas in the United States.

In the southwestern United States, joint firs (*Ephedra*) are grazed by livestock, and the leaves and stems are still brewed into "Mormon tea." To offset a slight bitterness, a teaspoon of sugar, honey, or jam is added to each cup of tea. Native Americans and pioneers used a concentrated version of the tea in the treatment of venereal diseases. They also ground the seeds into flour from which a bitter bread was made. The drug *ephedrine,* which is widely used in the treatment of asthma and other respiratory problems, still is extracted from a Chinese species, but most now in use is synthetically produced. The Chinese used ephedrine medicinally more than 4,700 years ago.

One species of *Gnetum* is cultivated in Java for its shoots, which are cooked in coconut milk and eaten for food. Fibers from the bark are made into a rope (see appendix 1 for scientific names of species discussed in this chapter).

Summary

Gymnosperms generally have exposed seeds produced on sporophylls arranged around an axis, thus forming a strobilus, or cone; the female (seed) cones and smaller male (pollen) cones are produced on the sporophytes. The female gametophyte develops within a nucellus that is enclosed in an integument. This integument later becomes the seedcoat of a seed after fertilization and development of an embryo has occurred. Four divisions of living gymnosperms are recognized: conifers (Division Coniferophyta), cycads (Division Cycadophyta), ginkgoes (Division Ginkgophyta), and gnetophytes (Division Gnetophyta).

Pines are the most numerous conifers; they have needlelike leaves arranged in clusters of two to five. The leaves have several modifications that permit the plants to withstand harsh environments. Modifications include thick-walled cells in a hypodermis, a thick cuticle, sunken stomata, and resin canals. Resin canals, which occur throughout the plants, are tubes lined with special cells that secrete resin, an aromatic substance that inhibits fungi and certain insect pests. Pine xylem contains no vessel elements or fibers, and is therefore relatively soft. The phloem lacks companion cells but has albuminous cells that apparently perform the same function. Pine roots are associated with mycorrhizal fungi, which are apparently essential to normal development of the plants.

Two kinds of spores are produced. Microspores are produced in papery male cones that, in turn, are produced in clusters toward the tips of lower branches. The microspores, which are produced as a result of spore mother cells undergoing meiosis in microsporangia that develop in pairs toward the base of the male cone scales, become four-celled pollen grains and are produced in huge numbers. Megaspores are produced in ovules at the bases of female cone scales.

The integument of the ovule has a pore called the micropyle. A spore mother cell within the nucellus undergoes meiosis, producing megaspores. One megaspore develops into a female gametophyte; the remainder degenerate. The mature female gametophyte contains archegonia; and during the spring before archegonia mature, pollen grains sift down between the cone scales and catch in a sticky fluid oozing from the micropyles. Each pollen grain produces a pollen tube that digests its way down to the vicinity of the developing archegonia and two of the original four cells in the pollen grain migrate into the tube as it grows. One of the two cells (the generative cell) divides and produces a sterile cell and a spermatogenous cell that itself divides, producing two sperms. The pollen tube reaches an archegonium about fifteen months after pollination, and one sperm unites with the egg, forming a zygote. The zygote develops into an embryo of a seed that has a membranous wing formed from a layer of the cone scale. Some conifers produce fleshy or berrylike "cones."

Cycads superficially resemble palm trees with unbranched trunks and crowns of large, pinnately divided leaves. They have strobili and life cycles similar to those of pine trees, but their sperms, unlike those of pines, have numerous flagella. Ginkgoes have small fan-shaped leaves with evenly forking veins; their life cycle also is similar to that of pines. Their seeds, which are enclosed in a fleshy covering, have a rank odor at maturity. Gnetophytes have diverse forms, but they all have vessels in their xylem. Half the species are in the genus *Ephedra*, whose members have jointed stems and somewhat resemble horsetails. *Gnetum* species have broad leaves and occur in the tropics, primarily in vinelike form. *Welwitschia* has only one species, which is confined to southwest African deserts. Its stem is in the form of a shallow cup with straplike leaves that extend from the rim; basal meristems on the leaves constantly add to their length.

The seeds and inner bark of pine trees are edible, and a tea has been made from the leaves.

Eastern white pine stems were favored for use as masts for sailing vessels and were used extensively for crates, furniture, flooring, paneling, and matchsticks. Western white pine is the source of most such lumber today. Resin from pines is the source of turpentine and rosin; turpentine is used as a solvent and rosin is used by musicians and by baseball players. Dammar from kauri pines is used in colorless varnishes; amber is fossilized kauri resin. Resin is also used in floor waxes, printer's ink, paper coatings, perfumes, and in the manufacture of menthol.

White spruce is the chief source of newsprint; it was also used for basketry and canoe lashing by Native Americans, with molasses or honey for treating scurvy, and in brewing a beer. Spruce resin was used for a type of chewing gum. The wood is used as soundboards for musical instruments and in the construction of aircraft. Larch and juniper woods are used for fence posts. Firs are used in the construction, paper, ornament, and Christmas tree industries. Douglas fir is probably the most desired timber tree in the world today. Coastal redwoods are also prized for their wood, which is resistant to fungi and insects. Bald cypress wood, used for coffins and shingles, is also resistant to decay. A dye and tannins are obtained from the eastern hemlock; Native Americans used parts of hemlocks for poultices and for food. Eastern white cedar's pliable wood was used for canoes, and that of the Atlantic cedar was used for construction of pipes for pipe organs. Yew wood is used for making bows. Podocarps of New Zealand have edible seeds. Incense cedar wood is used for cedar chests, cigar boxes, pencils, and fence posts. Juniper berries are used to flavor gin and were used by Native Americans for food and a beverage.

Arrowroot starch was once obtained from a cycad; *Ginkgo* seeds are edible; Mormon tea is brewed from the leaves and stems of joint firs (*Ephedra*), which, in the past, were also a source of the drug ephedrine and a venereal disease treatment. One *Gnetum* species is cultivated in Java for food.

Review Questions

1. What is a gymnosperm? How is it distinguished from any other kind of organism?

2. What is the difference between a seed and a spore?

3. How are the leaves of pine trees different from those of broadleaf flowering plants?

4. What is a resin canal, and what is its function? Where are resin canals found?

5. How do pine trees differ in their reproduction from ground pines and ferns?

6. How do pollen grains differ from spores or sperms?

7. Which conifers discussed in this chapter do not have woody female cones?

8. If you had samples of leaves of a pine tree, a cycad, a ginkgo, a joint fir, and a welwitschia all together, how could you tell them apart?

9. What parts of a pine tree are considered edible?

10. What is resin? Discuss some of its uses, past and present.

Discussion Questions

1. Ginkgoes and cycads have broad leaves, whereas those of pine trees are needlelike. Can you suggest any significance of this in terms of the climates and habitats involved?

2. If no distinction was made at the level of Kingdom between plants and animals, what would be the equivalent, if any, of sporophyte and gametophyte in humans?

3. Both bristlecone pines and redwoods live to be thousands of years old. What do you suppose makes this possible?

4. Most of the virgin stands of conifers in North America are now gone, and others will be gone soon. Much of what has been harvested is being replaced with new growth, sometimes with hybrids and nonnative plants. Our forests are essential to our economy as we know it. If you had the power to change the way we manage and exploit our natural forest resources, what would you do differently, assuming you did not want to damage the economy?

5. If money were no object and you wished to landscape your yard primarily with gymnosperms, realistically what would you include, taking into account your particular geographical area? Why?

Additional Reading

Bierhorst, D. W. *Morphology of vascular plants.* New York: Macmillan Co.; 1971.

Bold, H. C.; Alexopoulos, C. S.; Delevoryas, T. *Morphology of plants and fungi.* 4th ed. New York: Harper and Row, Publishers; 1980.

Foster, A. S.; Gifford, E.M., Jr. *Comparative morphology of vascular plants.* 2d ed. San Francisco: W.H. Freeman and Co.; 1974.

Marx, D. S.; Dugdale, C. B. *A modern American herbal.* Cranbury, NJ: A. S. Barnes and Co.; 1973.

Mirov, N. T.; Hasbrouck, J. *The story of pines.* Bloomington: Indiana University Press; 1976.

Sporne, K. R. *Morphology of gymnosperms. The structure and evolution of primitive seed plants.* Atlantic Highlands, NJ: Humanities Press, Inc.; 1967.

Weiner, M. A. *Earth medicine-earth foods: plant remedies, drugs, and natural foods of the North American Indians.* 2d ed. New York: Macmillan Co.; 1980.

Flowering Pl

Explorers in remote reaches o
tains in Bolivia occasiona
nary sight seen by few pe
years—a rare species
ure 20.1). There ar
but what makes
family so unu
magnitude
somewh
havin
alo

Overview

This chapter begins by calling attention to differences between angiosperms and gymnosperms, and then discusses the exceptional diversity of form and habit of the flowering plants. The parts of a flower are described and some modifications are noted. The chapter continues with a description of the parallel development of the gametophytes in the anthers and ovules, leading up to pollination, and then describes some details of fertilization and the development of a seed. Trends of specialization and classification in flowering plants are discussed, and a brief comparison between dicots and monocots is made.

A section on flower preservation deals with simple herbarium techniques and practice, and it ends with a discussion of the three-dimensional drying of flowers in a box. The chapter closes with a word of caution to plant collectors concerning unnecessary depletion of native floras.

Some Suggested Student Learning G

1. Describe the basic d between angiosperm gymnosperms.

2. Distinguish an annual from a biennial or a perennial.

3. Describe the parts of a typical flower and indicate the function of each.

4. Diagram the life cycle of a flowering plant, indicating shifts from haploid to diploid cells or tissues and vice versa.

5. Indicate the features that distinguish monocots from dicots.

6. List major trends of specialization in the flowering plants.

7. List the steps and equipment involved in the preparation of herbarium specimens.

...f the Andes Moun-
...ly view an extraordi-
...ople during the past 100
...of *puya* plant in flower (fig-
...e many rare plants, of course,
...this member of the bromeliad
...sual is the exceptional manner and
...of its flowering. These particular puyas
...t resemble *agaves* (century plants),
...a cluster of large basal leaves with spines
...g the margins. They grow for about 150 years,
...lowly producing new leaves and storing food, but
they do not flower until the last year of their long
existence. Then during the final months something unknown triggers a dramatic change.
Within the space of a few short weeks, an enormous flowering stalk develops. This ultimately
reaches a height of 10 to 12 meters (33 to 40 feet)
and becomes 2.5 meters (8 feet) wide, producing
a total of some eight thousand individual flowers.
When flowering and fruiting are completed, the
puya dies.

These rare puyas are but one of about 240
thousand known species of flowering plants that
comprise by far the largest and most diverse of
the divisions of the Plant Kingdom. These flowering plants are called **angiosperms.** They have
organs and tissues similar to those of the gymnosperms, but they differ in having their ovules
and seeds *enclosed* within ovaries and fruits instead of being exposed as they are in the gymnosperms.

All angiosperms are placed in the Division
Anthophyta (see chapter 13 for comments on
other classifications of flowering plants), which is
divided into two large classes, the Dicotyledonae
(dicots) and the Monocotyledonae (monocots).
Members of the two classes are distinguished
from one another on the basis of the features listed
in table 20.1.

Table 20.1 Some differences between dicots and monocots

Dicots	Monocots
1. Seed with two cotyledons	1. Seed with one cotyledon
2. Flower parts mostly in multiples of four or five	2. Flower parts in multiples of three
3. Leaf with a distinct network of veins	3. Leaf with more or less parallel veins
4. Vascular cambium, and frequently cork cambium, present	4. Vascular cambium and cork cambium absent
5. Vascular bundles in a ring	5. Vascular bundles scattered
6. Pollen grain mostly with three apertures	6. Pollen grain mostly with one aperture

Figure 20.2 Two species of duckweeds floating on water. The larger plants are about 4 millimeters (3/16 of an inch) in diameter.

Slightly less than three-fourths of all flowering plant species are dicots. Dicots include many annual plants and virtually all flowering trees and shrubs. Monocots, which are primarily herbaceous, include plants that produce bulbs (e.g., lilies), grasses and related plants, orchids, irises, and palms; they are believed to have developed from ancestors derived from primitive dicots.

Division Anthophyta— Flowering Plants

The plants of members of this division range in size from tiny duckweeds only 0.5 to 0.7 millimeter (0.02 to 0.03 inch) in diameter (figure 20.2) to the huge *Eucalyptus* trees of Tasmania, which rival the redwoods in total volume. The flowers themselves vary in their dimensions from the nearly invisible specks of the duckweeds to the enormous *Rafflesia* flowers of Indonesia, which are up to 1 meter (3 feet, 3 inches) in diameter and weigh up to 9 kilograms (20 pounds) each (figure 20.3). The *Rafflesia* flowers are produced on inconspicuous plants that lack chlorophyll and parasitize tropical vines. Other unrelated flowering plants are also parasitic. Dodders, for example, occasionally cause serious crop losses as they twine about their hosts and send haustoria into the xylem and phloem, intercepting food and water (see figure 5.12). Broomrapes also parasitize a variety of plants, as do mistletoes. The latter produce some chlorophyll and only partially depend on their hosts for food. Still others, such as the beautiful snowplant (plate 8A) and some of the orchids, are saprophytes. The vast majority of flowering plants, however, manufacture their food independently via the process of photosynthesis.

Figure 20.3 *Rafflesia* flower that is nearly one meter (3 feet 3 inches) in diameter. (Photo by W. H. Hodge)

The habitats of flowering plants are as varied as their form. They occur in both fresh and salt water, in the cracks and crevices of rocks, in deserts, jungles, in frigid arctic regions and in areas where the temperatures regularly soar to 45°C (113°F) in the shade—almost anywhere they can receive their basic needs of light, moisture, and a minimal supply of minerals. A species of chickweed survives at an altitude of 6,135 meters (over 20,000 feet) in the Himalaya Mountains, and fumitory plants in the same area flower even when the night temperatures plummet to −18°C (0°F). Some go through their life cycles dangling from wires or other plants (figure 20.4), and one Australian orchid species even flowers underground.

Flowering plants can complete their life cycles in less than a month, or, as noted, they may take as long as 150 years. Those that complete the cycle in a single season are called **annuals.**

Figure 20.4 Festoons of Spanish moss hanging from trees in a Florida swamp. Spanish moss is a nonparasitic flowering plant that goes through its life cycle suspended from other plants or objects such as wires. This plant should not be confused with lichens, which may also hang suspended from other objects. (USDA photo)

Biennials take two years to complete the cycle, while **perennials** may take several to many years to go from a germinated seed to a plant producing new seeds. Perennials may also produce flowers on new growth, which dies back each winter, while other parts of the plant persist indefinitely.

Structure and Reproduction of Flowering Plants

Regardless of form, flowers share certain basic features throughout the division. Let us look briefly at a peach flower to familiarize ourselves with the parts of a typical flower (figure 20.5).

Each flower occurs as a specialized branch at the tip of a stalk called a **peduncle.** The peduncle swells at its tip into a small pad known as the **receptacle.** The other parts of the flower, some of which are in whorls, are attached to this receptacle. The outermost whorl consists of five small green, somewhat leaflike **sepals.** Collectively, the sepals of a flower are referred to as a **calyx.** In many flowers, the calyx functions in protection, particularly while the flower is in the bud. The next whorl of flower parts consists of **petals,** which are collectively known as the **corolla.** When corollas are showy, they function in attracting pollinators such as bees, moths, or birds. Some may

Figure 20.5 Parts of a typical flower

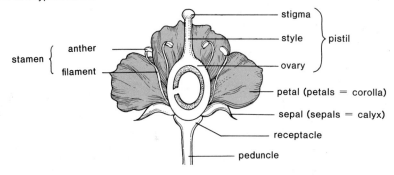

Figure 20.6 Catkins of a bur oak. (Courtesy Field Museum of Natural History, Chicago)

Figure 20.7 Jimson weed flower

have special markings that are invisible to humans but can be seen by bees whose vision functions in the ultraviolet range of the light spectrum (see figure 23.18). The corolla is often missing in wind-pollinated plants (see figure 22.24). Many trees, for example, are wind-pollinated and have very inconspicuous flowers (figure 20.6). In peach flowers, the petals are distinct separate units, but in other flowers, such as petunias or jimson weeds, the petals are fused together so that the corolla consists of a single, flared, trumpetlike sheet of tissue (figure 20.7). Sepals also may be fused together. While peach flowers are produced singly on their own peduncles, those of lilacs, grapes, bridal wreaths, and many others are produced in

clusters called **inflorescences** (figure 20.8). In an inflorescence, the peduncle has numerous little stalks serving individual flowers. These additional little stalks, one to each flower, are called **pedicels.**

In the center of the peach flower is a small greenish vaselike structure. This is the **pistil,** which consists of three regions that merge with one another. At the top is a slight swelling called the **stigma,** which is connected by a slender stalklike **style** to the swollen base called the **ovary,** which later develops into a fruit (see figure 20.14

PLATE 5. Eastern and midwestern wildflowers.
A. Swamp milkweed. *B.* Blazing star. *C.* Wild ginger.
D. Goldenrod. (*A*, Courtesy Perry J. Reynolds; *B, D,*
Courtesy Robert A. Schlising)

A

B

C

D

PLATE 6. Eastern and midwestern wildflowers.
A. Spring beauty. *B.* Wood betony. *C.* Pale corydalis.
D. Wood lily. (*B, C, D,* Courtesy Perry J. Reynolds)

A

B

C

D

PLATE 7. Western Wildflowers. *A.* Indian paintbrush. *B.* Gold violet. *C.* Red ribbons. *D.* Bridges' brodiaea. *E.* California poppy. (*A, B, C, D,* Courtesy Robert A. Schlising)

A

B

C

D

E

PLATE 8. Western wildflowers. *A.* Snowplant.
B. Baby blue eyes. *C.* Canchalagua. *D.* Tidy tips. (*A,*
B, C, D, Courtesy Robert A. Schlising)

A

B

C

D

Figure 20.8 Types of inflorescences. Each ball represents a flower.

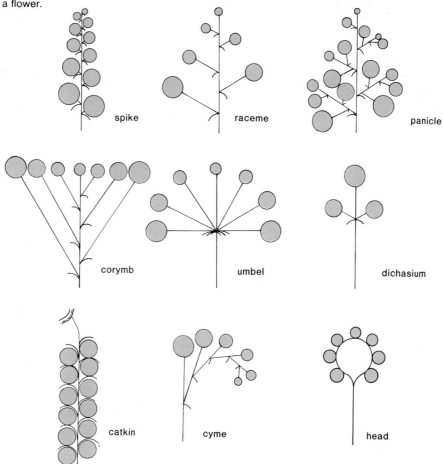

spike

raceme

panicle

corymb

umbel

dichasium

catkin

cyme

head

for variations in ovary positions). Within the ovary is a cavity containing an egg-shaped **ovule,** which is attached to the wall of the cavity by means of a short stalk. The ovule (in which a *megaspore* later gives rise to a female gametophyte) eventually becomes a **seed.**

Several to many **stamens** are attached to the receptacle around the base of the pistil. Each stamen consists of a semirigid but otherwise slender **filament** with a sac called an **anther** at the top. Within each anther numerous *microspores* are produced; these give rise to *pollen grains* (see figure 20.10), which later may develop male gametophytes. In most flowers, lengthwise slits that

develop on the anthers permit the release of pollen, but pores that function in the same way develop on the anthers of members of the heath family and those of a few other groups.

Development of Gametophytes

While the flower is developing in the bud, a diploid **megaspore mother cell** becomes differentiated from all the other cells in the ovule (figure 20.9). This cell undergoes meiosis, producing four haploid *megaspores*. Three of these megaspores degenerate and disappear not long after they are

Figure 20.9 Life cycle of a typical flowering plant

sporophyte

flower

ovary

ovule

megaspore mother cell

meiosis

microspore mother cells

anther

2N

young sporophyte

germination

seedcoat

embryo

seed

fruit

ovary develops into fruit; ovule develops into seed

zygote develops into embryo

3N endosperm tissue which develops from a cell formed when the two polar nuclei and one sperm unite.

fertilization

Chapter 20

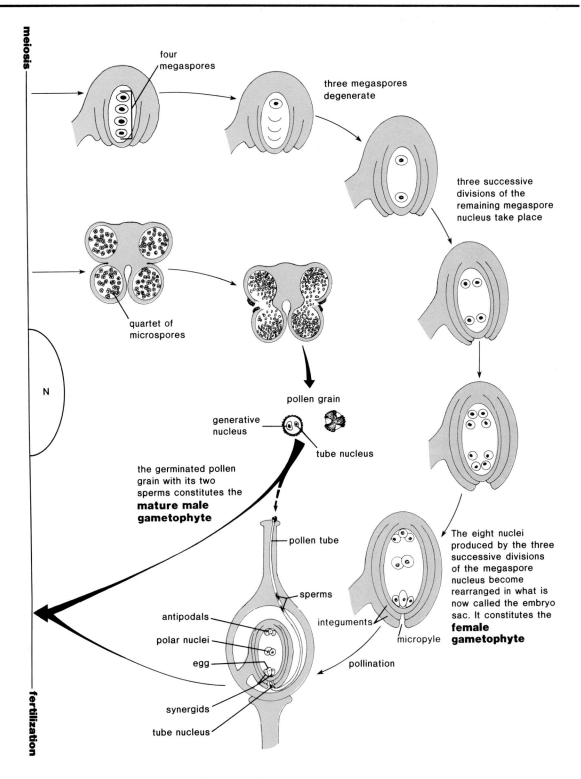

meiosis

four
megaspores

three megaspores
degenerate

three successive
divisions of the
remaining megaspore
nucleus take place

quartet of
microspores

N

pollen grain

generative
nucleus

tube nucleus

the germinated pollen
grain with its two
sperms constitutes the
**mature male
gametophyte**

pollen tube

sperms

antipodals

polar nuclei

egg

synergids

tube nucleus

integuments

micropyle

pollination

The eight nuclei
produced by the three
successive divisions
of the megaspore
nucleus become
rearranged in what is
now called the embryo
sac. It constitutes the
**female
gametophyte**

fertilization

Flowering Plants

produced, but the nucleus of the fourth undergoes mitosis and the cell enlarges. While the cell is growing bigger, its two haploid nuclei divide once more; the resulting four nuclei then divide yet another time. Consequently, eight nuclei in all are produced (without walls being formed between them). By the time these three successive divisions have been completed, the cell has grown to many times its original volume. At the same time, two outer cell layers of the ovule become differentiated. These layers, called **integuments,** later become the **seedcoat** of the seed. As they develop, they leave a pore, or gap, called the **micropyle** at one end.

At this stage, eight nuclei are in two groups of four toward the ends of the large cell. Now one nucleus from each group migrates toward the middle of the cell. These two **polar nuclei** may fuse together, forming a single diploid nucleus, or they may remain separate until fertilization occurs. Cell walls begin to form around the remaining nuclei. In the group closest to the micropyle, one of the cells functions as the female gamete, or *egg;* the other two cells, called *synergids,* either degenerate later or are destroyed during events that occur later. At the other end, the remaining three cells, called *antipodals,* apparently have no function and later they also degenerate. The large sac containing the seven or eight nuclei constitutes the *female gametophyte;* at maturity, it is called an **embryo sac.**

While the female gametophyte is being produced, a parallel process eventually leading to the formation of *male gametophytes* is taking place in the anthers. As an anther develops, four patches of tissue become distinguishable within the main mass of cells. These patches of tissue contain numerous diploid **microspore mother cells,** each of which undergoes meiosis, producing a quartet of *microspores.* Four chambers or cavities lined with nutritive cells are visible in cross section by the time the microspores have been produced. By the time the anther is mature, the walls between adjacent pairs of chambers have broken down so that only two larger sacs remain. After meiosis, the

Figure 20.10 Pollen grain of a poppy. The furrows are apertures through which a pollen tube may later emerge.

microspores in these chambers undergo several changes more or less simultaneously: (1) the nucleus in each microspore divides once; (2) the members of the quartet separate from one another (in some species this separation does not occur, but this is unusual); (3) a wall, which is often finely sculptured, develops around each unit. When these three events have been completed, the microspores have become **pollen grains** (figure 20.10). The outer part of the pollen grain wall, called the **exine,** contains substances that may later produce chemical reactions with the stigma of a flower. These reactions may result either in germination of the pollen grain or inhibition of further development, depending on whether or not it originated from the same plant, another plant of the same species, or a plant of a different species. Each pollen grain usually has three thin areas or *apertures* in the wall. One of the apertures functions in the completion of the development of the male gametophyte discussed in the next section. One of the pollen grain's two nuclei, the *generative nucleus,* will later divide, producing two sperms. The remaining *tube nucleus* is involved in events that occur after the pollen grain has left the anther.

Pollination

Frequently one reads in the press or popular magazines about flowers being pollinized, and the implication is made that the dusting of the pollen of one flower upon another is the equivalent of fertilization. Not so! **Pollination** is simply the transfer of pollen grains from an anther to a stigma, nothing more. **Fertilization** involves the union of egg and sperm; it may not occur until weeks or months after pollination has taken place or it may not follow pollination at all.

Most pollination is brought about by insects or wind, but water, birds, bats, other mammals, and gravity act as agents, or pollinators, in a number of species. The adaptations between the flower and its pollinators can be intricate and precise (see figure 23.21), and may even in certain instances involve force, drugs, or sexual enticement (see figure 23.23). A discussion of pollination ecology is given in chapter 23.

Fertilization and Development of the Seed

After pollination has occurred, further development of the male gametophyte may not take place unless the pollen grain is (1) from a different plant of the same species or (2) from a variety different from that of the flower receiving it. There are, however, some species or varieties (e.g., garden peas), that are naturally self-pollinating. Under suitable conditions, the dense cytoplasm of the pollen grain absorbs substances from the stigma and bulges out in the form of a tube through one of the apertures. This **pollen tube** then grows down between the cells of the stigma and style until it reaches the micropyle. In corn, it may have to grow more than 50 centimeters (20 inches) before it arrives at its destination, but in most plants, including peaches, the distance is considerably less. The pollen tube's journey may last less than twenty-four hours and usually does not take more than two days, although there are a few plants in which growth takes over a year. As the tube proceeds, the contents of the pollen grain are discharged into it. The tube nucleus stays at the tip,

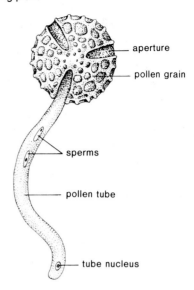

Figure 20.11 Mature male gametophyte of a flowering plant

- aperture
- pollen grain
- sperms
- pollen tube
- tube nucleus

while the generative nucleus (cell) lags behind and eventually divides by mitosis in the tube, producing two *sperms* that have no flagella. Sometimes the generative nucleus (cell) divides before the pollen tube has formed. The germinated pollen grain with its tube nucleus and two sperms constitutes the *mature male gametophyte* (figure 20.11).

When the pollen tube reaches the embryo sac, it usually discharges its contents into a degenerating synergid, or at least destroys the synergid upon entering. Then an event called **double fusion** (double fertilization), which is unique to angiosperms, occurs. One sperm migrates to and unites with the egg, forming a **zygote;** the other sperm migrates to and unites with the polar nuclei, producing a $3N$ (triploid) *endosperm nucleus*. The endosperm nucleus divides by mitosis, forming nutritive $3N$ tissue called **endosperm.** In some plants, such as corn and other grasses, the endosperm tissue becomes an extensive part of the seed, but in the legume family (e.g., peas, beans) the endosperm is used up by the **embryo** that develops from the zygote, so that the endosperm has disappeared by the time the seed is mature. The integuments harden, becoming a seedcoat, and the remaining haploid nuclei, or cells (antipodals, synergids, and tube nucleus), degenerate. At the conclusion of these various events, the ovule has become a seed, and at the same time the ovary matures into a fruit. Seed dispersal and fruits are discussed in chapter 21.

The process of embryo sac development described for a peach occurs in about 70% of the known flowering plants. The remaining 30% exhibit variations in which the embryo sac has from four to sixteen cells, or nuclei, at maturity, and the endosperm may be $5N$, $9N$, or even $15N$. One such variation occurs in lilies, which are favorites of manufacturers of microscope slides that show embryo sac development. When the megaspore mother cell undergoes meiosis in a lily, all four of the megaspores produced remain functional, with three of them uniting and forming a $3N$ nucleus. The embryo sac then develops in the same way

as that of a peach, but four of the ultimate eight nuclei are $3N$ and the other four are N. Thus, one of the polar nuclei is $3N$, while the other one is N. As a result, when fertilization occurs, a sperm (which is N) unites with both polar nuclei ($3N$ and N), forming a $5N$ endosperm nucleus.

Trends of Specialization and Classification in Flowering Plants

As indicated in chapter 13, various classifications of plants have been proposed ever since Theophrastus grouped them into trees, shrubs, and herbs in the fourth century B.C. Initially, classifications were merely for convenience and did not necessarily take into account natural relationships. Even today, plants may be lumped together in unnatural groupings in order to facilitate their identification. For example, some wildflower books arrange together all white-flowered species or all yellow-flowered species. There is nothing wrong with such arrangements as means of identification, but because such schemes do not reflect natural relationships (i.e., relationships based

Figure 20.12 Comparison between a primitive flower, magnolia (*left*), and an advanced flower, orchid (*right*).

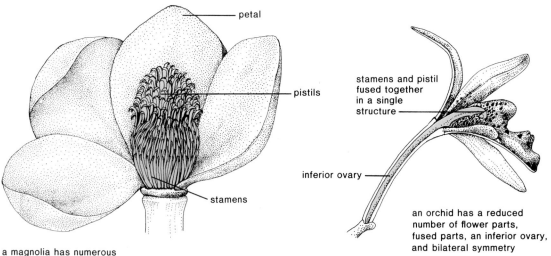

petal

pistils

stamens

a magnolia has numerous flower parts that are not fused together, superior ovaries, and radial symmetry

stamens and pistil fused together in a single structure

inferior ovary

an orchid has a reduced number of flower parts, fused parts, an inferior ovary, and bilateral symmetry

Chapter 20

on heredity), one does not recognize family characteristics or lineage. One should not infer that all blondes are more closely related to each other than they are to brunettes or that all long-haired dogs bear a closer kinship to each other than they do to short-haired dogs. Modern botanists therefore try to group plants according to their natural relationships, which are based on evidence gleaned from breeding experiments, forms and structures, chemistry, fossil records, and other features. There are, however, many interpretations of trends in the specialization of flowering plants and of other plants that are based primarily on inference from evidence currently available.

Although the information is very incomplete, the fossil record suggests that the flowering plants first appeared about 160 million years ago during the late Jurassic period (see table 18.1) and that they developed during the ensuing epochs of the Mesozoic and Cenozoic eras into the dominant elements of the flora they are today. Most botanists believe that primitive flowering plants, which they think appeared in the Triassic period, had the following features in common. Their leaves were simple. The flowers had numerous spirally arranged parts that were not fused to each other and were indefinite in number. The flowers were also radially symmetrical (i.e., the flowers could be divided into two equal halves along more than one lengthwise plane) and possessed both stamens and pistils.

While primitive flowers still abound today, many specializations and modifications have occurred since their initial appearance. Parts have become fewer and definite in number. Various parts have fused, and spiral arrangements have been compressed to whorls (figure 20.12). Originally a pistil was apparently formed by the blade of a leaflike structure with ovules along its edges rolling inward and fusing along the margins. Such a fertile blade was called a **carpel.** Eventually the separate carpels of primitive flowers fused together, forming the common compound pistil consisting of several carpels, which is found in many of today's angiosperms (figure 20.13). Each segment of an orange, for example, represents a

Figure 20.13 Ovaries in cross section. *Left:* The ovary of a simple pistil as found in a peach. *Right:* The ovary of a compound pistil as found in a lily.

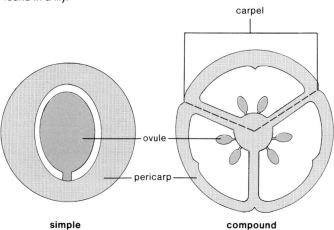

simple   compound

Figure 20.14 Ovary positions in flowers. Superior (e.g., peach). Half-inferior (e.g., cherry). Inferior (e.g., apple).

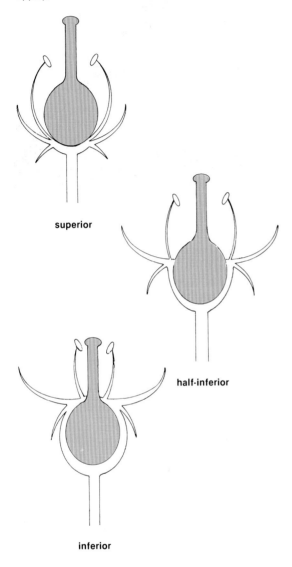

superior

half-inferior

inferior

Figure 20.15 Unisexual flowers of a wild gourd. *A.* Male. *B.* Female. Note the inferior ovary beneath the corolla.

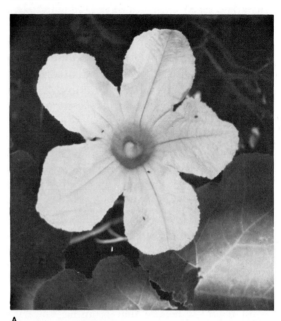

A.

B.

Chapter 20

single carpel, and if one cuts across a cucumber, three carpels are easily distinguishable. In advanced flowers, the receptacle or other flower parts have fused to and grown up around the ovary so that the calyx and the corolla appear to be attached at the top of it. When the ovary is embedded in this way, it is said to be an **inferior ovary,** in contrast to a primitive **superior ovary,** which sits on top of the receptacle with the other flower parts attached around its base. Some flowers have ovaries in an intermediate or *half-inferior* position (figure 20.14). Flowers have also tended to become *bilaterally symmetrical* (i.e., capable of being divided into two symmetrical halves only by a single lengthwise plane passing through the axis), as in sweet peas and orchids. In some families, the flowers have become **unisexual;** in other words, each flower has either stamens or a pistil but not both. Such flowers are seen in the pumpkin family (e.g., pumpkins, squashes, watermelons, cantaloupes, cucumbers) (figure 20.15). When both male and female unisexual flowers occur on the same plant, it is said to be **monoecious.** If male or female flowers occur only on separate plants, however, the species is said to be **dioecious.**

Within both the dicots and the monocots there has been reduction and fusion of parts and a shifting of the ovary from a superior to an inferior position as hypothetical progression is made from primitive families through intermediates to those that are advanced.

Observations on the human and ecological relevance of the flowering plants occupy literally thousands of volumes around the world. A brief overview of several aspects of this subject is given in chapters 22 and 23.

Herbaria and Plant Preservation

The botanical resources of many universities and other institutions include **herbaria** (singular: **herbarium**), which are something like libraries of dried and pressed plants arranged in such a manner that specific plants may be readily located.

Figure 20.16 Plant press

Specimens that are properly prepared and maintained can remain in excellent condition for three hundred or more years. Making one's own herbarium does not require formal training or experience. All that is needed are the materials, some simple equipment (which may be homemade), and the ability to follow a few relatively elementary procedures.

Flowers and other plant parts that are to be preserved need to have their moisture content reduced as rapidly as possible with a minimum of distortion. This is usually accomplished with the aid of a *plant press* (figure 20.16). This simple device consists of two pieces of plywood (or other wood materials, or thin metal plates) with dimensions of approximately 30 × 46 centimeters (12 × 18 inches) and a pair of webbing or leather straps to go around the boards. Between the boards are a number of felts or sheets of heavy blotting paper of similar dimensions. A folded page of newspaper is placed between each blotter, and a few sheets of stiff cardboard are interspersed between the blotters.

Any soil clinging to the roots of a specimen to be pressed is washed off, and the plant is laid out on one of the newspaper sheets. Leaves are

Figure 20.17 Mounted herbarium specimen

carefully straightened out, as are petals and other plant parts, so that they are not folded during pressing. Notes on where, when, and by whom the specimen was collected are penciled on the newspaper, or a number corresponding to such notes in a separate field notebook is used. The newspaper is then folded over the specimen and placed between two blotters. Many specimens may be placed in a press at one time or, if space permits, between the same sheets of newspaper. Only one species should be placed in a single fold of newspaper, however, as mixing species invariably leads to some being pressed better than others, due to the varying extents of woody tissues within the stems and other factors. After the newspaper and specimen are returned to the press, the straps are tightened around it as much as possible and it is placed in the sun or near a heater (but not close enough to scorch!) to dry for three or four days.

Unless the leaves were succulent or wet at the time they were placed in the press, they should be dry enough to mount on paper at this time. If a press is not available, plants may be pressed between newspapers and blotters by placing heavy weights on top of them.

If possible, 100% rag-content paper should be used for mounting the specimen, as pulp-content papers deteriorate with age. The paper should also be of heavy weight to lend some support to the now brittle specimen. The pressed plant(s) may be attached to the paper in one of several ways. Some spread a good white library type of glue on a glass plate and then place the specimen on the wet glue so that one side is covered with it. They then transfer the specimen to herbarium paper, which normally measures 29 × 42 centimeters (11.5 × 16.5 inches). Others place the specimen on the paper first and add glue or liquid plastic at strategic points. The bottom right-hand corner of the paper should be kept clear for a label indicating the scientific name of the plant, collection information, the collector's name, and the collection date (figure 20.17). Professional botanists also number their individual collections, giving a collection number after the collector's name. The label should measure about 7.5 × 12.5 centimeters (3 × 5 inches), and its paper should also be of 100% rag content. Specimens are then placed in manila folders and stored in some systematic fashion so that retrieval of individual specimens is easily accomplished. Major institutional herbaria usually arrange the plants in a presumed evolutionary sequence by families, and then the genera and species are alphabetically arranged within the families. Mothballs or similar material should be added to the cabinet or storage area to prevent insect damage during storage.

Some people are interested in pressing flowers for use in dry arrangements on cards, placemats, or other decorative items. In such cases, the pressing is done in the same manner as for the herbarium specimens, but the material is then further manipulated in one of several ways. For wall art, the specimens can be mounted on a piece

Figure 20.18 Pressed flowers framed as wall art

of smooth cardboard, covered with clear plastic or glass, matted, and framed (figure 20.18). Decorative notepaper can be made by placing pressed flowers on the paper and covering them with rice paper or facial tissue. A mixture of one part white glue to three parts water is then brushed over the tissue, causing it to become a permanent mount with the specimen clearly visible through the thin paper film. Clear Contact paper cut to the appropriate size makes a good mount for pressed flowers placed on cards or placemats. Pressed flowers can also be embedded in clear plastic poured into molds.

Flowers can be dried without pressing and with a little patience. Relatively lifelike three-dimensional preservations can be made, although flowers with petals that are not easily detached lend themselves more readily to three-dimensional drying than others. Most such drying is done in a shoebox, but almost any type of container can be used. The bottom of the box is covered with about 2 centimeters (0.8 inch) of sand, silica gel, or borax mixture. The fresh flower, with a little of the peduncle or stem still attached, is then gently laid on the surface. After this, more sand, silica gel, or borax mixture is slowly drizzled by hand into the box until the entire flower is buried, with care being taken not to create air pockets around any parts (figure 20.19). Each of the three drying agents mentioned has certain advantages and drawbacks. Sand must be thoroughly washed several times to be certain it is perfectly clean before use. The sand also needs to be completely dry, uniformly fine, and if possible, have individual grains that are relatively rounded rather than angular. Such sand with rounded grains is found around the Great Salt Lake in Utah and is available in arts and crafts stores. It takes about two weeks to dry most flowers with sand. Silica gel is also available commercially and is the quickest drying agent of the three mentioned, usually completing the job in four to five days (if the box is placed in a microwave oven, the drying process may be completed in as little as five minutes). Its granules tend to be of different sizes, however, giving the flower surface a slightly irregular texture. It also tends to darken certain colors, and it is expensive. When borax is used, two parts of it should be mixed with one part sand or cornmeal. It dries flowers in about three weeks, but is sometimes difficult to remove completely from the dried flower surfaces.

When the drying period has been completed, the container is tilted and the sand or other drying material is slowly and gently poured out in an uninterrupted motion. If any petals have come loose, they may be glued in place later. Wire may be inserted in the stem to add rigidity, and after any clinging granules have been carefully removed,

Figure 20.19 Three-dimensional drying. A flower placed in a box is embedded in sand, silica gel, or a borax mixture.

soft, powdered colored chalk may be dusted on to restore any fading of color. It is recommended that beginners use fairly large flowers whose petals are not easily detached (e.g., zinnias, roses) for their initial attempt at three-dimensional flower drying.

A word of caution: Dried plant collections of herbaria in particular have proved invaluable in botanical research in the past and will continue to do so in the future. Literally hundreds of plants native to North America are now on rare and endangered species lists, however, and thousands more are in similar predicaments on other continents. The day has come when persons, both professional and amateur, interested in plants must discipline themselves to exercise extreme caution in collecting native plants. Except for certain types of research, a good photograph of a plant may actually be preferable to a dried specimen and aesthetically more pleasing. It is sincerely hoped that each reader will confine a collection to photographs of native plants as much as possible.

Summary

Flowering plants or angiosperms are the largest and most diverse division (Anthophyta) of the Plant Kingdom. The angiosperms are divided into two classes, the Dicotyledonae (dicots), which has most of the species, and Monocotyledonae (monocots). Members of the two classes are distinguished from one another on the basis of differences in numbers of flower parts, numbers of cotyledons, venation, presence or absence of cambia, arrangement of vascular bundles, and number of apertures in the pollen grains. Angiosperms vary greatly in size, volume, habitat, nutrition, and length of life.

A peach flower consists of a stalk (peduncle) with a receptacle to which are attached sepals (calyx), petals (corolla), stamens, and pistil. Some flowers have no petals, in others the petals may be fused together, and in yet others many flowers may be aggregated in an inflorescence. The pistil has three regions: the stigma, the style, and the ovary. The ovary contains one or more ovules. Stamens consist of a pollen-bearing anther and a stalk or filament.

Gametophytes develop in separate structures. In the ovule a megaspore mother cell undergoes meiosis, producing four megaspores. Three megaspores degenerate and the fourth develops into an embryo sac containing eight nuclei in three groups. Closest to the micropyle is an egg flanked by two synergids; in the center are two polar nuclei which may fuse, and at the other end are three antipodals. Integuments, which later become a seedcoat, surround the embryo sac, or female gametophyte. Numerous microspore mother cells undergo meiosis in an anther, producing microspores. The microspores develop into pollen grains with two nuclei—the generative nucleus and the tube nucleus.

Pollination, which is simply the transfer of pollen grains from an anther to a stigma, is brought about by insects, wind, and other agents. After pollination a pollen tube may emerge through an aperture from a pollen grain and grow down the style toward the embryo sac. As it does so, the tube nucleus remains at its tip, and the generative nucleus divides, producing two sperms (without flagella), either before or after it enters the pollen tube. The contents of the pollen tube are discharged into the embryo sac. One sperm unites with the egg and forms a zygote, and the other sperm simultaneously unites with the polar nuclei, forming a $3N$ endosperm nucleus; this nucleus divides mitotically, becoming nutritive endosperm tissue. The endosperm may become an extensive part of the seed or it may be used up by the embryo by the time the seed is mature. Nuclei that have not participated in fertilization degenerate. Some flowering plants produce endosperm that is $5N$, $9N$, or $15N$ due to variations in the manner in which the embryo sac develops.

Some books group flowers on the basis of color to facilitate identification, but such grouping does not reflect natural relationships (i.e., relationships based on heredity). Accordingly, modern botanists try to group plants naturally on the basis of evidence gleaned from a variety of sources, including fossils that suggest the flowering plants first appeared about 160 million years ago. Primitive flowering plants had simple leaves and numerous spirally arranged flower parts that were not fused to each other, and possessed both stamens and pistils; they were also radially symmetrical (capable of being divided in half along more than one lengthwise plane).

Specializations that have occurred since primitive flowers first appeared include a reduction in the number of parts; fusion of parts; appearance of compound pistils composed of several individual pistils (referred to as carpels in their fused state); fusion of the receptacle to the ovary so that the ovary appears embedded, or inferior as opposed to superior, with some flowers having half-inferior ovaries; bilateral symmetry (capable of being divided in half along only one lengthwise plane), and unisexual flowers (flowers with either a pistil or stamens). Monoecious species have both male and female flowers on the same plants, whereas dioecious species have male and female flowers on separate plants.

Herbaria may be likened to libraries of dried and pressed plants arranged so that specific plants may be readily located. Properly pressed plants may last for hundreds of years. Initially a plant to be pressed is placed in a plant press between sheets of newspaper and absorbent material, after soil has been removed and parts carefully straightened. When the specimens are dry they may be affixed to sheets of high-quality paper with a label giving collection information. Pressing may also be done for nonacademic purposes such as the preparation of artwork, placemats, and so on. Flowers may be preserved in three dimensions by placing them in a box in drying materials such as sand, silica gel, borax, or cornmeal, but each agent has drawbacks.

Because so many plants are now on rare and endangered species lists, collectors should try to confine their future collecting as much as possible to photographs.

Review Questions

1. What are the basic differences between gymnosperms and angiosperms?

2. Indicate the features by which dicots are distinguished from monocots.

3. Define calyx, corolla, receptacle, peduncle, pedicel, pistil, filament, ovary, and megaspore.

4. What is the function of the integuments?

5. How do the pollen grains of flowering plants differ from those of pine trees?

6. What is the rough equivalent of an archegonium in a flower?

7. What is the function of endosperm, and how does it originate?

8. What differences are there between the embryo sacs of peaches and lilies?

9. Distinguish between dicots and monocots.

10. What distinguishes primitive flowers from those that are advanced?

Discussion Questions

1. The world's largest flowers occur on inconspicuous parasitic plants that resemble a fungus mycelium. If the flowers were not present, would it be possible to tell that the plant is not a fungus? How?

2. In most embryo sacs, the egg and the polar nuclei are the only ones of the eight original nuclei present that do not degenerate. What might be the consequences if the megaspore itself functioned as an egg and did not go on to produce other nuclei in an embryo sac?

3. It takes only one pollen grain to initiate the development of an ovule into a seed, yet a single flower may produce many thousands of pollen grains. Do you suppose such huge numbers of pollen grains are really necessary? Why?

4. Are such items as the name of the collector and the date significant on a herbarium specimen?

5. Is it really important that plants be classified? What would be the consequences if they were not?

Additional Reading

Bell, P.; Coombe, D.; translators of Denffer, D. von; Schumacher, W.; Magdefrau, K.; Ehrendorfer, F. *Strasburger's textbook of botany.* 30th ed. New York: Longman; 1976.

Bierhorst, D. W. *Morphology of vascular plants.* New York: Macmillan Co.; 1971.

Bold, H. C.; Alexopoulos, C. S.; Delevoryas, T. *Morphology of plants and fungi.* 4th ed. New York: Harper & Row, Publishers; 1980.

Condon, G. *The complete book of flower preservation.* Englewood Cliffs, NJ: Prentice-Hall; 1977.

Cronquist, A. *The evolution and classification of flowering plants.* Boston: Houghton Mifflin Co.; 1968.

Cronquist, A. *How to know the seed plants.* Dubuque, IA: Wm. C. Brown Company Publishers; 1979.

Foster, A. S.; Gifford, E. M., Jr. *Comparative morphology of vascular plants.* 2d ed. San Francisco: W. H. Freeman & Co.; 1974.

Holm, E. *The biology of flowers.* New York: Penguin Books; 1979.

uits and Seeds

ew

...pter fruits are introduced
...cussion of the nature of a
fruit and how it develops.
Parthenocarpy and fruit structure
are described. A listing and
discussion of the various types of
dry and fleshy fruits follow.

The agents of fruit and seed
dispersal (wind, animals, water,
etc.) are described. The chapter
concludes with a discussion of
seed structure and germination,
and some observations on seed
dormancy, longevity of seeds,
stratification, and vivipary.

Some Suggested Student Learning Goals

1. Distinguish a fruit from a vegetable.

2. Describe the regions of a mature fruit, such as a peach.

3. List five types of fleshy and dry fruits, and indicate how simple, aggregate, and multiple fruits are derived from the flowers.

4. Describe the adaptations of fruits and seeds to the agents by which they are dispersed.

5. Diagram and label a mature dicot seed (e.g., bean) and a monocot seed (e.g., corn) in section to show the parts and regions.

6. Describe the changes that occur when a seed germinates and note the environmental conditions essential to germination.

7. List the types of factors that control dormancy; describe how dormancy may be broken both naturally and artificially.

Introduction

In the 1930s, the United States Supreme Court ruled that a tomato was legally a vegetable rather than a fruit. This was in keeping with the general conception in the public's mind that fruits tend to be relatively sweet and in the nature of a dessert food, while vegetables tend to be more savory as salad or main course foods. Be that as it may, botanically speaking a fruit is any ovary that has developed and matured. It also usually contains seeds. By this definition, many so-called vegetables, including tomatoes, string beans, cucumbers, and squashes, are really fruits. On the other hand, vegetables can consist of leaves (e.g., lettuce, cabbage), stems (e.g., white potato), roots (e.g., sweet potato), stems and roots (e.g., beets), flower buds (e.g., globe artichoke), or other parts of the plant.

Fruits

All fruits arise from flowers and thus are found exclusively in the flowering plants. Fertilization usually indirectly determines whether or not a flower's ovary or ovaries will develop into a fruit. If at least a few of the ovules are not fertilized, the flower normally withers and drops without developing further. Pollen grains contain hormones that may initiate fruit development, and sometimes a quantity of dead pollen is all that is needed to produce the beginning of a fruit. It is the hormones produced by the developing seeds, however, that promote the greatest fruit growth. These hormones, in turn, stimulate the production of more fruit growth hormones by the ovary wall.

Some fruits can develop even without fertilization occurring. Examples of such **parthenocarpic** fruits, which are seedless, are found in navel oranges and certain varieties of figs and grapes. Occasionally, seed development occurs *apomictically* (see **apomixis** in chapter 12), without either meiosis or fertilization preceding it. This often happens in citrus fruits, where some normal seeds are interspersed with adventitious ones. To complicate matters further, not all seedless fruits are parthenocarpic. In Thompson seedless grapes, for example, fertilization does occur, but the ovules fail to develop with the fruit. Parthenocarpy can be induced artificially by the application of dilute hormone sprays to flowers. Seedless tomatoes are often produced in this way. Seedless watermelons are produced by crossing a diploid ($2N$) variety with a tetraploid ($4N$) variety. The hybrid that results from such a cross produces fruit, but the chromosomes cannot pair properly during meiosis. Fertilization and seed formation do not occur.

Fruit regions

By the time an ovary has matured into a fruit, the bulk of it is divided into three regions, which are, however, sometimes difficult to distinguish from one another (figure 21.1). The skin forms the **exocarp,** while the inner boundary around the seed(s) forms the **endocarp.** The endocarp may be hard and stony (as in a peach pit around the seed); it may be papery (as in apples); or it may not be distinct from the **mesocarp,** which is the name given to everything between the exocarp and the endocarp. The three regions are collectively called the **pericarp.** In dry fruits, the pericarp is often quite thin.

Some fruits consist of only the ovary and its seeds. Others have adjacent flower parts such as the receptacle or calyx fused to the ovary or different parts modified in various ways. Fruits may be either fleshy or dry at maturity, and they may or may not split, exposing the seeds. They may

Figure 21.1 Regions of a mature fruit

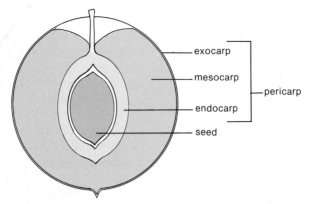

be derived from a single ovary or from more than one. Traditionally all these features have been used to classify fruits, but unfortunately not all fruits lend themselves to neat pigeonholing by such characteristics. Some of these problems are pointed out in the classification that follows.

Kinds of Fruits

Fleshy Fruits

Fruits whose mesocarp is at least partly fleshy at maturity are classified as fleshy fruits.

Simple Fleshy Fruits

Simple fleshy fruits are fruits that develop from a flower with a single pistil. The ovary may be superior or inferior, and it may consist of one or more carpels. The ovary alone may develop into the fruit, or other parts of the flower may develop with it.

Drupe

A drupe is a simple fleshy fruit with a single seed enclosed by a hard, stony endocarp, or pit (figure 21.2). It usually develops from flowers with a superior ovary containing a single ovule. The mesocarp is not always obviously fleshy, however. In coconuts, for example, the husk (consisting of the mesocarp and the exocarp), which is usually removed before the rest of the fruit is sold in markets, is very fibrous (the fibers are used in making mats and brushes). The seed ("meat") of the coconut is hollow and contains a watery endosperm referred to as "milk." It is surrounded by the thick, hard endocarp typical of drupes. Other examples of drupes include the stone fruits (e.g., apricots, cherries, peaches, plums), olives, and almonds. In almonds, the husk, which dries somewhat and splits at maturity, is removed before marketing, and it is the endocarp that we crack to obtain the seed.

Berry

Berries usually develop from a compound ovary and commonly contain more than one seed. The entire pericarp is fleshy, and it is difficult to distinguish between the mesocarp and the endocarp (figure 21.3). Four types of berries may be recognized.

A *true berry* is a fruit with a thin skin and one in which the entire pericarp is relatively soft at maturity. Although most contain more than one seed, notable exceptions are dates and avocados, which have only one seed. Typical examples of true berries include tomatoes, grapes, persimmons, peppers, and eggplants. Some fruits that popularly include the word *berry* in their common name (e.g., strawberry, raspberry, blackberry) botanically are not berries at all.

Chapter 21

Figure 21.2 Representative drupes. *A.* Peach.
B. Almond. *C.* Olive. (*A, C,* Photos by Bob Coyle)

A.

B.

C.

Figure 21.3 Representative berries. *A.* Grapes.
B. Tomatoes. (Photos by Bob Coyle)

A.

B.

Fruits and Seeds

Figure 21.4 Apples (representative of pomes). The bulk of the flesh is derived from the receptacle and the bases of the calyx, corolla, and stamens. (Photo by Bob Coyle)

False berries are similar to true berries outwardly, but they are derived from flowers with inferior ovaries so that other parts of the flower also contribute to the flesh. They can usually be distinguished from the true berries by the remnants of flower parts or their scars that persist at the tip. Examples of false berries include gooseberries, blueberries, cranberries, pomegranates, and bananas. In the cultivated banana, fruit development is parthenocarpic, so there are no seeds. Several other species of bananas produce an abundance of seeds.

The *pepo* has a relatively thick rind. Fruits of members of the pumpkin family, including pumpkins, cucumbers, watermelons, squashes, and cantaloupes, are pepos. Some authors do not think that the thick rind is particularly significant, and since pepos are also derived from flowers with inferior ovaries, they do not distinguish them from false berries.

The *hesperidium* is a berry with a leathery skin containing oils. Numerous outgrowths from the inner lining of the ovary wall become saclike and swollen with juice as the fruit develops. All members of the citrus family produce this type of fruit. Examples include oranges, lemons, limes, grapefruit, tangerines, and kumquats.

Pome

Pomes are simple fleshy fruits, the bulk of whose flesh comes from the enlarged receptacle that grows up around the ovary. The endocarp around the seeds is papery or leathery. Examples include apples, pears, and quinces. In an apple, the ovary consists of the core and a little adjacent tissue.

Figure 21.5 Blackberry. *A.* Flowers with many pistils. *B.* Aggregate fruits. (Courtesy Robert A. Schlising)

A.

B.

The remainder of the fruit has developed primarily from the receptacle (figure 21.4). Botany texts often refer to pomes, pepos, false berries, and other fruits derived from more than an ovary alone as *accessory fruits* or as fruits having *accessory tissue.*

Aggregate Fruits

An *aggregate fruit* is one that is derived from a single flower with several to many pistils. The individual pistils develop into tiny drupes or other fruitlets, but they mature as a clustered unit on a single receptacle (figure 21.5). Examples include raspberries, blackberries, and strawberries. In a strawberry, the cone-shaped receptacle becomes fleshy and red, while each pistil becomes a little *achene* (described in the following section

on dry fruits) on its surface. In other words, the strawberry, while being an aggregate fruit, is also partly composed of accessory tissue.

Multiple Fruits

Multiple fruits are derived from several to many individual flowers in a single inflorescence. Each flower has its own receptacle, but as the flowers mature separately into fruitlets they develop together into a single larger fruit, as in aggregate fruits. Examples of multiple fruits include mulberries (figure 21.6), osage oranges, pineapples, and figs. Pineapples, like bananas, usually develop parthenocarpically, and there are no seeds. The individual flowers are fused together on a fleshy axis, and the fruitlets coalesce into a single

Figure 21.6   Inflorescence and fruit of a mulberry

drupelets

Figure 21.7   Section through a developing fig

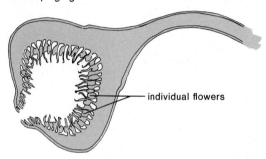

individual flowers

fruit. Figs mature from a unique "outside in" inflorescence. The individual flowers of the inflorescence are enclosed by the common receptacle, which has an opening to the outside at the tip (figure 21.7). Some varieties develop parthenocarpically, but others are pollinated by tiny wasps that crawl in and out through the opening. Some multiple fruits, such as those of the sweet gum, are dry at maturity.

Dry Fruits

Fruits whose mesocarp is definitely dry at maturity are classified as *dry fruits*.

Fruits That Split at Maturity

The fruits in this group are distinguished from one another by the manner in which they split.

Follicle

The *follicle* splits along one side or seam only, exposing the seeds within (figure 21.8). Examples include larkspur, columbine, milkweed, and peony.

Legume

The *legume* splits along two sides or seams (figure 21.9). Literally thousands of members of the legume family produce this type of fruit. Examples include peas, beans, garbanzo beans, lentils, carob, kudzu, and mesquite. Peanuts are also legumes, but they are atypical in that the fruits develop and mature underground. The seeds are usually released in nature by bacterial breakdown of the pericarp instead of through an active splitting action.

Figure 21.8 Follicles. *A.* Milkweed. *B.* Magnolia. The fruit of the magnolia is actually an aggregate fruit consisting of approximately forty to eighty individual one-seeded follicles on a common axis; the follicles and axis fall from the tree as a unit.

A.

B.

Figure 21.9 Legume of a garbanzo bean after it split. The two halves of the legume curled on drying, and the seeds were released with force. (Courtesy Field Museum of Natural History, Chicago)

Fruits and Seeds

Figure 21.10 *A.* Silique after it split open. The seeds are borne on a central, membranous partition. *B.* Silicles of a dollar plant. (*B,* Courtesy Field Museum of Natural History, Chicago)

A.

B.

Silique

Siliques also split along two sides or seams, but the seeds are borne on a central partition, which is exposed when the two halves of the fruit separate (figure 21.10A). Such fruits, which are less than three times as long as they are wide, are called *silicles* (figure 21.10B). Siliques and silicles are typically produced by members of the mustard family, which includes broccoli, cabbage, radish, shepherd's purse, and watercress.

Capsule

Capsules are the most common of the dry fruits that split (figure 21.11). They consist of at least two carpels and split in a variety of ways. Some split along the partitions between the carpels, while others split through the cavities (**locules**) in the carpels. Still others form a cap toward one end that pops off and permits release of the seeds, or they form a row of pores through which the seeds are shaken out as the capsule rattles in the wind. Examples include irises, orchids, lilies, poppies, violets, and snapdragons.

Fruits That Do Not Split at Maturity

In this type of dry fruit, the single seed is fused or attached to the pericarp to varying degrees.

Figure 21.11 Capsules. *A.* Butterfly iris. *B.* Bletilla orchid. *C.* Unicorn plant. *D.* Poppy. Note the row of pores through which the seeds are released near the top.

A.

C.

B.

D.

Fruits and Seeds

Figure 21.12 Dry fruits that do not split at maturity. *A.* Grain (caryopsis) of corn. *B.* Achene of sunflower. *C.* Nut (acorn) of oak.

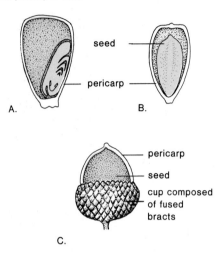

Figure 21.13 Samaras. *Top:* Maple. *Center:* Elm. *Bottom:* Ash.

Achene

The single seed of the *achene* is attached to its surrounding pericarp only at its base. Thus the husk (pericarp) is relatively easily separated from the seed. Examples include sunflower "seeds" (the edible kernel plus the husk constitute the achene) (figure 21.12), buttercup, and buckwheat.

Nut

Nuts are one-seeded fruits similar to achenes, but they are generally larger, and the pericarp is much harder and thicker. They develop with a cup, or cluster, of bracts at their base. Examples include acorns (figure 21.12), hazel nuts, hickory nuts, and chestnuts. Many nuts in the popular sense are not nuts, botanically speaking. We have already seen that peanuts are atypical legumes, and that coconuts and almonds are drupes. Walnuts and pecans are also drupes, whose "flesh" withers and dries after the seed matures. Brazil nuts are the seeds of a large capsule, and a cashew nut is the single seed of a peculiar drupe. It appears as a curved appendage at the end of a swollen pedicel, which is eaten raw in the tropics or made into preserves or wine. Pistachio nuts are also the seeds of drupes.

Grain (Caryopsis)

The pericarp of the grain is tightly fused to the seed and cannot be separated from it (figure 21.12). All members of the grass family, including corn, wheat, rice, oats, and barley, produce grains (also called *caryopses*).

Figure 21.14 Schizocarps of carrots

Samara

In *samaras,* the pericarp around the seed extends out into a wing or membrane, which aids in dispersal (figure 21.13). Samaras are produced in pairs in maples. In ashes, elms, and the tree of heaven they are produced singly.

Schizocarp

The twin fruit called a *schizocarp* (figure 21.14) is unique to the parsley family. Members of this family include parsley, carrots, anise, caraway, and dill. The twin fruits break into two one-seeded segments upon drying.

Fruit and Seed Dispersal

Why are so many species of orchids rare, while dandelions, shepherd's purse, and other weeds occur all over the world? Why are some plants confined to single continents, mountain ranges, or small niches occupying less than a hectare (2.47 acres) of land? The answers to these questions involve many different factors, including climate, soil, the adaptability of the plant, and its means of seed dispersal. How fruits and seeds are transported from one place to another is the subject of the following sections. Other factors are discussed in chapters 12 and 23.

Dispersal by Wind

Fruits and seeds have a variety of adaptations for wind dispersal (figure 21.15). The samara of a maple has a curved wing that causes the fruit to spin as it is released from the tree. In a brisk wind, samaras may be carried up to 10 kilometers (6

Figure 21.15 Types of seeds and fruits dispersed by wind

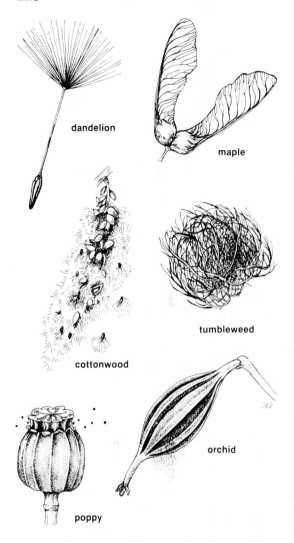

dandelion

maple

cottonwood

tumbleweed

poppy

orchid

miles) away from their source. In hop hornbeams, the seed is enclosed in an inflated sac that gives it some buoyancy in the wind. In some members of the buttercup and sunflower families, the fruits have plumes, and in the willow family the fruits are surrounded by cottony or woolly hairs that aid in wind dispersal. In button snakeroots and Jerusalem sage, the fruits are too large to be airborne, but they are spherical enough to be rolled along the ground by the wind. Seeds themselves may be so tiny and light that they can be blown great distances by the wind. Orchids and heaths, for example, produce seeds that are as fine as dust and equally light in weight. In trees such as catalpas and jacarandas, the seeds themselves are winged rather than the fruits, which remain on the branches and split, releasing their contents. Dandelion fruitlets have plumes that radiate out at the ends like tiny parachutes; these catch even a slight breeze. In tumble mustard and other tumbleweeds, the whole aboveground portion of the plant may break off and be blown away by the wind, releasing seeds as it bumps along.

Dispersal by Animals

The adaptations of fruits and seeds for animal dispersal are legion. Birds, mammals, and ants all act as disseminating agents (figure 21.16). Shore birds may carry seeds great distances in mud that adheres to their feet. Other birds and mammals eat fruits, whose seeds pass unharmed through their digestive tracts. In blackbirds, the seeds may remain in the tract as little as fifteen minutes, but in mammals the period is more commonly about twenty-four hours. In the giant tortoises of the Galápagos Islands, seeds do not pass through the tract for two weeks or more, and the seeds usually will not germinate unless they have been subjected to such treatment (see chapter 10). Some seeds and fruits are gathered and stored by rodents such as squirrels and mice and then are abandoned. Blue jays, woodpeckers, and other birds carry away nuts and other fruits, which they may drop in flight and abandon.

Many fruits and seeds catch in or adhere to the fur or feathers of animals and birds. Some, such as those of bedstraw and bur clover, are covered with small hooks that catch in fur (or a hiker's socks). The large capsules of unicorn plants have two giant curved extensions about 15 centimeters (6 inches) long. These catch in the fetlock of a deer or other animal that happens to step on the fruit, and the seeds are scattered as

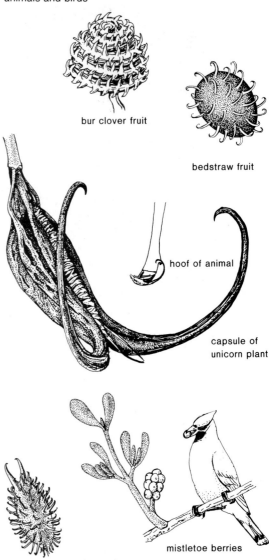

Figure 21.16 Types of seeds and fruit dispersed by animals and birds

bur clover fruit

bedstraw fruit

hoof of animal

capsule of unicorn plant

cocklebur

mistletoe berries

the animal moves along. Twinflowers and flax have fruits with sticky appendages that adhere to fur on contact, and those of the puncture vine penetrate the skin and stick by means of vicious little thorns (figure 21.17). Bleeding hearts, trilliums, and several dozen other plants have appendages on their seeds that contain oils attractive to ants (figure 21.18). The Scandinavian scientist

Figure 21.17 Fruits of a puncture vine clinging to a hiking boot. (Courtesy Robert A. Schlising)

Figure 21.18 Seeds of the Pacific bleeding heart. The glistening appendages are elaiosomes, which are removed from the seeds by ants and used for food.

Fruits and Seeds

Sernander once estimated that more than thirty-six thousand such seeds were carried by members of a single ant colony to their nest, where the ants stripped off the appendages for food but did not harm the seeds themselves.

Dispersal by Water

Some fruits are adapted to water dispersal by virtue of the fact that they contain trapped air. Many sedges, for example, have inflated sacs around the seeds (figure 21.19) that enable them to float. Others have waxy material on the surface of the seeds, which temporarily prevents them from absorbing water while they are floating. Sometimes a heavy downpour will create a torrent of water that dislodges masses of vegetation along a stream bank, carrying whole plants and their fruits to new locations. Large raindrops themselves may splash seeds out of their opened capsules. Seeds and fruits of a few plants have thick spongy pericarps that absorb water very slowly. Such fruits are adapted to dispersal by ocean currents, even though salt water eventually may penetrate enough to kill the delicate embryos. Enough fruits are beached before this occurs to ensure the survival of the species. The best known of ocean-dispersed plants is the coconut palm, whose large fruits apparently have been carried many hundreds of kilometers throughout the tropical seas of the world.

Other Dispersal Mechanisms

Fruits of many families mechanically eject seeds, sometimes with considerable force. For example, as the capsules of witch hazel dry, a splitting action occurs that may fling the seeds over 12 meters (40 feet) away. Seeds of some legumes and touch-me-nots also are explosively released, and those of dwarf mistletoes, when they are violently released in response to the heat of a warm-blooded animal coming close to the plants, can actually raise small welts on the skin. In manroots and a

Figure 21.19 Sedge adaptation to water dispersal. *Left:* Sedge fruit. The seed is enclosed within an inflated covering that enables it to float on water. *Right:* Sedge plant.

few other members of the pumpkin family, the seeds are squirted out of one end of the melonlike fruits like an erupting geyser. In filarees, each carpel of the fruit splits away and curls back from a central axis. Each fruitlet contains a single seed, which is pointed at its base. At the other end is a long, slender beak, which is sensitive to changes in humidity (figure 21.20). At night when the humidity increases, the beak is relatively straight, but in the sun it coils up like a corkscrew, literally drilling the pointed seed into the ground as it does so and effectively planting it in the process.

Humans are by far the most efficient transporters of fruits and seeds. Many noxious weeds and plant diseases, as well as valuable food and medicinal plants, have been carried from one continent to another by explorers and travellers over the past few hundred years in particular. Most countries now have strict regulations barring the importation of plant materials, except by

Figure 21.20 Filaree fruitlets. *A.* Under humid conditions. *B.* Under dry conditions.

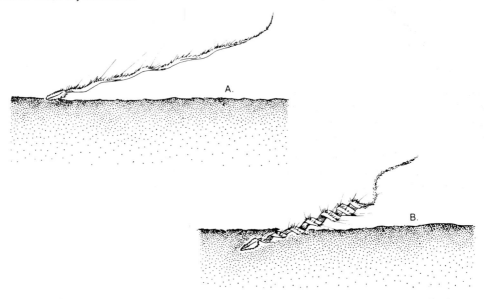

special permit, and some plants are not allowed across borders under any circumstances. Even within a country certain fruits and seeds that might carry diseases harmful to local agriculture may be barred from entry. In the United States, for example, Arizona, California, and Hawaii have border inspections in an attempt to prevent the importation of fruits such as citrus or popcorn across state lines, as such plants have in the past been carriers of diseases or pests that are presently under control.

Seed Structure

If an ordinary kidney bean (a dicot) is examined closely, one can see a small white scar called a *hilum* on the concave side. This marks the point at which the ovule was attached to the ovary wall. The *micropyle* may be visible as a small pore adjacent to the hilum. If this bean is placed in water for an hour or two, it may swell enough to split the seedcoat. Once the seedcoat is removed, the

bean can be seen to consist of two halves, or **cotyledons,** with a tiny rudimentary bean plant between them toward one edge (figure 21.21). The cotyledons are food storage organs, that also function as the first "seed leaves" of the new sporophyte. They and the tiny, rudimentary bean plant to which they are attached constitute the *embryo.* Some seeds (e.g., those of grasses and all other monocots) have only one cotyledon.

The tiny plant bears undeveloped leaves and a meristem at one end. This embryo shoot is called a **plumule.** The cotyledons are attached just below the plumule. The stem part of the axis above the cotyledons, which at this stage is almost nonexistent, is referred to as the **epicotyl,** whereas the part below the point of attachment is the **hypocotyl.** In an embryo it is often difficult to tell where the "stem" ends and the "root" begins, but the tip that will develop into a root is called a **radicle.**

Figure 21.21 Bean. *A.* Seed structure.
B. Germination and development of the seedling.

A.
- seedcoat
- plumule
- hypocotyl-radicle axis
- micropyle
- hilum
- cotyledon

B.

first true leaves
- cotyledon
- hypocotyl
- seedcoat

seedcoat — hypocotyl

primary root

epicotyl
- withered cotyledons
- hypocotyl

primary root
secondary root

When a kidney bean germinates, the hypocotyl lengthens below a crook so that the cotyledons are pulled above the ground, but in lima beans and peas the hypocotyl remains short so that the cotyledons do not emerge above the surface (see chapter 10).

In other seeds, a plumule–radicle axis may be in the center instead of to one side, and the cotyledon(s) may not play a significant food storage role. In corn, for example, the bulk of the food storage tissue is endosperm. Corn seeds (figure 21.22) also display other features not seen in beans. The plumule and the radicle are enclosed in tubular, sheathing structures called the **coleoptile** and the **coleorhiza,** respectively. These protect the delicate tissues within as the seeds germinate. Development of the coleoptile and coleorhiza ceases after they have attained several millimeters in length, and the plumule and radicle burst through the tips.

Germination

Germination of a seed depends on the interplay of a number of factors, both internal and external. Many seeds need to undergo a period of **dormancy** before they will germinate (see the discussion of dormancy in chapter 10). Dormancy is brought about by either mechanical or physiological circumstances or both. In the legume family and others, the seeds may have seedcoats so thick or tough that they prevent the

Figure 21.22 Corn. *A.* Grain structure.
B. Germination and development of the seedling.

A.
- pericarp
- endosperm
- cotyledon
- coleoptile
- plumule
- radicle
- coleorhiza

B.
- coleoptile
- first leaf
- coleoptile
- prop roots (adventitious roots)
- radicle
- coleorhiza
- primary root
- primary root

absorption of water or oxygen. Some seeds even have a one-way valve that lets moisture out but prevents its uptake (see figure 10.22). Dormancy in such seeds may sometimes be broken artificially by *scarification,* which involves nicking or slightly cracking the seeds or dipping them in a concentrated acid for a few seconds to a few minutes. In nature, such seeds may remain dormant until mechanical abrasion by rock particles in the soil, alternate thawing and freezing, or in some cases, bacterial action creates cracks in the seedcoat.

Dormancy may also be induced by growth-inhibiting substances present in the seedcoat, the interior of the seed, or tissues of the fruit surrounding it. Many desert plants have inhibitors in the seedcoat. These have to be washed away by soaking rains before germination will occur.

The inhibitors function in survival of the species by preventing germination unless there has been sufficient rainfall for a seedling to become established. Apples, pears, citrus fruits, tomatoes, and other fleshy fruits contain inhibitors that prevent germination of the seeds within the fruits. Once the seeds are removed and washed, they germinate readily. The embryos of some seeds, such as those of the American holly, consist of only a few unspecialized cells when the fruit ripens. The seeds will not germinate after the fruit has dropped until the embryo has developed fully with the aid of food materials stored in its endosperm. Such a process of development is called **after-ripening.**

Fruits and Seeds

449

In many woody plants of temperate areas, germination stimulators need to be present to initiate growth. These normally do not develop unless the seeds encounter a period of cold temperatures. Usually this period needs to be a minimum of four to six weeks. The dormancy of such seeds can be broken by placing them in a refrigerator, preferably in damp sand, for a few weeks. This technique is called *stratification.*

Even when mechanical and physiological barriers to germination are not present, a seed will not normally germinate unless environmental factors are favorable. Water and oxygen are essential to the completion of germination, and light or its absence also plays a role. Many seeds imbibe ten times or more their total weight in water before the radicle emerges. Some seeds, such as those of castor beans and certain spurges (figure 21.23), have appendages that function in water absorption, and thereby speed up the germination process. After water has been imbibed, enzymes begin to function in the protoplasm, which has now been rehydrated. Some enzymes convert stored proteins to amino acids, others convert fats and oils to soluble compounds, and still other enzymes aid in the conversion of starch to sugar. The soluble substances can then be conveyed to the embryo, and respiration, which in a dormant seed is almost imperceptible, can be greatly accelerated. Respiration releases the energy needed to initiate growth of the embryo, and a new plant begins to develop as mitosis and cell elongation take place.

Often the embryo obtains its energy initially through anaerobic respiration (see chapter 9), and then switches to aerobic respiration as soon as the splitting of the seedcoat admits oxygen. If seeds are kept waterlogged after planting, oxygen available to them is greatly reduced and germination then may fail to be completed. Most seeds require temperatures within certain ranges to germinate. These usually need to be above freezing but below 45°C (113°F). Germination percentages tend to be low approaching either

Figure 21.23 Seeds of a spurge. Note the water-absorbing appendage (caruncle) at the end of each seed. (Courtesy Field Museum of Natural History, Chicago)

extreme, however. Most crop plants have an optimum (ideal) germination temperature of between 20°C and 30°C (68°F to 86°F). The role of light in germination varies with the kinds of plants concerned. Seeds of some varieties of lettuce will not germinate in the dark (see discussion of phytochrome in chapter 10), while those of other seeds such as the California poppy germinate only in the dark. In lettuce seeds, the light apparently inactivates germination inhibitors, while in the California poppy it stimulates inhibitor formation.

Longevity of Seeds

From time to time one reads or hears of seeds of wheat or other edible plants germinating after lying dormant in Egyptian pyramids or Native American tombs and caves for thousands of years, but none of these reports has been confirmed. Reports of seeds of the aquatic lotus plant germinating after two thousand years, and another documenting the germination of Arctic tundra lupine seeds that were frozen for an estimated ten thousand years have been confirmed, however.

Seeds remain *viable* (retain the capacity to germinate) for periods that vary greatly, depending on the species and the conditions of storage. Some seeds such as those of certain willows, cottonwoods, orchids, and tea remain viable for only a few days or weeks, regardless of how they are stored, but the period of viability of most seeds is extended by months or even years when they are stored at low temperatures and kept dry. Packets of vegetable and flower seeds sold in stores are dated, giving the buyer a rough idea of how long a significant number of the seeds might be expected to remain viable. Generally, seeds of pumpkin family members (e.g., squash, cantaloupe, cucumber) retain a relatively high percentage of viability for several years, whereas those of members of the amaryllis family (e.g., onion, leek, chives) retain a good percentage of viability for only two or three years. Properly stored wheat seeds have been reported to retain better than 30% viability for more than thirty years, and some weed seeds stored under conditions of low oxygen, high humidity, and cool temperatures have remained viable for even longer periods.

In 1879 William J. Beal, a botanist who pioneered in the development of hybrid corn, buried twenty pint-sized bottles of weed seeds on the campus of what is now Michigan State University in East Lansing, Michigan. Each bottle contained 1,000 seeds of twenty different species of weeds. A bottle of seeds was dug up every five years until 1920, when the schedule was changed to every ten years, and the seeds were planted. When the first bottle was dug up in 1884, seeds of most of the weeds germinated; in 1960 seeds of evening primrose, curly dock, and moth mullein still germinated; and in 1980, 29 moth mullein seeds, 1 mullein seed, and 1 mallow seed germinated—101 years after they were placed in the bottles. It is of interest to note that a mallow seed previously had not germinated since 1899. Only six of the original bottles now remain; they are not scheduled to be unearthed until the year 2040.

A few species of both dicots and monocots produce seeds that have no period of dormancy at all. In some instances, the embryo that develops from the zygote continues to grow without pause in a phenomenon known as *vivipary*. In the red mangrove, a tropical tree associated with coastal waters and estuaries, each fruit contains a single seed in which the embryo continues to grow while the fruit is still on the tree, reaching a length of 25 centimeters (10 inches) or more before the seedling becomes detached and essentially plants itself in the mud below.

Summary

A fruit is a mature ovary; fruits are found exclusively in the flowering plants. Hormones produced by developing seeds promote the greatest fruit growth. Parthenocarpic fruits develop without fertilization occurring and are seedless, but not all seedless fruits are parthenocarpic. A mature fruit has an outer exocarp (skin), an endocarp around the seed(s), and a mesocarp ("flesh") between the exocarp and endocarp; the three regions may be fused together as a pericarp. At maturity fruits may be fleshy or dry, and the fruit parts may have been derived from the ovary alone or from adjacent flower parts as well.

Fleshy fruits may be simple fruits, which develop from a flower with a single pistil; aggregate fruits, which develop from a single flower having more than one pistil; or multiple fruits, which are derived from several to many individual flowers in a single inflorescence. Simple fleshy fruits include drupes, which have a single seed enclosed by a hard endocarp; berries, which develop from a compound ovary and have more than one seed (true berries have a thin skin and soft pericarp); false berries, which are derived from flowers with inferior ovaries so that other parts of the flower contribute to the flesh; pepos, which have a relatively thick rind present; hesperidiums, which have a leathery skin containing oils; and pomes, whose flesh for the most part is derived from the enlarged receptacle that grows up around the ovary. Fruits derived from more than the ovary alone are sometimes referred to as accessory fruits; some aggregate fruits (e.g., strawberries) are partly composed of accessory tissue. The individual fruitlets of a multiple fruit develop together into a single larger fruit.

Some fruits that are dry at maturity split as they mature; among such fruits are follicles, which split along one side only; legumes, which split along two sides; siliques or silicles, which not only split along two sides but bear their seeds on a central partition between the two halves; and capsules, which split in various ways. Dry fruits that do not split at maturity include achenes, in which the single seed is attached to its surrounding pericarp at the base only; nuts, which resemble achenes but are larger and have a harder pericarp; grains (caryopses), which have the pericarp tightly fused to the seed; samaras, which have the pericarp extended in a wing; and schizocarps, which are twin fruits that break into two one-seeded segments upon drying.

Some fruits and seeds have wings, plumes, and other adaptations for wind dispersal. Other fruits and seeds have appendages and other adaptations for animal and bird dispersal, and still others have adaptations for water dispersal. Some fruits eject seeds with force, and some have modifications that drill seeds into the ground. Humans disperse many seeds, and most countries and a few states have strict regulations governing the importation of plant materials, primarily to control the spread of pests and diseases.

A bean seed bears a scar called a hilum on the concave side and a pore called the micropyle adjacent to the hilum. Within the seedcoat are two cotyledons with a tiny rudimentary bean plant between them, the cotyledons and plant constituting the embryo. The embryo shoot is called a plumule; the part of the stem axis above the point of attachment of the cotyledons is called the epicotyl, and the part below is called the hypocotyl; the tip that develops into the root is called the radicle. In corn and other grain seeds the plumule and the radicle are protected by tubular sheathings called the coleoptile and the coleorhiza, respectively.

Germination of a seed depends on the interplay of a number of factors bringing about the cessation of dormancy. Dormancy may be sustained by mechanical circumstances such as tough seedcoats that need to be broken to admit water or valves that prevent the entry of water; scarification (nicking or softening the seedcoat) may break dormancy in such seeds. Dormancy may also be induced by growth or germination inhibitors, or immature embryos that must develop in a process of after-ripening before germination can occur. Germination stimulators that do not develop unless the seeds are subjected to a period of cold temperatures are necessary for the germination of seeds of some temperate woody plants; stratification (placing seeds in a moist cool environment for a few weeks) may break the dormancy of such seeds. Even when mechanical and physiological barriers to germination are not present, a seed will not germinate unless environmental factors including water, oxygen, light, and certain temperature ranges are favorable.

Seeds remain viable (capable of germinating) for periods that range from a few days to thousands of years, depending on the species and conditions of storage. The viability of most seeds is extended by storage at low temperatures under dry conditions, but some weed seeds have their viability extended by storage under humid, cool conditions that include little oxygen. William J. Beal buried bottles of weed seeds in 1879, and the bottles have been periodically dug up, one at a time, to check germination percentages. One hundred one years after the bottles were buried, seeds of three species still germinated. Some plants produce seeds that undergo no dormancy at all; the growth of the embryo while the seed and fruit are still on the plant is termed *vivipary*.

Review Questions

1. What is the difference between a fruit and a vegetable?

2. What causes an ovary to develop into a fruit?

3. What are the various parts of a fruit?

4. Define parthenocarpy and give examples.

5. Distinguish among simple, aggregate, and multiple fruits.

6. On what basis do some authors combine false fruits and pepos?

7. Which types of fruits are derived from flowers with inferior ovaries?

8. Distinguish among achenes, grains, samaras, and nuts.

9. What adaptations do seeds and fruits have for dispersal by water and animals?

10. Define plumule, radicle, coleoptile, coleorhiza, hypocotyl, after-ripening, stratification, vivipary.

Discussion Questions

1. Pine cones are sometimes referred to as "fruits." Would you call them fruits? Why?

2. Do you believe the botanical distinction between fruits and vegetables is a good one? If not, how would you change it?

3. In discussing pomes, it was observed that the bulk of the flesh in an apple comes from the receptacle. What could you do to prove or disprove this?

4. Seed and fruit dispersal is achieved with the aid of wind, water, animals, mechanical means, and humans. If you were "designing" a new plant, can you think of any new way in which it might be dispersed?

5. When volcanic cause new islands they eventually acqu Would you expect the t mechanisms for the flower these islands to be the same for ancient continents?

Additional Reading

Cutter, E. *Plant anatomy: experiment and interpretation.* Part II: *Organs.* Reading, MA: Addison-Wesley Publishing Co.; 1971.

Pijl, L. van der. *Principles of dispersal in higher plants.* 2d ed. Berlin: Springler-Verlag; 1972.

Ridley, H. N. *The dispersal of plants throughout the world.* Ashford, Kent, England: L. Reeve and Co.; 1930.

Scagel, R. F.; Bandoni, R. J.; Maze, J. R.; Rouse, G. E.; Schofield, W. B.; Stein, J. R. *Plants: an evolutionary survey.* Belmont, CA: Wadsworth Publishing Co.; 1984.

activity or coral polyps
to appear in the oceans,
ire some vegetation.
pes of dispersal
ing plants on
s they were

Flowering Plants and Civilization

Overview

This chapter begins with comments on some of the problems involved in distinguishing between fact and fancy in reported past uses of plants. It continues with a brief discussion of Vavilov's centers of origin of cultivated plants and a survey of sixteen well-known flowering plant families. Miscellaneous information given for the families, which are presented in phylogenetic sequence, includes brief comments on family characteristics and some past, present, or possible future uses of the following family members:

buttercup (Ranunculaceae)
laurel (Lauraceae)
poppy (Papaveraceae)
mustard (Brassicaceae)
rose (Rosaceae)
legume (Fabaceae)
spurge (Euphorbiaceae)
cactus (Cactaceae)
mint (Lamiaceae)
nightshade (Solanaceae)
carrot (Apiaceae)
pumpkin (Cucurbitaceae)
sunflower (Asteraceae)
grass (Poaceae)
lily (Liliaceae)
orchid (Orchidaceae)

Some Suggested Student Learning Goals

1. Give reasons for basing scientific evaluation on more than a single sampling.

2. List Vavilov's centers of distribution of cultivated plants, and identify several plants from each center.

3. List characteristics of ten flowering plant families.

4. Describe what constitutes a primitive family.

5. List five useful plants in the laurel, rose, legume, and spurge families.

6. Identify medicinal plants in the poppy and nightshade families.

7. Construct a simple, original key to five flowering plant families.

Introduction

One species of a genus of plants known as puccoons grows on dry plains and slopes throughout the western United States and British Columbia. The seeds of this puccoon, which has greenish yellow flowers, are so hard that in Nevada and California the plant is known by the common name of "stoneseed."

Native American women of the Shoshoni tribe of Nevada reportedly drank a cold water infusion of the roots of stoneseed every day for six months to ensure permanent sterility. A biologist by the name of Clellan Ford became curious about these reports and experimented by administering extracts of stoneseed plants to mice. He found that the extracts effectively eliminated the estrous cycle of the mice and decreased the weights of the ovaries, the thymus, and the pituitary glands.

Although in this instance a Native American reported use of a plant was demonstrated experimentally to have a basis in fact, distinguishing between fact and fancy in reported past uses of plants is often difficult, particularly if the plants have become rare or extinct. But this type of research is much needed today. Indeed, it is essential if we are to save potential sources of medicinal drugs and other useful plant products before they are permanently eliminated by the bulldozer and other symbols of "progress." For despite the abundance of superstition and folklore associated with the use of plants by primitive peoples, there is no question that some of their cures and treatments were effective, even though they had no knowledge of the scientific reasons for the results. Botanists who have recognized this have teamed with anthropologists, medical doctors, and interpreters to interview witch doctors and other medicine men in the tropics of the Americas and Africa. They are gleaning and sifting through as much of this information as possible before it is too late. It would be surprising if their work does not lead to useful new discoveries in the near future.

Origin of Cultivated Plants

In 1822 a Swiss botanist by the name of Alphonse de Candolle published a book entitled *Origin of Cultivated Plants*. After gathering data from a variety of sources, de Candolle deduced that cultivated plants probably originated in areas where their wild relatives grow. In 1916, ninety-four years after de Candolle's book was published, a Russian botanist named N. I. Vavilov began a followup of de Candolle's work, and in the ensuing twenty years he modified and expanded de Candolle's conclusions. Vavilov became persuaded, as a result of his research, that most cultivated plants differ appreciably from their wild relatives; he also concluded that dispersal centers of cultivated plants are characterized by the presence of dominant genes in plant populations (see chapter 12), with recessive genes becoming apparent toward the margins of a plant's distribution. Vavilov recognized eight centers of origin of cultivated plants, with some plants originating in more than one center. The centers (figure 22.1), some of which are subdivided, are as follows:

1. *Chinese Center*—the earliest independent center, consisting of mountainous areas and adjacent lowlands of western and central China. Cultivated plants believed to have originated here include millets, soybean and several other legumes, bamboo, radish, eggplant, cucumber, some citrus, peach, apricot, walnut, persimmon, tea, some sugar canes, hemp, and others.

2. *Indian Center*—second in importance only to the Chinese Center and consisting of Burma, and India (exclusive of the northwest portion). Included with the cultivated plants believed by Vavilov to have originated here are rice, sorghum, mung and several other beans, gourds, yam, mango, orange and other citrus fruits, sugar cane, bowstring hemp, black pepper, betel nut, cardamon, gum arabic, henna, senna, strychnine, rubber plant, and many others.

Figure 22.1 Major centers of origin of cultivated plants. (After N. I. Vavilov, *The origin, variation, immunity and breeding of cultivated plants.* Translated from the Russian by K. Starr Chester. New York: The Ronald Press Company, copyright 1951.)

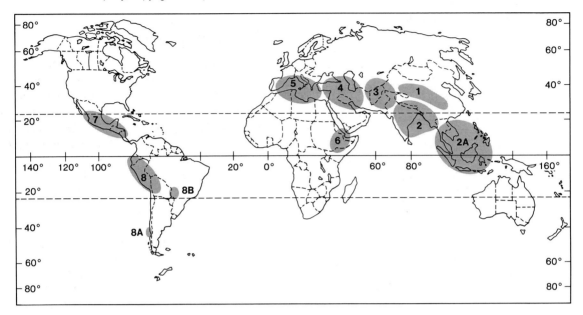

2A. *Indo-Malayan Center*—consists of Indochina, Malaysia, Java, Borneo, Sumatra, and the Philippines. Vavilov believed the giant bamboo, ginger, banana, breadfruit, candlenut, coconut, clove, nutmeg, Manila hemp, and many tropical fruits originated here.

3. *Central Asiatic Center*—consists of Northwest India, Afghanistan, and adjacent Soviet provinces. Wheat, garden pea, lentil, mustard, safflower, cotton, garlic, carrot, onion, basil, pear, almond, grape, apple, and other fruits and nuts are believed to have originated here.

4. *Near-Eastern Center*—consists of the interior of Asia Minor, Iran, Transcaucasia, and the highlands of Turkmenistan. Cultivated plants believed to have originated here include several types of wheat, rye, barley and oats, alfalfa, vetch, anise, poppy, cantaloupe, cabbage, lettuce, fig, pomegranate, cherry, and hazelnut.

5. *Mediterranean Center*—consists of areas bordering the Mediterranean ocean. Cultivated plants from this region include additional varieties of wheat, fava beans, clover, flax, black mustard, olive, carob, beet, parsley, leek, chive, savory, celery, parsnip, rhubarb, caraway, thyme, hyssop, lavender, peppermint, sage, rosemary, hop, and chufa.

6. *Abyssinian Center*—consists of Ethiopia (including the former country Eritrea) and Somaliland. Among the cultivated plants believed to have originated here are additional varieties of wheat, sesame, garden cress, coffee, okra, indigo, and fenugreek.

7. *South Mexican and Central American Center*—consists of southern Mexico, Guatemala, El Salvador, Honduras, Nicaragua, and Costa Rica. Maize (corn), common bean, lima bean, sweet potato, pepper, some cotton varieties, prickly pear, papaya, cashew, cacao (source of chocolate), cherry tomato, and annatto (a dye and spice plant) are believed to have originated here.

8. *South American Center*—consists of the highlands of Peru, Ecuador, and Bolivia. Several species of potatoes (other than the common white or Irish potato), tomato, pumpkin, marigold, cocaine plant, Egyptian cotton, guava, tobacco, and quinine tree are believed to have originated here.

8A. *The Chiloe Center*—confined to an island off the coast of southern Chile. The common white or Irish potato and the strawberry are believed to have originated here.

8B. *Brazilian–Paraguayan Center*—consists of Brazil and Paraguay. Despite the large diversity of species in this huge area, few cultivated plants are believed to have originated here. Manioc (source of cassava), peanut, passion fruit, pineapple, Brazil nut, and Pará rubber tree are among the cultivated plants from this area.

Each of these centers is isolated by mountain ranges, deserts, or oceans, and together they constitute less than 3% of the total land area of the earth. Agricultural development appears to have arisen independently in each center as primitive peoples investigated various members of the diverse local floras and in time came to make use of the plants available to them.

Selected Families of Flowering Plants

What follows is a survey of a few of the well-known plant families, indicating not only past uses of some of their members but also possible future uses in a few instances, along with biological notes and miscellaneous observations. *The reader is cautioned in particular against assuming that past medicinal uses were always effective or without harmful side effects and is urged to refrain from experimenting with such plants.*

More than three hundred families of flowering plants are recognized today. Flowers have proved to be considerably more reliable and stable indicators of heredity than leaves or other vegetative parts, and so flowering plant families are distinguished from one another primarily on the basis of flower parts and structure. Some families include a mere handful of species, while others are very large, their members numbering in the thousands. Space permits only brief discussion of a few of the larger, better-known families here, and the reader is referred to the additional readings for other sources of information. The families are taken up in a more or less phylogenetic sequence, beginning with those that have more primitive flower structure and progressing to those whose flowers are considered advanced. Primitive flowers generally have numerous parts that are not fused together and the ovaries are superior. In advanced flowers, the parts are reduced in number and often fused together; the ovary is inferior (see figure 20.12).

Buttercup Family (Ranunculaceae)

Flowers of the buttercup family have numerous stamens and pistils, and their petals are often irregular in number; the ovary is superior (figure 22.2). Most of the approximately fifteen hundred species, which are concentrated in north temperate and arctic regions, are herbaceous, often with dissected leaves that have no stipules and with petioles that are somewhat expanded at the base. A number of well-known representatives are ornamental plants (figure 22.3), including buttercup, columbine, larkspur, anemone, monkshood, and clematis. The columbine, which is the state flower of Colorado, receives its name from *columba*, the Latin word for dove. The name relates to the somewhat dovelike appearance of each of its five, spurred petals. Native Americans made a tea from boiled columbine roots for control of diarrhea, and members of at least two tribes believed the seeds to be an aphrodisiac. They would pulverize the seeds, rub them in the palms of their hands and would then try to shake hands with the woman of their choice. Others crushed and moistened the seeds and applied them to the scalp to repel lice.

Most members of the family are at least slightly poisonous, but the cooked leaves of cowslips have been used for food, and the well-cooked roots of the European bulbous buttercup are considered edible. In its natural state, the European buttercup does, however, cause blistering on the skin of sensitive individuals. Native Americans of the West gathered buttercup achenes, which they parched and ground into meal for bread. Others obtained a yellow dye from buttercup flower petals. Karok Indians made a blue stain for the shafts of their arrows from blue larkspurs and Oregon grape berries.

Figure 22.2 Buttercup flower

Goldenseal is a plant that was once abundant in the woods of temperate eastern North America but has become virtually extinct in the natural state because of relentless collecting by herb dealers. They sold the root for various medicinal uses, including remedies for inflamed throats, skin diseases, and sore eyes. At least one Native American tribe mixed the pounded root in animal fat and smeared it on the skin as an insect repellent. Another member of the buttercup family, monkshood, yields a drug complex called aconite, which was once used in the treatment of rheumatism and neuralgia. Although popular as garden flowers, the plants are very poisonous. Death follows within a few hours of consumption of any part of the plant. Most monkshood species have purplish to bluish or greenish flowers, but one Asian species called wolfsbane has yellow flowers. Wolf hunters in the past obtained a juice from wolfsbane root that they used to poison the animals.

Chapter 22

Figure 22.3 Representatives of the buttercup family.
A. Columbine (Courtesy Perry J. Reynolds).
B. Monkshood (Courtesy Robert A. Schlising).
C. Hepatica.

A.

C.

B.

Figure 22.4 Harvesting cinnamon in Sri Lanka. A pile of bark ("quills") stripped from young branches is in the foreground. The cinnamon of commerce is usually sold either as intact, dried strips of bark or in the form of bark that has been pulverized. (Photo by W. H. Hodge)

Laurel Family (Lauraceae)

The laurel family is also a primitive family whose flowers have no petals but whose six sepals are sometimes petal-like. The stamens, which occur in three or four whorls of three each, are a curiosity in that the anthers open by flaps that lift up. The ovary is superior. Most of the approximately one thousand species in this family are tropical evergreen shrubs and trees, many with aromatic leaves. The family received its name from the famous laurel cultivated in Europe for centuries. Its foliage was used by the ancient Greeks to crown victors in athletic events and later was employed in the conferring of academic honors.

Several important spices come from members of this family. Powdered cinnamon is the pulverized bark of a small tree (figure 22.4) that is native to India and Sri Lanka (formerly Ceylon), although it is also grown commercially elsewhere. Cassia, which is very similar, is often sold interchangeably with cinnamon today. Cinnamon oil is distilled from young leaves of the trees. Use of cinnamon and cassia dates back thousands of years. They were used in perfumes and anointing oils at the time of Moses, and other records indicate their use in Egypt at least thirty-five hundred years ago.

Camphor is another member of the laurel family that has been used since ancient times. This evergreen tree, which is native to China, Japan, and Taiwan, is the main source of camphor essence still used today in cold remedies and inhalants, insecticides, and perfumes. The essence is distilled from wood chips. Camphor is becoming popular as a street tree in cities and towns located in the milder climatic areas of the United States because of its ability to withstand smog.

Sassafras trees, which are native to the eastern United States and eastern Asia, also have spicy-aromatic wood. A flavoring widely used in toothpaste, chewing gum, mouthwashes, and soft drinks is obtained by distillation of wood chips and bark. Sassafras tea still is considered a refreshing beverage. It has in the past been used by country physicians for treating hypertension and for inducing a sweat in those with respiratory infections. Reports indicate it has a narcotic-stimulant effect in large doses. Sassafras is also an ingredient of some homemade root beers, and in the southern states an alcoholic beer has been made by adding molasses to boiled sassafras shoots and allowing the mixture to ferment. In Louisiana, powdered sassafras leaves (called *filé*) are used as a thickening and flavoring agent for gumbo.

The sweet bay, used as a flavoring agent in gravies, sauces, soups, and meat dishes, comes from the leaves of the laurel. Leaves of the related California bay are sometimes used as a substitute for sweet bay. The natural distribution of this California tree includes southwestern Oregon, where it is known as myrtle (true myrtles, however, belong to a different family). Its wood, which is hard, can be polished to a high luster and is used for making a variety of bowls, ornaments, and other smaller wooden articles. Early settlers in the West and Native Americans of the region used California bay in a number of ways. A bath for the relief of rheumatism was made by adding a quantity of leaves to hot water. The nutlike fruits (drupes) were roasted and used for winter food. A leaf was placed under a hat on the head to "cure" a headache (but even a small piece of leaf placed near the nostrils can produce an almost instant headache!). A few leaves placed on top of flour or grains in a canister will keep weevils away, and small branches have been used as chicken roosts to repel bird lice and fleas. A leaf or two rubbed on exposed skin functions as a mosquito repellent.

Avocados are also members of the laurel family. The fruits, which have more energy value by weight than red meats, are rich in vitamins and iron.

Poppy Family (Papaveraceae)

Members of the poppy family are mostly herbs distributed throughout temperate and subtropical regions north of the equator, but several poppies occur in the southern hemisphere, and a number are widely planted as ornamentals around the world. Poppies, like buttercups, tend to have numerous stamens, but most have a single pistil. Most also have milky or colored sap, and their sepals usually fall off as the flowers open. All members produce drugs.

A popular early flowering spring plant of eastern North American deciduous forests is called *bloodroot* (figure 22.5) because of the bright reddish sap produced in its rhizomes. This sap was used by some Native Americans as a facial dye, an insect repellent, and a cure for ringworm. Youngsters today still paint their nails with

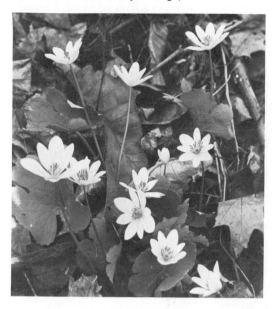

Figure 22.5 Bloodroot in flower. (Courtesy Field Museum of Natural History, Chicago)

it. Except when ingested in minute amounts, it has a very bitter taste. This latter feature made it effective in inducing vomiting, but members of one tribe counteracted the bitterness and used it for treating a sore throat by squeezing a few drops on a lump of maple sugar so that it could be held in the mouth.

Opium poppies have had a significant impact on societies of both the past and the present. Opium itself was described by Dioscorides in the first century A.D., and ancient Assyrian medical texts refer to both opium and opium poppies. Opium smoking, which does not extend back nearly as far as the use of the drug in other ways, became a major problem in China in the 1600s. Smoking opium has given way to other forms of use in recent years. The substance is obtained primarily by making small gashes in the green

Figure 22.6 Sectioned opium poppy and immature capsule that had been gashed with a razor blade. Note the opium-containing latex that oozed from the gashes. (Courtesy H. L. Dean)

capsules of the poppies (figure 22.6). The crude opium appears as a thick, whitish fluid oozing out of the gashes and is scraped off. It contains two groups of drugs. One group contains the narcotic and addicting drugs morphine and codeine, which are best known for their widespread medicinal use as painkillers and cough suppressants. Members of the other group are neither narcotic nor addictive. They include papaverine, which is used in the treatment of circulatory diseases, and noscapine, which is used as a codeine substitute because it functions like codeine in suppressing coughs but does not have its side effects.

Heroin, a scourge of modern societies, is a derivative of morphine. It is from four to eight times more powerful than morphine as a painkiller and less than one hundred years ago was advertised and marketed in the United States as a cough suppressant. It is now estimated that 75% of American drug addicts use heroin. The loss to society in terms of its economic, physical, and moral impact is enormous, since addicts frequently commit violent crimes to obtain the funds needed to support their habits (which often cost well over $100 per day).

The seeds of opium poppies contain virtually no opium and are widely used in the baking industry as a garnish. They also contain up to 50% edible oils, which are used in the manufacture of margarines and shortenings. Another type of oil obtained from the seeds after the edible oils have been extracted is used in soaps and paints.

Figure 22.7 Shepherd's purse

Mustard Family (Brassicaceae)

The original Latin name for the mustard family, still in widespread use today, was Cruciferae. The name describes the four petals of the flowers, which are arranged in the form of a cross. The flowers also have four sepals, usually four nectar glands, and six stamens, two of which are shorter than the other four. All members produce siliques or silicles, such fruits being unique to the family. All twenty-five hundred species of the family produce a pungent, watery juice, and nearly all are herbs distributed primarily throughout the temperate and cooler regions of the northern hemisphere.

Many edible plants are found in the mustard family. Some are widely cultivated, particularly in cooler temperate climates. Such plants include cabbage, cauliflower, brussels sprouts, broccoli, radish, kohlrabi, turnip, horseradish, watercress, and rutabaga. Some edible members are also widespread weeds. The leaves of shepherd's purse (figure 22.7), for example, can be cooked and eaten like cabbage, and the seeds can be used for bread meal. Other wild edible members include several cresses, peppergrass, sea rocket, toothwort, and wild mustard. The latter can become a weed problem in row crop planting. Their leaves are sometimes sold as vegetable greens in markets.

The seeds of wild mustard, shepherd's purse, and several other members of this family produce a sticky mucilage when wet. Biologists at the University of California at Riverside discovered a potential new use for these seeds. They were raising mosquitoes in water tanks for experiments and were feeding the larvae with alfalfa rabbit food pellets. They noticed that the larvae were sticking to the pellets and suffocating, since they had to come to the surface for air at frequent intervals. Curious, the workers examined the pellets under a microscope and found that they contained mustard seeds. Evidently the field where the alfalfa had been harvested had also contained mustard plants. The scientists then tried heating the mustard seeds to kill them and found that this did not affect the seeds' production of mucilage when wet. It was calculated that 0.45 kilogram (1 pound) of such seed could kill about twenty-five thousand mosquito larvae. Experiments are needed to determine if there is a practical way to control mosquitoes by such non-polluting means on a large scale.

Figure 22.8 Flower of wild rose

Native Americans mixed the tiny seeds of several members of this family with other seeds and grains for bread meal and gruel. Zuni Indians applied plants of western wallflower, ground and mixed with water, to the skin to prevent sunburn. Watercress, which is widely known as a salad plant, has had many medicinal uses ascribed to it. During the first century A.D., for example, Pliny listed more than forty medicinal uses. Native Americans of the west coast of the United States treated liver ailments with a diet consisting exclusively of large quantities of watercress for breakfast, abstinence from any further food until noon, and then resumption of an alcohol-free but otherwise normal diet for the remainder of the day. This was repeated until the disease, if curable, disappeared.

Dyer's woad, a European plant that has become naturalized and established in parts of North America, is the source of a blue dye that was used for body markings by the ancient Anglo-Saxons. Another member of the family, camelina, has been grown in the Netherlands for the oil that is obtained from its seeds. Camelina oil has been used in soaps and was once used as an illuminant for lamps.

Rose Family (Rosaceae)

The rose family comprises a large number of trees, shrubs, and herbs distributed throughout much of the world. The flowers characteristically have the basal parts fused into a cup, with petals, sepals, and numerous stamens being attached to the cup's rim (figure 22.8). The more than three thousand species of the family are divided into subfamilies on the basis of flower structure and

Figure 22.9 Harvesting damask roses for attar (otto) of roses perfume in Bulgaria. (Courtesy Field Museum of Natural History, Chicago)

Figure 22.9 Harvesting damask roses for attar (otto) of roses perfume in Bulgaria. (Courtesy Field Museum of Natural History, Chicago)

Figure 22.10 Mature rose hips. (Courtesy Field Museum of Natural History, Chicago)

fruits. The flowers of one group have inferior ovaries and produce pomes for fruits (see figure 21.4). Flowers of other groups have ovaries that are superior or partly inferior and produce follicles, achenes, drupes or clusters of drupelets (see figure 21.5).

Stone fruits (e.g., cherries, apricots, peaches, plums) and pome fruits (e.g., apples, pears) are all members of this family, as are aggregate fruits such as strawberries, blackberries, loganberries, and raspberries. The annual world production of peaches and plums is about 4.5 million metric tons (5 million tons) each, with 180,000 metric tons (200,000 tons) of plums, and an additional 135,000 metric tons (150,000 tons) of prunes being produced in the United States alone. A slightly greater world tonnage of pears is produced, and annual apple production is in the neighborhood of 18 million metric tons (20 million tons). United States production of strawberries exceeds 180,000 metric tons (200,000 tons). Obviously the economic impact of this family is enormous, particularly in temperate regions where these crops thrive.

Members of this family have been relevant to humans in many other ways in the past, however, and still continue to be so. Roses themselves, for example, have been favorite garden ornamentals of countless numbers of gardeners for centuries, and their elegant fragrance delights many. In Bulgaria and neighboring countries, a major perfume industry has grown up around the production of a perfume oil known as attar (or otto) of roses from damask roses (figure 22.9). In a valley near Sofia, more than two hundred thousand persons are involved in the industry, whose product brought more than $2,200 per kilogram ($1,000 per pound) during the 1970s. A considerable quantity of the oil is blended with less expensive substances in the perfume industry. It has been reported that perfume workers rarely develop respiratory disorders, thus suggesting that medicinal properties could be attributed to the plant extracts.

The fruits of wild roses, called *hips* (figure 22.10), are exceptionally rich in vitamin C. In fact, they may contain as much as sixty times the vitamin C of citrus fruit. Native Americans from coast to coast included rose hips in their diet (except for members of a British Columbia tribe, who believed they gave one an "itchy seat"), and it is believed that this practice contributed to the fact that scurvy was unknown among them. During

World War II when food supplies became scarce in some European countries, children in particular were kept healthy on diets that included wild rose hips. The hips also contain, in addition to vitamin C, significant amounts of iron, calcium, and phosphorus. Today many Europeans eat "Nyppon Sopa," a sweet, thick puree of rose hips, whenever they have a cold or influenza.

After giving birth, the women of one western Native American tribe drank western black chokecherry juice to staunch the bleeding. Other tribes frequently made a tea from blackberry roots to control diarrhea. Once it was reported that five hundred Oneida Indians cured themselves of dysentery with blackberry root tea, while many white settlers in the vicinity died from the disease. Men of certain tribes used older canes of roses for arrow shafts (presumably after removing the thorns!). Wild blackberries, raspberries, salmonberries, thimbleberries, dewberries, juneberries, and strawberries all provided food for Native Americans and early settlers, and they are still eaten today, either fresh or in pies, jams, and jellies. Wild strawberries are considered by many to be distinctly superior in flavor to cultivated varieties.

Legume Family (Fabaceae)

The legume family is the third largest of the approximately three hundred families of flowering plants, being exceeded in numbers of species only by the sunflower and orchid families. Its thirteen thousand members, which are cosmopolitan in distribution, include many important plants. The flowers range in symmetry from radial to bilateral. The latter have a characteristic *keel,* which is a boat-shaped fusion of two petals enclosing the pistil, two *wing petals,* and a larger *banner* petal

(figure 22.11). The stamens in such flowers are generally fused in the form of a tube around the ovary. The common feature that keeps the members together in one family is the legume fruit.

Important crop plants include peas, beans of all kinds (e.g., kidney, lima, garbanzo, broad, mung, tepary), lentils, peanuts, alfalfa, sweet clover, licorice, and wattle. The latter is an Australian tree that is grown commercially as a source of tannins for leather tanning. Carob, which is now widely used as a chocolate substitute, is also a member of this family. Several copals (hard resins used in varnishes and lacquers) are obtained from certain legume plants, as are gum arabic and gum tragacanth, which is used in mucilages, pastes, paints, and cloth printing.

Important dyes such as indigo, logwood (used in staining tissues for microscope slides and now scarce), and woadwaxen (a yellow dye) come from different legume plants. Locoweeds, which have been responsible for the death of many horses, cattle, and sheep, particularly in the southwestern United States, belong to a large genus of about sixteen hundred species. The poisonous principle in those species affecting livestock seems to vary in concentration according to the soil type in which the plants are growing. Other poisonous legumes include lupines, jequirity beans, black locusts, and mescal beans.

The sensitive plant, which grows as a weed in the tropics, is popular as a houseplant in temperate and colder areas because of the rapid manner in which its leaflets and leaves move in response to touch or other disturbances (see figure 10.18). About 90% of the members of the legume family exhibit leaf movements, but few are as rapid as those of the sensitive plants. Many of the movements are correlated more with daylength than with other factors.

Wild clovers have been widely used for food in the past by primitive peoples. The leaves are not readily digestible in quantity, but the rhizomes were gathered and usually roasted or

Figure 22.11 Parts of a sweet pea flower

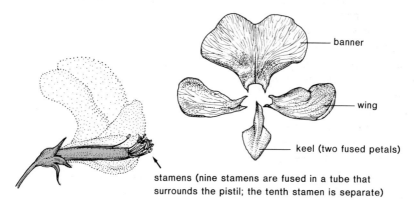

banner

wing

keel (two fused petals)

stamens (nine stamens are fused in a tube that
surrounds the pistil; the tenth stamen is separate)

steamed in salt water, then dipped in grease before being eaten. The seeds of both clovers and
vetches also were gathered and either ground for
meal or cooked in a little water and eaten as a
vegetable. Today seeds of several legumes, including alfalfa and mung beans, are popular for
their sprouts, which are widely used in salads and
Oriental dishes. A tropical bean called *winged
bean* has unusually high levels of protein, and all
parts of the plant are edible. It is presently being
grown in New Guinea, but it is believed to have
great potential for malnourished peoples of all
tropical areas.

Spurge Family (Euphorbiaceae)

Although many of the members of the spurge
family are tropical, a number are widespread in
temperate regions both north and south of the
equator. The stamens and pistils are produced in
separate flowers, and the flowers often lack a corolla, making them inconspicuous. In true spurges
(*Euphorbia*) the female flower is elevated on a
stalk called a *gynophore* and is surrounded by
several male flowers that each consist of little
more than an anther. Both the female and male
flowers are inserted on a cup composed of fused
bracts; the cup has distinctive glands on the rim.

This type of inflorescence is called a *cyathium*
(figure 22.12). Sometimes the inconspicuous
flowers are surrounded by brightly colored bracts
(e.g., poinsettia; see figure 7.14) that give the inflorescence the appearance of a single large flower.
Most members of this large family have milky
sap, and a number of species are poisonous.

Sooner or later many gardeners have "urges
to purge spurges," as some members of the family are exceptionally aggressive weeds that reproduce very rapidly. Other tropical spurges closely
resemble cacti and become very large. Several
economically important plants are cultivated,
particularly in frost-free regions. For example, an
estimated 90 million metric tons (100 million
tons) of cassava are harvested annually from
plants cultivated in the jungles of South America, Africa, and eastern Asia. The roots, which
develop thickened storage areas that resemble
large sweet potatoes, are the diet staple of tropical peoples, as white potatoes and cereals are of
the peoples of temperate areas. Poisonous principles are removed by boiling, fermenting, or by
squeezing out the juice. In dried and powdered

A.

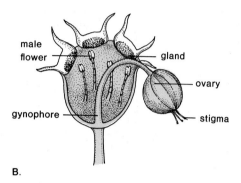

B.

form, the cassava is known as farinha. In Western countries, tapioca is prepared by forcing heated cassava pellets through a mesh while it is being agitated. Cassava starch is also used as a base for the production of alcohol, acetone, and other industrial chemicals.

Another cultivated spurge of the tropics is the Pará rubber tree, the source of the crude rubber from which most rubber products are made today. Although wild South American trees were the original commercial source of rubber, they have been widely planted in Indonesia, Africa, and adjacent areas. The trees, which vary in height from less than 5 to over 50 meters (16 to 164 feet), produce most of the latex in the inner bark. The

laticifers (see chapter 6) in which latex is secreted, spiral around inside the trunk at an angle of about 30 degrees. Accordingly, cuts are made at the same angle into the inner bark to obtain maximum yields of latex, which trickles down into collecting cups that are attached to the tree, in such a way as not to injure it. After collection, the latex is coagulated with the aid of either various chemicals or smoke and then shipped in sheet or crumbled form to processing plants. Sometimes an anticoagulant is mixed with it, and the liquid is transferred in tankers. Automobile and aircraft tires utilize a conspicuous share of the world's rubber, but other products manufactured from rubber are legion. The Pará rubber tree should not be confused with the broad-leaved ornamental known as rubber plant. The rubber plant, which is popular as a houseplant, also produces latex but is a member of the fig family (Moraceae).

The latex of other spurges may hold a key to future sources of fuel and lubricating oils.[1] In 1976, Melvin Calvin, a University of California Nobel prize winner, proposed the use of latex of gopher plants as a source of materials for oil. He estimated that such plants, which can grow in semidesert areas, would produce 10 to 50 barrels of oil per year on 0.4 hectare (1 acre) of land at a cost of $3 to $10 per barrel.

A spurge called *candelilla* occurs in remote areas of Mexico. It produces a wax on its stems that is collected and used in the making of candles and other wax products. Still another spurge produces seeds with a special oil used in plasticizers, and castor oil (from castor beans) is used in the manufacture of nylon, plastics, and soaps. Castor beans themselves are very poisonous—as few as one to three are sufficient to kill a child. The plants, which grow very rapidly, are popular as ornamentals but are possibly the leading natural cause of poisoning among American children.

A Mexican jumping bean is the seed of a certain spurge in which a small moth has laid an egg. When the egg hatches, the grub periodically changes position with a jerk, causing the seed to jump. The crown-of-thorns is an ornamental plant

Figure 22.13 Cactus flower. (Courtesy Field Museum of Natural History, Chicago)

with somewhat flexible twisting stems bearing vicious-looking spines. Some believe it to have been the plant from which the crown of thorns for the head of Christ was made. Poinsettias, or Christmas flowers (see figure 7.14), are favorite yuletide plants in various parts of the world. Tung oil, used in oil paints and varnishes, and Chinese vegetable tallow, a substance used in the manufacture of soap and candles, are two more commercially important products obtained from the seeds of cultivated members of the spurge family.

Cactus Family (Cactaceae)

Cacti are native only to the Americas but include many highly regarded ornamentals that have been exported around the globe. The flowers are usually showy (figure 22.13), with numerous stamens, petals, and sepals. The sepals are often colored like the petals; the inferior ovary develops into a berry. There are possibly more than fifteen hundred species, most occurring naturally in drier subtropical regions. The leaves of many are reduced in size or missing, with the fleshy, flattened or cylindrical, often fluted stems carrying

1. Jojoba, a member of the Box family (Buxaceae), is a desert shrub that has pea-sized seeds containing about 50% liquid wax. This high grade wax and another found in the seeds of meadow foam, a member of the meadow foam family (Limnanthaceae), are the only known natural substitutes for sperm whale oil, a vital ingredient in engine lubricants. The importing of sperm oil into the United States has been banned since 1970 to protect the nearly extinct large ocean mammals, and an expensive synthetic substitute is presently being used. Experiments and research are now in progress to find improved strains of jojoba and to test the feasibility of its being grown on a large scale. Two California counties have been using a mixture of petroleum oil and jojoba oil since 1981 in the transmissions of public transportation buses to see if the mixture will keep them from overheating. Preliminary results look promising and may reduce the need for transmission oil changes from once every 50,000 miles to once every 100,000 miles.

A.

B.

on the photosynthesis of the plants (figure 22.14). Many cacti can tolerate high temperatures, and some can withstand up to several years without moisture. They vary in size from pinheadlike forms to the giant saguaro, which can attain heights of 15 meters (50 feet) and weigh more than 4.5 metric tons (5 tons). They are generally exceptionally slow growing and make ideal houseplants for sunny windows because they need so little care.

In 1944, a marine pilot was forced to bail out of his aircraft over the desert near Yuma, Arizona. He survived the intense heat and low humidity of the area by chewing the juicy pulp of barrel cacti in the vicinity until he was rescued five days later. Since then, the use of cacti for emergency fluids and food has been recommended in most survival manuals. In fact, only

three cacti (peyote, living rock, and hedgehog cactus) are known to be inedible because of poisonous principles present; thus, the fruits of most cacti can be eaten. Those of prickly pears taste somewhat like pears. Prickly pear fruits also have seeds that Native Americans of the Southwest dried and ground for flour they used in *atole,* a staple food. A good syrup is obtained from boiling the fruits of prickly pears and also those of the giant saguaros. In the past, cactus candy has been made by partly drying strips of barrel cactus and boiling them in saguaro fruit syrup. Cane sugar has now largely been substituted for the saguaro syrup.

Native Americans of the Southwest used to scoop out barrel cacti, dry them, and use them for pots. They also mixed the sticky juice of prickly pear cacti in the mortar used in constructing their adobe huts. In Texas, a poultice of prickly pear stem was applied to spider bites. Hopi

C.

D.

Indians chewed raw cholla cactus as a treatment for diarrhea, and the skeletons of these cacti were used for flower arrangements.

In the middle of the nineteenth century, Australia imported a few prickly pear cacti, which were planted in the dry interior. These cacti found no natural enemies in their new environment and multiplied rapidly, infesting more than 24 million hectares (60 million acres) within seventy-five years. In an effort to control them, Australia introduced a moth from Argentina in 1925. The moth's caterpillars, which feed on prickly pear cacti, gradually brought the plants under control, and the land was made useable again.

Another cactus parasite, the cochineal insect (related to the mealybug, a common houseplant pest), feeds on prickly pear cacti in Mexico. At one time, the insects were collected for a crimson dye they produce, which was used in lipstick and rouge before aniline dyes were introduced.

Peyote cacti are small buttonlike plants, with roots resembling those of carrots and no spines. They contain several drugs, the best known of which is mescaline, a powerful hallucinogen. Dried slices of peyote have been used in native religious ceremonies in Mexico for centuries and more recently by at least thirty tribes of Native Americans. The drug gives the user a variety of hallucinations in vivid colors.

Mint Family (Lamiaceae)

The three thousand members of the mint family are relatively easy to distinguish from those of other families in having stems that are square in cross section, opposite leaves, and bilaterally symmetrical flowers (figure 22.15). They also generally produce aromatic oils in the leaves and

Figure 22.15 Flowers of henbit, a common weedy member of the mint family

Figure 22.16 Simple apparatus for distilling mint oil at home

stems. The superior ovary is four-parted, with each of the four divisions developing into a nutlet. Included in the family are such well-known plants as rosemary, thyme, sage (not to be confused with sagebrush, which is in the sunflower family), oregano, marjoram, basil, lavender, catnip, peppermint, and spearmint.

Mint oils can be distilled in the home with ordinary canning equipment. Whole plants (or at least the foliage) are loosely packed to a depth of about 10 centimeters (4 inches) or more in the bottom of a large canning pot. Then a wire rack or other support is also put in the pot, and a bowl is placed in the middle on the rack. Enough water is added to cover the vegetation, the pot is placed on a range, and the lid is inverted over it. The water is brought to a boil, and as it does so, ice is placed on the inverted lid. The oils vaporize, and condense when they contact the cold lid, dripping then from the low point into the bowl (figure 22.16). Of course, some moisture also condenses, but the oil, being lighter, floats on top.

Peppermint oil in particular is easy to collect this way and will keep indefinitely in a refrigerator.

Mint oils have been used medicinally and as an antiseptic in different parts of the world. Mohegan Indians used catnip tea for colds, and dairy farmers in parts of the midwestern United States use local mint oils to wash their milking equipment. As a result, mastitis, a common disease of dairy cattle, is seldom encountered in their herds. Horehound, a common mint weed of Europe, has become naturalized on other continents and is cultivated in France. A leaf extract is still used in horehound candy and cough medicines. In

Figure 22.17 Peppermint plants. Oil from the leaves is a source of menthol.

Figure 22.18 Chia. (Courtesy Robert A. Schlising)

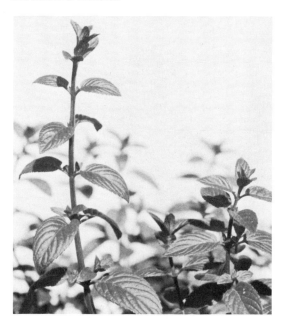

Figure 22.17 Peppermint plants. Oil from the leaves is a source of menthol.

England, it is a basic ingredient of horehound beer. Vinegar weed, also known as blue curls, is a common fall-flowering plant of western North America. Native Americans of the area used it in cold remedies, for the relief of toothaches, and in a bath for the treatment of smallpox. It was also used to stupefy fish.

Menthol, the most abundant ingredient of peppermint oil (figure 22.17), is widely used today in toothpaste, candles, chewing gum, liqueurs, and cigarettes. The primary American source is the Columbia River basin of Oregon and Washington, where it is grown commercially. Geese are sometimes used in the mint fields to control both insects and weeds, since they do not touch the mint plants themselves.

Ornamental mints include salvias and the popular variegated-leaf coleus plants, neither of which has typical mint oils in the foliage. Another relatively odorless mint is chia (figure 22.18), whose habitat is confined to the drier areas of western North America. Native Americans parched chia seeds and used them in gruel. The seeds, which become mucilaginous when wet, were also ground into a paste that was placed in the eye to aid in the removal of dirt particles. The paste was also used as a poultice for gunshot wounds, and Spanish Californians made a refreshing drink from ground chia seeds, lemon juice, and sugar. Chia seeds reportedly contain an unidentified substance that has effects similar to those of caffeine. Before the turn of the century, a physician by the name of Bard maintained that a tablespoon of chia seeds was sufficient to sustain a man on a twenty-four-hour endurance hike. Since that time, backpackers have experimented with the seeds, and results tend to support the earlier claim. A thorough scientific investigation of the matter is needed.

Nightshade Family (Solanaceae)

Flowers of the nightshade family, which is concentrated in the tropics of Central and South America, have their petals fused together, at least at the base, and the filaments of the stamens are fused to the corolla so that they appear to be arising from it (figure 22.19). The superior ovary develops into a berry or a capsule. The more than three thousand species of the family are widespread in distribution, have alternate leaves, and occur as herbs, shrubs, trees, or vines. Well-known representatives include tomato, white potato, eggplant, pepper, tobacco, and petunia.

Many nightshades produce poisonous drugs, some of which have medicinal uses. One of the best-known medicinal drug producers is the deadly nightshade of Europe. A drug complex called *belladonna* is extracted from its leaves. Belladonna was used in the "magic potions" of the past, but it is now the source of several widely used drugs, including atropine, scopolamine, and hyoscyamine. Atropine is used in shock treatment, for relief of pain, to dilate eyes, and to counteract muscle spasms. Scopolamine is used as a tranquilizer, and hyoscyamine has effects similar to those of atropine. Capsicum, obtained from a pepper, is used as a gastric stimulant.

Jimson weed is also a source of medicinal drugs that have been used in treatment of asthma and other ailments. The drugs can be fatal if ingested in sufficient quantities but have been much used in controlled amounts in Native American rituals of the past. Records indicate that users became temporarily insane but had no recollection of their activities when the effects of the drug wore off.

More than 800,000 hectares (2 million acres) of American farmland is allotted to the growing of tobacco, which in its dried form contains 1% to 3% of the drug nicotine. Nicotine is one of the

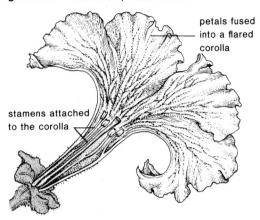

Figure 22.19 Sectioned petunia flower

petals fused into a flared corolla

stamens attached to the corolla

most important insecticides in use today, and it is also used for killing intestinal worms in farm livestock. It is, however, an addictive narcotic, and with the use of tobacco by humans increasingly being linked to lung and other cancers, the only benefit it may have to humans appears to be as a "leechicide." It is said that leeches attaching themselves to an average user of tobacco will drop off dead within five minutes from nicotine poisoning—a very dubious justification for continued human use!

Tomatoes are among the most popular of all "vegetables." About 18 million metric tons (20 million tons) are grown annually around the world. The plants are day-neutral (see chapter 10), and even though they require warm night temperatures (16°C or 60°F) to set fruit well, they are easily cultivated in greenhouses when natural conditions do not favor their growth. Most tomatoes grown commercially in the United States are processed into juice, tomato paste, and catsup. In Italy, a small amount of edible oil is extracted from the seeds after the pulp has been removed. Most American tomatoes are grown in California, where they are harvested with special machinery developed during the 1960s after inexpensive labor became no longer available (figure 22.20).

The white or Irish potato is one of the most important foods grown in temperate regions of the world, with annual production estimated at well over 270 million metric tons (300 million tons). The Soviet Union accounts for about 30% of the total, and a little over 4% is produced in the United States. It is believed that white potatoes originated on an island off the coast of Chile and were sent back to Europe by Spanish invaders of South America in the sixteenth century. In the 1840s, the late blight infestations that destroyed the potato crop in Ireland caused severe famine and subsequent immigration of Irish settlers to the United States. When potato tubers are exposed to the sun, they turn green at the surface. Poisonous drugs are produced in the green areas. These have proved fatal to both animals and humans and should never be eaten.

Carrot Family (Apiaceae)

Many members of the carrot family, which is abundantly represented in the northern hemisphere, have savory aromatic herbage. The flowers tend to be small and numerous and are arranged in umbels (see figure 20.8). The ovary is inferior, and the stigma is two-lobed. The petioles of the leaves, which are generally dissected, usually form sheaths around the stem at their bases. Included in the two thousand members of the family are dill, celery, carrot, parsley, caraway, coriander, fennel, anise, and parsnip. Anise is one of the earliest aromatics mentioned in literature. It is used for flavoring cakes, curries, pastries, and candy. Pocket gophers apparently are attracted by its aroma, and some poison baits are enhanced with anise. A liqueur known as anisette is flavored with it.

Figure 22.21   Water hemlock

Figure 22.22   Fruits and items associated with members of the pumpkin family (Cucurbitaceae). *A.* Hawaiian ceremonial gourd with feathers attached and gourds used in South America for drinking maté. A metal straw that strains out the maté leaves is resting in one gourd. *B.* Cantaloupes. *C.* Luffa (vegetable sponge).

A.

Another liqueur, called *kümmel,* is flavored with caraway seeds, which are well known for their use in rye and pumpernickel breads.

A few members of the carrot family are poisonous. Water hemlock (figure 22.21) and poison hemlock, which are common weeds in ditches and along streams, are deadly and have thus often been fatal to unwary wild food lovers. Socrates is believed to have died as a result of ingesting poison hemlock, which should not be confused with cone-bearing hemlock trees.

Several members of the carrot family, such as cow parsnip, squawroot, and hog fennel, have edible roots and were used for food by Native Americans. The reader is advised, however, to be absolutely certain of the identity of such plants before experimenting with them.

Pumpkin Family (Cucurbitaceae)

Although most species in the pumpkin family are tropical or subtropical, numbers occur in temperate areas of both the northern and southern hemispheres. Plants are prostrate or climbing herbaceous vines with tendrils. The flowers have fused petals and an inferior ovary with three carpels. All are unisexual. Some species have both

Figure 22.24 Flowers of grasses. *Left:* Expanded grass spikelet. *Right:* Enlargement of a single flower (floret).

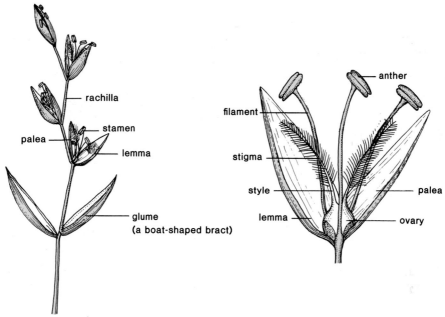

Many members of this family were widely used by Native Americans for various purposes. The seeds of balsamroot and mule ears were used for food. Balsamroot plants as a whole were eaten raw or cooked, and in the West extracts of both mule ears and tarweeds were used to treat poison oak inflammations. Salsify and thistle roots were also used for food.

Young leaves and roots of dandelions have long been on lists of wild edible plants, and the flowers have been used to make wine. Roasted dandelion and chicory roots have been used as a coffee substitute. During World War II chicory was grown as a crop specifically for use as a coffee adulterant. Recent crop failures and other problems in coffee-producing areas have produced a revival of interest in the use of chicory in coffee mixtures.

The dried and boiled leaves of American yarrow are said to make a nourishing broth. The European yarrow, which has become naturalized in North America, contains a drug that has been used in suppression of menstruation. This plant is believed to have been used by Achilles in treating the wounds of his soldiers.

Sunflowers themselves were widely used by Native Americans for their seeds, which were ground into a meal for bread. They are grown commercially today primarily for the edible oil extracted from the seeds, but their use in seed form by modern Americans is increasing.

Grass Family (Poaceae)

Although there are possibly only one-fourth as many species of grasses as there are members of the sunflower family, it is the largest family of flowering plants in terms of numbers of individuals and is the most widely distributed. The flowers of grasses are highly specialized in structure and have a terminology all their own (figure 22.24). The calyx and corolla are represented by tiny, inconspicuous scales, and the flowers are protected by boat-shaped bracts. The stigmas, when they are exposed, are feathery, and the leaves sheathe the stem at their bases.

All of the cereals, including wheat, barley, rye, oats, rice, and corn, belong in this family, and their economic importance is such that civilization as we know it would be vastly different without them. More than 900 million metric tons (1 billion tons) of cereals, feeding more than half of the world's population, are harvested each year, primarily in the Orient, North America, and Europe.

Sugar cane (figure 22.25), from which about 55 million metric tons (60 million tons) of sugar is extracted annually, is grown at lower elevations throughout humid tropical areas. It is a large grass, often growing to heights of 6 meters (20 feet). After the cane is harvested, the liquid portion is squeezed out. The solid waste is sometimes made into paper or particle board. It has also been converted to gasoline in South Africa and is now used in the production of electric power in Hawaii. Through a series of steps, sugar is crystallized out of the liquid portion and separated by centrifuging. The dark remnant, known as molasses, may be used as a base for rum or alcohol production.

Grasses have been widely used by primitive peoples for making mats and baskets and for thatching huts. One variety of sorghum is grown for its fibers, which are made into brooms. Some varieties are grown for their seeds, which are processed into cereal flours. Others are a source of silage and a carnaubalike wax. Citronella oil, once a common ingredient of mosquito repellents, is obtained from a grass grown in Indonesia. It is now used in cheaper soaps, cosmetics, and perfumes. Related grasses are the source of lemon grass oil, which is used in ways similar to those of citronella oil.

Juice squeezed from fresh young grass mowings has proved to have a high protein content and is being investigated as a source of protein for future human consumption.

Figure 22.25 Sugar cane plants

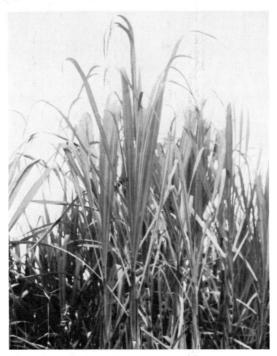

Lily Family (Liliaceae)

Although the two thousand species of lilies are particularly abundant in the tropics and subtropics, they occur in almost any area that supports vegetation. The flowers are often large, and their parts are all in multiples of three, with the sepals frequently colored the same as and resembling the petals (figure 22.26). The flowers of the related amaryllis family are similar. Lily flowers, however, generally have superior ovaries, whereas those of the amaryllis family are mostly inferior.

In addition to numerous types of lilies used widely as ornamentals, the family includes asparagus, sarsaparilla, squill, meadow saffron, bowstring hemp, and aloe. Sarsaparilla, which was at one time widely used for flavoring soft drinks and medicines, is obtained from the roots of a genus of woody vines whose stems are often covered with prickles. The bulbs of squills are the source of a rodent poison and also of a drug used as a heart stimulant. Meadow saffron is the source

Figure 22.26 Turk's cap lily flower. (Courtesy Perry J. Reynolds)

Figure 22.27 *Sansevieria* plant. (Courtesy Field Museum of Natural History, Chicago)

of colchicine, a drug once used to treat rheumatism and gout but now much more widely used in experimental agriculture to interfere with spindle formation in cells so that the chromosome number of plants may be artificially doubled. This doubling of the chromosome number can result in larger and more vigorous varieties of plants. (Meadow saffron should not be confused with true saffron, a member of the iris family and the source of the world's most expensive spice and a powerful yellow dye.) Bowstring hemps are related to the familiar, seemingly indestructible houseplants called *sansevierias* (figure 22.27) which have long, narrow, stiff leaves that stand upright from the base. The plants are cultivated in tropical Africa for their long fibers, which are used for string, rope, bowstrings, mats, and cloth. New Zealand flax, a larger plant, is grown in South America and New Zealand for similar purposes but is also widely used in ornamental plantings.

Several species of aloes produce juices used to treat X-ray and other burns. African aloes are prized as ornamentals in areas with milder climates. Their thick fleshy leaves have short spines along the margins; the spines were once used for phonograph needles.

Many lily bulbs are edible and were widely used for food by Native Americans (*Caution:* lily bulbs should not be confused with those of daffodils and other members of the amaryllis family. Daffodil and related bulbs are highly poisonous), but their use for food now should be discouraged, as such use to any great extent might render many species extinct. One member of the lily family confined to California and southern Oregon, the California soaproot (figure 22.28), had several uses in addition to being an important food item for Native Americans of the region. The large

Figure 22.28 California soaproot

Orchid Family (Orchidaceae)

Members of this very large family, which, according to some authorities has more than thirty-five thousand species, are very widely distributed although, like members of the lily family, they are especially abundant in the tropics. In many genera the number of individual plants at any one location may be quite small—sometimes limited to a single plant.

The flowers show exceptional variation in size and form, and the habitats of the plants are equally varied. The flowers of one Venezuelan species has a diameter of less than 1 millimeter (1 twenty-fifth of an inch), whereas those of a species native to Madagascar may exceed 45 centimeters (about 18 inches) in length. Many orchids are epiphytic on the bark of trees; others are aquatic or terrestrial, and a saprophytic species native to western Australia grows and flowers entirely underground.

Orchids commonly have three sepals and three petals, with one of the petals (the *lip petal*) differing in form from the other two. The stamens and pistil are united in a single structure (see figure 20.12), the *column,* which is unique to the family. The stigma often consists of a sticky depression on the column, and the anthers, which contain sacs of pollen called *pollinia,* are usually covered with a cap until they are removed by an insect or other pollinator. The specific adaptations between orchid flowers and their pollinators is extraordinary and sometimes bizarre (see figure 23.23).

Orchids have minute seeds that are often produced in prodigious numbers (e.g., a single fruit of certain orchid species may contain up to one million seeds). Each seed consists of only a few cells, and, in order for a seed to germinate it must become associated with a specific mycorrhizal fungus that produces substances necessary for its development. Once a seed has germinated, it may take from six to twelve years or more before the first flower appears.

bulbs are covered with coarse fibers, which were removed and tied to sticks to make small brooms. The bulbs themselves produce a lather in water and were used for soap. Sometimes numbers of bulbs were crushed and thrown in a small stream that had been dammed. Fish would be stupefied and float to the surface. The bulbs were generally eaten after being roasted in a stonelined pit in which a fire had been made. While they were roasting, a sticky juice would ooze out. This was used for gluing feathers to arrow shafts.

A resin used in stains and varnishes exudes from the stem of dragon's blood plants. Grass trees of Australia yield resins used in sealing waxes and varnishes.

Chapter 22

Contrary to popular belief, some orchids can be grown relatively easily on a windowsill that has bright light, but not direct sunlight (see appendix 4). Because orchids are among the most beautiful and prized of flowers, a large industry has grown up around their culture and propagation (see the section on mericloning and tissue culture in appendix 4). One species, the vanilla orchid, is grown commercially in the tropics for its fruits, which are the source of true vanilla flavoring (see appendix 1 for the scientific names of the plants mentioned in this chapter).

Summary

Distinguishing between fact and fancy in reported past uses of plants is often difficult because plants may have had medicinal or other values despite the fact that their use was based on superstition or reasoning devoid of scientific knowledge. At present, teams of professionals in several fields are interviewing witch doctors and medicine men and women of the tropics in an attempt to save potentially useful plants from extinction.

N. I. Vavilov, building on the earlier work of A. de Candolle, determined that there were eight major centers of distribution of cultivated plants. These centers are located in China, India, and Indo-Malaysia, Central Asia, the Near East, the Mediterranean, the Abyssinian area of Africa, Central America and adjacent South Mexico, and South America (with independent centers in Peru–Ecuador–Bolivia, Chiloe Island, and Brazil–Paraguay).

The flowering plant families surveyed in somewhat phylogenetic sequence are as follows: the buttercup family (Ranunculaceae)—its members include buttercup, columbine, larkspur, anemone, monkshood, clematis, goldenseal, and wolfsbane; the laurel family (Lauraceae)—its members include cinnamon, camphor, sassafras, sweet bay, California bay (or myrtle), avocado, and laurel; the poppy family (Papaveraceae)—its members include bloodroot and opium poppy (the source of medicinal drugs and heroin); the mustard family (Brassicaceae)—its members include cabbage, cauliflower, brussels sprouts, broccoli, radish, kohlrabi, turnip, horseradish, watercress, rutabaga, shepherd's purse, western wallflower, dyer's woad, and camelina; the rose family (Rosaceae)—its members include stone fruits (e.g., cherry, apricot, peach, plum), strawberry, raspberry, rose (whose fruits are a source of vitamin C), and related wild species; the legume family (Fabaceae)—its members include all legumes (e.g., pea, bean, lentil, peanut), alfalfa, clover, licorice, wattle (source of tannins), indigo, logwood, locoweed, sensitive plant, and winged bean (a tropical bean with high protein levels); the spurge family (Euphorbiaceae)—its members include spurge, poinsettia, cassava, Pará rubber, gopher plant, candelilla, castor bean, crown-of-thorns, Mexican jumping bean, and tung oil tree; the cactus family (Cactaceae)—its members include all cacti (e.g., prickly pear, cholla, barrel cactus, peyote); the mint family (Lamiaceae)—its members include rosemary, thyme, sage, oregano, marjoram, basil, lavender, catnip, peppermint, spearmint, horehound, salvia, coleus, and chia; the nightshade family (Solanaceae)—its members include tomato, white potato, eggplant, pepper, tobacco, belladonna (source of medicinal drugs), petunia, and jimson weed (source of hallucinogenic drugs); the carrot family (Apiaceae)—its members include dill, caraway, celery, carrot, parsley, coriander, fennel, anise, parsnip, water hemlock, poison hemlock, cow parsnip, squawroot, and hog fennel; the pumpkin family (Cucurbitaceae)—its members include pumpkin, squash, cucumber, cantaloupe, watermelon, vegetable sponge, gourd, melonette, and manroot; the sunflower family (Asteraceae)—its members include sunflower, dandelion, lettuce, endive, chicory, Jerusalem artichoke, dahlia, chrysanthemum, marigold, thistle, sagebrush, pyrethrum, balsamroot, tarweed, and yarrow; the grass family (Poaceae)—its members include all cereals (e.g., wheat, barley, rye, oats, rice, corn), sugar cane, sorghum, citronella, and lemon grass; the lily family (Liliaceae)—its members include lilies, asparagus, sarsaparilla, squill, meadow saffron, bowstring hemp, aloe, and New Zealand flax; the orchid family (Orchidaceae)—one of its members is the source of vanilla flavoring.

Review Questions

1. To which flowering plant family does each of the following belong: poinsettia, lupine, columbine, peach, pear, cinnamon, sarsaparilla, belladonna, peyote, horehound, rubber, gourd, jimson weed, parsley, sorghum, asparagus, broccoli, lettuce, tomato, opium?

2. Make a list of the poisonous plants in the families discussed.

3. Which plants mentioned are or have been used for medicines?

4. Which plants mentioned have been used for tools or utensils?

5. Native Americans made extensive use of plants for a wide variety of purposes in the past. List such uses for as many plants as possible.

Discussion Questions

1. Scientific investigations often take a great deal of time and money. Is it worth the effort to check out scientifically the past uses of plants?

2. A return to exclusively herbal medicines is being advocated in some quarters. Is this a good idea? What are the pros and cons?

3. Would you expect drugs produced naturally by plants to be more effective or better than drugs produced synthetically? Why?

4. If you were asked to single out the three most important families of the sixteen discussed, which would you choose? Why?

5. A number of wild edible plants were mentioned in this chapter. What would happen if a large portion of the population were to gather these wild plants as a major source of food?

Additional Reading

Advisory Committee on Technology Innovation. *Underexploited tropical plants with promising economic value.* Washington, DC: National Academy of Sciences; 1975.

Angier, B. *Field guide to edible wild plants.* Harrisburg, PA.: Stackpole Books; 1974.

———. *Field guide to medicinal wild plants.* Harrisburg, PA.: Stackpole Books; 1978.

Benson, L. *Plant classification.* 2d ed. Lexington, MA.: D. C. Heath and Company; 1979.

Brouk, B. *Plants consumed by man.* London: Academic Press; 1975.

Chrispeels, M. J.; Sadava, D. *Plants, food and people.* San Francisco: W. H. Freeman and Co.; 1977.

Fernald, M. L.; Kinsey, A. C.; Rollins, R. C. *Edible wild plants of eastern North America.* Rev. ed. New York: Harper & Row, Publishers; 1958.

Gibbons, E. *Stalking the healthful herbs*. New York: David McKay Co.; 1970.

Hardin, J. W.; Arena, J. M. *Human poisoning from native and cultivated plants*. Durham, NC: Duke University Press; 1974.

Harrington, H. D. *Edible native plants of the rocky mountains*. Albuquerque: University of New Mexico Press; 1974.

Kingsbury, J. M. *Poisonous plants of the United States and Canada*. 3d ed. Englewood Cliffs, NJ: Prentice-Hall; 1964.

Kirk, D. R. *Wild edible plants of the western United States*. Colored ed. Healdsburg, CA: Naturegraph Publishers; 1975.

Krochmal, A.; Krochmal, C. *A guide to the medicinal plants of the United States*. New York: Quadrangle/The New York Times Book Co.; 1973.

Lewis, W. H.; Elvin-Lewis, M. P. F. *Medical botany*. New York: John Wiley and Sons; 1977.

Murphey, E. V. A. *Indian uses of native plants*. Fort Bragg, CA: Mendocino County Historical Society; 1959.

Radford, A. E.; Dickison, W. C.; Massey, J. R.; Bell, C. R. *Vascular plant systematics*. New York: Harper and Row, Publishers; 1974.

Schery, R. W. *Plants for man*. 2d ed. Englewood Cliffs, NJ: Prentice-Hall; 1972.

Sweet, M. *Common edible and useful plants of the west*. Healdsburg, CA: Naturegraph Publishers; 1962.

Turner, N. J. *Food plants of British Columbia Indians,* Part 1: *Coastal peoples*. Victoria: British Columbia Provincial Museum; 1975.

Veninga, L.; Zaricor, B. R. *Goldenseal, etc*. Santa Cruz, CA: Ruka Publications; 1976.

Williams, K. *Eating wild plants*. Missoula, MT: Mountain Press Publishing Co.; 1977.

23

Ecology

Overview

This chapter explores some of the ecological topics not already extensively discussed elsewhere in the text.

Populations, communities, and ecosystems are discussed first. This is followed by a brief look at producers, primary and secondary consumers, decomposers, and food chains. Then energy flow through an ecosystem is considered, and human disruption of ecosystems is briefly explored. Next, the nitrogen and carbon cycles are introduced, and discussions of successions follow. The subject of fire ecology is then broached. Major biomes of North America are described, and a telescopic view of their plants and animals is given. The chapter concludes with a discussion of pollination ecology.

Some Suggested Student Learning Goals

1. Describe the functions of producers, primary consumers, secondary consumers, and decomposers in an ecosystem.

2. Describe the cycling of energy in an ecosystem.

3. List at least ten ways in which humans have disrupted ecosystems.

4. Describe how nitrogen and carbon are cycled.

5. Define succession and describe a succession that begins with bare rocks.

6. Define ecotype, secondary succession, eutrophication, climax vegetation.

7. List the major biomes of North America and describe the principal living members of each.

Introduction

The broad discipline of biology that deals not only with plant and animal relationships but with the relationship of organisms in general to one another and to their environment is known as **ecology.**

The word *ecology* was first proposed by the German biologist Ernst Haeckel in 1869, and ecology was recognized as a field of biological investigation at the beginning of this century. Its origins, however, date back to early civilizations when humans first learned to modify their environment through the use of tools and fire. Since the 1950s, ecology has become a household word, and it is now such a vast area of study, with so many aspects, that it is not possible in a text of this nature to do more than supply some signposts. In the past two decades alone, topics such as the effects of pollution on lands, waters, and peoples have filled literally thousands of volumes. During this time attempts to rectify or arrest ecological damage have resulted in the requirement that various construction projects file environmental impact reports before proceeding. These reports provide information that helps various agencies evaluate the possible effects a proposed project may have on the flora, fauna, and physical environment and determine what mitigations may be necessary before the project can be approved. The process has sometimes resulted in a great deal of controversy and emotional debate between persons involved in industry and those who feel that preservation of the environment should supersede material considerations.

Various aspects of ecology were treated in earlier chapters. Soils and organisms associated with them were discussed in chapter 5. Some of the adaptations and modifications of leaves associated with specific environments were discussed in chapter 7. The effects of light and temperature on plants were covered in chapters 9 and 10. Other important ecological topics comprise the bulk of this chapter, but the reader is referred to "Additional Reading" for a sampling of works covering a number of these themes in greater depth.

We recognize that plants, animals, and other organisms tend to be associated in various ways with one another and also with the physical environment. For example, the term *forest* is applied to *populations* (groups of individuals of the same species) of trees or other plants that form a **plant community** (unit composed of all the populations of plants occurring in a given area). The lichen and moss flora on a rock also constitutes a community, as do the various seaweeds in a tidepool (figure 23.1). These communities also invariably have animals and other living organisms associated with them; it is preferable, therefore, to refer to the unit composed of all the populations of living organisms in a given area as a **biotic community.** Considered together, the communities and their physical environments constitute **ecosystems,** which are interconnected by physical, chemical, and biological processes. Each of these terms may be applied to entities ranging in size from microscopic to global.

Figure 23.1 Tidepool community at low tide.
(Courtesy Field Museum of Natural History, Chicago)

Populations

Populations may vary in numbers, in density, and in the total mass of individuals. Depending on circumstances, a field biologist may investigate a population in various ways. If, for example, a conservation organization is concerned about the preservation of a rare or threatened species, the organization may simply count the *number* of individuals, although this may not always be feasible. If a count is not feasible, the organization may estimate population *density* (number of individuals per unit volume—e.g., five blueberry bushes per square meter). If the individuals in a population vary greatly in size or are unevenly scattered, a better estimate of the population's importance to the ecosystem may be calculated by determining the *biomass* (total mass of the individuals present).

Communities

Communities are composed of populations of one to many species of various types of organisms living together in the same location under the same conditions. A community is difficult to define precisely, however, because members of one community may also occur in other communities.

Chapter 23

Furthermore, members of one community may have specific genetic adaptations to that community, so that if individuals are transplanted to a second, different community where the same species occurs, the transplanted individuals may not necessarily be able to survive alongside their counterparts, which are themselves adapted to this second community. Individuals adapted to specific communities within their overall distribution are called *ecotypes*.

Ecosystems

Living organisms that are interacting with one another and with factors of the nonliving environment constitute an **ecosystem.** The nonliving factors of the environment include light, temperature, air circulation, precipitation, and soil type. The distribution of a plant species in an ecosystem is controlled mostly by temperature, precipitation, soil type, and the effects of other living organisms (biotic factors). In Mediterranean climates, for example, nearly all precipitation occurs during the winter months and the summers are dry. This type of climate favors spring annuals that complete their life cycles by summer. Forests may occur in areas where heavy winter snowfall soaks deeply into the soil, thus compensating to a certain extent for the lack of summer moisture.

Species of plants that occur naturally in areas of low precipitation and high temperatures (*xerophytes*) generally have modifications of leaves and other parts (see discussions under "Specialized Leaves" in chapter 7 and "Regulation of Transpiration" in chapter 8) that reduce transpiration and thus adapt them to their particular environment. Plants of arid areas may also have specialized forms of photosynthesis (see chapter 9). Similarly, plants that grow in water (*hydrophytes*) have modifications that adapt them to their aquatic environment (see chapter 7).

The mineral content of soils plays an important role in influencing the distribution of plant species. For example, serpentine soils—containing relatively high amounts of magnesium, iron, usually nickel and chromium, and low amounts of calcium and nitrogen—often support species that are not found on nearby nonserpentine soils. Biotic factors such as competition for light, the mineral nutrients and water available, and grazing by the animal members of the biotic community also influence the distribution of plant species.

Ideally, ecosystems sustain themselves entirely through photosynthetic activity and the recycling of nutrients. Organisms that are capable of carrying on photosynthesis capture light energy and convert it, along with carbon dioxide and water, to energy-storing molecules. Such organisms are called **producers.** Animals such as cows, caribou, and caterpillars feed directly on producers. They are called **primary consumers. Secondary consumers,** such as tigers, toads, and tsetse flies, feed on primary consumers. **Decomposers** break down organic materials to forms that can be reassimilated by the producers. The foremost decomposers in most ecosystems are bacteria and fungi.

In any ecosystem, the producers and consumers comprise **food chains,** which determine the flow of energy through the different levels (figure 23.2). Since most organisms have more than a single source of food and are themselves often consumed by a variety of consumers, there are considerable differences in the length and intricacy of food chains. This has led to the use of the term *food web* as being more appropriate to describe the interrelationships involved.

Energy itself, which enters at the producer level, is not recycled in an ecosystem. Only about 1% of the light energy falling on a community is actually converted to organic material. The energy then gradually escapes in the form of heat as it passes from one level to another. It has been estimated that when cattle graze only about 10% of the energy stored by the green plants they consume is converted to animal tissue. Most of the remaining energy dissipates into the atmosphere as heat. When we eat beef, only roughly 10% of

Figure 23.2 Food web

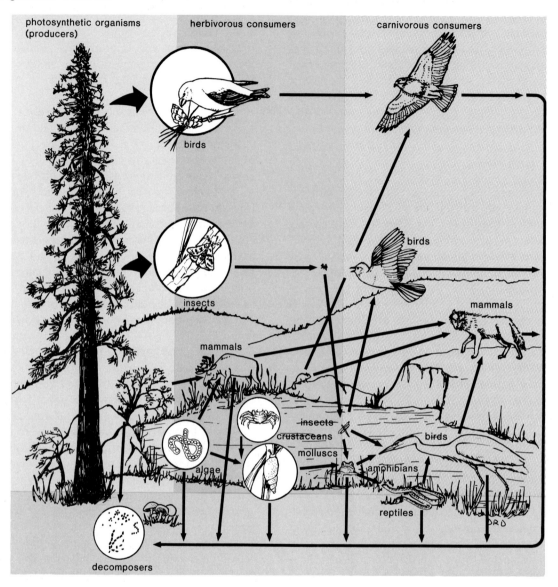

photosynthetic organisms (producers)

herbivorous consumers

carnivorous consumers

birds

insects

mammals

birds

mammals

insects
crustaceans
molluscs

algae

amphibians

birds

reptiles

decomposers

Chapter 23

its stored energy is utilized by our bodies to manufacture new blood cells and otherwise sustain life; the remaining energy is converted to heat. Obviously, then, in a long food chain the final consumer gains only a tiny fraction of the energy originally captured by the producer at the bottom of the chain. Conversely, there is proportionately much less loss of energy between levels in short food chains. Let us assume, for example, that for every 100 calories of light energy falling on lettuce plants each day, 10 calories are converted to lettuce tissue. Then suppose that the lettuce is fed to hogs. Again only 10%, or 1 calorie, of the original energy may be converted to animal tissue. If as secondary consumers we eat such pork, our bodies, in turn, may utilize 10% of the energy available in the pork, or only 0.1% of the original energy available. If, however, we eat the lettuce directly, one link in the food chain is eliminated, and as the primary consumers we receive 10% of the original energy available, a gain for us of approximately 90%. From this it follows that a vegetarian diet makes much more efficient use of solar energy than one that relies heavily on meats. It also follows that in terms of the numbers of individuals and the total mass there is a sharp reduction at each level of the food chain. In a given portion of ocean, for example, there may be billions of microscopic algal producers supporting millions of tiny crustacean consumers, which, in turn, support thousands of small fish, which meet the food needs of scores of medium-sized fish, which are finally consumed by one or two large fish (figure 23.3). In other words, one large fish may very well depend on a billion tiny algae to meet its energy needs every day.

The interrelationships and interactions among the components of an ecosystem can be quite complex, but all function together in a regulatory fashion. An increase in food made available by producers can result in an increase in consumers, but the increased number of consumers reduces the available food, which then inevitably leads to a reduction of consumers to earlier levels. The net result is sustained self-maintenance of the ecosystem.

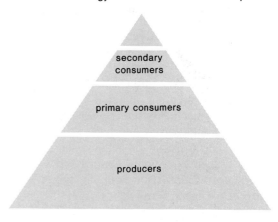

Figure 23.3 Energy pyramid of an ecosystem. There is much more energy at the bottom than at the top.

While it is easy to see that the total mass of consumers is largely determined by the total mass of food made available by the producers, the interactions among producers themselves and also between the decomposers and the other members of the ecosystem are usually more subtle. Many flowering plants produce substances that either inhibit or promote the growth of other flowering plants. Black walnut trees, for example, produce a substance that wilts tomatoes and potatoes and injures apple trees that come in contact with black walnut roots. Some bacteria and fungi limit higher plant growth through the production of various inhibitory chemical compounds. Other bacteria, fungi, and flowering plants limit population size through parasitism. The degree of parasitism varies considerably with the organisms involved. In the figwort family, some of the species have no chlorophyll and depend entirely on their flowering plant hosts for their energy and other nutritional needs. Other related species do have chlorophyll but apparently also require supplemental food from their hosts. Still other species often parasitize the roots of certain plants but are also capable of existing independently.

Large herbivorous animals such as deer and moose feed on a wide variety of plants, each differing in nutritional value. Each plant species also produces different combinations, types, and amounts of chemical compounds in addition to proteins, fats, and carbohydrates. Many of the chemical compounds are poisonous to the consumers, but the animals do not display symptoms of poisoning because their digestive systems are capable of breaking down the compounds and eliminating or excreting them to a limited degree. The limitations imposed by such compounds result in the consumers varying their diet, seeking familiar foods and being wary of new ones. If a plant species did not have some natural defense, such as chemical compounds or structural modifications (for example, spines), primary consumers of all kinds from insects to elephants would soon render that species extinct. In an ecosystem, the defenses that both producers and consumers have against each other have been developed through a process of natural selection and are maintained in delicate balance.

Humans in the Ecosystem

It has been estimated that the total human population of the world was less than 20 million in 6000 B.C. During the next 7,750 years it rose to 500 million; by 1850 it had doubled to 1 billion, and 70 years later it had doubled again to 2 billion. The 4 billion mark was reached during the early 1980s. The Earth remains constant in size, but humans obviously have occupied a great deal more of it over the past few centuries or at least have greatly increased in density of population. In feeding, clothing, and housing themselves humans have had a major impact on their environment. They have cleared natural vegetation from vast areas of land and drained wetlands; they have dumped wastes and other pollutants into rivers, oceans, lakes, and added pollutants to the atmosphere; and they have killed pests and plant disease organisms with poisonous substances.

These poisonous substances have also killed natural predators and other useful organisms, and in general have thoroughly disrupted the delicately balanced ecosystems that existed before the relentless onslaught of humans began.

If humans are to survive on this planet it is imperative that their growth rate be curbed drastically, and the many unwise agricultural and industrial practices that have accompanied the burgeoning of human populations must be supplanted with practices more in tune with restoring some ecological balance. Agricultural practices of the future will have to include the return of organic material to the soil after each harvest (see the section on composting in chapter 14), instead of adding only inorganic fertilizers. Harvesting of timber and other crops will have to be done in a manner that prevents topsoil erosion, and the practice of clearing brush with chemicals will have to be abolished. Industrial pollutants will have to be rendered harmless and recycled whenever possible. Many substances that now are still largely discarded (e.g., garbage, paper products, glass, metal cans) will also have to be recycled on a much larger scale. Biological controls (see appendix 2) will have to replace the use of poisonous controls whenever possible. Water and energy conservation will have to be universally practiced, and rare plant species, with their largely unknown gene potential for future crop plants, will need to be saved from extinction by preservation of their habitats and by other means. The general public will have to be made acutely aware of the urgency for wise land management and conservation—which will be needed especially when pressures are exerted by influential forces promoting unwise measures in the name of "progress"—before additional large segments of our natural resources are irreparably damaged or lost forever. The alternative appears to be nothing less than death from starvation, respiratory diseases, poisoning of our food and drink, and other catastrophic events ensuring the premature demise of large segments of the world's population.

The Nitrogen Cycle

As was noted in chapter 3, much of the proto-plasm of living cells consists of protein. The most abundant element in our atmosphere, nitrogen, constitutes about 18% of the protein. There are nearly 69,000 metric tons of nitrogen in the air over each hectare of land (35,000 tons per acre), but the total amount of nitrogen in the soil sel-dom exceeds 3.9 metric tons per hectare (2 tons per acre) and is usually considerably less. This discrepancy results from the nitrogen of the at-mosphere being chemically inert, which is an-other way of saying that it will not combine readily with other molecules. It is, therefore, largely unavailable to plants and animals for their use in building proteins and other substances containing nitrogen.

As noted in chapter 14, most of the nitrogen supply of plants (and indirectly, therefore, of an-imals) is derived from the soil in the form of in-organic compounds taken in by the roots. These compounds include those that contain nitrogen chemically combined with oxygen or hydrogen. They are produced by bacteria and fungi as these break down the more complex molecules of dead plant and animal tissues to simpler ones. Some nitrogen from the air is also *fixed,* that is, con-verted to compounds such as ammonia by various *nitrogen-fixing bacteria* (see figure 14.7). These organisms gain access to various plants, partic-ularly legumes (e.g., peas, beans, clover, alfalfa), through the root hairs.

As illustrated in figure 23.4, there is a con-stant flow of nitrogen from dead plant and animal tissues into the soil and from the soil back to the plants. Decay bacteria and fungi can reduce enormous quantities of dead leaves and other tis-sues to tiny fractions of their original volumes within a few days to a few months. If they were abruptly to cease their activities, the available ni-trogen compounds would be completely ex-hausted within a few decades, and the carbon dioxide supply needed for photosynthesis seri-ously depleted. Forests, jungles, and prairies would die as the accumulations of shed leaves, bodies, and debris buried the living plants and shielded their leaves from the light essential to photosynthetic activity. At present, even with the various bacteria involved in the nitrogen cycle functioning normally, the total amount of nitro-gen in the soil is not being increased by their ac-tivities but rather merely being recycled.

Further, significant amounts of nitrogen are continually being lost as water leaches it out or carries it away through erosion of topsoil. More is lost with each harvest, the average crop re-moving about 25 kilograms per hectare (25 pounds per acre) per year. This nitrogen loss from the harvesting of crops can be sharply reduced if vegetable and animal wastes are recycled and re-turned to the soil each year. While bacteria are decomposing tissues they use nitrogen, and little is available until they die and release their ac-cumulations into the soil. Accordingly, crops should not be planted in soils to which only par-tially decomposed materials have been added until bacteria have completed their breakdown of or-ganic matter. Likewise, when sawdust, straw, or other organic mulches are spread around plants in a garden to control weeds and conserve soil moisture, less nitrogen will be available to the growing plants until the mulches have been de-composed.

Weeds and stubble are frequently burned as a means of eradication. Fire causes serious loss of nitrogen, however, which has to be replaced. It has been estimated that the annual combined loss of nitrogen from the soil in the United States from fire, harvesting, and other causes exceeds 21 mil-lion metric tons (23 million tons), and only 15.5 million metric tons (17 million tons) are replaced by natural means. To offset the net loss, some 32 million metric tons (35 million tons) of chemical fertilizers are applied to the soils each year. If or-ganic matter is not added at the same time, how-ever, this application of chemical fertilizers, combined with the annual burning of stubble, may result eventually in the creation of a *hardpan* soil.

Figure 23.4 The nitrogen cycle

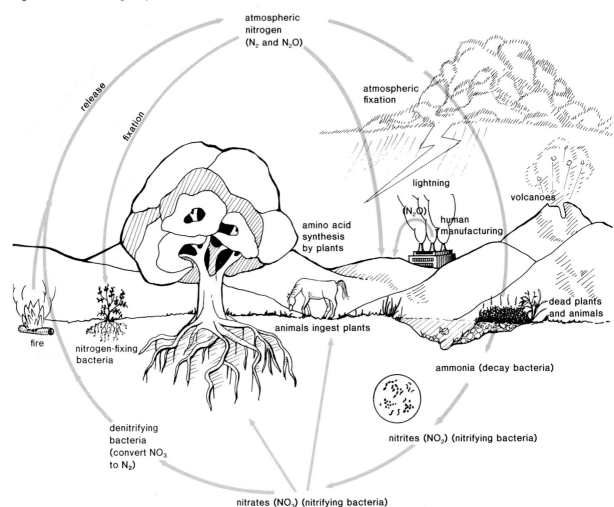

Hardpan develops through the gradual accumulation of salt residues, which dissolve humus and disrupt the structure of the soil, causing the clay particles to clump and also producing colloids that are impervious to moisture. In hardpan soils and others low in oxygen (e.g., flooded areas), *denitrifying bacteria* use nitrates instead of oxygen in their respiration, thus rapidly depleting the remaining soil nitrogen.

Precipitation returns a little nitrogen to the soil from the atmosphere, where it has accumulated as a result of the action of light on industrial pollutants, fixation by flashes of lightning, and diffusion of ammonia released through decay. The activities of nitrogen-fixing bacteria and volcanoes also contribute to the natural replenishment of nitrogen by converting it to forms that can be utilized by plants.

Chapter 23

Figure 23.5 The carbon cycle

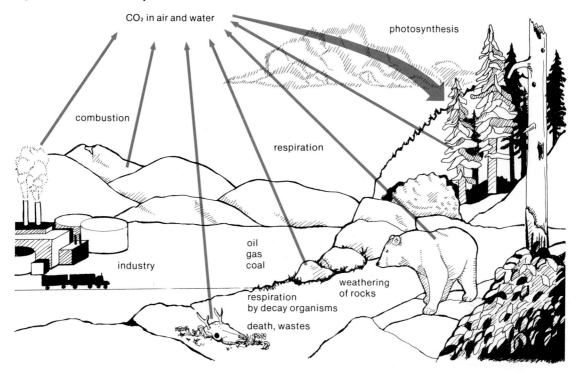

CO₂ in air and water

photosynthesis

combustion

respiration

industry

oil
gas
coal

weathering
of rocks

respiration
by decay organisms

death, wastes

The Carbon Cycle

In the process of recycling nitrogen, bacteria also play a major role in recycling carbon and many other substances (figure 23.5). As noted in chapter 9, one of the two raw materials of photosynthesis is carbon dioxide, which constitutes 0.03% of our atmosphere. It is estimated that the combined plant life of the oceans and the land masses uses about 14.5 billion metric tons (16 billion tons) of carbon obtained from carbon dioxide every year. This is replaced through the respiration of all living organisms, with perhaps as much as 90% or more being produced by the incredible numbers of decay bacteria and fungi as they decompose tissues. Lesser amounts of carbon dioxide are released into the air as a result of the burning of fossil fuels by the internal combustion engines of industry and transportation,

and a small amount originates with fires and volcanic activity. At the present rate of use by photosynthetic plants, it has been calculated that all of the carbon dioxide of our atmosphere would be used up in about twenty-two years if it were not constantly being replenished.

For aeons carbon, nitrogen, water, phosphorus, and other molecules have been passing through cycles. Some molecules that were a part of a primeval forest that became compressed and turned to coal may have become part of another plant after the coal burned. Then the new plant may have been eaten by an animal, which, in turn, contributed molecules to a part of yet another living organism. Just think of where the molecules in your own body may have been in the past few billion years. They may well have been a part of some prehistoric seaweed, a saber-toothed tiger, a mighty dinosaur, or even all three!

Succession

After a volcano spews lava over a landscape, or after an earthquake or a landslide exposes rocks for the first time, there is initially no sign of life on the lava or rock surfaces. Within a few years or sometimes within a few months, however, living organisms begin to appear and an orderly sequence of events known as **succession** takes place. During succession the species of plants and other organisms that first appear gradually alter their environment as they carry on their normal activities such as metabolism and reproduction. In time the accumulation of wastes, dead organic material, and inorganic debris, and other changes (such as of shade and water content in the habitat) favor different species, which may replace the original ones. These, in turn, modify the environment more so that yet other species become established.

Succession occurs whenever and wherever there has been a disturbance of natural areas on land or in water. It proceeds at varying rates, depending on the climate, the soils, and the animals or other organisms in the vicinity. Ecologists recognize a number of variations of two basic types of succession. *Primary succession* involves the actual formation of soil in the beginning stages, while *secondary succession* takes place in areas that had previously been covered with soil and vegetation.

Figure 23.6 Lichens in early stages of succession on a Hawaiian lava flow

Primary Succession: Xerosere

One of the most universal types of primary succession, called a *xerosere,* begins with bare rocks and lava (see discussion of soils in chapter 5) that have been exposed through glacial or volcanic activity or through landslides. Initially, the rocks are sometimes subjected to alternate thawing and freezing, at least in temperate to colder areas. Tiny cracks or flaking may occur on the surface as a result. Lichens become established on such surfaces (figure 23.6). They produce acids that very slowly etch the rocks, and as they die and contribute organic matter, they are replaced by other, larger lichens. Certain rock mosses

Figure 23.7 Live oak that grew from an acorn lodged in a small crack of the rock. The roots of the tree split the rock apart.

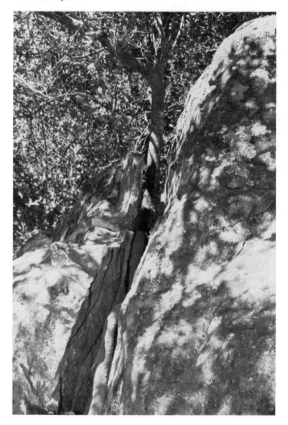

adapted to long periods of desiccation also become established, and a small amount of soil begins to build up. This is augmented by dust and debris blown in by the wind. Eventually enough of a mat of lichen and moss material is present to permit some ferns or even seed plants to become established, and the pace of soil buildup and rock breakdown accelerates. If deep cracks appear in the rocks, the larger seeds may widen them further as they germinate and the roots expand in girth. It has been calculated that germinating seedlings can exert a force of up to 31.635 kilograms per square centimeter (450 pounds per square inch). Indeed, there are known instances of seedlings splitting rocks that weigh several tons (figure 23.7).

As soil buildup continues, larger plants take over, and eventually the vegetation reaches an equilibrium in which the associations of plants and other organisms remain the same until another disturbance takes place or climatic changes occur. Such relatively stable plant associations are referred to as **climax vegetation.** The climax vegetation of deciduous forests in eastern North America are dominated by maples and beeches, oaks and hickories, hemlocks and white pines, or other combinations of trees. In desert regions, various cacti form a conspicuous part of the climax vegetation, while in the Pacific Northwest large coniferous trees predominate. In parts of the Midwest, prairie grasses and other herbaceous plants form the climax vegetation, and in wet tropical regions a complex association of jungle plants constitutes the climax.

Primary Succession: Hydrosere

Succession takes place in wet habitats as well as in drier ones and such succession is called a *hydrosere.* In the northern parts of midwestern states such as Michigan, Wisconsin, and Minnesota, ponds and lakes of various sizes abound. Many were left behind by retreating glaciers and often have no streams draining them. The water that evaporates from them is replaced annually by precipitation runoff. They also grow a tiny bit smaller each year as a result of succession (figure 23.8). This succession begins with algae either carried in by the wind or transported on the muddy feet of waterfowl and wading birds. The algae tend to multiply in shallow water near the margin, and with each reproductive cycle, the dead parts sink to the bottom. Floating plants such as duckweeds may then appear, often forming a band around the body of water just off-shore. Next, water lilies and other rooted aquatic plants with floating leaves become established, each group of plants contributing to the organic material on the bottom, which slowly turns to muck. Cattails and other flowering plants that produce their inflorescences above the water often take root in the muck around the edges, and the accumulation of organic material accelerates.

Meanwhile the algae, duckweeds, and other plants move farther out, and the surface area of exposed water gradually diminishes. Grasslike sedges become established along the damp margins and sometimes form floating mats as their roots interweave with one another. Dead organic material accumulates and fills in the area under the sedge mats, and herbaceous and shrubby plants then move in. As the margins become less marshy, coniferous trees whose roots can tolerate considerable moisture (e.g., tamaracks or eastern white cedars) gain a foothold, eventually growing across the entire site as the pond or lake disappears. The trees continue to aid in the formation of true soil, and in due course the climax vegetation takes over. No visible trace of the pond or lake now remains, and the only evidence of its having been there lies beneath the surface, where fossil pollen grains, bits of wood, and other material reveal the past history. Such succession may take thousands of years and has never been witnessed from start to finish. The evidence that it does occur, however, is extensive and compelling.

Under natural conditions stream-fed lakes and ponds eventually become filled with silt and debris, although this, too, may take thousands of years to occur. The streams that feed these lakes gradually bring in silt, and the nutrient content of the water slowly rises as dissolved organic and inorganic materials are brought in. This gradual enrichment, called **eutrophication,** permits the growth of algae and other organisms, which also add their debris to the bottom of the lake. When sewage and other pollutants are permitted to enter the lake, the process of eutrophication may be greatly accelerated through stimulation of the growth of aquatic organisms. Eutrophication may also be accelerated when trees are cleared from land surrounding lakes prior to the construction of summer homes and resorts. The cleared land erodes more readily, with precipitation runoff carrying soil into the water. Regardless of size, all bodies of water, including rivers, are subjected to these processes.

Secondary Succession

Secondary succession, which occurs more rapidly than primary succession, may take place if soil is already present. Many secondary successions follow man-made disturbances (e.g., land that was cleared when timber was harvested or land converted to farmland); other secondary successions follow fires. Grasses and other herbaceous plants become established on burned or logged land. These usually are followed by trees and shrubs that have widely dispersed seeds (e.g., aspen and sumac in the Midwest and East, and chaparral plants such as chamise and gooseberries in the West). After going through fewer stages than are typical of primary succession, the climax vegetation takes over, often within 100 years or less.

Fire Ecology

Natural fires started primarily by lightning and the activities of prehistoric humans occurred for thousands of years in North America and other continents. Trees such as the giant redwood and ponderosa pines, although scarred by certain types of fires, often survive, and the dates of fires can be determined by the proximity of the scars to specific annual rings (see figure 6.7). In the West checks of growth rings of ponderosa pines show that forests there used to burn on an average of approximately every six to seven years in the past. These fires and the climate, topography, and soil combined to have a profound effect on various biomes. But as humans became civilized, major and largely successful efforts to control fires were made and this, in turn, significantly altered vegetational patterns. As knowledge of the role of fires in the maintenance of ecosystems has accumulated, it has become apparent that trying to eliminate fires, at least in certain areas, disrupts natural habitats more in the long run than allowing them to occur, and agencies such as the U.S. National Park Service now allow fires at higher elevations to run their natural course.

Fires actually benefit grasslands, chaparral, and forests by converting accumulated dead organic material to mineral-rich ash, whose nutrients are recycled within the ecosystem. If the soil has been subjected to fire, some of its nutrients and organic matter may have been lost and the composition of microorganisms originally present is likely to have been altered. Losses are offset, however, by the fact that soil bacteria and blue-green algae, which are capable of fixing nitrogen from the air, increase in numbers after a fire, and there is a decrease in fungi that cause plant diseases.

Figure 23.9 Secondary succession on a burned area. Chamise and bracken ferns are resprouting the first season after a burn. (Courtesy Robert A. Schlising)

In some areas, such as the prairie states of the Midwest, grasses are better adapted to fire than woody plants. They produce seeds within a year or two after germination, and perennial grass buds, which are at the tips of rhizomes close to or beneath the surface where they escape the most intense heat of fires, usually survive, producing new growth the first season after a fire. Thus, a fire destroys only one season's growth of grass, often after growth has been completed. Shrubs, however, have much of their living tissue above ground, and a fire may destroy several years' growth. In addition, woody plants often do not produce seeds until several years after a seed germinates. Many shrubs do sprout from burned stumps, particularly in chaparral areas (figure 23.9), but repeated burning keeps them small, thus favoring grasses. There is evidence that at least some of the North American grasslands originated and were maintained because of fire. Since grassland fires have largely been controlled, many of the areas have now been invaded by shrubs that were once confined to water courses.

Fires also play a role in the composition of forests. In the mountains of east-central California, gooseberry and deerbrush appear in abundance after a fire, but their numbers stabilize within fifteen to thirty years when larger trees return to the area. Ponderosa, jack and southern longleaf pines, and Douglas firs (which do not tolerate shade) are among the species that repeatedly replace themselves after fires, and seeds of some species rarely germinate until they have been exposed to fire. The majority of chaparral species, both woody and herbaceous, are so adapted to fire that their seeds also will not germinate, as a rule, until fires remove accumulated litter and toxic wastes produced by the plants during growth.

Chapter 23

Figure 23.10 Major biomes of North America

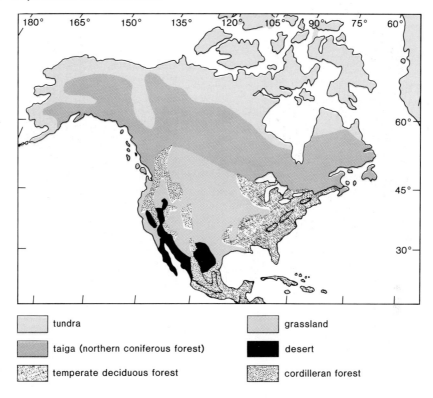

Legend:

- tundra
- taiga (northern coniferous forest)
- temperate deciduous forest
- grassland
- desert
- cordilleran forest

In view of the long-range beneficial effects of fires in some ecosystems, wise land management practices of the future undoubtedly will include guiding succession in many plant communities to the *sere* (a series of ecological communities) of greatest utility and safety.

Major Biomes of North America

Biotic communities considered on a global or at least on a continental scale are referred to as **biomes.** Several important biomes occupy the North American continent (figure 23.10), and most are also represented on other continents.

Tundra

Tundra (figure 23.11) occupies vast areas of the earth's land surfaces (about 20% in all), primarily above the Arctic circle. Patches of alpine tundra occur above timberline in mountains below the Arctic circle. It is a virtually treeless biome, although miniature willows only 2.5 centimeters (1 inch) tall do survive in some areas. Another characteristic of arctic tundra is the presence of *permafrost* (permanently frozen soil) at a depth of from a few centimeters (an inch or two) to about 1 meter (3 feet 3 inches) below the surface. The level of the permafrost determines the depth to which plant roots can penetrate. It is generally

Figure 23.11 Arctic tundra near Mt. McKinley. (Photo by Wm. E. Ferguson)

absent from alpine tundra areas. Precipitation averages less than 25 centimeters (10 inches) a year, but because of the permafrost, it is largely held at or near the surface of the land. There is a very short growing season of only two to three months, with frost possible on any day of the year; however, temperatures can soar to 27°C(81°F) during a long midsummer day. The vegetation of arctic tundra is dominated by lichens and grasses, with mosses and sedges abundant in some areas. Grasses and sedges tend to predominate in alpine tundra. During the brief growing season tiny perennial plants produce brightly colored flowers and form brilliant mats over the topsoil, which is largely organic and generally only 5.0 to 7.5 centimeters (2 to 3 inches) deep.

The biome is exceptionally fragile. A car driven across tundra compresses the soil sufficiently to kill plant roots, and the tracks are evident many years later. Occasionally sheep grazing tundra have been observed to pull up patches of the matted vegetation, leaving exposed edges. High winds then can catch the exposed edges and rip away larger segments of mat, leaving barren patches called *blowouts*.

Animals of the tundra include lemmings, the little rodents that reproduce prodigiously every three or four years and then decline in numbers. The arctic fox and snowy owl also fluctuate in numbers, the fluctuations being correlated with the size of the populations of lemmings, which constitute their principal food. Other animals of the tundra include caribou, polar bears (near the coast), shrews, ptarmigans, loons, plovers, jaegers, and arctic terns.

Figure 23.12 Taiga in northern Minnesota

Taiga

The taiga (figure 23.12), also referred to as northern coniferous or boreal forest, is found adjacent to and south of the arctic tundra. The vegetation is dominated by evergreen trees such as spruces, firs, hemlocks, and pines; tamaracks also occur in the taiga. Birches, aspens, and willows may be found in some of the wetter areas. Many perennials and a few shrubs occur in the taiga, but there is a dearth of annuals. Snow blankets the region during the winters, which are long and severe, with temperatures dropping to −50°C (−58°F) or lower during the coldest months. In summer, however, the temperatures often reach 27°C (81°F). Most precipitation occurs in the summer, ranging from about 25 centimeters (10 inches) to more than 100 centimeters (39 inches) in parts of western North America. Many lakes, ponds, and marshes dot the region, which is inhabited by a variety of birds, including jays, warblers, and nuthatches. Many rodents such as shrews and jumping mice, and larger mammals, notably moose, occur. Ermine and wolverines also make their home in the taiga.

Temperate Deciduous Forest

Deciduous trees are generally broad-leaved species that shed their leaves annually during the fall and remain dormant during the shorter days and colder temperatures of winter. Most temperate deciduous forests are found, like the taiga, on large continental masses in the northern hemisphere. In North America, this type of forest occurs from the Great Lakes region south to the Gulf of Mexico and extends from roughly the Mississippi river to the eastern seaboard, although some place the eastern coastal plain in a separate biome. Temperatures within the area vary a great deal but normally fall below 4°C (39°F) in midwinter and rise above 20°C (68°F) in summer. The climax trees of the forest are well adapted to subfreezing temperatures as long as the cold is accompanied by precipitation or snow cover. Annual precipitation totals between 50 and 165 centimeters (20 to 65 inches).

Some of the most beautiful of all the broad-leaved trees are found in this biome in a variety of associations. In the northwestern areas, sugar maples and American basswoods predominate (figure 23.13). Sugar maples are also found to the northeast, where they tend to be associated with the stately American beeches. In the west-central part of the forest, oaks and hickories abound. Oaks also are abundant along the eastern slopes of the Appalachian Mountains, where American chestnuts were once a conspicuous part of the flora. The chestnuts have now virtually disappeared, having succumbed to the ravages of chestnut blight disease, which was introduced and began taking its ugly toll during the early 1900s. Oak trees extend into the southeastern United States, where they become associated with pine trees.

Before the arrival of European immigrants, a mixture of large deciduous trees was found on the eastern slopes and valleys of the Appalachian Mountains that included maples, ashes, basswoods, beeches, buckeyes, hickories, oaks, tulip trees, and magnolias. Some of the trees grew over 30 meters (100 feet) tall and had trunks up to 3 meters (10 feet) in diameter. Except for a few protected pockets in the Great Smoky Mountains National Park, the largest trees of this rich forest

have been all but eliminated through logging. A smaller-treed extension of this forest is found in western Tennessee and Kentucky and in southern Illinois. American elms, also once a part of the forest, are rapidly disappearing as Dutch elm disease fells both wild trees and those planted along city streets. One small midwestern town, which had some six hundred elms planted along its streets, found itself left with only six live trees within a year or two after Dutch elm disease struck.

On the northern and southeastern borders of the forest is a mixture of deciduous trees and evergreens. Hemlocks and eastern white pines are found from New England west to Minnesota and south to the Appalachians. The once vast stands of eastern white pine are now largely gone, their valuable lumber having been used for construction and other purposes. Some have been lost to still another tree disease, white pine blister rust, but scattered trees remain, particularly where they have been protected. Various pines dominate the eastern coastal plain from New Jersey to Florida. Pitch pine is common in New Jersey, while some of the southern pines (e.g., long-leaf, slash, loblolly) are now cultivated in the southeastern United States for turpentine, lumber, and other commercially valuable products.

During the summer the trees of the deciduous forests form a relatively solid canopy that keeps most direct sunlight from reaching the floor. Many of the showiest spring flowers of the region (such as bloodroot, hepatica, trilliums, and violets) flower before the trees have leafed out fully and complete most of their growth within a few weeks. Other plants that can tolerate more shade, principally members of the sunflower family (e.g., asters and goldenrods), flower in succession in forest openings from midsummer through fall.

Animal life in the eastern deciduous forest includes red and gray foxes, raccoons, opossums, many rodents such as gray squirrels, snakes such as copperheads and black rat snakes, salamanders, and a wide variety of birds, including hawks, flickers, and mourning doves.

Chapter 23

Figure 23.14 Undisturbed western grassland in the spring

Grassland

Naturally occurring grasslands (figure 23.14) are found toward the interiors of continental masses. They tend to intergrade with forests, woodlands, or deserts at their margins, depending on precipitation patterns and amounts. Annually a grassland may receive as little as 25 centimeters (10 inches) or as much as 100 centimeters (39 inches) of rainfall. Temperatures can range from 50°C (122°F) in midsummer to −45°C (−50°F) in midwinter.

In North America huge herds of buffalo once grazed the prairies, but these disappeared as the settlers cultivated more and more of the land and hunters slaughtered more and more of the large animals. By 1889, there were only 551 left. Large areas are now used for growing cereal crops (particularly corn and wheat) and for grazing cattle. Before it was destroyed, the American prairie was a remarkable sight. In Illinois and Iowa, the grasses grew over 2 meters (6 feet 6 inches) tall during an average season and another meter taller during a wet one. A dazzling display of wildflowers began before the young perennial grasses emerged in the spring and continued throughout the growing season. I have counted over fifty species of flowering plants in bloom at the same time on 1 hectare (2.47 acres) of protected grassland in the middle of spring.

Grasslands occurring in areas with a Mediterranean climate (e.g., the Great Central Valley

Figure 23.15 Zonation in a vernal pool in northern California. (Courtesy Robert A. Schlising)

of California), where most of the precipitation takes place during the winters, usually include vernal pools. These are temporary accumulations of rainwater that evaporate after the rains come to an end. Their unique floras include an orderly succession of flowering plants, some appearing initially at the pool margins and each species forming a distinct zone or band until the water is gone (figure 23.15). Some species flower only in the damp soil and drying mud that remains. The seeds of many species germinate under water.

Grassland animals include cottontails, jackrabbits, gophers, mice, and pronghorns. As indicated earlier, buffalo were once abundant; some 15,000 now live in national parks, game preserves, and on ranches. Various sparrows (e.g., vesper and savannah sparrows) and other birds still find homes in uncultivated tracts of this biome.

Deserts

If you ask the average person to describe a desert, the response will probably include words like *sand, heat, mirage, oasis,* and *camels.* These are indeed found in some of the world's large deserts that are located in the vicinity of 30 degrees latitude both north and south of the equator, but deserts occur wherever precipitation is consistently low or the soil is too porous to retain water (figure 23.16). Most receive less than 12.5 centimeters (5 inches) of rain per year. The low humidity results in wide daily temperature ranges. On a summer day when the temperature has reached over 35°C (95°F), it will generally fall below 15°C (59°F) the same night. The light intensities reach higher peaks in the dry air than they do in areas where atmospheric water vapor filters out some of the sun's rays. Many desert plants have become adapted to these higher light intensities through the evolution of *crassulacean acid metabolism* (CAM) photosynthesis, a process by which certain organic acids accumulate in the chlorophyll-containing parts of the plants during the night and are converted to carbon dioxide during the day (see chapter 9). This permits much more photosynthesis to take place than would otherwise be possible, since most of the carbon dioxide of the atmosphere is excluded from the plants during the day by the stomata, which remain closed (thereby also retarding water loss).

Other adaptations of desert plants include thick cuticles, water storage tissues in stems and leaves, leaves with a leathery texture and/or reduced in size, or even total absence of leaves. In cacti and other succulents without functional leaves, the stems take over the photosynthetic activities of the plants. Desert perennials are adapted to the biome in various ways. Cacti and similar succulents have widespread shallow root systems that can rapidly absorb water from the infrequent rains. The water can then be stored for long periods in the interior of the stems, which

are modified and permit this function. Other perennials grow from bulbs that are dormant for much of the year. Annuals provide a spectacular display of color and variety, particularly during an occasional season when above-average precipitation has occurred. The seeds of the annuals often germinate after a fall or winter rain and then grow slowly or remain in rosette form for several months before producing flowers in the spring. Literally hundreds of different species of desert annuals may occur within a few square kilometers (one or two square miles) of typical desert in the southwestern United States.

Desert animals are adapted, in many instances, to foraging at night when lower air temperatures prevail. These include various mice and kangaroo rats, snakes (notably rattlesnakes), chuckwallas, and lizards (e.g., gila monster). Various thrashers, doves, and flycatchers are included in the bird life.

Cordilleran Forest

In the geologic past, deciduous forests extended to western North America. As the climate changed and summer rainfall was reduced, conifers largely replaced the deciduous trees, although some (e.g., maples, birches, aspens, oaks) still remain, particularly at the lower elevations. Today coniferous forests occupy vast areas of the Pacific Northwest and extend south along the Rocky Mountains and the Sierra Nevada and California coast ranges. Isolated pockets of this biome also occur in other parts of the West, particularly toward the southern limits of the mountains. The trees tend to be very large, particularly in and to the west of the Cascade Mountains of Oregon and Washington and on the western slopes of the Sierra Nevada. Part of the reason for the huge size of trees such as Douglas fir is the high annual rainfall, which exceeds 250 centimeters (100 inches) in some areas; however, the world's tallest trees, the coastal redwoods of California (figure 23.17), apparently depend more on fog than on large amounts of rain for their size and longevity.

One of the characteristics of the *cordilleran* (growing on a system of mountain ranges, or cordillera) forest and other forests in mountainous regions is a fairly conspicuous altitudinal zonation of species. In other words, one encounters different associations of plants as one proceeds from sea level up the mountainsides. At lower elevations in both the Rocky Mountains and the Sierra Nevada, the predominant conifer is ponderosa pine. At lower elevations in the northern part of the Cascades, Douglas firs, western red cedars, and western hemlocks are more common. At intermediate elevations in the Sierra Nevada, sugar pines, white firs, and Jeffrey pines take over, while at higher elevations other species of pines and firs predominate.

With the exception of the western side of the Olympic Mountains in Washington, most of the cordilleran forest has comparatively dry summers. This led to frequent forest fires, even before human carelessness became a factor. Several tree

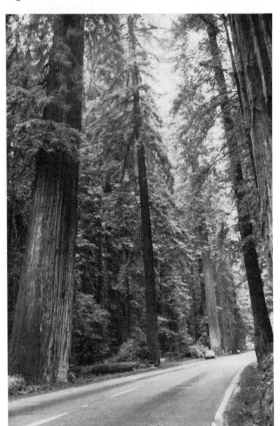

Figure 23.17 Coastal redwoods in northern California

species are well adapted to fires. The Douglas fir, for example, has a very thick protective bark that can be charred without transmitting sufficient heat to the interior to damage more delicate tissues, and its seedlings thrive in open areas after a fire. When the bark of the giant redwoods of the Sierra Nevada is burned, the trees are rarely killed. This has undoubtedly contributed to the great age and size many of the trees have attained. The cones of some of the pine trees (e.g., knobcone pine) remain closed and do not release their seeds until a fire causes them to open, while seeds of several other species germinate best after they have been exposed to fire. These attributes of members of the cordilleran forest biome have led to the practice in some of our national parks

Figure 23.18 Flower markings on a coneflower.
A. Under ordinary light. *B.* Under ultraviolet light.

A.

B.

of allowing fires at higher elevations to run their natural course. The higher-than-normal incidence of fires occurring since humans came in large numbers to the forest has made it necessary, however, for us to control them in most instances, even though in doing so we may be interfering with natural cycles that would otherwise occur in the biome.

Animal life in the cordilleran forest includes many different rodents (especially chipmunks and voles), bears, mountain lions, bobcats, mountain beavers, and mule deer. Large birds such as the golden eagle and many small birds, including mountain chickadees, warblers, and juncos, are an integral part of this biome.

Special Topic: Pollination Ecology

When certain consumers forage among producers for food, they often come in contact with flowers. Many insects and other animals become dusted with pollen, and in the course of their travels they unwittingly but effectively bring about pollination of the plants they visit.

Throughout the evolutionary history of the flowering plants, the pollinators have evidently coevolved with plants. In some instances, the relationship between the two has become highly specialized.

Included among pollinators of present-day flowering plants are some twenty thousand different species of bees. By far the best known of these are honeybees. Their chief source of nourishment is nectar, but they also gather pollen for their larvae. The flowers that bees visit are generally brightly colored and predominatly blue or yellow in color—rarely pure red. Pure red appears black to them, and therefore they generally overlook red flowers. Flowers often have lines or other distinctive markings that may function as honey guides that lead the bees to the nectar. Bees can see ultraviolet light (a part of the spectrum not visible to humans), and some flower markings are visible only in ultraviolet light, making patterns seen by bees sometimes different from those seen by humans (figure 23.18).

Many bee-pollinated flowers are delicately sweet and fragrant. By way of contrast, flowers

Figure 23.19 Stapelia (carrion flower). (Courtesy Field Museum of Natural History, Chicago)

Figure 23.20 Hummingbird visiting a flower. (Courtesy Vernon H. Oswald)

pollinated by beetles tend to have stronger yeasty, spicy, or fruity odors. They are usually white or dull in color, in keeping with the diminished visual sense of their pollinators. Some beetle-pollinated flowers secrete no nectar but either furnish the insects with pollen or have food available in special storage cells on the petals, which the beetles consume.

Certain flowers pollinated by short-tongued flies tend to be dull red or brown in color and often have foul odors resembling those of rotten meat. Such flowers include the stapelias of southern Africa (figure 23.19). These plants are related to our milkweeds, although superficially they do not resemble milkweeds at all. They are often called *carrion flowers* because of their foul odor and appearance. A number of bee-pollinated flowers are also pollinated by flies with longer tongues.

Moth- and butterfly-pollinated flowers are similar to bee-pollinated flowers in that they frequently have sweet fragrances. Many moths forage only at night, and the flowers they visit tend to be white or yellow—colors that stand out against dark backgrounds in starlight or moonlight. Some of the most specialized relationships between moths and flowers are found between certain small moths and members of the genus *Yucca*. This genus includes Joshua trees and Our Lord's Candle—the striking plants with the stiff trunks and sword-shaped leaves that stand out on the landscape of some southwestern desert areas of North America. The female moth gathers balls of sticky pollen from the stamens of one plant and carries them, with the aid of specialized mouth parts, to another plant. There the moth enters the flowers and pierces the ovary walls, laying eggs in the vicinity of the tiers of ovules. Then, depending on the species, the moth either packs the pollen balls into stigma cavities or rubs the pollen across the stigma surfaces. As the ovules develop into seeds, the eggs hatch, and the larvae consume some of the seeds for food. When the larvae are fully grown, they chew their way out of the ovaries and remain inactive in cocoons just below the surface of the ground until the next flowering season, when the adult moths emerge. The larvae seldom destroy more than a third of the seeds.

Chapter 23

Figure 23.21 Cluster of four pollen grains of California fuchsia, showing the threads that catch in the short bristle hairs at the base of a hummingbird's bill

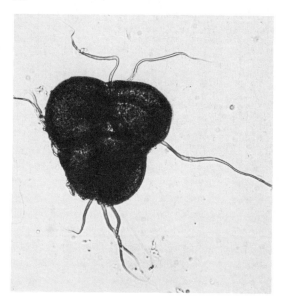

Figure 23.22 Bat visiting a flower of an organpipe cactus. (Photo by Donna J. Howell)

Some butterflies can detect red colors, and so red flowers are sometimes pollinated by them. The nectaries of these flowers are found at the bases of corolla tubes or spurs, where only the longer tongues of moths and butterflies can reach. Occasionally, however, an enterprising bumblebee will bypass convention and chew through the base of a spur to get at the nectar.

Birds—particularly the hummingbirds of the Americas and the sunbirds of Africa—and the flowers that they pollinate are also adapted to one another (figure 23.20). The birds do not have a highly developed sense of smell, but they have a keen sense of vision. Their flowers are thus frequently bright red or yellow and usually have little if any odor; they are typically large or are part of a large inflorescence. Birds are highly active pollinators and tend to burn up energy rapidly; therefore, they must feed frequently to sustain themselves. Many of the flowers they visit produce copious quantities of nectar, thus assuring repeated visits by the birds. The nectar is frequently produced in long floral tubes, which prevent most insects from gaining access to it. Some

native California fuchsias and their relatives have long threads extending from each pollen grain (figure 23.21). When a hummingbird inserts its long bill into such a flower, the pollen grain threads catch on some short stiff hairs located toward the base of its bill, and the bird unknowingly transfers pollen from one flower to another by this means.

Bat-pollinated flowers, found primarily in the tropics, tend to open only at night when the bats are foraging (figure 23.22). These flowers are dull in color, and like bird-pollinated flowers, they either are large enough for the animal to insert part of its head, or consist of ball-like inflorescences containing large numbers of small flowers whose stamens readily dust the visitor with pollen.

The very large orchid family, which has approximately thirty-five thousand species, has pollinators among all the types mentioned. Some of the adaptations between orchid flowers and their pollinators are extraordinary (figure 23.23).

Figure 23.23 Some pollinating mechanisms in orchids

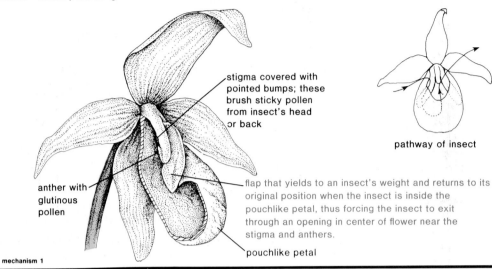

stigma covered with pointed bumps; these brush sticky pollen from insect's head or back

pathway of insect

anther with glutinous pollen

flap that yields to an insect's weight and returns to its original position when the insect is inside the pouchlike petal, thus forcing the insect to exit through an opening in center of flower near the stigma and anthers.

pouchlike petal

mechanism 1

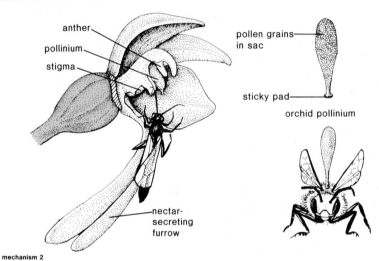

anther

pollinium

stigma

pollen grains in sac

sticky pad

orchid pollinium

nectar-secreting furrow

mechanism 2

An insect follows a nectar-secreting furrow on a petal until it reaches the center of the flower. Pollinia are removed when they adhere to the insect following an "explosion" initiated by the insect bumping a sensitive point below an anther. When the insect flies to another flower the pollinia on its head catch on the sticky surface of the stigma and are pulled off at the same time new pollinia are being attached.

Orchids pollinated by butterflies and moths may have a clamp at the base of the pollinia. When an insect inserts its tongue down the throat of the flower and touches a sensitive area, the clamp attaches the pollinia to the insect's tongue.

mechanism 3

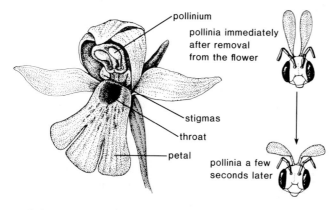

pollinium

pollinia immediately
after removal
from the flower

stigmas

throat

petal

pollinia a few
seconds later

In the showy orchids, there are two separate stigma patches. When an insect visits a flower the pollinia are attached to its head; then they twist outward and forward so that when the insect visits another flower the pollinia are simultaneously deposited in the separate stigmas while new pollinia are attached to the insect's head.

mechanism 4

One genus of orchids native to Europe and North Africa has a petal modified to resemble a female wasp. Male wasps try to copulate with the ''female'' and in doing so bump a sensitive area containing pollinia. The pollinia are ''glued'' to a wasp's head in the process, and carried to another flower. The wasps thus effectively bring about cross-pollination.

mechanism 5

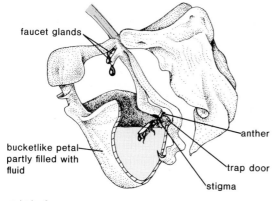

faucet glands

anther

bucketlike petal
partly filled with
fluid

trap door

stigma

The bucket orchids of South America have faucet glands that partially fill a bucketlike petal with fluid. In visiting the flower the insects may fall into the fluid. They escape by a trap door. The pathway of an insect assures contact with both the stigma and anthers containing pollinia, since they are located just in front of the trap door.

mechanism 6

Ecology

Many orchids produce their pollen grains in little sacs called **pollinia** (singular: **pollinium**), which typically have sticky pads at the bases. When a bee visits such a flower, the pollinia are usually deposited on its head. The "glue" of the sticky pads drys almost instantly, causing the pollinia to adhere tightly. In some orchids, the pollinia are forcibly "slapped" on the pollinator through a trigger mechanism within the flower. Members of one genus of orchids found in Europe and in North Africa have a petal modified so that it resembles a female bumblebee or wasp. Male bees or wasps emerge from their pupal stage a week or two before the females and apparently mistake the flowers for potential mates. They try to copulate with the flowers, and while they are doing so, pollinia are deposited on their heads. When they visit other flowers, the pollinia catch in sticky stigma cavities. The pollinia lost in the process are replaced with new ones, the switch taking place during a single visit. In orchids pollinated by moths and butterflies, the pollinia become attached to their long tongues by means of sticky clamps instead of pads. The pollinia of certain bog orchids become attached to the eyes of their pollinators, which happen to be female mosquitoes. After a few visits, the mosquitoes are blinded and unable to continue their normal activities—a striking example of biological controls within an ecosystem. Among the most bizarre of the orchid pollination mechanisms are those whose effects are to dunk the pollinator in a pool of watery fluid secreted by the orchid itself and permit the pollinator to escape underwater through a trapdoor. The route of the insect ensures contact with pollinia and stigma surfaces. In other orchids with powerful narcotic fragrances, pollinia are slowly attached to the drugged pollinator. When the transfer of pollinia has been completed, the fragrance abruptly disappears; the stupefied insect recovers and flies away.

Summary

Ecology deals with the relationships of organisms to one another and to their environment. It is a vast field of study with many facets. Groups of individuals of the same species constitute populations. Communities are units composed of all the populations of organisms occurring in a given area. Ecosystems are communities and their physical environment considered together.

Populations vary in numbers, in density, and in the total mass of individuals. Numbers may be counted whereas density and biomass (total mass of living organisms) may be calculated. A community is difficult to define precisely because given members of one community may also occur in another community, and variants of a given species (ecotypes) may be specifically adapted to a single community. The environmental factors playing the greatest roles in determining the distributions of plant species in an ecosystem include precipitation, temperature, soils, and biotic factors. Xerophytes (plants adapted to areas of low precipitation) and hydrophytes (plants adapted to aquatic environments) have modifications of leaves and other organs. The mineral content of soils influences species distribution, and biotic factors such as competition for light, nutrients, and water also play roles in this regard.

Ecosystems ideally sustain themselves entirely through photosynthetic activity and the recycling of nutrients. Producers are organisms that carry on photosynthesis, whereas consumers feed on producers. Primary consumers feed directly on producers and secondary consumers feed on primary consumers. Decomposers such as fungi and bacteria break down organic materials to forms that can be reassimilated by the producers. In any ecosystem, the producers and consumers comprise food chains, which determine the flow of energy through the different levels. Food chains vary in length and intricacy, and because of their interconnections, form what are often referred to as food webs.

Energy itself is not recycled in an ecosystem. At the producer level only about 1% of the light energy available is converted to organic material, and the energy gradually escapes in the form of heat as it passes from one level to another. In a long food chain the final consumer gains only a tiny fraction of the energy originally captured by the producer at the bottom of the chain. If producers increase the amount of food available in an ecosystem, there is a corresponding increase in consumers, but the increased number of consumers reduces the available food, with the result that the ecosystem is self-maintaining. Living components of an ecosystem may also influence its composition by the secretion of growth-inhibiting substances, growth-promoting substances, and by parasitism.

Human populations have increased dramatically in the past few centuries, and the disruption of ecosystems by the activities directly or indirectly associated with the feeding, clothing, and housing of billions of people threatens the survival of not only humans but many other living organisms as well. Some restoration of ecological balance by the curbing of pollution, wise agricultural and management practices, conservation of natural resources, and education of the public to the dangers of disrupting ecosystems is essential.

Some nitrogen from the air is fixed (converted to nitrogen-containing compounds) by the nitrogen-fixing bacteria found in legumes and other plants. In the nitrogen cycle there is a constant flow of nitrogen from dead plant and animal tissues into the soil and from the soil back to the plants. Water leaches nitrogen from the soil and carries it away when erosion occurs; other nitrogen is lost from harvesting crops, but the loss can be reduced if wastes are decomposed and annually returned to the soil; fire also causes nitrogen loss. Replacement of nitrogen loss by the application of chemical fertilizers can eventually create hardpan by altering the soil structure. Bacteria also recycle carbon and other substances such as water and phosphorus.

Succession (a directional change in the species composition of a given area over a period of time) occurs whenever there has been a disturbance of natural areas on land or in water. Primary succession involves the formation of soil in the beginning stages, whereas secondary succession takes place in areas previously covered with vegetation. Primary succession involving the conversion of bare rock or lava to soil through the activities of lichens, rock mosses, flowering and other plants, and physical forces, in an orderly progression of events over a period of time is called a xerosere. A stable vegetation (climax vegetation) becomes established at the conclusion of succession and remains until or unless a disturbance disrupts it. Succession initiated in a wet habitat (hydrosere) also proceeds in an orderly fashion and culminates in the establishment of a climax vegetation. As a lake or other body of water is filled in with silt and debris, eutrophication (the gradual enrichment of the nutrient content of the water) occurs; this permits the growth of algae and other organisms.

Secondary succession, which proceeds more rapidly than primary succession, may take place if soil is present. It may occur after fires, which alter the nutrient and organic composition of the soil and open up areas in which trees such as aspens may become established until slower growing climax species return.

Biotic communities considered on a global or at least a continental scale are called biomes. The following are the major biomes of North America. Tundra, which is found primarily above the Arctic circle, includes few trees and many lichens and grasses; it is characterized by the presence of permafrost (permanently frozen soil) below the surface. Taiga is dominated by coniferous trees, with birches, aspens, and willows in the wetter areas; many perennials and few annuals occur in taiga. Temperate deciduous forests are dominated by trees such as sugar maples, American basswoods, beeches, oaks, and hickories, with evergreens such as hemlocks and eastern white pines toward the northern and southeastern borders.

Grasslands are found toward the interiors of continental masses; those located in Mediterranean climatic zones usually include vernal pools with unique annual floras. Many grasslands have been converted to agricultural use. Deserts are characterized by low annual precipitation and wide daily temperature ranges, with plants adapted both in form and metabolism to the environment. Cordilleran forest occupies much of the Pacific Northwest and extends south along the Rocky Mountains and California mountain ranges. The primarily coniferous tree species tend to be in zones determined by altitude. This forest has mostly dry summers, and some of the species associated with it (e.g., Douglas fir, giant redwood) have thick bark that protects the trees from fires which occur frequently in this type of climate. Other species (e.g., knobcone pine) depend on fires for normal distribution and germination of seeds.

The relationship between pollinators and flowering plants has, in some instances, become highly specialized. Bee-pollinated flowers are delicately sweet and fragrant. Beetle-pollinated flowers tend to have stronger odors and are usually white or dull in color. Some fly-pollinated flowers emit foul odors. Moth-pollinated flowers tend to be white or yellow. Bird-pollinated flowers are usually bright red or yellow and have much nectar but little odor. Many orchids produce their pollen grains in sacs called *pollinia* that adhere or clamp on to parts of visiting insects. The flowers of some orchid species have developed bizarre pollination mechanisms (e.g., mechanisms that effect the dunking of the pollinator in a pool of fluid).

Review Questions

1. Distinguish among a plant community, a biotic community, an ecosystem, and a biome.

2. What is the difference between a primary consumer and a secondary consumer?

3. How are short food chains more efficient in the use of solar energy than long ones?

4. Why is the diet of larger herbivorous animals varied?

5. What types of checks and balances exist in an ecosystem?

6. What adaptations have developed between flowering plants and their pollinators?

7. How are nitrogen and carbon recycled in nature?

8. What is primary succession as opposed to secondary succession?

9. What are general characteristics of climax vegetation as compared with nonclimax species?

10. Characterize desert and tundra biomes.

11. What is the function of fire in grassland and cordilleran forest biomes?

Discussion Questions

1. Humans have disrupted ecosystems almost everywhere they populate, at least in industrialized countries. Do you believe that humans could also improve an ecosystem?

2. If a vegetarian diet makes more efficient use of solar energy, should we all strive to become vegetarians for this reason?

3. Besides the hypothetical example given (in which it was observed that eating lettuce directly was a more efficient use of solar energy than feeding it to hogs and then eating the pork), can you think of other ways in which food chains might be shortened?

4. Peanut butter has many of the nutrients needed in human nutrition. On the basis of what you have learned about the diet of animals in an ecosystem, do you think it would be a good idea to live on peanut butter and water as a means of saving money?

5. Criticize the following statements: "Flowers give nectar to bees so they will be pollinated." "When plants need to be pollinated by night-flying moths, they produce white or yellow flowers."

6. Could succession take place in an abandoned swimming pool?

7. Fire has been a natural phenomenon in several biomes for thousands of years, and most plant species are adapted to it in various ways in the ecosystem where it occurs regularly. Should we then *not* extinguish forest and grassland fires when they occur?

Additional Reading

Barbour, M.; Burk, J. H.; Pitts, W. D. *Terrestrial plant ecology.* Menlo Park, CA: Benjamin/Cummings Publishing Company, Inc.; 1980.

Darwin, C. *On the various contrivances by which British and foreign orchids are fertilized by insects, and on the good effects of intercrossing.* London: Murray; 1862.

Daubenmire, R. F. *Plant communities: a textbook of synecology.* New York: Harper and Row, Publishers; 1968.

————. *Plants and environment: a textbook of plant autecology.* 3d ed. New York: John Wiley and Sons; 1974.

Ehrlich, P. R.; Holdren, J. P.; Holm, R. W., eds. *Man and the ecosphere.* Scientific American Book. San Francisco: W. H. Freeman and Co.; 1971.

Faegri, K.; van der Pijl, L. *The principles of pollination ecology.* Elmsford, NY: Pergamon Press; 1972.

Gleason, H. A.; Cronquist, A. *The natural geography of plants.* New York: Columbia University Press; 1964.

Grant, V.; Grant, K. *Flower pollination in the phlox family.* New York: Columbia University Press; 1965.

Harborne, J. B., ed. *Phytochemical ecology.* New York: Academic Press; 1972.

Jensen, W. A.; Salisbury, F. B. *Botany: an ecological approach.* Belmont, CA: Wadsworth Publishing Co.; 1972.

Kormondy, E. J. *Concepts of ecology.* 2d ed. Englewood Cliffs, NJ: Prentice-Hall; 1976.

Kozlowski, T. T.; Ahlgren, C. E., eds *Fire and ecosystems.* New York: Academic Press; 1974.

Odum, E. P. *Fundamentals of ecology.* 3d ed. Philadelphia: W. B. Saunders Co.; 1971.

Powell, J. A.; Mackie, R. A. *Biological interrelationships of moths and Yucca whipplei.* Berkeley: University of California Press; 1966.

Proctor, M.; Yeo, P. *The pollination of flowers.* New York: Taplinger Publishing Co.; 1972.

Rice, E. L. *Allelopathy.* New York: Academic Press; 1974.

Scientific American. The biosphere. San Francisco: W. H. Freeman and Co.; 1970.

Whittaker, R. H. *Communities and ecosystems.* 2d ed. New York: Macmillan Co.; 1975.

Appendix **1**

Scientific Names of Organisms Mentioned in the Text

Common and scientific names of organisms mentioned in appendices 2 through 4 are provided within the respective appendices.

Common Names, Uses, and Scientific Names of Organisms			
Common Name or Use	**Scientific Name**	**Common Name or Use**	**Scientific Name**
Aardvark	*Orycteropus* spp.	**Algae, brown**	members of Division Phaeophyta, Kingdom Protista
Abrasives, horsetail source of	*Equisetum* spp.	**Algae, golden brown**	members of Class Chrysophyceae, Division Chrysophyta, Kingdom Protista
Absinthe liqueur, source of ingredients	*Pimpinella anisum, Artemisia absinthium,* and others		
Acetone-producing bacteria	*Clostridium acetobutylicum* and others	**Algae, green**	members of Division Chlorophyta, Kingdom Protista
Acidophilus bacteria	*Lactobacillus acidophilus*	**Algae, red**	members of Division Rhodophyta, Kingdom Protista
Aconite, source of	*Aconitum* spp.	**Algae, yellow-green**	members of Class Xanthophyceae, Division Chrysophyta, Kingdom Protista
Adder's tongue fern	*Ophioglossum* spp.		
Afghanistan pine	*Pinus eldarica*		
African sausage tree	*Kigelia pinnata*		
African violet	*Saintpaulia* spp.		
Agar, source of	*Gelidium* spp. and others	**Almond**	*Prunus amygdalus*
Agave	*Agave* spp.	**Aloe juice, source of**	*Aloe barbadensis, A. ferox, A. vera,* and others
Air plant	*Kalanchoë* spp.		
Alder	*Alnus* spp.	**Amaryllis**	*Amaryllis* spp.
Aleppo pine	*Pinus halepensis*	**Ammonifying bacteria**	*Clostridium* spp., *Micrococcus* spp., *Proteus* spp., *Pseudomonas* spp., and others
Alfalfa	*Medicago sativa*		
Alfalfa caterpillar	*Colias philodice*		
Algae	members of Kingdom Protista—all divisions exclusive of Division Myxomycota		
		Amoeba	*Amoeba proteus* and others
		Amoeba (fig. 3.11)	*Pelomyxa palustris*
Algae, blue-green	members of Class Cyanophyceae, Division Schizophyta, Kingdom Monera	**Amoeba, fungal internal parasites of**	*Cochlonema verrucosum* and others

Common Name or Use	Scientific Name	Common Name or Use	Scientific Name
Amoeba, fungal trappers of	*Dactylella* spp. and others	Astringent, horsetail source of	*Equisetum arvense, E. debile,* and others
Amphibious fern	*Marsilea* spp. and others	Athlete's foot fungus	*Trichophyton* spp.
Anabaena	*Anabaena* spp.	Atlantic white cedar	*Chamaecyparis thyoides*
Anemone	*Anemone* spp.	Australian baobab	*Adansonia gregorii*
Angelica	*Angelica archangelica*	Avocado	*Persea americana* and others
Anise	*Pimpinella anisum*		
Annatto	*Bixa orellana*	Azalea	*Rhododendron* spp.
Annual moss	*Acaulon* spp., *Ephemerum* spp., and others	Baby blue eyes	*Nemophila menziesii*
		Baby powder, ground pine source of	*Lycopodium clavatum*
Ant	*Formica* spp. and others	Bacteria, acetone-producing	*Clostridium acetobutylicum* and others
Ant and termite nest fungi	A *Leucoagaricus* sp. has been identified, but vast majority are unknown	Bacteria, acidophilus	*Lactobacillus acidophilus*
		Bacteria, ammonifying	*Clostridium* spp., *Micrococcus* spp., *Proteus* spp., *Pseudomonas* spp., and others
Anthrax bacteria	*Bacillus anthracis*		
Antibiotics, source of	*Penicillium* spp., *Cephalosporium* spp., and others		
Antler and bone mosses	*Tetraplodon* spp.	Bacteria, anthrax	*Bacillus anthracis*
Aphid	*Anuraphis* spp., *Aphis* spp., and others	Bacteria, botulism	*Clostridium botulinum*
		Bacteria, brucellosis	*Brucella abortus, B. suis, B. melitensis*
Apple	*Pyrus malus*	Bacteria, BT	*Bacillus thuringiensis*
Apple rust	*Gymnosporangium juniperi-virginianum*	Bacteria, bubonic plague	*Yersinia pestis*
Apricot	*Prunus armeniaca*	Bacteria, buttermilk	*Streptococcus lactis, S. cremoris, Leuconostoc citrovorum,* and others
Aquatic floating ferns	*Azolla* spp., *Salvinia* spp., and others		
Arbor-vitae	*Thuja occidentalis*	Bacteria, butyl alcohol	*Clostridium acetobutylicum* and others
Argentine moth	*Cactoblastosis cactorum*		
Arrowroot, Florida, source of	*Zamia floridana*	Bacteria, cheese—*see* Bacteria, buttermilk	
Artichoke, globe	*Cynaria scolymus*	Bacteria, cholera	*Vibrio cholerae*
Artichoke, Jerusalem	*Helianthus tuberosus*	Bacteria, decay	*Clostridium* spp., *Micrococcus* spp., *Proteus* spp., *Pseudomonas* spp., and others
Ash	*Fraxinus* spp.		
Ash (fig. 21.13)	*Fraxinus latifolia*		
Ash, blue	*Fraxinus quadrangulata*		
Asian climbing fern	*Lygodium salicifolium*		
Aspen	*Populus tremuloides* and others	Bacteria, denitrifying	*Micrococcus denitrificans, Thiobacillus denitrificans,* and others
Aster	*Aster* spp.		
Astringent, fern source of	*Actiniopteris radiata, Drynaria quercifolia, Pteridium aquilinum,* and others	Bacteria, dextran	*Leuconostoc mesenteroides*

Common Name or Use	Scientific Name	Common Name or Use	Scientific Name
Bacteria, diphtheria	*Corynebacterium diphtheriae*	Bacteria, paratyphoid fever	*Salmonella paratyphi*
Bacteria, ensilage	*Lactobacillus delbrueckii, L. plantarum,* and others	Bacteria, pneumonia (some forms)	*Streptococcus pneumoniae* and others
Bacteria, gas gangrene	*Clostridium novyi, C. perfringens, C. septicum*	Bacteria, PPLO	*Mycoplasma pneumoniae*
		Bacteria, pseudomonad	*Pseudomonas* spp.
Bacteria, glutamic acid-producing	*Arthrobacter* spp., *Brevibacterium* spp., *Micrococcus* spp.	Bacteria, purple nonsulfur	*Rhodomicrobium* spp., *Rhodopseudomonas* spp., *Rhodospirillum* spp.
Bacteria, gonorrhea	*Neisseria gonorrhoeae*	Bacteria, purple sulfur	*Amoebobacter* spp., *Lamprocystis* spp., *Rhodothece* spp., and others
Bacteria, green sulfur	*Chlorobium* spp., *Chloropseudomonas* spp., *Prosthecochloris* spp., and others	Bacteria, salmonella, ("food poisoning")	*Salmonella* spp.
Bacteria, hydrogen	*Hydrogenomonas* spp.	Bacteria, sauerkraut	*Leuconostoc* spp. and others
Bacteria, iron	*Gallionella* spp., *Sphaerotilus* spp.	Bacteria, sorbose	*Acetobacter suboxydans*
Bacteria, kefir	*Lactobacillus bulgaricus, Streptococcus lactis*	Bacteria, spotted fever	*Rickettsia rickettsii*
		Bacteria, "strep" throat	*Streptococcus* spp.
Bacteria, lactic acid	*Lactobacillus delbrueckii* and others	Bacteria, sulfur	*Desulfovibrio* spp., *Thiobacillus* spp., and others
Bacteria, legionnaire's disease	*Legionella pneumophilia*	Bacteria, syphilis	*Treponema pallidum*
		Bacteria, tetanus	*Clostridium tetani*
Bacteria, luminescent	*Achromobacter* spp., *Flavobacterium* spp., *Photobacterium* spp., *Pseudomonas* spp., *Vibrio* spp., and others	Bacteria, tularemia	*Francisella tularensis*
		Bacteria, typhoid fever	*Salmonella typhi*
		Bacteria, typhus fever	*Rickettsia prowazeki* and others
		Bacteria, vinegar	*Acetobacter* spp.
Bacteria, meningitis	*Neisseria meningitidis* and others	Bacteria, whooping cough	*Bordetella pertussis*
Bacteria, methane	*Methanobacterium* spp., *Methanococcus* spp., *Methanosarcina* spp., and others	Bacteria, yogurt	*Streptococcus thermophilus*
		Bald cypress	*Taxodium distichum*
		Balsa	*Ochroma lagopus*
Bacteria, milky spore disease	*Bacillus popilliae*	Balsam fir	*Abies balsamea*
		Balsamroot	*Balsamorhiza* spp.
Bacteria, mosquito-killing	*Bacillus thuringiensis* var. *israelensis*	Bamboo	*Bambusa* spp.
		Banana	*Musa paradisiaca* and others
Bacteria, nitrate (nitrifying)	*Nitrobacter* spp.	Banyan tree	*Ficus* spp.
Bacteria, nitrite (nitrosifying)	*Nitrosomonas* spp.	Baobab, Australian	*Adansonia gregorii*
		Barbasco	*Lonchocarpus nicou* var. *utilis* and others
Bacteria, nitrogen-fixing	*Azotobacter* spp., *Clostridium pasteurinum, Rhizobium* spp.	Barberry	*Berberis* spp.

Common Name or Use	Scientific Name	Common Name or Use	Scientific Name
Barberry, Common	*Berberis vulgaris*	Beetle	member of Order Coleoptera, Class Insecta, Phylum Arthropoda, Kingdom Animalia
Bark, green algae that inhabit	*Protococcus* spp.[1]		
Barley	*Hordeum* spp.		
Barn swallow	*Hirundo rustica erythrogaster*	Beetle, elm bark	*Hylurgopinus rufipes, Scolytus multistriatus*
Basil	*Ocimum basilicum*	Beetle, fungi used for food by ("ambrosia")	*Ambrosiella* spp., *Monilia* spp.
Basswood	*Tilia* spp.		
Bat	*Eidolon* spp., *Epomophorus* spp., and others		
		Begonia	*Begonia* spp.
Bat (fig. 23.22)	*Leptonycteris sanbornii*	Belladonna, source of	*Atropa belladonna*
Bay, California	*Umbellularia californica*	Bermuda grass	*Cynodon dactylon*
Bay, sweet	*Laurus nobilis*	Betel nut palm	*Areca catechu*
Bean, broad	*Vicia faba*	Betony, wood	*Pedicularis canadensis*
Bean, castor	*Ricinus communis*	Big tree	*Sequoiadendron giganteum*
Bean, garbanzo	*Cicer arietinum*		
Bean, garden	*Phaseolus vulgaris*	Birch	*Betula papyrifera* and others
Bean, jequirity	*Abrus precatorius*		
Bean, kidney	*Phaseolus vulgaris*	Bird's foot fern	*Pellaea mucronata*
Bean, lima	*Phaseolus lunatus*	Bird's nest fern	*Asplenium nidus*
Bean, Mexican jumping	*Sebastiana* spp., *Sapium* spp.	Bird's nest fungi	*Nidularia* spp. and others
Bean, mung	*Phaseolus aureus*	Bird's nest fungus (fig. 16.22)	*Crucibulum levis*
Bean, pinto	*Phaseolus vulgaris*		
Bean, scarlet runner	*Phaseolus coccineus*	Birth control pills, fungi used in manufacture of	*Rhizopus nigricans, R. arrhizus*
Bean, tepary	*Phaseolus acutifolius* var. *latifolius*		
		Bittersweet	*Celastrus scandens*
Bean, winged	*Psophocarpus tetragonolobus*	Blackberry	*Rubus argutus, R. laciniatus, R. procerus,* and others
Bear	*Ursus* spp., and others		
Bearberry	*Arctostaphylos uva-ursi*		
Bear, polar	*Thalarctos maritimus*	Blackbird	*Euphagus* spp. and others
Beaver, mountain	*Aplodontia rufa*		
Bedstraw	*Galium* spp.	Black bread mold	*Rhizopus stolonifer* and others
Beech	*Fagus grandifolia*		
Beech fern	*Thelypteris* spp.	Black Forest mushroom	*Lentinus edodes*
Beefsteak morel	*Helvella* spp.	Black locust	*Robinia pseudo-acacia*
Bee, honey	*Apis mellifera*	Black mangrove	*Avicennia nitida*
Beet, garden	*Beta vulgaris*	Black rat snake	*Elaphe obsoleta*
Beet, sugar	*Beta vulgaris* (horticulturally selected strains)	Black spruce	*Picea mariana*
		Black stem rust of wheat	*Puccinia graminis*
		Black walnut	*Juglans nigra*
		Bladderwort	*Utricularia* spp.
		Blazing star	*Liatris ligulistylis*
		Bleeding, ground pine used to arrest	*Lycopodium clavatum*
		Bleeding heart	*Dicentra* spp.

1. These algae are known under several names (*Desmococcus, Phytoconis, Pleurococcus, Protococcus*), and uncertainty exists as to which name has priority. The green algal component of certain lichens, *Trebouxia,* also occurs independently on bark.

Common Name or Use	Scientific Name	Common Name or Use	Scientific Name
Bleeding heart, Pacific	*Dicentra formosa*	Brown algae	members of Division Phaeophyta, Kingdom Protista
Bloodroot	*Sanguinaria canadensis*		
Blueberry	*Vaccinium* spp.	Brown fruit rot fungi	*Monolinia fruiticola* and others
Blue curls	*Trichostema* spp.		
Blue-green algae	members of Class Cyanophyceae, Division Schizophyta, Kingdom Monera	Brucellosis bacteria	*Brucella abortus, B. suis, B. melitensis*
		Brussels sprouts	*Brassica oleracea* var. *gemmifera*
Blue-green algae, Lake Chad edible	*Spirulina* sp.	Bryophyte (*see also* individual listings)	member of Class Anthocerotae, Class Hepaticae, or Class Musci; Division Bryophyta; Kingdom Plantae
Blue-green molds	members of Class Ascomycetes, Division Mycota, Kingdom Fungi		
Blue jay	*Cyanocitta cristata*	BT bacteria	*Bacillus thuringiensis*
Bobcat	*Lynx rufus*	BTI bacteria	*Bacillus thuringiensis* var. *israelensis*
Bollworm	*Pectinophora gossypiella*		
		Bubonic plague bacteria	*Yersinia pestis*
Boston fern	*Nephrolepis exaltata*		
Botulism bacteria	*Clostridium botulinum*	Bucket orchid	*Coryanthes* spp.
Bowstring fibers, source of	*Sansevieria metalaea*	Buckeye	*Aesculus* spp.
		Buckwheat	*Fagopyrum esculentum*
Bowstring hemp, source of	*Sansevieria* spp.	Buffalo	*Bison bison*
		Bur clover	*Medicago hispida*
Box elder	*Acer negundo*	Burn treatment, (ashes) horsetail source of	*Equisetum hyemale* and others
Boysenberry	*Rubus* hybrids, with *R. ursinus* as one parent		
		Burn treatment, fern source of	*Polystichum munitum*
Bracken fern	*Pteridium aquilinum*	Bur oak	*Quercus macrocarpa*
Bracket fungi	*Fomes* spp., *Daedalea* spp., and others	Butcher's broom	*Ruscus* spp.
		Buttercup	*Ranunculus* spp.
Bracket fungus (fig. 16.20)	*Pleurotus ostreatus*	Buttercup, European bulbous	*Ranunculus bulbosa*
Bracket fungus (plate 3C)	*Polyporus sulphureus*	Buttercup family	Ranunculaceae
		Butterfly	member of Superfamily Papilionoidea, Order Lepidoptera, Phylum Arthropoda, Kingdom Animalia
Brazil nut	*Bertholettia excelsa*		
Breadfruit	*Artocarpus* spp.		
Bridalwreath	*Spiraea vanhouttei* hybrids and others		
		Butterfly iris	*Moraea* sp.
Bridges' brodiaea	*Triteleia bridgesii*	Buttermilk bacteria	*Streptococcus lactis, S. cremoris, Leuconostoc citrovorum*, and others
Bristlecone pine	*Pinus longaeva*		
Bristlecone pine, Colorado	*Pinus aristata*		
Broad bean	*Vicia faba*	Butterwort	*Pinguicula* spp.
Broccoli	*Brassica oleracea* var. *botrytis*	Button snakeroot	*Eryngium* spp.
		Butyl alcohol bacteria	*Clostridium acetylbutylicum* and others
Bromeliad family	Bromeliaceae		
Broomrape	*Orobanche* spp.		

Common Name or Use	Scientific Name	Common Name or Use	Scientific Name
Cabbage	*Brassica oleracea*	Carpetweed family	Aizoaceae
Cabbage family	Brassicaceae (Cruciferae)	Carrot	*Daucus carota*
		Cashew	*Anacardium occidentale*
Cabbage looper	*Trichoplusia ni*	Cassava	*Manihot esculenta*
Cabbage worm	*Pieris rapae*	Cassia[2]	*Cinnamomum cassia*
Cacao	*Theobroma cacao*	Castor bean	*Ricinus communis*
Cactus (figure 22.13)	*Cereus hexagonus*	Catalpa	*Catalpa* spp.
Cactus, barrel	*Mamillaria* spp. and others	Caterpillar	larval stage of member of Order Lepidoptera, Phylum Arthropoda, Kingdom Animalia
Cactus, cholla	*Opuntia* spp. (cylindrical forms)		
Cactus, giant saguaro	*Carnegia gigantea*	Catnip	*Nepeta cataria*
Cactus, hedgehog	*Echinocereus* spp. and others	Cattail	*Typha* spp.
		Cattle—*see* Cow	
Cactus, living rock	*Ariocarpus fissuratus* and others	Cauliflower	*Brassica oleracea* var. *botrytis*[3]
Cactus, organpipe	*Lemaireocereus* spp.	Caussu wax, source of	*Calathea lutea*
Cactus, prickly pear	*Opuntia* spp.	Cedar, Atlantic white	*Chamaecyparis thyoides*
Cactus family	Cactaceae		
Cajeput, source of	*Melaleuca leucodendron*	Cedar, eastern red	*Juniperus virginiana*
Calabazilla	*Cucurbita foetidissima*	Cedar, eastern white	*Thuja occidentalis*
Calendula	*Calendula* spp.	Cedar, incense	*Calocedrus decurrens*
California bay	*Umbellularia californica*	Cedar, southern white	*Chamaecyparis thyoides*
California fuchsia	*Epilobium* spp.		
California nutmeg	*Torreya californica*	Cedar, western red	*Thuja plicata*
California poppy	*Eschscholzia californica*	Cedar waxwing	*Bombycilla cedorum*
California soaproot	*Chlorogalum pomeridianum*	Celery	*Apium graveolens*
		Cellular slime mold	member of Class Acrasiomycetes, Division Myxomycota, Kingdom Protista
Camel	*Camelus* spp.		
Camelina	*Camelina sativa*		
Camellia	*Camellia* spp.		
Camphor, source of	*Cinnamomum camphora*		
Canchalagua	*Centaurium venustum*	Century plant	*Agave* spp.
Candelilla	*Euphorbia antisyphilitca*	Chain fern	*Woodwardia fimbriata*
Candlenut	*Aleurites moluccana*	Chamise	*Adenostoma fasciculatum*
Cankerworm	*Alsophila pometaria* and others		
Cantaloupe	*Cucumis melo*	Chard	*Beta vulgaris* var. *cicla*
Cap-thrower fungus	*Pilobolus* spp.	Cheese bacteria—*see* Buttermilk bacteria	
Caraway	*Carum carvi*		
Cardamon	*Elettaria cardamomum*		
Caribou	*Rangifer* spp.		
Carnation	*Dianthus caryophyllus,* and others		
Carnaubalike wax, source of	*Stipa tenacissima*		
Carnauba wax, source of	*Copernicia cerifera*		
Carob	*Ceratonia siliqua*		

2. This should not be confused with the genus *Cassia,* the source of senna in the legume family, or cassie, a perfume oil whose source is another member of the legume family, *Acacia farnesiana.*

3. Broccoli and cauliflower are two different forms of the same variety.

Common Name or Use	Scientific Name	Common Name or Use	Scientific Name
Cheese fungi	*Penicillium camembertii* (for Camembert cheese), *P. roquefortii* (for blue, Gorgonzola, Roquefort, and Stilton cheeses)	Clover, bur	*Medicago hispida*
		Clover, sweet	*Melilotus* spp.
		Cloves	*Eugenia caryophyllata*
		Club fungus	member of Class Basidiomycetes, Division Mycota, Kingdom Fungi
Cherry	*Prunus avium* and others	Club moss	member of Division Lycophyta, Kingdom Plantae
Chestnut	*Castanea* spp.		
Chestnut, American	*Castanea americana*	Coastal redwood	*Sequoia sempervirens*
Chestnut blight	*Endothia parasitica*	Cobra plant	*Darlingtonia californica*
Chia	*Salvia columbariae*	Cocaine, source of	*Erythroxylon coca*
Chickadee, mountain	*Parus gambeli*	Cochineal insect	*Dactylopius coccus*
Chickweed (Himalayan)	*Stellaria decumbens*	Cocklebur	*Xanthium* spp.
Chicle, source of	*Achras sapota*	Cockroach	*Blatta orientalis, Blatella germanica,* and others
Chicory	*Cichorium intybus*		
Chilghoza pine	*Pinus gerardiana*		
Chinese vegetable tallow	*Sapium sebiferum*	Cockroach plant	*Haplophyton cimicidum*
		Cockscomb	*Celosia* spp.
Chipmunk	*Eutamias* spp. and others	Coconut palm	*Cocos nucifera*
		Codling moth	*Carpocapsa pomonella*
Chlamydomonas	*Chlamydomonas* spp.	Coffee	*Coffea arabica*
Chlorella	*Chlorella* spp.	Coleus	*Coleus blumei* and others
Chlorine-assimilating fungus	*Aspergillus terreus*		
		Columbine	*Aquilegia* spp.
Chocolate, source of	*Theobroma cacao*	Columbine (fig. 22.3A)	*Aquilegia canadensis*
Chokecherry	*Prunus virginiana* var. *melanocarpa*	Compass plant	*Lactuca serriola*
		Coneflower	*Rudbeckia* sp.
Cholera bacteria	*Vibrio cholerae*	Coneflower, Asian	*Strobilanthes* spp.
Cholla	*Opuntia* spp. (cylindrical forms)	Copal, sources of	*Agathis alba, Copaifera demeussei, Hymenea coubaril, Trachylobium verrucosum,* and others
Christmas flower	*Euphorbia pulcherrima*		
Chrysanthemum	*Chrysanthemum* spp.		
Chuckwalla	*Sauromalus obesus*		
Chufa	*Cyperus esculentus*		
Chytrids	member of Class Chytridiomycetes, Division Mycota, Kingdom Fungi	Copperhead	*Ancistrodon contortrix*
		Coralline algae	*Bossiella* spp., *Corallina* spp., *Lithothamnion* spp., and others
Cinnamon	*Cinnamomum zeylanicum*	Coral tree	*Erythrina crista-galli*
Cinnamon fern	*Osmunda cinnamonea*	Cordage fibers, source of	*Agave sisalina, A. heterocantha, A. lophantha, Phormium tenax,* and others
Citric acid, fungal producers of	*Aspergillus niger* and others		
Citronella oil, source of	*Cymbopogon nardus*		
Citrus	*Citrus* spp.	Coriander	*Coriandrum sativum*
Citrus family	Rutaceae	Cork oak	*Quercus suber*
Clematis	*Clematis* spp.	Corn	*Zea mays*
Climbing fern, Asian	*Lygodium salicifolium*	Corn rust	*Puccinia sorghi*
Clover	*Trifolium* spp.	Corn smut	*Ustilago maydis*

Common Name or Use	Scientific Name	Common Name or Use	Scientific Name
Corydalis, pale	*Corydalis sempervirens*	Dawn redwood	*Metasequoia glyptostroboides*
Cotton	*Gossypium* spp.		
Cottonwood	*Populus deltoides* and others	DDT-like compound, algal producers of	*Laurencia* spp. and others
Coughs, fern(s) used in treatment of	*Adiantum aethiopicum, A. lunulatum Polypodium glycyrrhiza,* and others	Death angel	*Amanita phalloides*
		Decay bacteria	*Clostridium* spp., *Micrococcus* spp., *Proteus* spp., *Pseudomonas* spp., and others
Coulter pine	*Pinus coulteri*		
Cow	*Bos* spp.	Deer	*Odocoileus* spp. and others
Cow parsnip	*Heracleum lanatum*		
Cowslip	*Caltha palustris*	Deer brush	*Ceanothus* spp.
Crabapple	*Crataegus* spp.	Deer, mule	*Odocoileus hemionus*
Cranberry	*Vaccinium macrocarpus*	Denitrifying bacteria	*Micrococcus dentrificans, Thiobacillus denitrificans,* and others
Cress, garden	*Lepidium sativum, Barbarea verna,* and others		
Crested wheat grass	*Agropyron cristatum*	Desmids	*Closterium* spp., *Cosmarium* spp., and others
Crocus	*Crocus* spp.		
Crocus, autumn	*Colchicum autumnale*		
Crown-of-thorns	*Euphorbia milii* var. *splendens* and others	Dewberry	*Rubus* hybrids with *R. ursinus* as one parent
Crustacean	member of Class Crustacea, Phylum Arthropoda, Kingdom Animalia	Dextran bacteria	*Leuconostoc mesenteroides*
		Diabetes, fern used in treatment of	*Adiantum caudatum*
Cucumber	*Cucumis sativus*	Diarrhea, fern(s) used in treatment of	*Botrychium lunaria, B. ternatum, Pteridium aquilinum,* and others
Cucumber, squirting	*Ecballium elaterium*		
Cup fungus	member of Class Ascomycetes, Division Mycota, Kingdom Fungi		
		Diatom	*Biddulphia* spp., *Cymbella* spp., *Navicula* spp., and others
Cup fungus (fig. 16.10)	*Caloscypha fulgens*		
Cycad (fig. 19.9)	*Dioon edule*	Dicot	member of Class Dicotyledonae, Division Anthophyta, Kingdom Plantae
Cyclamen	*Cyclamen* spp.		
Cypress, bald	*Taxodium distichum*		
Daffodil	*Narcissus* spp.		
Dahlia	*Dahlia* spp.	Digger pine	*Pinus sabiniana*
Daisy	*Dimorphotheca* spp., *Layia* spp., and others	Digitalis, source of	*Digitalis purpurea*
		Dill	*Anethum graveolens*
		Dinoflagellate	member of Division Pyrrophyta, Kingdom Protista
Damask rose	*Rosa damascena*		
Dandelion	*Taraxacum officinale*		
Dandruff, fern(s) used in treatment of	*Adiantum capillus-veneris, Polystichum munitum*	Dinoflagellate, midnight-phosphorescing	*Gonyaulax polyedra*
		Diphtheria bacteria	*Corynebacterium diphtheriae*
Date palm	*Phoenix dactylifera*		
		Dischidia	*Dischidia rafflesiana*

Common Name or Use	Scientific Name	Common Name or Use	Scientific Name
Diuretic, ferns used as	*Adiantum venustum, Lygodium japonicum*	Eelworms (nematodes), fungi that trap with constricting rings	*Dactylaria* spp., *Arthrobotrys dactyloides*
Dodder	*Cuscuta* spp.	Eelworms (nematodes), fungi that trap with passive rings	*Dactylella* spp.
Dogbane	*Apocynum* spp.		
Dogwood	*Cornus* spp.		
Douglas fir	*Pseudotsuga menziesii*	Eggplant	*Solanum melongena*
Dove	member of Family Columbidae, Class Aves, Phylum Vertebrata, Kingdom Animalia	Elephant	*Elephas* spp., *Loxodonta* spp.
		Elk	*Cervus canadensis*
		Elm	*Ulmus* spp.
Dove, mourning	*Zenaidura macroura*	Elm bark beetle	*Hylurgopinus rufipes, scolytus multistriatus*
Downy mildew of grape	*Plasmopora viticola*		
Dragon tree	*Dracaena draco*	Endive	*Cichorium endivia*
Dragon's blood	*Dracaena* spp., *Daemonorops* spp.	English ivy	*Hedera helix*
		Ensilage bacteria	*Lactobacillus delbrueckii, L. plantarum,* and others
Duckweed	*Lemna* spp., *Wolffia* spp., and others		
Dulse	*Rhodymenia* spp.		
Dung mosses (on dung of carnivores)	*Tayloria* spp.	Ergot	*Claviceps purpurea*
		Ermine	*Mustela erminea*
Dung mosses (on dung of herbivores)	*Splachnum* spp.	Eucalyptus, Tasmanian giant	*Eucalyptus globulus*
Dutch elm disease	*Ceratocystis ulmi*	Eucalyptus oil, source of	*Eucalyptus* spp.
Dutchman's breeches	*Dicentra cucullaria*		
Dwarf juniper	*Juniperus communis* and others	Euglenoids	members of Division Euglenophyta, Kingdom Protista
Dwarf mistletoe	*Arceuthobium* spp.		
Dyer's woad	*Isatis tinctoria*	European corn borer	*Pyrausta nubialis*
Dyes, fern source of	*Sadleria cyatheoides* (trunk) *Sphenomeria chusana* (fronds)	European larch	*Larix decidua*
		European spiderwort	*Tradescantia paludosa*
		European stone pine	*Pinus pinea*
Dyes, sources of	see listing in appendix 3	Extinguisher mosses	*Encalypta* spp.
Dysentery, fern(s) used in treatment of	*Botrychium lunaria, B. ternatum, Pteridium aquilinum,* and others	Eye diseases, fern used in treating	*Asplenium adiantum-nigrum*
		Fall webworm	*Hyphantria cunea*
		False morel	*Helvella* spp.
Eagle, golden	*Aguila chrysaëtos*	Fennel	*Foeniculum vulgare*
Earth star	*Geaster* spp. and others	Fenugreek	*Trigonella foenumgraecum*
Earthworm	*Lumbricus* spp. and others		
		Fern(s), adder's tongue	*Ophioglossum* spp.
Eastern hemlock	*Tsuga canadensis*	Fern(s), amphibious	*Marsilea* spp. and others
Eastern larch	*Larix laricina*		
Eastern white cedar	*Thuja occidentalis*	Fern(s), aquatic (floating)	*Azolla* spp., *Salvinia* spp.
Eastern white pine	*Pinus strobus*		
Ebony	*Diospyros ebenum*	Fern(s)—source of astringent	*Actiniopteris radiata, Drynaria quercifolia, Pteridium aquilinum,* and others
Eczema, fern used in treatment of	*Lygodium flexuosum*		
Eelworm (nematode)	member of Class Nematoda, Phylum Aschelminthes, Kingdom Animalia	Fern(s), beech	*Thelypteris* spp.
		Fern, bird's foot	*Pellaea mucronata*
		Fern, Boston	*Nephrolepis exaltata*

Common Name or Use	Scientific Name	Common Name or Use	Scientific Name
Fern, bracken	*Pteridium aquilinum*	Fern—used for treating insect stings and bites	*Adiantum capillus-veneris*
Fern—used in treating burns,	*Polystichum munitum*	Fern—used for easing labor pains	*Athyrium filix-femina*
Fern, chain	*Woodwardia fimbriata*		
Fern, cinnamon	*Osmunda cinnamomea*	Fern, lady	*Athyrium filix-femina*
Fern, climbing (Asian)	*Lygodium salicifolium*	Fern(s)—used as laxative	*Asplenium trichomanes, Polypodium vulgare*
Fern(s)—used in treating coughs	*Adiantum aethiopicum, A. lunulatum, Polypodium glycyrrhiza*	Fern—used in treating leprosy	*Marsilea quadrifolia*
Fern(s)—used in treating dandruff	*Adiantum capillus-veneris, Polystichum munitum*	Fern, licorice	*Polypodium glycyrrhiza*
		Fern(s)—poisonous to livestock	*Onoclea sensibilis, Pteridium aquilinum*
Fern—used in treating diabetes	*Adiantum caudatum*	Fern, maidenhair (plate 4B)	*Adiantum pedatum*
Fern(s)—used in treating diarrhea	*Botrychium lunaria, B. ternatum, Pteridium aquilinum,* and others	Fern—edible Malaysian (relative of lady fern)	*Athyrium esculentum*
Fern(s)—used as diuretic	*Adiantum venustum, Lygodium japonicum*	Fern, male	*Dryopteris filix-mas*
Fern—sources of dyes	*Sadleria cyatheoides* (trunk), *Sphenomeris chusana* (fronds)	Fern, mosquito	*Azolla caroliniana*
		Fern, nest	*Asplenium nidus*
		Fern—used to arrest nosebleeds	*Pellaea mucronata*
Fern(s)—used in treating dysentery	*Botrychium lunaria, B. ternatum, Pteridium aquilinum,* and others	Fern(s)—used for orchid bark	*Cibotium* spp., *Osmunda* spp.
		Fern, Oriental water	*Ceratopteris thalictroides*
Fern—used in treating eczema	*Lygodium flexuosum*	Fern—used as poison antidote	*Polystichum squarrosum*
Fern—used in treating eye diseases	*Asplenium adiantum-nigrum*	Fern(s)—used in treating rickets	*Asplenium ruta-muraria, Osmunda regalis*
Fern—used to reduce fevers	*Marsilea quadrifolia*	Fern(s), seed (Pteridosperms)	*Lyginopteris* spp., *Medullosa* spp., and others
Fern, five-finger	*Adiantum pedatum*		
Fern(s)—used as food	*Athyrium filix-femina, Dryopteris austriaca, D. filix-mas, Polystichum munitum,* and others	Fern(s)—used for stuffing mattresses, pillows, upholstery	*Cibotium* spp., *Sadleria* spp.
		Fern, sword	*Polystichum munitum*
		Fern—used in treating toothache	*Pityrogramma triangularis*
Fern(s), fossil	*Psaronius* spp., *Thamnopteris* spp., and others	Fern(s), Hawaiian tree	*Cibotium* spp., *Sadleria* spp.
Fern, goldback	*Pityrogramma triangularis*	Fern(s), tree	*Cyathea* spp., *Ctenitis* spp., *Dicksonia* spp., *Sphaeropteris* spp., and others
Fern, holly	*Polystichum lonchitis*		
Fern—used for hairwash	*Dryopteris dilatata*		
		Fern, tropical tree (fig. 18.18)	*Cibotium glaucum*
Fern(s)—used by hummingbirds	*Cyathea arborea, Lophosoria quadripinnata, Nephelea mexicana*	Fern, tropical tree (fig. 18.25)	*Sphaeropteris glauca*
		Fern—used for expelling worms (intestinal tract)	*Dryopteris filix-mas*

Common Name or Use	Scientific Name	Common Name or Use	Scientific Name
Fern(s)—used for treating wounds	*Lygodium circinatum, Ophioglossum vulgatum*	Fossil ferns	*Psaronius* spp., *Thamnopteris* spp., and others
Fevers, fern used to reduce	*Marsilea quadrifolia*	Fossil horsetails	*Equisetites* spp., *Hyenia* spp., *Sphenophyllum* spp., and others
Fevers, ground pine used to reduce	*Lycopodium clavatum*	Fossil horsetails, treelike	*Calamites* spp.
Field horsetail	*Equisetum arvense*	Fox, arctic	*Alopex lagopus*
Fig, common	*Ficus carica*	Fox, gray	*Urocyon cinereoargentus*
Fig, tropical	*Ficus* spp.		
Fig, tropical (fig. 5.11)	*Ficus macrophyllus*	Fox, red	*Vulpes fulva*
Figwort family	Scrophulariaceae	Foxglove	*Digitalis* spp.
Filaree	*Erodium* spp.	Frangipanni	*Plumeria* spp.
Fir, balsam	*Abies balsamea*	Frog	*Rana* spp., and others
Fir, Douglas	*Pseudotsuga menziesii*	Fruit fly	*Drosophila melanogaster* and others
Fir, white	*Abies concolor*		
Fish	member of Class Pisces, Phylum Vertebrata, Kingdom Animalia	Fuchsia, California	*Epilobium* spp.
		Fumitory, Himalayan	*Corydalis gerdae*
Fish molds	*Saprolegnia* spp. and others	Fungi that produce antibiotics	*Penicillium* spp., *Cephalosporium* spp., and others
Five-finger fern	*Adiantum pedatum*		
Flashlight fish	*Anomalops katoptron, Photoblepharon palpebratus*	Fungi that cause aspergilloses	*Aspergillus fumigatus, Candida albicans, Coccidiodes immitis,* and others
Flashlight powder, ground pine source of	*Lycopodium* spp.		
		Fungi that cause athlete's foot	*Trichophyton* spp.
Flatworm	*Convoluta roscoffensis*	Fungi used by beetles for food	*Ambrosiella* spp., *Monilia* spp.
Flatworm algae	*Platymonas* spp.		
Flax	*Linum* spp.	Fungi, bird's-nest	*Nidularia* spp.
Flax, New Zealand	*Phormium tenax*	Fungi—used in manufacturing birth control pills	*Rhizopus nigricans, R. arrhizus*
Flea	member of Order Siphonaptera, Phylum Arthropoda, Kingdom Animalia		
		Fungi, bracket	*Fomes* spp., *Daedalea* spp., and others
Flicker	*Colaptes* spp.	Fungi, brown fruit rot	*Monolinia fruticola* and others
Florida arrowroot	*Zamia floridana*		
Flour, Hopi Indian horsetail source of	*Equisetum laevigatum*	Fungi, cap-thrower	*Pilobolus* spp.
		Fungi, cheese	*Penicillium camembertii* (for Camembert cheese), *P. roquefortii* (for blue, Gorgonzola, Roquefort, and Stilton cheeses)
Flowerpot leaf plant	*Dischidia rafflesiana*		
Fly	member of Order Diptera, Phylum Arthropoda, Kingdom Animalia		
		Fungi, citric acid-producing	*Aspergillus niger* and others
Fly agaric	*Amanita muscaria*		
Flycatcher	*Empidonax* spp., *Myiarchus* spp., and others	Fungi, flavor-producing	*Aspergillus* spp.
Fossil, compression (fig. 18.26)	*Annularia radiata*		

Common Name or Use	Scientific Name	Common Name or Use	Scientific Name
Fungi, hallucinogenic	*Amanita muscaria, Conocybe* spp., *Panaeolus* spp., *Psilocybe* spp., and others	Funori, source of	*Gloiopeltis* spp.
		Fur, green algae that inhabit animal	*Trentepohlia* spp.
Fungi, horse dung	*Pilobolus* spp.	Garbanzo bean	*Cicer arietinum*
Fungi, industrial alcohol-producing	*Aspergillus* spp.	Garden bean	*Phaseolus vulgaris*
		Gas gangrene bacteria	*Clostridium novyi, C. perfringens, C. septicum*
Fungi, insect-parasitizing	members of Order Laboulbeniales, Class Ascomycetes, Division Mycophyta, Kingdom Fungi and others	Gentian, source of	*Gentiana* spp.
		Geranium	*Geranium* spp., *Pelargonium* spp.
		Geranium family	Geraniaceae
Fungi, jelly	*Auricularia* spp., *Exidia* spp., *Tremella* spp., and others	Giant Galápagos tortoise	*Testudo elephantopus porteri*
		Giant horsetail	*Equisetum telmateia*
Fungi, meat-tenderizing	*Thamnidium* spp.	Giant kelp	*Macrocystis pyrifera*
Fungi, shelf—*see* Fungi, bracket		Giant redwood	*Sequoiadendron giganteum*
Fungi, *shoyu*	*Aspergillus oryzae, A. soyae*	Giant saguaro cactus	*Carnegia gigantea*
		Giant waterlily	*Victoria amazonica*
Fungi—used in silvering of mirrors	*Aspergillus* spp.	Gila monster	*Heloderma suspectum*
Fungi—used in manufacturing soap	*Penicillium* spp.	Ginger	*Zingiber officinale* and others
		Ginger, wild (plate 5C)	*Asarum canadense*
Fungi, soil	*Fusarium* spp. and others	Ginseng, source of	*Panax quinquefolium*
		Giraffe	*Giraffa camelopardalis*
Fungi, soy sauce	*Aspergillus oryzae, A. soyae*	Gladiola	*Gladiolus* spp.
		Gloeocapsa	*Gloeocapsa* spp.
Fungi, *sufu*	*Actinomucor elegans, Mucor* spp.	Glutamic acid-producing bacteria	*Arthrobacter* spp., *Brevibacterium* spp., *Micrococcus* spp.
Fungi, *teonanacatl* (sacred)	*Conocybe* spp., *Panaeolus* spp., *Psilocybe* spp., and others	Goat	*Capra* spp.
		Goldback fern	*Pityrogamma triangularis*
Fungus, chlorine-assimilating	*Aspergillus terreus*	Golden-brown algae	members of Class Chrysophyceae, Division Chrysophyta, Kingdom Protista
Fungus, "foolish seedling"	*Gibberella fujikuroi*		
Fungus, *miso*	*Aspergillus oryzae*	Golden chain tree	*Laburnum anagyroides*
Fungus—used in producing plastics	*Aspergillus terreus*	Goldenrod	*Solidago* spp.
Fungus, *tempeh*	*Rhizopus oligosporus*	Goldenrod, woodland (plate 5D)	*Solidago nemoralis*
Fungus—used in manufacturing toothpaste	*Aspergillus niger*	Goldenseal	*Hydrastis canadensis*
		Goldenweed	*Haplopappus gracilis*[4]
Fungus—used in manufacturing yellow food-coloring agent	*Blakeslea trispora*		

4. Species with four chromosomes per cell.

Common Name or Use	Scientific Name	Common Name or Use	Scientific Name
Gold violet	*Viola douglasii*	Gum arabic, source of	*Acacia senegal* and others
Gonorrhea bacteria	*Neisseria gonorrhoeae*	Gum tragacanth, source of	*Astragalus echidenaeformis, A. gossypinus, A. gummifer,* and others
Goose	*Branta* spp., and others		
Gooseberry	*Ribes* spp.		
Gopher	*Geomys* spp., *Thomomys* spp.		
Gopher, pocket	*Geomys bursarius,* and others	Guppy	*Lebistes reticulatus*
Gopher plant	*Euphorbia lathyrus*	Gypsy moth	*Porthetria dispar*
Gourd	*Lagenaria siceraria,* and others	Hadeda ibis	*Hagedashia hagedash hagedash*
Grape	*Vitis* spp.	Haircap moss	*Polytrichum* spp.
Grapefruit	*Citrus paradisi*	Hairwash, fern used as	*Dryopteris dilatata*
Grass	*Bromus* spp., and others[5]	Hairwash, horsetail used as	*Equisetum hyemale*
Grass, Bermuda	*Cynodon dactylon*	Hallucinogenic fungi	*Amanita muscaria, Conocybe* spp., *Panaeolus* spp., *Psilocybe* spp., and others
Grass, crested wheat	*Agropyron cristatum*		
Grass, Indian	*Sorghastrum nutans*		
Grass family	Poaceae (Gramineae)		
Grass tree	*Xanthorrhea* spp.		
Gray fox	*Urocyon cinereoargentus*	Hawk	*Buteo* spp., *Falco* spp., and others
Greenbrier	*Smilax* spp.	Hazel nut	*Corylus* spp.
Green rope algae	*Codium* spp.	Heath	*Erica* spp. and others
Green sulfur bacteria	*Chlorobium* spp., *Chloropseudomonas* spp., *Prosthecochloris* spp., and others	Heath family	Ericaceae
		Hedgehog cactus	*Echinocereus* spp. and others
		Hemlock, eastern	*Tsuga canadensis*
		Hemlock, mountain	*Tsuga mertensiana*
Ground pine	*Lycopodium* spp.	Hemlock, poison	*Conium maculatum*
Ground pine, fossil relative of	*Baragwanathia* spp., *Drephanophycus* spp., *Protolepidodendron* spp., and others	Hemlock, water	*Cicuta* spp.
		Hemlock, western	*Tsuga heterophylla*
		Hemp	*Cannabis sativa*
		Henbit	*Lamium amplexicaule*
		Henna	*Lawsonia inermis*
Ground pine—used to arrest bleeding	*Lycopodium clavatum*	Hepatica	*Hepatica* spp.
		Hepatica (fig. 22.3C)	*Hepatica americana*
Ground pine—used for baby powder	*Lycopodium clavatum*	Hickory	*Carya* spp.
		Himalayan fumitory	*Corydalis gerdae*
Ground pine—used as intoxicant	*Lycopodium selago*	Hog	*Sus scrofa* and others
Ground pines—used for ornaments	*Lycopodium clavatum, L. complanatum,* and others	Hog fennel	*Lomatium* spp.
		Holly, American	*Ilex opaca*
		Holly fern	*Polystichum lonchitis*
Ground pine—used to reduce fevers	*Lycopodium clavatum*	Honeybee	*Apis mellifera*
		Honey locust	*Gleditsia triacanthos*
Guava	*Psidium guajava*	Hoopoe	*Upupa africana*

5. About 4,500 species of the grass family, Poaceae (Gramineae), in all.

Common Name or Use	Scientific Name	Common Name or Use	Scientific Name
Hop hornbeam	*Ostrya virginiana*	Hyssop	*Hyssopus officinalis*
Hops	*Humulus lupulus*	Ice plant	*Mesembryanthemum* spp.
Horehound	*Marrubium vulgare*		
Hornwort	*Anthoceros* spp.	Incense cedar	*Calocedrus decurrens*
Horse	*Equus caballus*	India, toxic blue-green algae of	*Lyngbya majuscula*
Horseradish	*Rorippa armoracia*		
Horsetail	*Equisetum* spp.	Indian paintbrush (plate 7A)	*Castilleja latifolia*
Horsetail(s)—used as abrasive	*Equisetum* (all spp.)		
		Indian pipe	*Monotropa* spp.
Horsetail(s)—used as astringent	*Equisetum arvense, E. debile,* and others	Indian warrior	*Pedicularis densiflora*
		Indicator mosses for absence of calcium	*Andraea* spp., *Rhacomitrium lanuginosum*
Horsetail(s)—ashes used for treating burns	*Equisetum hyemale* and others		
		Indicator mosses for presence of calcium	*Didymodon* spp., *Desmatodon* spp., and others
Horsetail—used for treating diarrhea	*Equisetum hyemale*		
		Indicator mosses for saline (salty) soil	*Pottia* spp.
Horsetail(s)—used as diuretic	*Equisetum arvense, E. debile,* and others		
		Indicator mosses for seasonal running water	*Fontinalis* spp.
Horsetail—used for treating dysentery	*Equisetum hyemale*		
Horsetail, field	*Equisetum arvense*	Indigo	*Indigofera* spp.
Horsetail(s), fossil	*Equisetites* spp., *Hyenia* spp., *Sphenophyllum* spp., and others	Industrial alcohol, fungal producers of	*Aspergillus* spp.
		Inky cap mushrooms	*Coprinus* spp.
Horsetail, Hopi Indian flour source	*Equisetum laevigatum*	Insects—*see* individual entries	
		Insects, fern used for treating stings and bites of	*Adiantum capillus-veneris*
Horsetail, treelike fossil	*Calamites* spp.		
Horsetail, giant	*Equisetum telmateia*		
Horsetail—used as hairwash	*Equisetum hyemale*	Intoxicant, ground pine used as	*Lycopodium selago*
Horsetail—used as water source	*Equisetum telmateia*	Ipecac, source of	*Cephaelis ipecacuanha*
		Iris	*Iris* spp.
Hot springs, blue-green algae of	*Bacillosiphon induratus, Synechococcus* spp., and others	Iris, butterfly	*Moraea,* sp.
		Iris family	Iridaceae
		Irish moss	*Chondrus crispus*
"Human hair" slime mold	*Stemonitis* spp.	Iron bacteria	*Gallionella* spp., *Sphaerotilus* spp.
Hummingbird	*Archilocus* spp. and others	Ironwood, South American	*Krugiodendron ferreum*
Hummingbird, ruby-throated (fig. 23.20)	*Archilocus colubris*	Ivy, Boston	*Parthenocissus tricuspidata*
Hummingbirds, ferns used by (for nest material)	*Cyathea arborea, Lophosoria quadripinnata, Nephelea mexicana*	Ivy, English	*Hedera helix*
		Jacaranda	*Jacaranda* spp.
		Jack pine	*Pinus banksiana*
Hummingbirds, tropical	*Chlorostilbon maugaeus* and others	Jaeger	*Stercorarius* spp.
		Jeffrey pine	*Pinus jeffreyi*
Hyacinth	*Hyacinthus* spp.	Jelly fungi	*Auricularia* spp., *Exidia* spp., *Tremella* spp., and others
Hyacinth, grape	*Muscari* spp.		
Hydrogen bacteria	*Hydrogenomonas* spp.		
		Jequirity bean	*Abrus precatorius*

Common Name or Use	Scientific Name	Common Name or Use	Scientific Name
Jerusalem sage	*Phlomis fruticosa*	Laurel	*Laurus nobilis*
Jimson weed	*Datura* spp.	Laurel family	Lauraceae
Jimson weed (fig. 20.7)	*Datura stramonium*	Lavender	*Lavandula officinalis*
Jojoba	*Simmondsia californica*	Laxative, ferns used as	*Asplenium trichomanes, Polypodium vulgare*
Joshua tree	*Yucca brevifolia*		
Jumping mouse	*Zapus hudsonius, Napaeozapus insignis*	Leaf miner	*Agromyza* spp. and others
Junco	*Junco* spp.	Leaf roller	*Archips argyrospila* and others
Junco, slate-colored	*Junco hyemalis*	Leafy liverwort—*see* Liverwort, leafy	
Juneberry	*Amelanchier* spp.	Legionnaire's disease bacteria	*Legionella pneumophilia*
Juniper	*Juniperus* spp.		
Juniper, dwarf	*Juniperus communis,* and others	Legume family	Leguminosae (Fabaceae)
Jute	*Corchorus* spp.	Lemming	*Lemmus* spp., *Dicrostonyx groenlandicus*
Kaffir lily	*Clivia* sp.		
Kauri pine	*Agathis australis, A. robusta*	Lemon	*Citrus limon*
Kefir bacteria	*Lactobacillus bulgaricus, Streptococcus lactis*	Lemongrass oil, source of	*Cymbopogon citratus, C. flexuosus*
		Lentil	*Lens esculenta*
Kelp	*Alaria* spp., *Dictyoneurum* spp., *Egregia* spp., *Laminaria* spp., *Lessoniopsis* spp., *Nereocystis* spp., and others	Leprosy, fern used for treating	*Marsilea quadrifolia*
		Lettuce	*Lactuca sativa*
		Lichen (symbiotic association of an alga and a fungus)	Class Lichens, Kingdom Fungi[6]
Kelp, giant	*Macrocystis pyrifera*	Lichen, foliose (fig. 16.34B)	*Lobaria pulmonaria*
Kidney bean	*Phaseolus vulgaris*		
Knobcone pine	*Pinus attenuata*	Lichen, foliose (fig. 16.33)	*Physcia* sp.
Knotweed	*Polygonum aviculare*		
Kohlrabi	*Brassica oleracea* var. *caulorapa*	Lichen, fruticose (fig. 16.34C)	*Ramalina menziesii*
Kudzu	*Puerania thunbergiana*	Lichen, grazed by North African sheep	*Lecanora* spp.
Kumquat	*Fortunella japonica*		
Labor pains, fern used to ease	*Athyrium filix-femina*	Lichen, litmus	*Rocella* spp.
		Lichen, natural dye	*Parmelia* spp., *Usnea* spp., and others
Lactic acid bacteria	*Lactobacillus delbrueckii* and others	Lichen, perfume stabilizer	*Evernia* spp.
Lady fern	*Athyrium filix-femina*		
Lake Chad region, edible blue-green alga of	*Spirulina maxima*		
Larch, eastern	*Larix laricina*		
Larch, European	*Larix decidua*		
Larch, western	*Larix occidentalis*		
Larkspur, blue	*Delphinium* spp.		
Larkspur, red	*Delphinium nudicaule*		
Late blight of potato	*Phytophthora infestans*		

6. The lichens are arbitrarily treated as a class within the Kingdom Fungi on the basis of the fact that the fungal component of each species of lichen is unique to the species, whereas the algal component may be common to more than one species of lichen.

Common Name or Use	Scientific Name	Common Name or Use	Scientific Name
Lichen, reindeer ("reindeer moss")	*Cladonia* spp., *Cetraria islandica*	Locoweed	*Astragalus* spp.
		Loganberry	*Rubus* hybrids, with *R. ursinus* as one parent
Lichens, crustose (plate 3A; fig. 16.34A)		Longleaf pine	*Pinus palustris*
black	*Rinodina* sp.	Loon	*Gavia* spp.
chartreuse	*Acarospora citrina*	Louse	Order Mallophaga, Order Anaplura, Class Insecta, Phylum Arthropoda, Kingdom Animalia
gray	*Psora* sp.		
orange-red	*Caloplaca elegans*		
yellow	*Candelariella vitellina*		
Lichens used as miniature trees and shrubs	*Cladonia* spp. and others	Luffa	*Luffa cylindrica*
		Luminescent bacteria	*Achromobacter* spp., *Flavobacterium* spp., *Photobacterium* spp., *Pseudomonas* spp., *Vibrio* spp., and others
Licorice, source of	*Glycyrrhiza glabra*		
Licorice fern	*Polypodium glycyrrhiza*		
Lignum vitae	*Guaiacum officinale*		
Lilac	*Syringa vulgaris*		
Lily	*Lilium* spp. and others	Luminous moss	*Schistostega pennata*
Lily, tiger	*Lilium pardalinum*	Lupine	*Lupinus* spp.
Lily, Turk's cap	*Lilium superbum*	Lupine, tree (with seed valves)	*Lupinus arboreus*
Lily, wood	*Lilium superbum*		
Lily family	Liliaceae	Madder family	Rubiaceae
Lima bean	*Phaseolus lunatus*	Magnolia	*Magnolia* spp.
Lime	*Citrus aurantifolia*	Maidenhair fern	*Adiantum* spp.
Litmus indicator dye, source of	*Rocella* spp.	Malaysian edible fern (relative of lady fern)	*Athyrium esculentum*
Liverwort	member of Class Hepaticae, Division Bryophyta, Kingdom Plantae	Male fern	*Dryopteris filix-mas*
		Mallow	*Malva* spp.
		Mango	*Mangifera indica*
Liverwort, leafy (fig. 17.8)	*Porella navicularis*	Mangrove	*Rhizophora mangle* and others
Liverworts, leafy	*Frullania* spp., *Jungermannia* spp., *Porella* spp., and others[7]	Mangrove, tropical (fig. 5.8)	*Sonneratia alba*
		Manila hemp	*Musa textilis*
		Manioc—*see* Cassava	
Liverworts, thalloid	*Marchantia* spp., and others	Manroot	*Marah* spp.
Livestock, ferns poisonous to	*Onoclea sensibilis*, *Pteridium aquilinum*	Maple	*Acer* spp.
		Maple	*Acer saccharinum*
		Maple, hard	*Acer saccharum*
Living rock cactus	*Ariocarpus fissuratus* and others	Maple, sugar	*Acer saccharum*
		Marigold	*Tagetes* spp.
Lizard	*Sceloporus* spp. and others	Marijuana	*Cannabis sativa*
		Marjoram	*Majorana hortensis*
Lobeline sulphate, source of	*Lobelia inflata*	Maté	*Ilex paraguariensis*
		Meadow foam	*Limnanthes* spp.
Loblolly pine	*Pinus taeda*	Meadow saffron	*Colchicum autumnale*
		Mealy bugs	*Pseudococcus* spp.
		Meat-tenderizing fungi	*Thamnidium* spp.

---

7. There are thousands of species of leafy liverworts assigned to about 200 genera.

Common Name or Use	Scientific Name	Common Name or Use	Scientific Name
Melon	*Cucumis melo*	Moose	*Alces americana*
Melon, honeydew	*Cucumis melo* (variety)	Morel	*Morchella esculenta* and others
Melonette	*Melothria pendula*		
Meningitis bacteria	*Neisseria meningitidis* and others	Morning glory	*Ipomoea violacea* and others
Merkus pine	*Pinus merkusii*	Mosquito	*Anopheles* spp., *Culex* spp., and others
Mermaid's wineglass	*Acetabularia* spp.	Mosquito fern	*Azolla caroliniana*
Mesquite	*Prosopis glandulosa*	Mosquito-killing bacteria	*Bacillus thuringiensis* var. *israelensis*
Methane bacteria	*Methanobacterium* spp., *Methanococcus* spp., *Methanosarcina* spp., and others	Moss	member of Class Musci, Division Bryophyta, Kingdom Plantae
Mexican jumping bean	*Sebastiana* spp., *Sapium* spp.	Moss, luminous	*Schistostega pennata*
Mexican jumping bean moth	*Carpocaps asaltitans*	Mosses, annual (bare soil)	*Acaulon* spp., *Ephemerum* spp., and others
Mexican poppy	*Hunnemania* spp.	Mosses, antler and bone	*Tetraplodon* spp.
Mexican stone pine	*Pinus cembroides*		
Milkweed	*Asclepias* spp.	Mosses, dung (on dung of carnivores)	*Tayloria* spp.
Milkweed, swamp	*Asclepias incarnata*		
Milky spore disease bacteria	*Bacillus popilliae*	Mosses, dung (on dung of herbivores)	*Splachnum* spp.
Millet	*Pennisetum glaucum*, *Setaria italica*, and others	Mosses, extinguisher	*Encalypta* spp.
		Mosses, haircap	*Polytrichum* spp.
Millipede	member of Class Diplopoda, Phylum Arthropoda, Kingdom Animalia	Mosses, indicator for absence of calcium	*Andreaea* spp., *Rhacomitrium lanuginosum*
Mint—*see* Peppermint, Spearmint, etc.		Mosses, indicator for presence of calcium	*Didymodon* spp., *Desmatodon* spp., and others
Mint family	Lamiaceae (Labiatae)	Mosses, indicator for saline (salty) soil	*Pottia* spp.
Miso fungi	*Aspergillus oryzae*	Mosses, indicator for seasonal running water	*Fontinalis* spp.
Mistletoe	*Phoradendron* spp.		
Mistletoe, dwarf	*Arceuthobium* spp.		
Mite	member of Order Acarina, Phylum Arthropoda, Kingdom Animalia	Mosses, peat	*Sphagnum* spp.
		Mosses, pygmy—*see* Mosses, annual	
Mollusc	member of Phylum Mollusca, Kingdom Animalia	Mosses, rock	*Grimmia* spp. and others
		Mosses, sphagnum	*Sphagnum* spp.
Monkey	*Ateles dariensis* and others	Mosses used with splints	*Philonotis* spp., *Fontinalis* spp., and others
Monkey flower	*Mimulus* spp.	Moss rose	*Portulaca grandliflora*
Monkshood	*Aconitum* spp.	Moth	member of Order Lepidoptera, Class Insecta, Phylum Arthropoda, Kingdom Animalia
Monocot	member of Class Monocotyledonae, Division Anthophyta, Kingdom Plantae		
Monterey pine	*Pinus radiata*		

Common Name or Use	Scientific Name	Common Name or Use	Scientific Name
Moth, Argentine	*Cactoblastis cactorum*	Myrtle[9]	*Umbellularia californica*[9]
Moth, Mexican jumping bean	*Carpocaps asaltitans*	Nasturtium (garden)	*Tropaeolum majus*
Moth, Yucca	*Pronuba* spp., *Tegeticula* spp.	Natural dye lichens	*Parmelia* spp., *Usnea* spp., and others
Moth mullein	*Verbascum blattaria*	Nematode	member of Class Nematoda, Phylum Aschelminthes, Kingdom Animalia
Mountain beaver	*Aplodontia rufa*		
Mountain hemlock	*Tsuga mertensiana*		
Mouse	*Mus musculus, Peromyscus* spp., and others	Nest fern	*Asplenium nidus*
		Nettle	*Urtica* spp.
Mouse, jumping	*Zapus hudsonius, Napaeozapus insignis*	Nicotine relative (nornicotine), source of	*Duboisia hopwoodii, Nicotiana tabacum*
Mulberry	*Morus* spp.	Nightshade, deadly	*Atropa belladonna*
Mulberry, red	*Morus rubra*	Nightshade family	Solanaceae
Mule ears	*Wyethia* spp.	Nitrate (nitrifying) bacteria	*Nitrobacter* spp.
Mullein	*Verbascum thapsus*		
Mung bean	*Phaseolus aureus*	Nitrite (nitrosifying) bacteria	*Nitrosomonas* spp.
Mushroom[8]	*Agaricus* spp., and others		
		Nitrogen-fixing bacteria	*Azotobacter* spp., *Clostridium pasteurinum*
Mushroom, Black Forest	*Lentinus edodes*		
Mushroom, pore (fig. 16.19)	*Boletus* spp.	Nosebleeds, fern used to arrest	*Pellaea mucronata*
Mushroom, shaggy mane	*Coprinus comatus*	Nostoc	*Nostoc* spp.
		Nutmeg	*Myristica fragrans*
Mushrooms, common woods (plate 3D)	*Russula* sp.	Nutmeg, California	*Torreya californica*
		Oak	*Quercus* spp.
Mushrooms, inky cap	*Coprinus* spp.	Oak, bur	*Quercus macrocarpa*
Mushrooms, oyster (fig. 16.20)	*Pleurotus ostreatus*	Oak, cork	*Quercus suber*
		Oak, Hooker	*Quercus lobata*
Mushrooms, *teonanacatl* (sacred)	*Conocybe* spp., *Panaeolus* spp., *Psilocybe* spp., and others	Oak, live (fig. 23.7)	*Quercus wislizenii*
		Oak, white	*Quercus alba*
		Oats	*Avena* spp.
		Obedient plant	*Physostegia* sp.
Muskrat	*Ondatra zibesthicus*	Oedogonium	*Oedogonium* spp.
Mustard	*Brassica campestris, B. nigra,* and others	Olibanum tree	*Boswellia* spp.
		Olive	*Olea europaea*
Mustard, cultivated	*Brassica alba, B. juncea,* and others	Onion	*Allium cepa*
		Orange	*Citrus sinensis*
Mustard family	Brassicaceae (Cruciferae)	Orchid	*Cattleya* spp. and others[10]
Myrrh tree	*Commiphora* spp.		

---

9. This plant, also known as the California bay, is in the laurel family. True myrtles are in the myrtle family (Myrtaceae).

---

8. *Mushroom* is a term generally applied to the fruiting bodies with stalked, caplike structures produced by members of the Class Basidiomycetes, Division Mycota, Kingdom Fungi; the term is also loosely applied to some of the fruiting bodies of members of other classes of true fungi. There are thousands of known species.

---

10. Depending on which authorities are followed, the number of orchid species (all in the family Orchidaceae) may exceed 30,000.

Appendix 1

Common Name or Use	Scientific Name	Common Name or Use	Scientific Name
Orchid (fig. 18.24)	*Oncidium* sp.	Peanut	*Arachis hypogaea*
Orchid, bletilla (fig. 21.11B)	*Bletilla* sp.	Pear	*Pyrus communis*
		Peat moss	*Sphagnum* spp.
Orchid, underground-flowering	*Rhizanthella gardneri*	Pecan	*Carya illinoensis*
		Penicillin mold[11]	*Penicillium* spp.[11]
Orchid bark, fern sources of	*Cibotium* spp., *Osmunda* spp.	Pennyroyal	*Hedeoma pulegioides*
		Peony	*Paeonia* spp.
Orchid with cladophylls	*Epidendrum* spp.	Peperomia	*Peperomia* spp.
Orchid family	Orchidaceae	Pepper	*Capsicum anuum,* *C. frutescens*[12]
Orchids, saprophytic	*Corallorhiza* spp., *Eburophyton austinae,* and others	Peppergrass	*Lepidium* spp.
		Peppermint	*Mentha piperita*
Orchids, vanilla	*Vanilla* spp.	Perfume stabilizer lichen	*Evernia* spp.
Oregano	*Origanum vulgare* and others	Persimmon	*Diospyros* spp.
Oregon grape	*Berberis aquifolium* and others	Petitgrain oil, source of	*Citrus aurantium* var. *amara*
Organpipe cactus	*Lemaireocereus* spp.	Petunia	*Petunia* spp.
Oriental water fern	*Ceratopteris thalictroides*	Peyote	*Lophophora williamsii*
		Phoebe	*Sayornis phoebe*
Ornaments, ground pines used for	*Lycopodium complanatum, L. clavatum,* and others	Pillbugs	*Cylisticus convexus* and others
		Pine	*Pinus* spp.
Osage orange	*Maclura pomifera*	Pine, Afghanistan	*Pinus eldarica*
Oscillatoria	*Oscillatoria* spp.	Pine, Aleppo	*Pinus halepensis*
Our Lord's Candle	*Yucca whipplei*	Pine, bristlecone	*Pinus longaeva*
Owl, snowy	*Nyctea scandiaca*	Pine, Chilghoza	*Pinus gerardiana*
Oyster mushrooms (fig. 16.20)	*Pleurotus ostreatus*	Pine, Colorado bristlecone	*Pinus aristata*
Painted lady	*Echeveria derenbergii*	Pine, Coulter	*Pinus coulteri*
Palm, panama hat	*Carludovica palmata*	Pine, digger	*Pinus sabiniana*
Palm, Seychelles Island	*Lodoicea maldivica*	Pine, eastern white	*Pinus strobus*
Palm, wax	*Copernicia cerifera*	Pine, European stone	*Pinus pinea*
Palm family	Palmaceae	Pine, jack	*Pinus banksiana*
Panama hat palm	*Carludovica palmata*		
Pansy	*Viola tricolor*		
Papaya	*Carica papaya*		
Pará rubber	*Hevea brasiliensis*		
Paratyphoid fever bacteria	*Salmonella paratyphi*		
Parsley	*Petroselinum sativum*		
Parsley family	Apiaceae (Umbelliferae)		
Parsnip	*Pastinaca sativa*		
Passion fruit	*Passiflora* spp.		
Patchouli oil, source of	*Pogostemon cablin* and others		
Pea	*Pisum sativum*		
Pea, sweet	*Lathyrus odoratus*		
Peach	*Prunus persica*		
Peach leaf curl	*Taphrina deformans*		

---

11. The original penicillin producer discovered by Sir Alexander Fleming was *Penicillium notatum*; current commercially used penicillin producers are strains of *Penicillium chrysogenum.*

---

12. The drug capsicum, whose active ingredient is the oleoresin capsaicin, is derived from these species, and garden peppers include these and other species of *Capsicum.* Condiment or black pepper is derived from the unrelated *Piper nigrum.*

Common Name or Use	Scientific Name	Common Name or Use	Scientific Name
Pine, jeffrey	*Pinus jeffreyi*	Plover	*Charadrius* spp. and others
Pine, kauri	*Agathis australis, A. robusta*	Plum	*Prunus domestica* and others
Pine, knobcone	*Pinus attenuata*	Pneumonia bacteria	*Streptococcus pneumoniae* and others
Pine, loblolly	*Pinus taeda*		
Pine, longleaf	*Pinus palustris*		
Pine, Merkus	*Pinus merkusii*	Podocarps, New Zealand timber	*Podocarpus dacrydoides, P. totara*
Pine, Mexican pinyon	*Pinus cembroides*		
Pine, Mexican stone	*Pinus cembroides*		
Pine, Monterey	*Pinus radiata*	Podocarps used as ornamentals	*Podocarpus macrophylla, P. nagi, and others*
Pine, pinyon	*Pinus edulis, P. monophylla, P. quadrifolia*		
		Poinsettia	*Euphorbia pulcherrima*
Pine, pitch	*Pinus rigida*	Poison antidote, fern used for	*Polystichum squarrosum*
Pine, ponderosa	*Pinus ponderosa*		
Pine, shortleaf	*Pinus echinata*	Poison hemlock	*Conium maculatum*
Pine, Siberian white	*Pinus sibirica*	Poison ivy	*Toxicodendron radicans*
Pine, slash	*Pinus caribaea, P. elliottii*	Poison oak	*Toxicodendron diversilobum*
Pine, southern yellow— *see* Pine, loblolly; Pine, longleaf; Pine, shortleaf; and Pine, slash		Polyanthus	*Primula polyanthus* and hybrids
		Pomegranate	*Punica granatum*
		Ponderosa pine	*Pinus ponderosa*
Pine, stone—*see* Pine, European stone and Pine, Mexican stone		Poor man's pepper	*Lepidium virginicum*
		Popcorn	*Zea mays* (horticultural variety)
Pine, sugar	*Pinus lambertiana*	Poplar	*Populus* spp.
Pine, western white	*Pinus monticola*	Poplar leaf spot rust	*Melampsora medusae*
Pine, western yellow	*Pinus ponderosa*	Poppy	*Papaver* spp. and others
Pineapple	*Ananas comosus*		
Pinedrops	*Pterospora* spp.	Poppy, bush	*Dendromecon rigida*
Pinto bean	*Phaseolus vulgaris*	Poppy, California	*Eschscholzia californica*
Pinyon pines	*Pinus edulis, P. monophylla, P. quadrifolia*	Poppy, Mexican	*Hunnemannia* spp.
		Poppy, opium	*Papaver somniferum*
		Poppy, Oriental	*Papaver orientale*
Pistachio	*Pistacia vera*	Poppy family	Papaveraceae
Pitcher plants	*Sarracenia* spp. and others	Porcupine	*Erethizon* spp., *Hystrix* spp.[13]
Pitcher plants, Asian	*Nepenthes* spp. and others	Potato, Irish	*Solanum tuberosum*
		Potato, sweet	*Ipomea batatas*
Pitch pine	*Pinus rigida*	Potato, white—*see* Potato, Irish	
Plantain	*Plantago* spp.		
Plastic, fungus used in production of	*Aspergillus terreus*	Potato vine	*Solanum jasminoides*
		Powdery mildew	*Erysiphe* spp. and others
Plasticizers, source of oil for	*Euphorbia agascae*		
Platelike colonial green algae	*Pediastrum* spp., *Chaetopeltis* spp., and others	PPLO pneumonia bacteria	*Mycoplasma pneumoniae*

13. *Hystrix* is also a name for a genus of grasses.

Appendix 1

Common Name or Use	Scientific Name	Common Name or Use	Scientific Name
Prayer plant	*Maranta* spp.	Raspberry, red	*Rubus idaeus, R. strigosus,* and their hybrids
Preferns	*Cladoxylon* spp., *Protopteridium* spp., and others	Rat	*Rattus norvegicus, R. rattus,* and others
Prickly pear cactus	*Opuntia* spp.	Rat, kangaroo	*Dipodomys* spp.
Primrose	*Primula* spp.	Rat snake, black	*Elaphe obsoleta*
Prochlorophyte	member of Division Prochlorophyta, Kingdom Monera	Rattlesnake	*Crotalus* spp.
		Red algae	members of Division Rhodophyta, Kingdom Protista
Pronghorn	*Antilocarpa americana*		
Pseudomonad bacteria	*Pseudomonas* spp.	Red cedar	*Juniperus virginiana*
Ptarmigan	*Lagopus* spp.	Red cedar, western	*Thuja plicata*
Pteridosperms	*Lyginopteris* spp., *Medullosa* spp., and others	Red fox	*Vulpes fulva*
		Red mangrove	*Rhizophora mangle*
Puffballs	*Calvatia* spp., *Lycoperdon* spp.	Red raspberry	*Rubus idaeus, R. strigosus* and their hybrids
Pulque, source of	*Agave* spp.		
Pumpkin	*Cucurbita pepo*	Red ribbons	*Clarkia concinna*
Pumpkin family	Cucurbitaceae	Red Sea, blue-green alga of the	*Trichodesmium erythraeum*
Puncture vine	*Tribulus terrestris*		
Purple laver	*Porphyra* spp.	Red spruce	*Picea rubens*
Purple laver (fig. 15.23)	*Porphyra tenera*	Redwood, coastal	*Sequoia sempervirens*
Purple nonsulfur bacteria	*Rhodomicrobium* spp., *Rhodopseudomonas* spp., *Rhodospirillum* spp.	Redwood, giant	*Sequoiadendron giganteum*
		Reindeer	*Rangifer* spp.
Purple sulfur bacteria	*Amoebobacter* spp., *Lamprocystis* spp., *Rhodothece* spp., and others	Reindeer lichens	*Cladonia* spp., *Cetraria islandica*
		Reserpine, source of	*Rauwolfia serpentina*
		Resurrection plant	*Selaginella lepidophylla*
Puya (rare)	*Puya raimondii*	Rhododendron	*Rhododendron* spp.
Pygmy mosses	*Acaulon* spp., *Ephemerum* spp., and others	Rhubarb	*Rheum rhaponticum*
		Rice	*Oryza sativa*
		Ricepaper plant	*Tetrapanax papyrifera*
Quillwort	*Isoetes* spp.	Rickets, ferns used in treating	*Asplenium ruta-maria, Osmunda regalis*
Quillwort, fossil relatives of	*Isoetites* spp.		
Quince	*Cydonia oblonga*	Ringworm fungi	*Epidermophyton* spp., *Microsporium* spp., *Trichophyton* spp.
Quinine, source of	*Cinchona* spp.		
Rabbit	*Oryctolagus cuniculus*	Robin	*Turdus migratorius*
Rabbit, cottontail	*Sylvilagus* spp.	Rock cress	*Arabis* sp.
Rabbit, jack	*Lepus* spp.	Rock mosses	*Grimmia* spp. and others
Raccoon	*Procyon lotor*		
Radish	*Raphanus sativus*	Rock-rose, European	*Helianthemum vulgare*
Rafflesia	*Rafflesia* spp.	Rock tripe	*Umbilicaria* spp.
Rafflesia (fig. 20.3)	*Rafflesia rochussenii*	Rockweeds	*Fucus* spp., *Pelvetia* spp., and others
Ragweed	*Ambrosia* spp.		
Raspberry	*Rubus* spp.		

Scientific Names of Organisms

Common Name or Use	Scientific Name	Common Name or Use	Scientific Name
Rose	*Rosa* spp.	Sea anemone	*Stephanauge* spp. and others
Rose, damask	*Rosa damascena*	Sea hare	*Aplysia californica*
Rose family	Rosaceae	Sea lettuce	*Ulva* spp.
Rosemary	*Rosmarinus officinalis*	Sea palm	*Postelsia palmaeformis*
Rotenone relative, source of	*Tephrosia vogelii*	Sea rocket	*Cakile edentula*
Rubber, Pará	*Hevea brasiliensis*	Sedge	*Carex* spp. and others
Rubber plant	*Ficus elastica*	Seed ferns (Pteridosperms)	*Lyginopteris* spp., *Medullosa* spp., and others
Ruellia	*Ruellia portellae* and others		
Rutabaga	*Brassica campestris* var. *napobrassica*	Senna	*Cassia senna* and others
Rye	*Secale cereale*	Sensitive plant	*Mimosa pudica*
Ryegrass	*Lolium* spp.	Sesame	*Sesamum indicum*
Safflower	*Carthamus tinctorius*	Seychelles Island palm	*Lodoicea maldivica*
Saffron, meadow	*Colchicum autumnale*	Shaggy mane mushroom	*Coprinus comatus*
Saffron (true)	*Crocus sativus*		
Sage	*Salvia officinalis*[14]	Sheep	*Ovis* spp.
Saguaro	*Carnegia gigantea*	Shelf fungi	*Fomes* spp., *Daedalea* spp., and others
Salmon	*Oncorhynchus* spp., *salmo salar* and others	Shepherd's purse	*Capsella bursa-pastoris*
		Shortleaf pine	*Pinus echinata*
Salmonberry	*Rubus spectabilis*	Showy orchid	*Orchis* spp.
Salmonella bacteria	*Salmonella* spp.	Shrimp	*Crago* spp. and others
Salsify	*Tragopogon* spp.	Siberian white pine	*Pinus sibirica*
Salvia	*Salvia* spp.	Silvering of mirrors, fungi used in	*Aspergillus* spp.
Sansevieria	*Sansevieria* spp.		
Santonin, source of	*Artemisia cina*	Sisal	*Agave sisalina*
Saprophytic orchids	*Corallorhiza* spp., *Eburophyton austinae*, and others	Sitka spruce	*Picea sitchensis*
		Skunk	*Mephitis* spp.
		Slash pine	*Pinus caribaea*, *P. elliottii*
Sargassum (fig. 15.18)	*Sargassum echinocarpum*	Slate-colored junco	*Junco hyemalis*
Sarsaparilla, source of	*Smilax* spp.	Slime mold	member of Division Myxomycota, Kingdom Protista
Sassafras	*Sassafras albidum*		
Sauerkraut bacteria	*Leuconostoc* spp. and others		
		Slime mold (fig. 15.26A)	*Arcyria versicolor*
Sausage tree, African	*Kigelia pinnata*	Slime mold (fig. 15.26B)	*Lycogala epidendrum*
Savory	*Satureia hortensis*	Smut	*Ustilago* spp. and others
Saxifrage	*Saxifraga* spp.		
Scarlet runner bean	*Phaseolus coccineus*	Snail	*Haplotrema concava* and others
Screw pine	*Pandanus* spp.	Snapdragon	*Antirrhinum majus*
		Snowbank algae	*Chlamydomonas nivale* and others

14. This sage, which is in the mint family, should not be confused with sagebrush, which is in the sunflower family (Asteraceae).

		Snowplant	*Sarcodes sanguinea*
		Snowy owl	*Nyctea scandiaca*
		Soap, fungi used in manufacturing	*Penicillium* spp.

Common Name or Use	Scientific Name	Common Name or Use	Scientific Name
Soaproot, California	*Chlorogalum pomeridianum*	Squills	*Urginea maritima*
Sorbose bacteria	*Acetobacter suboxydans*	Squirrel	*Citellus* spp., *Sciuris* spp., and others
Sorghum	*Sorghum* spp.	Squirrel, gray	*Sciurus carolinensis*
Sorrel	*Oxalis* spp.	Squirrel corn	*Dicentra canadensis*
Southern white cedar	*Chamaecyparis thyoides*	Squirting cucumber	*Ecballium elaterium*
		Stapelia	*Stapelia* spp.
Southern yellow pine— *see* Pine, loblolly; Pine, longleaf; Pine, shortleaf; and Pine, slash		Stinkhorn	*Mutinus* spp., *Phallus* spp., and others
		Stinkhorn, common (fig. 16.14)	*Mutinus caninus*
		Stonecrop	*Sedum* spp. and others
Soybean	*Glycine max*	Stone pine, European	*Pinus pinea*
Soy sauce fungi	*Aspergillus oryzae, A. soyae*	Stone pine, Mexican	*Pinus cembroides*
		Stoneseed	*Lithospermum ruderale*
Spanish moss	*Tillandsia* spp.	Stonewort	*Chara* spp., *Nitella* spp.
Sparrow, savannah	*Passerculus sandwichensis*	Strawberry	*Fragaria* spp.
		"Strep" throat bacteria	*Streptococcus* spp.
Sparrow, vesper	*Pooecetes gramineus*	String-of-pearls	*Senecio rowellianus*
Spearmint	*Mentha spicata*	Strychnine, source of	*Strychnos* spp.
Sperm whale	*Physeter catodon*	Stuffing, ferns used for mattresses, pillows, upholstery	*Cibotium* spp., *Sadleria* spp., and others
Spiderwort	*Tradescantia* spp.		
Spiderwort, European	*Tradescantia paludosa*		
Spike moss	*Selaginella* spp.	Sugar cane	*Saccharum officinarum* and others
Spike moss, fossil relatives of	*Lepidodendron* spp., *Sigillaria* spp., and others		
		Sugar pine	*Pinus lambertiana*
		Sulfur bacteria	*Desulfovibrio* spp., *Thiobacillus* spp., and others
Spinach	*Spinacia oleracea*		
Spirogyra	*Spirogyra* spp.		
Sponge	*Spongilla* spp. and others	Sumac	*Rhus* spp.
		Sunbird	*Anthodiaeta* spp., *Notiocinnyris* spp. and others
Sponge, vegetable	*Luffa cylindrica*		
Sponge algae	*Chlorella* spp., *Zoochlorella* spp.	Sundew	*Drosera* spp.
		Sundew relative used for flypaper	*Drosophyllum lusitanicum*
Spotted fever bacteria	*Rickettsia rickettsii*		
Spring beauty	*Claytonia virginica*	Sunflower	*Helianthus annuus*
Spruce	*Picea* spp.	Sunflower family	Asteraceae (Compositae)
Spruce, black	*Picea mariana*		
Spruce, red	*Picea rubens*	Sweet pea	*Lathyrus odoratus*
Spruce, sitka	*Picea sitchensis*	Sword fern	*Polystichum munitum*
Spruce, white	*Picea glauca*	Sycamore	*Platanus* spp.
Spurge	*Euphorbia* spp.	Syphilis bacteria	*Treponema pallidum*
Spurge (fig. 21.23)	*Euphorbia montana*	Tamarack	*Larix* spp.
Spurge (fig. 22.12)	*Euphorbia cyparissias*	Tamarisk	*Tamarix* spp.
Spurge family	Euphorbiaceae	Tangerine	*Citrus reticulata*
Squash	*Cucurbita mixta, C. pepo,* and others	Taro	*Colocasia* spp.
Squawroot	*Perideridia* spp.		
Squill	*Scilla* spp.		

Scientific Names of Organisms

Common Name or Use	Scientific Name	Common Name or Use	Scientific Name
Tarragon	*Artemisia dracunculus*	Toothpaste, fungus used in manufacturing	*Aspergillus niger*
Tarweed	*Grindelia* spp.		
Tarweed, western (fig. 4.3)	*Calycadenia* sp.	Toothwort	*Dentaria* spp.
Tea	*Camellia sinensis*	Tortoise, giant Galápagos	*Testudo elephantopus porteri*
Tempeh fungus	*Rhizopus oligosporus*	Touch-me-not	*Impatiens* spp.
Tent caterpillar	*Malacosoma americanum* and others	Tree fern	*Cyathea* spp., *Ctenitis* spp., *Dicksonia* spp., *Sphaeropteris* spp., and others
Teonanacatl (sacred) mushrooms	*Conocybe* spp., *Panaeolous* spp., *Psilocybe* spp., and others		
		Tree fern, Hawaiian	*Cibotium* spp., *Sadleria* spp.
Tepary bean	*Phaseolus acutifolius* var. *latifolius*	Tree fern, small Hawaiian (plate 4D)	*Sadleria cyatheoides*
Tequila, source of	*Agave* spp.	Tree fern, tropical (fig. 18.25)	*Sphaeropteris glauca*
Termite	*Odontotermes* spp., *Reticulitermes* spp., and others	Tree lupine	*Lupinus arboreus*
		Tree-of-heaven	*Ailanthus altissima*
Tetanus bacteria	*Clostridium tetani*	Trillium	*Trillium* spp.
Thalloid liverworts	*Marchantia* spp., and others	Truffles	*Tuber* spp.
		Tsetse fly	*Glossina morsitans*, *G. palpalis*
Thermal blue-green algae	*Bacillosiphon induratus*, *Synechococcus* spp., and others		
		Tularemia bacteria	*Francisella tularensis*
Thimbleberry	*Rubus parviflorus*	Tulip	*Tulipa* spp.
Thistle	*Cirsium* spp., and others	Tulip tree	*Liriodendron tulipifera*
		Tumble mustard	*Sisymbrium altissimum*
Thistle, Canada	*Cirsium arvense*	Tumbleweeds	*Amaranthus albus*, *Salsola pestifera* and others
Thrasher	*Toxostoma* spp.		
Thyme	*Thymus vulgaris* and others		
		Tung oil, source of	*Aleurites* spp.
Tidy tips	*Layia fremontii*	Turk's cap lily	*Lilium superbum*
Tiger	*Panthera tigris*	Turmeric, source of	*Curcuma longa*
Ti plant	*Cordyline* spp.	Turnip	*Brassica rapa*
Toad	*Bufo americanus*	Turtle	*Chelydra* spp., *Chrysemys* spp., and others
Tobacco	*Nicotiana tabaccum*		
Tomato	*Lycopersicon esculentum*	Turtleback algae	*Basicladia* spp., *Dermatophyton* spp.
Tomato, Galápagos	*Lycopersicon esculentum* var. *minor*, *L. pimpinellifolium*		
		Twinflower	*Linnaea borealis*
		Typhoid bacteria	*Salmonella typhi*
		Typhus fever bacteria	*Rickettsia prowazeki* and others
Tomato fruitworm	*Heliothis armigera*	Ulothrix	*Ulothrix* spp.
Tomato hornworm	*Protoparce quinquemaculata*	Ultraviolet light, flowers seen in (fig. 23.18)	*Rudbeckia* sp.
Toothache, fern chewed for	*Pityrogramma triangularis*	Unicorn plant	*Proboscoidea* spp.
		Vanilla orchid	*Vanilla* spp.
		Vegetable sponge	*Luffa cylindrica*
		Venus flytrap	*Dionaea muscipula*
		Vetch	*Vicia* spp.

Common Name or Use	Scientific Name	Common Name or Use	Scientific Name
Vetchling, yellow	*Lathyrus aphaca*	Weaver birds	*Anaplectes* spp., *Hyphantoris* spp., and others
Vinegar bacteria	*Acetobacter* spp.		
Vinegar weed	*Trichostema* spp.		
Violet	*Viola* spp.	Western hemlock	*Tsuga heterophylla*
Violet, African	*Saintpaulia* spp.	Western larch	*Larix occidentalis*
Violet, gold	*Viola douglasii*	Western red cedar	*Thuja plicata*
Virginia creeper	*Parthenocissus quinquefolia*	Western wallflower	*Erysimum capitatum*
		Western white pine	*Pinus monticola*
Virus[15]		Whale, sperm	*Physeter catodon*
Vole	*Microtus* spp. and others	Wheat	*Triticum* spp. and their hybrids
Wahoo	*Euonymus alata* and others	Wheel tree	*Trochodendron aralioides*
Wake robin	*Trillium* spp.	Whisk fern	*Psilotum* spp.
Wallflower, western	*Erysimum capitatum*	Whisk fern, fossil relatives of	*Asteroxylon* spp., *Psilophyton* spp., *Rhynia* spp., and others
Walnut	*Juglans* spp.		
Walnut, black	*Juglans nigra*		
Warbler	*Dendroica* spp. and others		
Watercress	*Nasturtium officinale*	Whisk fern, living relatives of	*Tmesipteris* spp.
Water fern, oriental	*Ceratopteris thalictroides*	White fir	*Abies concolor*
		White fly	*Aleurocanthus woglumi* and others
Water hemlock	*Cicuta* spp.		
Water hyacinth	*Eichhornia crassipes*	White piedra fungus	*Trichosporon beigeli*
Water lily	*Nymphaea* spp. and others	White pine, eastern	*Pinus strobus*
		White pine, Siberian	*Pinus sibirica*
Water lily, giant	*Victoria amazonica*	White pine, western	*Pinus monticola*
Watermelon	*Citrullus vulgaris*	White pine blister rust	*Cronartium ribicola*
Water molds	member of Class Oomycetes, Division Mycophyta, Kingdom Fungi	White spruce	*Picea glauca*
		Willow	*Salix* spp.
		Willow family	Salicaceae
		Window leaves, plants with	*Fenestraria* spp. and others
Water net	*Hydrodictyon* spp.	Winged bean	*Psophocarpus tetragonolobus*
Water silk	*Spirogyra* spp.		
Water source, horsetails used as	*Equisetum telmateia*	Wintergreen oil, sources of	*Gaultheria procumbens* and others
Water weed	*Elodea* spp.	Wisteria	*Wisteria sinensis* and others
Water weed, yellow	*Ludwigia repens*		
Wattle	*Acacia decurrens* and others	Witch hazel	*Hamamelis virginiana*
		Woad, dyer's	*Isatis tinctoria*
		Woadwaxen	*Genista tinctoria*
		Wolfsbane	*Aconitum vulparia*
		Wolverine	*Gulo luscus*
		Woodpecker	*Dendrocopus* spp. and others

15. Depending on the classification used, viruses may not have a scientific name. Many are named after the disease they cause; e.g., tobacco mosaic virus causes tobacco mosaic disease. One classification attempts to give them at least a Latin prefix, so that the virus for warts is *Papavovirus;* for smallpox, *Poxvirus;* for polio, *Picornavirus;* for measles and mumps, *Paramyxovirus.*

Common Name or Use	Scientific Name	Common Name or Use	Scientific Name
Worms, fern used in expelling from intestinal tract	*Dryopteris filix-mas*	Yellow-green algae	members of Class Xanthophyceae, Division Chrysophyta, Kingdom Protista
Wormwood	*Artemisia absinthium*		
Wounds, ferns used for treating	*Lygodium circinatum, Ophioglossum vulgatum*	Yellow vetchling	*Lathyrus aphaca*
		Yellow water weed	*Ludwigia repens*
Yam	*Dioscorea* spp.	Yew	*Taxus* spp.
Yareta	*Azorella yareta*	Yogurt bacteria	*Streptococcus thermophilus*
Yarrow, American	*Achillea lanulosum*		
Yarrow, European	*Achillea millefolium*	Yucca	*Yucca* spp.
Yeast	*Saccharomyces* spp.	Zamia	*Zamia* spp.
		Zebra	*Equus zebra* and others
		Zinnia	*Zinnia elegans* and others

Appendix 1

Appendix 2
Biological Controls

If you were to ask the average farmer or backyard gardener how to control a particular insect or plant pest, you might be given the name of some poisonous spray or bait that has proved "effective" in the past. Evidence has been mounting for many years, however, that spraying with such substances yields only temporary results and frequently causes larger invasions of pests than before. In addition, the residues of poisonous sprays often accumulate in the soil and disrupt the microscopic living flora and fauna essential to the soil's health. The problem is compounded and the ecology further upset when large amounts of chemical fertilizers are added. As increasing numbers of people develop an awareness of the devastating effects on the environment of pesticides and herbicides, they have been turning to *biological controls* as an alternative to the use of poisonous sprays. To the surprise of some, such controls are often more effective than traditional controls.

One reason why poisonous sprays actually promote pest invasions is that the sprays frequently kill numerous beneficial insects along with the undesirable ones. In addition, the pests often become resistant to the sprays through mutations. In undisturbed natural areas, one rarely sees pests destroying the community, even though such pests are present, and weeds are never a problem. Why is this so? As was observed in chapter 23, all members of a community are in ecological balance with one another. The plants produce a variety of substances that may either repel or attract insects, inhibit or promote the growth of other plants, and generally contribute to the "health" of the community as a whole.

Virtually all insects have their own pests and diseases, as do most living organisms, and each pest functions, at least indirectly, to ensure that the various species of a community are perpetuated. To a certain extent this principle of nature can be applied to farming and gardening. The following are some general and specific biological controls either now in widespread use or showing promise for the future in various tests.

General Controls

Establishment of Beneficial Insects

Ladybugs (Family Coccinellidae)
The larval stages, in particular, of the small and often colorful beetles called *ladybugs* consume large numbers of aphids, thrips, insect eggs, weevils, and other pests. They are obtainable from various commercial sources (e.g., Bio-Control Co., P.O. Box 247, Cedar Ridge, CA 95924; Unique Insect Control, P.O. Box 15376, Sacramento, CA 95852), but they probably will establish themselves without being imported if given a chance. When obtained from other sources, they should be placed in groups at the bases of plants on which pests are present, preferably after watering in the early evening.

Lacewings (Families Chrysopidae and Hemerobiidae)
Lacewings are relatively slow-flying and delicate-winged insects that consume large numbers of aphids, mealybugs, and other pests. They lay their eggs on the undersides of leaves, each egg

being borne at the tip of a slender stalk. The larvae consume the immature stages of leafhoppers, bollworms, caterpillar eggs, mites, scale insects, thrips, aphids, and other destructive pests. Commercial sources include California Green Lacewing, P.O. Box 2495, Merced, CA 95340; All Pest Control, 6030 Grenville Lane, Lansing, MI 48910.

Praying Mantis (Family Mantidae)

About twenty species of praying mantis are now established in the United States. These are voracious feeders that prey somewhat indiscriminately on flying insects, and sometimes even on other mantises. They can be established by tying their egg cases to tree branches or at other locations above the ground. The egg cases, which form compact, somewhat oblongoid masses about 2.5 to 5.0 centimeters (1 to 2 inches) long, are obtainable from various commercial sources, including Mincemoyer's Nursery, Rt. 7, Box 379, Jackson, NJ 08527; Bio-Control Co., P.O. Box 247, Cedar Ridge, CA 95924.

Trichogramma Wasps (Family Trichogrammatidae)

Trichogramma wasps are minute insects, mostly less than 1 millimeter (1/25 inch) long; they parasitize insect eggs and are known to have significantly reduced populations of well over one hundred different insect pests, including alfalfa caterpillars, armyworms, cabbage loopers, cutworms, hornworms, tent caterpillars, and the larvae of numerous species of moths. As with other insects used as biological controls, care should be exercised in the release of trichogramma wasps, as lack of sufficient pest eggs in the vicinity can result in the wasps parasitizing eggs of beneficial butterflies and other useful insects. They are available from commercial sources such as California Green Lacewing, P.O. Box 2495, Merced, CA 95340; Gothard, Inc., Box 370, Canutillo, TX 79835.

Ichneumon Wasps (Family Ichneumonidae)

The ichneumon wasps belong to a very large family of primarily stingless wasps that are mostly slender and have long ovipositors that sometimes exceed the length of the body. Most insects are parasitized by at least one species of ichneumon; many species parasitize the larval stages of insects, consuming the host internally after hatching from eggs deposited on the body; alternatively, they may complete development in a later stage. Ichneumons will usually appear naturally in a backyard or farm population of pests if toxic sprays and other unnatural conditions have not interfered with their normal activities.

Tachinid Flies (Family Tachinidae)

Many members of the large family of tachinid flies resemble houseflies or bumblebees. All parasitize other insects, including a large variety of caterpillars, Japanese beetles, European earwigs, grasshoppers, gypsy moths, tomato worms, sawflies, and various beetles. Write to Insect Pest Advisory Service, 762 S. First St., Kerman, CA 93630 for further information.

Use of Pathogenic Bacteria

Bacillus thuringiensis (BT) is one of two pathogenic bacteria registered for use on edible plants in the United States. It reproduces only in the digestive tracts of caterpillars and is completely harmless to all other wildlife, including earthworms, birds, and mammals. It is exceptionally effective against a wide range of caterpillars such as tomato hornworms and fruitworms, cabbage worms and loopers, grape leaf rollers, corn borers, cutworms, fall webworms, and tent caterpillars. It is mass-produced and sold in a powdered spore form at nurseries and garden supply stores under the trade names of Dipel, Biotrol, and Thuricide.

Establishment of Toads and Frogs

It has been estimated that a single adult toad will consume about ten thousand insects and slugs in a single growing season. Toads and frogs feed at night when snails, slugs, sowbugs, earwigs, and other common pests are active.

Use of Beneficial Nematodes

These exceedingly numerous microscopic round-worms have gained a widespread negative reputation because of the damage several species inflict on economically important crops when they invade plant roots and other underground organs. Most species, however, are either harmless or beneficial to plants. They have been used successfully in parasitizing cabbage worm caterpillars, codling moth larvae, Japanese beetle grubs, and tobacco budworms and have shown considerable potential against other pests. One species that has been particularly effective in controlling ants, beetles, bugs, flies, wasps, and numerous other insects is the caterpillar nematode (*Neoaplectana carpocapsae*). It carries a symbiotic bacterium (*Xenorhabdus nematophilus*), which multiplies rapidly in the host, killing most insects within twenty-four hours after initial contact. It may be obtained from Nematode Farm, Inc., 2617 San Pablo Avenue, Berkeley, CA 94702.

Use of Limonoid Sprays

Limonoids are bitter substances found in the rinds, seeds and juice of citrus fruits (especially grapefruit). If seeds and rinds are ground up, soaked overnight in water, strained, and then sprayed on plants, the bitter principle apparently stops or reduces the feeding of larvae on the foliage. In experiments, limonoid sprays have proved effective against corn earworm, fall armyworm, tobacco budworm, and pink bollworm, but undoubtedly will deter many other pests as well.

Use of Liquefied Pest Sprays

Jeff Cox, an editor of *Organic Gardening and Farming* magazine, called attention to this method of pest control in 1976 (*OGF,* October) and again in 1977 (*OGF,* May). Insect pests or slugs are gathered in small quantities and liquefied with a little water in a blender. The material is then further diluted with water and sprayed throughout the infested area. It is not known why spraying with "bug juice" is effective against pests; however, it is known that virtually all organisms are subject to viral infections, and it has been theorized that even the healthy insects and slugs may be carriers of viruses that are somehow activated in the process of liquefaction. The viruses would be spread throughout an entire yard or farm if all parts of the area were sprayed. Most viruses are highly specific, generally attacking a single species of organism.

A resident pest control entomologist in Florida, M. Sipe, who has recommended the "bug juice" technique, has also suggested the possibility that the odor of the liquefied insects attracts their predators and parasites or that the insects' distress *pheromones* (naturally produced insect chemicals that influence sexual or other behavior) are released by the blender. The *pheromones* could act as an insect repellent; possibly the observed effects of spraying "bug juice" are the result of a combination of viruses, predator attraction, and repellent pheromones. Sipe warns that if one tries this method of pest control, care should be taken to use only pest species and only those that are doing significant damage. Failure to heed this warning could disrupt the activities of natural predators and other natural controls present.

This approach still needs further testing and investigation of its safety for use by humans, but preliminary results in various areas of North America have thus far yielded impressive results with no evidence of harm to humans or beneficial organisms.

Use of Resistant Varieties

Many plants may kill or inhibit disease fungi or bacteria with chemicals known as *phytoalexins.* Phytoalexins are synthesized at the point of attack or invasion by the pathogen and are toxic to the fungus or bacterium. In selecting for improved fruit quality, vigor of growth, or other desirable characteristics, horticulturists in the past have sometimes unknowingly bred out a plant's capacity to produce certain phytoalexins, although general vigor is usually accompanied by disease resistance. Now that this aspect of a plant's defense mechanisms has become known,

Table A2.1 Specific weeds and agents involved in their biological control

Weed	Agent(s) of Biological Control
Alligator weed (*Alternanthera philoxeroides*)	Flea beetles (*Agasicles hygrophila*)
Bladder campion (*Silene cucubalus*)	Tortoise beetle (*Cassida hemisphaerica*)
Brazil peppertree (*Schinus terebinthifolius*)	Weevil (*Bruchus atronotatus*) and others
Brushweed (*Cassia surattensis*)	Imperfect fungus (*Cephalosporium* sp.)
Curly dock (*Rumex crispus*)	Rust (*Uromyces rumicis*)
Curse (*Clidemia hirta*)	Thrip (*Liothrips urichi*) and others
Cypress spurge (*Euphorbia cyparissias*)	Sphinx moth (*Hyles euphorbiae*)
Dalmatian toadflax (*Linaria dalmatica*)	Leaf miner (*Stagmatophora serratella*) and others
Emex (*Emex australis*)	Seed weevils (*Apion antiquum*) and others
Gorse (*Ulex europaeus*)	Seed weevils (*Apion ulicis*) and others
Halogeton (*Halogeton glomeratus*)	Casebearer (*Coleophora parthenica*) and others
Hawaiian blackberry (*Rubus penetrans*)	Sawflies (*Pamphilius sitkensis; Priophorus morio*) and others
Jamaica feverplant (*Tribulus terrestris*)	Weevils (*Microlarinus* spp.)
Joint vetch (*Aeschynomene virginica*)	Imperfect fungus (*Colletotrichum gloeosporioides*)
Klamath weed (*Hypericum perforatum*)	Leaf beetles (*Chrysolina* spp.); buprestid beetle (*Agrilus hyperici*)
Lantana (*Lantana camara*)	Seed weevil (*Apion* sp.); ghost moth (*Hepialus* sp.); plume moth (*Platyptilia pusillidactyla*); hairstreaks (*Strymon* spp.); and others
Leafy spurge (*Euphorbia esula*)	Wood-boring beetle (*Oberea* sp.) and others
Mediterranean sage (*Salvia aethiopis*)	Snout beetles (*Phrydiuchus* spp.)
Milkweed vine (*Morrenia odorata*)	Oomycete fungus (*Phytophthora citrophthora*); rust (*Aecidium asclepiadinum*)
Prickly pear (*Opuntia* spp.)	Moth (*Cactoblastis cactorum*); cochineal insects (*Dactylopius* spp.); and others
Puncture vine (*Tribulus terrestris*)	Weevils (*Microlarinus* spp.)

breeders are concentrating on developing varieties capable of producing phytoalexins against various fungi, bacteria, and even nematodes. Several tomato varieties, for example, are listed as being *VFN*. The letters *V* and *F* indicate a resistance to *Verticillium* and *Fusarium* (common pathogenic fungi), while the letter *N* denotes a resistance to root-knot nematodes. Other aspects of plant disease resistance include thick cuticles; the secretion of gums, resins, and other metabolic products that may interfere with fungal and bacterial spore germination; and the presence within all the cells of the plant of chemical compounds toxic to pathogens.

Interplanting with Plants That Produce Natural Insecticides or Substances Offensive to Pests

Although no single species of plant produces substances that repel all pests, many produce substances that repel a significant number. The best known include marigolds, garlic, and members of the mint family such as pennyroyal, peppermint, basil, and lavender. See appendix 3 for an expanded discussion of the subject.

Weed	Agent(s) of Biological Control
Scotch broom (*Cytisus scoparius*)	Seed weevil (*Apion fuscirostre*) and others
Skeleton weed (*Chondrilla juncea*)	Gall mite (*Aceria chondrillae*); root moth (*Bradyrrhoa gilveolella*); rust (*Puccinia chondrillina*); powdery mildews (*Erysiphe cichoracearum, Leveillula taurica*)
Spiny emex (*Emex spinosa*)	Seed weevil (*Apion antiquum*)
Tansy ragwort (*Senecio jacobaea*)	Seed fly (*Hylemya seneciella*); cinnabar moth (*Tyria jacobaeae*); leaf beetle (*Longitarsus jacobaeae*)
Thistles:	
Bull thistle (*Cirsium vulgare*)	Weevil (*Ceuthorrhynchidius horridus*); tortoise beetle (*Cassida rubiginosa*)
Canada thistle (*Cirsium arvense*)	Weevil (*Ceutorhynchus litura*); flea beetle (*Altica carduorum*); stem gall fly (*Urophora cardui*)
Diffuse knapweed (*Centaurea diffusa*)	Seed fly (*Urophora affinis*)
Italian thistle (*Carduus pycnocephalus*)	Flea beetles (*Rhinocyllus conicus; Psylliodes chalcomera*); weevil (*Ceutorhynchus trimaculatus*)
Milk thistle (*Silybum marianum*)	Flea beetle (*Rhinocyllus conicus*)
Musk thistle (*Carduus nutans*)	Weevils (*Ceutorhynchus trimaculatus; Ceuthorrhynchidius horridus; Rhinocyllus conicus*); flea beetle (*Psylliodes chalcomera*)
Perennial sowthistle (*Sonchus arvensis*)	Peacock fly (*Tephritis dilacerata*)
Plumeless thistle (*Carduus acanthoides*)	Tortoise beetle (*Cassida rubiginosa*); seed weevil (*Rhinocyllus conicus*); weevil (*Ceuthorrhynchidius horridus*)
Russian thistle (*Salsola kali* var. *tenuifolia*)	Casebearer (*Coleophora parthenica*) and others
Slenderflower thistle (*Carduus tenuiflorus*)	Weevil (*Ceutorhynchus trimaculatus*)
Spotted knapweed (*Centaurea maculosa*)	Seed fly (*Urophora affinis*) and others
Star thistle (*Centaurea nigrescens*)	Weevil (*Ceuthorrhynchidius horridus*)
Yellow star thistle (*Centaurea solstitialis*)	Seed fly (*Urophora siruna-seva*)
Water hyacinth (*Eichhornia crassipes*)	Weevils (*Neochetina bruchi; N. eichhorniae*); moth (*Sameodes albiguttalis*)
Water purslane (*Ludwigia palustris*)	Snout beetle (*Nanophyes* sp.)

Specific Controls

Weeds

In 1974 the Weed Science Society of America published a special committee report (*Weed Science* 22:490–95) on the biological control of weeds, summarizing the status of projects on the biological control of weeds with insects and plant pathogens in the United States and Canada. Table A2.1 is condensed from that report and supplemented with additional information. Many other biological controls for these and other weeds are currently under investigation.

Insects

The maintenance of ecological balance in nature includes a vast array of predator–prey relationships between animals, birds, insects, and other organisms. Specific biological controls for several types of insect pests, in addition to the general controls previously discussed, are given in table A2.2.

Table **A2.2** Specific biological controls for several types of insect pests	
Insect	**Control**
Ants (about 8,000 spp. within the superfamily Formicoidea)	Ants that carry aphids into trees and consume ripening fruits can be prevented from getting farther than the trunk by applying a band of sticky material around the trunk. A commercial preparation sold under the trade name of Tanglefoot is particularly effective. A water suspension of ground hot peppers (*Capsicum* spp.) used as a spray can act as an ant deterrent. *Caution:* Many ants are beneficial to a balanced ecology; they should not be decimated indiscriminately.
Grasshoppers (there are several families of grasshoppers, but the insects that usually constitute the most serious pests are species of *Melanoplus,* **family Acrididae)**	In 1980 the Environmental Protection Agency permitted private companies to begin the mass culture of a protozoan, *Nosema locustae,* for use in controlling rangeland grasshoppers. Tests have shown that properly timed applications of spores mixed with wheat bran can reduce grasshopper populations by up to 50%.
Gypsy moths (*Porthetria dispar*)	Parasitic wasps (*Apanteles flavicoxis, A. indiensis*) imported from India lay their eggs in gypsy moth caterpillars and kill large numbers.
Japanese beetles (*Popillia japonica*)	The pathogenic bacterium *Bacillus popillae,* which is sold commercially, is specific for Japanese beetle larvae. It causes what is known as "milky spore disease" in the grubs while they are still in the soil, and it is very destructive.
Mealybugs (*Pseudococcus* spp.)	The small brown beetles called *crypts* (*Cryptolaemus montrouzieri*) effectively control mealybugs in greenhouses and also outdoors on apple, pear, peach, and citrus trees. Order from Rincon-Vitova Insectaries, Inc., P.O. Box 95, Oakview, CA 93022.
Mosquitoes (*Culex* spp.; *Anopheles* spp., and others)	The bacterium *Bacillus thuringiensis* var. *israelensis* has proved to be very effective in destroying mosquito larvae. A fungus (*Lagenidium giganteum*) has also proved highly effective against mosquito larvae if the temperature is above 20°C (67°F). The bacterium is available from several sources, including Abbott Laboratories, Dept. 95-M, 1400 Sheridan Rd., N. Chicago, IL 60064; Sandoz, Inc., 480 Camino del Rio S., San Diego, CA 92108.
Red spider mites (*Tetranychus telarius*)	Predatory mites (*Phytolesius persimilis*, which works best when weather is not hot, and *Amblyseius californicus,* which is more effective in hot weather) effectively control populations of red spider mites.
White flies (*Trialeurodes vaporariorum*)	A minute wasp *Encarsia formosa,* parasitizes white flies exclusively. The wasps have been known to be very effective in greenhouses. They are obtainable from White Fly Control Co., Box 986, Milpitas, CA 95035; Rincon-Vitova Insectaries, Inc., P.O. Box 95, Oakview, CA 93022.

Companion Planting

An examination of the Additional Reading list will reveal that the literature on the chemical interactions between plants and also between plants and their consumers is already extensive. Despite the scientific evidence on the subject to date, however, a significant amount of the "backyard biological control" that is practiced today is based primarily on *empirical* information. Such information has been obtained from thousands of gardeners and farmers who have tried various techniques with their plantings and pest controls. As a result, they have come to conclusions that certain things work while others do not, but they have not deliberately set up controlled experiments, nor have they necessarily understood the scientific basis for what they have observed. This does not mean that their observations are not useful or that they are invalid. In fact, such empirical observations have often been the inspiration for investigations and experiments by scientists. The scientific investigations have sometimes revealed that the empirical observations were biased or not carefully made or that erroneous conclusions had been drawn, but frequently sound scientific bases for these observations have been revealed.

Further insights into the means by which plants inhibit or enhance the growth of others and into the nature of their resistance to disease or insect-repelling mechanisms continue to be discovered. Observations of such phenomena in the past have led organic gardeners and others to the practice of *companion planting,* which involves the interplanting of various crops and certain other plants in such a way that each species derives some benefit from the arrangement. The following companion planting list, based primarily on empirical information, appeared in the February 1977 issue of *Organic Gardening and Farming* magazine. It is included here with the permission of Rodale Press, Inc.

Table A2.3 is a list of combinations of vegetables, herbs, flowers, and weeds that are mutually beneficial, according to current reports of organic gardeners and to companion planting traditions.

Table A2.3	Companion plants
Plant	**Companions and Effects**
Asparagus	Tomatoes, parsley, basil
Basil	Tomatoes (improves growth and flavor); said to dislike rue; repels flies and mosquitoes.
Beans	Potatoes, carrots, cucumbers, cauliflower, cabbage, summer savory, most other vegetables and herbs; around houseplants when set outside.
Beans (bush)	Sunflowers (beans like partial shade, sunflowers attract birds and bees), cucumbers (combination of heavy and light feeders), potatoes, corn, celery, summer savory.
Beets	Onions, kohlrabi.
Borage	Tomatoes (attracts bees, deters tomato worm, improves growth and flavor), squash, strawberries.
Cabbage family	Potatoes, celery, dill, chamomile, sage, thyme, mint, pennyroyal, rosemary, lavender, beets, onions. Aromatic plants deter cabbage worms.
Carrots	Peas, lettuce, chives, onions, leeks, rosemary, sage, tomatoes.
Catnip	Plant in borders; protects against flea beetles.
Celery	Leeks, tomatoes, bush beans, cauliflower, cabbage.
Chamomile	Cabbage, onions.
Chervil	Radishes (improves growth and flavor).
Chives	Carrots; plant around base of fruit trees to discourage insects from climbing trunk.
Corn	Potatoes, peas, beans, cucumbers, pumpkin, squash.
Cucumbers	Beans, corn, peas, radishes, sunflowers.
Dill	Cabbage (improves growth and health), carrots.

Plant	Companions and Effects
Eggplant	Beans
Fennel	Most plants are supposed to dislike it.
Flax	Carrots, potatoes.
Garlic	Roses and raspberries (deters Japanese beetle); with herbs to enhance their production of essential oils; plant liberally throughout garden to deter pests.
Horseradish	Potatoes (deters potato beetles); around plum trees to discourage curculios.
Lamb's quarters	Nutritious edible weed; allow to grow in modest amounts in the corn.
Leek	Onions, celery, carrots.
Lettuce	Carrots and radishes (lettuce, carrots, and radishes make a strong companion team), strawberries, cucumbers.
Lovage	Plant here and there in garden.
Marigolds	The workhorse of pest deterrents. Keeps soil free of nematodes; discourages many insects. Plant freely throughout garden.
Marjoram	Here and there in garden.
Mint	Cabbage family; tomatoes: deters cabbage moth.
Mole plant	Deters moles and mice if planted here and there throughout the garden.
Nasturtium	Tomatoes, radishes, cabbage, cucumbers; plant under fruit trees. Deters aphids and pests of cucurbits.
Onion	Beets, strawberries, tomato, lettuce (protects against slugs), beans (protects against ants), summer savory.
Parsley	Tomato, asparagus.
Peas	Squash (when squash follows peas up trellis), plus grows well with almost any vegetable; adds nitrogen to the soil.

Plant	Companions and Effects
Petunia	Protects beans; beneficial throughout garden.
Pigweed	Brings nutrients to topsoil; beneficial growing with potatoes, onions, and corn; keep well thinned.
Potato	Horseradish, beans, corn, cabbage, marigold, limas, eggplant (as trap crop for potato beetle).
Pot marigold	Helps tomato, but plant throughout garden as deterrent to asparagus beetle, tomato worm, and many other garden pests.
Pumpkin	Corn
Radish	Peas, nasturtium, lettuce, cucumbers; a general aid in repelling insects.
Rosemary	Carrots, beans, cabbage, sage; deters cabbage moth, bean beetles, and carrot fly.
Rue	Roses and raspberries; deters Japanese beetle. Keep it away from basil.
Sage	Rosemary, carrots, cabbage, peas, beans; deters some insects.
Southernwood	Cabbage; plant here and there in garden.
Soybeans	Grows with anything helps everything.
Spinach	Strawberries
Squash	Nasturtium, corn.
Strawberries	Bush beans, spinach, borage, lettuce (as a border).
Summer savory	Beans, onions. Deters bean beetles.
Sunflower	Cucumbers
Tansy	Plant under fruit trees; deters pests of roses and raspberries; deters flying insects; also Japanese beetles, striped cucumber beetles, squash bugs, deters ants.
Tarragon	Good throughout garden.
Thyme	Here and there in garden; deters cabbage worm.
Tomato	Chives, onion, parsley, asparagus, marigold, nasturtium, carrot, limas.

Appendix 2

Plant	Companions and Effects
Turnip	Peas
Valerian	Good anywhere in garden.
Wormwood	As a border, keeps animals from the garden.
Yarrow	Plant along borders, near paths, near aromatic herbs; enhances essential oil production of herbs.

Some Sources of Herb Plants and Seeds

ABC Nursery and Greenhouse, Rt. 1, Box 313, Lecoma, MO 65540

Borchelt Herb Gardens, 474 Carriage Shop Road, East Falmouth, MA 02356

Capriland's Herb Farm, Silver Street, Coventry, CT 06238

Carroll Gardens, 444 East Main Street, Westminster, MD 21157

Casa Yerba, Star Route 2, Box 21, Days Creek, OR 97429

Fox Hill Farm, Box 70G, Parma, MI 49269

Golden Sun Herb Gardens, 5318 North Gravenstein Highway, Sebastopol, CA 95472

Gourmet Garden, West Danville, VT 05873

Helix, Box 1808-G, Boulder, CO 80306

Hickory Hollow, Route 1, Box 52, Peterstown, WV 24963

Howe Hill Herbs, Camden, ME 04843

Mincemoyers, Dept. 00, Route 526, Jackson, NJ 08527

Nichols Garden Nursery, 1190 North Pacific Highway, Albany, OR 97321

Otto Richter and Sons, Box 260, Goodwood, Ontario, Canada LOC 1A0

Sandy Mush Herbs, Route 2, Dept. OG, Leicester, NC 28748

Sunnybrook Farms Nursery, Box 6, 9448 Mayfield Road, Chesterland, OH 44026

Taylor's Herb Garden, 1535 Lone Oak Road, Vista, CA 92083

Yankee Peddler Herb Farm, Route 4, Box 77, Brenham, TX 77833

Additional Reading

Baker, K.F.; Cook, R.J. *Biological control of plant pathogens.* San Francisco: W.H. Freeman and Co.; 1974.

Bosch, R. van den; Messenger, P.S. *Biological control.* New York: Intext Educational Publishers; 1973.

Burges, H.D.; Hussey, N.W., eds. *Microbial control of insects and mites.* New York: Academic Press; 1971.

Carson, R. *Silent spring.* Boston: Houghton Mifflin Co.; 1962.

Debach, P. *Biological control by natural enemies.* London and New York: Cambridge University Press; 1974.

Goeden, R.D.; Andres, L.A.; Freeman, T.E.; Harris, P.; Pienkowski, R.L.; Walker, C.R. Present status of projects on the biological control of weeds with insects and plant pathogens in the United States and Canada. *Weed Science* 22: 490–95; 1974.

Henry, J.E. Natural and applied control of insects by protozoa. *Annual Review of Entomology* 26: 49–73; 1981.

Huffaker, C.B.; Messenger, P.S. *Theory and practice of biological control.* New York: Academic Press; 1977.

Hunter, B.D. *Gardening without poisons.* 2d ed. Boston: Houghton Mifflin Co.; 1971.

Rice, E.L. *Allelopathy.* New York: Academic Press; 1974.

Whittaker, R.H.; Feeny, P. P. Allelochemics: chemical interactions between species. *Science* 171: 757–70; 1971.

Yepsen, R.B., Jr., ed. *Organic plant protection.* Emmaus, PA: Rodale Press; 1976.

Wild Edible Plants

Words of Caution

Literally thousands of native and naturalized plants, or at least parts of them, have been used for food and other purposes by Native Americans and the immigrants who came later from other quarters of the globe. Table A3.1 has been compiled from a variety of sources; I have had opportunities to sample only a fraction of these plants myself and thus cannot confirm the edibility of all of the plants listed. *The reader is cautioned to be certain of the identity of a plant before consuming any part of it.* Cow parsnip (*Heracleum lanatum*) and water hemlocks (*Cicuta* spp.), for example, resemble each other in general appearance, but although cooked roots of cow parsnip have been used for food for perhaps many centuries, those of water hemlocks are very poisonous and have caused many human fatalities.

As was indicated in chapter 20, many species of organisms are now on rare and endangered species lists, and a number of them are doomed to extinction within the next few years. Although the wild edible plants discussed here are not included on such lists it might not take much indiscriminate gathering to endanger their existence as well. Because of this, one should exercise the following rule of thumb: *Never reduce a population of plants by more than 10% when collecting them for any purpose!*

Table A3.1 Wild edible plants		
Plant	**Scientific Name**	**Uses**
Amaranth	*Amaranthus* spp.	young leaves used like spinach; seeds ground with others for flour
Arrow grass	*Triglochin maritima*	seeds parched or roasted (**Caution:** *The plant is otherwise poisonous.*)
Arrowhead	*Sagittaria latifolia*	tubers used like potatoes
Balsamroot	*Balsamorhiza* spp.	whole plant edible, especially when young, either raw or cooked
Basswood	*Tilia* spp.	fruits and flowers ground together to make a paste that can serve as a chocolate substitute; winter buds edible raw; dried flowers used for tea
Bedstraw	*Galium aparine*	roasted and ground seeds make good coffee substitute
Beechnuts	*Fagus grandifolia*	seeds used like nuts; oil extracted from seeds for table use
Biscuit root	*Lomatium* spp.	roots eaten raw or dried and ground into flour; seeds edible raw or roasted
Bitterroot	*Lewisia rediviva*	outer coat of the bulbs should be removed to eliminate the bitter principle; bulbs are then boiled or roasted

Plant	Scientific Name	Uses
Blackberry (wild)	*Rubus* spp.	fruits edible raw, in pies, jams, and jellies
Black walnut	*Juglans nigra*	nut meats highly edible
Bladder campion	*Silene cucubalus*	young shoots (less than 5 cm tall) cooked as a vegetable
Blueberry	*Vaccinium* spp.	fruits edible raw, frozen, and in pies, jams, and jellies
Bracken fern	*Pteridium aquilinum*	young uncoiling leaves ("fiddleheads") cooked like asparagus; rhizomes also edible but usually tough (**Caution:** *Recent evidence indicates that frequent consumption of bracken fern can cause cancer of the intestinal tract.*)
Broomrape	*Orobanche* spp.	entire plant eaten raw or roasted
Bulrush (Tule)	*Scirpus* spp.	roots and young shoot tips edible raw or cooked; pollen and seeds also edible
Butternut	*Juglans cinerea*	nut meats edible
Caraway	*Carum carvi*	young leaves in salads; seeds for flavoring baked goods and cheeses
Cattail	*Typha* spp.	copious pollen produced by flowers in early summer is rich in vitamins and can be gathered and mixed with flour for baking; rhizomes can be cooked and eaten like potatoes
Chicory	*Cichorium intybus*	leaves eaten raw or cooked; dried, ground roots make good coffee substitute
Chokecherry	*Prunus virginiana*	fruits make excellent jelly or can be cooked with sugar for pies and cobblers
"Coffee" (wild)	*Triosteum* spp.	berries dried and roasted make good coffee substitute
Common chickweed	*Stellaria media*	plants cooked as a vegetable
Corn lily	*Clintonia borealis*	youngest leaves can be used as a cooked vegetable
Cow parsnip	*Heracleum lanatum*	roots and young stems cooked
Cowpea	*Vigna sinensis*	"peas" and young pods cooked as a vegetable (plant naturalized in southern U.S.)
Crab apple	*Pyrus* spp.	jelly made from fruits
Crowberry	*Empetrum* spp.	fruits should first be frozen then cooked with sugar
Dandelion	*Taraxacum officinale*	leaves rich in vitamin A; dried roots make good coffee substitute; wine made from young flowers
Dock	*Rumex* spp.	leaves cooked like spinach; tartness of leaves varies from species to species and sometimes from plant to plant—tart forms should be cooked in two or three changes of water
Douglas fir	*Pseudotsuga menziesii*	cambium and young phloem edible; tea made from fresh leaves
Elderberry	*Sambucus* spp.	fresh flowers used to flavor batters; fruits used in pies, jellies, wine (**Caution:** *Other parts of the plant are poisonous.*)
Evening primrose	*Oenothera hookeri, O. biennis,* and others	young roots cooked

Plant	Scientific Name	Uses
Fairy bells	*Disporum trachycarpum*	berries can be eaten raw
Fennel	*Foeniculum vulgare*	leaf petioles eaten raw or cooked
Fireweed	*Epilobium angustifolium*	young shoots and leaves boiled as a vegetable
Ginger (wild)	*Asarum* spp.	rhizomes can be used as substitute for true ginger
Gooseberry	*Ribes* spp.	berries eaten cooked, dried, or raw; make excellent jelly
Grape (wild)	*Vitis* spp.	berries usually tart but can be eaten raw; make good jams and jellies
Grass	Many genera and species	seeds of most can be made into flour; rhizomes of many perennial species can be dried and ground for flour
Greenbrier	*Smilax* spp.	roots dried and ground; refreshing drink made with ground roots, sugar, and water
Groundnut	*Apios americana*	tubers cooked like potatoes
Hawthorn	*Crataegus* spp.	fruits edible raw and in jams and jellies
Hazel nut	*Corylus* spp.	nuts eaten raw or roasted
Hickory	*Carya* spp.	nuts edible
Highbush cranberry	*Viburnum trilobum*	fruits make excellent jellies and jams
Huckleberry	*Vaccinium* spp.	berries eaten raw or in jams and jellies
Indian paintbrush	*Castilleja* spp.	flowers of many species edible (**Caution:** *On certain soils, plants absorb toxic quantities of selenium.*)
Indian pipe	*Monotropa* spp.	whole plant edible raw or cooked
Juniper	*Juniperus* spp.	"berries" dried, ground, and made into cakes
Labrador tea	*Ledum* spp.	tea made from young leaves
Lamb's quarters	*Chenopodium album*	leaves and young stems used as cooked vegetable
Licorice	*Glycyrrhiza lepidota*	roots edible raw or cooked
Mallow	*Malva* spp.	leaves and young stems used as vegetable (use only small amounts at one time)
Manzanita	*Arctostaphylos* spp.	berries eaten raw, in jellies or pies, or made into "cider"
Maple	*Acer* spp.	sugar maples (*Acer saccharum*) well known for the sugar content of the early spring sap; other species (e.g., box elder—*A. negundo*, bigleaf maple—*A. macrophyllum*) also contain usable sugars in their early spring sap
Mariposa lily	*Calochortus* spp.	bulbs edible raw or cooked
Mayapple	*Podophyllum peltatum*	fruit good raw or cooked (**Caution:** *Other parts of the plant are poisonous.*)
Maypops	*Passiflora incarnata*	fruits edible raw or cooked
Miner's lettuce	*Montia perfoliata*	leaves eaten raw as a salad green
Mint	*Mentha arvensis* and others	leaves of several mints used for teas
Mormon tea	*Ephedra* spp.	tea from fresh or dried leaves (add sugar to offset bitterness); seeds for bitter meal
Mulberry	*Morus* spp.	fruits of the red mulberry (*M. rubra*) are used raw and in pies and jellies; fruits of white mulberry (*M. alba*) edible but insipid

Useful Plants and Poisonous Plants

Plant	Scientific Name	Uses
Mushrooms	Many genera and species	*Utmost caution* should be exercised in identifying mushrooms before consuming them. Although poisonous species are in the minority, they are common enough. Edible forms that are relatively easy to identify include morels (*Morchella esculenta*), most puffballs (*Lycoperdon* spp.), and inky cap mushrooms (*Coprinus* spp.)
Mustard	*Brassica* spp.	leaves used as vegetable; condiment made from ground seeds
Nettles	*Urtica* spp.	leaves and young stems boiled like spinach
New Jersey tea	*Ceanothus americanus*	tea from leaves
Nutgrass	*Cyperus esculentus* and others	tubers can be eaten raw
Oak	*Quercus* spp.	acorns were widely used ground for flour by native North Americans; all contain bitter tannins that must be leached out before use
Onion (wild)	*Allium* spp.	bulbs edible raw or cooked
Orach	*Atriplex patula* and others	leaves and young stems cooked as a vegetable
Pawpaw	*Asimina triloba*	fruit edible raw or cooked
Pennycress	*Thlaspi arvense*	young leaves are edible raw
Peppergrass	*Lepidium* spp.	immature fruits add zest to salads; seeds spice up meat dressings
Persimmon	*Diospyros virginiana*	fully ripened fruits can be eaten raw or cooked
Pickerel weed	*Pontederia cordata*	fruits edible raw or dried
Pigweed (*see* Amaranth)		
Pines	*Pinus* spp.	cambium and young phloem edible; tea from fresh needles rich in vitamin C
Pipsissewa	*Chimaphila umbellata*	drink made from boiled roots and leaves (cool after boiling)
Plantain	*Plantago* spp.	young leaves eaten in salads or as cooked vegetable
Poke	*Phytolacca americana*	fresh young shoots boiled like asparagus (older parts of plant poisonous)
Prairie turnip	*Psoralea esculenta*	turniplike roots cooked like potatoes
Prickly pear	*Opuntia* spp.	fruits and young stems peeled and eaten raw or cooked
Purple avens	*Geum rivale*	liquid from boiled root has flavor similar to chocolate
Purslane	*Portulaca oleracea*	leaves and stems cooked like spinach
Raspberry (wild)	*Rubus* spp.	fruits edible raw or in pies, jams, and jellies
Redbud	*Cercis* spp.	flowers used in salads; cooked young pods edible
Rose (wild)	*Rosa* spp.	fruits ("hips") exceptionally rich in vitamin C; hips can be eaten raw, pureed, or candied
Salsify	*Tragopogon* spp.	roots edible raw or cooked
Saltbush	*Atriplex* spp.	seeds nutritious (*Caution:* On certain soils, plants can absorb toxic amounts of selenium.)
Sassafras	*Sassafras albidum*	tea from roots (*Caution:* Large quantities have narcotic effect); leaves and pith used for Louisiana filé

Appendix 3

Plant	Scientific Name	Uses
Serviceberry	*Amelanchier* spp.	all fruits edible (mostly bland)
Shepherd's purse	*Capsella bursa-pastoris*	leaves cooked as vegetable; seeds eaten parched or ground for flour
Showy milkweed	*Asclepias speciosa*	flowers eaten raw or cooked; young shoots cooked
Soap plant	*Chlorogalum pomeridianum*	bulbs slow-baked and eaten like potatoes after fibrous outer coats are removed
Solomon's seal	*Polygonatum* spp.	rootstocks dried and ground for bread flour
Sorrel	*Oxalis* spp.	leaves mixed in salads
Spatterdock	*Nuphar polysepalum*	seeds placed on hot stove burst like popcorn and are edible as such; peeled tubers eaten boiled or roasted
Speedwell	*Veronica americana* and others	leaves and stems used in salads
Spring beauty	*Claytonia* spp.	bulbs edible raw or roasted
Strawberry (wild)	*Fragaria* spp.	fruits superior in flavor to cultivated varieties
Sunflower	*Helianthus annuus*	seeds eaten raw or roasted; seeds yield cooking oil
Sweet cicely	*Osmorhiza* spp.	roots have aniselike flavor
Sweet flag	*Acorus calamus*	young shoots used in salads; roots candied
Thistle	*Cirsium* spp.	peeled stems edible; roots edible raw or roasted
Vetch	*Vicia* spp.	tender green pods edible baked or boiled
Watercress	*Nasturtium officinale*	leaves edible raw in salads or cooked as a vegetable
Waterleaf	*Hydrophyllum* spp.	young shoots raw in salads; shoots and roots cooked as vegetable
Water plantain	*Alisma* spp.	the bulblike base of the plant is dried and then cooked
Water shield	*Brasenia schreberi*	tuberlike roots are peeled and then dried to be ground for flour or boiled
Winter cress	*Barbarea* spp.	leaves and young stem edible as cooked vegetable
Yarrow	*Achillea lanulosa*	plant dried and made into broth (**Caution:** *The closely related and widespread European yarrow—A. millefolium—is somewhat poisonous.*)
Yellow pond lily (*see* Spatterdock)		
Yew	*Taxus* spp.	bright red pulpy part of berries edible (**Caution:** *Seeds and leaves are poisonous.*)

Poisonous Plants

Literally thousands of plants contain varying amounts of poisonous substances. In many instances, the poisons are not present in sufficient quantities to cause adverse effects in humans when only moderate contact or consumption is involved and cooking may destroy or dissipate the substance. Some plants have substances that produce toxic effects in some organisms but not in others. Ordinary onions (*Allium cepa*), for example, occasionally poison horses or cattle yet are widely used for human food, and poison ivy (*Toxicodendron radicans*) or poison oak (*Toxicodendron diversilobum*) produce dermatitis in some individuals but not in others. Table A3.2 and table A3.3 include plants that are native to, or cultivated in, the United States and Canada.

Table A3.2 Plants known to have caused human fatalities

Plant	Scientific Name	Poisonous Parts
Angel's trumpet	*Datura suaveolens*	all parts, especially seeds and leaves
Azalea	*Rhododendron* spp.	leaves and flowers (however, poisoning is rare)
Baneberry	*Actaea* spp.	berries and roots
Belladonna	*Atropa belladonna*	all parts, especially fruits and roots
Black cherry	*Prunus serotina*	bark, seeds, leaves (***Caution:*** *Seeds of most cherries, plums, and peaches contain a poisonous principle.*)
Black locust	*Robinia pseudo-acacia*	seeds, leaves, inner bark
Black snakeroot	*Zigadenus* spp.	bulbs
Buckeye	*Aesculus* spp.	seeds, leaves, shoots, flowers
Caladium	*Caladium* spp.	all parts
Carolina jessamine	*Gelsemium sempervirens*	all parts
Castor bean	*Ricinus communis*	seeds
Chinaberry	*Melia azedarach*	fruits and leaves
Daphne	*Daphne mezereum*	all parts
Death camas (*see* Black snakeroot)		
Dieffenbachia	*Dieffenbachia* spp.	all parts
Duranta	*Duranta repens*	berries
English ivy	*Hedera helix*	berries and leaves
False hellebore	*Veratrum* spp.	all parts
Foxglove	*Digitalis purpurea*	all parts
Golden chain	*Laburnum anagyroides*	seeds and flowers
Jequirity bean	*Abrus precatorius*	seeds
Jimson weed	*Datura stramonium* and other *Datura* spp.	all parts, especially seeds
Lantana	*Lantana camara*	unripe fruits
Lily of the valley	*Convallaria majalis*	all parts
Lobelia	*Lobelia* spp.	all parts
Mistletoe	*Phoradendron* spp.	berries
Monkshood	*Aconitum* spp.	all parts
Moonseed	*Menispermum canadense*	fruits
Mountain laurel	*Kalmia latifolia*	leaves, shoots, flowers
Mushrooms	Many genera and species, especially *Amanita* spp.	all parts
Nightshade	*Solanum* spp.	unripened fruits (***Caution:*** *Common potatoes are* Solanum tuberosum; *a poisonous principle is produced in tubers exposed to light.*)
Oleander	*Nerium oleander*	all parts
Poke	*Phytolacca americana*	roots and mature stems
Rhododendron (*see* Azalea)		
Rhubarb	*Rheum rhaponticum*	leaf blades (***Caution:*** *Although young petioles are widely eaten, dangerous accumulations of a poisonous substance can occur in leaf blades.*)
Rubber vine	*Cryptostegia grandiflora*	all parts

Plant	Scientific Name	Poisonous Parts
Sandbox tree	*Hura crepitans*	milky sap and seeds
Tansy	*Tanacetum vulgare*	leaves, flowers
Tung tree	*Aleurites fordii*	all parts, especially seeds
Water hemlock	*Cicuta* spp.	roots
White snakeroot	*Eupatorium rugosum*	all parts
Yellow oleander	*Thevetia peruviana*	all parts, especially fruits
Yew	*Taxus* spp.	all parts except "berry" pulp

Table A3.3   Other plants producing significant quantities of poisonous substances

Plant	Scientific Name	Poisonous Parts
Amaryllis	*Amaryllis* spp.	bulbs
Autumn crocus	*Colchicum autumnale*	all parts
Bittersweet	*Celastrus scandens*	seeds
Bleeding hearts	*Dicentra* spp.	all parts
Bloodroot	*Sanguinaria canadensis*	all parts
Blue cohosh	*Caulophyllum thalictroides*	fruits, leaves
Boxwood	*Buxus sempervirens*	leaves
Buckthorn	*Rhamnus* spp.	fruits
Bushman's poison	*Acokanthera* spp.	all parts
Buttercup	*Ranunculus* spp.	all parts; toxicity varies from species to species; mostly cause blistering
Buttonbush	*Cephalanthus occidentalis*	leaves
Chincherinchee	*Ornithogalum thyrsoides*	all parts
Crown of thorns	*Euphorbia milii*	milky latex
Culver's root	*Veronicastrum virginicum*	root
Daffodil	*Narcissus* spp.	bulbs
Desert marigold	*Baileya radiata*	all parts
Dutchman's breeches	*Dicentra cucullaria*	all parts
Fly poison	*Amianthemum muscaetoxicum*	leaves, roots
Four-o'clock	*Mirabilis jalapa*	seeds, roots
Gloriosa lily	*Gloriosa* spp.	all parts
Goldenseal	*Hydrastis canadensis*	rhizomes, leaves
Holly	*Ilex aquifolium*	berries
Horse chestnut	*Aesculus hippocastanum*	seeds, flowers, leaves
Hyacinth	*Hyacinthus* spp.	bulbs
Hydrangea	*Hydrangea* spp.	buds, leaves
Jack-in-the-pulpit	*Arisaema triphyllum*	roots, leaves
Jessamine	*Cestrum* spp.	leaves, young stems
Jonquil (*see* Daffodil)		
Karaka nut	*Corynocarpus laevigata*	seeds
Kentucky coffee tree	*Gymnocladus dioica*	fruits
Larkspur	*Delphinium* spp.	young plants, seeds

Plant	Scientific Name	Poisonous Parts
Lignum vitae	*Guaiacum officinale*	fruits
Locoweed	*Astragalus* spp.	location of poisonous principles varies from species to species; plants more of a problem for livestock than for humans
Lupine	*Lupinus* spp.	location of poisonous principles varies from species to species, primarily in pods and seeds
Marijuana	*Cannabis sativa*	resins secreted by glandular hairs among flowers
Mayapple	*Podophyllum peltatum*	all parts except ripe fruits
Mescal bean	*Sophora secundiflora*	seeds
Narcissus (*see* Daffodil)		
Ngaio	*Myoporum laetum*	leaves
Opium poppy	*Papaver somniferum*	unripe fruits
Philodendron	*Philodendron* spp.	leaves, stems
Pittosporum	*Pittosporum* spp.	fruits, leaves, stems
Poinsettia	*Euphorbia pulcherrima*	milky latex
Poison ivy	*Toxicodendron radicans*	leaves
Poison oak	*Toxicodendron diversilobum*	leaves
Poison sumac	*Toxicodendron vernix*	leaves
Prickly poppy	*Argemone* spp.	seeds, leaves
Privet	*Ligustrum vulgare*	fruits
Sneezeweed	*Helenium* spp.	all parts
Snow-on-the-mountain	*Euphorbia marginata*	milky latex
Squirrel corn	*Dicentra canadensis*	all parts
Star-of-Bethlehem	*Ornithogalum umbellatum*	all parts
Sweet pea	*Lathyrus* spp.	seeds
Tobacco	*Nicotiana tabacum*	leaves (when eaten)

Medicinal Plants

Although numerous modern medicines include synthetic substances, many still contain drugs naturally produced by plants, and as recently as fifty years ago the vast majority of medicines used in the treatment of human diseases and ailments were plant-produced. Table A3.4 includes a sampling of plants associated with medicinal uses in the past. Some of the drugs concerned are still prescribed for specific ailments by modern physicians, while others are a part of folk medicine still practiced in rural areas. *Caution: Do not use any of the plants listed here for medicinal purposes without consulting a physician.*

Appendix 3

Table A3.4 Plants associated with medicinal uses

Plant	Scientific Name	Uses
Aloe	*Aloe* spp. (especially *Aloe vera*)	juice from leaves used on burns, especially X-ray burns
American mountain ash	*Pyrus americana*	liquid made from steeping inner bark in water used as an astringent; tea of berries used as a wash for piles; berries eaten to prevent or cure scurvy
Anemone	*Anemone canadensis*	pounded boiled root applied to wounds as an antiseptic
Arnica	*Arnica* spp.	plants applied as a poultice to bruises and sprains
Balm of Gilead	*Populus* × *gileadensis*	buds used as an ingredient in cough syrups
Balsam poplar	*Populus balsamifera*	buds made into ointment, which was placed in nostrils by Native Americans for relief of congestion
Blackberry	*Rubus* spp.	tea of roots used by northern California Native Americans to cure dysentery
Black cohosh	*Cimicifuga racemosa*	dried roots used in cough medicines and for rheumatism
Black haw	*Viburnum prunifolium*	bark used in treatment of asthma and for relieving menstrual irregularities
Bloodroot	*Sanguinaria canadensis*	Native Americans used rhizome for ringworm, as an insect repellent and for sore throat
Blue cohosh	*Caulophyllum thalictroides*	tea of root drunk by Native Americans and early settlers a week or two before giving birth to promote rapid parturition
Boneset	*Eupatorium perfoliatum*	water infusion of dried plant tops widely used to treat fevers and colds
Broom snakeweed	*Gutierrezia sarothrae*	Navajo Indians applied chewed plant to insect stings and bites of all kinds
Burdock	*Arctium lappa*	used as an insulin substitute in folklore; root extract used in seventeenth century for venereal diseases
Button snakeroot	*Eryngium* spp.	Natchez Indians inserted chewed stem in nostrils to arrest nosebleed
California bay	*Umbellularia californica*	Yuki Indians put leaves in bath of hot water and bathed for relief of rheumatism; leaves used as an insect repellent
Camphor	*Cinnamomum camphora*	oil from leaves and wood used in cold remedies and liniments
Camptotheca	*Camptotheca acuminata*	extracts from flowers and immature fruits yield camptothecin, which has given evidence of being effective against certain forms of cancer
Cascara	*Rhamnus purshiana*	bark extract widely used in laxatives
Catnip	*Nepeta cataria*	leaf tea used for treatment of colds and to relieve infantile colic
Cayenne pepper	*Capsicum frutescens*	used to reduce mucous drainage (recent evidence suggests it may be carcinogenic)
Cherry (wild)	*Prunus serotina*	tea brewed from bark used for coughs and colds

Plant	Scientific Name	Uses
Chia	*Salvia columbariae*	mucilaginous seeds used by Spanish Californians to make a refreshing drink; seeds contain a caffeinelike principle that enabled Native Americans to perform unusual feats of endurance; seed paste used in eyes irritated by foreign matter
Cinchona	*Cinchona* spp.	bark yields quinine drugs used in treating malaria
Club moss	*Lycopodium clavatum*	spores dusted on wounds or inhaled by Native Americans to arrest nosebleeds
Coca	*Erythroxylon coca*	cocaine from leaves used as a local anesthetic; South American laborers use it as a stimulant
Cotton	*Gossypium* spp.	cotton root bark used by black slaves and Native Americans to induce abortions
Creosote bush	*Larrea divaricata*	decoction from leaves used as a cure-all by Native Americans but especially for respiratory problems
Cubebs	*Piper cubeba*	dried fruit best known as a condiment but is also used in treatment of asthma
Deadly nightshade	*Atropa belladonna*	belladonna, a drug complex extracted from leaves, contains the drugs atropine, hyoscyamine, and scopolamine; these are used as an opium antidote, in shock treatments, and for dilation of pupils; scopolamine is also used as a tranquilizer and for "twilight sleep" in childbirth
Dogbane	*Apocynum androsaemifolium*	roots boiled in water and resulting liquid used as a heart medication (contains a drug similar in action to digitalis)
Dogwood	*Cornus* spp.	inner bark boiled in water and resulting liquid drunk to reduce fevers
Ephedra	*Ephedra* spp.	drug ephedrine, widely used to relieve nasal congestion and low blood pressure, obtained from stems (most ephedrine now in use is synthetic)
Ergot	(Source: *Claviceps purpurea* on cereal grains)	used to treat migraine headaches and to control bleeding after childbirth
Flowering ash	*Chionanthus virginicus*	bark used as a laxative and in treatment of liver ailments
Foxglove	*Digitalis purpurea*	drug digitalis, widely used as a heart stimulant, obtained from leaves
Gentian	*Gentiana catesbaei*	Catawba Indians applied hot-water extract of roots to sore backs; liquid drunk as a remedy for stomachaches
Geranium	*Geranium maculatum*	dried root used for dysentery, diarrhea, and hemorrhoids
Ginger (wild)	*Asarum* spp.	extract of rhizome used as a broad-spectrum antibiotic
Ginseng	*Panax quinquefolium* and other *Panax* spp.	considered a general panacea, especially by the Chinese
Goldenseal	*Hydrastis canadensis*	rhizome source of alkaloidal drugs used in treatment of inflamed mucous membranes; also used as a tonic

Plant	Scientific Name	Uses
Goldthread	*Coptis groenlandica*	Native Americans boiled plant and gargled resulting liquid for treatment of sore or ulcerated mouths
Green hellebore	*Helleborus viridis*	extract of plant used in treatment of hypertension (drug now largely synthesized); Thompson Indians used it in small amounts for treatment of syphilis
Hemlock	*Tsuga* spp.	Native Americans made tea of inner bark for treatment of colds and fevers
Horehound	*Marrubium vulgare*	extract of dried tops of plants used in lozenges for relief of sore throat and colds
Horsetail	*Equisetum* spp.	plants boiled in water and resulting liquid used as a hairwash; liquid gargled for treatment of mouth ulcers
Indigo (wild)	*Baptisia tinctoria*	Native Americans boiled plant and used resulting liquid as antiseptic for skin sores
Ipecac	*Cephaelis ipecacuana*	drug from roots and rhizome used to treat amoebic dysentery; also used as an emetic
Jimson weed	*Datura* spp.	drugs atropine, hyoscyamine, and scopolamine obtained from seeds, flowers, and leaves; drug stramonium used for knockout drops and in treatment of asthma (see Deadly nightshade)
Joe-pye weed	*Eupatorium purpureum*	dried root said to prevent formation of gallstones
Joshua tree	*Yucca brevifolia*	cortisone and estrogenic hormones made from sapogenins produced in roots
Juniper	*Juniperus* spp.	tea of "berries" drunk by Zuni Indian women to relax muscles following childbirth
Kansas snakeroot	*Echinacea angustifolia*	dried roots used as antiseptic for treatment of sores and boils, periodontal disease, and sinus drainage problems
Lily of the valley	*Convallaria majalis*	all parts of plant contain a heart stimulant similar to digitalis
Lobelia	*Lobelia inflata*	drug lobeline sulphate obtained from dried leaves; drug used in antismoking preparations and in treatment of respiratory disorders
Mandrake	*Mandragora officinarum*	extracts of plant used as painkiller in folk medicine (drugs hyoscyamine, podophyllin, and mandragorin have been isolated; podophyllin experimentally used in treatment of paralysis)
Manroot	*Marah* spp.	Native Americans used oil from seeds to treat scalp problems and the crushed roots for relief from saddle sores
Marginal fern	*Dryopteris marginalis*	rhizomes contain oleoresin used in expulsion of tapeworms from intestinal tract
Marijuana	*Cannabis sativa*	tetrahydrocannabinol obtained from resinous hairs in inflorescences; ancient medicinal drug of China
Mayapple	*Podophyllum peltatum*	podophyllin obtained from roots used experimentally in treatment of paralysis; dried root powder used on warts

Useful Plants and Poisonous Plants

Plant	Scientific Name	Uses
Maypop	*Passiflora incarnata*	dried leaves used as a sedative; Native Americans used juice for treatment of sore eyes
Mesquite	*Prosopsis juliflora*	Native Americans mixed powdered leaves with water to make liquid used to treat sore eyes
Milkweed	*Asclepias syriaca*	Quebec Indians promoted temporary sterility by drinking infusion of pounded roots
Mistletoe	*Phoradendron flavescens*	Native Americans of California drank tea of leaves to induce abortions or as a contraceptive
Monkshood	*Aconitum napellus*	source of aconite once used in treatment of rheumatism and neuralgia
Mulberry	*Morus rubra*	Rappahannock Indians applied milky latex of leaf petioles to scalp for ringworm
Mullein	*Verbascum thapsus*	Native Americans smoked leaves for respiratory ailments and asthma; flowers once widely used in cough medicines
Onion (wild)	*Allium* spp.	bulbs eaten to prevent scurvy; Cheyenne Indians applied bulbs in poultice to boils
Opium poppy	*Papaver somniferum*	morphine and codeine obtained from latex of immature fruits
Oregon grape	*Berberis aquifolium*	bark tea drunk by Native Americans to settle upset stomach; used in strong doses as a remedy for venereal diseases
Pansy (wild)	*Viola tricolor*	plants ground up and applied to skin sores
Pennyroyal	*Hedeomia pulegioides*	Native Americans used leaf tea for relief of headaches and flatulence and to repel chiggers
Persimmon	*Diospyros virginiana*	liquid from boiled fruit used as an astringent; leaves said to be rich in vitamin C.
Peyote	*Lophophora williamsii*	alcoholic extract of plant used as an antibiotic
Pinkroot	*Spigelia marilandica*	powdered root very effective in expulsion of roundworms from intestinal tract
Pipsissewa	*Chimaphila umbellata*	Native Americans steeped plant in water and applied resulting liquid to draw out blisters
Pitcher plant	*Sarracenia purpurea*	Native Americans used root widely as smallpox cure (records indicate it was effective)
Plantain	*Plantago ovata* and others	seeds used in laxatives
Pleurisy root	*Asclepias tuberosa*	liquid resulting from roots being boiled in water taken for respiratory problems
Prickly ash	*Zanthoxylum americanum*	bark and berries widely used by Native Americans for toothache (pieces inserted in cavities) and liquid infusion drunk for venereal diseases
Quassia	*Picraea excelsa; Quassia amara*	wood extract used as pinworm remedy and as an insecticide
Rauwolfia	*Rauwolfia serpentina*	reserpine obtained from roots; drug used in treatment of mental illness and in counteracting effects of LSD

Plant	Scientific Name	Uses
Saffron (meadow)	*Colchicum autumnale*	drug colchicine from corms once used in treatment of gout and back ailments but now largely used for experimental doubling of chromosome numbers in plants
Sarsaparilla	*Aralia nudicaulis*	cough medicines made from roots
Sassafras	*Sassafras albidum*	tea of root bark used to induce sweating; used externally as a liniment
Self-heal	*Prunella vulgaris*	Native Americans applied plants in poultices to boils
Seneca snakeroot	*Polygala senega*	bark boiled in water; liquid applied to snakebites and taken internally as an abortifacient; also used as a remedy for coughs
Senna	*Cassia senna* and others	extract of leaves used as a purgative
Skeleton weed	*Lygodesmia juncea*	widely used by Native American women to increase flow of milk
Skullcap	*Scutellaria laterifolia*	dried plant used as an anticonvulsive in treatment of epilepsy and as a sedative
Slippery elm	*Ulmus fulva*	dried inner bark contains an aspirinlike substance that soothes inflamed membranes
Spicebush	*Lindera benzoin*	berries, buds, bark, and leaves brewed for tea used to reduce fevers
Spruce	*Picea* spp.	Cree Indians ate small cones for treatment of sore throat
Squills	*Urginea maritima*	bulbs of red variety are the source of a heart stimulant; bulbs of white variety are the source of a widely used rodent killer
Stoneseed	*Lithospermum ruderale*	Shoshoni Indian women reported to have drunk water infusion of roots daily for six months to produce permanent sterility (experiments with mice suggest substance to the reports)
Strophanthus	*Strophanthus sarmentosus* and others	seeds are a major source of cortisone and also the source of a heart stimulant
Strychnine plant	*Strychnos nox-vomica*	strychnine extracted from seeds widely used as an animal and insect poison and is the principal active ingredient in blowgun darts and poison arrows of primitive South American aborigines; minute amounts stimulate the central nervous system and relieve paralysis
Sumac	*Rhus glabra* and others	Native Americans applied leaf decoction as a remedy for frostbite; fruits and liquid made from leaves applied to poison ivy rash and gonorrhea sores; root chewed for treatment of mouth ulcers
Sweet flag	*Acorus calamus*	boiled root applied to burns; root chewed for relief of colds and toothache
Sweet gum	*Liquidambar styraciflua*	bud balsam used to treat chigger bites; balsam also used in insect fumigating powders

Plant	Scientific Name	Uses
Sword fern	*Polystichum munitum*	boiled rhizome used by Native Americans to treat dandruff; sporangia applied to burns
Tamarind	*Tamarindus indica*	fruit pulp used as a laxative
Valerian	*Valeriana septentrionalis*	pulverized plants applied to wounds
Velvet bean	*Mucuna* spp.	seeds contain L-dopa used in treatment of Parkinson's disease
Virginia snakeroot	*Aristolochia serpentaria*	Native Americans used tea of plant for reducing high fevers
Wahoo	*Euonymus atropurpureus*	bark steeped in water; resulting liquid has digitalislike effect on heart
Western wallflower	*Erysimum capitatum*	Zuni Indians ground plant with water and applied it to skin to prevent sunburn
Willow	*Salix* spp.	Chickasaw Indians snuffed infusion of roots as a remedy for nosebleed; Pomo Indians boiled bark in water and applied resulting liquid for relief of skin itches; fresh inner bark contains salicin, an aspirinlike compound used to reduce fevers
Wintergreen	*Gaultheria procumbens*	oil from leaves used as a folk remedy for body aches and pains
Witch hazel	*Hamamelis virginiana*	oil distilled from twigs used as an external medicine
Wormseed	*Chenopodium ambrosioides*	oil from seeds used to expel intestinal worms
Wormwood	*Artemisia* spp.	Yokia Indians made tea from leaves to treat bronchitis; other Native American tribes used tea as a remedy for colds
Yarrow	*Achillea millefolium*	Native Americans used infusion of plant for treatment of wounds, burns, and earaches
Yellow lady's slipper	*Cypripedium calceolus*	dried root used for relief of insomnia or as a sedative
Yellow nut grass	*Cyperus esculentus*	Paiute Indians pounded tubers with tobacco leaves and applied mass in wet dressing for treatment of athlete's foot
Yerba santa	*Eriodictyon californicum*	Native Americans smoked leaves or drank tea of leaves for treatment of colds or asthma

Hallucinogenic Plants

Although a few hallucinogenic substances produced by animals have been isolated and some have been synthesized, the majority of known hallucinogens are produced by plants. Table A3.5 is not a complete list but it includes the better known sources. The reader is referred to the Additional Reading for further information.

Table A3.5 Hallucinogenic substances produced by plants			
Plant	**Scientific Name**	**Part Used**	**Principal Active Substance**
Ajuca	Mimosa hostilis	roots	nigerine
Belladonna	Atropa belladonna	leaves	hyoscyamine, scopolamine
Caapi	Banisteriopsis caapi	wood	harmine
Canary broom	Cytisus canariensis	seeds	cytisine
Catnip	Nepeta cataria	leaves	unknown
Cohoba	Piptadena peregrina	seeds (snuff)	tryptamines
Coral bean	Erythrina spp.	seeds	unknown
Cubbra borrachera	Methysticodendron amesianum	leaves	scopolamine
Ergot fungus	Claviceps purpurea	rhizomorph	ergine (LSD)
Fly agaric	Amanita muscaria	mushroom cap	ibotenic acid, muscimol
Henbane	Hyoscyamus spp.	leaves	hyoscyamine, scopolamine
Iboga	Tabernanthe iboga	root bark	ibogaine
Jimson weed	Datura spp.	all parts	scopolamine
Kava kava	Piper methysticum	root (large amounts of beverage produce hallucinations)	myristicinlike compound
Mace	Myristica fragrans	aril of seed	myristicin
Mescal bean	Sophora secundiflora	seeds	cytisine
Morning glory	Ipomoea violacea	seeds	ergine
Nutmeg	Myristica fragrans	seeds	myristicin
Ololiuqui	Rivea corymbosa	seeds	turbicoryn
Peyote	Lophophora williamsii	stems	mescaline
Psilocybe mushrooms	Psilocybe spp., Conocybe spp., Panaeolus spp., and others	all parts	psilocybin, psilocin
Rape dos Indios	Maquira sclerophylla	dried plant (snuff)	unknown
San Pedro	Trichocereus pachanoi	stems	mescaline
Sassafras	Sassafras albidum	root bark (large amounts of tea)	safrole
Sweet flag	Acorus calamus	dried root	asarone, β-asarone
Syrian rue	Peganum harmala	seeds	harmine
Vygie	Mesembryanthemum expansum	all parts	mesebrine
Wood rose	Argyreia nervosa	seeds	ergoline alkaloids
Yakee (Parica)	Virola spp.	resin from inner surface of freshly removed bark (snuff)	tryptamine
Yohimbehe	Corynanthe spp.	bark	yohimbine

Spice Plants

The word *spice* describes any aromatic plant or part of a plant used to flavor or season food; spices are also used to add scent or flavor to manufactured products (table A3.6). Although spices have no nutritional value, they add a pleasurable zest to meals, and before food preservation was possible they helped make palatable food that was still edible but unappealing.

The value placed on spice plants was responsible for changing the course of Western civilization as a principal motive behind the voyages of discovery.

Table **A3.6** Plants used to season or flavor			
Spice	**Scientific Name of Plant**	**Parts Used; Remarks**	**Principal Source**
Allspice	*Pimento officinalis*	powdered dried fruit	Jamaica
Almond	*Prunus amygdalus*	oil from seed used for flavoring baked goods	Mediterranean; U.S.
Angelica	*Angelica archangelica*	stems candied; oil from seeds and roots used in liqueurs	Europe; Asia
Anise	*Pimpinella anisum*	oil distilled from fruits used for flavoring	widely cultivated
Arrowroot	*Maranta arundinacea*	powdered root used in milk puddings, baked goods	South America
Asafoetida	*Ferula asafoetida*	powdered gum from stems and roots used in minute quantities with fish	Middle East
Balm (Melissa)	*Melissa officinalis*	oil from leaves used in beverages; leaves used as food flavoring	U.S., Mediterranean
Basil	*Ocimum basilicum*	leaves used in meat dishes, soups, sauces	Mediterranean
Bay	*Laurus nobilis*	leaves used in soups, sauces	Europe
Bell pepper	*Capsicum frutescens*	dried diced fruit used in chip dips, salad dressings	widely cultivated
Bergamot	*Monarda didyma*	leaves used with pork (*Note:* A perfume oil obtained from a variety of orange—*Citrus aurantium* var. *bergamia*—is also called bergamot.)	North America (*Monarda*); Italy (*Citrus*)
Black pepper	*Piper nigrum*	dried fruits used as a condiment	India; Indonesia
Borage	*Borago officinalis*	leaves used as a beverage flavoring	England
Burnet	*Sanguisorba minor*	used in soups and casseroles	Eurasia
Calamus	*Acorus calamus*	powdered rhizome used for flavoring	Europe; Asia; North America
Capers	*Capparis spinosa*	flower buds used for flavoring relishes, pickles, sauces	Mediterranean
Caraway	*Carum carvi*	seeds used in breads, cheeses; seed oil used in the liqueur kümmel	North America; Europe
Cardamon	*Elletaria cardamomum*	dried fruit and seeds used for flavoring baked goods (*Note:* several false cardamons—*Amomum* spp.—are sold commercially.)	India; Sri Lanka; Central America
Cassia	*Cinnamomum cassia*	powdered bark used as cinnamon substitute	Southeast Asia
Cayenne pepper	*Capsicum* spp.	powdered dried fruits used in chili powder, Tabasco sauce	American tropics

Spice	Scientific Name of Plant	Parts Used; Remarks	Principal Source
Celery	Apium graveolens	seeds used in celery salt, soups	Europe; U.S.
Chervil	Anthriscus cerefolium	used as a parsley substitute	Europe; Near East
Chives	Allium schoenoprasum	leaves, bulbs used with sour cream, butter	widely cultivated
Chocolate	Theobroma cacao	ground seeds used for flavoring	Africa; South America
Cilantro	Coriandrum sativum	leaves used in avocado dip and with poultry	Europe
Cinnamon	Cinnamomum zeylanicum	ground bark used for flavoring baked goods; oil from leaves used as flavoring, clearing agent	Seychelles; Sri Lanka
Citrus	Citrus spp.	fruits, especially rinds, source of flavoring oil	Mediterranean; South Africa; U.S.
Coffee	Coffea arabica	roasted seeds source of mocha-coffee flavoring	Tropics
Coriander	Coriandrum sativum	ground seed used in German frankfurters, curry powders	Mediterranean
Cubebs	Piper cubeba	dried fruits used as seasoning	East Indies
Cumin	Cuminum cyminum	ground seed used with meats, pickles, cheeses, curry	Mediterranean
Curry	—	a spicy condiment containing several ingredients, such as turmeric, cumin, fenugreek, and zedoary	India
Dill	Anethum graveolens	seeds used in pickling brines; leaves used for seasoning meat loaves, sauces	Europe; Asia
Dittany	Origanum dictamnus	leaves used as seasoning for poultry, meats	Crete
Eucalyptus	Eucalyptus spp.	oil from leaves used in toothpastes, flavoring agents	Australia
Fennel	Foeniculum vulgare	seeds used in baked goods	Europe
Fenugreek	Trigonella foenumgraecum	oil distilled from seeds used in pickle, chutney, curry powders, imitation maple flavoring	widely cultivated
Filé (see Sassafras)			
Garlic	Allium sativum	fresh or dry bulbs used for meat seasonings	widely cultivated
Ginger	Zingiber officinale and others	dried rhizomes used for flavoring many foods and drinks	India; Taiwan
Grains of paradise	Afromomum melegueta	seeds used to flavor beverages and medicines	West Africa
Hops	Humulus lupulus	dried inflorescences of female plants used in brewing beer	Europe; North America
Horseradish	Rorippa armoracia	grated fresh root used as a condiment	Europe; North America
Juniper	Juniperus spp.	"berries" used to season beef roasts, poultry, sauces	North America
Licorice	Glycyrrhiza glabra	dried rhizome and root used to flavor pontefract cakes, candies	Middle East
Lovage	Ligusticum scoticum	stems candied; seeds used in pickling sauces; celery substitute	Europe
Mace	Myristica fragrans	aril of seed used for flavoring beverages, foods	Grenada; Indonesia; Sri Lanka

Useful Plants and Poisonous Plants

Spice	Scientific Name of Plant	Parts Used; Remarks	Principal Source
Marigold	*Tagetes* spp.	petals substituted for saffron in rice dishes, stews	widely cultivated
Marjoram	*Majorana hortensis*	leaves used in stews, dressings, sauces	Mediterranean
Mustard	*Brassica* spp.	ground seeds used in meat condiment	Europe; China
Nasturtium	*Tropaeolum majus*	flowers, seeds, leaves used in salads	widely cultivated
Nutmeg	*Myristica fragrans*	seeds used for flavoring foods, beverages	Grenada; Indonesia; Sri Lanka
Oregano	*Origanum vulgare* and others	leaves used as seasoning with poultry, meats	Europe
Paprika (*see* Cayenne pepper)			
Parsley	*Petroselinum sativum*	leaves used as meat garnish and flavoring in sauces	widely cultivated
Peppermint	*Mentha piperita*	oil from leaves used for food, drink, dentifrice flavoring (much commercial menthol is derived from *Mentha arvensis* grown in Japan)	U.S.; Russia
Pimiento	*Capsicum* spp.	bright red fruits of a cultivated variety of pepper used in stuffing olives and in cold meats, cheeses	Central and South America
Poppy	*Papaver somniferum*	seeds used in baking	widely cultivated
Rosemary	*Rosmarinus officinalis*	oil from leaves used in perfumes, soaps	Mediterranean
Rue	*Ruta graveolens*	flavoring for fruit cups, salads	Europe
Saffron	*Crocus sativus*	dried stigmas used to flavor oriental-style dishes	Spain; India
Sage	*Salvia officinalis*	leaves used in poultry and meat dressings	Yugoslavia
Sarsaparilla	*Smilax* spp.	roots are source of flavoring for beverages, medicines	American Tropics
Sassafras	*Sassafras albidum*	bark and wood yield flavoring for beverages, toothpaste, gumbo	U.S.
Savory (summer)	*Satureia hortensis*	leaves used in green bean and bean salads, lentil soup, with fish	Mediterranean
Savory (winter)	*Satureia montana*	leaves used as seasoning in stuffings, meat loaf, stews	Europe
Scallion	*Allium fistulosum*	leaves used in wine cookery, soups	widely cultivated
Sesame	*Sesamum indicum*	seeds used in baking	Asia
Shallot	*Allium ascalonicum*	bulbs, leaves used in Colbert butter, wine cookery	widely cultivated
Southernwood	*Artemisia abrotanum*	leaves used to flavor cakes	Europe
Star anise	*Illicium verum*	fruits used in candy and cough drops	China
Stonecrop	*Sedum acre*	dried leaves (ground) used as pepper substitute	Europe
Tarragon	*Artemisia dracunculus*	leaves and flowering tops used in pickling sauces	Europe
Thyme	*Thymus vulgaris*	leaves used in meat and poultry dishes, soups, sauces	widely cultivated

Spice	Scientific Name of Plant	Parts Used; Remarks	Principal Source
Tonka bean	*Dipteryx* spp.	seeds source of flavoring for tobacco; vanilla substitute (now largely synthesized)	American tropics
Turmeric	*Curcuma longa*	rhizomes powdered and used in curry powders, meat flavoring	India; China
Vanilla	*Vanilla planifolia*	flavoring extracted from fruits; used in foods, drinks	Malagasay Republic
Wintergreen	*Gaultheria procumbens*	oil from leaves, bark used as flavoring for confections, toothpaste	U.S.
Zedoary	*Curcuma zedoaria*	dried rhizome used in liqueurs, curry powders	India

Dye Plants

In the recent and the ancient past, dyes from many different plants were used to color cotton, linen, and other fabrics. Since the middle of the nineteenth century, however, natural dyes have been almost completely replaced in industry by synthetic dyes, and today the use of natural dyes is largely confined to individual hobbyists.

Any reader interested in experimenting with natural dyes is encouraged to choose not only those plant materials included in table A3.7 but to try any local plants available. The experimenter will soon find that quite unexpected colors may be derived from plants, as the colors of fresh flowers, bark, or leaves often bear little relationship to the colors of the dyes. For methods of dyeing, see the footnote given under the heading of Lichens in chapter 16 and references in the Additional Reading list.

Table A3.7	Plant sources of natural dyes	
Plant or Dye	**Scientific Name of Plant Source**	**Remarks**
Acacia	*Acacia* spp.	brown dyes from bark and fruits
Alder	*Alnus* spp.	brownish dyes from bark
Alkanet	*Alkanna tinctoria*	red dye from roots
Annatto	*Bixa orellana*	yellow or red dye from pulp surrounding seeds
Bamboo	*Bambusa* spp.	light green dye from leaves
Barberry	*Berberis vulgaris*	grayish dye from leaves
Barwood	*Baphia nitida*	purplish dyes from wood
Bearberry	*Arctostaphylos uva-ursi*	yellowish dye from leaves
Bedstraw	*Galium* spp.	light reddish brown dyes from roots
Birch	*Betula* spp.	light brown to black dyes from bark
Black cherry	*Prunus serotina*	red dye from bark; gray to green dyes from leaves
Black walnut	*Juglans nigra*	rich brown dye from bark; brown dye from walnut hulls
Bloodroot	*Sanguinaria canadensis*	red dye from rhizomes
Blueberry	*Vaccinium* spp.	blue to gray dye from mature fruits (tends to fade)
Bougainvillea	*Bougainvillea* spp.	light brownish dyes from floral bracts
Brazilwood	*Caesalpinia* spp.	reddish dyes from wood
Buckthorn	*Rhamnus* spp.	green dyes from fruits
Buckwheat	*Fagopyrum esculentum*	blue dye from stems

Plant or Dye	Scientific Name of Plant Source	Remarks
Buckwheat (wild)	*Eriogonum* spp.	dark gold, pale yellow, and beige dyes from stems and flowers
Buffaloberry	*Shepherdia argentea*	red dye from fruit
Butternut	*Juglans cinerea*	yellow to grayish brown dyes from fruit hulls
Cocklebur	*Xanthium strumarium*	dark green dye from stems and leaves
Coffee	*Coffea arabica*	light brown dye from ground roasted seeds
Cudbear (Archil)	*Rocella* spp. (lichen)	red dye obtained by fermentation of thallus
Cutch	*Acacia* spp.; *Uncaria gambir*	brown to drab green dyes from stem gums
Dock	*Rumex* spp.	light brown dyes from stems and leaves
Dogwood	*Cornus florida*	red dye from bark; purplish dye from root
Doveweed	*Eremocarpus setigerus*	light to olive green dye from entire plant
Dyer's rocket	*Reseda luteola*	orangish dye from all parts
Elderberry	*Sambucus* spp.	blackish dye from bark; purple, blue, or dark brown dyes from fruits
Eucalyptus	*Eucalyptus* spp.	beige dyes from bark
Fennel	*Foeniculum vulgare*	yellow dyes from shoots
Fig	*Ficus carica*	green dyes from leaves and fruits
Fustic	*Chlorophora tinctoria*	yellow, bright orange, and greenish dyes from heartwood
Gamboge	*Garcinia* spp.	yellow dye from resins that ooze from cuts made on stems
Giant reed	*Arundo donax*	pale yellow dye from leaves
Grape	*Vitis* spp.	bright yellow to olive green dyes from leaves
Hawthorn	*Crataegus* spp.	pink dye from ripe fruits
Hemlock	*Tsuga* spp.	reddish brown dye from bark
Henna	*Lawsonia inermis*	orange dye from shoots and leaves
Hickory	*Carya tomentosa*	yellow dye from bark
Hollyhock	*Althaea rosea*	purplish black dye from flower petals
Horsetail	*Equisetum* spp.	tan dyes from all green parts
Indigo	*Indigofera tinctoria*	bright blue dyes from leaves
Kendall green (*see* Woadwaxen)		
Larkspur	*Delphinium* spp.	blue dyes from petals
Lichens	Many genera and species	many lichens yield brilliant shades of yellows, golds, and browns with various mordants
Litmus	*Rocella tinctoria*	famous pink-to-blue pH indicator dye from thallus
Logwood	*Haematoxylon campechianum*	dark blue purple dye from heartwood
Lokao	*Rhamnus* spp.	green dye from wood
Lupine	*Lupinus* spp.	greenish dyes from flowers
Madder	*Rubia tinctorium*	bright red dye from roots
Madrone	*Arbutus menziesii*	brown dye from bark
Manzanita	*Arctostaphylos* spp.	beige to dull yellow dyes from dried fruits
Maple	*Acer* spp.	pink dye from bark
Marsh marigold	*Caltha palustris*	yellow dye from petals
Milkweed	*Asclepias speciosa*	pale yellow dyes from leaves
Morning glory	*Ipomoea violacea*	gray green dye from blue flowers
Mullein	*Verbascum thapsus*	gold dyes from leaves
Oak	*Quercus* spp.	yellow dye from bark
Onion	*Allium cepa*	reddish brown dyes from dry outer bulb scales of red onions; yellow dyes from similar parts of yellow onions

Appendix 3

Plant or Dye	Scientific Name of Plant Source	Remarks
Oregon grape	Berberis aquifolium	yellow dyes from roots
Osage orange	Maclura pomifera	yellow, gray, and green dyes from fruits; yellow orange dye from wood
Peach	Prunus persica	green dyes from leaves
Poke	Phytolacca americana	red dyes from mature fruits
Pomegranate	Punica granatum	dark gold dye from fruit rinds
Prickly lettuce	Lactuca serriola	green dye from leaves
Privet	Ligustrum vulgare	yellow green dye from leaves; deep gray dye from berries
Quercitron	Quercus velutina	bright yellow dye from bark
Rhododendron	Rhododendron spp.	tan dyes from leaves
Safflower	Carthamnus tinctorius	reddish dye from flower heads
Saffron	Crocus sativus	powerful yellow dye from stigmas
Sage	Salvia officinalis	yellow dye from shoots
Sandalwood	Pterocarpus santalinus	red dye from wood
Sappanwood	Caesalpinia sappan	red dye from heartwood
Sassafras	Sassafras albidum	orange brown dye from bark
Scotch broom	Cytisus scoparius	yellow dye from all parts of plant
Smoke tree	Cotinus coggyria	orange yellow dye from wood (dye sometimes called "young fustic")
Smooth sumac	Rhus glabra	grayish brown dye from bark
St. John's wort	Hypericum spp.	light brownish dyes from leaves
Tansy	Tanacetum spp.	yellow, green dyes from leaves
Toyon	Heteromeles arbutifolia	reddish brown dyes from leaves
Turmeric	Curcuma longa	orangish dye from rhizome
Woad	Isatis tinctoria	blue dye from leaves
Woadwaxen	Genista tinctoria	yellow dye from all parts
Yerba santa	Eriodictyon californicum	rich dark brown dyes from leaves

Additional Reading

Adrosko, R. J. *Natural dyes in the United States.* National Museum Bulletin 281. Washington, DC: Smithsonian Institution; 1968.

Emboden, W. *Narcotic plants.* New York: Macmillan Co.; 1980.

Hoffer, A.; Osmond, H. *The hallucinogens.* New York: Academic Press; 1967.

Lesch, A. *Vegetable dyeing.* New York: Watson Guptill; 1970.

Loewenfeld, C.; Back, P. *The complete book of herbs and spices.* Pomfret, VT: David and Charles, Inc.; 1978.

Merory, J. *Food flavorings: composition, manufacture and use.* 2d ed. New York: Chemical Publishing Co.; 1968.

Schleiffer, H. *Sacred narcotic plants of the New World Indians.* New York: Hafner Press; 1973.

Schultes, R. E.; Hoffman, A. *The botany and chemistry of hallucinogens.* 2d ed. Springfield, IL: Charles C. Thomas, Publishers; 1980.

Shetky, J., ed. *Dye plants and dyeing.* Plants and Gardens Handbook. New York: Brooklyn Botanic Garden; 1964.

Usdin, E.; Efron, D. H. *Psychoactive drugs and related compounds.* Washington, DC: U.S. Public Health Publication No. 1589; 1967.

Usher, G. *A dictionary of plants used by man.* New York: Hafner Press; 1974.

Vogel, V. J. *American Indian medicine.* Norman: University of Oklahoma Press; 1977.

See also the Additional Reading entries in chapter 22.

Vegetative Propagation

The value of being able to take a piece of stem of a desirable plant from one locality to another and induce it to grow into a new plant identical to its "parent," or of being able to divide rhizomes, tubers, and other parts of plants and have them each develop into new plants has been recognized since ancient times. Grafting, which involves the permanent union of parts of two plants, also has its origins in antiquity. Plants with desirable features such as superior fruits or flowers or attractive leaves may be grafted onto related plants whose chief attraction may lie in a vigorous root system or one that is resistant to disease. The Chinese were apparently practicing grafting by the year 1000 B.C., and Theophrastus discussed grafting and other forms of vegetative propagation in his book *Causes of Plants,* written in the third century B.C. During the fourteenth and fifteenth centuries A.D. large numbers of plants were imported into European gardens and maintained by grafting. Even though the internal nature of the graft union was not fully understood, many forms of grafting were developed; no fewer than 119 different methods of grafting were described in 1821 by A. Thouin. Today the practice of various forms of vegetative propagation is almost universal and is an important part of the economies of the world. Some of the more widespread techniques of vegetative propagation are discussed in the following sections.

Cuttings

Occasionally a basic change may occur in the chromosomes of a cell, and the result is a bud or a whole plant that develops differently from the rest of the tree or variety. Such a **mutation,** as it is called (see chapter 12), is likely to be undesirable or to result in the death of the cell or tissue. In a few instances, however, the mutation may produce superior characteristics such as better flowers or fruit, dwarf forms, or disease resistance. If you were to plant seeds from such a mutant form, the plant that would develop as a result could differ as much from the mutant parent as you do from either your mother or your father. Since pollen from one parent has to be transferred to a flower of another parent for fruits and seeds to form, the mutant plant carries characteristics from both parents and thus may pass on to future generations a variety of combinations of characteristics. When the mutation is desirable, the element of chance inheritance can be eliminated by cultivating tissue directly from the mutant plant, so that each of the offspring has characteristics identical to the single parent; the desirable feature can thus be perpetuated indefinitely. Most of the Washington navel oranges, for example, have come from a mutation that appeared on a sweet orange tree growing near Bahia, Brazil, around 1820. The grower, whose identity is not known, apparently recognized the superior nature of the fruit on a branch of one of the trees and either by bud grafting or by making *cuttings* (pieces of a plant that are induced to produce roots and are then planted to grow on their own) multiplied the new form. Fifty years later, a missionary sent a dozen budded navel orange trees from Brazil to the United States Commissioner of Agriculture in Washington, D.C., and in 1873 two trees were shipped from Washington to Riverside, California. From these California trees, one of which is still living, almost the entire navel orange industry of the world has been developed through vegetative propagation.

A large number of plants can be started with cuttings from a few trees or shrubs in a relatively small amount of space and time, and little skill is required. Cuttings also do not have the problem of incompatibility, which is found with grafting. It is not always desirable to propagate by cuttings, however, particularly if the variety concerned does not have a sturdy, disease-resistant root system. Some plants are also very difficult to start from cuttings because they do not readily form roots. Nevertheless, these problems have not been sufficient to prevent this form of propagation from becoming the most widely used means, next to planting seeds, of multiplying plants in the world today.

Most cuttings are made from stems that are 0.6 to 2.5 centimeters (0.25 to 1 inch) thick and contain at least two nodes. They may vary in length from 7.5 centimeters to 6 decimeters (3 inches to 2 feet). With deciduous plants, the cuttings are made—while the buds are dormant—from healthy wood of the previous season that has been growing in the sun. Pieces may be cut transversely, or a small part of wood from an older branch may be left attached at the bottom end. If a number of cuttings are being made at one time, they may be tied together with rubber bands, placed bottom ends up in cool, damp sand or sawdust from which water will drain easily, covered, and stored until spring. They should be checked from time to time to see if buds are developing. If they are, the cuttings should either be planted or refrigerated, as development of buds without corresponding root growth will kill the material. In the spring, the bottom ends of the cuttings can then be inserted several inches in damp sand (to minimize problems that may occur with soil fungi) or directly into the soil. Here adventitious roots develop, and normal growth should occur. Evergreen plants tend to produce adventitious roots much more slowly, taking from several months to a year to produce sufficient growth to become established. They need to be misted with water frequently to keep the cuttings from drying out, and they also need to be kept out of the sun and handled rapidly while they are being prepared. Leaves from the lower part of the cutting should be removed. If bottom heating is available, the additional warmth usually accelerates the development of the roots as long as high humidity can be maintained.

Leaf Cuttings

Houseplants such as African violets, peperomias, begonias, sansevierias, and others with somewhat succulent leaves can be propagated in various ways from the leaves (figure A4.1). One commonly used method is to cover a jar or pan of water with foil or plastic film, punch small holes in the covering, and insert the petioles so that they contact the water, which should be changed occasionally. After a few weeks, adventitious roots develop at the base of the petiole. The leaves may then be planted, leaving the blade exposed, resulting in the gradual appearance of new plants from beneath the soil.

Rex begonias are sometimes propagated by punching out disks from the leaf blades with an instrument that resembles a cookie cutter; each disk should be about 1.9 centimeters (0.75 inch) in diameter. The disks are placed on damp filter paper in Petri dishes, where they develop both roots and shoots. Rex begonia leaves may also be induced to form new plants by making cuts about 3 millimeters (0.16 inch) deep across the main veins on the lower surface. Leaves are then placed lower side down on moist potting medium, and the edges of the leaf are pinned down with toothpicks. Eventually, new plants develop at the vein cuts, and the original leaf withers.

Other houseplants, most of the raspberry–blackberry group (excluding red raspberries), citrus plants, rhododendrons, and camellias may be propagated by cutting a short piece of stem, containing a single node with a leaf and its axillary bud, and inserting it into a mixture of sand and peat moss with the bud 1.3 centimeters (0.5 inch) or more below the surface. As with all evergreen or herbaceous plants, it is essential that the humidity be kept high during propagation. If a misting device is not available, the area should be covered with glass or polyethylene sheeting—allowing, however, for some air circulation—and the leaf surfaces should be occasionally sprayed with water.

Root Cuttings

A variety of plants may be propagated from root cuttings. The roots should be obtained after a growing season has been completed and the plants are dormant. The roots are cut into pieces from 2.5 to 7.5 centimeters (1 to 3 inches) long and placed in a potting medium or damp sand, either in a horizontal position or with the bottom end down. Since it is often difficult to tell the top end from the bottom once a root has been divided, some growers make transverse cuts at the top end and slanting cuts at the bottom end to avoid confusion.

The production of adventitious roots by cuttings is apparently brought about through the stimulation of auxin produced within the cells. Thus, it is common practice to add to the natural auxin of the plant by the application of either additional IAA or synthetic growth-promoting substances, particularly indole butyric acid (IBA) or naphthalene acetic acid (NAA), which often seem

to be more effective than natural auxin in stimulating root growth. These substances are available at nurseries and florists in powder or paste form, and their use not only significantly increases the number of successful cuttings but also initiates more vigorous root growth. In the propagation of leaf disks, cytokinins have also been used. Since fungi are frequently a problem in vegetative propagation, fungicides have been found useful in inhibiting their growth.

Layering

Tip Layering

In blackberries, boysenberries, and other plants with somewhat flexible stems, the tips of the canes are bent over so that they touch the ground and they are covered with a small mound of soil. Roots form on the portion of buried stem, and eventually shoots appear also. These new plants can then be separated from the parent stems. Variations of tip layering involve forcing a stem to lie horizontally and covering it with small mounds of soil at intervals, or heaping soil around the base of a plant so that the individual stems produce roots

Figure A4.2 Tip layering. The tips of canes are bent over and covered with a small mound of soil. When a new plant has developed at the tip, it can be cut from the parent plant and planted independently.

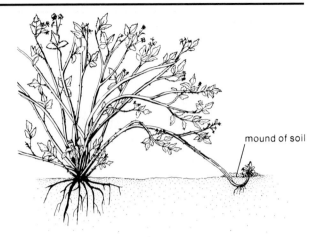

mound of soil

Figure A4.3 Steps in air layering. *A.* Nearly vertical cuts are made on the branch. *B.* Damp sphagnum moss is placed around the cut area. *C.* Polyethylene film is wrapped around the moss. *D.* The ends of the film are taped shut. Adventitious roots should develop in the cut area during ensuing weeks. (USDA Photos)

A.

B.

C.

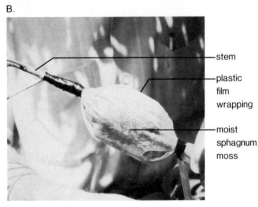

stem

plastic film wrapping

moist sphagnum moss

D.

Appendix 4

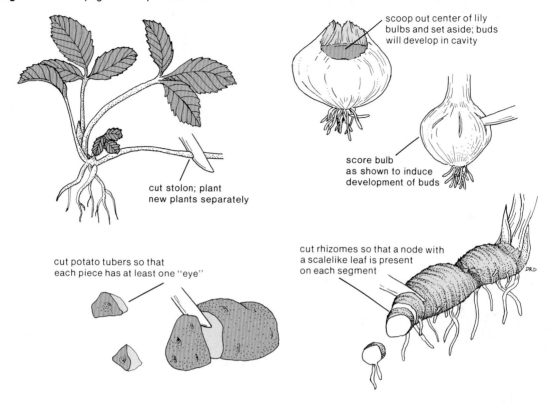

scoop out center of lily bulbs and set aside; buds will develop in cavity

score bulb as shown to induce development of buds

cut stolon; plant new plants separately

cut potato tubers so that each piece has at least one "eye"

cut rhizomes so that a node with a scalelike leaf is present on each segment

there. Once roots have been established, the individual plantlets or pieces of stem can be cut off from the original parent and grown independently (figure A4.2).

Air Layering

Tropical trees and shrubs are sometimes propagated by wrapping sphagnum moss around a branch or main stem, dampening the moss with water, and then covering the area with polyethylene film. Adventitious roots eventually form on the portion of the stem covered by the moss. Sometimes the stem is girdled by removing about 2.5 centimeters (1 inch) of the bark all around the stem before the moss is applied. The moss should not be kept too wet or disease organisms may rot the stem (figure A4.3).

Propagation from Specialized Stems and Roots

In chapters 5 and 6, modified plant organs such as rhizomes, tubers, and corms were discussed. Virtually all of these may be divided, and each piece is capable of developing into a new plant (figure A4.4). In rhizomes and tubers, any piece containing at least one node may be planted and will develop into an independent plant. Strawberry and saxifrage runners and the stolons of mints root at intervals and develop shoots above the roots. As soon as the new plants appear, they may be separated and grown independently. The scales of lily bulbs are quite fleshy and may be used in vegetative propagation. In late summer, while the weather is still warm, the scales are separated and planted in moist soil about 2.5 to 5 centimeters (1 to 2 inches) deep. Three to five bulblets will form at the base of each scale in a few weeks; these can be transplanted the following spring. Bulbs of hyacinth, grape hyacinth, squills, and similar plants can be induced to form bulblets by inverting them and cutting down about a third of the way as though to section an apple or by scooping out the center with a melon baller. The razor, knife, or melon baller needs to be sterile. After they are cut, the bulbs are placed in dry sand in a warm place for about two weeks. During this time *callus* tissue (made up of undifferentiated cells) forms along the cuts. After the callus has formed, the bulbs are removed from the sand and stored in trays at room temperature under humid but well-aerated conditions. Small bulblets form along the cuts during the ensuing two or three months. These are left attached, and the whole original bulb is planted. After a year's growth, the bulblets are separated and replanted independently. In addition to using sterile instruments, it may be necessary to use fungicides to prevent fungal diseases from damaging the materials.

Some bulbs such as those of daffodils, tulips, and amaryllis have dry scales on the outside and fleshy scales on the inside. Unlike those of the lilies, these bulbs do not respond to being separated. These plants and plants with corms (e.g., gladioli) naturally produce bulblets or cormlets, however, which may be separated and grown independently. The number of bulblets produced appears to be correlative with depth in the soil. Most develop when the original bulb is about 7.5 centimeters (3 inches) deep. Few, if any, are produced when the bulb is more than 17 centimeters (8 inches) deep.

Plants with fleshy storage roots such as sweet potatoes or dahlias can be propagated by dividing the root so that each section has a shoot bud. This is done in early spring, just before planting, when shoot buds have appeared. Sweet potatoes can also be cut in pieces and induced to form new plants by suspending each segment with toothpicks from the top of a jar partially filled with water. The segments form adventitious shoots and additional roots in a relatively short time.

Mericloning and Tissue Culture

Of all cultivated plants, orchids are among the most prized. They also produce the smallest seeds known, and it commonly takes seven years or more from seed until the first flower appears. Normally the seeds have to become associated with a specific fungus before they will germinate, and such fungi may not occur outside of the orchid's natural habitat. Orchid breeders have partially overcome the problem by substituting certain hormones and other substances for the fungus in artificial growing media. When they cross plants to produce hybrids (see chapter 12), they do not know if they have been successful in developing plants with superior characteristics until the orchids flower several years later. This is because of the variables inherited from the parents. Several techniques, however, have been developed that reduce by two years or more the time involved in producing mature orchid plants and that eliminate the often unpredictable inherited characteristics in plants grown from seeds. The techniques, which are applied to a number of plants besides orchids, all involve aseptic practices (i.e., practices that seek to eliminate bacteria and other disease organisms) designed to isolate unique tissues with the potential to become new disease-free plants and to permit rapid multiplication of the materials.

The best results are obtained when the techniques are employed in a room or chamber that has been sterilized with ultraviolet light or has been steam heated before work begins. All instruments, glassware, and media need to be sterilized in advance. An *autoclave,* which is something like a large pressure cooker, is generally used for this purpose. The ingredients for the media are measured very precisely. They usually include various inorganic salts, organic substances such as yeast extracts, vitamins and sugar; and distilled water.

Figure A4.5 Orchid shoot meristem culture. The work is done in a chamber designed to minimize contamination (Courtesy Rod McLellan Co., South San Francisco, California)

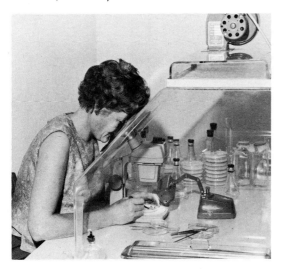

Figure A4.6 Closeup view of cultured orchid tissue being divided. (Courtesy Rod McLellan Co., South San Francisco, California)

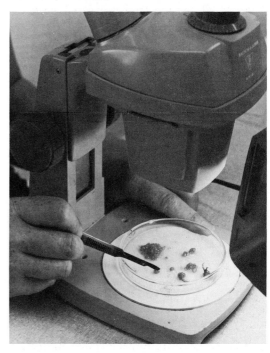

Embryo Culture

In embryo culture, a seed is carefully opened up under aseptic conditions, and the embryo is removed and placed on solidified agar in which nutrients have been dissolved. When shoots and roots appear, the seedlings are transferred to a sterilized growing medium.

Shoot Meristem Culture

The shoot meristem technique, also called **mericloning,** involves the careful removal of the apical meristem and the youngest one or two leaf primordia under aseptic conditions. A dissecting microscope is usually needed (figures A4.5 and A4.6). The meristem is then placed on a sterile medium and kept at room temperature under lights (figure A4.7). Adventitious roots should develop within a month. In orchid culture, the tissue that develops prior to the appearance of shoots and roots is subdivided as it grows, and each subdivision is then capable of becoming a new plant. Commercial growers produce thousands of plants in a relatively short time by this process, all genetically identical to the plant from which the meristem originated.

Figure A4.7 Modern laboratory for mericloning orchids. Each stoppered flask contains a piece of orchid meristem that is being cultured into a new plant. (Courtesy Rod McLellan Co., South San Francisco, California)

Tissue Culture

A tissue culture is essentially a mass of callus tissue growing on an artificial medium. It can be started from almost any part of the plant, although tissues taken from the vicinity of meristems usually produce the best results. With the proper media, the callus tissue eventually differentiates into shoots and roots. The technique has been used in the culture of tissues from carrot, tobacco, citrus, and other plants.

Grafting

Simple grafting basically involves the insertion of a short portion of stem, called a **scion,** into another stem with a root system, the **stock** (figure A4.8). A segment of stem, called an **interstock,** is sometimes grafted between the stock and the scion, or a single bud with a little surrounding tissue may be grafted on to a stock. In all grafts, the cambia of the stock and scion must be placed tightly in contact with each other, and all parts being grafted must be related to one another (e.g., different varieties of apples may be grafted together, but it is not possible to graft elm or maple wood to an apple tree). After the graft is made, the fit may be tight enough for the graft not to need additional support, but it is customary to tie flat rubber strips, raffia, plastic tape, or waxed string around the union and cover it with grafting wax. The grafting wax is needed to prevent the tissues from drying and also to reduce the possibility of disease organisms gaining access to the interior of the stem. If the grafts are buried beneath the surface of the ground, they should be checked after growth begins to see if the wrapping is breaking down. If they are not underground, the wrapping should be removed as growth begins to prevent it from hindering the increase in girth that occurs as the cambium adds new tissue.

Figure A4.8 Simple graft. The stock (rooted portion) and scion (portion to be grafted onto the scion) are cut so that the two parts will fit together with the cambium of each part in close contact.

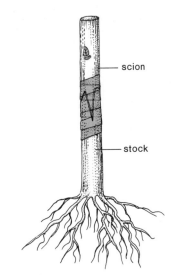

Major Types of Grafting

Whip or Tongue Grafting

Whip, or *tongue, grafting* is widely used for relatively small material, that is, wood between 0.7 and 1.25 centimeters (0.25 and 0.5 inch) in diameter. The stock and scion (the rooted portion and the aerial portion of the graft, respectively) should be as nearly identical in diameter as possible to bring about maximum contact between the cambia. The scion should contain two or three buds, and the cuts on both the stock and scion should be made in an internode. As shown in figure A4.9, a smooth tangential cut about 5 centimeters (2 inches) long is made with a *sharp* sterilized knife at the bottom of the scion and at the top of the stock. The angles of both cuts should be as nearly identical as possible, and there should be no irregularities or undulations in the surfaces (such as those that can be produced by a dull cutting instrument). A second cut is then made in

Figure A4.9 Stages in whip grafting. *A.* A smooth tangential cut is made at the bottom of the scion and at the top of the stock. *B.* Cuts are made back into the wood of the stock and the scion. *C.* The cuts are widened to form a little tongue on each portion. *D.* The scion is inserted into the stock as tightly as possible without forcing a split; the graft is bound with rubber strips and sealed with grafting wax.

A. B. C. D.

both the stock and scion about one-third of the distance from the tip of the cut surfaces. This cut is made back into the wood, nearly parallel to the first cut so that it forms a little tongue. The scion is then inserted into the stock as tightly as possible, taking care, however, not to force a split. In addition, the bottom edge of the scion should not protrude past the bottom of the cut of the stock. The process is completed by binding the materials and adding grafting wax.

If the stock and scion are not identical in diameter, it is still possible to obtain a graft if care is taken to bring the cambia in close contact along one edge of the cuts (figure A4.10).

Splice Grafting

Splice grafting is sometimes used with plants where the pith is extensive. It is essentially the same as whip grafting, except that no second cut is made to form a tongue.

Figure A4.10 Whip graft in which stock and scion are of different diameters

Cleft Grafting

Cleft grafting is a widespread method of grafting used routinely when the stock has a considerably greater diameter than that of the scion or scions. The stock branch or trunk is first cut at right angles, making sure the bark is not torn. If some bark is pulled loose by the saw, a new cut should be made. Commercial growers often minimize detachment of the bark by making a cut one-third of the way in on one side and then making a slightly lower cut on the opposite side. This usually leaves a surface with clean edges. Next, a meat cleaver or heavy knife is hammered 5.0 to 7.5 centimeters (2 to 3 inches) into the wood to make a vertical cut or shallow split. A wedge is then temporarily inserted into the cut to keep it open. Scions, usually 7.5 to 10 centimeters (3 to 4 inches) long, are inserted on each side of the cut toward the outer parts so that as much of the cambia as possible are in contact. When the wedge is removed, a tight enough fit should result that the scions cannot easily be pulled out by hand. Finally, any exposed surface is sealed with grafting wax (figure A4.11).

Side Grafting

Side grafting is often used with stocks that are about 2.5 centimeters (1 inch) in diameter. A cut about 2.5 centimeters (1 inch) deep is made with a heavy knife or chisel at an angle of 20 to 30 degrees to the surface of the stock. The bottom end of the scion, which should be 7.5 to 10.0 centimeters (3 to 4 inches) long and about 0.75 centimeter (0.25 inch) in diameter, is cut into a smooth wedge about 2.5 centimeters (1 inch) long. The stock is then bent slightly to open up the cut, and the scion is inserted, making certain that the maximum contact between each cambium is obtained; the pressure is then released to ensure a tight fit (figure A4.12). This is followed by sealing with grafting wax. Side grafting may be used to replace limbs lost through storm or other damage or for cosmetic purposes such as improving the symmetry of the plant.

A variation of this graft, the *side tongue graft,* is often used with small broad-leaved evergreen plants. This graft involves slicing about 2.5 to 3.25 centimeters (1 to 1.5 inches) of stem out of the side toward the base to a depth of 0.75 centimeter (0.25 inch), followed by a second, smaller cut to form a tongue within the original cut. A scion, prepared in similar fashion to the side-grafted scion, is then inserted, and the graft is tied and sealed.

Approach Grafting

If two related plants tend not to form grafts very well by other means, *approach grafting* can be tried. Two independently growing plants, at least one of which is usually in a container, are prepared by making smooth cuts identical in length

Figure A4.11 Stages in cleft grafting. *A.* The stock is cut transversely and a meat cleaver or a heavy knife is hammered into the wood to make a vertical cut, or split. *B.* A wedge is temporarily inserted into the vertical cut to keep it open. *C.* Scions are inserted into the vertical cut in the vicinity of the cambium and the wedge is removed. *D.* Exposed surfaces are sealed with grafting wax.

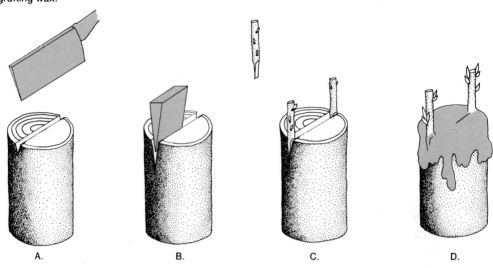

A.   B.   C.   D.

Figure A4.12 Stages in side grafting

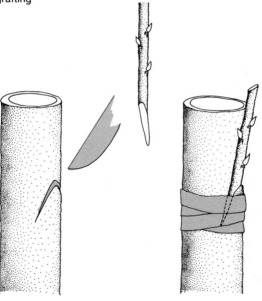

Home Gardening and Houseplants

Figure A4.13 Approach graft

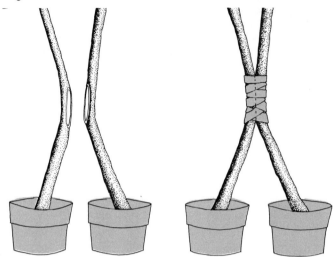

and depth at the same height on both stems. A tongue sometimes also is cut in both exposed areas, and then the two parts are fitted together, tied, and sealed (figure A4.13). This can be done at any time of the year but is most likely to be successful during periods of active growth. After union is achieved, the top of one plant may be cut off above the graft and the bottom of the other removed below the graft.

Inarching is a variation of approach grafting sometimes used to save a valuable tree whose root system has been damaged. Young seedlings or trees of the same kind are planted around the base of the tree. When they have become established, the tip of each seedling, which should be 0.75 to 1.25 centimeters (0.25 to 0.5 inch) thick, is cut vertically for about 1.5 decimeters (6 inches) on the side nearest the main tree. Then vertical cuts of similar length should be made on the tree to the exact width of the prepared seedling tip and deep enough to expose the cambium; a small flap of bark should be left at the top to cover the tip of the seedling. Next, the prepared seedling tips are fit into the slots and nailed in with four to six flat-headed nails, and the entire area is sealed with grafting wax. If any side shoots develop from the

seedlings after the grafting union has developed, they should be pruned off. Because the larger tree will be producing a considerable amount of food, the seedlings often grow very rapidly after successful inarching (figure A4.14).

Bridge Grafting

Sometimes in temperate and colder regions, a particularly deep snowpack may prevent rodents and other animals from reaching their usual winter food. When this occurs, they may turn to the bark of trees and gnaw off a band of tissue, sometimes a decimeter or two wide. The damage usually extends through the phloem and cambium, as these tissues are the most palatable to the animals. This stripping of tissue is frequently referred to as **girdling** a tree and if left untreated will probably result in its death through starvation of the roots, since the phloem cannot conduct food past the damaged area. The tree can often be saved, however, through *bridge grafting,* particularly if it is done in the early spring just as new growth is beginning. The scions should be cut from dormant twigs of the same tree and kept in a refrigerator until needed. The damaged area should be cleaned and prepared by cutting out any remaining dead or tattered tissues. The scions are

Figure A4.14 Inarching. *A.* Established seedlings, which had been planted around the base of the tree, are cut vertically at their tips, and vertical slots of similar length are made in the tree adjacent to the seedling tips. *B.* The seedling tips are nailed into the slots. *C.* Growth of the grafted seedling bases may be very rapid if the inarching is successful.

A.

B.

C.

then inserted above and below the girdle, about 5.0 to 7.5 centimeters (2 to 3 inches) apart, with the natural bottom ends facing down and the tip ends up (figure A4.15). The graft will not succeed if a scion is put in upside down.

Bud Grafting

Bud grafting, or **budding,** is a form of grafting that utilizes a single bud. The method is widely used in commercial nurseries, partly because a single team of two to three workers can produce over a thousand such grafts a day and also because frequently better than 95% of the grafts are successful.

Budding is usually done in the summer while the season's axillary buds are mature and while the sap of the stocks is flowing freely. Budding is generally most successful when plump leaf buds (not flower or mixed buds) of the current season's growth are grafted to healthy stocks that are two to three years old.

The stock is prepared by removing all the leaves and side branches below and in the vicinity of the point at which the graft is to be made. Then

Figure A4.15 Bridge grafting

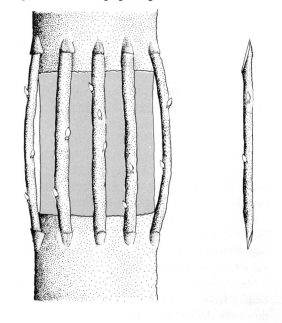

Home Gardening and Houseplants

Figure A4.16 Budding (bud grafting). Leaves and side branches are removed from the stock below the point at which the graft is to be made. A "bud stick" is prepared by removing leaf material while leaving a short portion of each petiole. *A.* A bud is cut from the "bud stick" so that an oval patch of tissue surrounds it. *B.* A T-shaped incision is made in the bark of the stock. *C.* The bark is raised slightly at the corners. *D.* The bud and its surrounding tissue are inserted into the T-shaped incision on the stock. *E.* The flaps of bark are folded over the tissue, leaving the bud exposed, and the T tied shut with flat rubber strips. *F.* After growth begins, the stock is cut off just above the bud.

A.   B.   C.   D.   E.   F.

a T-shaped incision is made through the bark with the aid of a sterilized, razor-sharp knife. The transverse cut should be roughly 1.25 centimeters (0.5 inch) wide and the vertical cut about 2.5 to 3.0 centimeters (1 to 1.2 inches) long. Both cuts should be no deeper than the cambium, and the bark should peel back easily at the junction of the two cuts.

A *bud stick* is prepared by removing all leaf material except for 1.25 centimeters (0.5 inch) of the petioles, which are left to serve as handles. Then a bud is carefully cut from the stick so that an oval piece of tissue about 1.75 centimeters (0.75 inch) in diameter surrounds it. If the bud separates from the tissue, it should be discarded and another one cut. The bud and its oval shield is next inserted into the T-shaped cut of the stock, and the flaps of bark are folded over the shield, leaving the bud exposed. Flat rubber strips are used to tie the T shut so that only the bud remains visible. After growth begins, the stock is cut off just above the bud. The stock should not be cut off any earlier because the bud derives benefit from the transport of substances up and down the stock (figure A4.16).

Bud grafting can also be done in the spring, as soon as possible after growth of the stock begins but before the bud sticks become active. The bud material is frequently cut and stored in a refrigerator before growth begins. In areas with long growing seasons, bud grafting may also be done in early summer if the bark still peels back easily, but it should not be done any later because a young tree needs to produce sufficient growth before fall to be healthy and vigorous the following season.

In thick-barked trees, such as the Pará rubber tree and some of the nut trees, a rectangular patch of bark is cut out of the stock and replaced with a similar patch containing a bud. Normally the patch is not more than 5 centimeters (2 inches) in diameter. This method is slower than the other budding methods described but generally gives much better results in species with thick bark.

Root Grafting

Whip or tongue grafts with roots or pieces of roots used for stocks are sometimes used in the propagation of apple, pear, and other fruit trees. After the grafts have been made, they are usually stored

Appendix 4

in a cool, damp place for about two months and then refrigerated until early spring, when they are planted before growth starts. After growth begins, they are checked to make sure the scion is not producing its own roots, or the advantages of the original rootstock will be lost. These advantages may include disease resistance or dwarfing.

Double-Working

In *double-working* grafts, an *interstock* consisting of a stem segment varying in length from 2.5 centimeters to 3 decimeters (1 to 12 inches) or more is grafted between the stock and scion (i.e., three sections of stem are used for two grafts). Double-working grafts are used for special purposes. One such purpose is dwarfing, usually achieved by using special combinations of materials but sometimes by this method of grafting. A young tree is cut off above the ground, an additional segment is cut to serve as the interstock, and the stock, interstock, and top (scion) are grafted together with the interstock *inverted*. The method will work only with certain varieties, as inversion normally effectively blocks the flow of materials up and down the stem. Other purposes of double-working grafts—in which the interstock is not inverted—include the influencing of growth so as to promote greater flower production than would otherwise occur, providing a disease- or cold-resistant trunk, or circumventing graft incompatibility (the failure of grafts to form permanent unions). With regard to this last purpose, if scions of certain varieties will not produce permanent unions when grafted to stocks of another variety but will form good grafts with a third variety, it may be possible to graft the third variety to the stock so that it can function as an interstock, and thus the problem can be circumvented.

Pruning

A good gardener or orchardist makes a habit of pruning trees, shrubs, and other plants regularly for a variety of reasons. He or she may wish to improve the quality and size of the fruits and flowers, restrict the size of the plants, keep the plants healthy, shape the shrubbery, or generally get more from the plants.

Except for spring-flowering ornamental shrubs, which should be pruned right after flowering, most maintenance pruning is done in the winter when active growth is not taking place. It usually involves removal of portions of stems, but it can also involve roots. When a terminal bud is removed, the axillary or lateral bud just below the cut will usually develop into a branch, and a bushier growth will result. Some gardeners pinch off terminal buds routinely to encourage such growth. The following sections provide a few generalities and specifics pertaining to several types of plants.

Fruit Trees

When young fruit trees are first planted, all except four or five stems and any damaged roots should be pruned off. The remaining stems should be cut back so that there is one central leader about 1 meter (3 feet) tall, with shorter side branches facing in different directions (figure A4.17). When the stems are cut, one should be careful to cut in such a way that the axillary or lateral bud just below the cut is facing outward. Each succeeding year, new growth should be pruned back to within a few centimeters of the previous year's cut. Any dead or diseased branches should be removed, and stems that have grown so that they are rubbing against each other should be pruned. The central leader should be cut out the second year in peach trees so that the interior of the tree is left relatively open. Any sucker shoots that develop from the base or along the trunk of the tree should be regularly removed.

Grapevines

There are several methods of pruning grapevines, depending on the type of vine and the circumstances under which the vines are being grown. In general, grapevines should be pruned heavily in late winter for best fruit production. After a

Figure A4.17 Young peach tree. *A.* Before pruning.
B. After pruning.

central "trunk" has been allowed to develop, each shoot, regardless of its length, should be cut back so that no more than three axillary buds remain. Exceptions to this rule involve cases in which the vines are trained on arbors or wires, when the shoots initially may be allowed to grow longer. Even then, however, after the desired training has taken place (figure A4.18), pruning should be heavy for best results.

Roses

Rose bushes should be pruned heavily—they will recover! In general, new stems should be cut back to within 1 to 2 decimeters (4 to 8 inches) of their point of origin, with care taken that the top remaining axillary bud of each stem is pointing outward. This promotes growth that leaves the center open for better air circulation. Any dead or diseased canes should be removed and the number of remaining canes limited to three or four per plant.

Raspberries, Blackberries, and Their Relatives

Berry canes are biennial. They are produced from the base the first year, branch during the summer, and usually produce fruit on the branches the second year, although in milder climates they may also produce fruit on the first year's growth. The canes die after the second year.

Old, dead canes should be removed and all but three or four canes developing from each crown should be pulled out when the ground is soft. New canes should be cut back to lengths of 1 meter (3 feet) or less in the spring. Branches of

Appendix 4

Figure A4.18 Grapevine. *A.* Before pruning. *B.* After pruning.

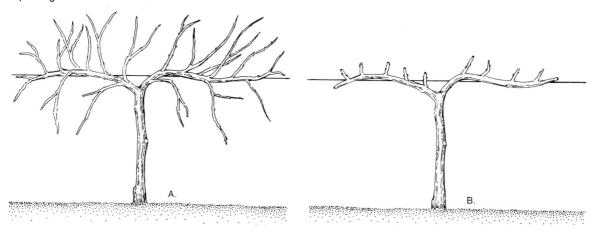

Figure A4.19 Bonsai plant. This sitka spruce is less than 3 decimeters (1 foot) tall and is over 40 years old. (Courtesy Mr. and Mrs. George F. Petersen)

one-year-old canes should be cut back in early spring to lengths of about 3 decimeters (12 inches) for larger fruits.

Bonsai

Container-grown trees that are dwarfed through the constant careful pruning of both roots and stems, the manipulation of soil mixtures, and the weighting of branches are called **bonsai** (figure A4.19). Some of these dwarfed trees attain ages of fifty to seventy-five years or more and may be less than 1 meter (3 feet) tall. Bonsai is an art that requires knowledge of the environmental requirements and tolerances of individual species.

In general, growers involved in bonsai pinch out new growth above a bud every few days during a growing season but never prune when a plant is dormant. The reader is referred to the Additional Reading for more information on the subject.

Growing Vegetables

General Tips on Vegetable Growing

Seed Germination

Many gardeners germinate larger seeds (e.g., squash, pumpkin) in damp newspaper. A few sheets of newspaper are soaked in water for a minute and then hung over a support for about fifteen minutes or until the water stops dripping. The seeds are then lined up in a row on the newspaper, wrapped, and the damp mass is placed in a plastic bag. The bag is tied off or sealed and placed in a *warm* (not hot!), shaded location. Depending on the species, germination should occur within two to several days.

Tiny seeds (e.g., carrots, lettuce) may be mixed with clean sand before sowing to bring about a more even distribution of the seed in the rows.

Transplanting

Roots should be disturbed as little as possible when seedlings or larger plants are transplanted. Even a few seconds' exposure to air will kill root hairs and smaller roots. They should be shaded (e.g., with newspaper) from the sun and transplanted late in the day or on a cool, cloudy day if at all possible. To minimize the effects of transplanting, immediately water the seedlings or plants in their new location with a dilute solution of vitamin B/hormone preparation (e.g., Superthrive).

Bulb or Other Plants with Food Storage Organs

All such plants (e.g., beets, carrots, onions) develop much better in soil that is free of lumps and rocks. If possible, the areas where these plants are to be grown should be dug to a minimum depth of 3 decimeters (12 inches) and the soil sifted through a 0.7 centimeter (approximately 0.25 inch) mesh before planting. Obviously such a procedure is not always practical, but it can yield dramatic results.

Cutworms

Cutworms forage at or just beneath the surface of the soil. Their damage to young seedlings can be minimized by a collar placed around each plant. Tuna cans with both ends removed make effective collars when pressed into the ground to a depth of about 1 to 2 centimeters (0.2 to 0.4 inch).

Protection Against Cold

Some seedlings can be given an earlier start outdoors if plastic-topped coffee cans with the bottoms removed are placed over them. The plastic lids can be taken off during the day and replaced at night during cool weather. Conical paper frost caps can serve the same purpose.

Watering

Proper watering promotes healthy growth. It is much better to water an area thoroughly (e.g., for twenty to thirty minutes) every few days than to wet the surface for a minute or two daily. Shallow daily watering promotes root development near the surface, where midsummer heat can damage the root system. Conversely, too much watering can leach minerals out of the topsoil.

Fertilizers

Manures, bone and blood meals, and other organic fertilizers, which release the nutrients slowly and do not "burn" young plants, are preferred. Plants will utilize minerals from any available source, but in the long run the plants will be healthier and subject to fewer problems if they are not given sudden boosts with liquid "chemical" fertilizers.

Pests and Diseases

Biological controls are listed in appendix 2. If sprays must be used, biodegradable substances such as rotenone and pyrethrum should be used.

Common Vegetables and Their Nutritional Values

Note: The nutritional values (NV) given are per 100 grams (3.5 ounces), edible portion, as determined by the United States Department of Agriculture.

Asparagus

NV (spears cooked in water): 20 calories; protein 2.2 gm; fats 0.2 gm; vit. A 900 I.U.; vit. B_1 0.16 mg; vit. B_2 0.18 mg; niacin 1.5 mg; vit. C 26 mg; fiber 0.7 gm; calcium 21 mg; phosphorus 50 mg; iron 0.6 mg; sodium 1.0 mg; potassium 208 mg.

Asparagus can be started from seed, but time until the first harvest can be reduced by a year or two if planting begins with one-year-old root clusters of healthy, disease-resistant varieties (e.g., "Mary Washington"). Asparagus requires little care if appropriate preparations are made before planting in the permanent location. Seeds should be sown sparsely, and the seedlings thinned to about 7.5 centimeters (3 inches) apart. Before transplantation the following spring, dig a trench 3 to 6 decimeters (12 to 24 inches) deep and about 5 decimeters (20 inches) wide in an area that receives full sun, usually along one edge of the garden. If the soil is heavy, place crushed rock or gravel on the bottom of the trench to provide good drainage. Add a layer of steer manure about 1 decimeter (4 inches) thick, followed by about 7.5 centimeters (3 inches) of rich soil that has been prepared by thorough mixing with generous quantities of steer manure and bone meal. Place root clusters about 4.5 decimeters (18 inches) apart in the trench and cover with about 1.5 decimeter (6 inches) of prepared soil (be sure not to allow root clusters to dry out). As the plants grow, gradually fill in the trench. If one-year-old roots are planted, wait for two years before harvesting tips; if two-year-old roots are planted, some asparagus may be harvested the following year. In all cases, no harvesting should be done after June, so that the plants may build up reserves for the following season. Cut shoots below the surface but well above the crown before the buds begin to expand. Cut all stems to the ground after they have turned yellow later in the season.

Beans

String or Snap Beans

NV (young pods cooked in water): 25 calories; protein 1.6 gm; fats 0.2 gm; vit. A 540 I.U.; vit. B_1 0.07 mg; vit. B_2 0.09 mg; niacin 0.5 mg; vit. C 12 mg; fiber 1.0 gm; calcium 50 mg; phosphorus 37 mg; iron 0.6 mg; sodium 4 mg; potassium 151 mg.

String or snap beans are warm-weather plants, although they can be grown almost anywhere in the United States. Wait until all danger of frost has passed and the soil is warm. Prepare the soil, preferably the previous winter, by digging to a depth of 3 decimeters (12 inches) and mixing in aged manure and bone meal. Pulverize the soil just before sowing; if soil has a low pH, add lime. Plant seeds thinly in rows about 4 to 5 decimeters (16 to 20 inches) apart; thin plants to 1 decimeter (4 inches) apart when the first true leaves have developed. Beans respond unfavorably to a very wet soil—do not overwater! In areas with hot summers, beans also prefer some light shade, particularly in midafternoon. As the bean plants grow, nitrogen-fixing bacteria invade the roots and supplement the nitrogen supply. Early vigorous growth can be enhanced by inoculating the seeds with such bacteria, which are available commercially in a powdered form. To maintain a continuous supply of green beans, plant a new row every two to three weeks during the growing season until two months before the first predicted frost. Cultivate regularly to control weeds, taking care not to damage root systems. Do not harvest or work with beans while the plants are wet, as this may invite disease problems.

Pole Beans

NV similar to those of string beans.

Soil preparation and cultivation are the same as for string beans. Plant beans in hills around poles that are not less than 5 centimeters (2 inches) in diameter and at least 2 meters (6.5 feet) tall. As beans twine around their supports, it helps to tie them to the support with plastic tape as they grow. If harvested before the pods are mature, pole beans will produce over a longer period of time than bush varieties.

Lima Beans

NV (immature seeds cooked in water): 111 calories; protein 7.6 gm; fats 0.5 gm; vit. A 280 I.U.; vit. B$_1$ 0.18 mg; vit. B$_2$ 0.10 mg; niacin 1.3 mg; vit. C 17 mg; fiber 1.8 gm; calcium 47 mg; phosphorus 121 mg; iron 2.5 mg; sodium 1.0 mg; potassium 422 mg.

Lima beans take longer to mature than other beans and are more sensitive to wet or cool weather. They definitely need warm weather to do well. Prepare soil and cultivate as for string beans.

Soybeans

NV (dry, mature seeds): 403 calories; protein 34.1 gm; fats 17.7 gm; vit. A 80 I.U.; vit. B$_1$ 1.10 mg; vit. B$_2$ 0.31 mg; niacin 2.2 mg; vit. C—[1]; fiber 4.9 gm; calcium 226 mg; phosphorus 554 mg; iron 8.4 mg; sodium 5.0 mg; potassium 1677 mg.

Prepare soil and cultivate as for string beans.

Broad (Fava) Beans

NV (dry mature seeds): 338 calories; protein 25.1 gm; fats 1.7 gm; vit. A 70 I.U.; vit. B$_1$ 0.5 mg; vit. B$_2$ 0.3 mg; niacin 2.5 mg; vit. C—[1]; fiber 6.7 gm; calcium 47 mg; phosphorus 121 mg; iron 2.5 mg; sodium 1.0 mg; potassium 422 mg.

Unlike other beans, broad beans need cool weather for their development. Sow as early as possible (in mild climates they may be sown in the fall, as they can withstand light frosts). Since the plants occupy a little more space than bush beans, plant in rows about 9 to 10 decimeters (3 feet or more) apart and thin to about 2 decimeters (8 inches) apart in the rows. After the first pods mature, pinch out the tips to promote bushier development. To most palates, broad beans do not taste as good as other types of beans.

Beets

NV (cooked in water): 32 calories; protein 1.1 gm; fats 0.1 gm; vit. A 20 I.U.; vit. B$_1$ 0.03 mg; vit. B$_2$ 0.04 mg; vit. C 6 mg; fiber 0.8 gm; calcium 14 mg; phosphorus 23 mg; iron 0.5 mg; sodium 43 mg; potassium 208 mg.

Beets will grow in a variety of climates but do best in cooler weather. They can tolerate light frosts and can be grown on a variety of soil types, although they prefer a sandy loam supplemented with well-aged organic matter. As with any bulb or root crop, they develop best in soil that is free of rocks and lumps.

Beet "seeds" are really fruits containing several tiny seeds. Plant them in rows 4 to 6 decimeters (16 to 24 inches) or more apart and thin to about 1 decimeter (4 inches) apart in the rows after germination. After harvesting, the beets will keep in cold storage for up to several months. The leaves, if used when first picked, make a good substitute for spinach.

Broccoli

NV (spears cooked in water): 26 calories; protein 3.1 gm; fats 0.3 gm; vit. A 2,500 I.U.; vit. B$_1$ 0.09 mg; vit. B$_2$ 0.2 mg; niacin 0.8 mg; fiber 1.5 gm; calcium 88 mg; phosphorus 62 mg; iron 0.8 mg; sodium 10 mg; potassium 267 mg.

Broccoli is a cool-weather plant that will thrive in any good prepared soil, providing that it has not been heavily fertilized just prior to planting (fresh fertilizer promotes rank growth). The plants can stand light frosts and are planted in both the spring and fall in areas with mild climate. Although broccoli will continue to produce during the summer, most growers prefer not to keep the plants going during warm seasons because of the large numbers of pest insects they may attract.

---

1. Values not available

Sow seeds indoors and transplant outdoors after danger of killing frosts has passed. Place plants about 9 to 10 decimeters (3 feet or more) apart and keep well watered. Keep area weeded and pests under control. Harvest heads (bundles of spears) while they are still compact. Smaller heads will develop very shortly after the first harvest; if these are removed regularly, the plants will continue to produce for some time, although the heads become smaller as the plants age.

Cabbage

NV (raw): 24 calories; protein 1.3 gm; fats 0.2 gm; vit. A 130 I.U.; vit. B$_1$ 0.05 mg; vit. B$_2$ 0.05 mg; niacin 0.3 mg; vit. C 47 mg; fiber 1.0 gm; calcium 49 mg; phosphorus 29 mg; iron 0.4 mg; sodium 20 mg; potassium 233 mg.

Growth requirements of cabbage are similar to those of broccoli.

Carrots

NV (raw): 42 calories; protein 1.1 gm; fats 0.2 gm; vit. A 11,000 I.U.; vit. B$_1$ 0.6 mg; vit. B$_2$ 0.5 mg; niacin 0.6 mg; vit. C 8 mg; fiber 1.0 gm; calcium 37 mg; phosphorus 36 mg; iron 0.7 mg; sodium 47 mg; potassium 341 mg.

Carrots are hardy plants that can tolerate a wide range of climate and soils, but the soil must be well prepared, free of rocks and lumps, and preferably not too acid. The seeds are slow to germinate. Plant in rows a little more than 3 decimeters (12 inches) apart and thin seedlings to about 5 centimeters (2 inches) apart in the rows. Weed the rows regularly until harvest. Carrots keep well in below-ground storage containers when freezing weather arrives.

Cauliflower

NV (cooked in water): 22 calories; protein 2.3 gm; fats 0.2 gm; vit. A 60 I.U.; vit. B$_1$ 0.09 mg; vit. B$_2$ 0.08 mg; niacin 0.6 mg; vit. C 55 mg; fiber 1.0 gm; calcium 21 mg; phosphorus 42 mg; iron 0.7 mg; sodium 9 mg; potassium 206 mg.

Growth requirements of cauliflower are similar to those of broccoli except that heavier fertilizing is required. As cauliflower heads develop,

protect them from the sun by tying the larger leaves over the tender heads. Harvest while the heads are still solid.

Corn

NV (cooked sweet corn kernels): 83 calories; protein 3.2 gm; fats 1.0 gm; vit. A 400 I.U. (yellow varieties; white varieties have negligible vit. A content); vit. B$_1$ 0.11 mg; vit. B$_2$ 0.10 mg; niacin 1.3 mg; vit. C 7 mg; fiber 0.7 gm; calcium 3 mg; phosphorus 89 mg; iron 0.6 mg; sodium trace; potassium 165 mg.

There are several types of corn (e.g., popcorn, flint corn, dent corn), but sweet corn is the only type grown to any extent by home gardeners. It can be grown in any location where there is at least a ten-week growing season and warm summer weather.

Corn prefers a fertile soil, which should be prepared by mixing with compost and liberal amounts of chicken manure or fish meal. Since corn is wind-pollinated, it can be helpful to grow the plants in several short rows at right angles to the prevailing winds rather than in a single long row. For best results, use only fresh seeds and plant in rows 6 decimeters (24 inches) apart for dwarf varieties and 9 decimeters (36 inches) apart for standard varieties. Thin to 2 to 3 decimeters (8 to 12 inches) apart in the rows after the plants have produced three to four leaves. Cultivate frequently to control weeds. The corn is ready to harvest when the silks begin to wither.

Cucumber

NV (raw, with skin): 15 calories; protein 0.9 gm; fats 0.1 gm; vit. A 250 I.U.; vit. B$_1$ 0.03 mg; vit. B$_2$ 0.04 mg; niacin 0.2 mg; vit. C 11 mg; fiber 0.6 mg; calcium 25 mg; phosphorus 27 mg; iron 1.1 mg; sodium 6 mg; potassium 160 mg.

Until drought- and disease-resistant varieties were developed in recent years, cucumbers were considered rather "temperamental" plants to grow. The newer varieties are no more difficult to raise than those of most other common vegetables.

The soil should be a light loam—neither too heavy nor too sandy. It should be mixed with well-aged manure and compost and heaped into small mounds about 2 meters (6.5 feet) apart. Five to six seeds should be planted in each mound about 2.5 centimeters (1 inch) below the surface in the middle of the spring. When the plants are about 1 decimeter (4 inches) tall, thin to three plants per mound. Cultivate regularly, and to promote continued production, pick all cucumbers as soon as they attain eating size.

Eggplant

NV (cooked in water): 19 calories; protein 1.0 gm; fats 0.2 gm; vit. A 10 I.U.; vit. B_1 0.05 mg; vit. B_2 0.04 mg; niacin 0.5 mg; vit. C 3 mg; fiber 0.9 gm; calcium 11 mg; phosphorus 21 mg; iron 0.6 mg; sodium 1 mg; potassium 150 mg.

Eggplant is strictly a hot-weather plant that is sensitive to cold weather or dry periods and needs heavy fertilizing. Since seedling development is initially slow, plant the seeds indoors about two months before the plants will be set out, which should be about five to six weeks after the last average date of frost.

Eggplants do best in enriched sandy soils that are supplemented with additional fertilizer once a month. Never permit them to dry out. Place the seedlings about 7 to 8 decimeters (28 to 32 inches) apart in rows 9 to 10 decimeters (3 feet or more) apart. Some staking of the plants may be desirable. The fruits are ready to harvest when they have a high gloss. They are still edible after greenish streaks appear and the gloss diminishes, but they are not as tender at this stage.

Lettuce

NV (crisp, cabbage-head varieties): 13 calories; protein 0.9 gm; fats 0.1 gm; vit. A 330 I.U.; vit. B_1 0.06 mg; vit. B_2 0.06 mg; niacin 0.3 mg; vit. C 6 mg; fiber 0.5 gm; calcium 20 mg; phosphorus 22 mg; iron 0.5 mg; sodium 9 mg; potassium 175 mg.

NV (leaf varieties): 18 calories; protein 1.3 gm; fats 0.3 gm; vit. A 1,900 I.U.; vit. B_1 0.05 mg; vit. B_2 0.08 mg; niacin 0.4 mg; vit. C 18 mg; calcium 68 mg; phosphorus 25 mg; iron 1.4 mg; sodium 9 mg; potassium 264 mg.

This favorite salad plant comes in a wide variety of types and forms, all of which prefer cooler weather, although a few of the leaf types (e.g., oak leaf) can tolerate some hot periods. As long as the individual plants are given room to develop and the soil is not too acid, most varieties can be grown on a wide range of soil types.

Since lettuce can stand some frost, sow the seeds outdoors as early in the spring as the ground can be cultivated. Do not cover the seeds with more than a millimeter or two of soil—they need light to germinate. Mix the soil with a well-aged manure and a general-purpose fertilizer a week or two before sowing. Plant seedlings about 3 decimeters (12 inches) apart in rows 3 to 4 decimeters (12 to 16 inches) apart. For best results, do not allow the soil to dry out, and plant only varieties suited to local conditions. The most common crisp, cabbage-head varieties found in produce markets will not form heads in hot weather, and many others will bolt (begin to flower) during hot weather and longer days. Cultivate weekly between rows to promote rapid growth and to control weeds.

Onion

NV (raw): 38 calories; protein 1.5 gm; fats 0.1 gm; vit. A 40 I.U. (yellow varieties only); vit. B_1 0.03 mg; vit. B_2 0.04 mg; niacin 0.2 mg; vit. C 10 mg; fiber 0.6 gm; calcium 27 mg; phosphorus 36 mg; iron 0.5 mg; sodium 10 mg; potassium 157 mg.

These easy-to-grow vegetables do best in fertile soils that are free of rocks and lumps, are well drained, and are not too acid or sandy.

Onions may take several months to mature from seed. The viability of the seed decreases rapidly after the first year. Bulb formation is determined by daylength rather than by the total number of hours in the ground. Because of these characteristics of onions, most gardeners prefer to purchase "sets" (young plants that already

Appendix 4

have a small bulb) from commercial growers, although green or bunching onions are still easily grown from seed.

Plant the sets upright 6 to 7.5 centimeters (2.5 to 3 inches) apart in rows and firm in place. Except for weeding, watering, and occasional shallow cultivation, they will need little care until harvest about fourteen weeks later. The onions are mature when the tops fall over. After they are pulled from the ground, allow them to dry in the shade for two days. Then remove the tops 2 to 3 centimeters (about 1 inch) above the bulbs, and spread out the bulbs to continue "curing" for two to three more weeks. After this, store them in sacks or other containers that permit air circulation until needed.

Peas

NV (green, cooked in water): 71 calories; protein 5.4 gm; fats 0.4 gm; vit. A 540 I.U.; vit. B_1 0.28 mg; vit. B_2 0.11 mg; niacin 2.3 mg; vit. C 20 mg; fiber 2.0 gm; calcium 23 mg; phosphorus 99 mg; iron 1.8 mg; sodium 1 mg; potassium 196 mg.

Peas are strictly cool-weather plants that generally produce poorly when the soil becomes too warm. Seeds should be planted in the fall or very early spring. As is the case with beans, peas receive a better start if the seeds are inoculated with nitrogen-fixing bacteria (see discussion of beans) at planting time. Prepare the soil by mixing thoroughly with liberal amounts of aged manure and bone meal. Plant the seeds about 2.5 centimeters (1 inch) deep in heavy soil or 5 centimeters (2 inches) deep in light, sandy soil, about 2.5 centimeters (1 inch) apart in single rows for dwarf bush varieties or in double files 1.5 decimeters (6 inches) apart for standard varieties, with intervals of 8 to 9 decimeters (32 to 36 inches) between the rows. After germination, thin the plants to 1 decimeter apart. Place support wires, strings, or chicken wire between the rows at the time of planting; peas will not do well without such support.

Green peas should be picked while still young and cooked or frozen immediately, as the sugars that make them sweet are converted to starch within two to three hours after harvest.

Peppers

NV (raw sweet or bell peppers): 22 calories; protein 1.2 gm; fats 0.2 gm; vit. A 420 I.U.; vit. B_1 0.08 mg; vit. B_2 0.08 mg; niacin 0.5 mg; vit. C 128 mg; fiber 1.4 gm; calcium 9 mg; phosphorus 22 mg; iron 0.7 mg; sodium 13 mg; potassium 213 mg.

Peppers, like eggplants, are strictly hot-weather plants for most of their growing season, but unlike eggplants they actually do better toward the end of their season if temperatures have moderated somewhat. Sweet or bell peppers are closely related and have similar cultural requirements.

Plant seeds indoors eight to ten weeks before the outdoor planting date, which is generally after the soil has become thoroughly warm. They will grow in almost any sunny location in a wide variety of soils, but to obtain the large fruits seen in produce markets the plants need to be fertilized heavily and watered regularly. Plant seedlings 5 to 6 decimeters (20 to 24 inches) apart in rows that are 6 to 9 decimeters (24 to 36 inches) apart. Sweet peppers can be harvested at almost any stage and are still perfectly edible after they have turned red.

Potatoes

NV (baked in skin): 93 calories; protein 2.6 gm; fats 0.1 gm; vit. A trace; vit. B_1 0.10 mg; vit. B_2 0.04 mg; niacin 1.7 mg; vit. C 20 mg; fiber 0.6 gm; calcium 9 mg; phosphorus 65 mg; iron 0.7 mg; sodium 4 mg; potassium 503 mg.

Potatoes grow best in a rich, somewhat acid, well-drained soil that has had compost or well-aged manure added to it. They are subject to several diseases, and it is advisable to use disease-free seed potatoes purchased from a reputable dealer. Two to three weeks before the average date of the last spring frost, plant the seed potatoes whole or cut into several pieces, making sure that each piece has at least one eye. Place the potato pieces about 3 decimeters (12 inches) or more apart at a depth of about 12 to 15 centimeters (5 to 6 inches) in rows 9 to 10 decimeters apart.

Later plantings are feasible. Spread a thick mulch (e.g., straw) over the area after planting to keep soil temperatures down and to retain soil moisture.

Potatoes are ready for harvest when the tops start turning yellow, but they may be left in the ground for several weeks after that if the soil is not too wet. After harvest, wash the potatoes immediately and place them in a dry, cool, dark place until needed. If left exposed to light, the outer parts of the potato turn green; poisonous substances are produced in these tissues, and such potatoes should be discarded.

Spinach

NV (raw): 26 calories; protein 3.2 gm; fats 0.3 gm; vit. A 8,100 I.U.; vit. B$_1$ 0.10 mg; vit. B$_2$ 0.20 mg; niacin 0.6 mg; vit. C 51 mg; fiber 0.6 gm; calcium 93 mg; phosphorus 51 mg; iron 3.1 mg; sodium 71 mg; potassium 470 mg.

Spinach is a cool season crop that goes to seed as soon as the days become long and warm. It should be planted in the fall or early spring. If protected by straw or other mulches, it will over-winter in the ground in most areas and be ready for use early in the spring. Spinach has a high nitrogen requirement and reacts negatively to acid soils. It is otherwise easy to grow. Mix the soil thoroughly with aged manure and bone meal before planting. Plant seedlings 3 to 5 centimeters (1 to 2 inches) apart in rows 4 to 5 decimeters (16 to 20 inches) apart. Keep the plants supplied with adequate moisture, and their growing area free of weeds. Harvest the whole plant when a healthy crown of leaves develops.

Squash

NV (cooked zucchini): 12 calories; protein 1.0 gm; fats 0.1 gm; vit. A 300 I.U.; vit. B$_1$ 0.05 mg; vit. B$_2$ 0.08 mg; niacin 0.8 mg; vit. C 9 mg; fiber 0.6 gm; calcium 25 mg; phosphorus 25 mg; iron 0.4 mg; sodium 1 mg; potassium 141 mg.

All varieties of squash are warm-weather plants, and all are targets of a variety of pests. Thorough preparation of the soil before planting pays dividends in production and in the health of the plants. Mix compost and aged manure with the soil and heap the soil in small hills about 12 decimeters (4 feet) apart from one another. Plant four to five seeds in each hill and thin the seedlings to three after they are about 1 decimeter (4 inches) tall. Summer squashes (e.g., zucchini) mature in about two months, while winter squashes (e.g., acorn) can take twice as long to mature. Summer squashes should be harvested while very young—they can balloon seemingly overnight into huge fruits. Winter squashes should be harvested before the first frost; only clean, undamaged fruits will store well. Keep such squashes laid out in a cool, dry place until use and not piled on top of one another. Check them occasionally for the development of surface fungi.

Tomatoes

NV (raw, ripe): 22 calories; protein 1.1 gm; fats 0.2 gm; vit. A 900 I.U.; vit. B$_1$ 0.06 mg; vit. B$_2$ 0.04 mg; niacin 0.7 mg; vit. C 23 mg; fiber 0.5 gm; calcium 13 mg; phosphorus 27 mg; iron 0.5 mg; sodium 3 mg; potassium 244 mg.

These almost universally used fruits are easy to grow providing one understands a few basic aspects of their cultural requirements: (1) Many tomato plants normally will not initiate fruit development from their flowers when night temperatures drop below 14°C (57°F) or day temperatures climb above 40°C (104°F). For the earliest yields, seeds may be germinated indoors several weeks before the plants are to be placed outside, but little is accomplished by transplanting before the night temperatures begin to remain above 14°C (57°F); in addition, some growers insist that plants given an early start indoors do not always do as well later as those germinated outdoors. (2) Tomatoes require considerably more phosphorus than nitrogen from any fertilizers added to the soil where they are to be grown. Many inexperienced gardeners make the mistake of giving the plants lawn or general-purpose fertilizers that are proportionately high in nitrogen. As a result, the plants may grow vigorously but produce very few tomatoes. Give tomatoes bone meal, "tomato food" (Magamp is an excellent commercial slow-release preparation), or steer manure mixed with superphosphate. (3) Tomato plants seem more susceptible than most to soil fungi and to root knot nematodes. The damage caused by these organisms may not become evident until the plants begin to bear. Then the lower leaves begin to wither, and yellowing progresses up the plant or there seems to be a general loss of vigor and productivity. Using disease- and nematode-resistant varieties (usually indicated by the letters *V, F,* and *N* on seed packets) is by far the simplest method of controlling these problems. Another effective control involves dipping the seedling roots in an emulsion of 0.25% corn oil in water when transplanting; experiments have shown that the corn oil greatly reduces root-knot nematode infestation. (4) Many garden varieties of tomatoes need to be staked to keep fruits off the ground where snails and other organisms can gain easy access to them. When using wooden stakes, be sure they have a diameter of 5 centimeters (2 inches) or more, and tie the plants securely to the stakes with plastic tape. Thinner stakes are likely to break or collapse when the plants grow to a height of 2 meters (6.5 feet) or more. Some growers prefer to use heavy wire tomato "towers" instead of stakes. (5) Hornworms and tomato worms almost invariably appear on tomato plants during the growing season. They can virtually strip a plant and ruin the fruits if not controlled. Fortunately control is simple and highly effective with the use of *Bacillus thuringiensis,* which was discussed in appendix 2. (6) The "eating season" for garden tomatoes can be extended for about two months past the first frost if all the green tomatoes are picked before frost occurs. Place the tomatoes on sheets of newspaper on a flat surface in a cool, dry place, where they will ripen slowly a few at a time. Generally the taste of tomatoes ripened this way is superior to that of hothouse tomatoes sold in produce markets. Be sure when picking the green tomatoes to handle them very gently, as they bruise very easily and molds quickly develop in the bruised areas.

Growing Houseplants

If sales volume is an indication, houseplants have never been more popular in the United States than they are now. Many are easy to grow and will brighten windowsills, planters, and other indoor spots for years if a few simple steps are followed to ensure their health and vigor.

Water

Houseplants are commonly overwatered, resulting in the unnecessary development of rots and diseases (see table A4.1). As a rule, the surface of the potting medium should be dry to the touch before watering, but the medium should not be allowed to dry out completely unless the plant is dormant. Care should be taken, particularly during the winter, that the water is at room temperature. If rainwater is available, it is to be preferred over tap water, particularly if the water has a high mineral content or is chlorinated. Broad-leaved plants should periodically have house dust removed with a damp sponge (never use detergents to clean surfaces—they remove protective waxes). Many plants benefit from a daily misting with water, particularly in heated rooms.

Containers

In time plants may develop too extensive a root system for the pots in which they are growing (commonly called *becoming root bound*). Nutrients in the potting medium may become exhausted, salts and other residues from fertilizers and water may build up to the point of inhibiting growth, or the plants themselves may produce substances that accumulate until they interfere with the plant's growth. To resolve these problems, the plants should be periodically repotted and divided, if necessary, at the time of repotting.

Temperatures

Most houseplants do not thrive where the temperatures are either too high or too low (see table A4.1). They generally tend to prefer minimum temperatures of about 13°C (55°F) and maximum temperatures of about 29°C (84°F). Many that prefer warmer temperatures while actively growing also benefit from a "rest" period at lower temperatures after flowering.

Light

Next to overwatering, the most common contributor to the demise of houseplants is a lack of sufficient light (see table A4.1). This does not mean that houseplants prefer direct sunlight—such light frequently damages them, but filtered sunlight (as, for example, through a muslin curtain) is usually better for the plant than the light available in the middle of the room. Plants can also thrive in artificial light of appropriate quality. Ordinary incandescent bulbs have too little light of blue wavelengths, and ordinary fluorescent tubes emit too little red light. A combination of the two, however, works very well. Generally the wattage of the incandescent bulbs should be only one-fourth that of the fluorescent tubes in such a combination. Several types of fluorescent tubes specially balanced to imitate sunlight are also available.

Humidity

Dry air is hard on most houseplants. The level of humidity around the plants can be raised by standing the pots in dishes containing gravel or crushed rock to which water has been added. The humidity level can also be raised through the use of humidifiers, which come in a variety of sizes and capacities. Daily misting, as mentioned previously, can also help.

Table A4.1 Common ailments of houseplants	
Problem Symptoms	**Possible Causes**
Wilting or collapse of whole plant	lack of water; too much heat; too much water or poor drainage resulting in root rot
Yellowish or pale leaves	insufficient light; too much light; microscopic pests (especially spider mites); too much or too little fertilizer
Brown, dry leaves	humidity too low; too much heat; poor air circulation; lack of water
Tips and margins of leaves brown	mineral content of water; drafts; too much sun or heat; too much or too little water
Ringed spots on leaves	water too cold
Leaves falling off	improper watering or water too cold; excessive use of fertilizer or wrong fertilizer; too much sun or, if lower leaf drop only, too little light
Stringy growth	needs more light; too much fertilizer
Base of plant soft or rotting	overwatering
No flowers or flower buds drop	too much or too little light; night temperatures too high
Water does not drain	drain hole plugged; potting mixture has too much clay
Mildew present	fungi present—arrest with sulphur dust
Common Pests	**Controls**
Aphids	wash off under faucet or spray with soapy (*not* detergent) water; pyrethrum or rotenone sprays also effective
Mealybugs	remove with cotton swabs dipped in alcohol; spray with Volck oil
Scale insects	remove by hand; spray with Volck oil
Spider mites	use sprays containing small amounts of xylene (act as soon as possible—spider mites multiply exceedingly rapidly)
Thrips	spray with pyrethrum/rotenone or Volck oil sprays
White flies	spray with soapy water or pyrethrum/rotenone sprays every four days for two weeks (only the adults are susceptible to the sprays)

For additional controls, see the biological controls in appendix 2.

Common Houseplants

Explanation of Symbols Given
with Each Plant

Temperature

= cool. Maximum 13–16°C
(55–61°F); minimum 5–7°C
(41–45°F)

= cool to medium. Maximum
18–21°C (65–70°F); minimum
10–13°C (50–55°F)

= medium. Maximum 30°C
(86°F)

= medium to warm

= warm

Many houseplants are native to the tropics, where they thrive under year-around warm temperatures, while cacti and succulents prefer cool winters. The closer one is able to imitate a plant's natural environment, the better the plant will grow (see table A4.2).

Light

= needs shading or indirect daylight

= prefers bright light but needs to be screened from direct sunlight

= prefers direct sunlight

= shading to bright

= bright to direct

As mentioned previously, improper lighting is second only to overwatering as a cause of problems for houseplants; generally they are given too little light or occasionally too much. A south-facing windowsill may be ideal for certain plants in midwinter but excessively bright in midsummer; conversely, a north-facing windowsill may have enough light for certain plants in midsummer but not in midwinter. Adjustable screens permit manipulation of daylight to suit the plants involved.

Water

 = needs little water (applies primarily to cacti and succulents; these plants store water in such a way that the soil can be completely dry for a week or two without their being adversely affected)

= water regularly but not excessively; wait until the potting medium surface is dry to the touch before watering

= immerse pot in water for a few minutes each week and water frequently, never allowing the potting medium to become dry; do not, however, leave the base of the pot standing in water

 = little to regular

= regular to frequent

Minerals in hard water are taxing on houseplants, and commercial water softeners do not improve water for the plants. Use rainwater or filtered water if at all possible; otherwise repot more often.

Humidity

 = will tolerate dry air

 = dry to regular

 = will tolerate the air in most houses provided it is mist-sprayed occasionally

 = regular to humid

 = needs high humidity; use a humidifier if possible

Virtually all plants with or symbols benefit from having a pan of gravel with water beneath the pot.

Potting Medium

 = requires a porous, slightly acid medium that drains immediately

 = requires a loam that is slightly alkaline (e.g., a mixture of sand and standard commercial potting medium)

 = requires a peaty potting mixture and acid fertilizer

Table A4.2	Environments suitable for common houseplants		
Plant	**Scientific Name**	**Environmental Requirements**	**Remarks**
Aechmea	Aechmea fasciata		see Bromeliad; produces side shoots that should be propagated as main plant dies after flowering
African lily	Agapanthus spp.		do not repot until pot is full; keep cool in winter
African violet	Saintpaulia spp.		let rest under cooler conditions after flowering; dislikes cold water
Agave	Agave spp.		keep cool and dry in winter
Algerian ivy	Hedera canariensis		resembles a variegated English ivy
Aloe	Aloe spp.		keep cool and dry in winter
Aluminum plant	Pilea cadierei		plants do not usually survive long in houses
Amaryllis	Amaryllis spp.		let leaves die back in fall; put bulb in cool, dark place until early spring; then repot, water, and fertilize weekly

Plant	Scientific Name	Environmental Requirements	Remarks
Anthurium	*Anthurium* spp.		if it grows without flowering, try putting it in a cooler location for a few weeks
Aphelandra	*Aphelandra squarrosa*		mist-spray frequently; fertilize regularly
Aralia (*see* Fatsia)			
Asparagus fern	*Asparagus plumosus*		not a true fern; repot annually; fertilize weekly
Aspidistra	*Aspidistra* spp.		sometimes called ''cast iron plant'' because it can stand neglect
Aucuba	*Aucuba japonica*		must be kept cool in winter
Avocado	*Persea* spp.		easily propagated from seed; provides good greenery but will not produce fruit indoors
Azalea	*Rhododendron* spp.		needs acid fertilizer; avoid warm locations
Bamboo (dwarf)	*Bambusa angulata*		needs good air circulation, bright light
Begonia	*Begonia* spp.		easily propagated from leaves; repot regularly
Bilbergia	*Bilbergia* spp.		*see* Bromeliad; tough plant that can tolerate some neglect
Birdcatcher plant	*Pisonia umbellifera*		sticky exudate on fruits attracts birds in the plant's native habitat of New Zealand; strictly a foliage plant in houses
Bird of paradise plant	*Strelitzia reginae*		can be grown outdoors in milder climates
Bird's nest fern	*Asplenium nidus*		produces a spongelike mass of roots at base; requires much water and regular fertilizing
Black eyed susan	*Thunbergia alata*		annual climbing vine; grow from seed
Bloodleaf	*Iresine herbstii*		easily propagated
Boston fern	*Nephrolepis exaltata*		needs regular watering and fertilizing

Plant	Scientific Name	Environmental Requirements	Remarks
Bromeliads	many species		these plants absorb virtually all their water and nutrients through their leaves; they should not be placed in regular potting soil, nor should they be watered with high calcium content water; they produce offshoots that should be propagated as main plant dies after flowering
Cacti	many species		contrary to popular belief, these slow-growing plants should not be grown in pure sand; add some humus to potting mixture and withhold water in winter; keep cool in winter
Caladium	*Caladium* spp.		must have high humidity; keep root at 18°C in pot during winter
Calceolaria	*Calceolaria herbeohybrida*		discard after flowering
Calla lily	*Zantedeschia aethiopica*		after flowering, allow plant to dry up; repot in fall and start over
Cape jasmine	*Gardenia jasminoides*		night temperatures below 22°C needed to initiate flowering; needs cool temperatures in winter
Carrion flower	*Stapelia* spp.		cactuslike plants with foul-smelling flowers
Century plant (*see* Agave)			
Chinese evergreen	*Aglaonema costatum*		needs warm temperatures and much water all year
Chrysanthemum	*Chrysanthemum* spp.		plants may be artificially dwarfed through use of chemicals; flowering initiated by short days
Cineraria	*Senecio cruentus*		needs cool temperatures; discard after flowering
Cliff brake	*Pellaea rotundifolia*		hanging basket fern; needs minimum temperature above 10°C

Appendix 4

Plant	Scientific Name	Environmental Requirements	Remarks
Coffee	*Coffea arabica*		handsome foliage plant that will produce fruit if self-pollinating variety is obtained
Coleus	*Coleus blumei*		to control size, restart plants from cuttings annually
Copperleaf	*Acalypha wilkesiana*		seldom survives average house environment for long
Corn plant	*Dracaena massangeana*		uses much water when large; easy to grow
Croton	*Codiaeum* spp.		needs constant high humidity and bright light
Crown of thorns	*Euphorbia milii* and *E. splendens*		deviation from watering routine may result in loss of leaves, but plant generally recovers
Cyclamen	*Cyclamen* spp.		fertilize weekly; keep cool; withhold water after flowering for few weeks, then start over
Donkey tail	*Sedum morganianum*		keep cool and dry in winter
Dracaena	*Dracaena* spp.		many kinds—all easy to grow and tolerant of some neglect
Dumbcane	*Dieffenbachia* spp.		needs regular fertilizing; keep away from small children (poisonous)
Dwarf banana	*Musa cavendishii*		will produce small edible bananas if given enough light, water, and humidity
Dwarf cocos palm	*Microcoleum weddelianum*		keep temperature above 18°C at all times
Echeveria	*Echeveria* spp.		keep cool in winter
English ivy	*Hedera helix*		needs cool temperatures to grow at its best
False aralia	*Dizygotheca elegantissima*		benefits from frequent mist-spraying
Fatshedera	*Fatshedera lizei*		climbing plant
Fatsia	*Fatsia japonica*		also called *Aralia*
Ferns	many species		water regularly; propagate from spores or runners

Plant	Scientific Name	Environmental Requirements	Remarks
Figs:			
Climbing fig	*Ficus pumila*		damp sponge leaves regularly
Fiddleleaf fig	*Ficus lyrata*		damp sponge leaves regularly
Weeping fig	*Ficus benjamina*		damp sponge leaves regularly
Fingernail plant	*Neoregelia* spp.		*see* Bromeliad; name from red tips of leaves
Fittonia	*Fittonia* spp.		strictly terrarium plants—humidity too low elsewhere
Flame violet	*Episcia cupreata*		add charcoal and peat to potting medium
Flowering maple	*Abutilon striatum*		needs bright light to flower
Fuchsia	*Fuchsia* spp.		soil must be alkaline
Gardenia (*see* Cape jasmine)			
Geranium	*Pelargonium* spp.		make cuttings annually and discard parent plants each fall; keep cool through winter; available with scents of orange, rose, or coconut
Gloxinia	*Sinningia speciosa*		fertilize heavily and water frequently; after flowering, withhold water and keep bulb cold for few weeks
Goldfish plant	*Columnea* spp.		pot in mixture of leaf mold, fern bark, peat moss, and charcoal; use only rainwater or filtered water
Grape ivy	*Rhoicissus rhomboidea*		tolerates low light better than most plants
Haworthia	*Haworthia* spp.		aloelike plants that need minimum temperatures above 10°C in winter
Hen and chickens	*Sempervivum tectorum*		keep cool in winter
Hibiscus	*Hibiscus rosa-sinensis*		fertilize weekly
Hippeastrum (*see* Amaryllis)			
Holly fern	*Cyrtomium falcatum*		relatively tough fern; keep cool in winter
Houseleek	*Sempervivum* spp.		keep cool in winter

Plant	Scientific Name	Environmental Requirements	Remarks
Hydrangea	*Hydrangea* spp.	(icons)	prune after flowering; keep cool in winter; pink-flowering plant can be converted to blue by changing the soil to acid and vice versa
Impatiens	*Impatiens* spp.	(icons)	exceptionally easy to propagate from cuttings
Ivy-arum	*Rhaphidophora aurea*	(icons)	can tolerate some neglect
Jade plant	*Crassula argenta* and others	(icons)	keep cool in winter
Kaffir lily	*Clivia miniata*	(icons)	save the plant's energy by removing flowers as they wither
Kalanchoë	*Kalanchoë* spp.	(icons)	withhold water and fertilizer for a few weeks after flowering, then repot and start over
Lantana	*Lantana camara*	(icons)	fertilize twice a month; can be espaliered
Madagascar jasmine (*see* Stephanotis)			
Maidenhair fern	*Adiantum* spp.	(icons)	mist-spray regularly
Meyer Fern	*Asparagus densiflora* var. *meyeri*	(icons)	fertilize regularly; repot annually
Moneywort	*Lysimachia nummularia*	(icons)	hanging pot plant; needs bright light
Moonstones	*Pachyphytum* spp.	(icons)	keep cool in winter
Moses in the cradle	*Rhoeo* spp.	(icons)	can tolerate some neglect
Mother in law's tongue (*see* Sansevieria)			
Mother of thousands	*Saxifraga* spp.	(icons)	also called *Saxifrage*; hanging plant; plantlets formed on runners can be removed and grown separately
Neanthe palm	*Chamaedorea elegans*	(icons)	sometimes also called *Parlor palm*; stays less than 1 meter tall
Norfolk Island Pine	*Araucaria* spp.	(icons)	needs cold temperatures (2–3°C) in winter
Octopus tree	*Schefflera arboricola*	(icons)	does best under cool conditions

Plant	Scientific Name	Environmental Requirements	Remarks
Oleander	_Nerium oleander_		keep pot cool in winter for better flowering; keep away from small children (poisonous)
Orchid	thousands of species	no single set of environmental conditions applies	contrary to popular belief, the common _Cattleya_ and related orchids do not need high temperatures and humidity; most can get along with a minimum temperature of 13°C at night and a minimum humidity of 40 percent. Most need bright light. They should never be placed in soil; pot them in sterilized pots with chips of fir bark or shreds of tree fern bark. _See_ Additional Reading for culture references
Oxtongue	_Gasteria_ spp.		can tolerate some neglect; needs a cool and relatively dry winter to flower
Palms	many species		use deep pots; fertilize regularly
Parlor palm	_Howea fosteriana_		one of the easiest palms to grow
Peperomia	_Peperomia_ spp.		many kinds; keep warm, humid; fertilize regularly
Persian violet	_Exacum affine_		needs good air circulation
Philodendron	_Philodendron_ spp.		relatively tough plants; repot each spring
Piggyback plant	_Tolmiea menziesii_		plantlets formed on leaves can be separated and propagated
Pineapple	_Ananas comosus_		_see_ Bromeliad; easily grown from the top of a pineapple; if plant has not flowered after one year, enclose it in a plastic bag with a ripe apple for a few days (ethylene from apple should initiate flowering); no temperature below 15°C

Plant	Scientific Name	Environmental Requirements	Remarks
Pink polka dot plant	*Hypoestes sanguinolenta*		susceptible to diseases and pests
Pittosporum	*Pittosporum* spp.		put several cuttings in one pot for bushy appearance
Poinsettia	*Euphorbia pulcherrima*		after flowering, let plant dry under cool conditions until it loses its leaves, then restart
Prayer plant	*Maranta leuconeura*		name derived from fact that leaves fold together in evening
Primrose	*Primula* spp.		needs much water; does well outside in cool weather
Purple tiger	*Calathea amabilis*		use pots that are broader than deep
Rosary plant	*Ceropegia woodii*		hanging pot plant whose potting medium must drain well or plant will not survive
Rubber plant	*Ficus elastica*		do not overwater!
Sago palm	*Cycas revoluta*		very slow growing (not a palm but a gymnosperm); never allow to dry out, but be certain water drains
Sansevieria	*Sansevieria* spp.		perhaps the toughest of all houseplants— nearly indestructible
Satin pothos	*Scindapsus pictus*		basket or pot plant
Screw pine	*Pandanus* spp.		if given space can become large; mist-spray often
Selaginella (*see* Spike moss)			
Sensitive plant	*Mimosa pudica*		leaves fold when touched; does not usually last more than a few months in most houses
Shrimp plant	*Beloperone guttata*		winter temperatures should be above 15°C
Spathe flower	*Spathiphyllum wallisii*		prefers warm winters and even warmer summers
Spider plant	*Chlorophytum comosum*		plantlets formed at tips of stems can be propagated separately

Plant	Scientific Name	Environmental Requirements	Remarks
Spiderwort	*Tradescantia* spp.		easy to grow; do not overwater
Spike moss	*Selaginella* spp.		can become a weed in greenhouses
Splitleaf philodendron	*Monstera deliciosa*		plant adapts to various indoor locations quite well
Sprenger fern	*Asparagus densiflora* var. *sprengeri*		fertilize weekly; repot annually
Staghorn fern	*Platycerium* spp.		tough plant; immerse in water weekly
Stephanotis	*Stephanotis floribunda*		use very little fertilizer; keep cool in winter but water sparingly
Stonecrop	*Sedum* spp.; *Crassula* spp.		keep cool in winter
Stove fern	*Pteris cretica*		water and fertilize regularly
String-of-pearls	*Senecio rowellianus*		keep cool in winter
Sundew	*Drosera* spp.		sterilize pots; grow only on sphagnum moss
Syngonium	*Syngonium podophyllum*		repot annually in spring
Tillandsia	*Tillandsia* spp.		*see* Bromeliad; best known species is called *Spanish moss*
Ti plant	*Cordyline terminalis*		needs high humidity; seems to do better with other plants in pot
Treebine	*Cissus antarctica*		plant dislikes acid potting medium
Umbrella plant	*Cyperus* spp.		one of very few plants that needs to stand in water
Velvetleaf	*Gynura sarmentosa*		gets "stringy" but is easily restarted from cuttings
Venus flytrap	*Dionaea muscipula*		sterilize pots; grow only in sphagnum moss; repot annually

Plant	Scientific Name	Environmental Requirements	Remarks
Vriesia	*Vriesia* spp.		*see* Bromeliad
Wandering Jew	*Zebrina pendula*		easy to grow; do not overwater
Wax plant	*Hoya* spp.		climber; leave pot in one place—does not like to be moved

Additional Reading

Arthurs, K., ed. *How to grow house plants.* A Sunset Book. Menlo Park, CA: Lane Publishing Co.; 1976.

————. *How to grow orchids.* A Sunset Book. Menlo Park, CA: Lane Publishing Co.; 1977

Bailey Hortorium Staff. *Hortus third: a concise dictionary of plants cultivated in the United States and Canada.* New York: Macmillan Co.; 1976.

Baumgardt, J. P. *How to prune almost everything.* New York: William Morrow and Co.; 1968.

Davidson, W.; Rochford, T. C. *The complete all-color guide to house plants, cacti and succulents.* New York: Galahad Books; 1976.

Edinger, P., ed. *How to grow herbs.* A Sunset Book. Menlo Park, CA: Lane Publishing Co.; 1972.

Free, M. *Plant propagation in pictures.* Rev. ed., Dietz, M. J., ed. Garden City, NY: Doubleday and Co.; 1979.

Kester, D. E.; Hartmann, H. T. *Plant propagation: principles and practices.* 3rd ed. Englewood Cliffs, NJ: Prentice-Hall; 1975.

Kramer, J. *The complete book of flowering house plants.* New York: St. Martin's Press, Inc.; 1980.

McDowell, J., ed. *Bonsai: culture and care of miniature trees.* A Sunset Book. Menlo Park, CA: Lane Publishing Co.; 1965.

Stowell, J. P. *Bonsai: indoors and out.* New York: Van Nostrand Reinhold Co.; 1966.

Tiedjens, V. A. *The vegetable encyclopedia and gardener's guide.* New York: Avenel Books; 1973.

Watt, B. K., et al., eds. *Handbook of the nutritional contents of foods* (unabridged republication of Agriculture Handbook No. 8, *Composition of foods,* published in 1963 by the United States Department of Agriculture). New York: Dover Publications; 1975.

Wyman, D. *Wyman's gardening encyclopedia.* Rev. ed. New York: Macmillan Co.; 1977.

Appendix 5

Metric Equivalents

Metric System of Measurement		
Application	**International System of Units**	**English System Equivalents**
Length	kilometer	0.62137 miles
	meter	39.37 inches
	centimeter	0.3937 inch
	millimeter	0.03937 inch
	micrometer	0.00003937 inch
	nanometer	0.00000003937 inch
	angstrom	0.0000000003937 inch
Mass (Weight)	metric ton	2,200 pounds
	kilogram	2.2 pounds
	gram	0.03527 ounce
	milligram	0.00003527 ounce
Volume	liter	1.06 quart
	milliliter	0.00106 quart
	cubic meter	35.314 cubic feet
	cubic centimeter	0.061 cubic inch
Area	hectare	2.471 acres
Temperature	To convert Celsius to Fahrenheit, multiply the Celsius figure by 9, divide the total by 5, and add 32.	

Glossary

a

Abscisic acid (ab-siz'ik as'id) (ABA) a growth-inhibiting hormone of plants; it is involved with other hormones in dormancy

Abscission (ab-sizh'un) the separation of leaves, flowers, and fruits from plants after the formation of an abscission zone at the base of their petioles, peduncles, and pedicels

Accessory pigments (ak-ses'uh-ree pig'mentz) pigments that capture light energy and transfer it to chlorophyll *a*

Acid (as'id) a substance that dissociates in water, releasing hydrogen ions

Active transport (ak'tiv trans'port) the expenditure of energy by a cell in moving a substance across a plasma membrane against a diffusion gradient

Adventitious (ad-ven-tish'uss) said of buds developing in internodes or on roots, or of roots developing along stems or on leaves

Aerobic respiration (air-oh'bik res-puh-ray'shun) respiration that requires free oxygen

After-ripening (af'tur-ry'pen-ing) metabolic changes that must take place in a dormant seed before germination can occur

Agar (ah'gur) a gelatinous substance produced by certain red algae; it is often used as a culture medium, particularly for bacteria

Albuminous cell (al-byu'min-uss sel) a gymnosperm phloem parenchyma cell closely associated with an adjacent sieve cell

Algin (al'jin) a gelatinous substance produced by certain brown algae; it is used in a wide variety of food substances and in pharmaceutical, industrial, and household products

Alternation of generations (ol-tur-nay'shun uv jen-ur-ay'shunz) alternation between a haploid gametophyte phase and a diploid sporophyte phase in the life cycle of sexually reproducing organisms

Amino acid (ah-mee'noh as'id) one of the organic, nitrogen-containing units from which proteins are synthesized

Anaerobic respiration (an-air-oh'bik res-puh-ray'shun) respiration in which the hydrogen removed from the glucose during glycolysis is combined with an organic ion (instead of oxygen)

Angiosperm (an'jee-oh-spurm) a plant whose seeds develop within ovaries that mature into fruits

Annual (an'you-ul) a plant that completes its entire life cycle in a single growing season

Annual ring (an'you-ul ring) a single season's production of xylem (wood) by the vascular cambium

Annulus (an'yu-luss) a specialized layer of cells around a fern sporangium; it aids in spore dispersal through a springlike action; also a membranous ring around the stipe of a mushroom

Anther (an'thur) the pollen-bearing part of a stamen

Antheridiophore (an-thur-id'ee-oh-for) a stalk that bears an antheridium

Antheridium (pl. antheridia) (an-thur-id'ee-um; pl. an-thur-id'ee-ah) the male gametangium of certain algae, fungi, bryophytes, and vascular plants other than gymnosperms and angiosperms

Anthocyanin (an-thoh-sy'ah-nin) a water-soluble pigment found in cell sap; anthocyanins vary in color from red to blue

Antibiotic (an-tee-by-ot'ik) a substance produced by a living organism that interferes with the normal metabolism of another living organism

Apical dominance (ay'pi-kul dom'i-nunts) suppression of growth of lateral buds by hormones

Apical meristem (ay'pi-kul mair'i-stem) a meristem at the tip of a shoot or root

Apomixis (ap-uh-mik'sis) reproduction without fusion of gametes or meiosis in otherwise normal sexual structures

Archegoniophore (ahr-kuh-goh'nee-oh-for) a stalk bearing an archegonium

Archegonium (pl. archegonia) (ahr-kuh-goh'nee-um; pl. ahr-kuh-goh'nee-ah) the multicellular female gametangium of bryophytes and most vascular plants other than angiosperms

Aril (air'il) an often brightly colored appendage surrounding the seed of certain plants

Ascus (pl. asci) (as'kus; pl. as'eye) one of often numerous, frequently fingerlike hollow structures in which the fusion of two haploid nuclei is followed by meiosis; a row of ascospores (usually eight) is ultimately produced in each ascus on or within the sexually initiated reproductive bodies of cup (sac) fungi

Asexual reproduction (ay-seksh'yule ree-proh-duk'shun) any form of reproduction not involving the union of gametes

Assimilation (uh-sim-i-lay'shun) cellular conversion of raw materials into protoplasm and cell walls

Atom (at'um) the smallest individual unit of an element that retains the properties of the element

Autotrophic (aw-toh-troh'fik) descriptive of an organism capable of sustaining itself through conversion of inorganic substances to organic material

Auxin (awk'sin) a growth-regulating substance produced either naturally by plants or synthetically

Axil (ak'sil) the angle formed between a twig and the petiole of a leaf; normally the site of an *axillary bud* (also called *lateral bud*)

b

Backcross (bak'kross) a cross involving a hybrid and one of its parents

Bacteriophage (bak-teer'ee-oh-fayj) a virus whose host is a bacterium

Bark (bahrk) tissues of a woody stem between the vascular cambium and the exterior

Basidium (pl. basidia) (buh-sid'ee-um; pl. buh-sid'ee-ah) one of usually numerous, frequently club-shaped hollow structures in which the fusion of two haploid nuclei is followed by meiosis, the four resulting nuclei becoming externally borne basidiospores; basidia are produced on or within sexually initiated reproductive bodies of the club fungi (e.g., mushrooms, puffballs)

Biennial (by-en'ee-ul) a plant that normally requires two seasons to complete its life cycle, the first season's growth being strictly vegetative

Biological controls (by-oh-loj'i-kull kun-trohlz') the use of natural enemies and inhibitors in combating insect pests and other destructive organisms

Biome (by'ohm) similar biotic communities considered on a worldwide or at least continental scale (e.g., desert biome, grassland biome)

Biotic community (by-ot'ik kuh-myu'nit-ee) an association of plants, animals, and other organisms (e.g., woodland)

Bivalent (biv'uh-lint or by-vay'lent) an associated pair of homologous chromosomes as seen in prophase 1 of meiosis

Blade (blayd) the conspicuous, flattened part of a leaf (also called *lamina*)

Bond (bonnd) a force that holds atoms together

Bonsai (bon-sy') container-grown plants (usually trees) that have been dwarfed artificially through skillful pruning and manipulation of the growing medium

Botany (bot'an-ee) science involving the study of plants

Botulism (bot'yu-lizm) poisoning from consumption of food infected by botulism bacteria

Bract (brakt) a structure that is usually leaflike and modified in size, shape, or color

Bryophyte (bry'oh-fyt) a photosynthetic, terrestrial, aquatic, or epiphytic plant without xylem and phloem (e.g., mosses, liverworts, hornworts)

Budding (bud'ing) a form of grafting in which the scion is a bud surrounded by a small shield of tissue; also a form of asexual reproduction in which a new cell develops to full size from a protuberance arising from a mature cell, as in yeasts

Bulb (buhlb) an underground food storage organ that is essentially a modified bud consisting of fleshy leaves that surround and are attached to a small stem

Bundle scar (bun'dul skahr) a small scar left by a vascular bundle within a leaf scar when the leaf separates from its stem through abscission

Bundle sheath (bun'dul sheeth) the parenchyma and/or sclerenchyma cells surrounding a vascular bundle

c

Callose (kal'ohs) a complex carbohydrate that develops in sieve tubes following an injury; it is commonly associated with the sieve areas of sieve tube elements

Callus (kal'uss) undifferentiated tissue that develops around injured areas of stems and roots; also the undifferentiated tissue that develops during tissue culture

Calvin cycle (kal'vin sy'kuhl) see *dark reactions*

Calyptra (kuh-lip'truh) tissue from the enlarged archegonial wall of many mosses that forms a partial or complete cap over the capsule

Calyx (kay'liks) collective term for the sepals of a flower

Cambium (kam'bee-um) a meristem producing secondary tissues; see *vascular cambium, cork cambium*

Capillary water (kap'i-lair-ee waw'tur) water held in the soil against the force of gravity

Carbohydrate (kahr-boh-hy'drayt) an organic compound containing carbon, hydrogen, and oxygen, with twice as many hydrogen as oxygen atoms per molecule

Carpel (kahr'pul) an ovule-bearing unit that is a part of a pistil

Casparian strip (kass-pair'ee-un strip) the band of suberin around the radial and transverse walls of an endodermal cell

Cell (sel) the basic structural and functional unit of living organisms; in plants it consists of protoplasm usually surrounded by a cell wall

Cell biology (sel by-ol'uh-jee) the biological discipline involving the study of cells and their functions

Cell cycle (sel sy'-kul) sequence of events involved in the division of a cell

Cell membrane (sel mem'brayn) see *plasma membrane*

Cell plate (sel playt) the precursor of the middle lamella; it forms at the equator during telophase

Cell sap (sel sap) the liquid contents of a vacuole

Cellulose (sel'yu-lohss) a complex, insoluble carbohydrate that constitutes the principal component of plant cell walls

Cell wall (sel wawl) the relatively rigid boundary of cells of plants and certain other organisms

Centromere (sen'truh-meer) the dense constricted portion of a chromosome to which a spindle fiber is attached (also called *kinetochore*)

Chiasma (pl. chiasmata) (ky-az'mah; pl. ky-az'mah-tah) the X-shaped configuration formed by two chromatids of homologous chromosomes as they remain attached to each other during prophase 1 of meiosis

Chlorenchyma (klor-en'kuh-mah) tissue composed of parenchyma cells that contain chloroplasts

Chlorophyll (klor'uh-fil) green pigments essential to photosynthesis

Chloroplast (klor'uh-plast) an organelle containing chlorophyll, found in cells of most photosynthetic organisms

Chromatid (kroh'muh-tid) one of the two strands of a chromosome; they are united by a centromere

Chromatin (kroh'muh-tin) a readily staining complex of DNA and proteins found in chromosomes

Chromoplast (kroh'muh-plast) a plastid containing pigments other than chlorophyll; the pigments are usually yellow to orange

Chromosome (kroh'muh-sohm) a body consisting of a linear sequence of genes and composed of DNA and proteins; chromosomes are found in cell nuclei and appear in contracted form during mitosis and meiosis

Cilium (pl. cilia) (sil'ee-um; pl. sil'ee-uh) a short hairlike structure usually found on the cells of unicellular aquatic organisms, normally in large numbers and arranged in rows; the most common function of cilia is propulsion of the cell

Circadian rhythm (sur-kay'dee-an rith'um) a daily rhythm of growth and activity found in living organisms

Cladophyll (klad'uh-fil) a flattened stem that resembles a leaf

Clamp connection (klamp kuh-nek'shun) a looplike connection between adjacent cells of a dikaryotic hypha of a club fungus

Class (klas) a category of classification between a division and an order

Climax vegetation (kly'maks vej-uh-tay'shun) vegetational association that perpetuates itself indefinitely at the culmination of ecological succession

Coenocytic (see-nuh-sit'ik) having many nuclei not separated from one another by crosswalls, as in the hyphae of water molds

Cohesion-tension-transpiration theory (koh-hee'zhun - ten'shun - trans-puh-ray'shun thee'uh-ree) theory that explains the rise of water in plants through a combination of cohesion of water molecules in capillaries and tension on the water columns brought about by transpiration

Coleoptile (koh-lee-op'tul) a protective sheath surrounding the emerging shoot of seedlings of the grass family (Poaceae) (e.g., corn, wheat)

Coleorhiza (koh-lee-uh-ry'zuh) a protective sheath surrounding the emerging radicle (immature root) of members of the grass family (Poaceae) (e.g., corn, wheat)

Collenchyma (kuh-len'kuh-muh) tissue composed of cells with unevenly thickened walls

Colloid (kol'oyd) substance consisting of a medium in which fine particles are permanently dispersed

Community (kuh-myu'nit-ee) collective term for all the living organisms sharing a common environment and interacting with one another

Companion cell (kum-pan'yun sel) a specialized cell derived from the same parent cell as the closely associated sieve tube element immediately adjacent to it (in angiosperm phloem)

Compound (kom'pownd) a substance whose molecules are composed of two or more elements

Compound leaf (kom'pownd leef) a leaf whose blade is divided into distinct leaflets

Conidium (pl. conidia) (kuh-nid'ee-um; pl. kuh-nid'ee-uh) an asexually produced fungal spore formed outside of a sporangium

Conifer (kon'i-fur) a cone-bearing tree or shrub

Conjugation (kon-juh-gay'shun) a process leading to the fusion of isogametes in algae, fungi, and protozoa; also, the means by which certain bacteria exchange DNA

Conjugation tube (kon-juh-gay'shun t(y)oob) a tube permitting transfer of a gamete or gametes between adjacent cells, as in *Spirogyra* or desmids

Cork (kork) tissue composed of cells whose walls are impregnated with suberin at maturity; the outer layer of tissue of an older woody stem; cork is produced by the cork cambium

Cork cambium (kork kam'bee-um) a narrow cylindrical sheath of cells between the exterior of a woody root or stem and the central vascular tissue; it produces *cork* to its exterior and *phelloderm* to its interior; it is also called *phellogen*

Corm (korm) a vertically oriented, thickened food storage stem that is usually enveloped by a few papery nonfunctional leaves

Corolla (kuh-rahl'uh) collective term for the petals of a flower

Cortex (kor'teks) a primary tissue composed primarily of parenchyma; the tissue usually extends between the epidermis and the vascular tissue

Cotyledon (kot-uh-lee'dun) an embryo leaf ("seed leaf") that usually either stores or absorbs food

Crossing over (kross'ing oh'vur) the exchange of corresponding segments of chromatids between homologous chromosomes during prophase 1 of meiosis

Crozier (kroh'zhur) the spirally coiled "fiddlehead" of an immature fern frond; also, a hook-shaped hyphal tip that precedes the development of an ascus in sac fungi

Cuticle (kyut'i-kul) a waxy or fatty layer of varying thickness on the outer walls of epidermal cells

Cutin (kyu'tin) the waxy or fatty substance of which a cuticle is composed

Cyclosis (sy-kloh'sis) the flowing or streaming of cytoplasm within a cell

Cytokinesis (sy-toh-kuh-nee'sis) cell division

Cytokinin (syt-uh-ky'nin) a growth hormone involved in cell division and several other metabolic activities of cells

Cytology (sy-tol'uh-jee) see *cell biology*

Cytoplasm (sy'tuh-plazm) the protoplasm of a cell exclusive of the nucleus

Cytoplasmic streaming (sy-tuh-plaz'mik streem'ing) see *cyclosis*

d

Dark reactions (dahrk ree-ak'shunz) a cyclical series of chemical reactions that utilizes carbon dioxide and energy generated during the light reactions of photosynthesis, psoducing sugars, some of which are stored as insoluble carbohydrates while others are recycled; the reactions are independent of light and occur in the stroma of chloroplasts

Day-neutral plant (day new'trul plant) a plant that is not dependent on specific daylengths for the initiation of flowering

Deciduous (di-sij'yu-wuss) shedding leaves annually

Decomposer (dee-kuhm-poh'zur) organism (e.g., bacterium, fungus) that breaks down organic material to forms capable of being recycled

Development (di-vel'up-ment) changes in the form of a plant resulting from growth and differentiation of its cells into tissues and organs

Dicotyledon (dy-kot-uh-lee'dun) a class of angiosperms whose seeds commonly have two cotyledons; the term is frequently abbreviated to *dicot*

Dictyosome (dik'tee-uh-sohm) see *Golgi body*

Differentially permeable membrane (dif-uh-rensh'uh-lee pur'mee-uh-bul mem'brayn) a membrane that permits the passage of certain materials through it but restricts or prevents the flow of others; also called *semipermeable*

Differentiation (dif-uh-ren-shee-ay'shun) the change of a relatively unspecialized cell to a more specialized one (e.g., the change of a cell just produced by a meristem to a vessel element or fiber)

Diffusion (dif-fyu'zhin) the random movement of molecules or particles from a region of higher concentration to a region of lower concentration, ultimately resulting in uniform distribution

Digestion (dy-jes'jin) enzyme-controlled conversion of complex, usually insoluble substances to simpler, usually soluble substances

Dihybrid (dy-hy'brid) a cross involving two different pairs of chromosomes in the parents

Dikaryotic (dy-kair-ee-ot'ik) having a pair of nuclei in each cell or unit of the mycelium in club fungi

Dioecious (dy-ee'shuss) having unisexual flowers or cones, with the male flowers or cones confined to certain plants and the female flowers or cones of the same species confined to other different plants

Diploid (dip'loyd) having two sets of chromosomes in each cell; the 2N chromosome number characteristic of the sporophyte generation

Diuretic (dy-yu-ret'ik) a substance tending to increase the flow of urine

Division (duh-vizh'un) the largest category of classification of organisms within a kingdom

DNA standard abbreviation of deoxyribonucleic acid, the carrier of genetic information in cells and viruses

Dominance (dom'uh-nintz) the Mendelian principle that one member of a pair of genes may mask or inhibit the expression of the other

Dominant (dom'uh-nint) the member of a pair of genes that masks or suppresses the phenotypic expression of the recessive gene

Dormancy (dor'man-see) a period of growth inactivity in seeds and plant organs that requires certain environmental stimuli or conditions to bring it to an end

Double fusion (dub'ul fu'shun) the more or less simultaneous union of one sperm and egg (forming a zygote) and union of another sperm and polar nuclei (forming a primary endosperm nucleus) that occurs in the embryo sac of flowering plants

e

Ecology (ee-kol'uh-jee) the biological discipline involving the study of the relationship of organisms to their environment

Ecosystem (ee'koh-sis-tim) a system involving interactions of living organisms with one another and with their nonliving environment

Egg (eg) a nonmotile female gamete

Elater (el'uh-tur) straplike appendage (usually occurring in pairs) attached to a horsetail (*Equisetum*) spore; also, a somewhat spindle-shaped sterile cell occurring in large numbers in liverwort sporangia; both types of elaters facilitate spore dispersal

Electron (i-lek'tron) a negatively charged particle of an atom

Element (el'uh-mint) one of 105 types of matter, most existing naturally but some man-made, each of which is composed of one kind of atom

Embryo (em'bree-oh) immature sporophyte that develops from a zygote within an ovule or archegonium after fertilization

Embryo sac (em'bree-oh sak) the female gametophyte of angiosperms, which, in approximately 70 percent of the species investigated, contains eight nuclei

Enation (i-nay'shun) one of the tiny green leaflike outgrowths on the stems of whisk ferns (*Psilotum*)

Endocarp (en'doh-kahrp) the innermost layer of a fruit wall

Endodermis (en-doh-dur'mis) a single layer of cells surrounding the vascular tissue (stele) in roots and some stems; many of the cells have Casparian strips

Endoplasmic reticulum (en-doh-plaz'mik ruh-tik'yu-lum) a complex system of interlinked double membrane channels, subdividing the cytoplasm of a cell into compartments; parts of it are lined with ribosomes

Endosperm (en'doh-spurm) a food storage tissue that develops through divisions of the primary endosperm nucleus; it is digested by the sporophyte after germination in some species (e.g., corn) or before maturation of the seed in other species (e.g., beans)

Energy (en'ur-jee) the capacity to do work; various forms of energy include heat, light, and kinetic

Enzyme (en'zym) one of numerous complex proteins that speeds up a chemical reaction in living cells without being used up in the reaction (i.e., it catalyzes the reaction)

Epicotyl (ep'uh-kaht-ul) the part of an embryo or seedling above the attachment point of the cotyledon(s)

Epidermis (ep-uh-dur'mis) the exterior tissue, usually one cell thick, of leaves and young stems and roots

Epiphyte (ep'uh-fyt) an organism that is attached to and grows on another organism without parasitizing it

Ergotism (ur'got-izm) a disease resulting from consumption of goods made with flour containing ergot fungus

Essential element (uh-sen'shul el'uh-mint) one of eighteen elements generally considered essential to the normal growth, development, and reproduction of most plants

Ethnobotany (eth-no-bot'an-ee) the biological discipline involving the study of the various uses of plants by humans

Ethylene (eth'uh-leen) a simple gaseous substance that inhibits plant growth and promotes the ripening of fruit

Etiolation (ee-tee-oh-lay'shun) a condition characterized by long internodes, poor leaf development, and pale, weak appearance, due to a plant's having been deprived of light

Eukaryotic (yu-kair-ee-ot'ik) pertaining to cells having distinct membrane-bounded organelles, including a nucleus with chromosomes

Eutrophication (yu-troh-fuh-kay'shun) the gradual enrichment of a body of water through the accumulation of nutrients, resulting in a corresponding increase in algae and other organisms

Exine (ek'seen or ek'syne) the outer layer of the wall of a pollen grain or spore

Exocarp (ek'soh-kahrp) the outermost layer of a fruit wall

Eyespot (eye'spot) a small, often reddish structure within a motile unicellular organism; it appears to be sensitive to light (also called *stigma*)

f

F₁ first filial generation; the offspring of a cross between two parent plants

F₂ second filial generation; the offspring of F₁ plants

Family (fam'uh-lee) a category of classification between an order and a genus

Fat (fat) an organic compound containing carbon, hydrogen, and oxygen but with proportionately much less oxygen than is present in a carbohydrate molecule

Fermentation (fur-men-tay'shun) respiration in which the hydrogen removed from the glucose during glycolysis is transferred back to pyruvic acid, creating substances such as ethyl alcohol or lactic acid

Fertilization (fur-til-i-zay'shun) formation of a zygote through the fusion of two gametes

Fiber (fy'bur) a long thick-walled cell whose protoplasm often is dead at maturity

Filament (fil'uh-mint) threadlike body of certain algae and fungi; also the stalk portion of a stamen

Fission (fish'un) the division of cells of bacteria, blue-green algae, and related organisms into two new cells

Flagellum (pl. flagella) (fluh-jel'um; pl. fluh-jel'uh) a fine threadlike structure protruding from a motile unicellular organism or the motile cells produced by multicellular organisms; it functions primarily in locomotion

Floret (flor'et) a small flower that is a part of the inflorescence of members of the sunflower family (Asteraceae) and the grass family (Poaceae)

Food chain (food chayn) a natural chain of organisms of a community wherein each member of the chain feeds on members below it and is consumed by members above it, with autotrophic organisms (producers) being at the bottom; interconnected food chains are referred to as *food webs*

Foot (foot) the basal part of the embryo of bryophytes and other plants; it is attached to and absorbs food from the gametophyte

Fossil (fos'ul) the remains or impressions of any natural object that has been preserved in the earth's crust

Frond (frond) a fern leaf

Fruit (froot) a mature ovary usually containing seeds; term also somewhat loosely applied to the reproductive structures of groups of plants other than angiosperms

Fucoxanthin (fyu-koh-zan'thin) a brownish pigment occurring in brown and other algae

Fundamental tissue (fun-duh-ment'ul tish'yu) parenchyma tissue of monocots (equivalent of cortex and pith in dicots)

g

Gametangium (pl. gametangia) (gam-uh-tan'jee-um; pl. gam-uh-tan'jee-ah) any cell or structure in which gametes are produced

Gamete (gam'eet) a sex cell; one of two cells that unite, forming a zygote

Gametophore (guh-me'toh-for) a stalk on which a gametangium is borne

Gametophyte (guh-me'toh-fyte) the haploid (*N*) gamete-producing phase of the life cycle of an organism that exhibits alternation of generations

Gemma (pl. gemmae) (jem'uh; pl. jem'ee) a small outgrowth of tissue that becomes detached from the parent body and is capable of developing into a complete new plant or other organism; gemmae are produced in cuplike structures on liverwort thalli, and are also produced by certain fungi

Gene (jeen) a unit of heredity; part of a linear sequence of such units occurring in the DNA of chromosomes

Generative cell (jen'uh-ray-tiv sel) the cell of the male gametophyte of angiosperms that divides, producing two *sperms;* also, the cell of the male gametophyte of gymnosperms that divides, producing a *sterile cell* and a *spermatogenous cell*

Genetics (juh-net'iks) the biological discipline involving the study of heredity

Genotype (jeen'oh-typ) the genetic constitution of an organism; it may or may not be visibly expressed, as contrasted with *phenotype*

Genus (pl. genera) (jee'nus; pl. je'ner ah) a category of classification between a family and a species

Gibberellin (jib-uh-rel'in) one of a group of plant hormones that have a variety of effects on growth; they are particularly known for promoting elongation of stems

Gill (gil) one of the flattened plates of compact mycelium that radiate out from the stalk on the underside of the caps of most mushrooms

Girdling (gurd'ling) the removal of a band of tissues extending inward to the vascular cambium on the stem of a woody plant

Gland (gland) a small body of variable shape and size that may secrete certain substances but that also may be functionless

Glycolysis (gly-kol'uh-sis) the initial phase of all types of respiration in which glucose is converted in the absence of free oxygen to pyruvic acid

Golgi body (gohl'jee bod'ee) an organelle consisting of disc-shaped, often branching hollow tubules that apparently function in accumulating and packaging substances used in the synthesis of materials by the cell; also called *dictyosome;* collectively, the Golgi bodies of a cell may be referred to as the *Golgi apparatus*

Graft (graft) the union of a segment of a plant, the *scion,* with a rooted portion, the *stock*

Granum (pl. grana) (gra'num; pl. gra'nuh) a series of stacked thylakoids within a chloroplast

Gravitational water (grav-uh-tay'shun-ul waw'tur) water that drains out of the pore spaces of a soil after a rain

Ground meristem (grownd mair'i-stem) meristem that produces all the primary tissues other than the epidermis and stele (e.g., cortex, pith)

Growth (grohth) progressive increase in size and volume through natural development

Guard cell (gahrd sel) one of a pair of specialized cells forming a stoma

Guttation (guh-tay'shun) the exudation of water in liquid form from leaves due to root pressure

Gymnosperm (jim'nuh-spurm) a plant whose seeds are not enclosed within an ovary during their development (e.g., pine tree)

h

Haploid (hap'loyd) having one set of chromosomes per cell, as in gametophytes; also referred to as having *N* chromosomes (as contrasted with 2*N* chromosomes in the *diploid* cells of sporophytes)

Haustorium (pl. haustoria) (haw-stor'ee-um; pl. haw-stor'ee-uh) a protuberance of a fungal hypha or plant organ such as a root that functions as a penetrating and absorbing structure

Heartwood (hahrt'wood) nonliving, usually darker-colored wood whose cells have ceased to function in water conduction

Herbaceous (hur-bay'shuss *or* ur-bay'shuss) referring to nonwoody plants

Herbal (hur'bul *or* ur'bul) a sixteenth- and seventeenth-century botany book that emphasized medicinal uses, edibility, and other utilitarian functions of plants

Herbarium (pl. herbaria) (hur-bair'ee-um *or* ur-bair'ee-um; pl. hur-bair'ee-uh) a collection of dried pressed specimens, usually mounted on paper and provided with a label that gives collection information and an identification

Heterocyst (het'uh-roh-sist) a transparent, thick-walled, slightly enlarged cell occurring in the filaments of certain blue-green algae

Heterozygous (het-uh-roh-zy'guss) having a pair of genes with contrasting characters at the same location on homologous chromosomes

Holdfast (hold'fast) attachment organ or cell at the base of the thallus or filament of certain algae

Homologous chromosomes (hoh-mol'uh-guss kroh'muh-sohmz) pairs of chromosomes that associate together in prophase 1 of meiosis; each member of a pair is derived from a different parent

Homozygous (hoh-moh-zy'guss) having a pair of genes with identical characters at the same location on a pair of homologous chromosomes

Hormone (hor'mohn) an organic substance generally produced in minute amounts in one part of an organism and transported to another part of the organism where it controls or affects specific metabolic processes

Hybrid (hy'brid) offspring of two parents that differ in one or more genes

Hydrolysis (hy-drol'uh-sis) the breakdown of complex molecules to simpler ones as a result of the union of water with the compound; the process is usually controlled by enzymes

Hygroscopic water (hy-gruh-skop'ik waw'tur) water that is chemically bound to soil particles and therefore unavailable to plants

Hypha (pl. hyphae) (hy'fuh; pl. hy'fee) a single, usually tubular, threadlike filament of a fungus; *mycelium* is a collective term for many hyphae

Hypocotyl (hy-poh-kot'ul) the portion of an embryo or seedling between the radicle and the cotyledon(s)

Hypodermis (hy-poh-dur'mis) a layer of cells immediately beneath the epidermis and distinct from the parenchyma cells of the cortex in certain plants

Hypothesis (hy-poth'uh-sis) a postulated explanation for some observed facts that must be tested experimentally before it can be accepted as valid or discarded if it proves to be incorrect

i

Imbibition (im-buh-bish'un) adsorption of water by nonliving materials and subsequent swelling because of the adhesion of the water molecules to the internal surfaces

Indusium (pl. indusia) (in-dew'zee-um; pl. in-dew'zee-uh) small, membranous, sometimes umbrellalike covering of a developing fern *sorus*

Inferior ovary (in-feer'ee-or oh'vuh-ree) an ovary to which parts of the calyx, corolla, and stamens have become more or less united so that they appear to be attached at the top of it

Inflorescence (in-fluh-res'ints) collective term for a group of flowers attached to a common axis in a specific arrangement

Inorganic (in-or-gan'ik) descriptive of compounds having no carbon

Integument (in-teg'yu-mint) the outermost layer of an ovule; it usually develops into a seedcoat; a gymnosperm ovule usually has a single integument, and an angiosperm ovule usually has two integuments

Intermediate-day plant (in-tur-me'dee-ut - day plant) a plant that has two critical photoperiods so that it will not flower if the days are either too short or too long

Internode (in'tur-nohd) a stem region between nodes

Interstock (in'tur-stok)   a segment of stem that is grafted between a stock and a scion

Ion (eye'on)   a molecule or atom that has become electrically charged through the loss or gain of one or more electrons

Isogamy (eye-sog'uh-me)   sexual reproduction in certain algae and fungi having gametes that are alike in size

Isotope (eye'suh-tohp)   one of two or more forms of an element that have the same chemical properties but differ in the number of neutrons in the nuclei of their atoms

k

Kinetochore (kih-net'uh-kor)   see *centromere*

Kingdom (king'dom)   the highest category of classification (e.g., Plant Kingdom, Animal Kingdom)

Knot (not)   a portion of the base of a branch enclosed within wood

Krebs cycle (krebz' sy'kul)   a complex series of reactions following glycolysis in aerobic respiration that ultimately results in the conversion of pyruvic acid to hydrogen, carbon dioxide, and electrons

l

Lamina (lam'uh-nuh)   see *blade*

Lateral bud (lat'uh-rul bud)   see *axil*

Laticifer (luh-tis'uh-fur)   specialized cells or ducts resembling vessels; they form branched networks of *latex*-secreting cells in the phloem and other parts of plants

Leaf gap (leef gap)   a parenchyma-filled interruption in a stem's cylinder of vascular tissue immediately above the point at which a branch of vascular tissue (*leaf trace*) leading to a leaf occurs

Leaflet (leef'lit)   one of the subdivisions of a compound leaf

Leaf scar (leef skahr)   the suberin-covered scar left on a twig when a leaf separates from it through abscission

Leaf trace (leef trays)   see *leaf gap*

Lenticel (lent'uh-sel)   one of usually numerous, slightly raised, somewhat spongy groups of cells in the bark of woody plants; lenticels permit gas exchange between the interior of a plant and the external atmosphere

Leucoplast (loo'kuh-plast)   a colorless plastid commonly associated with starch accumulation

Light reactions (lyt ree-ak'shunz)   a series of chemical reactions through which light energy is converted to chemical energy with the aid of chlorophyll molecules; in the process water molecules are split, with hydrogen ions and electrons being produced and oxygen gas being released; ATP and $NADPH_2$ also are created

Lignin (lig'nin)   an organic hardening substance with which certain cell walls (e.g., those of wood) become impregnated

Ligule (lig'yool)   the tiny tonguelike appendage at the base of a spike moss (*Selaginella*) or quillwort (*Isoetes*) leaf; also, the outgrowth from the upper and inner side of a grass leaf at the point where it joins the sheath; also, the conspicuous straplike portion of the corolla of an outer floret in the flower head of a member of the sunflower family (Asteraceae)

Linkage (link'ij)   the tendency of two or more genes located on the same chromosome to be inherited together

Lipid (lip'id)   a general term for fats, fatty substances, and oils

Locule (lok'yool)   a cavity within an ovary or a sporangium

Long-day plant (long-day plant)   a plant in which flowering is not initiated unless exposure to more than a critical daylength occurs

Lysosome (ly'suh-sohm)   an organelle containing enzymes that regulate digestion in a cell

m

Macrophyll (mak'roh-fil)   a leaf having branching veins; it is associated with a leaf gap

Megaspore (meg'uh-spor)   a spore that develops into a female gametophyte

Megaspore mother cell (meg'uh-spor muth'ur sel)   a diploid cell that produces megaspores upon undergoing meiosis

Megasporophyll (meg-uh-spor'uh-fil)   a leaf, usually reduced in size, on or within which megaspores are produced

Meiosis (my-oh'sis)   the process of two successive nuclear divisions through which segregation of genes occurs and a single diploid ($2N$) cell becomes four haploid (N) cells

Mericloning (mair'i-kloh-ning)   multiplication of plants through culturing and artificial dividing of shoot meristems

Meristem (mair'i-stem)   a region in which undifferentiated cells divide

Mesocarp (mez'uh-karp)   the middle region of the fruit wall that lies between the exocarp and the endocarp

Mesophyll (mez'uh-fil)   parenchyma (chlorenchyma) tissue between the upper and lower epidermis of a leaf

Metabolism (muh-tab'uh-lizm)   the sum of all the interrelated chemical processes occurring in a living organism

Microphyll (my'kroh-fil)   a leaf having a single unbranched vein not associated with a leaf gap

Micropyle (my'kroh-pyl)   a pore or opening in the integuments of an ovule through which a pollen tube gains access to an embryo sac or archegonium of a seed plant

Microspore (my'kroh-spor)   a spore that develops into a male gametophyte

Microspore mother cell (my'kroh-spor muth'ur sel)   a diploid cell that produces microspores upon undergoing meiosis

Microsporophyll (my-kroh-spor'uh-fil)   a leaf, usually reduced in size, on or within which microspores are produced

Middle lamella (mid'ul luh-mel'uh)   a layer of pectic material that cements two adjacent cell walls together

Midrib (mid'rib)   the central (main) vein of a pinnately veined leaf or leaflet

Mitochondrion (pl. mitochondria) (my-toh-kon'dree-un; pl. my-toh-kon'dree-uh)   an organelle containing enzymes that function in the Krebs cycle and the electron transport chain of aerobic respiration

Mitosis (my-toh′sis)   nuclear division, usually accompanied by cytokinesis, during which the chromatids of the chromosomes separate and two genetically identical daughter nuclei are produced

Mixture (miks′chur)   a substance containing two or more ingredients, the atoms and molecules of which retain a separate identity and are not in a fixed proportion to one another

Molecule (mol′uh-kyul)   the smallest unit of an element or compound retaining its own identity; it consists of two or more atoms

Monocotyledon (mon-oh-kot-uh-lee′dun)   a class of angiosperms whose seeds have a single cotyledon; the term is commonly abbreviated to *monocot*

Monoecious (moh-nee′shuss)   having unisexual male flowers or cones and unisexual female flowers or cones both on the same plant

Monohybrid (mon-oh-hy′brid)   a cross involving a single pair of genes with contrasting characters in the parents

Monokaryotic (mon-oh-kair-ee-ot′ik)   having a single nucleus in each cell or unit of the mycelium in club fungi

Motile (moh′tul)   capable of independent movement

Mulch (mulch)   any substance (e.g., straw, polyethylene sheeting) placed on the ground to protect plants from the environment, to conserve soil moisture, or to keep fruit clean

Mutation (myu-tay′shun)   an inheritable change in a gene or chromosome

Mycelium (my-see′lee-um)   a mass of fungal hyphae

Mycorrhiza (pl. mycorrhizae) (my-kuh-ry′zuh; pl. my-kuh-ry′zee)   a symbiotic association between fungal hyphae and a plant root

n

Net venation (net ve-nay′shun)   network of veins characteristic of a dicot leaf

Neutron (new′tron)   an uncharged particle in the nucleus of an atom

Node (nohd)   region of a stem where one or more leaves are attached

Nucellus (new-sel′us)   ovule tissue within which an embryo sac develops

Nuclear envelope (new′klee-ur en′vuh-lohp)   a porous double membrane enclosing a nucleus

Nucleic acid (new-klee′ik as′id)   see *DNA, RNA*

Nucleolar organizer (new-klee′oh-lurr or′gun-eye-zurr)   specialized area of certain chromosomes; it is associated with the formation of a nucleolus

Nucleolus (pl. nucleoli) (new-klee′oh-luss; pl. new-klee′oh-ly)   a somewhat spherical body within a nucleus; it contains primarily RNA and protein; there may be more than one nucleolus per nucleus

Nucleotide (new′klee-oh-tyd)   the structural unit of DNA and RNA

Nucleus (new′klee-uss)   the organelle of a living cell that contains chromosomes and is essential to the regulation and control of all the cell's functions; also, the core of an atom

o

Oil (oyl)   a fat in a liquid state

Oogamy (oh-og′uh-mee)   sexual reproduction in which the female gamete or egg is nonmotile and larger than the male gamete or sperm, which is motile

Oogonium (pl. oogonia) (oh-oh-goh′nee-um; pl. oh-oh-goh′nee-ah)   a female sex organ of certain algae and fungi; it consists of a single cell that contains one to several eggs

Operculum (oh-per′kyu-lum)   the lid or cap that protects the peristome of a moss sporangium

Order (or′dur)   a category of classification between a class and a family

Organelle (or-guh-nel′)   a membrane-bounded body in the cytoplasm of a cell; there are several kinds, each with a specific function (e.g., mitochondrion, chloroplast)

Organic (or-gan′ik)   pertaining to or derived from living organisms, and to the chemistry of carbon-containing compounds

Osmosis (oz-moh′sis)   the diffusion of water or other solvents through a differentially permeable membrane from a region of higher concentration to a region of lower concentration

Osmotic potential (oz-mot′ik puh-ten′shul)   potential pressure that can be developed by a solution separated from pure water by a differentially permeable membrane (the pressure that must be applied to the solution to stop the movement of water)

Osmotic pressure (oz-mot′ik presh′ur)   see *osmotic potential*

Ovary (oh′vuh-ree)   the enlarged basal portion of a pistil that contains an ovule or ovules and develops into a fruit

Ovule (oh′vyool)   a structure of seed plants that contains a female gametophyte and has the potential to develop into a seed

p

Palisade mesophyll (pal-uh-sayd′ mez′uh-fil)   mesophyll having relatively uniform rows of tightly packed parenchyma (chlorenchyma) cells located beneath the upper epidermis of a leaf

Palmately compound; palmately veined (pahl′mayt-lee kom′pownd; pahl′mayt-lee vaynd)   having leaflets or principal veins radiating out from a common point

Papilla (pl. papillae) (puh-pil′uh; pl. puh-pill-ay)   a small, usually rounded or conical protuberance

Parallel venation (pair′uh-lel ve-nay′shun)   venation in which the principal veins are more or less parallel to one another; characteristic of monocot leaves

Parasite (pair′i-syt)   an organism that derives its nourishment from the living cells or tissues of another organism, which is referred to as the *host*

Parenchyma (puh-ren′kuh-muh)   thin-walled cells varying in size, shape, and function; the most common type of plant cell

Parthenocarpic (par-thuh-noh-kar′pik)   developing fruits from unfertilized ovaries; the resulting fruit is, therefore, usually seedless

Passage cell (pas'ij sel) a thin-walled cell of an endodermis

Pectin (pek'tin) a water-soluble organic compound occurring primarily in the middle lamella; when combined with organic acids and sugar it becomes a jelly

Pedicel (ped'i-sel) the individual stalk of a flower that is part of an inflorescence

Peduncle (pee'dun-kul) the stalk of a solitary flower or the main stalk of an inflorescence

Peptide bond (pep'tyd bond) the type of chemical bond formed when two amino acids link together

Perennial (puh-ren'ee-ul) a plant that continues to live indefinitely after flowering

Pericarp (per'uh-karp) collective term for all the layers of a fruit wall

Pericycle (per'uh-sy-kul) tissue sandwiched between the endodermis and phloem of a root; it is often only one or two cells wide in transverse section and is the site of origin of branch roots

Peristome (per'uh-stohm) a series of flattened, often ornamented structures (teeth) arranged around the margin of the open end of a moss sporangium; the teeth are sensitive to changes in humidity and facilitate the release of spores

Petal (pet'ul) a frequently flattened, often colored unit of a corolla

Petiole (pet'ee-ohl) the stalk of a leaf

Phage (fayj) see *bacteriophage*

Phellogen (fel'uh-jun) see *cork cambium*

Phenotype (fee'noh-typ) the physical appearance of an organism

Pheromone (fer'uh-mohn) a substance produced by an organism that facilitates chemical communication with another organism

Phloem (flohm) the food-conducting tissue of a vascular plant

Photon (foh'ton) a unit of light energy

Photoperiodism (foh-toh-pir'ee-ud-izm) the initiation of flowering and certain vegetative activities of plants in response to relative lengths of day and night

Photosynthesis (foh-toh-sin'thuh-sis) the conversion of light energy, water, and carbon dioxide in the presence of chlorophyll to carbohydrate, with oxygen being released as a by-product

Photosynthetic unit (foh-toh-sin-thet'ik yew'nit) one of two groups of about 250 to 400 pigment molecules each that function together in chloroplasts in the light reactions of photosynthesis; the units are exceedingly numerous in each chloroplast

Photosystem (foh'toh-sis-tum) collective term for a specific functional aggregation of photosynthetic units

Phragmoplast (frag'moh-plast) set of microfibrils forming a ring at the margin of the developing cell plate at the close of mitosis

Phytochrome (fy'tuh-krohm) pigment associated with the absorption of light; it is found in the cytoplasm of cells of green plants

Pigment (pig'mint) coloring matter; substance that absorbs light

Pilus (pl. pili) (py'lis; pl. py'lee) the equivalent of a conjugation tube in bacteria

Pinna (pl. pinnae) (pin'uh; pl. pin'ee) primary subdivision of a fern frond; the term is also applied to a leaflet of a compound leaf

Pinnately compound; pinnately veined (pin'ayt-lee kom'pownd; pin'ayt-lee vaynd) having leaflets or veins on both sides of a common axis (e.g., rachis, midrib) to which they are attached

Pistil (pis'tul) the central unit of a flower; it is composed of one or more carpels and consists of an ovary, style, and stigma

Pit (pit) a depression or cavity in a cell wall

Pith (pith) central tissue of a dicot stem and certain roots; it usually consists of parenchyma cells that become crushed in woody plants as cambial activity increases the organ's girth

Plankton (plank'ton) free-floating aquatic organisms that are mostly microscopic

Plant anatomy (plant ah-nat'uh-mee) the botanical discipline that pertains to the internal structure of plants

Plant community (plant kuh-myu'nuh-tee) an association of plants inhabiting a common environment and interacting with one another

Plant geography (plant jee-og'ruh-fee) the botanical discipline that pertains to the broader aspects of the space relations of plants and their distribution over the surface of the earth

Plant morphology (plant mor-fol'uh-jee) the botanical discipline that pertains to plant form and development

Plant physiology (plant fiz-ee-ol'uh-jee) the botanical discipline that pertains to the metabolic activities and processes of plants

Plant taxonomy (plant tak-son'uh-mee) the botanical discipline that pertains to the classification, naming, and identification of plants

Plasma membrane (plaz'muh mem'brayn) the outer boundary of the protoplasm of a cell; also called *cell membrane,* particularly in animal cells

Plasmodesma (pl. plasmodesmata) (plaz-muh-dez'muh; pl. plaz-muh-dez'mah-tah) minute strands of cytoplasm that extend between adjacent cells through pores in the walls

Plasmolysis (plaz-mol'uh-sis) the shrinking in volume of the protoplasm of a cell and the separation of the cytoplasm from the cell wall due to loss of water via osmosis

Plastid (plas'tid) an organelle associated primarily with the storage or manufacture of carbohydrates (e.g., *leucoplast, chloroplast)*

Plumule (ploo'myool) the terminal bud of the embryo of a seed plant

Polar nuclei (poh'lur new'klee-eye) nuclei, frequently two in number, that unite with a sperm in an embryo sac, forming a primary endosperm nucleus

Pollen grain (pahl'un grayn) a structure derived from the microspore of seed plants that develops into a male gametophyte

Pollen tube (pahl'un t(y)oob) a tube that develops from a pollen grain and conveys the sperms to the female gametophyte

Pollination (pahl-uh-nay'shun) the transfer of pollen from an anther to a stigma

Pollinium (pl. pollinia) (pah-lin'ee-um; pl. pah-lin'ee-ah) a cohesive mass of pollen grains commonly found in members of the orchid family (Orchidaceae) and the milkweed family (Asclepiadaceae)

Polyploidy (pahl'i-ploy-dee) having more than two complete sets of chromosomes per cell

Pressure-flow hypothesis (presh'ur - floh hy-poth'uh-sis) the theory that food substances in solution in plants flow along concentration gradients between the sources of the food and *sinks* (places where the food is utilized)

Primary consumer (pry'-mer-ree kon-soo'-mur) animals that feed directly on producers

Primary endosperm nucleus (pry'mer-ee en'doh-spurm new'klee-uss) the nucleus produced by the fusion of a sperm and polar nuclei in the embryo sac of angiosperms; it gives rise to *endosperm* tissue

Primordium (pry-mord'ee-um) an organ or structure (e.g., leaf, bud) at its earliest stage of development

Procambium (proh-kam'bee-um) the primary meristem that gives rise to primary xylem and phloem

Producer (pruh-dew'sur) an organism that manufactures food through the process of photosynthesis

Progametangium (proh-gam-uh-tan'jee-um) a cell or structure that becomes a gametangium

Prokaryotic (proh-kair-ee-ot'ik) having a cell or cells that lack a distinct nucleus and other membrane-bounded organelles (e.g., bacteria, blue-green algae)

Proplastid (proh-plas'tid) a tiny cytoplasmic body from which a plastid develops

Protein (proht'ee-in *or* proh'teen) an organic compound containing carbon, hydrogen, oxygen, nitrogen, and frequently sulphur in complex molecules composed of numerous amino acids linked together by peptide bonds

Prothallus (pl. prothalli) (proh-thal'us; pl. proh-thal'eye) the gametophyte of ferns and their relatives; also called *prothallium*

Protoderm (proh'tuh-durm) the primary meristem that gives rise to the epidermis

Proton (proh'ton) a positively charged particle in the nucleus of an atom

Protonema (proh-tuh-nee'muh) a green, usually branched, threadlike or sometimes platelike growth from a moss spore; it gives rise to "leafy" gametophytes

Protoplasm (proh'tuh-plazm) the living substance of a cell (includes the cytoplasm and nucleus)

Pruning (proon'ing) removal of portions of plants for aesthetic purposes, for improving quality and size of fruits or flowers, or for elimination of diseased tissues

Pyrenoid (py'ruh-noyd) a small body found on the chloroplasts of certain green algae and hornworts; pyrenoids are associated with starch accumulation; they may occur singly on a chloroplast or they may be numerous

Pyruvic acid (py-roo'vik as'id) the organic compound that is the end product of the glycolysis phase of respiration

q

Quartersawed (kwor-tur-sawd') wood that has been cut radially, i.e., along the rays

r

Rachis (ray'kiss) the axis of a pinnately compound leaf or frond extending between the lowermost leaflets or pinnae and the terminal leaflet or pinna (corresponds with the midrib of a simple leaf)

Radicle (rad'i-kuhl) the part of an embryo in a seed that develops into a root

Ray (ray) radially oriented tiers of cells that conduct food, water, and other materials laterally in the stems and roots of woody plants; they are generally continuous across the vascular cambium between the xylem and the phloem; the portion within the wood is called a *xylem ray,* whereas the extension of the same ray in the phloem is called a *phloem ray*

Receptacle (ree-sep'tuh-kul) the commonly expanded tip of a peduncle or pedicel to which the various parts of a flower (e.g., calyx, corolla) are attached

Recessive (ree-ses'iv) descriptive of a member of a pair of genes whose phenotypic expression is masked or suppressed by the dominant gene

Red tide (red tyd) the marine phenomenon that results in the water becoming temporarily tinged with red due to the sudden proliferation of certain dinoflagellates that produce substances poisonous to animal life and humans; the occurrence of the phenomenon is unpredictable

Reproduction (ree-proh-duk'shun) the development of new individual organisms through either sexual or asexual means

Resin canal (rez'in kuh-nal') a tubular duct of many conifers and some angiosperms that is lined with resin-secreting cells

Respiration (res-puh-ray'shun) cellular breakdown of sugar and other foods, accompanied by release of energy; in aerobic respiration oxygen is utilized

Rhizoid (ry'zoyd) delicate root- or root-hairlike structures of algae, fungi, the gametophytes of bryophytes, and certain structures of a few vascular plants; they function in anchorage and absorption but have no xylem or phloem

Rhizome (ry'zohm) an underground stem, usually horizontally oriented, which may be superficially rootlike in appearance but which has definite nodes and internodes

RNA the standard abbreviation for ribonucleic acid, an important cellular substance that occurs in three forms, all involved in the synthesis of proteins

Root cap (root kap) a thimble-shaped mass of cells at the tip of a growing root; it functions primarily in protection

Root hair (root hair) a delicate protuberance that is part of an epidermal cell of a root; root hairs occur in a zone behind the growing tip

Runner (run'ur) see *stolon*

Glossary

Saprophyte (sap'ruh-fyt)   an organism that obtains its food directly from nonliving organic matter

Sapwood (sap'wood)   outer, usually functional layers of wood in a tree trunk; sapwood is usually lighter in color than heartwood

Science (sy'ints)   a branch of study involved with the observation, recording, organization, and classification of facts, and the establishment of verifiable laws, chiefly through induction and hypotheses

Scion (sy'un)   a segment of plant that is grafted onto a stock

Sclereid (sklair'id)   a sclerenchyma cell with pits; it may vary in shape but usually one axis is not conspicuously longer than the other

Sclerenchyma (skluh-ren'kuh-muh)   tissue composed of cells with thick walls; the tissue functions primarily in strengthening and support

Secondary consumer (sek'-on-dair-ee kon-soo'-mer)   an animal that feeds on other consumers

Secondary tissues (sek'un-der-ee tish'yuz)   tissues produced by the vascular cambium and the cork cambium (e.g., virtually all the xylem and phloem in a tree trunk)

Secretory cell, tissue (see'kruh-tor-ee sel, tish'yu)   cell or tissue producing a substance or substances that are moved outside the cells

Seed (seed)   a mature ovule containing an embryo and bounded by a protective seedcoat

Seedcoat (seed'koht)   the outer boundary layer of a seed; it is developed from the integument(s)

Sepal (see'puhl)   a unit of the calyx that frequently resembles a reduced leaf; sepals often function in protecting the unopened flower bud

Seta (see'tuh)   the stalk of a moss sporophyte

Sexual reproduction (seksh'yule ree-proh-duk'shun)   reproduction involving the union of gametes

Shoot (shoot)   a young twig or branch; also, the exposed portion of a plant

Short-day plant (short-day plant)   a plant in which flowering is initiated when the days are shorter than its critical photoperiod

Sieve cell (siv sel)   a food-conducting cell found in the phloem of gymnosperms, ferns, and other lower vascular plants

Sieve plate (siv playt)   area of the wall of a sieve tube element that contains several to many perforations that permit cytoplasmic connections between similar adjacent cells, the cytoplasmic strands being larger than plasmodesmata

Sieve tube (siv t(y)oob)   column of sieve tube elements arranged end to end; food is conducted from cell to cell through sieve plates

Sieve tube element (siv t(y)oob el'uh-mint)   a single cell of a sieve tube

Simple leaf (sim'pul leef)   a leaf with the blade undivided into leaflets

Solute (sol'yute)   a dissolved substance

Solution (suh-lew'shun)   typically a liquid in which the molecules of a solute are interspersed between the molecules of a solvent

Soredium (pl. soredia) (sor-ee'dee-um; pl. sor-ee'dee-uh)   an asexual reproductive structure of lichens; it consists of several algal cells surrounded by fungal hyphae

Sorus (pl. sori) (sor'uss; pl. sor'eye)   a cluster of sporangia; the term is most frequently applied to clusters of fern sporangia

Species (spee'seez)   the basic unit of classification; a population of individuals capable of interbreeding freely with one another but not with members of another species

Sperm (spurm)   a male gamete; it is usually motile and smaller than the corresponding female gamete

Spice (spyss)   an aromatic organic substance used to season or flavor food or drink

Spindle (spin'dul)   an aggregation of fiberlike threads (microtubules) that appears in cells during mitosis and meiosis; some threads are attached to the centromeres of chromosomes, whereas other threads extend directly or in arcs between two invisible points designated as *poles* (see also *phragmoplast*)

Spine (spyn)   a relatively strong, sharp-pointed woody structure usually located on a stem; it is frequently a modified leaf or stipule

Spongy mesophyll (spun'jee mez'uh-fil)   mesophyll having loosely arranged cells and numerous air spaces; it is generally confined to the lower part of the interior of a leaf just above the lower epidermis

Sporangiophore (spuh-ran'jee-uh-for)   the stalk on which a sporangium is produced

Sporangium (pl. sporangia) (spuh-ran'jee-um; pl. spuh-ran'jee-uh)   a structure in which spores are produced; it may be either unicellular or multicellular

Spore (spor)   a reproductive cell or aggregation of cells capable of developing directly into a gametophyte or other body without uniting with another cell (*Note:* a bacterial spore is not a reproductive cell but is an inactive phase that enables the cell to survive under adverse conditions); *sexual spores* formed as a result of meiosis occurring are often called *meiospores;* spores produced by mitosis may be referred to as *vegetative spores*

Spore mother cell (spor muth'ur sel)   a diploid cell that becomes four haploid spores or nuclei as a result of undergoing meiosis

Sporophyll (spor'uh-fil)   a modified leaf that bears a sporangium or sporangia

Sporophyte (spor'uh-fyt)   the diploid (2*N*) spore-producing phase of the life cycle of an organism exhibiting alternation of generations

Stamen (stay'min)   a pollen-producing structure of a flower; it consists of an anther and usually also a filament

Starch (starch)   an insoluble carbohydrate composed of numerous glucose units; it is the principal food storage substance of plants

Stele (steel)   the central cylinder of tissues in a stem or root; it usually consists primarily of xylem and phloem

Stigma (stig'muh)   the pollen receptive area of a pistil; also, the eyespot of certain motile algae

Stipe (styp) the supporting stalk of mushrooms and other stationary organisms

Stipules (stip′yoolz) paired appendages of varying size, shape, and texture present at the base of the leaves of some plants

Stock (stok) the rooted portion of a plant to which a scion is grafted

Stolon (stoh′lun) a stem that grows horizontally along the surface of the ground; it typically has relatively long internodes; also called a *runner*

Stoma (pl. stomata) (stoh′muh; pl. stoh′mah-tuh) a minute pore or opening in the epidermis of leaves, herbaceous stems, and the sporophytes of hornworts (*Anthoceros*); it is flanked by two guard cells that regulate its opening and closing and thus regulate gas exchange and transpiration; the guard cells and pore are also collectively referred to as a stoma

Strobilus (pl. strobili) (stroh′buh-luss; pl. stroh′buh-leye) an aggregation of sporophylls on a common axis; it usually resembles a cone or is somewhat conelike in appearance

Stroma (stroh′muh) a colorless fluid substance constituting the bulk of the volume of a chloroplast or other plastid; it contains enzymes that in chloroplasts play a key role in photosynthetic reactions

Style (styl) the tissue that connects a stigma and an ovary

Suberin (soo′buh-rin) fatty substance found primarily in the cell walls of cork and the Casparian strips of endodermal cells

Succession (suk-sesh′un) an orderly progression of changes in the composition of a community from the initial development of vegetation to the establishment of a climax community

Superior ovary (soo-peer′ee-or oh′vuh-ree) an ovary that is free from the calyx, corolla, and other floral parts so that the sepals and petals appear to be attached at its base

Syngamy (sin′gam-mee) union of gametes; fertilization

t

Tendril (ten′dril) a slender structure that coils on contact with a support of suitable diameter; it usually is a modified leaf or leaflet, and aids the plant in climbing

Thallus (pl. thalli) (thal′uss; pl. thal′eye) a multicellular plant body that is usually flattened and not organized into roots, stems, or leaves

Thylakoid (thy′luh-koyd) coin-shaped membranes containing chlorophyll and arranged in stacks that form the *grana* of chloroplasts

Tissue (tish′yu) an aggregation of cells having a common function

Tracheid (tray′kee-id) a xylem cell that is tapered at the ends and has thick walls containing pits

Transpiration (trans-puh-ray′shun) loss of water in vapor form; most transpiration takes place through the stomata

Tropism (troh′pizm) response of a plant organ or part to an external stimulus, usually in the direction of the stimulus

Tuber (t(y)oo′bur) a swollen, fleshy underground stem (e.g., white potato)

Turgid (tur′jid) firm or swollen because of internal water pressures resulting from osmosis

Turgor pressure (tur′gur presh′ur) pressure within a cell resulting from the uptake of water

Tylosis (pl. tyloses) (ty-loh′sis; pl. ty-loh′seez) a ballooning of material through a pit into a vessel, thus preventing the vessel from functioning; tyloses are common in heartwood

U

Unisexual (yu-nih-seksh'yu-ul) a term usually applied to a flower lacking either stamens or a pistil

V

Vacuolar membrane (vak-yu-oh'lur mem'brayn) a membrane between the cytoplasm and a vacuole of a cell

Vacuole (vak'yu-ohl) a pocket of cell sap that is separated from the cytoplasm of a cell by a membrane; it may occupy more than 90 percent of a cell's volume in plants; also, food storage or contractile pockets within the cytoplasm of unicellular organisms

Vascular bundle (vas'kyu-lur bun'dul) a strand of tissue composed mostly of xylem and phloem and usually enveloped by a bundle sheath

Vascular cambium (vas'kyu-lur kam'bee-um) a narrow cylindrical sheath of cells that produces secondary xylem and phloem in stems and roots

Vascular plant (vas'kyu-lur plant) a plant having xylem and phloem

Vein (vayn) a term applied to any of the vascular bundles that form a branching network within leaves

Venter (ven'tur) the site of the egg in the enlarged basal portion of an archegonium

Vessel (ves'uhl) one of usually very numerous cylindrical tubes occurring in the xylem of most angiosperms and a few other vascular plants; each vessel is composed of *vessel elements* laid end to end; the perforated or open-ended walls of the vessel elements permit water to pass through freely

Vessel element (ves'uhl el'uh-mint) a single cell of a vessel

Vitamin (vyt'uh-min) a complex organic substance produced primarily by photosynthetic organisms; various vitamins are essential in minute amounts for normal metabolism and growth

Volva (vol'vuh) a cuplike or egglike structure at the base of the stalk of certain mushrooms

W

Water potential (waw'tur puh-ten'shul) the potential of cells to take up water until they are *turgid* (term usually expressed mathematically)

Whorled (wirld) having three or more leaves at a node

Wilting percentage (wilt'ing pur-sent'ij) percentage of soil water not available to the roots; plants wilt when soil moisture is reduced to the wilting percentage

X

Xylem (zy'lim) the tissue through which most of the water and dissolved minerals utilized by a plant are conducted; it consists of several types of cells

Z

Zoospore (zoh'uh-spor) a motile spore occurring in algae and fungi

Zygospore (zy'goh-spor) a zygote that develops a thick protective wall

Zygote (zy'goht) the product of the union of two gametes

Index

a

Aardvark, 24, 218, A1
ABA. *See* Abscisic acid
Abscisic acid, 167, 171
Abscission, 125–26, 171–72, 187, 398
Abscission zone, 125
Absence of dominance. *See* Incomplete dominance
Absinthe, 128, A1
Absorption spectrum, 149
Acetabularia, 284
Acceptor end (of RNA), 44
Acceptor molecule electron, 148, 150
Q, 150
Accessory fruit, 437
Accessory pigments, 147, 237, 265
Acetabularia, 284
Acetic acid, 106, 297
Acetone, 258, 468, A1
Acetyl CoA, 154–55, 158–59
Achene, 437, 442, 458, 465
Achilles, 479
Acid (definition), 17
Acidophilus bacteria, 258, A1
Aconite, 77, 458, A1, A48
Acorn, 315, 442, 497
Active transport, 134, 138, 178
Adaptation, 15, 421, 482, 487, 489, 508, 511
Adder's tongue fern, 35, 380, A1
Adenine, 41, 42
Adenosine diphosphate, 147, 155, 158
Adenosine triphosphate, 147, 150, 153–58, 330
Adirondack Mountains, 401
ADP. *See* Adenosine diphosphate
Advanced flower, 422–25, 457
Adventitious, 433
  buds, 71, 100
  roots, 72, 73, 98, 100, 121, 166, 364, 371, 376, 377, 449, A60–A64
Aechmea, A86
Aecia, 327
Aeciospores, 327
Aerenchyma, 56

Aerobic respiration, 153–56, 450
Aerotaxy, 182
Aerotropism, 179
Afghanistan pine, 403, A1
African lily, A86
African sausage tree, 182, A1
African violet, A1, A60, A86
After-ripening, 449
Agar, 167, 295, 298, A1
Agaricus campestris, 330
Agave, 126, 127, 412, A1, A86
Agent Orange, 169
Age of Ferns, 381
Aggregate fruit, 437, 439, 465
Ahnfeltia, 292
Air layering, A62, A63
Air plant, 121, A1
Ajuca, A51
Akinete, 262, 263, 281
Albrightia, 260
Albuminous cell, 60, 392
Alcohol, 5, 19, 25, 77, 295, 315, 317, 318, 328, 333, 468, 480. *See also* Ethyl alcohol, Industrial alcohol
Alder, 76, A1, A55
Ale, 383
Aleppo pine, 403, A1
Alfalfa, 63, 128, 296, 456, 463, 466, 467, 493, A1
Alfalfa caterpillars, 257, A1, A28
Algae, 167, 180, 183, 233–37, 275–99, 335, 384, 491, 498, A1. *See also* Blue-green algae, Brown algae, *etc.*
Algae, comparison of divisions of, 295
Algae, human and ecological relevance of, 296–99
Algal bloom, 263, 285
Algerian ivy, A86
Algin, 290, 296, 297
Alkaline compound. *See* Base
Alkaloids (alkaloidal drugs), 96, 220, A51
Alkanet, A55
Alligator weed, A30
Allium cepa, 235
Allspice, A52
Almond, 434, 435, 442, 456, A1, A52

Aloe, 127, 481, A1, A45, A86
Alternation of generations, 201, 202, 346, 356
Aluminum plant, A86
Amaranth, A37
Amaryllis, A1, A43, A64
Amaryllis family (Amaryllidaceae), 451, 480, 481, A86
Ama'uma'u, plate, 4D
Amazon Indians, 96
Amber, 403, 404
Ambrose, Howe, 356
American holly, 449
American mountain ash, A45
Amino acid, 20, 44–46, 144, 152, 258, 450, 494
Amino group, 20
Ammonia, 251, 252, 494, A1
Amoeba, 232, 261, 300, A1
Amoeboid movement, 275, 300, 308
Amphibian, 490
Amphibious fern, 380, A2
Amphibious liverwort, 350
Amylase, 159
Amyloplast, 34
Anabaena, 260, 262, 263, A2
Anaerobic respiration, 155, 156, 254, 450
Anaphase, 37, 40, 198–201
Anemone, 458, A2, A45
Anemone, sea, 261
Anesthetic, 315, A46
Angelica, 77, A2, A52
Angel's trumpet, A42
Angiosperms, 412–85
Angstrom (unit), A97
Aniline dyes, 458
Anise, 128, 443, 456, 475, A52
Anisette, 475
Annatto, 457, A2, A55
Annual, 91, 414, 503, 507
Annual mosses. *See* Pygmy mosses
Annual ring, plate 1, 69, 92–94, 392, 499
Annulus, 321, 322, 380
Ant, 78, 79, 121, 331, 444–46, A2, A29, A32
Anther, 416–18, 420, 460
Antheridiophore, 348, 349
Antheridium, 281, 394
  of ferns, 377, 380, 381
  of *Fucus*, 291
  of fungi, 316

of ground pines (*Lycopodium*), 364, 365
of hornworts, 351
of horsetails (*Equisetum*), 372, 373
of liverworts, 348, 349
of mosses, 353–55
of *Oedogonium*, 281, 282
of *Psilotum*, 363
of quillworts (*Isoetes*), 368
of spike mosses (*Selaginella*), 366, 367
of stoneworts, 285
of water molds, 310, 311
of whisk ferns (*Psilotum*), 363
Anthocerotae, 351, 352
Anthocyanin, 36, 126
Anthophyta, 238, 412–85
Anthrax, 255, 256, A2
Anthurium, 382, A87
Antibiotic, 255, 269, 296, 299, 332, 338, A2, A46, A48
Anticodon, 44, 45
Anticodon loop, 44
Antidote. *See* Poison antidote
Antipodals, 419, 420, 422
Antler and bone mosses, A2
Apertures (pollen grain), 412, 420, 421
Aphanocapsa, 262
Aphelandra, A87
Aphid, 113, 137, 381, A2, A27, A28, A32, A84
Aphrodisiac, A50
Apiaceae. *See* Carrot family
Apical dominance, 170
Apical meristem, 38, 51, 52, 58, 65, 66, 87, 88, 98
Aplanospore, 279, 283, 290
Apomixis, 220, 433
Apple, 14, 24, 172, 390, 424, 436, 449, 456, 465, 491, A2, A32, A66
Apple rust, 328, A2
Apricot, 434, 455, 465, A2
  silver, 398
Aquatic (floating) ferns, A2
Aralia, A87
Arborvitae, 407, A2
Archegoniophore, 348, 349

Archegonium
  of ferns, 377, 380
  of ground pines
    (*Lycopodium*),
    364, 365
  of gymnosperms, 393–95
  of hornworts, 351
  of horsetails (*Equisetum*),
    372, 373
  of liverworts, 348, 349
  of mosses, 353–55
  of *Psilotum*, 363
  of quillworts (*Isoetes*),
    368
  of spike mosses
    (*Selaginella*), 366,
    367
  of whisk ferns (*Psilotum*),
    363
Arctic fox, 502
Arctic tern, 502
Arcyria, 301
Argentine moth, 470, A2
Aril, 396, 407, A53
Aristotle, 7
Army worm, A28
Arnica, A45
Arrow grass, A37
Arrowhead, A37
Arrowroot, 407, A2, A52
Arroyo, Padre, 403
Arthropod, 308
Arthrospira, 260
Artichoke, globe, 433, A2
Artichoke, Jerusalem, 186,
  478, A2
Asafoetida, A52
Asarone, β-asarone, A51
Ascocarp, 316, 317, 338
Ascogonium, 316
Ascomycetes, 315–19. *See
  also* Cup fungi
Ascorbic acid, 258
Ascospores, 316, 317
Ascus, 316, 317
Asexual reproduction, 193,
  245. *See also specific
  organisms*
Ash, 95, 126, 170, 442, 443,
  504, A2
Ash, American mountain, A45
Ash, blue, 126, A2
Ash, flowering, A46
Ash, white, 102
Asian climbing fern, 383, A2
Asian coneflower, 182
Asparagus, 100, 254, 480,
  A33, A77
Asparagus fern, 254, A87
Aspartate, 152
Aspen, 126, 499, 503, 508,
  A2
Aspergilloses, 333, 334
Aspergillus, 333, 334
Aspidistra, A87
Assimilation, 15, 159, 160
Aster, 184, 504, A2
Aster (of animal cells), 39

Asteraceae. *See* Sunflower
  family
Asthma, 318, 407, A45–A50
Astringent, 375, A2, A45
Athlete's foot, 333, A2, A50
Atlantic white cedar, 407, A2
Atole, 470
Atom, 16, 17, 18, 19, 20, 21,
  22
Atomic numbers, 16
ATP. *See* Adenosine
  triphosphate
Atropine, 127, 474, A46, A47
Attar of roses, 465
Aucuba, A87
Australian baobab, 95, A2
Author (of scientific name),
  235
Autoclave, A64
Autotrophic, 249
Autotrophic bacteria, 249
Autumn crocus, 220, A43.
  See Crocus
Auxins, 167–69, 176, 178,
  188, A61
  functions of, 167–69
Auxospores, 287
Available water, 81
Avocado, 434, 461, A2, A53,
  A87
Axil, 86, 100, 351, 366, 400
Axillary bud, 69, 86, 87, 98,
  99, 100, 170
Azalea, A2, A42, A87

b

Baby blue eyes, plate 8B, A2
Bacillariophyceae, 286
Bacilli, 244, 245
Bacillosiphon, 260
Bacillus, 260, 268
Bacillus popilliae, 258, A32
Bacillus thuringiensis, 257,
  A28, A83
Bacillus thuringiensis var.
  israelensis, 258, A32
Backcross, 212
Bacteria, 15, 25, 26, 76, 78,
  79, 81, 124, 125, 143,
  156, 157, 242–59,
  283, 298, 300, 303,
  308, 332, 346, 384,
  406, 438, 449, 489,
  491, 493, 495, 499,
  A2, A3, A29, A30. *For
  comprehensive list of
  individual bacteria see
  also pp. A2-A3*
  anaerobic, 254, 258
  autotrophic, 249
  and blue-green algae
    compared, 260
  chemoautotrophic, 249
  chemosynthetic, 242
  chlorophyll in, 249
  conjugation, 245–47
  decay, 252, 493, 494
  as decomposers, 251–53
  denitrifying, 494

diphtheria, 253
and disease, 243, 253–56
DNA of, 243, 245–49
green sulfur, 249
heterotrophic, 249
hydrogen, 251
iron, 251
luminescent, 242, 243
nitrifying, 251, 494
nitrogen-fixing, 250, 251,
  263, 269, 493,
  494, A18, A77
nutrition and metabolism,
  249–51
pathogenic, 243, 253–56,
  A32
photosynthetic, 249
plasmid, 245, 248
pseudomonad, 319
purple nonsulfur, 249
purple sulfur, 249
reproduction, 245–47
saprobic, 249
scientific names of, 243,
  A2, A3
size, 243
spores, 245, 254
sulfur, 156, 251
toxins of, 254
transduction, 247
transformation, 247
tuberculosis, 253
useful, 257–59, 493, 494
Bacteriochlorophyll, 249
Bacteriophage, 267, 268, 269
Bald cypress, 71, 102, 171,
  406, A3
Balm (melissa), A52
Balm of Gilead, A45
Balsa, 101, A3
Balsam, 406, A3
Balsam poplar, A45
Balsamroot, 479, A3, A37
Baltic amber, 404
Bamboo, 298, 455, A3, A55
Bamboo, dwarf, A87
Bamboo, giant, 181, 456
Banana, 126, 172, 437, 456,
  A3
Baneberry, A42
Banner (legume flower), 466,
  467
Banyan tree, 72, 73, 177, A3
Baobab, Australian, 95, A3
Barbasco, 77, A3
Barberry, 118, 182, 183, A3,
  A55
Barberry, common, 119, 326,
  327, A4
Bard, 473
Bark, 51, 96, 382, 392, 402,
  406, 407, 460
  inner, 96, 401, 407
  outer, 51, 96
Barley, 6, 64, 98, 442, 456,
  480, A4
Barn swallow, 344, A4
Barrel cactus, 470
Barwood, A55
Base (definition), 17

Basidiocarp, 320–24
Basidiomycetes, 320–32. *See
  also* Club fungi
Basidiospores, 320–28
Basidium, 320–22
Basil, 126, 456, 472, A4, A30,
  A33, A52
Basswood, plate 1, 57, 94,
  101, 504, A4, A37
Bat, 187, 421, 511, A4
Bateson, W. and R. C.
  Punnett, 214
Bay, A52
Bay, California, 460, A4
Bay, sweet, 460, A4
Beagle, H. M. S., 217
Beal, William J., 451
Beans, 76, 169, 254, 421,
  438, 455, 457, 466,
  467, 493, A33, A77.
  See also Garden
  bean, Lima bean, *etc.*
Bear, 374, 509, A4
Bearberry, 126, A4, A55
Bear, polar, 502, A4
Beaver, mountain, 509, A4
Bedrock, 78
Bedstraw, 112, 444, A4, A37,
  A55
Bee, 206, 221, 415, 416, 509.
  See also Honeybee
Beech, 95, 497, 504, A4
Beech fern, 380, A4
Beechnuts, A37
Beefsteak morel, 317, A4
Beer, 6, 296, 297, 318, 405,
  460, 473, A53
Beetle, 309, 328, A4, A28,
  A29
  buprestid, A30
  crypts, A32
  elm bark, 319
  flea, A30, A31
  flowers, 510
  Japanese, 258, A28, A29,
    A32
  leaf, A30
  snout, A31
  tortoise, A30, A31
  wood-boring, 331, A30
Beets, 70, 76, 113, 254, 433,
  456, A33, A78
  garden, 185, A4
  sugar, 76, 296, A4
Begonia, A4, A60, A87
Begonia, Rex, A60
Belladonna, 127, 474, A4,
  A42, A46, A51
Bell pepper, A52
Beneficial insects, A27, A28
Beneficial nematodes, A29
Bergamot, A52
Bermuda grass, 178, A4
Berry, 407, 434, 435, 469, 474
  false, 436, 437
  true, 434
Betacyanins, 126
Betel nut, 455, A4
Biennial, 76, 187, 415

Index

Big leaf maple, A39
Big tree. *See* Giant redwood
Bilateral symmetry (of
    flowers), 422, 425,
    466, 470
Bilbergia, A87
Binomials, 231, 232, 234
Binomial System of
    Nomenclature,
    230–34
Biological clocks, 180
Biological controls, 71, 257,
    258, 473, 492,
    A27–A35
Biomass, 487, 491
Biome, 501–9
Biotic community, 487, 489,
    501
Birch, 90, 135, 503, 508, A4,
    A55
Birdcatcher plant, A87
Bird of paradise plant, A87
Birds, 79, 187, 253, 257, 344,
    346, 367, 407, 415,
    421, 444, 490, 498,
    503, 504, 506, 507,
    509, 511. *See also*
    Barn swallow, Junco,
    etc.
Bird's foot fern, 383, A4
Bird's nest fern, 376, A4, A87
Bird's-nest fungus, 320, 324,
    325, A4
Birth control pills, 315
Biscuit root, A37
Bitterroot, A37
Bittersweet, 91, A4, A43
Bivalent, 196
Blackberry, 220, 434, 437,
    465, 466, A4, A38,
    A45, A60, A61, A74
Blackbird, 444, A4
Black bread mold (*Rhizopus*),
    312, 313, A4
Black cherry, A42, A55
Black cohosh, A45
Black Death. *See* Bubonic
    plague
Black-eyed Susan, A87
Black Forest mushroom, 329,
    330, A4
Black haw, A45
Black locust, 100, 101, 466,
    A4, A42
Black mangrove, 71, A4
Black mustard, 456
Black pepper, 455, A19, A52
Black rat snake, 504, A4
Black snakeroot, A42
Black spruce, 405, A4
Black stem rust of wheat,
    326, 327, A4
Black walnut, 102, 112, 491,
    A4, A38, A55
Bladder
    of bladderwort, 123, 124
    of kelp, 275, 289, 290
Bladder campion, A30, A38

Bladderwort, 15, 123, 124, A4
Blade
    algal, 275, 284, 285, 289,
      290
    leaf, 86, 87, 114, 378, 379
Blazing star, plate 5A, A4
Bleeding heart, 444, A4
Bleeding heart, Pacific, 445,
    A5, A43
Bletilla orchid, 441, A19
Blight
    chestnut, 319
    late, of potato, 297, 312
Bloodleaf, A87
Bloodroot, *cover*, 461, 504,
    A5, A43, A45, A55
Bloom, algal, 263, 285
Blowouts, 502
Blue ash, 126, A2
Blueberry, 436, 487, A5, A38,
    A55
Blue cohosh, A43, A45
Blue curls, 473, A5
Blue-green algae, 15, 183,
    259–63, 296, 335,
    351, 384, 499, A5
    Anabaena, 260, 262, 263
    and bacteria compared,
      260
    human and ecological
      relevance of, 263
    Nostoc, 262, 263
    Oscillatoria, 183, 262, 263
Blue jay, 444, A5
Bobcat, 509, A5
Bog orchid, 514
Boletus, 323
Bollworms, 257, A5, A28, A29
Bolting, 185
Bond, 16, 17
Bond, peptide, 20
Bonding, 16, 17
Boneset, A45
Bonsai, A75
Booza, 6
Borage, A33, A52
Bordeaux mixture, 310, 312
Bordered pit, 46
Boreal forest. *See* Taiga
Boston fern, 381, A5, A87
Boston ivy, 73, 100
Botanical garden, 6
Botany, 6
Botryocladia, 292
Botulism, 254, 265, A5
Bougainvillea, A55
Bowstring fibers, 126, A5
Bowstring hemp, 455, 480,
    481, A5
Box elder, A4, A39
Box family (Buxaceae), 469
Boxwood, A43
Boysenberry, A5, A63
Bracken fern, 382, 383, 500,
    A5, A38
Bracket fungus, plate 3C,
    300, 320, 324, A5
Bract, 54, 121, 122, 442, 467
Brassicaceae. *See* Mustard
    family

Brazil nut, 442, 457, A5
Brazil peppertree, A30
Brazilwood, A55
Breadfruit, 456, A5
Bread molds, 312–14
Breil, David, 344
Bridalwreath, 416, A5
Bridges' brodiaea, plate 7D,
    A5
Bristlecone pine, 391, A5
Bristlecone pine, Colorado,
    A5
British Columbia Indians, 407
Broad bean, 456, 466, A5,
    A78
Broad-leaved plantain, 229
Broccoli, 440, 463, A5, A6,
    A78
Bromeliad, A88
Bromeliad family
    (Bromeliaceae), 412,
    A5
Broomrape, 74, 413, A5, A38
Broom snakeweed, A45
Brown, Robert, 25
Brown algae, 288–91, 295,
    344, A5
Brown fruit rot, 315, A5
Brownian movement, 131
Brucellosis, 255, A5
Brunfels, Otto, 8
Brushweed, A30
Brussels sprouts, 463, A5
Bryophyta, 236, 237, 343–59,
    A5
Bryophytes, 343–59
    human and ecological
      relevance of, 356,
      357
BTI. *See Bacillus
    thuringiensis* var.
    israelensis
Bubonic plague, 255, 256, A5
Buchner, Eduard, 25
Bucket orchid, 513, A5
Buckeye, 112, 504, A5, A42
Buckthorn, A43, A55
Buckwheat, 442, A5, A55
Buckwheat, wild, A56
Bud
    axillary, 69, 86, 87,
      98–100, 170, 171
    development, 86
    dormancy, 169, 171, 186
    flower, 415, 433
    gap, 88
    primordia, 88, 171
    scale, 86, 87
    terminal, 86, 87
    yeast, 315
Budding, 315, A71. *See also*
    Bud grafting
Bud grafting, A59, A71, A72
Buffalo, 505, A5
Buffaloberry, A56
Bulb, 73, 99, 119, 177, 413,
    480–82, 507, A63,
    A64, A76
Bulbil, 364
Bulblet, 100, 221, A64

Bulliform cells, 116, 181
Bull thistle, A31
Bulrush (tule), A38
Bumblebee, 511, 514, A28
Bundle, vascular, 90, 91, 97,
    98
Bundle scar, 86, 87
Bundle sheath, 97, 116, 152
Buprestid beetle, A30
Bur clover, 444, A5
Burdock, A45
Burnet, A52
Bur oak, 416, A5
Bushman's poison, A43
Butcher's broom, 100, A5
Buttercup, 68, 442, 458, A5,
    A43
Buttercup, European
    bulbous, 458, A5
Buttercup family
    (Ranunculaceae), 443,
    458, 459, A5
Butterfly, 257, 510–12, A5,
    A28
Butterfly iris, A5
Buttermilk, 258, A5
Butternut, A38, A56
Butterwort, 124, A5
Button (stage of mushroom),
    321, 322, 331
Buttonbush, A43
Button snakeroot, 443, A5,
    A45
Buttress roots, 73, 74
Butyl alcohol, 258, A5
Buxaceae. *See* Box family

C

C₃ pathway (of
    photosynthesis),
    149–52
C₄ pathway (of
    photosynthesis), 152
Caapi, A51
Cabbage, 126, 169, 206, 253,
    433, 440, 456, 463,
    A6, A33, A79
Cabbage family. *See* Mustard
    family—Brassicaceae
Cabbage loopers, 257, A6,
    A28
Cabbage moth, A34
Cabbage worms, 257, A6,
    A29, A34
Cacao, 457, A6
Cactaceae. *See* Cactus family
Cactus, 56, 63, 100, 116, 126,
    152, 467, 469, 470,
    497, 507, 511, A6,
    A88. *See also* Barrel
    cactus, Cholla cactus,
    etc.
Cactus family (Cactaceae),
    469, A6
Caffeine, 473
Cajeput, 126, A6
Calabazilla, 70, A6

Caladium, A42, A88
Calamites, 375
Calamus, A52
Calceolaria, A88
Calcium propionate, 312
Calendula, 185, A6
California bay, 460, A6, A18, A45
California fuchsia, 511, A6
California nutmeg, 396, A6
California poppy, plate 7E, 450, A6
California soaproot, 481, 482, A6
Calla lily, A88
Callose, 14
Callus, 15, A64, A66
Calvin, Melvin, 147, 469
Calvin cycle, 147, 151, 152
Calyptra, 349, 350, 354, 355
Calyx, 415, 416, 425, 433, 437
Cambium, 38, 51, 91, 94, 394, 401, 407, A66. *See also* Vascular cambium, Cork cambium
Camel, 506, A6
Camelina, 464, A6
Camellia, 57, A6, A60
Camphor, 126, 460, A6, A45
CAM photosynthesis. *See* Crassulacean acid metabolism photosynthesis
Camptotheca, A45
Camptothecin, A45
Canada balsam, 406
Canada thistle, 71, A31
Canary broom, A51
Cancer, 265, A45
Canchalagua, plate 8C, A6
Candelilla, 469, A6
Candlenut, 456, A6
Candolle, A. P. de, 186, 455
Cankerworm, 257, A6
Canned fruit fungi, 315
Cannel, 375
Cantaloupe, 332, 425, 436, 451, 456, 476, 477, A6
Cap (mushroom), 320–24
Cape jasmine, A88
Capers, A52
Capillarity, 136
Capillary water, 81
Capillitium, 302
Capsaicin, A19
Capsicum, 474, A19
Capsule
    fruit, 440, 441–46, 462, 474
    of liverworts, 349, 350
    of mosses, 354, 355
Cap-thrower fungus. *See* Pilobolus
Caraway, 443, 456, 475, 476, A6, A38, A52

Carbohydrate (defined), 18, G3
Carbon cycle, 495
Carbon dioxide, role in photosynthesis, 145
Carboniferous period, 369, 374, 381, 391
Carbon monoxide, 283, 290
Carboxyl group, 20
Cardamon, 455, A6, A52
Caribou, 338, 489, 502, A6
Carinal canals, 371, 375
Carnation, 185, A6
Carnaubalike wax, 480, A6
Carnauba wax, 52, 128, A6
Carnivores, 345, 490
Carob, 438, 456, 466, A6
Carolina jessamine, A42
Carotene, 125, 126, 295. *See also* Carotenoids
Carotenoids, 34, 36, 125, 150, 261, 344
Carpel, 423, 425, 434, 440, 446, 476
Carpetweed family (Aizoaceae), 121, A6
Carpogonium, 293–95
Carposporangium, 294, 295
Carpospores, 294, 295
Carrageenin, 298
Carrion flower, 510, A88
Carrot, 70, 76, 133, 253, 443, 456, 475, A6, A33, A66, A76, A79
Carrot family (Apiaceae), 443, 475, 476
Caruncle, 450
Caryopsis, 442
Cascara, A45
Casebearer, A30
Cashew, 442, 457, A6
Casparian strip, 67
Cassava, 76, 77, 457, 467, 468, A6
Cassia, 460, A6, A52
Cassia, A6
Cassie, A6
Cast, 385
Castor bean, 450, 469, A6, A42
Cat, 253, 328
Catalpa, 101, 444, A6
Catalyst, 20, 25, 165
Catawba Indians, A46
Caterpillar nematode, A29
Caterpillars, 92, 257, 331, 470, 489, A6, A28
Catkin, 416, 417
Catnip, 472, A6, A33, A45, A51
Cattail, 498, A6, A38
Cattle, 263, 466, 472, 489
Cattleya, A92
Cauliflower, 463, A6, A79
Caussu wax, 128, A6
Cayenne pepper, A45, A52
Cedar, 101, 328, 391. *See also* Atlantic white cedar, Incense cedar, etc.

Cedar waxwing, 344, A6
Celery, 126, 133, 456, 475, A6, A33, A53
Cell, 14, 24–47, 165, 170, 249
    asters, 39
    centrioles, 36, 39
    communication, 40, 41
    companion, 59, 60, 97
    cycle, 36–38
    differentiation, 165, 170
    division, 39, 40, 170
    elongation, 450
    equator, 39, 40
    eukaryotic, 36, 233, 236, 245
    lumen, 56, 57
    membrane, 19, 36, 165
    meristematic, 51
    nucleus, 25, 34, 35, 51
    organelles, 26
    phragmoplast, 40, 344
    plant vs. animal, 36
    plate, 40, 344
    poles, 39, 40
    prokaryotic, 36, 233, 236, 241, 259
    sap, 35, 36
    size, 26
    spindle fibers, 34, 39, 40
    structure, 26–29
    theory, 25
    transfer, 56
    wall, primary, 28, 29, 40, 46, 56
    wall, secondary, 29, 40, 46, 56
Cell biology, 10, 26, 40
Cell membrane, 35, 175. *See also* Plasma membrane
Cells, differences between plant and animal, 36
Cellular reproduction, 36–40
Cellular slime mold, 303, A6
Cellulose, 18, 29, 40, 101, 134, 152, 159, 276, 278, 308, 344
Cenozoic era, 423
Central canal (of horsetail stems), 371
Centrioles, 36, 39
Centromere, 39, 40, 194–201
Century plant. *See* Agave
Ceratium, 288
Cesalpino, Andrea, 330
Chain fern, 383, A6
Chamaesiphon, 262
Chamberland, Charles, 265
Chamise, 499, 500, A6
Chamomile, A33
Chaparral, 499
Chard, 126, A6
Chargraff, E., 42
Charophyta, 237, 285
Cheese, 332, A6, A7
Cheilanthes, 379
Chelates, iron, 175
Chemical and physical bases of life, 16
Chemoautotrophic bacteria, 249

Chemosynthetic bacteria, 242
Chemotaxy, 182
Chemotropism, 179
Cherry, 71, 90, 187, 424, 434, 456, 465, A7
Cherry, wild, A45
Cherry tomato, 457
Chervil, A33, A53
Chestnut, 442, A7
Chestnut, American, 319, 504, A7
Chestnut blight, 319, 504, A7
Cheyenne Indians, A48
Chia, 473, A7, A46
Chiasma, 197
Chickadee, mountain, 509, A7
Chickasaw Indians, A50
Chicken pox, 265, 269
Chickweed, Himalayan, 414, A7
Chicle, 96, 333, A7
Chicory, 478, 479, A7, A38
Chiggers, A48, A49
Chilghoza pine, 401, A7
Chinaberry, A42
Chincherinchee, A43
Chinese agriculture, 6
Chinese evergreen, A88
Chinese vegetable tallow, 469, A7
Chipmunk, 509, A7
Chitin, 308
Chives, 451, 456, A33, A53
Chlamydias, 253
Chlamydomonas, 276–78, A7
Chlorella, 283, 297, A7
Chlorenchyma, 56, 115, 135, 348, 371
Chlorine, 173, 333, A7
Chlorobium chlorophyll, 249
Chlorophyll, 33, 74, 125, 146, 147, 491
    a, 147, 149, 150, 249, 260, 261, 288, 290, 295, 344
    and photosynthesis, 146, 147, 149–53
    b, 147, 150, 295, 344
    bacterio-, 249
    c, 147, 287, 288, 290, 295
    chlorobium, 249
    d, 147, 295
    e, 147, 295
    pigments, 146, 147
Chlorophyta, 237, 275–85, 295. *See also* Green algae
Chloroplast, 27, 28, 32, 33, 144, 145, 147, 152, 236, 261, 265, 276, 277, 280, 288, 346, 348, 351, 356
Chocolate, 319, 457, 466, A7, A53
Chokecherry, western black, 466, A7, A38
Cholera, 254, A7
Cholesterol, 330

Cholla cactus, 470, A7
Christmas flower, 121, 122, 186, 469, A7
Chromatid, 39, 40, 196
Chromatin, 35
Chromoplast, 32
Chromosome, 35, 39, 40, 165, 193–202, 215, 219, 220, 245, 248, 433, 481, A48, A59
  centromeres, 39
  chromatids, 39, 40, 196
  deletion, 219
  homologous, 194–201
  inversion, 219
  mapping, 214, 215
  satellites, 39
  translocation, 219
Chrysanthemum, 5, 170, 185, A7, A88
Chrysolaminarin, 295
Chrysophyta, 286, 287, 295. See also Golden-brown algae
Chuckwalla, 507, A7
Chufa, 456, A7
Chytridiomycetes, 309, 310
Chytrids, 309, 310, A7
Cibotium, 379
Cider, 318
Cilantro, A53
Cilia, 175
Cinchona, A46
Cineraria, A88
Cinnabar moth, A31
Cinnamon, 460, A7, A52, A53
Cinnamon fern, plate 4A, A7
Circadian rhythm, 180
Citric acid, 17, 158, 258, 333, A7
Citric acid cycle. See Krebs cycle
Citronella oil, 126, 480, A7
Citrus, 57, 172, 173, 253, 433, 449, 455, A7, A29, A32, A52, A53, A60, A66
Citrus family (Rutaceae), 436, A7
Cladophyll, 99, 100
Clamp connection, 321
Class, 231, 234
Classification, 228–40, 422–25
Clay, 80
Clematis, 118, 458, A7
Cliff brake, A88
Climax vegetation, 497–99, 503
Climbing fern, Asian, 383, A7
Climbing plants, 73, 100, 118
Clinostat, 179
Clock, Linnaeus' garden, 180
Closed system, 5, 283
Closterium, 283
Clostridium botulinum, 254
Clover, 188, 250, 456, 466, 467, 493

Clover, bur, 444, A7
Clover, sweet, 466, A7
Cloves, 456, A7
Club fungi, 320–32, 335, A7
  human and ecological relevance of, 328–32
Club mosses, 352, 361, 362–67, 369, 375, A7, A46
  human and ecological relevance of, 353
CoA. See Coenzyme A
Coal, 4, 98, 153, 375, 495
Coal Age forest, 374
Coastal redwood, plate 2D, 95, 370, 397, 406, 508, A7
Cobra plant, 124, A7
Coca, 127, A46
Cocaine, 127, 457, A7, A46
Cocci, 244, 245
Cochineal insects, 470, A7, A30
Cocklebur, 135, 188, 444, A7, A56
Cockroach, 128, A7
Cockroach plant, 128, A7
Cockscomb, 185, A7
Cocoa beans, 258
Coconut, 170, 407, 434, 442, 456
Coconut palm, 390, 446, A7
Codeine, 462, A48
Codling moth, A7, A29
Codon, 44
Coenocytic, 310–14
Coenzyme A, 154, 155, 158
Coevolution, 509
Coffee, 77, 188, 258, 330, 457, 479, A7, A53, A56, A89
Coffee, wild, A38, A53
Cohesion-tension-transpiration theory, 136, 137
Cohoba, A51
Colchicine, 220, 481, A48
Cold, common, 265, 269, 466, 473, A45–A50
Coleoptile, 166, 167, 179, 448, 449
Coleorhiza, 166, 448, 449
Coleus, 88, 178, 473, A7, A89
Collenchyma, 56
Colloid, 80, 134
Colony, 242, 261, 275, 276
Colorado bristlecone pine, A5
Columbine, 438, 458, 459, A7
Columbus, 9
Column, 482
Common chickweed, A38
Common name, 229–40
Communication, cellular, 40, 41
Communication between cells, 46
Community, 487–91, 507
Companion cell, 59, 60, 97, 392

Companion planting, A33–A35
Compass plant, 117, A7
Compost, 251–53, 331
Compost pile, 251–53, 383
Compound (definition), 17
Compound leaf, 111–13, 407
Compound ovary, pistil, 423
Compression fossil, 385
Conceptacle, 291
Condiments, 4, A19. See also specific entries
Conducting tissue. See Phloem, Xylem
Cone, 390, 393–97, 401
Coneflower, 509, A7
Coneflower, Asian, 182, A7
Conidiophore, 315, 334
Conidium, 315, 317, 320, 331
Conifer, 68, 384, 390–97, 401–7, 497, 498, 503, 508
Coniferophyta, 238, 390–97. See also Gymnosperms
Conjugation, 280, 283
Conjugation tube, 279, 280
Consumers, 490, 491, 509
  primary, 489–91
  secondary, 489–91
Contact movements, 181, 182
Contractile roots, 73, 177
Contraction movements, 177
Cooperage, 106
Copal, 466, A7
Copeland, H. F., 233
Copperhead, 504, A7
Copperleaf, A89
Coprolites, 386
Coral bean, A51
Coralline algae, 292, A7
Coral tree, A7
Cordage, 98, 126, A7
Cordilleran forest, 501, 508, 509
Coriander, 475, A7, A53
Cork, 24, 53, 55, 69, 89, 91, 392
Cork cambium, 38, 51, 53, 68, 89, 91, 412
Cork oak, 53, A7
Corm, cormlet, 99, 100, 220, 367, 370
Corn, 4, 9, 31, 66, 72, 98, 135, 170, 215, 254, 401, 421, 442, 448, 449, 451, 457, 477, 480, 505, A7, A33, A79
Corn borers, 257, A28
Corn earworm, A29
Corn lily, A38
Corn plant (houseplant), A89
Corn rust, 328, A7
Corn smut, 326, A7
Corolla, 415, 416, 424, 437
Correns, Carl, 206
Cortex, 65, 67, 76, 88, 180, 335, 336, 362, 371, 407

Cortisone, A47, A49
Corydalis, pale, plate 6C, A8
Corymb, 417
Cosmic radiation, 219
Costa, 352
Cotton, 185, 253, 456, 457, A8, A46
Cottonseed oil, 401
Cottontail, 506
Cottonwood, 101, 443, 451, A8
Cotyledon, 90, 165, 166, 412, 447–49
Coulter pine, 393, A8
Cousteau, Jacques, 242
Covalent bond, 17
Cow, 333, 374, 489, A8. See also Cattle
Cow parsnip, 476, A8, A38
Cowpea, A38
Cowpox, 265
Cowslip, 458, A8
Cox, Jeff, A29
Crab, 308
Crabapple, 7, 100, A8, A38
Cranberry, 436, A8
Crassulacean acid metabolism photosynthesis, 139, 152, 153, 506, 507
Cree Indians, A48
Creosote bush, A46
Cress, 457, 463, A8
Crested wheat grass, 198, 199, A8
Cretaceous period, 369
Crick, Francis H. C., 41–43
Cristae, 30, 31
Crocus, 100, 367, A8
Crocus, autumn, 220, A8, A43
Cross
  dihybrid, 212–14
  monohybrid, 211, 212
Crossing over, 197, 214, 215
Crossovers, 197, 198, 214, 215
Croton, A89
Crowberry, A38
Crown-of-thorns, 469, A8, A43, A89
Crozier, 376, 378
Cruciferae. See Brassicaceae
Crustacean, 263, 490, 491, A8
Crustose lichen. See Lichens, crustose
Crypts, A32
Cubbra borrachera, A51
Cubebs, A46, A53
Cucumber, 100, 118, 425, 433, 436, 451, 455, 477, A8, A33, A79, A80
Cucumber, squirting, 183, A8
Cucurbitaceae. See Pumpkin family

Cudbear (archil), A56
Cultivated plants, origin of, 455–57
Culver's root, A43
Cumin, A53
Cup fungi, 315–19, 335, A8
  human and ecological relevance of, 317–19
Curculios, A34
Curly dock, 451, A30
Currant, 327
Curry, A53, A55
Curse, A30
Cutch, A56
Cuticle, 52, 53, 67, 87, 113, 114, 121, 344, 391, 507, A29
Cutin, 52
Cuttings, 168, A59–A61
  leaf, A60, A61
  root, A61
Cutworms, A28, A76
Cyanophyceae, 259–63
Cyanophycean starch, 261
Cyathea, 379
Cyathium, 467, 468
Cycad, 261, 391, 397, 407, A8
Cycadeoid, 397
Cycadophyta, 238, 391
Cyclamen, 185, A8, A89
Cycle
  carbon, 495
  cell, 36–38
  nitrogen, 76, 493, 494
Cyclic photophosphorylation, 151
Cyclosis, 36, 137
Cyme, 417
Cypress, bald, 71, 102, 171, 406, A8
Cypress spurge, A30
Cystocarp, 294, 295
Cystopteris, 379
Cytisine, A51
Cytochrome c, 21
Cytochromes, 150, 155, 158
Cytokinesis, 37, 38
Cytokinins, 167, 170, 171, 188, A61
Cytology, 10
Cytoplasm, 27, 28, 145, 165, 421
Cytoplasmic streaming. See Cyclosis
Cytosine, 41, 42

d

Daffodil, 481, A8, A43, A64
Dahlia, 185, 478, A8, A64
Daisy, 220, 478, A8
Dalmatian toadflax, A30
Damask rose, 465, A8
Dammar, 403
Dandelion, 64, 70, 73, 96, 169, 185, 220, 443, 444, 478, 479, A8, A38

Dandruff, 383, A8, A50
Daphne, A42
Dark reactions, 147, 148
Darwin, Charles, 10, 167, 176, 216–18
Darwin, Francis, 167
Darwin, Robert, 217
Date, 6, 434
Date palm, 6, A8
Davallia, 379
Dawn redwood, 406, A8
Day-neutral plant, 185
DDT, 144, 299, 319
DDT-like compound, 299, A8
Deadly nightshade, 127, A46
Death angel, 321, 323, 328, A8
Death camas. See Black snakeroot
Decay bacteria, 252, A8
Deciduous plants, 86, 125, 319, 503, 504, 507
Decomposer, 251, 489, 490, 494, 495
Deer, 444, A8
Deer, mule, 509
Deerbrush, 500, A8
Deerfly, 255
Deficiency symptoms, 174, 175
Dehydration movements, 183
Deletion, chromosomal, 219
Denitrifying bacteria, 494, A8
Deoxyribonucleic acid, 35, 37, 41–47, 197, 247, 267, 268
  bacterial, 241, 242
  recombinant, 248, 249
  replication of, 37, 42, 43, 242
  structure, 41, 42
  viral, 267, 268
Desert, 79, 118, 121, 139, 187, 497, 501, 506, 507
Desert marigold, A43
Desmids, 283, A8
Desmococcus, A4
Deuteromycetes, 332–34. See also Imperfect fungi
Development, 165
Devonian period, 362, 381, 390
Dewberry, 466, A8
Dextran, 258, A8
Diabetes, 383, A8
Diarrhea, 298, 375, 383, 458, 466, 470, A8
Diatom (Bacillariophyceae), 183, 261, 286, 287, 296, A8
Diatomaceous earth, 296, 297
Dicentra, 222, 229, 230
Dichasium, 417
Dichotomous key, 236–38
Dichotomous venation, 112, 113, 238, 398
Dicot, 63, 65, 68, 412, 413, 425, 447, 451, A8

Dicots and monocots compared, 90, 91, 412
Dicotyledon, 90
Dicotyledonous stems, 90–96
Dictyosome. See Golgi body
Dieffenbachia, A42. See also Dumbcane
Differentially permeable membrane, 132–34
Differentiation, 165, 170, 417, 420
Diffuse knapweed, A31
Diffusion, 131–33, 137, 14
Diffusion gradient, 131, 134, 178
Digester, organic, 259
Digestion, 15, 19, 124, 159, 160
Digger pine, 401, A8
Digitalis, 127, A8, A46
Dihybrid cross, 212–14
Dikaryotic, 321–23
Dill, 126, 258, 443, 475, A8, A33, A53
Dimethyl sulfoxide, 29
Dinobryon, 286
Dinoflagellates, 180, 236, 288, 295, A8
Dinosaurs, 495
Dioecious, 425
Dioscorides, 461
Dioxin, 169
Diphtheria, 253, 269, A8
Diploid (definition), 201
Disaccharide, 18, 144
Dischidia, 120, 121, A8
Disease,
  bacteria and, 253–56
  fungal, 253, 310, 311, 319, 333
  resistance to, A29, A30
  viruses and, 265–69
Disk floret, 478
Dispersal, seed, 394, 443–47
Dittany, A53
Diuretic, 375, A9
Divisions (plant or other organismal), 234–38
DMSO. See Dimethyl sulfoxide
DNA. See Deoxyribonucleic acid
Dock, A38, A56
Dock, curly, 451
Doctrine of Signatures, 8, 346
Dodder, 74, 75, 413, A9
Dog, 263, 328, 422
Dogbane, 96, A9, A46
Dogwood, 112, 121, A9, A46, A56
Dollar plant, 440
Dominance
  incomplete, 209, 210
  in heredity, 209, 210, 455
Dominant, 209–16, 220, 455
Donkey tail, A89
Dormancy, 187, 188, 448–51, A60
Double fusion, 421
Double-working, A73

Douglas fir, 96, 406, 508, A9, A38
Dove, 458, 507, A9
Dove, mourning, 504, A9
Doveweed, A56
Downy mildew, 310, A9
Dracaena, 98, A89
Dragon's blood, 482, A9
Drugs, 77, 96, 127, 212, 383, 421, 461, 462, 470, 474, 478, 479, 481, A44–A50
Drupe, drupelet, 434, 437, 442, 461, 465
Dry fruits, 433, 438–43
Duckweed, 63, 110, 383, 413, 498, A9
Dulse, 297, A9
Dumbcane, A89
Dung mosses, 356, A9
Duranta, A42
Dutch elm disease, 319, 504, A9
Dutchman's breeches, 221, 222, 229, 230, A9, A43. See also Dicentra
Dutrochet, R. J. H., 24
Dwarf banana, A89
Dwarf cocos palm, A89
Dwarfing, A73
Dwarf juniper, 407, A9
Dwarf mistletoe, 183, 446, A9
Dye, 77, 101, 106, 126, 131, 333, 334, 338, 382, 406, 407, 457, 458, 461, 464, 466, 470, 481, A9, A55–A57
Dyeing, natural, 334, 338, 458, 461, 464, A55–A57
Dyer's rocket, A56
Dyer's woad, 464, A9
Dysentery, 254, 383, 466, A9, A45–A50

e

Eagle, golden, 509, A9
Earth star, plate 3B, 320, 324, A9
Earthworm, 78, 79, 257, 376, A9
Earwig, A28
Earwig, European, A28
Eastern hemlock, 407, 497, A9
Eastern larch, 498, 503, A9
Eastern white cedar, 407, 498, A9
Eastern white pine, 102, 401, 403, 497, 504, A9
Ebony, 91, A9
Echeveria, A89
Ecology, 486–517
Ecosystem, 487–92, 499, 501
Ecotype, 489
Eczema, 383, A9

Edible plants, wild, 401, 407, 476, 479, 481, A37–A41. *See also* chapter 22
Eelworm, 312, 314, A9. *See also* Nematodes
Egg. *See* Gametes
Egg (bird's-nest fungus), 324, 325
Eggplant, 434, 455, 474, A9, A34, A80
Einstein, Albert, 149
Elaioplast, 34
Elaiosome, 445
Elaters
 hornwort, 351
 horsetail, 373
 liverwort, 349, 350
Elderberry, 24, 55, A38, A56
Electromagnetic spectrum, 149
Electron, 16
 acceptor molecule, 150
 donor, 150
 flow, 150
 hole, 150
 microscope, 26
 transport chain, 150, 155, 156, 158
Electrotropism, 179
Element, 16
 deficiency symptoms, 174
 essential, 173, 174
Elephant, 29, 492, A9
Elk, A9
Elm, 170, 319, 442, 443, 504, A9, A66
Elm bark beetle, 319, A9
Elodea, 133
Embryo, 344, 348, 364, 390
 angiosperm, 418, 422, 446
 bryophyte, 348, 354
 culture, A65
 gymnosperm, 394, 395
 of seed, 63, 165, 166, 446, 447, 449–50
Embryophyta, 344
Embryo sac, 419–22
Emex, A30
Empirical information, A33
Enation, 237, 361, 363
Endive, 478, A9
Endocarp, 433, 434, 436
Endodermis, 67, 68, 117, 132, 371, 391
Endoplasmic reticulum, 27, 28, 31, 40
 rough, 28, 31
 smooth, 28, 31
Endosperm, 418, 421, 422, 434, 448, 449
Endosperm nucleus, 418, 421
Endospores, 245
Energy, 18, 29, 134, 143–59
 ecosystems and, 489–92
 forms of, 17
 living organisms and, 134, 489–92

photosynthesis and, 143–53
pyramid, 491
respiration and, 153–59
English ivy, 73, 100, A9, A42, A89
Ensilage, 258, 480, A9
Enzyme, 20, 25, 31, 32, 35, 43, 46, 96, 124, 135, 144, 152–60, 165, 187, 248, 308, 314, 319, 333, 450
Enzyme, repair, 248
Enzymes, restriction, 248
Ephedra, 399, 407, A46
Ephedrine, 318, 407, A46
Epicotyl, 166, 188, 447, 448
Epidermis, 52, 53, 67, 72, 87, 90, 121, 188, 351, 362
 of leaf, 52–54, 113–15, 138, 391
 of root, 67, 72
 of stem, 90, 370, 371
Epilepsy, A48
Epiphyte, 281, 482
Equator (of cell), 39, 40
Equatorial division, 196
Equilibrium, 131
Equisetum, 370–75
Ergine, A51
Ergoline, A51
Ergosterol, 330
Ergot, 315, 318, A9, A46, A51
Ergotism, 318
Ermine, 503, A9
Escherichia coli, 44
Essential elements, 173–75
Estrous cycle, 455
Ethyl alcohol, 154, 318, 319
Ethylene, 167, 171, 172, 177, 184, 188
Etiolation, 184
Eucalyptus, 114, 252, A9, A53, A56
Eucalyptus, Tasmanian giant, 413, A9
Eucalyptus oil, 126, A9
Eucapsis, 260
Euglena, 285
Euglenoids, 232, 236, 275, 285, 295, A9
Euglenophyta, 236, 285, 286, 295
Eukaryotic cells, 36, 233, 236, 245
 eukaryotic vs. prokaryotic, 35
Euphorbia, 467
Euphorbiaceae. *See* Spurge family
European bulbous buttercup, 458
European corn borers, 257, A9
European larch, 406, A9
European rock rose, 178
European stone pine, 401, A9
Eutrophication, 499
Evening primrose, 451, A38
Evolution, 216–23

Evolution, mechanisms of, 219–21
Exine, 420
Exocarp, 433, 434
Extinguisher mosses, 354, A9
Eye, potato, 99, 100
Eye disease, 383, A9
Eyespot, 276–78, 285

f

F. *See* First filial generation
F. *See* Second filial generation
Fabaceae. *See* Legume family
Factor (genetic), 209
FAD. *See* Flavin adenine dinucleotide
Fairy bells, A39
Fairy ring, 323
Fall armyworm, A29
Fall webworms, 257, A9, A28
False aralia, A89
False berry, 436, 437
False hellebore, A42
False morel, 317, A9
Family (category of classification), 234, 427, 457–84
Farinha, 468
Far-red light, 183, 184
Fat, 19, 160, 295, 353, 450
 polyunsaturated, 19
 saturated, 19
Fatshedera, A89
Fatsia, A89
Fatty acid, 19, 30, 160
Faucet gland, 513
Fava bean. *See* Broadbean
Fennel, 475, A9, A34, A39, A53, A56
Fenugreek, 457, A9, A53
Fermentation, 25, 154–56, 318
Fern, edible Malaysian, 383, A10
Fern allies, 361–75
Ferns, 86, 110, 173, 182, 345, 346, 360, 361, 376–84, 390, 497, A9–A11, A89. *See also* Adder's tongue fern, Bracken fern, *etc.*
 human and ecological relevance of, 381–84, A9–A11
Ferrodoxin-reducing substance, 151
Fertilization, 202, 421, 422, 433
Fertilizer, 81, 252, 297, 492, 493, A27, A76, A77, A84
Fever blister, 265
Fevers, 254, 256, 369, 383, A11, A45–A50
Fibers, plate 1, plate 2c, 57–59, 89, 94, 98, 126, 391, 407, 434, 480–82

Fibril, 40, 183, 267, 268
Fiddlehead, 377, 378, 382
Field capacity, 81
Field horsetail, 374, A11
Fig, 6, 433, 437, 438, 456, A11, A56
 climbing, A90
 fiddleleaf, A90
 tropical, 52, 72, 73, A11
 weeping, A90
Fig family (Moraceae), 469
Figwort family (Scrophulariaceae), 491, A11
Filament, 242, 261
 of algae, 275, 277–83, 292
 of stamen, 416, 417
Filaree, 183, 446, 447, A11
Filé, 460
Filterable viruses, 265
Final oxidation (aerobic respiration), 155, 156
Fingernail plant, A90
Fir, 101, 170, 391, 406, 503
 balsam, 406, A11
 Douglas, 96, 406, 508, A11, A38
 white, 508, A11
Fire ecology, 499–501, 508, 509
Fireweed, A34
First filial generation, 208–14
Fish, 77, 258, 263, 275, 288, 290, 296, 310, 473, 478, 482, 491, A11
Fish mold, 310, A11
Fish poison, 77
Fission, 242, 245, 246
Fittonia, A90
Five-finger fern, 383, A11
Flagella, 175, 232, 242, 245, 276, 277, 278–83, 285, 288, 290, 295, 302, 303, 308, 310, 353, 394, 397, 398
Flame violet, A90
Flashlight fish, 242, 243, A11
Flash powder, 369, A11
Flatworm, 275, A11
Flatworm algae, 275, A11
Flavin adenine dinucleotide, 158
Flavoprotein acceptor molecule, 158
Flax, 98, 444, 456, A11, A34
Flax, New Zealand, 481, A11
Flea, 128, 255, 256, 375, 461, A11
Flea beetle, A30, A31
Fleming, Sir Alexander, 332, A19
Flicker, 504, A11
Flora, 10
Floral leaves (bracts), 54, 121, 122
Floret, 478
 disk, 478
 ray, 478

Florida arrowroot starch, 407, A11
Floridean starch, 237, 295
Florigen, 186
Flour, 375, 460, 470, 480, A11
Flower, 57, 309, 425, 433, 437
  bracts, 54, 121, 122
  development of, 417–22
  essential parts, 415–17
  inflorescences, 416, 417, 437, 438
  preservation, 425–28
  primitive vs. advanced, 422–25, 457, 460
  specialization trends, 422–25
  structure and reproduction, 415–25
  symmetry, 422–25
Flowering ash, A46
Flowering maple, A90
Flowering plants, 95, 234, 361, 384, 390, 391, 399, 411–85, 491
  structure and reproduction, 415–25
  trends of specialization and classification, 422–25
Flower pot leaves. See Dischidia
Fluorescence, 149
Fly, 128, 320, 356, A29
  carrot, A34
  deer, 255
  fruit, 214, 215
  house, 254, A28
  long-tongued, 510
  peacock, A31
  seed, A31
  short-tongued, 510
  tachinid, A28
Fly agaric mushroom, 329, A11, A51
Flycatcher, 507, A11
Fly poison, A43
Foliose lichen. See Lichens, foliose
Follicle, 438, 439, 465
Food, translocation of, 137, 138
Food chain, 126, 275, 296, 489–92
Food poisoning, 254
Food storage, 70, 137, 467
Food web, 489, 490
Foot, 349–52, 354, 364
Footcandle, 146
Ford, Clellan, 455
Formaldehyde, 288
Fossil, 52, 221, 223, 344, 361, 362, 369, 375, 384–86, 390, 391, 398, 423, 495, 499, A11

Fossil fern, 223, 381, A11
Fossil horsetails, 375
Fossil lycopod, 369
4-carbon pathway (of photosynthesis), 152
Four-o'clock, A43
Fox
  arctic, 502, A11
  gray, 504, A11
  red, 504, A11
Foxglove, 91, 127, A11, A42, A46
Fragmentation, 263, 280, 290, 320
Frangipanni, 116, A11
Frankincense, 95
Franklin, Rosalind and M. H. F. Williams, 42
French bean, 249
Frets, 32
Frog, 125, A11, A28
Frond, 376–79, 381–83
FRS. See Ferrodoxin-reducing substance
Fructose, 154, 155, 478
Fructose, 1, 6-biphosphate, 158
Fruit, 183, 390, 412, 415, 416, 418, 432–47, 451
  accessory, 437
  aggregate, 437
  development of, 422, 433
  dispersal, 443–47
  distinction between vegetables and, 433, 434
  dry, 433, 438–43
  fleshy, 433–38
  kinds of, 434–43
  multiple, 437, 438
  regions of, 433, 434
  ripening of, 433
  seedless, 433
  simple, 434, 436
Fruit fly, 214, 215, A11
Fruit trees, pruning of, A73
Fruticose lichen. See Lichens, fruticose
Fuchs, Leonard, 8
Fuchsia, A90
Fuchsia, California, 511, A11
Fucoxanthin, 287, 290, 295
Fucus, 290, 291
Fumitory, Himalayan, 414, A11
Fundamental tissue, 97
Fungi, 101, 157, 169, 217, 231, 261, 308–42, 346, 391, 406, 489, 491, 493, 495, 499, A11, A12, A29, A30, A32, A64
  bird's nest, 320, 324, 325
  bracket, plate 3C, 300, 320, 324, A11
  club, 320–32
  cup, 315–19
  diseases, 253, 310, 311, 332–34, A11
  earth stars, plate 3B, 320, 324, A9

gardens, 331, A2
gill, 320–24, A11
hallucinogenic, 328, 329, A12
human and ecological relevance of, 314, 315, 317–19, 328–33, A11, A12
jelly, 320, 330
mycorrhizal, 74, 75, 332, 350, 362, 364, 392, 482, A64
nematode-trapping, 312–14
puffballs, 308, 320, 324, 325, A21
rust, 308, 320, 326–28, 378, A30
sac, 315–19, A22
smut, 308, 320, 326, A22
stinkhorn, 308, 320, A23
Fungicides, 310, A61, A64
Fungi Imperfecti, 332–34, A30
Funori, 298, A12
Fusarium, A30
Fustic, A56

g

Galileo, 8
Gallic acid, 333
Gall mite, A31
Gamboge, A56
Gametangium, 291, 312, 313, 344, 353, 361. See also Antheridium, Archegonium, Oogonium
Gametes, 193, 210
  of algae, 278–94
  of angiosperms, 418–22
  of bryophytes, 346–56
  of ferns, 377, 380, 381
  of fungi, 309–27
  of gymnosperms, 394, 395
Gametophore, 347–49
Gametophyte, 193, 201
  of algae, 278–95
  of angiosperms, 417–22
  of ferns, 377, 380, 381
  of ground pines (Lycopodium), 364, 365
  of gymnosperms, 390–95
  of hornworts, 351
  of horsetails (Equisetum), 372, 373
  of liverworts, 347–49
  of mosses, 352–56
  of quillworts (Isoetes), 368
  of spike mosses (Selaginella), 364–67
  of whisk ferns (Psilotum), 361–63
Gametophyte generation, 201, 202, 346
Gane, R., 171
Gangrene, 318

Garbanzo bean, 183, 438, 439, 466, A12
Garden bean, 27, 32, 165, 185, A12
Gardenia, 168
Garlic, 456, A30, A34. A53
Garner, W. W. and H. A. Allard, 184
Gas gangrene, 255, A12
Geese. See Goose
Gelidium, 298
Gemmae, 348, 349
Gemmae cup, 348, 349
Gene, 35, 165, 176, 209, 219, 248, 455
Gene, mutator, 219
Genera. See Genus
General Sherman tree, 397
Generative cell, 394, 395, 421
Generative nucleus, 419–21
Gene splicing, 248, 249
Genetic engineering, 248, 249
Genetic recombination, 245
Genetics, 10, 205–16
Genotype, 209–16
Gentian, 77, A12, A46
Genus, 230, 231, 427, 455
Geologic time scale, 384
Geomagnetropism, 179
Geotropism, 179
Geranium, 14, A12, A46, A90
Geranium family (Geraniaceae), 183, A12
German measles, 269
Germination, seed, 134, 135, 165, 166, 172, 173, 177, 187, 188, 448–50
Ghost moth, A30
Giant bamboo, 181, 456
Giant Galápagos tortoise, 187, 444, A12
Giant horsetail, 375, A12
Giant kelp, 288–90, 296, 297, A12
Giant puffball, 325
Giant redwood, 392, 397, 406, 499, 508, A12
Giant reed, A56
Giant saguaro cactus, 470, 471, A12
Giant sequoia. See Giant redwood
Giant water lily, 110, A12
Gibberellins, 167, 169, 170, 187
Gibbons, Euell, 401
Gigartina, 292
Gila monster, 507, A12
Gill, 321–24
Gill fungi, 320–24
Ginger, 456, A12, A46, A53
Ginger, wild, plate 5C, A12, A39
Ginkgo, 112, 384, 391, 397, 398, 407
Ginkgophyta, 238, 391, 398
Ginseng, 77, A12
Giraffe, 218, A12

Girdling, A70
Gladiola, 100, 367, A12, A64
Gland, 48, 49, 113, 122, 124, 215, 463, 467, 468
Glandular hairs, 57, 113, 122–24
Gliding movement, 183, 242, 263
Globe artichoke, 433
Gloeocapsa, 262, A12
Gloriosa lily, A43
Gloxinia, A90
Glucose, 18, 148, 152, 155–59
Glume, 479
Glutamic acid, 258, A12
Glycerol, 18, 19, 160, 319
Glycine, 19
Glycogen, 261, 308
Glycolysis, 154, 155, 157, 158
Gnetophyta, 238, 391, 399, 400
Gnetophytes, 391, 397, 399, 400
Gnetum, 399, 407
Goat, 333, 374, A12
Goiter, 297
Goldback fern, 383, A12
Golden-brown algae (Chrysophyta; Chrysophyceae), 286, 287, 295, A12
Golden chain tree, 95, A12, A42
Golden eagle, 509, A9
Goldenrod, plate 5D, 185, 504, A12
Goldenseal, 458, A12, A43, A46
Goldenweed, 35, A12
Goldfish plant, A90
Goldthread, A47
Gold violet, plate 7B, A13
Golgi body (dictyosome), 28, 32, 40, 167
Gomphosphaeria, 262
Gonorrhea, 255, 375, 407, A13, A49
Gonyaulax, 288
Goose, 374, 473, A13
Gooseberry, 327, 436, 499, 500, A13, A39
Gopher, 506, A13
Gopher, pocket, 475, A13
Gopher plant, 469, A13
Gordon, J., 249
Gorse, A30
Gourd, 424, 455, 476, 477, A13
Grafting, 86, A59, A66–A73
    approach, A68–A70
    bridge, A70, A71
    bud, A71, A72
    cleft, A68, A69
    double-working, A73
    inarching, A70, A71
    natural, 69
    root, A72, A73

side, A68, A69
side tongue, A68
splice, A67
tongue, A66, A68, A72
whip, A66, A67, A68, A72
Grain, 442
Grains of paradise, A53
Gram, Christian, 245
Gram-negative, 245
Gram-positive, 245
Grana. *See* Granum
Granum, 33, 147, 152
Grape, 6, 100, 118, 310, 416, 433–35, 456, A13, A39, A56, A73
Grape, wild, A39
Grapefruit, 436, A13, A29
Grape hyacinth, A14, A64
Grape ivy, A90
Grape leaf rollers, A28
Grapes, seedless, 170
Grapevine, A73, A75
Grass, 65, 86, 96, 116, 126, 169, 178, 180, 261, 312, 314, 344, 384, 413, 421, 447, 479, 480, 499, 500, 502, 505, A13, A20, A39
    flower structure, 479
Grass, Bermuda, 178, A13
Grass, crested wheat, 198–99, A13
Grass, Indian, 185, A13
Grass family (Poaceae), 166, 442, 479, A13
Grasshoppers, A28, A32
Grasslands, 499–501, 505, 506
Grass tree, 482, A13
Gravitational water, 81
Gray fox, 504, A13
Gray squirrel, 504
Green algae (Chlorophyta), 275–85, 295, 335, 344
Greenbrier, 100, 118, A13, A39
Green hellebore, A47
Green sulfur bacteria, 249, A13
Grew, Nehemiah, 9, 24, 136
Grimmia, 345
Ground meristem, 38, 46, 65, 87, 88
Groundnut, A39
Ground pine, 364, 365, A13
Ground squirrel, 256
Ground tissue, 97
Growth, 14, 164–90
Growth, mineral requirements for, 173, 174
Growth movements, 175, 176
Growth regulators, 165, 167–73
Guanine, 41, 42
Guard cells, 52, 138, 171
Guava, 457, A13
Gullet (*Euglena*), 232, 285
Gum, chewing, 96, 333, 405, 460, 473
Gum arabic, 455, 466, A13

Gums, 96, 101, A30
Gum tragacanth, 466, A13
Guppy, 14, A13
Guttation, 139
Gymnosperms, 389–410, 412
    human and ecological relevance of, 401–7
Gynophore, 467, 468
Gypsy moth, 257, A13, A28, A32

h

Hadeda ibis, 344, A13
Haeckel, Ernst, 233, 487
Haircap moss, 346, 354, A13
Hairs, plate 2B, 52, 54, 57, 122–25, 379, 382, 397, 443
Hairstreaks, A30
Hairwash, A13
Hall-inferior ovary, 424, 425
Hallucinogenic fungi, 328, 329, A13
Hallucinogenic plants, 470, 474, A51
Halogeton, A30
Haploid (definition), 201, 278
Hardpan, 493, 494
Hardwood, 94, 135, 391
Hardy, G. H., 216
Hardy-Weinberg law, 216
Harmine, A51
Haustorium, 74, 75, 314
Hawaiian blackberry, A30
Hawaiian tree ferns, plate 4D, 376, 378, 381–83, 390, 397, A24
Hawk, 504, A13
Haworthia, A90
Hawthorn, A39, A56
Hazel nut, 442, 456, A13, A39
Head (inflorescence), 417
Heartwood, 69, 95
Heath, 444, A13
Heath family (Ericaceae), 417, A13
Hedgehog cactus, 470, A13
Helmont, J. B. van, 9
Hemlock, 103, 391, 476, 497, 503, 504, A47, A56
    eastern, 407, A13
    mountain, 407, A13
    poison, 476, A13
    water, 70, 476, A13
    western, 407, 508, A13
Hemp, 98, 455, A13. *See also* Marijuana
Hen and chickens, A90
Henbane, A51
Henbit, 471, A13
Henna, 126, 455, A13, A56
Hepatica, 459, 504, A13
Hepaticae, 346–51
Hepatitis, infectious, 269
Herbaceous, 91, 96, 146
Herbal, 8
Herbalist, 8, 346
Herbal medicine, 8

Herbarium, 425–27
Herbarium specimen, 425–27
Herbicide, 52, 170, A27
Herbivore, 345, 490, 491
Herbs, A33–A35
    sources, A35
Heredity, 205–16, 423, 457
Heroin, 96, 462
Herring, 290
Hesperidium, 436
Heterocyst, 262, 263
Heterozygous, 210–14
Hibiscus, A90
Hickory, 103, 442, 497, 504, A13, A39, A56
Highbush cranberry, A39
Hilum, 188, 447, 448
Himalayan fumitory, 414, A13
Hippeastrum, A90
Hips, rose, 465, 466
Hog, 259, 263, 315, 491, A13
Hog fennel, 476, A13
Hogg, J., 233
Holdfast, 278, 279, 284, 285, 289
Holly, 127, 172, A43
Holly, American, 449, A13
Holly fern, plate 4C, A13, A90
Hollyhock, A56
Homologous chromosomes, 194–201
Homozygous, 210–14
Honey, 405
Honeybee, 509, A13
Honeydew, 137
Honeydew melon, 173
Honey guides, 509
Honey locust, 100, A13
Hooke, Robert, 24
Hooker, Sir Joseph D., 10
Hooker Oak, 94
Hoopoe, 229, 344, A13
Hop hornbeam, 442, A14
Hopi Indians, 374, 470
Hops, 170, 456, A14, A53
Horehound, 472, A14, A47
Horizons, soil, 78
Hormones, 57, 125, 165, 167–73, 186, 281, 433
Hornworm. *See* Tomato hornworm
Hornwort, 351, 352, A14
Horse, 259, 263, 331, 375, 466, A14
Horse chestnut, A43
Horse dung fungi. *See* *Pilobolus*
Horseradish, 71, 76, 463, A14, A34, A53
Horsetail, 361, 370–75, 384, 399, A14, A47, A56
    human and ecological relevance of, 374, 375, A14
Horsetail, field, 374, A14
Hot springs, 261, A14, A24
Housefly. *See* Fly
Houseleek, A90

Houseplants, 381, A83–A95
diseases and pests, A84
environmental
requirements,
A83–A95
Huckleberry, A39
Humans in the ecosystem,
492
Humboldt, Alexander von, 10
Hummingbirds, 344, 382, 510,
511, A14
Humus, 78, 376, 494
Hyacinth, 100, A14, A43, A64
Hyacinth, grape, A14, A64
Hyacinth, water, 69
Hybrid, 5, 211–14, 220, 327,
433, 451
Hybridization, role of in
evolution, 220
Hydathode, 110, 139
Hydrangea, A43, A91
Hydrodictyon, 284
Hydrogen, 17
Hydrogen bacteria, 251, A14
Hydrogen bond, 17
Hydrogen sulfide, 249
Hydrolysis, 19, 159, 160
Hydrophyte, 489
Hydrosere, 497
Hydrotropism, 179
Hydroxyl ion, 17
Hygroscopic water, 81
Hymenium, 316
Hyoscyamine, 474, A46, A47,
A51
Hypha, 76, 308–38
Hypocotyl, 166, 176, 184,
447, 448
Hypodermis, 117, 391
Hypoglycemia, 328
Hypothesis, 6, 137
Hyssop, 456, A14

i

IAA. *See* Indoleacetic acid
IBA. *See* Indolebutyric acid
Iboga, A51
Ibogaine, A51
Ibotenic acid, A51
Ice plant, 183, A14
Ichneumon wasps, A28
Igneous rocks, 79
Imbibition, 134, 135, 157, 450
Impatiens, A91
Imperfect fungi, 332–34, A30
Imprint, 385
Inarching, A70, A71
Incas of Peru, 81
Incense cedar, 407, A14
Incomplete dominance, 210
Increment borer, 94
Independent assortment, law
of, 212
Indian grass, 185
Indian paintbrush, plate 7A,
A14, A39

Indian pipe, A39
Indians. *See* Catawba
Indians, Cree Indians,
etc.
Indian warrior, 74, A14
Indicator mosses, 356, A14
Indigo, 457, 466, A14, A56
Indigo, wild, A47
Indoleacetic acid, 168, A61
Indolebutyric acid, A61
Indusium, 377–79
Indusium, false, 379
Industrial alcohol, 315, A14
Inferior ovary, 422, 424, 425,
434, 436, 457, 465,
469, 475, 476, 480
Inflorescence, 416, 417, 467,
468, 479, 511
Influenza, 265, 268, 269, 330,
466
Inheritance, 205–16
of acquired
characteristics,
218
Mendelian laws of,
209–12
Inhibitors, growth, 170, 173,
449, 491
Initials (cambial), 51
Inky cap mushrooms, 328,
A14
Inorganic, 18
Insecticide, 77, 299, 460, 474
Insects, 122–25, 182, 207,
258, 269, 308, 315,
345, 356, 383, 403,
421, 427, 473, 478,
490, 492, 509,
511–14, A14
beneficial, A27, A28, A32
pests, 269, A27, A28, A32
pheromones, A29
pollination, 509–14
prehistoric, 404
Insect-trapping leaves, 57,
122–25
Insomnia, A50
Insulin, A45
Integument, 390, 399, 407
of flowering plants, 419,
420, 422
of pine, 393, 394
Intercalary meristem, 52, 98
Interferon, 268, 269, 330
Intermediate-day plant, 185
International Code of
Botanical
Nomenclature, 232,
234
Internode, 69, 86, 87, 98, 184,
371
Interphase, 36, 37
Interstock, A73
Intoxicant, 127, 329, 369, A14
Introgression, 220
Inversion, chromosomal, 219
Invertase, 319
Iodine, 173, 297
Ion, 16, 138
Ionic bonding, 16

Ipecac, 77, A14, A47
Iris, 413, 440, A14
Iris, butterfly, 441, A14
Iris family (Iridaceae), A14
Irish moss, 298, 352, A14
Irish potato. *See* Potato, Irish
Iron bacteria, 251, A14
Iron chelates, 175
Ironwood, South American,
101, A14
Isocitric acid, 152
Isogamy, 280
Isoetes. See Quillworts
Isogamy, 280
Isotope, 16, 43, 145
Italian thistle, A31
Ivy, Boston, 73, 100, A14
Ivy, English, 73, 100, A14,
A42, A89
Ivy-arum, A91

j

Jacaranda, 444, A14
Jack-in-the-pulpit, A43
Jack pine, 394, 500, A14
Jackrabbit, 506, A21
Jade plant, A91
Jaeger, 502, A14
Jamaica feverplant. *See*
Puncture vine
Janssen, Hans, 8
Janssen, Zacharias, 8, 24
Japanese beetle, 258, A28,
A29, A32
Jay, 503
Jeffrey pine, 508, A14
Jelly fungi, 320, 330, A14
Jenner, Edward, 265
Jequirity bean, 466, A14, A42
Jerusalem artichoke, 186,
478
Jerusalem sage, 443, A15
Jessamine, A43
Jesty, Benjamin, 265
Jimson weed, 416, 474, A15,
A42, A47, A51
Joe-pye weed, A47
Joint fir, 399, 407
Joint vetch, A30
Jojoba, 4, 469, A15
Jonquil. *See* Daffodil
Joshua tree, 510, A15, A47
Jumping mouse, 503, A15
Junco, 509, A15
Junco, slate-colored, 344,
A15
Juneberry, 466, A15
Juniper, 397, 406, 407, A15,
A39, A47, A53
Juniper, dwarf, 407, A15
Jute, 98, A15

k

Kaffir lily, 52, A15, A91
Kalanchoë, A91
Kangaroo rat, 507, A21
Kansas snakeroot, A47
Karaka nut, A43
Karok Indians, 458

Karpechenko, G. D., 206
Kauri pine, 403, A15
Kava kava, A51
Keel (of legume flower), 466,
467
Kefir, 258, A15
Kelp, 275, 289, A15
Kelp, giant, 288–90, 296, 297,
A15
Kelp harvesting, 298
Kendall green, A57
Kentucky coffee tree, A43
Key, synoptic, 236–38
Kidney bean, 446, 448, 466,
A15
Kinetic energy, 17
Kinetochore, 39
Kingdom, 232–40
Animalia, 233, 234
Fungi, 202, 233, 234,
308–42, A15
Monera, 233, 241–65, 308
Plantae, 202, 233, 234,
343–485
Protista, 202, 233, 234,
274–306, 308
Klamath weed, A30
Knee, bald cypress, 71, 406
Knobcone pine, 394, 508,
A15
Knot, 101, 103, 406, 407
Knot, root, 76
Koch, Robert, 256
Koch's postulates, 256
Kohlrabi, 463, A15
Krebs, Hans, 154
Krebs cycle, 154, 155, 158,
159
Kudzu, 438, A15
Kümmel, 476, A52
Kumquat, 436, A15
Kurosawa, E., 169

l

Labrador tea, A39
Lacewings, A27
Lactic acid, 154–56, 258, A15
Ladybugs, A27
Lady fern, 383, A15
Lake Chad, 263, A5, A15
Lamarck, Jean Baptiste, 24,
25, 218
Lamb's quarters, A34, A39
Lamiaceae. *See* Mint family
Lamina, 111
Laminarin, 237, 290, 295
Lantana, A30, A42, A91
Larch, 328, 406, 498. *See
also* Tamarack
Eastern, 498, 503, A15
European, 406, A15
Western, 406, A15
Larkspur, 185, 438, 458, A43,
A56
blue, 458
Late blight of potato, 297,
312, 475, A15

Lateral bud. *See* Axillary bud
Latex, 57, 58, 96, 468, 469, A48
Laticifers, 96, 469
Lauraceae. *See* Laurel family
Laurel, 330, 460, A15
Laurel family (Lauraceae), 460, 461, A15
Lavender, 126, 456, 472, A15, A30
Law of dominance, 209
Law of independent assortment, 212
Law of segregation, 211
Law of unit characters, 209
Laxative, 298, 338, 362
Layering, A61–A63
    air-layering, A62, A63
    tip-layering, A61–A63
L-dopa, A50
Leaf beetle, A30, A31
Leaf gap, 88, 89, 111
Leaf hoppers, A28
Leaflet, 113, 180
Leaf miners, 257, A15, A30
Leaf primordia, 87, 88, 171
Leaf rollers, 257, A15
Leaf scar, 86, 87
Leaf trace, 88, 89
Leafy liverwort, 346, 350, 351, A15
Leafy spurge, A30
Leaves, 52, 109–29, 309, 376, 433
    abscission of, 125, 126
    arrangement of, on stem, 86, 110–13
    coloration, 35, 125, 126
    compound, 111–13, 407
    fleshy, 99, 100, 119
    floral (bract), 54, 121, 122, 442
    flower pot, 120, 121, 122
    human and ecological relevance of, 126–28
    insect-trapping, 57, 122–25
    mesophyll of, 26, 113–16, 136, 145
    midrib of, 113
    pine, 117, 391
    primordium of, 87, 88, 111, 171, A65
    rachis of, 113
    reproductive, 121
    scalelike, 99, 100, 237, 309, 371
    simple, 111–13
    specialized, 116–25, 466
    stipules, 86, 87, 111
    storage, 119
    structure of, 113–16
    succulent, 112, 121
    sun and shade, 117
    veins, 112, 113–16, 136
    venation, 112, 113, 412
    window, 120, 121

Leech, 474
Leechicide, 474
Leek, 451, 456, A34
Leeuwenhoek, Anton van, 8, 24
Legionnaire's disease, 254, A15
Legume family (Fabaceae), 95, 176, 180, 421, 448, 466, 467, A6, A15
Legumes, 76, 188, 251, 263, 269, 438, 442, 446, 455, 466, 467, 493
    fruits, 438
    nitrogen fixation and, 76, 251, 263, 493, 494
Lemma, 479
Lemming, 502, A15
Lemon, 436, 473, A15
Lemongrass oil, 126, 480, A15
Lenticel, 53, 55, 87, 90
Lentil, 438, 456, 466, A15
Lentinacin, 330
Lepidodendron, 369
Leprosy, 383, A15
Lettuce, 126, 133, 171, 184, 185, 433, 450, 456, 478, 491, A15, A34, A76, A80
Leucoplast, 32
Lewin, R. A., 265
Lice. *See* Louse
Lichens, 234, 261, 275, 334–38, 356, 415, 487, 497, 502, A15, A56
    crustose, plate 3A, 336, 337, 338, A16
    foliose, 336, 337, A15
    fruticose, 336, 337, 338, A15
    human and ecological relevance of, 338, A15, A16
    litmus, 338, A15
    perfume stabilizer, use in, 338, A15
    reindeer, 338, A16
    structure, 335, 336
Licorice, 77, 466, A16, A39, A53
Licorice fern, 383, A16
Life, nature of, 13–20
Light
    far-red, 183, 184, 314
    growth and, 177–79
    intensity, photosynthesis and, 146, 152, 506, 507
    properties of, 146, 149
    reactions (photosynthesis), 147–51
    red, 183, 184
    ultraviolet, 416, 509, A24
Lignin, 29, 56, 101, 346
Lignum vitae, 101, A16, A44
Ligule, 366–68
Lilac, 416, A16

Liliaceae. *See* Lily family
Lily, 73, 96, 100, 110, 119, 138, 177, 186, 196, 412, 422, 440, 480, 481, A16, A64. *See also* African lily, Calla lily, *etc.*
Lily family (Liliaceae), 480–82, A16
Lily of the valley, A42, A47
Lima bean, 166, 448, 457, 466, A16, A78
Lime, 310, 436, A16
Liming, 81
Liminoid sprays, A29
Limnanthaceae. *See* Meadow foam family
Linden. *See* Basswood
Linkage, 214, 215
Linnaeus, 7, 9, 10, 180, 230, 231, 234, 363
Linnaeus' garden clock, 180
Lion, mountain, 509
Lip cells, 380
Lipid, 19, 29, 146, 152, 159, 261
Lip petal, 482
Liquefied pest sprays, 269, A29
Liqueur, 128, 473, 475, 476, A52, A55
Litmus, 338, A16, A56
Live oak, 497
Liverwort, 346–51, A16
    leafy, 346, 350, 351, A15, A16
    thalloid, 346–49, A16
Living organisms, characteristics of, 14, 15
Living rock cactus, 470, A16
Lizard, 507, A16
Loam, 78
Lobelia, 127, A42, A47
Lobeline sulphate, 127, A16, A47
Loblolly pine, 504, A16
Lockjaw, 255
Locoweed, 466, A16, A44
Locule, 440
Locus, 214
Locust, 92
Loganberry, 465, A16
Log cutting, 101–6
Logwood, 466, A56
Lokao, A56
Long-day plant, 185
Longevity of seeds, 157, 450, 451
Longleaf pine, 402, 500, 504, A16
Loon, 502, A16
Lotus, 450
Louse, 128, 255, 256, 375, 458, 461, A16
Lovage, A34, A53
LSD. *See* Lysergic acid diethylamide
Luffa. *See* Vegetable sponge
Luminescent bacteria, 242, 243, A16

Luminous moss, 346, A16
Lupine, 230, 451, 466, A16, A44, A56
Lupine, tree, 188, A16
Lupinus, 230
Lycogala, 301
Lycophyta, 237, 361–70, 384
Lycopodium, 364, 365, 370
Lyngbya, 262
Lysergic acid diethylamide, 318, A48
Lysosome, 35

m

Mace, A51, A53
Macrocystis, 289
Macronutrients, 173
Macrophyll, 237
Madagascar jasmine, A85
Madder, A56
Madder family (Rubiaceae), 77, A16
Madrone, A56
Magnolia, 422, 439, 504
Maidenhair fern, plate 4B, 398, A16, A91
Maidenhair tree, 398
Maize. *See* Corn
Malaria, A46
Malate. *See* Malic acid
Malaysian edible fern, 383, A16
Male fern, 383, A16
Malic acid, 152
Mallow, 451, A16, A39
Malpighi, Marcello, 9
Maltase, 159
Malthus, 217
Maltose, 159
Mammals, 14, 356, 421, 444, 469, 490, 503
Mandragorin, A47
Mandrake, A47
Mango, 172, 455, A16
Mangrove, 71, A16
Mangrove, black, 71, A4
Mangrove, red, 451, A21
Manila hemp, 126, 456, A16
Manioc, 457
Mannitol, 295
Manroot, 70, 446, 478, A16, A47
Manzanita, A39, A56
Maple, 95, 112, 126, 217, 442, 443, 461, 497, 504, 508, A16, A39, A56, A66
Maple, bigleaf, A39
Maple, hard, 102, A16
Maple, sugar, 503, 504, A16, A39
Marchantia, 347–50
Marginal fern, A47
Marigold, 457, 478, A16, A30, A34, A54
Marijuana, plate 2B, 127, 310, A16, A44, A47

Mariposa lily, A39
Marjoram, 126, 472, A16, A34, A54
Marsh marigold, A56
Mass, 16, 29, 487
Mastitis, 472
Maté, 127, 476, A16
Matter, 16
Maxwell, James, 149
Mayapple, A39, A44, A47
Maypop, A39, A48
Meadow foam, 469, A16
Meadow foam family (Limnanthaceae), 469
Meadow saffron, 480, 481, A16
Mealy bugs, 381, 470, A16, A27, A32, A84
Measles, 265, 269, A25
Measles, German, 269
Measurement, units of microscopic, A97
Mechanisms of evolution, 219–21
Medicinal drugs. See Atropine, Belladonna, etc.
Medicinal plants, 6, 9, 407, 455, 458, 462, 464, 465, 472, 474, 480, 481, A44–A50
Mediterranean sage, A30
Medulla, 336
Megaphyll, 361
Megasporangium, 368
Megaspore, 367, 368, 380, 393–95, 417, 419, 422
Megaspore mother cell, 367, 368, 395, 417, 418, 422
Megasporophyll, 366–68
Meiosis, 192–204. For meiosis in individual organisms see also chapters 15–20
Meiospore, 201
Melon, 118, 253, A17
Melon, honeydew, 173, A17
Melonette, 478, A17
Membranes, 28, 132–34, 165, 242
    differentially permeable, 132–34
    of organelles, 27–34
    plasma, 28, 29, 30, 32, 36, 132
    semipermeable. See Membranes, differentially permeable
    vacuolar, 28, 35, 132
Mendel, Gregor, 10, 206, 207, 208–12
Mendel's laws, 209–12
Meningitis, 253, A17
Meniscus, 136
Mentha, 230, 231

Menthol, 403, 473
Mericloning, 298, 483, A64, A65
Merismopedia, 262
Meristem, 38, 51, 52, 289, 351, 400, 447, A65
    apical, 38, 51, 52, 58, 65, 66, 87, 88, 98
    ground, 38, 51, 65
    intercalary, 52, 98
Meristematic tissue, 51, 52, 183, 347
Merkus pine, 391, A17
Mermaid's wineglass, 26, 284, A17. See also Acetabularia
Mescal bean, 466, A44, A51
Mescaline, 470, A51
Mesebrine, A51
Mesocarp, 433, 434, 438
Mesophyll, 26, 113–17, 136, 145, 391
Mesozoic era, 397, 423
Mesquite, 63, 119, 438, A17, A48
Messenger RNA, 44–46
Metabolism, 15, 165, 266. See also Assimilation, Digestion, Photosynthesis, Respiration
Metamorphic rocks, 79
Metaphase, 37, 40, 193, 194, 197–201
Meter, A97
Methane, 5, 259, A17
Methane digester, 259
Mexican jumping bean, 469, A17
Mexican jumping bean moth, 469, A17
Mexican poppy, 183, A17
Mexican stone pine, 401, A17
Meyer fern, A91
Mice. See Mouse
Micelle, 80
Microbiologists, 243
Microbodies, 35
Microcladia, 292
Microcystis, 260
Microfilament, 36
Micrometer, A97
Micronutrients, 173
Microphyll, 237, 361, 364, 367, 368, 369, 371
Micropyle, 393–95, 399, 419–21, 447
Microscope, 8, 24, 25, 26
    electron, 26
Microsporangium, 368, 392, 393
Microspore, 367, 368, 380, 392, 393, 417, 419, 420
Microspore mother cell, 366, 368, 395, 418, 420
Microsporophyll, 366
Microtubules, 34, 39
Middle lamella, 26, 28, 29, 46, 56

Midrib, 113, 124, 350, 378, 398
Mildews, 308, A84
    downy, 310, A9
    powdery, 315, A20
Milk thistle, A31
Milkweed, 96, 438, 439, 510, A17, A48, A56
Milkweed, showy, A41
Milkweed, swamp, plate 5A, A17
Milkweed vine, A30
Milky spore disease, 258, A17
Miller, Norton, 356
Millet, 6, 455, A17
Millimeter, A97
Millipede, 78, A17
Mineral requirements for growth, 173–75
Minerals, 173, 375
Miner's lettuce, A39
Mint, 57, 58, 230, A34, A39
Mint family (Lamiaceae), 403, 470, A17, A22, A30
Miso, 333, A17
Mistletoe, 74, 413, 444, A17, A42, A48
Mistletoe, dwarf, 183, 446, A17
Mites, 375, 381, A17, A28, A32
Mitochondrion, 27, 28, 29, 30, 31, 153
Mitosis, 36–40, 450
Mixture (definition), 17
Mnium, 353, 354
Mohawk Indian, 401
Mohegan Indians, 472
Molasses, 405, 460, 480
Mold, 157, 308, 332, 333
    bread, 312–14
    slime, 183, 232–34, A22
    water, 308, 310–12, A25
Mold (fossil), 385
Molecule (definition), 15, 16, 131
Mole plant, A34
Moles, A34
Mollusc, 490, A17
Monera, Kingdom, 241–65
Moneywort, A91
Monkey, 477, A17
Monkey flower, 182, A17
Monkshood, 458, 459, A17, A42, A48
Monocot, 65, 66, 68, 90, 412, 413, 425, 447, 451, A17
Monocotyledon, 90
Monocotyledonous stems, 96, 97, 98
Monoecious, 425
Monohybrid cross, 211, 212
Monokaryotic, 321, 322
Monterey pine, 403, A17
Moonseed, A42
Moonstones, A91
Moose, 503, A17
Mordant, 334
Morel, 315, 317, A17

Morel, beefsteak, 317
Morel, false, 317
Morgan, T. H., 214
Mormon tea, 407, A39
Morning glory, 176, A17, A51, A56
Morphine, 96, 462, A48
Moses in the cradle, A91
Mosquito, 126, 258, 383, 461, 463, 480, 514, A17, A32
Mosquito fern, 383, A17
Mossamedes desert, 399
Mosses, 231, 345, 352–57, 384, 487, 497, 502, A17. See also Annual mosses, Antler and bone mosses, etc.
Moss rose, 182, A17
Moth, 257, 415, 510, A28, A30
    Argentine, 470, A2, A18, A30
    cabbage, A34
    cinnabar, A31
    codling, A7
    ghost, A30
    gypsy, 257, A13, A32
    Mexican jumping bean, A17
    plume, A30
    root, A31
    sphinx, A30
    yucca, 510, A18
Mother-in-law's tongue. See Sansevieria
Mother of thousands, A91
Moth mullein, 451, A18
Motile, 237, 290
Mountain beaver, 509, A18
Mountain chickadee, 509, A7
Mountain hemlock, 407, A18
Mountain laurel, A42
Mountain lion, 509
Mourning dove, 504
Mouse, 125, 219, 249, 444, 455, 506, 507, A18
Mouse, jumping, 503, A18
Movement, gliding, 183, 242, 263
Movement, molecular, 131, 132
Movement of water through a plant, 135–37
Movements, plant, 15, 175–83
Moyle, Susan, 344
mRNA. See Ribonucleic acid, messenger RNA
Mucilage, 57, 64, 351, 463, 466, 473
Mulberry, 437, 438, A18, A39, A48
Mulberry, red, 101, A18
Mulberry, white, A39
Mule deer, 509, A8
Mule ears, 479, A18
Mullein, 451, A18, A48, A56

Index

Mullein, moth, 451, A18
Multiple fruit, 437, 438
Mumps, 265, 269, A25
Mung bean, 455, 466, 467, A18
Musci. See Mosses
Mushroom, 171, 178, 300, 308, 320–32, A18, A40, A42. See also Black Forest mushrooms; Inky cap mushrooms, etc.
Mushroom spawn, 331
Muskrat, 367, 374, A18
Musk thistle, A31
Mussel, 375
Mustard, 77, 456, 463, A18, A40, A54
Mustard, black, 456
Mustard, cultivated, A18
Mustard family (Brassicaceae), 440, 463, A18
Mutation, 216, 219, 221, 268, 327, A59
Mutator gene, 219
Mutualism, 74
Mycelium, 308–38
Mycologist, 308, 322
Mycophagist, 308
Mycorrhizae, 74, 75
mycorrhizal fungus, 74, 75, 332, 350, 362, 364, 392, 482
mycorrhizal roots, 74–76, 332, 392
Mycota, 308–42
Myristicin, A51
Myrrh, 96, A18
Myrrh tree, 96, A18
Myrtle, 461, A18
Myrtle family (Myrtaceae), A18
Myxamoeba, 302, 303
Myxomycota, 237, 299–303

n

NAA. See Napthalene acetic acid
NAD. See Nicotinamide adenine dinucleotide
NADH. See Nicotinamide adenine dinucleotide
NADP. See Nicotinamide adenine dinucleotide phosphate
NADPH. See Nicotinamide adenine dinucleotide phosphate
Names, plant, 228–40
Namib desert, 399
Nanometer, A97
Naphthalene acetic acid, A61
Narcissus, A44
Narcotics, 127, 460–62, 514
Nastic movements, 177
Nasturtium, 77, 118, 185, A18, A34, A54

Natchez Indians, A45
Native Americans. See Catawba Indians; Cree Indians, etc.
Natural dyes, 334, 338, 458, A18
Natural selection, 217–19
Navajo Indians, A45
Naval stores, 403
Navel orange, 170, 433
Neanthe palm, A91
Neck canal, 353
Nectar, 57, 122, 509–14
Nectaries, 463, 511
Nematode, caterpillar, A29
Nematodes, 76, 312, 314, 478, A18, A30. See also Eelworms
beneficial, A29
root-knot, 76
Nereocystis, 289
Nest fern, A18
Nettle, 26, A18, A40
Net venation, 116, 412
Neutron, 16
New Jersey tea, A40
Newton, Sir Isaac, 149
New Zealand flax, 481
Ngaio, A44
Nicandros, 330
Nicotinamide adenine dinucleotide, 154, 158
Nicotinamide adenine dinucleotide phosphate, 147, 151
Nicotine, 128, 474, A18
Nigerine, A51
Nightshade, 126, A18, A42
Nightshade, deadly, 127, A18
Nightshade family (Solanaceae), 126, 474, 475, A18
Nitrate, 251, 494, A18
Nitrifying bacteria, 251, A18
Nitrite, 251, 494, A18
Nitrogen
in atmosphere, 251, 493, 494
in soil, 79, 493, 494
Nitrogen cycle, 76, 493, 494
Nitrogen fixation, 251, 263, 493, 494
Nitrogen-fixing bacteria, 81, 250, 251, 269, 493, 494, 499, A18, A77
Nitrogen-fixing blue-green algae, 251, 263, 351, 384, 499
Noah's ark, 403
Nodding movements, 176
Node, 52, 69, 86, 87, 98, 371, 399
Noncyclic photophosphorylation, 151
Norfolk Island pine, A91
Nori, 297, 299
Northern coniferous forest. See Taiga
Noscapine, 462
Nostoc, 262, 263, A18

Nucellus, 390, 393, 394
Nuclear envelope, 34, 35, 39, 40, 196, 197, 201
Nucleolus, 25, 27, 28, 34, 39, 40, 196, 197, 201
Nucleoplasm, 34
Nucleotide, 41–44,
Nucleus
atomic, 16
cell, 24, 27, 28, 31, 34, 35, 40
Nut, nutlet, 442, 456, 472
Nut grass, A40
Nuthatch, 503
Nutmeg, 456, A18, A51, A54
Nutmeg, California, 396, A18
Nutritional values of vegetables, A77–A82
Nyppon sopa, 466

o

Oak, 112, 315, 330, 442, 497, 504, 508, A18, A40, A56
bur, 416, A18
cork, 53, A18
Hooker, 94, A18
live, 497, A18
white, 102, A18
Oats, 98, 167, 188, 442, 456, 480, A18
Octopus tree, A91
Oedogonium, 281, 282, A18
Oil, 4, 19, 56, 57, 58, 101, 126, 295, 296, 350, 401, 407, 436, 444, 450, 460, 462, 464, 465, 469–74, 478, 479
Okra, 457
Oleander, A42, A92
Olibanum tree, 96, A18
Olive, 6, 330, 396, 434, 435, 456, A18
Ololiuqui, A51
Oneida Indians, 466
Onion, 100, 119, 235, 451, 456, A18, A34, A41, A56, A76, A80, A81
Onion, wild, A40, A48
Oogamy, 283, 285
Oogonium, 281, 282, 291, 310, 311
Oomycetes, 308, 310–12, A30
Operculum, 354, 355
Opium, 58, 461, 462, A46
Opium poppy, 96, 461, 462, A44, A48
Opossum, 504
Orach, A40
Orange, 126, 333, 423, 436, 455, A18, A52, A59
Orange, navel, 170, 433, A59
Orchid, 52, 63, 72, 73, 75, 100, 182, 298, 382, 413, 414, 422, 440, 441, 443, 444, 451, 482, 483, 511–14, A18, A64, A65, A92. See also Bletilla orchid, Bucket orchid, etc.

Orchidaceae. See Orchid family
Orchid family (Orchidaceae), 482, 483, 511–14, A18, A19
Order, 234
Oregano, 126, 472, A19, A54
Oregon grape, 458, A19, A48, A57
Organelle, 27, 29–35, 242, 276
Organic, 18
acid, 35, 152, 154
acid cycle, 154
compounds, 18, 153
digester, 259
evolution, 216–23
farming and gardening, A33–A35
Organpipe cactus, 511, A19
Organs, plant. See Leaves, Roots, Stems
Oriental water fern, 383, A19
Origin of cultivated plants, 455–57
Osage orange, 101, 437, A19, A57
Oscillatoria, 183, 262, 263, A19
Osmometer, 132
Osmosis, 29, 132, 133, 136, 137
Osmotic potential, 132
Osmotic pressure, 132
Otto of roses, 465
Our Lord's Candle, 510, A19
Ovary, 412, 416, 417–19, 422–25, 433–43, 447, 510
Ovary position, 424, 425, 434, 436, 457, 460, 465, 469, 471, 474–76
Ovule, 390, 393–96, 399, 412, 417, 418, 422, 423, 433, 447
Owl, snowy, 502, A19
Oxaloacetic acid, 152, 155, 158
Oxidation, 148
Oxidation, final, 155, 156
Oxidation-reduction reactions, 148
Oxidative phosphorylation, 158
Oxtongue, A92
Oxygen, 17
generator, 5
Oyster mushrooms, 324, A19

p

P$_{682}$, 150
P$_{700}$, 150, 151
Painted lady, 177, A19
Paiute Indians, A50
Palea, 479
Pale corydalis, plate 6C, A8
Paleobotanists, 386

Palisade mesophyll, 114–16
Palm, 63, 98, 126, 390, 397, 413, A92. *See also* Neanthe palm, Panama hat palm, *etc.*
Palmate compounding, venation, 112, 113
Palm family (Arecaceae), A19
Panama hat palm, 126, A19
Panicle, 417
Pansy, A19
Pansy, wild, A48
Pantothenic acid, 158
Papaveraceae. *See* Poppy family
Papaverine, 462
Papavovirus, 266
Papaya, 457, A19
Papilla, 280
Paprika, A49, A52
Parallel venation, 112, 113, 412
Paramylon, 236, 285, 295
Paraphysis, 353, 354
Pará rubber, 96, 457, 468, A19, A72
Parasites, 76, 249, 253, 263, 309, 312, 314, 315, 470
   bacterial, 267, 268
   energy, 253
   flowering plant, 74, 413
   fungal, 308, 326, 327, 335, 350
Parasitism, 74, 491
Paratyphoid fever, 254, A19
Parenchyma, 53–55, 59, 65, 73, 88, 97
Parental generation, 208–14
Parkinson's disease, A50
Parlor palm, A92
Parsley, 126, 456, 475, A19, A33, A34, A53, A54
Parsley family, A19. *See also* Carrot family— (Apiaceae)
Parsnip, 76, 456, 475, A19
Parthenocarpy, 433, 436–38
Passage cell, 67, 68
Passion fruit, 457, A19
Pasteur, Louis, 25, 265
Patchouli oil, 126, A19
Pauling, Linus, 42
Pawpaw, A40
Pea, 76, 118, 166, 170, 172, 206–14, 421, 438, 448, 456, 466, 493, A19, A34, A81
Pea, sweet, 214, 425, 467, A19, A23, A44
Peach, 173, 187, 415, 416, 421, 422, 424, 434, 435, 455, 465, A19, A32, A57, A73, A74
Peach leaf curl, 319, A19
Peach tree borers, 257
Peacock fly, A31

Peanut, 188, 229, 438, 442, 457, 466, A19
Pear, 26, 71, 173, 253, 436, 449, 456, 465, 470, A19, A32
Peat moss, 345, 352, 356, 357, A19
Pecan, 442, A19
Pectin, 29, 40, 125, 286
Pediastrum, 277
Pedicel, 178, 416, 442
Peduncle, 415, 416
Peel (fossil), 386
Pellicle, 285
Penicillin, 256, 297, 332, 333, A19
Penicillium mold, 332, 333
Pennycress, A40
Pennyroyal, 126, 127, A19, A29, A48
Peony, 438, A19
PEP. *See* Phosphoenol pyruvate
Peperomia, A19, A60, A92
Pepo, 436, 437
Pepper, 186, 434, 457, 474, A19, A32, A81
Pepper, black, 455, A19
Pepper, red, 34
Pepper family, 52
Peppergrass, 463, A19, A40
Peppermint, 126, 231, 456, 472, 473, A19, A30, A54
Peptide bond, 20
Perennial, 415, 502, 503, 507
Perennial sowthistle, A31
Perfume, 333, 403, 460, 465, A6
Perfume stabilizer, 333, A19
Pericarp, 423, 433, 434, 438, 440, 442, 446
Pericycle, 68, 69
Periderm, 53, 54, 55, 96
Peristome, 355, 356
Permafrost, 501, 502
Permanent tissue, 52
Permanent wilting point, 81
Persian violet, A92
Persimmon, 434, 455, A19, A40, A48
Pesticides, A27
Petal, 121, 415, 416, 422, 426, 427, 455
Peter the Great, 318
Petiole, 86, 87, 100, 110, 114, 122, 126, 378, 383, 458
Petitgrain oil, 126, A19
Petrifaction, 385, 386
Petunia, 14, 474, A19, A34
Peyote, 470, 471, A19, A48, A51
PGA. *See* Phosphoglyceric acid
PGAL. *See* Phosphoglyceraldehyde
Phaeophyta, 237, 288–91, 295. *See also* Brown algae

Phage. *See* Bacteriophage
Pharmacopoeia, 6
Phelloderm, 90, 91
Phellogen. *See* Cork cambium
Phenotype, 209–16
Pheromones, A29
Philodendron, A44, A92
Phloem, plate 1, 59, 60, 91, 97, 137, 346, 401
   components of, 59, 60
   function of, 60
   parenchyma, 59
   primary, plate 1, 65, 68, 88
   ray cells, plate 1, 59
   secondary, plate 1, 89, 392
   transport of food substances, 137, 138
Phoebe, 344, A19
Phosphoenol pyruvate, 152
Phosphoenol pyruvate carboxylase, 152
Phosphoglyceraldehyde, 151, 155, 158
Phosphoglyceric acid, 151
Phospholipid, 19
Phosphorescence, 149, 180, 288
Photoelectric effect, 149
Photon, 149
Photoperiod, 184–86
Photoperiodism, 184–87
Photophosphorylation
   cyclic, 151
   noncyclic, 151
Photorespiration, 146, 152
Photosynthesis, 15, 35, 56, 139, 143–53, 335, 346, 413, 489, 495, 507
   CAM, 139, 506, 507
   C_3, 149–52
   C_4, 152
   dark reactions, 147, 148
   details of, 147–53
   essence of, 144
   light reactions, 147–51
   major steps, 147–53
Photosynthesis and respiration compared, 153
Photosynthetic units, 147, 149
Photosystems, 149–51
Phototaxy, 182
Phototorsion, 178
Phototropism, 177–79
Phragmoplast, 40, 344
Phycobilins, 147, 236, 261, 295
Phycocyanin, 261
Phycoerythrin, 261
Phylum, 234, 303
Physarum, 301
Physiology, plant, 8
Phytoalexins, A29, A30
Phytochrome, 183, 184, 186

Phytoconis, A4
Pickerel weed, A40
Pig. *See* Hog
Piggyback plant, A92
Pigment, 125, 126, 146, 147, 149, 184, 245, 261, 265, 275, 286–95, 344, 382
Pignolias, 401
Pigweed, A34
Pillbugs, 78, A19
Pilobolus, 179, 314, 315
Pilus (pili), 242, 247
Pimiento, A54
Pine, 58, 90, 102, 110, 117, 139, 170, 327, 390–95, 397, 398, 401–3, 503, 504, A19, A40. *See also* Afghanistan pine, Aleppo pine, *etc.*
Pineapple, 169, 173, 187, 437, 457, A20, A92
Pinedrops, 74, A20
Pine leaf, 117, 391, 401
Pine needle. *See* Pine leaf
Pink bollworm, A29
Pink polka dot plant, A93
Pinkroot, A48
Pinna, 377–79, 398
Pinnate compounding, venation, 112, 113
Pinto bean, 134, A20
Pinworm, A48
Pinyon, 401, A20
Pipsissewa, A40, A48
Pistachio, 442, A20
Pistil, 416, 417, 422, 423–25, 434, 437, 458
Pit, 46, 47, 58, 59, 95, 434
Pitcher plants, 122–24, 350, A20, A48
Pitcher plants, Asian, 122, A20
Pitch pine, 403, 504, A20
Pith, 65, 88, 90, 91, 170, 371, 392, 407, A67
Pittosporum, A44, A92
Plankton, 275, 296, 297
Plantago major, 229
Plantain, 169, 229, A20, A40, A48
Plant anatomy, 8
Plant cells vs. animal cells, 36
Plant community, 487
Plant ecology, 10. *See also* chapter 23
Plant geography, 10
Plant morphology, 10
Plant movements, 15, 174–83, 466
Plant physiology, 9
Plant press, 425–27
Plant systematics, 9
Plant taxonomist, 9, 236
Plant taxonomy, 9, 236
Plasma membrane, 28, 29, 30, 32, 36, 285

Plasmids, 245, 248
Plasmodesma, 27, 28, 36, 40, 46
Plasmodium, 300–303, 308
Plasmolysis, 133
Plastid, 32–34, 36, 233. *See also* Chloroplast, Chromoplast
Plato, 7, 230
Pleurisy root, A48
Pleurococcus, A4
Pleuropneumonia-like organisms, 256
Pliny, 464
Plover, 502, A20
Plum, 187, 398, 434, 465, A20
Plumeless thistle, A31
Plume moth, A30
Plumule, 166, 447–49
Pneumatophore, 71
Pneumonia, 243, 253, 256, A20
Poaceae. *See* Grass family
Pocket gopher, 475, A13
Podocarps, 396, 407, A20
Podophyllin, A47
Poi, 139
Poinsettia, 121, 122, 185, 186, 467, 469, A20, A44, A92
Poison, 6, 127, 328
Poison antidote, 6, 328, 383, A20
Poison hemlock, 476, A20
Poison ivy, A20, A41, A44, A49
Poison oak, 479, A20, A41, A44
Poisonous plants, 70, 382, 407, 455, 466, 467, 469, 470, 474–76, 481, A41–A44
Poison sumac, A44
Poke, A40, A42, A57
Polar bear, 502, A4
Polar nuclei, 418–22
Pole beans, 169, A78
Poles (cellular), 39, 40, 194–97
Polio, 265, 330, A25
Pollen, 182, 206, 207, 396, 397, 421, 509–14
Pollen chamber, 393
Pollen cone, 393, 395
Pollen grains, 386, 390, 393–95, 412, 417, 419–21, 433, 499, 511
Pollen tube, 394, 395, 419, 421
Pollination, 169, 206, 207, 394, 416, 421, 482, 509
Pollination ecology, 421, 509–14
Pollinium, 482, 512–14
Pollution, 4
    effects of, 114, 263, 335, 487, 492

Polyanthus, 14, A20
Polypeptide chain, 20, 45
Polyploid, 220
Polyploidy, 220
Polypodium, 379
Polysaccharide, 18, 144, 148, 261, 308
Polysiphonia, 293–95
Pome, 436, 437, 465
Pomegranate, 6, 436, 456, A20, A57
Pomo Indians, A50
Ponderosa pine, 401, 403, 499, 500, 508, A20
Poor man's pepper, 187, A20
Popcorn, 447, A20
Poplar, 230, 330, A20
Poplar leaf spot rust, 328, A20
Poppy, 6, 420, 440, 441, 443, 456, 461, A20, A54. *See also* California poppy, Mexican poppy, *etc.*
Poppy family (Papaveraceae), 461, A20
Population, 487, 488
Populus, 230
Porcupine, 367, A20
Pore fungi, 323
Porifera, 234, 303
Postelsia, 289
Potato, 457, 491, A34
Potato, Irish, 4, 5, 6, 9, 99, 100, 173, 185, 253, 433, 457, 467, 474, 475, A20, A63, A81, A82
Potato, sweet, 70, 76, 77, 457, 467, A20, A64
Potato, white. *See* Potato, Irish
Potato blight. *See* Late blight of potato
Potato vine, 118, A20
Potential pressure, 132
Pot marigold, A34
Powdery mildew, 315, A20, A31
PPLO pneumonia, 256, A20
PPLO's. *See* Pleuropneumonia-like organisms
Prairie, 497, 505
Prairie turnip, A40
Prayer plant, 177, 180, A21, A93
Praying mantis, A28
Preferns, 381, A21
Preservation of flowers, 425–27
Pressure-flow hypothesis, 137, 138
Prickles, 100, 480
Prickly ash, A48
Prickly lettuce, A57
Prickly pear cactus, 347, 457, 470, A21, A30, A40
Prickly poppy, A44
Primary consumer, 489–91

Primary meristem, 51, 65
Primary phloem, 65, 68, 87, 88
Primary succession, 496–99
Primary tissue, 66, 88
Primary wall, 28, 29, 40, 46, 56
Primary xylem, 65, 68, 87, 88
Primitive flower, 422, 423, 457
Primordium, 87, 88, 111, 171, A65
Primrose, 6, A21, A93
Privet, A44, A57
Procambium, 38, 51, 65, 87
Prochloron, 264
Prochlorophyta, 236, 265
Prochlorophytes, 259, 264, 265, A21
Producer, 489–91
Progametangium, 312, 313
Prokaryotic, 36, 233, 241, 259, 308
Pronghorn, 506, A21
Propagation. *See* Vegetative propagation
Prophase, 37, 39, 194, 196, 201
Proplastid, 34
Prop roots, 72, 73, 166, 449
Protective layer (abscission), 125
Protein, 20, 21, 29, 96, 144, 159, 261, 319, 329, 354, 401, 450, 467, 480, 493
    synthesis, 44–47
Prothallus, 377, 380, 381
Protista, 233, 259, 274–306
Protococcus, A4
Protoctista, 233
Protoderm, 38, 51, 65, 87, 88
Proton, 16, 150
Protonema, 346, 351, 355, 356
Protoplasm, 14, 18, 27, 29–36, 145, 275, 493
Protozoans, 233, 234, 261, 275, 303, 312, A32
Prunes, 465
Pruning, A73–A76
Pseudomonad bacteria, 319, A21
Pseudoplasmodium, 303
Psilocin, A51
Psilocybe mushrooms. *See* Teonanacatl mushrooms
Psilocybin, A51
Psilophyta, psilophytes, 237, 361–63
Psilophyton, 381
Psilotum, 361–63
Psittacosis, 253
Ptarmigan, 502, A21
Pteridosperms, 390, A21
Pterophyta, 237, 376–84
Puccoon, 455
Puffballs, 308, 320, 324, A21
Pulp (wood), 103, 104, 404

Pulque, 127, A21
Pulvinus, 180–82
Pumpkin, 425, 436, 457, 477, A21, A34, A76
Pumpkin family (Cucurbitaceae), 70, 118, 173, 425, 436, 446, 451, 476–78, A21
Puncture vine, 444, 445, A21, A30
Punnett square, 212
Pun-tsao, 6
Purine, 41, 42
Purple avens, A40
Purple laver, 297, 299, A21
Purple nonsulfur bacteria, 249, A21
Purple sulfur bacteria, 249, A21
Purple tiger, A93
Purslane, A40
Puya, 412, A21
Pygmy mosses, 345, A21
Pyrenoid, 276–81, 351
Pyrethrum, 478, A77, A84
Pyrimidine, 41, 42, 44
Pyrrophyta, 236, 288, 295. *See also* Dinoflagellates
Pyruvic acid, 152, 154–58

q

Q (acceptor molecule), 150
Quartersawed, 101, 103
Quassia, A48
Quebec Indians, A48
Quercitron, A57
Quill, cinnamon, 460
Quillworts, 361, 363, 367–70, A21
    human and ecological relevance of, 369, 370
Quince, 6, 436, A21
Quinine, 457, A21, A46

r

Rabbit, 263, A21
    cottontail, 506, A21
    jack, 506, A21
Rabies, 265
Raccoon, 504, A21
Raceme, 417
Rachilla, 479
Rachis, 113, 118, 378, 383
Radial symmetry (of flowers), 422–24, 466
Radicle, 63, 165, 188, 447–49
Radioactive tracer, 137
Radish, 35, 70, 76, 206, 440, 455, 463, A21, A34
Rafflesia, 413, 414, A21
Ragweed, 185, A21
Rambler, 100
Ranunculaceae. *See* Buttercup family

Rape dos Indios, A51
Rappahannock Indians, A48
Raspberry, 100, 119, 434, 437, 465, A21, A60, A74
Raspberry, wild, 466, A40
Rat, 255, 507, A21
Rat snake, black, 504, A21
Rattlesnake, 507, A21
Rauwolfia, A48
Ray, 58, 94, 95
    parenchyma, 58, 95
    phloem, plate 1, 94
    xylem, plate 1, 94
Ray floret, flower, 478
Reactive molecule, 150, 151
Receptacle, 290, 291, 415, 416, 417, 425, 433, 436–38, 478
Receptive hyphae, 327
Recessive, 209–16, 455
Red algae, 292–96, A21
Redbud, A40
Red cedar, 407, A21
Red cedar, western, 508, A21
Red fox, 504, A21
Red light, 183, 184
Red mangrove, 451, A21
Red mulberry, 101, A18
Red ribbons, plate 7C, A21
Red Sea, 261, A21
Red spider mites, A32
Red spruce, 103, 405, A21
Red tide, 288
Reduction division, 196
Redwood, 26, 29, 391, 397, 406, 413
Redwood, coastal, plate 2D, 95, 370, 397, 406, 508, A21
Redwood, dawn, 406, A8
Redwood, giant, 392, 397, 406, 499, 508, A21
Reed, 344
Reindeer, 329, 338, 345, A21
Reindeer lichens, 338, A21
Reindeer moss, 352
Reproduction, 14. See also Asexual reproduction, Sexual reproduction
Reproductive isolation, 220, 221
Reproductive leaves, 121
Reptiles, 14, 490
Reserpine, 77, A21, A48
Reservoir (of Euglena), 285
Resin, 58, 101, 391, 402, 403, 406, 466, 482, A30
Resin canal, 95, 96, 117, 391, 392, 403
Resistant varieties, A29
Respiration, 15, 31, 153–59, 167, 450, 495
    aerobic, 153–56, 450
    anaerobic, 156, 450
    details of, 154–59
    electron transport chain, 155, 156, 158

essence of, 153
    factors affecting rates of, 156, 157
    fermentation, 154–56
    final oxidation, 155, 156
    glycolysis, 154, 155, 157, 158
    Krebs cycle, 154, 155, 158, 159
    major steps, 154–59
Response to stimuli, 14
Resurrection plant, 369, A21
Retting, 98
Rex begonia, A54
Rhizine, 336
Rhizoid, 309, 347, 350, 352, 354, 355, 363, 373, 377, 380
Rhizome, 98, 99, 361, 363–66, 371, 372, 373, 376, 377, 383, 461, 466, 500, A63
Rhizopus, 312, 313
Rhododendron, A21, A42, A57, A60
Rhodophyta, 237, 292–96. See also Red algae
Rhubarb, 126, 456, A21, A42
Rhynia, 362, 381
Rib (Equisetum), 354, 355
Ribonucleic acid, 35, 44, 267, 268, 330
    messenger RNA, 44, 45
    ribosomal RNA, 45, 46
    and synthesis of proteins, 44–47
    transfer RNA, 44, 45, 46
Ribosomes, 27, 28, 31, 32, 44–46, 242
Ribulose biphosphate, 147, 148, 151, 153
Ribulose I, 5-biphosphate. See Ribulose biphosphate
Rice, 6, 98, 169, 333, 383, 442, 445, 480, A21
Ricepaper plant, 71, A21
Rickets, 383, A21
Rickettsias, 256, A21
Ringworm, 333, 461, A21, A45, A48
RNA. See Ribonucleic acid
Robin, 344, A21
Rock cress, A21
Rock mosses, 345, 352, 356, 497, A21
Rockrose, European, 178, A21
Rock tripe, 338, A21
Rockweed, 290, 291, A21
Root beer, 460
Root cap, 64, 65, 66, 179
    functions of, 64
Root grafting, A66
Root hair, 52, 63, 66, 67, 332, 366, 493
Root knot, 76
Root-knot nematodes, 76
Root moth, A31
Root pressure, 136

Roots, 62–76, 120, 121, 392, 404, 407, 425, 433, 447–49, 455, 458, 466, 493, 497
    adventitious, 63, 72, 73, 98, 100, 121, 166, 364, 371, 376, 377, 449
    aerial, 52, 63, 72, 73, 177, 382
    branch, origin of, 68, 69
    buttress, 73, 74
    contractile, 73
    fibrous, 63, 64
    food storage, 70, 467
    functions of, 63
    human relevance of, 76, 77
    mineral absorption by, 63
    mycorrhizal, 74–76, 332, 392
    nodules, 76, 250
    parasitic, 74, 75
    photosynthetic, 72, 73
    pneumatophores, 71
    prop, 72, 73, 166
    propagation by, A61
    propagative, 71
    radicle, 63, 165
    regions of, 38, 65–69
    specialized, 69–76
    structure, 38, 64–69
    taproot, 63, 64, 76, 400
    velamen, 52, 72
    water storage, 70, 478
Rosaceae. See Rose family
Rosary plant, A93
Rose, 119, 185, 465, 466, A22, A74
Rose, damask, 465, A22
Rose, wild, 464, 465, 466, A22, A40
Rose family (Rosaceae), 464, 465, A22
Rose hips, 465, 466
Rosemary, 456, 472, A22, A34, A54
Rosin, 403
Rotenone, 77, 128, A22, A77, A84
Roundworms, A29, A48
rRNA. See Ribonucleic acid, ribosomal RNA
Rubber, 58, 96, 319, 468
Rubber, Pará, 96, 457, 468, A22
Rubber plant, 455, 468, A22, A93
Rubber vine, A42
RuDP. See Ribulose diphosphate
Rue, A54
Ruellia, 139, A22
Rum, 480
Runner, 98
Russian thistle, A31
Rust, 308, 320, 326–28, 378, A30
Rutabaga, 76, 463, A22
Rye, 98, 456, 480, A22
Ryegrass, 63, 67, A22

S

Sac fungi. See Cup fungi
Safflower, 456, A57
Saffron, meadow, 480, 481, A22, A49
Saffron, true, 481, A22, A54, A57
Safrole, A51
Sage, 126, 456, 472, A22, A34, A54, A57
Sagebrush, 128, 472, 478, A22
Sago palm, A93
Saguaro. See Giant saguaro cactus
St. Anthony's fire, 318
St. John's wort, A57
Sake, 318
Salamander, 504
Salicin, A50
Salmon, 310, A22
Salmonberry, 466, A22
Salmonella (food poisoning), 254, A22
Salsify, 70, 479, A22, A40
Salt, 17, 35, 96, 139, 251, 333, 385
Saltbush, A40
Salvia, 185, 473, A22
Samara, 442, 443
Sandalwood, A57
Sandbox tree, A43
San Pedro, A51
Sansevieria, 98, 126, 481, A22, A60, A93
Santonin, 478, A22
Sappanwood, A57
Saprobe, 249, 308
Saprolegnia, 311
Saprophyte, 309, 362, 413
Saprophytic orchids, 482, A22
Sapwood, 69, 95
Sarcina, 260
Sargassum, 290, A22
Sarsaparilla, 77, 480, A22, A49, A54
Sassafras, 77, 460, A22, A40, A49, A51, A54, A57
Satellites (chromosomal), 39
Satin pothos, A93
Sauerkraut, 258
Sausage tree, African, 182, A22
Saussure, Theodore de, 148
Savory, 126, 456, A22
Savory, summer, A54
Savory, winter, A54
Sawflies, A28, A30
Saxifrage, 100, A22, A63
Scale, 237, 348, 393–97
Scale insects, A28, A84
Scale leaves, 99, 100, 237, 309, 371
Scallion, A54
Scarification, 449
Scarlet runner bean, 166, A22

Scenedesmus, 277
Schizocarp, 443
Schizomycetes, 242–59
Schizophyta, 236–63
Schleiden, Matthias, 25
Schwann, Theodor, 25
Science, 6, 221
Scientific method, 6
Scientific names, 229–40,
    A1–A32, A37–A58,
    A86–A95
Scientific procedure, 6
Scion, A66–A73
Sclereid, plate 2A, 56, 57
Sclerenchyma, plate 1, 56,
    57, 97, 98, 118, 371
Scopolamine, 127, 474, A46,
    A47, A51
Scotch broom, A31, A57
Scouring rushes. *See*
    Horsetails
Screw pine, 72, A22, A93
Scurvy, 401, 405, 465, A45,
    A48
Sea anemone, 261, A22
Sea hare, 299, A22
Sea lettuce, 284, 285, 297,
    A22
Sea palm, 289, A22
Sea rocket, 463, A22
Sea squirt, 264, 265
Seaweed, 275, 284, 288–99,
    487, 495
Seaweeds, preservation of,
    292
Secondary consumer, 489–91
Secondary phloem, plate 1,
    89, 392
Secondary succession, 496,
    499, 500
Secondary tissues, 66
Secondary wall, 29, 40, 46,
    56
Secondary xylem, plate 1, 89
Second filial generation,
    208–14
Secretory cells, tissue, 53,
    54, 56, 57
Sedge, 446, 498, 502, A22
Sedimentary rocks, 79
Seedcoat, 134, 166, 187, 390,
    394, 418, 420, 422,
    447, 448, 449
Seed cone, 393
Seed dispersal, 394, 443–47,
    499
Seed ferns, 384, 390, A22
Seed fly, A31
Seeds, 75, 135, 184, 193,
    206, 208, 217, 390,
    401, 407, 412, 415,
    417, 418, 420, 421,
    433–53, 455, 458,
    462, 497, 500, A76
cotyledons of, 447–49
development of, 394,
    418–22

dispersal of, 394, 443–47
dormancy, 187, 188,
    448–51
germination of, 134, 135,
    165, 166, 172,
    173, 177, 187,
    188, 448–50,
    A76–A82
longevity, 157, 450, 451
structure of, 165, 422,
    447–49
Seed weevil, A30, A31
Selaginella, 364–67, A93
Selection, 216
Selenium, 173
Self-heal, A49
Self-pollination, 206, 421
Seneca snakeroot, A49
Senna, 455, A22, A49
Sensitive plant, 15, 182, 466,
    A22, A93
Sepal, 415, 416, 460
Separation layer (abscission),
    125
Sere, 501
Sernander, 446
Serpentine soil, 489
Serviceberry, A41
Sesame, 457, A22, A54
Seta, 348, 349, 351, 354, 355
Sexual reproduction,
    193–202. *See also
    specific organisms*
Seychelles Island palm, 110,
    A22
Shade leaves, 117
Shaggy mane mushroom,
    328, A22
Shallot, A54
Sheep, 262, 333, 338, 466,
    502, A22
Shelf fungi, 300, 320, 324,
    A22
Shellfish, 288, 356
Shen Nung, 6
Shepherd's purse, 440, 463,
    A22, A41
Shiitake mushroom. *See*
    Black Forest
    mushroom
Shoot, 38, 165
Shoot meristem culture, A65
Shoot system, 86
Short-day plant, 185
Shortleaf pine, A22
Shoshoni Indians, 455, A49
Showy milkweed, A41
Showy orchid, 513, A22
Shoyu, 333
Shrew, 502, 503
Shrimp, 258, 283, A22
Shrimp plant, A93
Siberian white pine, 401, A22
Sieve cell, 60
Sieve plate, 59, 60
Sieve tube, 59, 137, 138
Sieve tube element, 59, 60,
    94, 97
Silage. *See* Ensilage
Silica, 370–72, 374, 375, 385

Silicle, 440, 463
Silique, 440, 463
Silt, 80, 498
Silurian, 362
Simple fruit, 434–37
Simple leaf, 111–13, 423
Simple pistil, ovary, 423
Simple pit, 46, 47
Sink, 137
Sipe, M., A29
Sisal, 98, A22
Sitka spruce, 405, A22, A75
Skeleton weed, A31, A49
Skullcap, A49
Skunk, 131, A22
Slash pine, 403, 504, A22
Slate-colored junco, 344, A22
Sleep movements, 180, 181
Slenderflower thistle, A31
Slime mold, 183, 232–34,
    299–303, 308, A22
Slippery elm, A49
Slugs, 303, A28, A29
Smallpox, 265, 473, A25, A48
Smog, 113, 403, 460
Smoke tree, A57
Smooth sumac, A57
Smut, 308, 320, 326, A22
Smut, corn, 326
Snails, 258, 261, 287, 328,
    478, A22, A28
Snake, 318, 504, 507. *See
    also* Black rat snake,
    Copperhead, *etc.*
Snap beans, A77
Snapdragon, 185, 210, 440,
    A22
Sneezeweed, A44
Snout beetles, A30, A31
Snowbank algae, 275, A22
Snow-on-the mountain, A44
Snowplant, plate 8A, 413,
    A22
Snowy owl, 502, A22
Soap, 338, 482, A22
Soap plant, A41
Soaproot, California, 481,
    482, A23
Socrates, 476
Softwood, 391
Soils, 77–81, 121, 253, 425,
    489, 493, 497, 499,
    502
aggregates, 81
air in, 78
bedrock, 78
cultivation of, A76–A82
erosion, 79
essential elements,
    173–75
horizons, 78
mineral components, 80
organic composition of,
    79
organisms in, 78, 79, 493,
    494, 499
parent material, 78, 79
profile, 78
serpentine, 469
structure of, 80, 81

texture of, 80
topsoil, 78, 79
water in, 80, 81
Solanaceae. *See* Nightshade
    family
Solomon's seal, A41
Solution, 132, 133
Solvent, 132, 258, 319, 403
Sorbose, 258, A23
Soredium, 336
Sorghum, 6, 185, 455, 480,
    A23
Sorrel, 77, 177, 328, A23, A41
Sorus, 377–79
Southern white cedar, 407,
    A23
Southernwood, A34, A54
Southern yellow pines. *See*
    Loblolly pine,
    Longleaf pine,
    Shortleaf pine, Slash
    pine
Sowbugs, A28
Soybean, 6, 184, 185, 314,
    333, 455, A23, A34,
    A78
Soy sauce, 333, A23
Spanish moss, 352, 415, A23
    savannah, 506, A23
    vesper, 506
Spathe flower, A93
Spatterdock, A41
Spawn (mushroom), 331
Spearmint, 126, 472, A23
Specialized leaves, 116–25,
    466
Specialized roots, 69–76,
    A63, A64
Specialized stems, 98–100,
    A63, A64
Species, 231, 234, 243, 412,
    427, 455, 457
Species Plantarum, 10, 231,
    232, 234
Specific epithet, 235
Spectrum, 146, 149
Speedwell, A41
Sperm. *See* Gametes
Spermagonium, 327
Spermatangium, 293, 294
Spermatium, 293, 294, 327
Spermatogenous cell, 394,
    395
Sperm whale, 469, A23
Sphagnum moss, 352, A63.
    See also Peat moss
Sphenophyta, 237, 361,
    370–75
Sphinx moth, A30
Spice, seasoning and
    flavoring plants, 77,
    126, 457, 478, 481,
    A52–A55
Spicebush, A49
Spider, 470
Spider mite, A32, A84
Spider plant, A93
Spiderwort, A23, A94

Spike, 230, 417
Spikelet, grass, 479
Spike moss *(Selaginella)*,
    364–67, 369, A23,
    A94
Spinach, 126, 185, A23, A34,
    A82
Spindle (and fibers), 34, 39,
    40, 194–201, 481
Spine, 118, 119, 407, 469,
    481, 492
Spiny emex, A31
Spiraling movements, 176
Spirilli, 244, 245
Spirillum, 260
Spirogyra, 32, 280, 281, A23
Spirulina, 262
Splitleaf philodendron, A94
Sponge, 233, 234, 303, A23
Sponge, vegetable, 476, 477,
    A24
Sponge algae, 275
Spongy mesophyll, 114–16,
    391
Sporangiophore, 312, 313,
    372–74
Sporangium, 300–303,
    312–15, 344, 349,
    350, 354, 355, 362,
    363, 366, 372, 373,
    374, 377–80
Spore, 193, 201, 202, 217,
    300–303, 310–38,
    347–56, 362, 363,
    372, 373, 375, 377,
    380, 390
Spore mother cell, 202,
    349–52, 354, 355,
    363–66, 372, 373,
    377, 380, 394
Spore print, 322, 323
Sporophyll, 366–68, 390
Sporophyte, 201, 447
    of algae, 278–95
    of angiosperms, 412–22,
        447
    of ferns, 376–84
    of ground pines
        (Lycopodium),
        364, 365
    of gymnosperms,
        390–400
    of hornworts, 351
    of horsetails *(Equisetum)*,
        370–74
    of liverworts, 347–49
    of mosses, 352–56
    of quillworts *(Isoetes)*,
        367, 368
    of spike mosses
        (Selaginella),
        364–67
    of whisk ferns *(Psilotum)*,
        361–63
Sporophyte generation, 201,
    202, 346
Spotted fever, 256, A23
Spotted knapweed, A31

Sprenger fern, A94
Spring beauty, plate 6A, A23,
    A41
Spruce, 170, 391, 503, A23,
    A49
    black, 405, A23
    red, 103, 405, A23
    Sitka, 405, A23, A75
    white, 404, 405, A23
Spurge, 100, 450, 467, 468,
    469, A23
Spurge family
    (Euphorbiaceae), 58,
    96, 467–69, A23
Squash, 118, 176, 253, 425,
    433, 436, 451, 477,
    A23, A34, A76, A82
Squawroot, 476, A23
Squills, 480, A23, A49, A64
Squirrel, 394, 444, A23
Squirrel, gray, 504, A23
Squirrel, ground, 256
Squirrel corn, 221, 229, A23,
    A44. *See also*
    Dicentra
Squirting cucumber, 183, A23
Staghorn fern, A94
Stamen, 231, 416, 417, 422,
    423, 425, 437, 458
Stanley, Wendell, 266
Stapelia, 510, A23
Staphylococcus, 260
Star anise, A54
Starch, 18, 34, 56, 70, 76,
    101, 152, 159, 179,
    275, 295, 344, 407,
    450
Star-of-Bethlehem, A44
Star thistle, A31
Statocytes, 179
Statospore, 286, 287
Stem gall fly, A31
Stemonitis, 300
Stems, 85–108, 433
    aerial, 86
    applied uses of, 101–6
    development of, 87–90
    dicot herbaceous, 90, 91
    dicot woody, 91–96
    growth, 87–90
    herbaceous, 91
    monocot, 96, 97, 98
    origin of, 87–90
    pine, 392
    specialized, 98–100, A63
    tissue patterns in, 90–98
Stephanotis, A94
Sterigma, 321, 322
Sterile cell, 394, 395
Stigma, 182, 206, 416, 420,
    421
Stinkhorn, 308, 320, A23
Stipe (of kelps), 289
Stipitococcus, 286
Stipules, 86, 87, 100, 118,
    119, 458
Stipule scars, 86
Stock, A66–A73
Stolon, 98, 99, 100, A63

Stoma(ta), 52, 90, 110, 114,
    115, 121, 135, 137,
    138, 139, 171, 351,
    354, 507
    distribution, 115
    function of, 115
    guard cells, 52, 53, 115,
        116, 139
    mechanism of opening
        and closing, 115,
        139
    sunken, 117, 391
Stone cell. *See* Sclereid
Stonecrop, 152, 177, 183,
    A23, A54, A94
Stone pine. *See* European
    stone pine, Mexican
    stone pine
Stoneseed, 455, A23, A49
Stonewort, 285, A23
Storage tissue. *See*
    Parenchyma
Stove fern, A94
Stramonium, A47
Stratification, 187, 450
Strawberry, 100, 185, 193,
    434, 434, 437, 457,
    466, A23, A41, A63
"Strep" throat, 253, A23
Streptobacillus, 260
Streptococcus pneumoniae,
    243
String beans, 433, A77
String-of-pearls, 112, A23,
    A94
Strobel, G., 319
Strobilus, 364, 372–75, 390,
    393–97, 399, 400
Stroma, 33, 147
Strophanthus, A49
Strychnine, 256, 455, A23,
    A49
Strychnine plant, A49
Style, 416, 421
Stylet, 137
Suberin, 53, 67, 89, 125
Subsoil, 78
Succession, 496–99
    after a fire, 499–501
    primary, 496–99
    secondary, 496, 499
Succulent, 152, 507
Sucker, 71
Sucrose, 18, 144, 159
Sudden infant death, 254
Suez canal, 63
Sufu, 314
Sugar, 18, 25, 29, 35, 41, 46,
    96, 144, 159, 261,
    296, 333, 354, 401,
    405, 450, 470, 473,
    480. *See also*
    Fructose, Glucose,
    etc.
Sugar beets, 76, 296
Sugar cane, 170, 455, 480,
    A23
Sugar maple, 503, 504, A39
Sugar pine, 393, 508
Sulfur bacteria, 143, 156, A23

Sulphur dioxide, 335
Sulphuric acid, 81
Sumac, 126, 499, A23, A49
Summer savory, A34
Sunbean, 249
Sunbird, 344, 511, A23
Sundew, 57, 123, 124, A23,
    A94
Sunflower, 90, 185, 249, 442,
    479, A23, A34, A41
Sunflower family
    (Asteraceae), 443,
    472, 478, 479, 504,
    A22, A23
Superior ovary, 422, 424,
    425, 434, 457, 458,
    460, 465, 472, 474,
    480
Support tissue. *See*
    Sclerenchyma
Surface tissue. *See*
    Epidermis
Swamp milkweed, plate 5B,
    A17
Swarm cell, 302, 303
Sweet bay, 460
Sweet cicely, A41
Sweet clover, 466
Sweet flag, A41, A49, A51
Sweet gum, 438, A49
Sweet pea, 214, 425, 467,
    A19, A23, A44
Sweet potato, 70, 76, 77, 457,
    467, A64
Sword fern, 383, A23, A50
Sycamore, A23
Symbiosis, 74, 251, 261, 335,
    362, 384, A29
Synergids, 419–22
Syngamy, 202
Syngonium, A94
Synoptic key, 236–38
Syphilis, 255, 407, A23, A47
Syrian rue, A51

t

Tachinid flies, A28
Taiga, 501, 503
Tamarack, 328, 498, 503, A23
Tamarind, A50
Tamarisk, 63, A23
Tangerine, 436, A23
Tannins, 56, 101, 330, 466
Tansy, A34, A43, A57
Tansy ragwort, A31
Tapeworm, 383, A47
Tapioca, 76, 468
Taro, 139, A23
Tarragon, 126, 478, A24, A34,
    A54
Tarweed, 479, A24
Tarweed, western, 59, A24
Taxis, 182
Taxonomy, taxonomists, 9,
    236
TCDD. *See* 2, 3, 7, 8-tetra-
    chlorodibenzoparadioxin

Tea, 127, 258, 401, 451, 455, 458, 466, A24
Telia, 327
Teliospores, 327
Telophase, 37, 40, 200
Tempeh, 314, A24
Temperate deciduous forest, 319, 501, 503, 504
Temperature
  growth and, 186, 187
  respiration and, 156
Tendril, 99, 100, 118, 176, 476
Tent caterpillars, 257, A24, A28
Teonanacatl mushrooms, 328, A24, A51
Tepary bean, 466, A24
Tequila, 127, A24
Terminal bud, 86, 87, 171
Termite, 331, 344, A1, A24
Tetanus, 255, A24
Tetrahydrocannabinol, 127, A47
Tetrasporangium, 294, 295
Tetraspore, 294, 295
Tetrasporophyte, 293–95
Thalloid liverworts, 346–50, A24
Thallophyta, 344
Thallus, 290, 293–95, 335, 336, 346–51
THC. See Tetrahydrocannabinol
Theophrastus, 7, 230, 422, A59
Thermoperiod, 186, 187
Thermoperiodism, 186
Thermotropism, 179
Thigmotropism, 179
Thimbleberry, 466, A24
Thioctic acid, 328
Thistles, 478, 479, A24, A31, A41. See also Bull thistle, Canada thistle, etc.
Thompson Indians, A47
Thompson seedless grapes, 433
Thoreau, 375
Thorn, 100, 119, 444. See also Spine
Thouin, A., A59
Thrasher, 507, A24
3-carbon pathway (of photosynthesis), 147–52
Thrips, A27, A28, A30, A84
Thylakoid, 33, 34, 147
Thyme, 126, 456, 472, A24, A34, A54
Thymine, 41, 42
Tick, 255, 256, 269
Tidepool, 487, 488
Tidy tips, plate 8D, A24
Tiger, 489, A24
Tiger, sabre-toothed, 495
Tiger lily, 100

Tilia. See Basswood
Tillandsia, A94
Ti plant, 98, A24, A94
Tip layering, A61–A63
Tissue, 24, 29, 50–60
  complex, 52, 58, 233
  fundamental, 97
  ground, 97
  patterns in stems, 90–98
  primary, 51, 88
  secondary, 51, 89
  simple, 52
  types of, 50–60
Tissue culture, 298, 483, A64, A65, A66
Tmesipteris, 362
Toad, 489, A24, A28
Toadstool, 320
Tobacco, 59, 127, 174, 184, 457, 474, A24, A44, A50, A66
Tobacco budworm, A29
Tobacco mosaic, 266, A25
Tomato, 34, 76, 139, 140, 185, 253, 433–35, 449, 457, 474, 491, A24, A29, A33, A34, A82, A83
Tomato, cherry, 457, A24
Tomato, Galápagos, 187, A24
Tomato fruitworm, 257, A24, A28, A83
Tomato harvester, 474
Tomato hornworm, 257, A24, A28, A83
Tonka bean, A55
Toothwort, 463, A24
Tortoise, giant Galápagos, 188, 444, A24
Tortoise beetle, A30, A31
Touch-me-not, 446, A24
Toxin, 257
Toxins (bacterial), 254, 288
Toyon, A57
Trace, 89
Tracheid, 58, 59, 89, 94, 136, 188, 391, 405
Tracheophyta, 234, 236
Transduction, bacterial, 247
Transfer cell, 56
Transformation, bacterial, 247
Translocation
  chromosomal, 219
  food, 137, 138
Transpiration, 110, 116, 132, 135, 139, 489, A24, A64
  rates, 139
  regulation of, 138, 139
Transplanting, 67, 489, A76
Trap molecule, 150, 151
Traumotropism, 179
Travertine, 260
Trebouxia, A4
Treebine, A94
Tree ferns, 376, 378, 381–83, 390, 397, A24. See also Hawaiian tree ferns
Tree lupine, 188, A24
Tree of heaven, 443, A24
Tree snake, 344

Triassic period, 384, 423
Trichogramma wasps, A28
Trichogyne, 293, 294
Trigger hairs, 124
Trillium, 444, 504, A24
Triploid (tissue), 421
tRNA. See Ribonucleic acid, transfer RNA
Tropisms, 177–79. See also Aerotropism, Chemotropism, etc.
Truffles, 308, 315, 316, A24
Tryptamine, A51
Tschermak, Eric von, 206
Tsetse fly, 489, A4
Tube nucleus, 395, 419–22
Tuber, 99, 100, 186, 351, A63
Tuberculosis, 256, 338, 375
Tularemia, 255, 256, A24
Tulip, 100, A24, A64
Tulip tree, 112, 504, A24
Tumble mustard, 444, A24
Tumbleweeds, 443, 444, A24
Tundra, 450, 501, 502
Tung oil, 469, A24
Tung tree, A43
Turbicoryn, A51
Turgor, 133, 135, 136, 139, 180–82
Turgor movements, 180–82
Turgor pressure, 133, 137, 138, 174
Turk's cap lily, 481, A24
Turmeric, A24, A55, A57
Turnip, 70, 76, 463, A24, A35
Turpentine, 403, 406, 504
Turtle, 261, 275, A24
Turtleback algae, 275, A24
Twig, woody, 86, 87
Twinflower, 444, A24
Twining movements, 176
2, 4-D, 169
2, 4, 5-T, 169
2, 4, 5-TP, 169
2, 3, 7, 8-tetrachlorodi-benzoparadioxin, 169
Tyloses, 95
Typhoid fever, 254, A24
Typhus fever, 256, A24

u

Ulothrix, 278–80, A24
Ultraviolet light, 416, 509, A24, A64
Ulva. See Sea lettuce
Umbel, 417, 475
Umbrella plant, A94
Unaltered fossils, 386
Underground-flowering orchid, 414
Underleaves, 350
Undulant fever, 255
Unicorn plant, 441, 444, A24
Unisexual flowers, 424, 425, 476
Unit characters, law of, 209
Uracil, 44
Uredinia, 327
Uredospores, 327

V

Vaccination, vaccine, 265, 269
Vacuolar membrane, 28, 35, 132
Vacuole, 27, 28, 35, 36, 66, 133, 165, 276, 277, 278, 285, 300
Valence, 17
Valerian, A35, A50
Vallecular canals, 371
Vanadium, 173
Vanilla, 73, 258, 483, A55
Vanilla orchid, 73, 483, A24
Vascular bundle, 90, 91, 97, 98, 111, 113–16, 412, 436
Vascular cambium, 38, 51, 68, 89, 91, 367, 412
Vascular plants, introduction to, 361
Vascular tissue, 58
Vaucheria, 286
Vavilov, N. I., 455, 456
Vegetables
  culture of, A76–A82
  distinguished from fruits, 433
  nutritional values of, A77–A82
Vegetable sponge, 476, 477, A24
Vegetative propagation, 193, A59–A75
Vein, 111, 136, 398
Velamen roots, 52, 72
Velvet bean, A50
Velvet leaf, A94
Venation, 112, 113, 412. See also Palmate venation, Pinnate venation
Veneer, 103
Venereal diseases. See Gonorrhea, Syphilis
Venter, 353, 354
Venus flytrap, 123, 124, 181, A24, A94
Vernal pools, 187, 506
Verticillium, A30
Vessel, 58, 94, 136, 391, 399
Vessel element, 58, 89, 97, 391
Vestigial organs, 223
Vetch, 183, 456, 467, A24, A41
Vetchling, yellow, 118, A25
Viability of seeds, 157, 450, 451
Vinegar, 258, A25
Vinegar weed, 473, A25
Violet, 185, 440, 504
  African, A1, A25
  Gold, plate 7B, A25
Virchow, Rudolf, 25
Virginia creeper, 73, A25
Virginia snakeroot, A50

Viruses, 234, 247, 248, 265–69, 330, A25
human and ecological relevance of, 269
replication of, 268
Vitamin, 158, 165, 256, 296, 297, 298, 314, 318, 330, 401, 461, 465, 466, A48, A77–A82
Vivipary, 451
Volck oil, A84
Vole, 509, A25
Volva, 321, 323
Volvox, 276, 284
Vries, Hugo de, 206, 219
Vriesia, A95
Vygie, A51

w

Wahoo, 126, A25, A50
Wake robin, 188, A25
Walking fern, 121
Wallace, Alfred, 216, 217
Wallflower, western, 464, A25
Walnut, 135, 252, 442, 455, A25
Walnut, black, 102, 112, 491, A25, A38, A55
Wandering jew, A95
Warbler, 503, 509, A25
Warts, 265, 266, A25, A47
Wasp, 438, 513, 514, A28, A29, A32
Water, 17
absorption of, 132–34, 450
available, 81
capillary, 81
conduction of, 58, 59, 74, 86, 135–37
gravitational, 81
hygroscopic, 81
movement through plant, 135–37
photosynthesis and, 145
respiration and, 157
role of in plants, 130–39
soil, 80, 81
Water conservation movements, 180
Watercress, 440, 463, 464, A25, A41
Water fern, floating, 383, A9
Water hemlock, 70, 476, A25, A43
Water hyacinth, 69, A25, A31
Waterleaf, A41
Water lily, 56, 498, A25
Water lily, giant, 110, A25
Watermelon, 350, 425, 436, 477, A25
Water molds, 308, 310–12, A25
Water net, 284, A25
Water plantain, A41
Water pollution, 4
Water potential, 136

Water purslane, A31
Water shield, A41
Water silk, 280, 281, A25
Water storage, 70, 352, 478, 507
Water table, 81
Water weed, 133, A25
Water weed, yellow, 71, A25
Watson, J. D. and Francis H. C. Crick, 42, 43
Wattle, 466, A25
Wax, 19, 52, 113, 114, 124, 128, 406, 446, 469, 482. *See also* Carnauba wax, Carnaubalike wax, *etc.*
Wax palm, 52
Wax plant, A95
Weaver birds, 344, A25
Weed controls, A30, A31
Weed killer, 169, 175
Weeds, 229, 446, 451, 463, 467, 473, 493, A30, A31. *See also* individual listings
Weevils, 461, A27, A30, A31
Weevils, seed, A30, A31
Weinberg, W., 216
Wellensiek, 10
Welwitschia, plate 2C, 399, 400
Went, Frits, 167, 186
Western black chokecherry, 466
Western hemlock, 407, 508, A25
Western larch, 406, A25
Western red cedar, 508, A25
Western tarweed, 54, A24
Western wallflower, 464, A25, A50
Western white pine, 102, 403, A25
Whale, sperm, 469, A23
Wheat, 6, 98, 171, 185, 326, 327, 331, 442, 450, 451, 456, 457, 480, 505, A25
Wheel tree, plate 2A, A25
Whisk fern (*Psilotum*), 361–63, 381, A25
White ash, 102
White cedar, eastern, 407, 498, A6
White fir, 508, A25
White fly, 478, A25, A32, A84
White mulberry, A39
White oak, 102
White piedra, 333, A25
White pine, eastern, 102, 401, 403, 497, 504, A25
White pine, Siberian, 401, A25
White pine, western, 102, 403, A25
White pine blister rust, 327, 403, 504, A25
White snakeroot, A43
White spruce, 404, 405, A25

Whittaker, R. H., 233, 234
Whooping cough, 253
Whorl, 86, 112, 364, 371, 374, 399, 415, 423
Wild edible plants, 401, 407. *See also chapter 22*
Willdenow, Carl, 10
Willow, 9, 101, 170, 451, 501, 503, A25, A50
Willow family (Salicaceae), 443, A25
Wilting point, permanent, 81
Window leaves, 120, 121, A25
Window plant, 120, 121
Wind pollination, 393, 416, 421
Wine, 156, 310, 318, 403, 442, 479
Winged bean, 467, A25
Wing petals (of legume flower), 466, 467
Winter cress, A41
Wintergreen, 126, A25, A50, A55
Witch hazel, 183, 446, A25, A50
Woad, dyer's, 464, A25, A57
Woadwaxen, 466, A25, A57
Wolf, 458
Wolfsbane, 458, A25
Wolverine, 503, A25
Wood, 91–96, 135, 153
alcohol, 106
annual ring, plate 1, 69, 92–94, 392
applied uses of, 101–6
density, 101
durability, 101
hard, 94, 135, 391
heartwood, 69, 95
kilns, 101
knots, 101, 103, 406
plain-sawed, 101, 103
properties of, 101
pulp, 103
quartersawed, 101, 103
rays, 101
sapwood, 69, 95
seasoning of, 101
slab cut, 101
slabs, 101
soft, 391
specific gravity of, 101
spring, 92
structure, 91–96
summer, 92
tangentially cut, 101
veneer, 103
Wood betony, plate 6
Wood-boring beetles, 332, A30
Wood lily, plate 6D
Woodpecker, 444, A25
Wood rose, A51
Worms, intestinal, 298, 383, 478, A26, A47, A48, A50
Wormseed, A50
Wormwood, 128, A26, A35, A50

x

Xanthophylls, 125, 126, 295
Xerophyte, 489
Xerosere, 497
Xylem, 58, 59, 69, 73, 91, 97, 136, 137, 346, 361, 391, 392
components of, 58
function of, 58, 136, 137
parenchyma, 58
primary, 65, 68, 88
ray cells, plate 1, 58
secondary, 68, 89

y

Yakee (Parica), A51
Yam, 70, 76, 455, A26
Yareta, 126
Yarrow, American, 479, A26, A35, A41, A50
Yarrow, European, 479, A26, A35, A41, A50
Yeast, 25, 156, 265, 269, 315, 318–20, A26
Yellow fever, 265, 269
Yellow-green algae (Xanthophyceae), 286, 295, A26
Yellow lady's slipper, A50
Yellow nutgrass, A50
Yellow oleander, A43
Yellow pond lily. *See* Spatterdock
Yellow star thistle, A31
Yellow vetchling, 118, A26
Yellow water weed, 71, A26
Yerba santa, A50, A57
Yew, 112, 396, 407, A26, A41, A43
Yogurt, 258, A26
Yohimbehe, A51
Yohimbine, A51
Yokia Indians, A50
Yucca, 510, A26
Yucca moth, 510
Yuki Indians, A45

z

Zamia, 24, A26
Zebra, 24, A26
Zedoary, A55
Zinnia, 218, A26
Zoologist, 36, 232
Zoosporangium, 311
Zoospore, 278–83, 287, 310, 311, 312
Zosterophyllum, 362
Zuni Indians, 464, A47
Zygnema, 32, 277
Zygomycetes, 312–14
Zygospore, 312, 313
Zygote, 193, 210. *See also* individual organisms

Index